Principles of Physical Biochemistry

Second Edition

Kensal E. van Holde
Professor Emeritus of Biochemistry and Biophysics
Department of Biochemistry and Biophysics
Oregon State University

W. Curtis Johnson
Professor Emeritus of Biochemistry and Biophysics
Department of Biochemistry and Biophysics
Oregon State University

P. Shing Ho
Professor and Chair, Biochemistry and Biophysics
Department of Biochemistry and Biophysics
Oregon State University

PEARSON
Prentice
Hall

Pearson Education International

Executive Editor: Gary Carlson
Marketing Manager: Andrew Gilfillan
Art Editors: Eric Day and Connie Long
Production Supervision/Composition: Progressive Publishing Alternatives/Laserwords
Art Studio: Laserwords
Art Director: Jayne Conte
Cover Designer: Bruice Kenselaar
Manufacturing Buyer: Alan Fischer
Editorial Assistant: Jennifer Hart

© 2006, 1998 by Pearson Education, Inc.
Pearson Prentice Hall
Pearson Education, Inc.
Upper Saddle River, NJ 07458

ISBN 0-13-201744-X

Pearson Education Ltd. *London*
Pearson Education Australia Pty. Ltd., *Sydney*
Pearson Education Singapore, Pte. Ltd.
Pearson Education North Asia Ltd., *Hong Kong*
Pearson Education Canada, Inc., *Toronto*
Pearson Educacíon de Mexico, S.A. de C.V.
Pearson Education—Japan, *Tokyo*
Pearson Education Malaysia, Pte. Ltd.
Pearson Education, *Upper Saddle River, New Jersey*

Contents

Preface   **xiii**

Chapter 1   Biological Macromolecules   **1**

1.1   General Principles   1
    1.1.1   Macromolecules   2
    1.1.2   Configuration and Conformation   5
1.2   Molecular Interactions in Macromolecular Structures   8
    1.2.1   Weak Interactions   8
1.3   The Environment in the Cell   10
    1.3.1   Water Structure   11
    1.3.2   The Interaction of Molecules with Water   15
    1.3.3   Nonaqueous Environment of Biological Molecules 16
1.4   Symmetry Relationships of Molecules   19
    1.4.1   Mirror Symmetry   21
    1.4.2   Rotational Symmetry   22
    1.4.3   Multiple Symmetry Relationships and Point Groups 25
    1.4.4   Screw Symmetry   26
1.5   The Structure of Proteins   27
    1.5.1   Amino Acids   27
    1.5.2   The Unique Protein Sequence   31
    Application 1.1: Musical Sequences   33
    1.5.3   Secondary Structures of Proteins   34
    Application 1.2: Engineering a New Fold   35
    1.5.4   Helical Symmetry   36
    1.5.5   Effect of the Peptide Bond on Protein Conformations 40
    1.5.6   The Structure of Globular Proteins   42
1.6   The Structure of Nucleic Acids   52
    1.6.1   Torsion Angles in the Polynucleotide Chain   54
    1.6.2   The Helical Structures of Polynucleic Acids   55
    1.6.3   Higher-Order Structures in Polynucleotides   61
    Application 1.3: Embracing RNA Differences   64
    Exercises   68
    References   70

Chapter 2   Thermodynamics and Biochemistry   **72**

2.1   Heat, Work, and Energy—First Law of Thermodynamics   73
2.2   Molecular Interpretation of Thermodynamic Quantities   76
2.3   Entropy, Free Energy, and Equilibrium—Second Law
         of Thermodynamics   80
2.4   The Standard State   91
2.5   Experimental Thermochemistry   93
         2.5.1   The van't Hoff Relationship   93
         2.5.2   Calorimetry   94
         Application 2.1: Competition Is a Good Thing   102
         Exercises   104
         References   105

Chapter 3   Molecular Thermodynamics   **107**

3.1   Complexities in Modeling Macromolecular Structure   107
         3.1.1   Simplifying Assumptions   108
3.2   Molecular Mechanics   109
         3.2.1   Basic Principles   109
         3.2.2   Molecular Potentials   111
         3.2.3   Bonding Potentials   112
         3.2.4   Nonbonding Potentials   115
         3.2.5   Electrostatic Interactions   115
         3.2.6   Dipole-Dipole Interactions   117
         3.2.7   van der Waals Interactions   118
         3.2.8   Hydrogen Bonds   120
3.3   Stabilizing Interactions in Macromolecules   124
         3.3.1   Protein Structure   125
         3.3.2   Dipole Interactions   129
         3.3.3   Side Chain Interactions   131
         3.3.4   Electrostatic Interactions   131
         3.3.5   Nucleic Acid Structure   133
         3.3.6   Base-Pairing   137
         3.3.7   Base-Stacking   139
         3.3.8   Electrostatic Interactions   141
3.4   Simulating Macromolecular Structure   145
         3.4.1   Energy Minimization   146
         3.4.2   Molecular Dynamics   147
         3.4.3   Entropy   149
         3.4.4   Hydration and the Hydrophobic Effect   153
         3.4.5   Free Energy Methods   159
         Exercises   161
         References   163

Chapter 4   Statistical Thermodynamics   166

4.1   General Principles   166
         4.1.1   Statistical Weights and the Partition Function   167
         4.1.2   Models for Structural Transitions in Biopolymers   169
4.2   Structural Transitions in Polypeptides and Proteins   175
         4.2.1   Coil-Helix Transitions   175
         4.2.2   Statistical Methods for Predicting Protein
                    Secondary Structures   181
4.3   Structural Transitions in Polynucleic Acids and DNA   184
         4.3.1   Melting and Annealing of Polynucleotide Duplexes   184
         4.3.2   Helical Transitions in Double-Stranded DNA   189
         4.3.3   Supercoil-Dependent DNA Transitions   190
         4.3.4   Predicting Helical Structures in Genomic DNA   197
4.4   Nonregular Structures   198
         4.4.1   Random Walk   199
         4.4.2   Average Linear Dimension of a Biopolymer   201
         Application 4.1: LINUS: A Hierarchic Procedure to
         Predict the Fold of a Protein   202
         4.4.3   Simple Exact Models for Compact Structures   204
         Application 4.2: Folding Funnels: Focusing Down to the Essentials   208
         Exercises   209
         References   211

**Chapter 5   Methods for the Separation and Characterization
                of Macromolecules   213**

5.1   General Principles   213
5.2   Diffusion   214
         5.2.1   Description of Diffusion   215
         5.2.2   The Diffusion Coefficient and the Frictional Coefficient   220
         5.2.3   Diffusion Within Cells   221
         Application 5.1: Measuring Diffusion of Small DNA Molecules in Cells   222
5.3   Sedimentation   223
         5.3.1   Moving Boundary Sedimentation   225
         5.3.2   Zonal Sedimentation   237
         5.3.3   Sedimentation Equilibrium   241
         5.3.4   Sedimentation Equilibrium in a Density Gradient   246
5.4   Electrophoresis and Isoelectric Focusing   248
         5.4.1   Electrophoresis: General Principles   249
         5.4.2   Electrophoresis of Nucleic Acids   253
         Application 5.2: Locating Bends in DNA by Gel Electrophoresis   257
         5.4.3   SDS-Gel Electrophoresis of Proteins   259
         5.4.4   Methods for Detecting and Analyzing Components on Gels   264

5.4.5   Capillary Electrophoresis   266
5.4.6   Isoelectric Focusing   266
Exercises   270
References   274

Chapter 6   X-Ray Diffraction   276

6.1   Structures at Atomic Resolution   277
6.2   Crystals   279
       6.2.1   What Is a Crystal?   279
       6.2.2   Growing Crystals   285
       6.2.3   Conditions for Macromolecular Crystallization   286
       Application 6.1: Crystals in Space!   289
6.3   Theory of X-Ray Diffraction   290
       6.3.1   Bragg's Law   292
       6.3.2   von Laue Conditions for Diffraction   294
       6.3.3   Reciprocal Space and Diffraction Patterns   299
6.4   Determining the Crystal Morphology   304
6.5   Solving Macromolecular Structures by X-Ray Diffraction   308
       6.5.1   The Structure Factor   309
       6.5.2   The Phase Problem   317
       Application 6.2: The Crystal Structure of an Old
       and Distinguished Enzyme   327
       6.5.3   Resolution in X-Ray Diffraction   334
6.6   Fiber Diffraction   338
       6.6.1   The Fiber Unit Cell   338
       6.6.2   Fiber Diffraction of Continuous Helices   340
       6.6.3   Fiber Diffraction of Discontinuous Helices   343
       Exercises   347
       References   349

Chapter 7   Scattering from Solutions of Macromolecules   351

7.1   Light Scattering   351
       7.1.1   Fundamental Concepts   351
       7.1.2   Scattering from a Number of Small Particles:
                Rayleigh Scattering   355
       7.1.3   Scattering from Particles That Are Not Small
                Compared to Wavelength of Radiation   358
7.2   Dynamic Light Scattering: Measurements of Diffusion   363
7.3   Small-Angle X-Ray Scattering   365
7.4   Small-Angle Neutron Scattering   370
       Application 7.1: Using a Combination of Physical Methods
       to Determine the Conformation of the Nucleosome   372
7.5   Summary   376

Exercises 376
References 379

Chapter 8   Quantum Mechanics and Spectroscopy **380**

8.1   Light and Transitions 381
8.2   Postulate Approach to Quantum Mechanics 382
8.3   Transition Energies 386
    8.3.1   The Quantum Mechanics of Simple Systems 386
    8.3.2   Approximating Solutions to Quantum Chemistry Problems 392
    8.3.3   The Hydrogen Molecule as the Model for a Bond 400
8.4   Transition Intensities 408
8.5   Transition Dipole Directions 415
    Exercises 418
    References 419

Chapter 9   Absorption Spectroscopy **421**

9.1   Electronic Absorption 421
    9.1.1   Energy of Electronic Absorption Bands 422
    9.1.2   Transition Dipoles 433
    9.1.3   Proteins 435
    9.1.4   Nucleic Acids 443
    9.1.5   Applications of Electronic Absorption Spectroscopy 447
9.2   Vibrational Absorption 449
    9.2.1   Energy of Vibrational Absorption Bands 450
    9.2.2   Transition Dipoles 451
    9.2.3   Instrumentation for Vibrational Spectroscopy 453
    9.2.4   Applications to Biological Molecules 453
    Application 9.1: Analyzing IR Spectra of Proteins for Secondary Structure 456
9.3   Raman Scattering 457
    Application 9.2: Using Resonance Raman Spectroscopy
    to Determine the Mode of Oxygen Binding to Oxygen-Transport Proteins 461
    Exercises 463
    References 464

Chapter 10   Linear and Circular Dichroism **465**

10.1   Linear Dichroism of Biological Polymers 466
    Application 10.1   Measuring the Base Inclinations
    in dAdT Polynucleotides 471
10.2   Circular Dichroism of Biological Molecules 471
    10.2.1   Electronic CD of Nucleic Acids 476
    Application 10.2: The First Observation of Z-form
    DNA Was by Use of CD 478

10.2.2   Electronic CD of Proteins   481
10.2.3   Singular Value Decomposition and Analyzing the
          CD of Proteins for Secondary Structure   485
10.2.4   Vibrational CD   496
Exercises   498
References   499

Chapter 11   Emission Spectroscopy   **501**

11.1   The Phenomenon   501
11.2   Emission Lifetime   502
11.3   Fluorescence Spectroscopy   504
11.4   Fluorescence Instrumentation   506
11.5   Analytical Applications   507
11.6   Solvent Effects   509
11.7   Fluorescence Decay   513
11.8   Fluorescence Resonance Energy Transfer   516
11.9   Linear Polarization of Fluorescence   517
         Application 11.1: Visualizing c-AMP with Fluorescence   517
11.10   Fluorescence Applied to Protein   524
         Application 11.2: Investigation of the Polymerization of G-Actin   528
11.11   Fluorescence Applied to Nucleic Acids   530
         Application 11.3: The Helical Geometry of Double-Stranded
         DNA in Solution   532
         Exercises   533
         References   534

Chapter 12   Nuclear Magnetic Resonance Spectroscopy   **535**

12.1   The Phenomenon   535
12.2   The Measurable   537
12.3   Spin-Spin Interaction   540
12.4   Relaxation and the Nuclear Overhauser Effect   542
12.5   Measuring the Spectrum   544
12.6   One-Dimensional NMR of Macromolecules   549
         Application 12.1: Investigating Base Stacking with NMR   553
12.7   Two-Dimensional Fourier Transform NMR   555
12.8   Two-Dimensional FT NMR Applied to Macromolecules   560
         Exercises   575
         References   577

Chapter 13   Macromolecules in Solution: Thermodynamics and Equilibria **579**

13.1   Some Fundamentals of Solution Thermodynamics   580
         13.1.1   Partial Molar Quantities: The Chemical Potential   580

13.1.2   The Chemical Potential and Concentration:
Ideal and Nonideal Solutions   584
13.2   Applications of the Chemical Potential to Physical Equilibria   589
13.2.1   Membrane Equilibria   589
13.2.2   Sedimentation Equilibrium   597
13.2.3   Steady-State Electrophoresis   598
Exercises   600
References   603

Chapter 14   Chemical Equilibria Involving Macromolecules   605

14.1   Thermodynamics of Chemical Reactions in Solution: A Review   605
14.2   Interactions Between Macromolecules   610
14.3   Binding of Small Ligands by Macromolecules   615
14.3.1   General Principles and Methods   615
14.3.2   Multiple Equilibria   622
Application 14.1: Thermodynamic Analysis of the
Binding of Oxygen by Hemoglobin   641
14.3.3   Ion Binding to Macromolecules   644
14.4   Binding to Nucleic Acids   648
14.4.1   General Principles   648
14.4.2   Special Aspects of Nonspecific Binding   648
14.4.3   Electrostatic Effects on Binding to Nucleic Acids   651
Exercises   654
References   658

Chapter 15   Mass Spectrometry of Macromolecules   660

15.1   General Principles: The Problem   661
15.2   Resolving Molecular Weights by Mass Spectrometry   664
15.3   Determining Molecular Weights of Biomolecules   670
15.4   Identification of Biomolecules by Molecular Weights   673
15.5   Sequencing by Mass Spectrometry   676
15.6   Probing Three-Dimensional Structure by Mass Spectrometry   684
Application 15.1: Finding Disorder in Order   686
Application 15.2: When a Crystal Structure Is Not Enough   687
Exercises   690
References   691

Chapter 16   Single-Molecule Methods   693

16.1   Why Study Single Molecules?   693
Application 16.1: RNA Folding and Unfolding Observed at
the Single-Molecule Level   694
16.2   Observation of Single Macromolecules by Fluorescence   695

16.3   Atomic Force Microscopy   699
        Application 16.2: Single-Molecule Studies of Active Transcription
        by RNA Polymerase   701
16.4   Optical Tweezers   703
16.5   Magnetic Beads   707
        Exercises   708
        References   709

Answers to Odd-Numbered Problems   **A-1**

Index   **I-1**

Preface

What criteria justify revision of a successful text? It seemed to us, as authors, that there were several factors dictating the production of a second edition of "Principles of Physical Biochemistry". Foremost is the fact that the field has changed—new methods have become of major importance in the study of biopolymers; examples are mass spectrometry of macromolecules and single-molecule studies. These must be included if we are to educate today's students properly. Some older techniques see little use today, and warrant more limited treatment or even elimination. Second, we have realized that some reorganization of the text would increase its usefulness and readability. Finally, it is almost always possible to say things more clearly, and we have benefited much from the comments of teachers, students and reviewers over the last several years. We thank them all, with special thanks to the reviewers, J. Ellis Bell, University of Richmond; Michael Bruist, University of the Sciences—Philadelphia; Lukas Buehler, University of California—San Diego; Dale Edmondson, Emory University; Adrian H. Elcock, University of Iowa; David Gross, University of Massachusetts; Marion Hackert, University of Texas at Austin; Diane W. Husic, East Stroudsburg University; Themis Lazaridis, City University of New York; Jed Macosko, Wake Forest University; Dr. Kenneth Murphy, University of Iowa; Glenn Sauer, Fairfield University; Gary Siuzdak, Scripps Research University; Ann Smith, University of Missouri K.C.; Catherine Southern, College of the Holy Cross; John M. Toedt, Eastern Connecticut State University; Pearl Tsang, University of Cincinnati; Steven B. Vik, Southern Methodist University; Kylie Walters, University of Minnesota; David Worcester, University of Missouri.

We realize that there are some important areas of biophysical chemistry we still do not cover—electron spin resonance is one, chemical kinetics another. We regret not treating these, but have held to the principle that we only discuss areas in which we authors have had hands-on experience.

Biochemistry and molecular biology are today in a major transition state, largely driven by new techniques that allow dissection of macromolecular structures with precision and ease, and are beginning to allow the study of these molecules within living cells. We hope that the text will continue to be of use to students and researchers in the exciting years to come.

Kensal E. van Holde
W. Curtis Johnson
P. Shing Ho

CHAPTER 1

Biological Macromolecules

1.1 GENERAL PRINCIPLES

In physical biochemistry, we are interested in studying the physical properties of biological macromolecules, including proteins, RNA and DNA, and other biological polymers (or *biopolymers*). These physical properties provide a description of their structures at various levels, from the atomic level to large multisubunit assemblies. To measure these properties, the physical biochemist will study the interaction of molecules with different kinds of radiation, and their behavior in electric, magnetic, or centrifugal fields. This text emphasizes the basic principles that underlie these methodologies.

In this introductory chapter, we briefly review some of the basic principles of structure and structural complexity found in biological macromolecules. Most readers will have already learned about the structure of biological macromolecules in great detail from a course in general biochemistry. We take a different point of view; the discussion here focuses on familiarizing students with the quantitative aspects of structure. In addition, this discussion includes the symmetry found at nearly all levels of macromolecular structure. This approach accomplishes two specific goals: to illustrate that the structures of macromolecules are very well defined and, in many ways, are highly regular (and therefore can be generated mathematically); and to introduce the concepts of symmetry that help to simplify the study and determination of molecular structure, particularly by diffraction methods (Chapters 6 and 7). This discussion focuses primarily on the structures of proteins and nucleic acids, but the general principles presented apply to other macromolecules as well, including polysaccharides and membrane systems.

1.1.1   Macromolecules

As a basic review of molecular structure, perhaps the place to start is to ask the question, What is a *molecule*? Here, the definition of a biological molecule differs slightly from the definition learned in chemistry. In organic chemistry, a molecule consists of two or more atoms that are covalently bonded in specific proportions according to weight or stoichiometry, and with a unique geometry. Both stoichiometry (the chemical formula) and geometry (the chemical structure) are important. Dichloroethylene, for example, has the specific chemical formula $C_2H_2Cl_2$. This, however, does not describe a unique molecule, but rather three different molecules. The geometry for one such molecule is defined by the arrangements of the chlorine atoms, as in *cis*-1,2-dichloroethylene (Figure 1.1). Now, the identity of the molecule is unambiguous.

In biochemistry, a single molecule is considered to be a component that has well-defined stoichiometry and geometry, and is not readily dissociated. Thus, to a biochemist, a molecule may not necessarily have all the parts covalently bonded, but may be an assembly of noncovalently associated polymers. An obvious example of this is hemoglobin. This is considered to be a single molecule, but it consists of four distinct polypeptides, each with its own heme group for oxygen binding. One of these polypeptide-heme complexes is a *subunit* of the molecule. The heme groups are noncovalently attached to the polypeptide of the subunit, and the subunits are noncovalently interacting with each other. The stoichiometry of the molecule can also be described by a chemical formula, but is more conveniently expressed as the

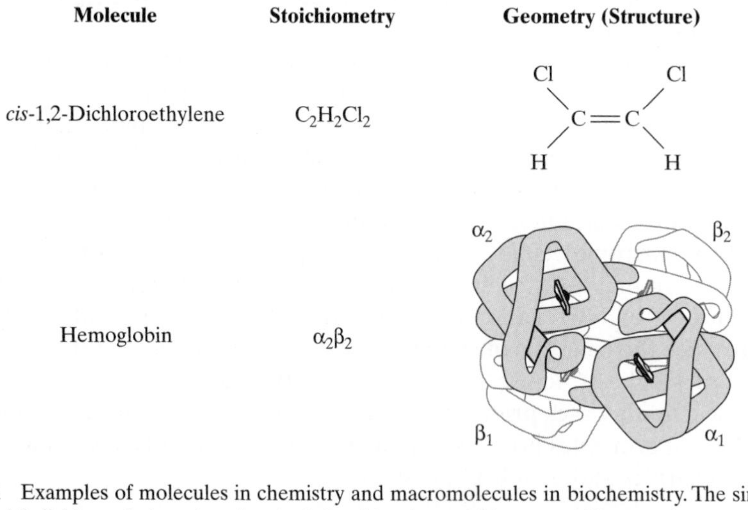

Molecule	**Stoichiometry**	**Geometry (Structure)**
cis-1,2-Dichloroethylene	$C_2H_2Cl_2$	
Hemoglobin	$\alpha_2\beta_2$	

Figure 1.1 Examples of molecules in chemistry and macromolecules in biochemistry. The simple compound *cis*-1,2-dichloroethylene is uniquely defined by the stoichiometry of its atomic components and the geometry of the atoms. Similarly, the structure of a biological macromolecule such as hemoglobin is defined by the proportions of the two subunits (the α and β-polypeptide chains) and the geometry by the relative positions of the subunits in the functional complex.

composition of *monomer* units. The stoichiometry of a protein therefore is its amino acid composition. The geometry of a biological molecule is again the unique linear and three-dimensional (3D) arrangements of these components. This is the *structure* of a biochemical molecule.

A *macro*molecule is literally a large molecule. A biological macromolecule or *biopolymer* is typically defined as a large and complex molecule with biological function. We will take a chemical perspective when dealing with macromolecules, so, for this discussion, size will be judged in terms of the number of components (atoms, functional groups, monomers, and so on) incorporated into the macromolecule. Complexity generally refers to the organization of the three-dimensional structure of the molecule. We will treat size and structural complexity separately.

What is considered large? It is very easy to distinguish between molecules at the two extremes of size. Small molecules are the diatomic to multiple-atom molecules we encounter in organic chemistry. At the upper end of large molecules is the DNA of a human chromosome, which contains tens of billions of atoms in a single molecule. At what point do we decide to call something a macromolecule? Since these are biopolymers, their size can be defined by the terms used in polymer chemistry, that is, according to the number of sugar or amino acid or nucleic acid residues that polymerize to form a single molecule. Molecules composed of up to 25 residues are called *oligomers*, while polymers typically contain more than 25 residues. This is an arbitrary distinction, since some fully functional molecules, such as the DNA-condensing J-protein of the virus G4, contain 24 residues.

The structure of biological macromolecules is hierarchical, with distinct levels of structure (Figure 1.2). These represent increasing levels of complexity, and are defined below.

> *Monomers* are the simple *building blocks* that, when polymerized, yield a macromolecule. These include sugars, amino acids, and nucleic acid residues of the polymers described above.
>
> *Primary structure* (abbreviated as 1°) is the linear arrangement (or sequence) of residues in the covalently linked polymer.
>
> *Secondary structure* (abbreviated as 2°) is the local regular structure of a macromolecule or specific regions of the molecule. These are the *helical* structures.
>
> *Tertiary structure* (abbreviated as 3°) describes the global 3D fold or *topology* of the molecule, relating the positions of each atom and residue in 3D space. For macromolecules with a single subunit, the functional tertiary structure is its *native structure.*
>
> *Quaternary structure* (abbreviated as 4°) is the spatial arrangement of multiple distinct polymers (or subunits) that form a functional complex.

Not all levels of structure are required or represented in all biological macromolecules. Quaternary structure would obviously not be relevent to a protein such as myoglobin that consists of a single polypeptide. In general, however, all biological macromolecules require a level of structure up to and including 2°, and typically 3°

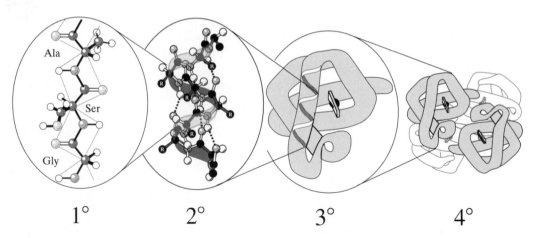

Figure 1.2 Hierarchical organization of macromolecular structure. The structures of macromolecules are organized starting with the simple monomers to form the sequence in the primary structure, which folds into the local regular helices of secondary structure, the global tertiary structure, and the association of folded chains to form complexes in the quaternary structure.

for biological function. The relationship between these levels of structure is often presented in sequential order as $1°$, followed by $2°$, which is followed by $3°$, and finally $4°$ (if present). This sequential relationship is a convenient means of presenting the increasing complexity of macromolecular structure; however, it is not clear that this is how a molecule folds into its functional form. The most recent models for protein folding suggest that a less compact form of $3°$ (often called a *molten globule* state, see Section 4.4.3) must occur first in order to form the environment to stabilize helices ($2°$). One of the goals in physical biochemistry is to understand the rules that relate these levels of structural complexity. This is often presented as the problem of predicting 3D structure ($2°$ to $3°$) from the sequence ($1°$) of the building blocks. The problem of predicting the complete 3D structure of a protein from its polypeptide sequence is the *protein-folding problem*. We can define a similar folding problem for all classes of macromolecules.

We will see how this hierarchical organization of structure applies to the structures of proteins and nucleic acids, but first we need to discuss some general principles that will be used throughout this chapter for describing molecular structure. It should be emphasized that we cannot directly see the structure of a molecule, but can only measure its properties. Thus, a picture of a molecule, such as that in Figure 1.2, is really only a model described by the types of atoms and the positions of the atoms in 3D space. This model is correct only when it conforms to the properties measured. Thus, methods for determining the structure of a molecule in physical biochemistry measure its interactions with light, or with a magnetic or electric field, or against a gradient. In all cases, we must remember that these are models of the structure, and the figures of molecules presented in this book are nothing more than representations of atoms in 3D space. It is just as accurate (and often more useful) to represent the structure as a list of these atoms and their atomic coordinates (x, y, z) in a standard Cartesian axis system.

1.1.2   Configuration and Conformation

The arrangement of atoms or groups of atoms in a molecule is described by the terms *configuration* and *conformation*. These terms are not identical. The configuration of a molecule defines the position of groups around one or more nonrotating bonds or around *chiral* centers, defined as an atom having no plane or center of symmetry. For example, the configuration of *cis*-1,2-dichloroethylene has the two chlorine atoms on the same side of the nonrotating double bond (Figure 1.3). To change the configuration of a molecule, chemical bonds must be broken and remade. A conversion from the *cis*- to *trans*-configuration of 1,2-dichloroethylene requires that we first break the carbon-carbon double bond, rotate the resulting single bond, then remake the double bond. In biological macromolecules, configuration is most important in describing the stereochemistry of a chiral molecule. A simple chiral molecule

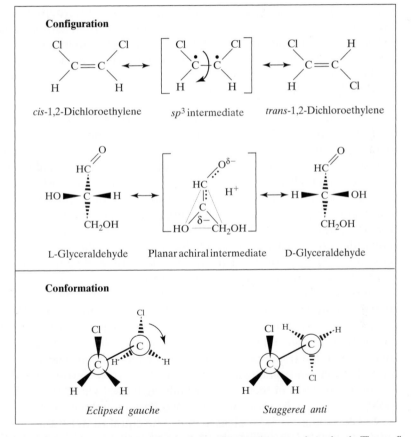

Figure 1.3 Configuration and conformation both describe the geometry of a molecule. The configuration of a molecule can be changed only by breaking and remaking chemical bonds, as in the conversion of a *cis*-double bond to one that is in the *trans*-configuration, or in converting from the L- to the D-stereoisomer of a chiral molecule. Conformations can be changed by simple rotations about a single bond.

has four unique chemical groups arranged around a tetrahedral atom (usually a carbon atom with sp^3 hybridization). To change the configuration or chirality of this molecule, we must break one bond to form a planar *achiral* intermediate, and re-form the bond on the opposite side of the plane. The resulting molecule is the *stereoisomer* or *enantiomer* of the starting structure. The stereoisomers of a molecule, even though they are identical in chemical composition, are completely different molecules with distinct properties, particularly their biological properties. Sugars that have more than one chiral center have more complex stereochemistry.

The conformation of a molecule, on the other hand, describes the spatial arrangement of groups about one or more freely rotating bonds. For example, 1,2-dichloroethane, the saturated version of dichloroethylene, has no restrictions to rotation about the chemical bonds to prevent the chlorine atoms from sitting on the same or opposite sides of the central carbon-carbon bond. These positions define the *gauche* and *anti* structural isomers, respectively. In addition, the conformation can be *eclipsed* or *staggered,* depending on whether the groups are aligned or misaligned relative to each other on either side of the carbon-carbon bond. The conformation of a molecule thus describes the structural isomers generated by rotations about single bonds (Figure 1.3). A molecule does not require any changes in chemical bonding to adopt a new conformation, but may acquire a new set of properties that are specific for that conformation.

The stereochemistry of monomers.   The monomer building blocks of biological macromolecules are *chiral* molecules, with only a few exceptions. There are many conventions for describing the stereochemistry of chiral molecules. The stereochemistry of the building blocks in biochemistry has traditionally been assigned according to their absolute configurations. This provides a consistent definition for the configuration of all monomers in a particular class of biopolymer. For example, the configurations of sugar, amino acid, and nucleic acid residues are assigned relative to the structures of L- and D-glyceraldehyde (Figure 1.4). In a standard projection formula, the functional groups of D-glyceraldehyde rotate in a clockwise direction around the chiral carbon, starting at the aldehyde, and going to the hydroxyl, then the hydroxymethyl, and finally the hydrogen groups. The configuration of the building blocks are therefore assigned according to the arrangement of the analogous functional groups around their chiral centers. Since glyceraldehyde is a sugar, it is easy to see how the configurations of the carbohydrate building blocks in polysaccharides are assigned directly from comparison to this structure. Similarly, the configuration of the ribose and deoxyribose sugars of the nucleic acids can be assigned directly from glyceraldehyde. Biopolymers are typically constructed from only one enantiomeric form of the monomer building blocks. These are the L-amino acids in polypeptides and the D-sugars in polysaccharides and polynucleotides.

For an amino acid such as alanine, the chiral center is the C_α carbon directly adjacent to the carboxylic acid. The functional groups around the C_α carbon are analogous but not identical to those around the chiral center of glyceraldehyde. The L-configuration of an amino acid has the carboxylic acid, the amino group, the α-hydrogen and the methyl side chain arranged around the C_α carbon in a

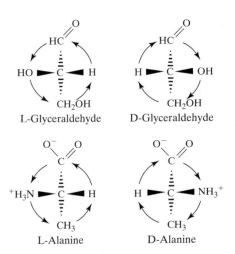

L-Glyceraldehyde   D-Glyceraldehyde

L-Alanine   D-Alanine

Figure 1.4 Absolute configuration of monomer building blocks. The stereochemistry of the monomers in biopolymers are assigned relative to L- and D-glyceraldehyde. Carbohydrates and the sugars of nucleic acids are assigned directly according to the rotation starting at the carbonyl group. For amino acids, the stereochemistry is defined according to the rotation starting at the analogous carboxyl group.

manner analogous to the aldehyde, hydroxyl, hydrogen, and hydroxymethyl groups in L-glyceraldehyde.

Conformation of molecules.   Unlike the configuration of a macromolecule, the number of possible conformations of a macromolecule can be enormous because of the large number of freely rotating bonds. It is thus extremely cumbersome to describe the conformation of a macromolecule in terms of the alignment of each group using the *gauche/anti* and *eclipsed/staggered* distinctions. It is much more convenient and accurate to describe the *torsion angle* θ about each freely rotating bond. The torsion angle is the angle between two groups on either side of a freely rotating chemical bond. The convention for defining the torsion angle is to start with two nonhydrogen groups (A and D) in the *staggered anti* conformation with $\theta = -180°$. Looking down the bond to be rotated (as in Figure 1.5) with atom A closest to you, rotation of D about the B—C bond in a clockwise direction gives a positive rotation of the bond. Thus, the values for θ are defined as 0° for the

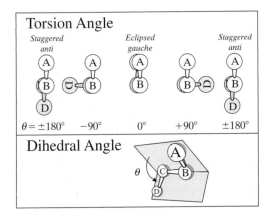

Figure 1.5   Torsion angles and dihedral angles (θ). The rotation around a single bond is described by the torsion angle of the four atoms around the bond (A—B—C—D) and the dihedral angle θ relating the planes defined by atoms A—B—C and by B—C—D.

eclipsed gauche conformation to +180° for the *staggered anti* conformation. Notice that the start and end points ($\theta = \pm 180°$) are identical.

The angle between the two groups of atoms can also be defined by the *dihedral angle*. Mathematically, the dihedral angle is defined as the angle between two planes. Any three atoms about a freely rotating bond (two atoms in the bond, plus one extending from that bond, as in A—B—C and B—C—D in Figure 1.5) defines a plane. Thus, we can see from this definition that the torsion and dihedral angles are identical.

Changing the conformation of a molecule does not make a new molecule, but can change its properties. The properly folded conformation of a protein, referred to as the *native* conformation, is its functional form, while the unfolded or *denatured* conformation is nonfunctional and often targeted for proteolysis by the cell. Thus, both the configuration and conformation of a molecule are important for its shape and function, but these represent distinct characteristics of the molecule and are not interchangeable terms. The conformations of polypeptides and polynucleic acids will be treated in greater detail in later sections.

1.2 MOLECULAR INTERACTIONS IN MACROMOLECULAR STRUCTURES

The configurations of macromolecules in a cell are fixed by covalent bonding. The conformations, however, are highly variable and dependent on a number of factors. The sequence-dependent folding of macromolecules into secondary, tertiary, and quaternary structures depends on a number of specific interactions. This includes the interactions between atoms in the molecule and between the molecule and its environment. How these interactions affect the overall stability of a molecule and how they can be used to construct models of macromolecules are discussed in greater detail in Chapter 3. In this introductory chapter, we define some of the characteristics of these interactions, so that we can have some understanding for how the various conformations of proteins and polynucleic acids are held together.

1.2.1 Weak Interactions

The covalent bonds that hold the atoms of a molecule together are difficult to break, releasing large amounts of energy during their formation and concomitantly requiring large amounts of energy to break (Figure 1.6). For a stable macromolecule, they can be treated as invariant. The conformation of a macromolecule, however, is stabilized by weak interactions, with energies of formation that are at least one order of magnitude less than that of a covalent bond. The weak interactions describe how atoms or groups of atoms are attracted or repelled to minimize the energy of a conformation.

These are, in general, distance-dependent interactions, with the energies being inversely proportional to the distance r or some power of the distance (r^2, r^3, etc.) separating the two interacting groups (Table 1.1). As the power of the inverse

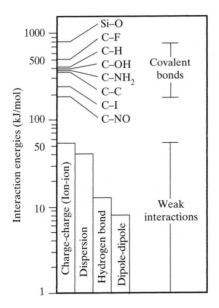

Figure 1.6 Energies of molecular interactions. The interactions that define the structure of a molecule range from the strong interactions of covalent bonds (200 to 800 kJ/mol) to the weak charge-charge (or ion-ion), dipole-dipole, dispersion, and hydrogen-bonding interactions (0 to 60 kJ/mol).

distance dependency increases, the interaction approaches zero more rapidly as r increases, and thus becomes a shorter range interaction. The interaction energy between two charges varies as $1/r$; this is a long-range interaction. At the other extreme are the induced dipole-induced dipole (or *dispersion*) interactions. These interactions describe the natural tendency of atoms to attract, regardless of charge and polarity, because of the polarizability of the electron clouds. Its dependence on $1/r^6$ defines this as a very short-range interaction, having a negligible interaction energy at about 1 nm or greater. Directly opposing this attraction, however, is steric repulsion, which does not allow two atoms to occupy the same space at the same time. This repulsion occurs at even shorter distances and is dependent on $1/r^{12}$. Together, the attractive dispersion and repulsive exclusion interactions define an optimal distance separating any two neutral atoms at which the energy of interaction is a minimum. This optimal distance thus defines an effective radius (the *van der Waals radius*, or r_{vdw}) for each type of atom. The potential energy functions for

Table 1.1 Relationship of Noncovalent Interactions to the Distance Separating the Interacting Molecules, r

Type of Interaction	Distance Relationship
Charge-charge	$1/r$
Charge-dipole	$1/r^2$
Dipole-dipole	$1/r^3$
Charge-induced dipole	$1/r^4$
Dispersion	$1/r^6$
Repulsion	$1/r^{12}$

each interaction and their application to simulating the thermodynamic properties of macromolecules are treated in detail in Chapter 3.

The energies associated with long-range interactions (charge-charge, charge-dipole, and dipole-dipole) are dependent on the intervening medium. The interaction between two charged atoms, for example, becomes shielded in a polar medium and is therefore weakened. The least polarizable medium is a vacuum, with a dielectric constant of $\kappa\epsilon_O = 4\pi 8.85 \times 10^{-12} \, C^2 \, J \cdot m$, where $\epsilon_O = 8.85 \times 10^{-12} \, C^2 \, J \cdot m$ and $\kappa = 4\pi$ for a point charge. The polarizability of a medium is defined as its dielectric constant D relative to that of a vacuum. The expressions for the energy of long-range interactions are all inversely related to the dielectric of the medium and are therefore weakened in a highly polarizable medium such as water.

With the dielectric constant, we introduce the environment as a factor in stabilizing the conformation of a macromolecule. How the environment affects the weak interactions is discussed in the next section. In the process, two additional interactions (hydrogen bonds and hydrophobicity) are introduced that are important for the structure and properties of molecules.

1.3   THE ENVIRONMENT IN THE CELL

The structures of macromolecules are strongly influenced by their surrounding environment. For biopolymers, the relevant environment is basically the solvent within the cell. Because the mass of a cell is typically more than 70% water, there is a tendency to think of biological systems primarily as aqueous solutions. Indeed, a large majority of studies on the properties of biological macromolecules are measured with the molecule dissolved in dilute aqueous solutions. This, however, does not present a complete picture of the conditions for molecules in a cell. First, a solution that is 70% water is in fact highly concentrated. In addition, the cell contains a very large surface of membranes, which presents a very different environment for macromolecules, particularly for proteins that are integral parts of the bilayer of the membranes. The interface between interacting molecules also represents an important nonaqueous environment. For example, the recognition site of the TATA-binding protein involves an important aromatic interaction between a phenylalanyl residue of the protein and the nucleotide bases of the bound DNA.

In cases where solvent molecules are observed at the molecular interfaces (for example, between the protein and its bound DNA), the water often helps to mediate interactions, but is often treated as part of the macromolecule rather than as part of the bulk solvent. In support of this, a well-defined network of water molecules has been observed to reside in the minor groove of all single-crystal structures of DNA duplex. Results from studies using nuclear magnetic resonance (NMR) spectroscopy indicate that the waters in this spine do not readily exchange with the bulk solvent and thus can be considered to be an integral part of the molecule. We start by briefly discussing the nature of the aqueous environment because it is the dominant solvent system, but we must also discuss in some detail the nonaqueous environments that are also relevant in the cell.

1.3.1   Water Structure

Water plays a dominant role in defining the structures and functions of many molecules in the cell. This is a highly polar environment that greatly affects the interactions within molecules (*intramolecular interactions*) and between molecules (*intermolecular interactions*). It is useful, therefore, to start with a detailed description of the structure of water.

A single H_2O molecule in liquid water is basically tetrahedral. The sp^3 oxygen atom is at the center of the tetrahedron, with the hydrogens forming two of the apices, and the two pairs of nonbonding electrons forming the other two apices (Figure 1.7). In the gas phase, the nonbonding orbitals are not identical (the spectroscopic properties of water are discussed in Chapter 9), but in the hydrogen-bonded network they are. The oxygen is more electronegative than are the hydrogens (Table 1.2), leaving the electrons localized primarily around the oxygen. The O—H bond is therefore polarized and has a permanent dipole moment directed from the hydrogen (the positive end) to the oxygen (the negative end). A dipole moment also develops with the positive end at the nucleus of the oxygen, pointing toward each of the nonbonding pairs of electrons. The magnitude of these dipoles becomes exaggerated in the presence of other charged molecules or other polar molecules. The magnitude of the dipole moment increases from 1.855 debye (debye $= 3.336 \times 10^{-30}$ C/m) for an isolated water molecule to 2.6 debye in a cluster of six or more molecules to 3 debye in ice. Water is therefore highly polarizable, as well as being polar. It has a very high dielectric constant relative to a vacuum ($D = 78.5\kappa\epsilon_O$).

The interaction between two water molecules is an interaction between polar compounds. This is dominated by the dipoles, which align the O—H bonds with the

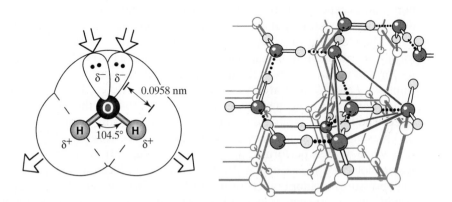

Figure 1.7   The structure of water. Each H_2O molecule has two hydrogens and two lone pairs of unbonded electrons at each oxygen. In ice, the hydrogens act as hydrogen-bond donors to the lone pairs of the oxygens, which act as hydrogen-bond acceptors. This results in a hexagonal lattice of hydrogen-bonded water molecules, with each H_2O molecule having four neighbors arranged in a tetrahedron. [Adapted from Mathews and van Holde (1996), *Biochemistry*, 2nd ed., 33. Benjamin-Cummings Co., Menlo Park, CA.]

Table 1.2 Electronegativities of Elements
Typically Found in Biological Molecules

Element	Electronegativity
O	3.5
Cl	3.0
N	3.0
S	2.5
C	2.5
P	2.1
H	2.1
Cu^{2+}	1.9
Fe^{2+}	1.8
Co^{2+}	1.8
Mg^{2+}	1.2
Ca^{2+}	1.0
Na^{+}	0.9
K^{+}	0.8

Higher values indicate a higher electron
affinity.

dipole moment of nonbonding electrons of the oxygen. These dipole-dipole interactions bring the oxygen and hydrogen atoms closer than the sum of their van der Waal's radii, and thus are classified as weak bonds. This is the water-water *hydrogen bond.* In this case, the O—H donates the hydrogen to the bond and is the *hydrogen-bond donor.* The oxygen, or more precisely the nonbonding electrons of the oxygen, acts as the *hydrogen-bond acceptor.* Water molecules therefore form a hydrogen-bonded network, with each H_2O potentially donating hydrogen bonds to two neighbors and accepting hydrogen bonds from two neighbors. Other hydrogen-bond donors and acceptors that are important in biopolymers are listed in Table 1.3.

The structure in which the H_2O molecules are exactly tetrahedral and uniformly distributed into hexagonal arrays (Figure 1.7) is found only in the crystalline ice form of water. However, the hydrogens of the ice observed under normal conditions (*ice I* at 0°C and 1 atm pressure) remain disordered. They cannot be assigned to any particular oxygen at any given time, even though the oxygen atoms remain fixed. The hydrogens can be ordered precisely, but only at pressures greater than 20 kbars at temperatures less than 0°C (Figure 1.8). This *ice VIII,* therefore, forms only when work is performed against the inherent entropy in the hydrogens of the water molecules (see Chapter 2). Water can be induced at low temperatures and high pressures to adopt other forms or *phases* of ice that are unstable under normal conditions. The molecules in *ice IX,* for example, are arranged as pentagonal arrays. This arrangement is similar to many of the faces in the clathrate-like structures observed around ions and alkyl carbons under standard conditions (Figure 1.9).

The structure of liquid water is very similar to that of ice I. This liquid form, which we will now simply refer to as water, is also a hydrogen-bonded network. The average stretching frequency of the O—H bond in water is more similar to that of

Table 1.3 Hydrogen-Bond Donors and Acceptors
in Macromolecules

Donor	Acceptor	r (nm)

	0.29
	0.29
	0.31
	0.37
	0.28
	0.28

ice than to H_2O molecules that do not form hydrogen bonds (Figure 1.10). At the air-water interface, the water molecules are well organized, much like that of ice, and form a highly cohesive network. A similar ordered structure is found at the interface between water and the surface of molecules dissolved in water, which we describe in the next section.

However, the structure of water is more dynamic than that of ice, with the pattern of hydrogen bonds changing about every picosecond. The redistribution of protons results in a constant concentration of hydronium ions (a hydrate proton) and hydroxide ions in aqueous solutions, as defined by the equilibrium constant (K_{eq})

$$K_{eq} = [H_3O^+][OH^-]/[H_2O] \tag{1.1}$$

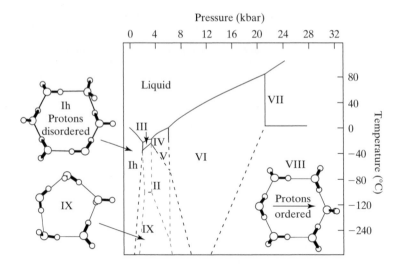

Figure 1.8 Phase diagram for water. Liquid water freezes in different ice forms, depending on the temperature and pressure. Under normal conditions, ice is a hexagonal network in which the protons of the hydrogen bonds are equally shared and cannot be assigned to a specific oxygen center (ice Ih). More compact forms (e.g., ice IX) or more ordered forms (e.g., ice VIII) are observed at low temperatures and high pressures. [Adapted from H. Savage and A. Wlodawer (1986), *Meth. Enzymol.* **127**; 162–183.]

Equation 1.1 is reduced to the standard equation for self-dissociation of water

$$K_W = K_{eq}[H_2O] = [H^+][OH^-] = 10^{-14}\,M^2 \tag{1.2}$$

with the concentration of H_3O^+ represented by $[H^+]$. Alternatively, this is given as

$$pK_W = pH + pOH = 14 \tag{1.3}$$

with p{anything} $= -\log_{10}$\{anything\}.

    Free protons do not exist in aqueous solution but are complexed with a local aggregate of water molecules. This is also true for the resulting hydroxide ion. The two ions are indeed distinct and have different properties, even in terms

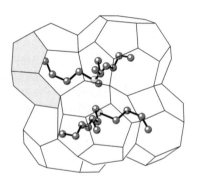

Figure 1.9 Clathrate structure of waters in the hydrated complex $(nC_4H_9)_3S^+F^- \cdot 23\,H_2O$. The solvent structure is composed of regular hexagonal and pentagonal faces (one of each is highlighted), similar to those found in ice structures. [Adapted from G. L. Zubay, W. W. Parson, and D. E. Vance (1995), *Principles of Biochemistry*, 14. Wm. C. Brown, Dubuque, IA.]

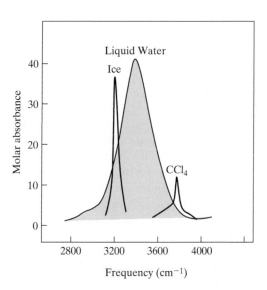

Figure 1.10 Vibrational frequency of $O-H$ bond of H_2O in ice, in liquid water, and in CCl_4. The vibration in CCl_4 is very similar to that of the bond in water vapor. [Adapted from C. Tanford (1980), *The Hydrophobic Effect: Formation of Micelles and Biological Membranes*, 2nd ed., 36. John Wiley & Sons, New York.]

of the distribution of protons around each water molecule. A proton in $H_5O_2^+$ sits at an average position between oxygens. In $H_3O_2^-$, the average distance between oxygens is increased, leaving the shared proton distributed toward one or the other oxygen atom. The difference in the chemical properties of the two ionic forms of water may be responsible for the differences observed in how acids and bases affect biochemical reactions, particularly in the effects of deuterium or tritium on the kinetics of enzyme-catalyzed reactions that require proton transfers.

1.3.2   The Interaction of Molecules with Water

A molecule dissolved into water must interact with water. The polarizability of the aqueous medium affects the interactions between charged groups of atoms, polar but uncharged groups, and uncharged and nonpolar groups in macromolecules. These interactions are discussed in greater detail in Chapter 3. At this point, we will provide a general picture of how molecules interact with water, and how this affects the properties of the molecule as well as the properties of the solvent.

When any molecule is placed in water, the solvent must form an envelope that is similar in many respects to the air-water interface. This is true whether the compound is an ion or a hydrocarbon. Water molecules form a cage-like clathrate structure around ions (Figure 1.9). Compounds that can overcome the inherently low entropy of this envelope by interacting strongly with the water will be soluble. These *hydrophilic* compounds are water loving. Salts such as NaCl are highly soluble in water because they dissociate into two ions, Na^+ and Cl^-. The strong interaction between the charged ions and the polar water molecules is highly favorable, so that the net interaction is favorable, even with the unfavorable entropy contribution from the structured waters.

Hydrocarbons such as methane are neither charged nor polar, and thus are left with an inherently unfavorable cage of highly structured surrounding waters. This cage is ice-like, often with pentagonal faces similar to ice IX. These rigid ice-like cage structures are low in entropy and this is the primary reason that hydrocarbons are insoluble in water. These compounds are thus *hydrophobic* or water hating. The pentagonal arrays help to provide a curved surface around a hydrophobic atom, much like that seen at the air-water interface. The waters around hydrophilic atoms typically form arrays of six and seven water molecules.

In contrast, hydrophobic molecules are highly soluble in organic solvents. Methane, for example, is highly soluble in chloroform. The favorable interactions between nonpolar molecules come from van der Waals attraction. Thus, polar and charged compounds are soluble in polar solvents such as water, and nonpolar compounds are soluble in nonpolar organic solvents, such as chloroform. This is the basis for the general chemical principle that *like dissolves like*.

Molecules that are both hydrophilic and hydrophobic are *amphipathic*. For example, a phospholipid has a charged phosphoric acid head group that is soluble in water, and two long hydrocarbon tails that are soluble in organic solvents (Figure 1.11). In water, the different parts of amphipathic molecules sequester themselves into distinct environments. The hydrophilic head groups interact with water, while the hydrophobic tails extend and interact with themselves to form an oil drop–like hydrophobic environment. The form of the structures depends on the type of molecules that are interacting and the physical properties of the system. In the case of phospholipids, the types of structures that form include micelles (formed by dilute dispersions), monolayers (at the air-water interface), and bilayers. A bilayer is particularly useful in biology as a membrane barrier to distinguish between, for example, the interior and exterior environment of a cell or organelle.

Proteins and nucleic acids are also amphipathic. Proteins consist of both polar and nonpolar amino acids, while nucleic acids are composed of hydrophobic bases and negatively charged phosphates. These biopolymers will fold into structures that resemble the structures of micelles. In general, molecules or residues of a macromolecule that are hydrophilic will prefer to interact with water, while hydrophobic molecules or residues will avoid water. This is the basic principle of the *hydrophobic effect* that directs the folding of macromolecules (such as proteins and nucleic acids) into compact structures in water. The basis for the hydrophobic effect and its role in stabilizing macromolecular structures is discussed in Chapter 3.

1.3.3   Nonaqueous Environment of Biological Molecules

Several biologically important molecules exist in nonaqueous environments. These are predominantly molecules (primarily proteins) found in the phospholipid bilayers of cellular membranes. There are a number of significant differences between a cell membrane and the aqueous solution in a cell. The most obvious is that those parts of a molecule residing within the hydrocarbon tails of the membrane bilayer must be

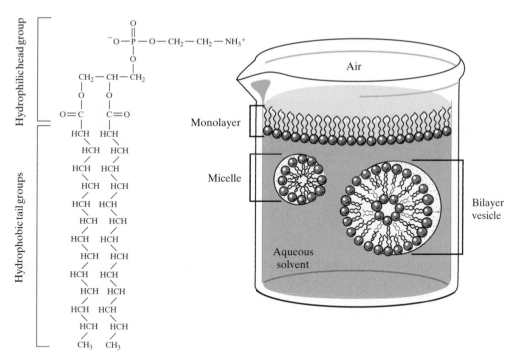

Figure 1.11   Structures formed by amphipathic molecules in water. An amphipathic molecule, such as phosphatidyl choline, has a head group that is hydrophilic and long hydrophobic tails. In water, these compounds form monolayers at the water-air interface, globular micelles, or bilayer vesicles. [Adapted from Mathews and van Holde (1996), *Biochemistry*, 2nd ed., 37. Benjamin-Cummings, Menlo Park, CA.]

hydrophobic. The structure of an integral membrane protein can be thought of as being inverted relative to the structure of a water-soluble protein, with the hydrophobic groups now exposed to the solvent, while the hydrophilic atoms form the internalized core. An example of this inverted topology is an ion channel (Figure 1.12). The polar groups that line the internal surface of the channel mimic the polar water solvent, thus allowing charged ions to pass readily through an otherwise impenetrable bilayer.

In addition to affecting the solubility of molecules, the organic nature of the hydrocarbon tails in a membrane bilayer makes them significantly less polar than water. Thus, the dielectric constant is approximately 40-fold lower than an aqueous solution. The effect is to enhance the magnitude of interactions dependent on D by a factor of about 40. One consequence of this dramatically lower dielectric constant is that the energy of singly charged ions in the lipid bilayer is significantly higher than that in aqueous solution. A measure of the energy of single charges in a particular medium is its self-energy E_S. This can be thought of as the energy of a charge in the absence of its counterion and thus defined by an expression similar to that of a charge-charge interaction.

$$E_S = q^2/2DR_S \tag{1.4}$$

Figure 1.12 (a) The crystal structures of the ion channel gramicidin and a calcium-binding ionophore A23187. Gramicidin is a left-handed antiparallel double helix in the crystal. In this structure, the central pore is filled with cesium and chloride ions. [Adapted from B. A. Wallace and K. Ravikumar (1988), *Science,* **241**; 182–187.] (b) The structure of A23187 binds a calcium by coordination to oxygen and nitrogen atoms. [Adapted from Chaney (1976), *J. Antibiotics* **29**, 4.]

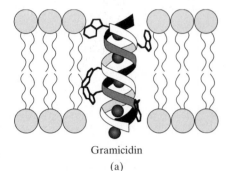

Gramicidin

(a)

Calcium-binding ionophore A23187

(b)

In this case, E_S is dependent on the square of the single charge q^2, and thus the self-energy is always positive for single cations or anions. In this relationship, R_S is the Stokes' radius of the molecule (the effective molecular radius, see Section 5.2.1). The inverse relationship to the Stokes' radius R_S indicates that a charge that is distributed over a larger ion or molecule has a lower self-energy than an ideal point charge. The dependence of E_S on $1/D$ means that the self-energy of an ion in water is 40 times lower than in a lipid bilayer. This translates into a probability that the ion will reside in the membrane is $\sim 10^{-18}$ times that in water, thus making membranes highly efficient barriers against the passage of charged molecules. The movement of ions and other polar molecules through a cellular membrane requires the help of ion carriers, or ionophores, that form water-filled channels through the membrane or transport ions directly across the membranes (Figure 1.12).

Membranes are also distinguished from an aqueous environment in that membranes are essentially two-dimensional (2D) surfaces. With the exception of very small molecules such as the ionophores, molecules travel mostly in two dimensions in membranes. The concepts of concentration and diffusion-controlled kinetics must be defined in terms of this 2D surface, as opposed to a 3D volume. In solution, the concentration of a molecule is given in units of moles/dm^3 (moles/l = M). The concentration of a molecule in a membrane is defined as the number of molecules per given surface area (moles/dm^2). For example, the concentration of molecules at the surface of a sphere will be diluted by a factor of 4 if the radius of that sphere is doubled, while molecules within the volume of the sphere will be diluted by a factor of 8.

The diffusional rates of molecules in aqueous solution and in membranes show these volume versus surface area relationships.

Finally, it is not necessary for a molecule to be imbedded in a membrane to experience a nonaqueous environment. The interior of a globular protein consists primarily of hydrophobic amino acids, and the polarizability of this environment is often compared to that of an organic solvent such as octanol. The consequence is that it is very difficult to bury a single charge in the interior of a large globular protein. This is reflected in a lower pK_a for the side chains of the basic amino acids lysine and arginine, or a higher pK_a for the acidic amino acids aspartic acid and glutamic acid when buried in the interior of a protein. We can estimate the effect of the *self-energy* on the pK_a of these amino acids in solution as opposed to being buried in the interior of a protein. We should reemphasize that these are estimates. A lysine with a $pK_a = 9.0$ for the side chain would be protonated and positively charged in water. If we transfer this charged amino acid (with a Stokes' radius ≈ 0.6 nm) into a protein interior ($D \approx 3.5\kappa\epsilon_O$), the difference in self-energy in the protein versus water ΔE_S is about 40 kJ/mol. We can treat this energy as a perturbation to the dissociation constant by

$$\Delta pK_a = \Delta E_S/2.303 k_B T \tag{1.5}$$

and predict that the $pK_a < 1$ for a lysine buried in the hydrophobic core of a globular protein and therefore would be uncharged unless it is paired with a counterion such as an aspartic acid residue.

1.4   SYMMETRY RELATIONSHIPS OF MOLECULES

Biological systems tend to have symmetry, from the shape of an organism to the structure of the molecules in that organism. This is true despite the fact that the monomer building blocks (amino acids, for example) are always asymmetric. Yet, these often combine to form elegantly symmetric structures. In this section, we describe the symmetry relationships of biological macromolecules, both conceptually and mathematically. This mathematical formalism provides a means to precisely overlay or *map* two symmetry-related objects on top of each other, or to construct a set of symmetry-related objects from a starting model. This is useful in that it simplifies many problems in physical biochemistry, including structure prediction, structure determination using techniques such as X-ray diffraction and electron diffraction, and image reconstruction that improves the results obtained from electron microscopy or atomic microscopies.

Symmetry is the correspondence in composition, shape, and relative position of parts that are on opposite sides of a dividing line or median plane or that are distributed about a center or axis (Figure 1.13). It is obvious that two or more objects are required in order to have a symmetric relationship. The unique object in a symmetry-related group is the *motif*. A motif *m* is repeated by applying a symmetry element or symmetry operator $\hat{O}$ to give a related motif *m'*.

$$\hat{O}(m) = m' \tag{1.6}$$

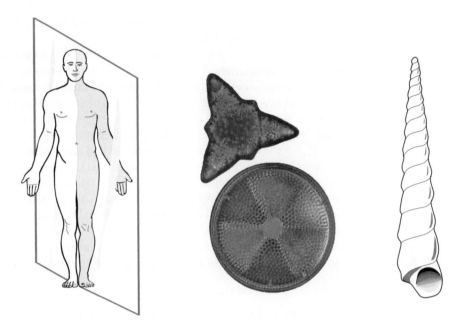

Figure 1.13   Examples of mirror, rotational, and screw symmetry. The human body shows mirror symmetry through a plane, diatoms show rotational symmetry about an axis, and a spiral shell shows screw symmetry about an axis.

It is also clear that the symmetry described here is centered at a point or line or plane that passes through the center of mass of the motifs. Thus, this definition of symmetry is for *point symmetry,* and motifs that are related by the same symmetric relationships are said to belong to the same *point group*.

There are two types of point symmetry that are relevant in biology and to molecular structure. Simple single-cell and multicellular organisms seem to represent nearly all possible forms of symmetry. *Mirror symmetry* relates two motifs on opposite sides of a dividing line or plane. Both simple and higher organisms show mirror symmetry, at least superficially. The human body, when bisected vertically by a plane, results in two halves that are, to the first approximation, mirror images of each other (left hand to right hand, left foot to right foot, and so on). In this case, the motif is half the body, and the two halves are related by mirror symmetry. The second type of symmetry from this definition is *rotational symmetry,* which relates motifs distributed about a point or axis. This includes radial symmetry about a single axis or multiple axes, and helical or screw symmetry, which is rotation and translation along an axis.

To express symmetry relationships mathematically for macromolecules, we must define some conventions. First, a structure will be described by a model in which each atom of a molecule is placed at some unique set of coordinates (x, y, z) in space. These coordinates will always be placed in a right-handed Cartesian coordinate system (Figure 1.14). The fingers of the right hand point from x to y and the thumb is along z. All rotations in this coordinate system will be right-handed. With

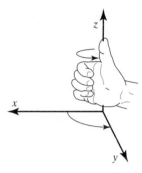

Figure 1.14 Right-handed Cartesian coordinates and right-handed rotations. In a right-handed axis system, the fingers point from the x-axis toward the y-axis when the thumb is aligned along the z-axis. This same system describes a right-handed rotation, where the fingers of the right hand represent a positive rotation about a particular axis, in this case the z-axis.

the thumb of the right hand pointed in the direction of the rotation axis, the fingers point in the direction of a positive rotation.

1.4.1   Mirror Symmetry

To the first approximation, the left and right hands of the human body are related by *mirror symmetry* (Figure 1.15), the left hand being a reflection of the right hand through a *mirror* plane that bisects the body. A simple symmetry operator is derived for mirror symmetry by first placing the hands onto a three-dimensional axis system. We start by defining the xz plane (formed by the x- and z-axes) as the mirror plane, with the y-axis perpendicular to the xz plane. In this axis system, the fingers of each hand are assigned a unique set of coordinates. The thumb of the right hand has the coordinates (x, y, z), while the thumb of the left hand has (x', y', z'). For the thumbs of the two hands, we can see that $x' = x, z' = z$, but $y' = -y$. The right hand is thus inverted through the plane to generate the left. The two sets of coordinates are related to each other by a symmetry operator $\hat{i}$ such that

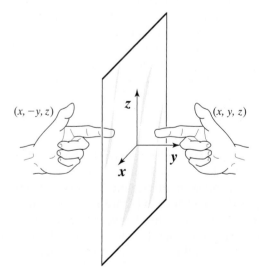

Figure 1.15 Mirror symmetry of left and right hands. The left and right hands are related by mirror symmetry through a plane. In this axis system, the two hands are related by an inversion of the y coordinate through the xz plane.

$\hat{\imath}(x, y, z) = (x', y', z') = (x, -y, z)$. For a 3D coordinate system, a symmetry operator can be represented by the three simultaneous equations:

$$a_1 x + b_1 y + c_1 z = x'$$
$$a_2 x + b_2 y + c_2 z = y'$$
$$a_3 x + b_3 y + c_3 z = z'$$

This can be rewritten in the matrix form

$$\begin{vmatrix} a_1 & b_1 & c_1 \\ a_2 & b_2 & c_2 \\ a_3 & b_3 & c_3 \end{vmatrix} \times \begin{vmatrix} x \\ y \\ z \end{vmatrix} = \begin{vmatrix} x' \\ y' \\ z' \end{vmatrix} \tag{1.7}$$

The operator $\hat{\imath}$ would thus have the following coefficients for the three simultaneous equations:

$$1 \times x + 0 \times y + 0 \times z = x'$$
$$0 \times x - 1 \times y + 0 \times z = y'$$
$$0 \times x + 0 \times y + 1 \times z = z'$$

Alternatively, we can write the matrix form of $\hat{\imath}$ as

$$\hat{\imath} = \begin{vmatrix} 1 & 0 & 0 \\ 0 & -1 & 0 \\ 0 & 0 & 1 \end{vmatrix} \tag{1.8}$$

This is the mathematical relationship between any pair of motifs that are exact mirror images of each other, including the stereoisomers of the chiral monomer building blocks in biological macromolecules. If the structure of a molecule is available, an exact mirror image can be generated by the mirror operator.

On a larger scale, the symmetry of the human body is superficial and does not show true mirror symmetry. Although the left and right hands of the body are apparently mirror images, the inside of the body is not. The heart is slightly displaced to the left side in the body, and there is no corresponding heart on the right side. Motifs that appear symmetric, but that are not truly symmetric, show *pseudosymmetry*.

1.4.2 Rotational Symmetry

The symmetry around a point or axis is rotational symmetry. In this case, there is not an inversion of a motif, but a reorientation in space about the center of mass. Again, we can start with a motif and generate all symmetry-related motifs with a rotational symmetry operator $\hat{c}$. For example, the two hands in Figure 1.16 are related by a rotation of 180° (or 1π radian) about an axis that lies between the hands and is perpendicular to the page. We can derive the operator that relates the two rotated hands by following the same procedure as the one for mirror symmetry. The motif

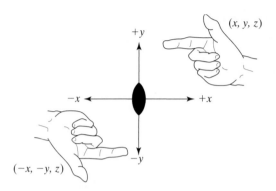

Figure 1.16 Rotational symmetry. The two hands in this figure are related by two-fold rotational symmetry about the z-axis. In this example, both the x and y coordinates are inverted.

(one hand) is placed in the xy plane of a standard Cartesian coordinate system such that the z-axis is perpendicular to the page of the figure. We place the first hand at (x, y, z) and the second hand at $(-x, -y, z)$. The operator $\hat{c}$ that relates one hand to the other can thus be written as

$$\hat{c} = \begin{vmatrix} -1 & 0 & 0 \\ 0 & -1 & 0 \\ 0 & 0 & 1 \end{vmatrix} \tag{1.9}$$

which defines an operator that rotates a hand 180° about the z-axis.

We notice that this specific rotational operator, when used twice, rotates the motif a full 360°, so that $(\hat{c}(\hat{c}(x, y, z)) = \hat{c}(-x, -y, z) = (x, y, z))$. In other words, it takes two applications of this rotational operator to return to the original starting point. Thus, this operator is identified as giving a *two-fold rotation*. The z-axis about which the hand was rotated is called the *two-fold rotational axis* and the set of motifs is said to have *two-fold symmetry*. This is also known as *dyad symmetry,* and the axis of rotation is the *dyad axis.*

Two-fold rotation about the z-axis is only one very specific example of rotational symmetry. Analogous operators can be derived for two-fold axes along the x-axis and along the y-axis. There are also rotational matrices for any set of motifs related by any rotational angle. For rotation about the z-axis by any angle θ, the general operator in matrix form is

$$\begin{vmatrix} \cos\theta & -\sin\theta & 0 \\ \sin\theta & \cos\theta & 0 \\ 0 & 0 & 1 \end{vmatrix} \tag{1.10}$$

For $\theta = 180°$, $\cos(180°) = -1$ and $\sin(180°) = 0$, and this general matrix reduces to that for two-fold symmetry in the matrix (Figure 1.9). For the rotational angle θ, the symmetry is said to be related by n-fold rotation or have C_n symmetry, where n is the number of times the operator must be applied to return to the starting point $(n = 360°/\theta)$. The symbols for various symmetry axes are listed in Table 1.4.

Table 1.4   Symbols for Symmetry

Symbol	Symmetry	Motif
	C_2 (two-fold)	Monomer
	2_1 (two-fold screw)	Monomer
	C_3 (three-fold)	Monomer
	3_1 (right-handed three-fold screw)	Monomer
	3_2 (left-handed three-fold screw)	Monomer
	C_4 (four-fold)	Monomer
	4_1 (right-handed four-fold screw)	Monomer
	4_2 (four-fold screw)	Dimer
	4_3 (left-handed four-fold screw)	Monomer
	C_6 (six-fold)	Monomer
	6_1 (right-handed six-fold screw)	Monomer
	6_2 (right-handed six-fold screw)	Dimer
	6_3 (six-fold screw)	Trimer
	6_4 (left-handed six-fold screw)	Dimer
	6_5 (left-handed six-fold screw)	Monomer

1.4.3   Multiple Symmetry Relationships and Point Groups

In describing rotational symmetry, we showed that two applications of a two-fold rotation brings the motif back to its starting point. Indeed, this is true for n-applications of any n-fold rotation operator. Symmetry operators, however, need not be restricted to identical n-fold rotations, or even to rotations about the same axis or point. Multiple symmetry elements may be applied to produce a higher level of symmetry between motifs.

In biological macromolecules, multiple sets of symmetry elements relate identical subunits that associate at the level of quaternary structure. The typical point groups found in biological molecules are rotational symmetry about a central point (C), dihedral symmetry (D), tetragonal symmetry (T), octahedral symmetry (O) and icosahedral symmetry (I) (Figure 1.17). We have described the point group for simple rotational symmetry. We can define a motif or set of motifs having n-fold rotational symmetry as being in the point group C_n. Thus, two-fold symmetry falls in the point group C_2. The n-fold rotational axis is the C_n axis. The value n also refers to the number of motifs that are related by the C_n axis. Thus, C_1 symmetry describes a single motif that has no rotational relationships. This can be found only in asymmetric molecules that have no plane or center of symmetry (i.e., chiral molecules).

At the next higher level of symmetry, molecules that are related by a C_n rotational axis, and n perpendicular C_2 axes are related by dihedral symmetry D_n. The C_n axis generates a set of motifs rotated by an angle $360°/n$, all pointing in one direction around the axis. The perpendicular C_2 axes generate a second set below the first in which all the motifs point in the opposite direction. Thus, the number of motifs related by D_n symmetry is $2n$. Alternatively, if we can think of each pair of molecules

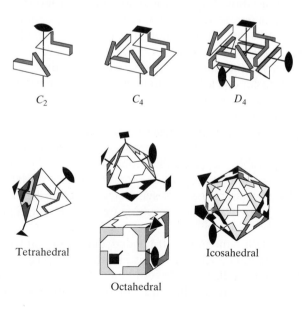

C_2 C_4 D_4

Tetrahedral   Icosahedral

Octahedral

Figure 1.17 Point groups. The repeating motif in each figure is represented by an arrow. The C_2 point group shows two-fold rotational symmetry, C_4 shows four-fold symmetry, and D_4 shows both four-fold and two-fold symmetry. In the D_4 point group, two of the four two-fold axes exactly overlap, leaving only two unique two-fold axes perpendicular to the four-fold axis. Tetrahedral symmetry has two-fold axes at the edges and three-fold axes at the corners and faces of the four sides. Octahedral symmetry has two-fold, three-fold, and four-fold symmetry. This defines both a standard octahedron and a cube. Icosahedral symmetry is defined by two-fold, three-fold, and five-fold symmetry axes.

related by the C_2 axes as symmetric dimers, then D_n symmetry is simply the rotational relationship between n number of dimers.

Finally, tetrahedral, octahedral, and icosahedral symmetry are point groups that combine multiple rotational axes. The symmetry increases going from T to O to I, but these are related point groups. For any shape with m-hedral symmetry, there are m faces on the solid shape. In terms of the symmetry of these shapes, we will see that m is best defined by the total number of C_3 symmetry axes in each point group. A true octahedron, for example, has eight triangular faces, each with a C_3 axis. However, a cube, which has eight corners, each with a C_3 axis, also has octahedral symmetry.

The number of repeating motifs N required for each point group is $N = 3m$. For icosahedral symmetry, the number of repeating motifs is $N = 3 \times 20 = 60$. Since N must be a constant, and must be related by symmetry, the types and numbers of the other symmetry operators that contribute to these point groups are very well defined. The parameter N will always be the product of $m \times n$ for m number of C_n symmetry axes. Thus, the number of C_2 axes in icosahedral symmetry is $60/2 = 30$. Clearly, m and n must be roots of N. Thus, icosahedral symmetry also has 12 C_5 axes. Not all factors, however, are represented: There are no true C_6 axes in this point group. Examples of these higher levels of symmetry are described in our discussion of quaternary structures in proteins.

1.4.4   Screw Symmetry

Motifs that are symmetric about an axis are not always radially symmetric. For example, a spiral staircase is symmetric about an axis. However, this spiral does not bring us to the starting position after a full 360° rotation unless there is no rise (that is, the spiral is a circle). To describe this type of symmetry, we need to introduce another symmetry element, translation. Translation simply moves a motif from one point to another, without changing its orientation. A translational operator T can shift the coordinates of a motif some distance along the x-axis, y-axis, and/or z-axis. This can be written as $(x, y, z) + T = (x + T_x, y + T_y, z + T_z)$ where T_x, T_y, and T_z are the x, y, and z components of the translational operator, respectively.

The combination of translations and rotations in a spiral staircase no longer shows point symmetry but has *screw symmetry* or *helical symmetry,* which describes the helical structures in macromolecules. The root for this name is self-evident (Figure 1.18). A screw operates by inducing translational motion (the driving of the screw into a board) using a rotating motion (the turning of the screwdriver). These concerted motions are inextricably linked to the properties of the threads of the screw. The properties of screw symmetry rely on the rotational and translational elements of the threads. The operator for screw symmetry therefore has the form $C_n(x, y, z) + T = (x', y', z')$. The translation resulting from a 360° rotation of the screw is its *pitch.*

A screw can either be right-handed or left-handed, depending on whether it must be turned clockwise or counterclockwise to drive it into a board (Figure 1.18). Looking down the screw, if we turn the screwdriver in the direction that the fingers of the right hand take to drive the screw in the direction of the thumb of that hand,

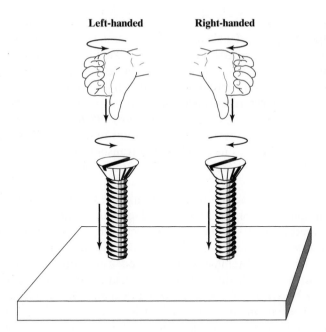

Left-handed   **Right-handed**

Figure 1.18   Screw symmetry. The rotation of a screw becomes translational motion by screw symmetry. Screws can be either left-handed or right-handed, depending on which direction the screw must be turned to drive it into the board.

then the screw is right-handed. The nomenclature for screw symmetry comes from the two elements of the operator. A helix can be said to have *n-fold screw symmetry* if the rotation component of the operator must be applied *n* times to give one complete turn of the helix. This definition will be used to describe the helical structures that define the secondary structures of macromolecules.

1.5   THE STRUCTURE OF PROTEINS

Proteins are the functional forms of polypeptides. We start with a detailed description of proteins because their structures represent all levels of the heirarchy of macromolecular structure. We will summarize each of these levels of structure as they apply to the native form of proteins. The structures of proteins are defined as much by the environment as by the chemical properties of the polypeptide chain. Thus, there are distinctive categories of proteins: the water-soluble globular proteins; water-insoluble fibrous proteins; and proteins that associate with the hydrophobic environment of membrane bilayers. Each of these is characterized by distinct amino acid compositions and sequences, but all of their structures can be described using an identical set of basic principles.

1.5.1   Amino Acids

Proteins are polymers built from amino acids. There are 20 amino acids that are common to all living organisms on Earth. These are all α-amino acids, with the amino and carboxylic acid groups separated by a single C_α carbon. All common amino acids are L-amino acids, with the exception of glycine, which is achiral. Each amino acid is distinguished by the chemistry of the side chains that are attached to the C_α carbon (Figure 1.19). The common amino acids do not, however, represent the only components of proteins. D-amino acids are found in the antiviral proteins valinomycin and gramicidin, produced by bacteria. Amino acids in proteins of eukaryotes are often covalently modified following synthesis. Finally, proteins often require additional simple organic and inorganic *cofactors* or *prosthetic groups* to function. This includes the heme group of the cytochromes and the oxygen-transport proteins myoglobin and hemoglobin, and many of the vitamin-derived cofactors involved in enzyme-catalyzed reactions (Table 1.5).

To demonstrate the significance of the configuration of building blocks, we can ask a very basic and simple question: Will an inverted structure result from a protein constructed from the stereoisomers of the naturally occurring amino acids? The intuitive answer would be yes, but this was not proven until the structures were determined for molecules that were chemically synthesized using unnatural D-amino acids (Figure 1.20). Both rubredoxin and the protease from the human immunodeficiency virus (HIV) have been synthesized using D-amino acids. The structures of native and the inverted rubredoxin were determined by X-ray crystallography and were found to be exact mirror images of each other. The crystal used to solve the structure was a *racemic mixture,* containing equal proportions of the two stereoisomers with the structures related by a mirror symmetry operator. The stereoisomers of HIV protease were found to recognize the mirror images of the enzyme substrate. The inverted protease did not catalyze lysis of the natural substrate, and the native enzyme was inactive against the inverted substrate. Thus, the stereochemistry of the amino acids is important for both the structure and function of proteins.

The amino acids in proteins are the most varied of the monomers found in biopolymers. Amino acids are distinguished by the chemical properties of their side chains. They can be hydrophobic or hydrophilic. Most proteins are amphipathic; that is, they include both hydrophobic and hydrophilic amino acids. The hydrophobic amino acids tend to avoid water by residing in the interior of a globular protein or within the lipid bilayer for membrane-bound proteins; the hydrophilic residues prefer to remain hydrated. This partitioning of amino acids between aqueous and nonaqueous solvent environments leads to the hydrophobic effect that drives the folding of proteins into compact globular tertiary structures.

Table 1.5 Enzyme Cofactors and Their Dietary Precursors

Coenzyme	Precursor
Thiamine pyrophosphate	Thiamine (vitamin B_1)
Flavin adenine dinucleotide	Riboflavin (vitamin B_2)
Pyridoxal phosphate	Pyridoxine (vitamin B_6)
5′-Deoxyadenosylcobalamine	Vitamin B_{12}

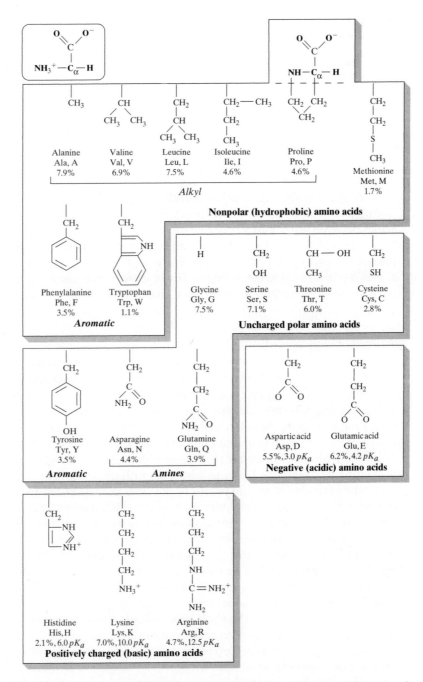

Figure 1.19 Twenty common amino acids. The amino acids that are common to all living organisms are these α-amino acids in the L-configuration. The side chains can be categorized as being nonpolar hydrophobic, uncharged polar, or charged at neutral pHs. The amino acids are listed along with their frequency of occurrence in typical proteins.

Figure 1.20 Mirror images of rubredoxin. The native protein (shaded) and its mirror image were solved from the same crystal. The two proteins are related by mirror symmetry and rotational symmetry. [Adapted from Zawadzke and Berg (1993), *Proteins* **16**; 301.]

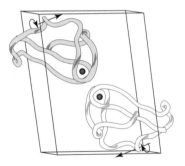

To quantify the contribution of each amino acid to the hydrophobic effect (the *hydropathy* of an amino acid), we can explicitly measure the partitioning of the molecules between water and an organic solvent such as octanol. For this measurement, an amino acid is placed in a two-solvent system with separate aqueous and organic phases. The hydropathy of the amino acid is represented by the partition coefficient P, which is measured as the mole fraction of molecules in the aqueous phase χ_{aq} relative to the mole fraction in the organic phase at equilibrium χ_{nonaq}.

$$P = \chi_{aq}/\chi_{nonaq} \tag{1.11}$$

The hydropathy is expressed in Table 1.6 for each type of amino acid as $-\log(P)$. Although the hydropathy index is consistent with what we would generally expect for the amino acids (positive hydropathy for nonpolar side chains and negative hydropathy for polar and charged side chains), we should note that the hydrophobic effect was originally defined only for nonpolar compounds. Still, the coefficients are useful for describing the relative behavior of the amino acids in proteins and their effect on the thermodynamic stability of proteins (see Chapter 3).

The hydrophobic amino acids have either alkyl or aromatic side chains. The aromatic side chains are bulky and interact with other aromatic side chains. These interactions tend to place the planes of the rings perpendicular to each other when buried in the interior of a globular protein, or within the hydrophobic region of the membrane bilayer in membrane proteins. This orientation of aromatic groups is similar to the perpendicular arrangement of benzene molecules in solution. There are more ways to arrange planar molecules such as benzene with their planes lying perpendicular to each other than with their faces stacked parallel. Thus, this arrangement of aromatic molecules is entropically favored although this does not preclude additional short-range interactions between π-systems that could favor these arrangements. In contrast, the nucleotide bases in the structures of DNA and RNA form parallel stacks to reduce their exposure to the solvent. The perpendicular arrangement therefore is preferred by aromatic rings that do not need to minimize their exposed surfaces to water.

The side chains of the hydrophilic amino acids are polar, and may be charged under physiological pHs. The charged amino acids include the acidic amino acids aspartic acid and glutamic acid, and the basic amino acids histidine, lysine, and arginine.

Table 1.6 Hydropathy Index
of Amino Acids

Amino Acid	Hydropathy
Ile	4.5
Val	4.2
Leu	3.8
Phe	2.8
Cys	2.5
Met	1.9
Ala	1.8
Gly	−0.4
Thr	−0.7
Ser	−0.8
Trp	−0.9
Tyr	−1.3
Pro	−1.6
His	−3.2
Asn	−3.5
Gln	−3.5
Asp	−3.5
Glu	−3.5
Lys	−3.9
Arg	−4.5

Source: From J. Kyte and R. F. Doolittle,
(1982), *J. Mol. Biol.* **157**; 105–132.

The overall charge of a protein is dependent on the number of acidic and basic amino acids that are charged at a particular pH. The pH at which the total charge is zero (that is, all the negative charges are balanced by positive charges) is the *isoelectric point* or pI of the protein. The charge density of a protein ρ_C is estimated as the ratio of the effective charge of a protein at any pH relative to its molecular weight (MW).

$$\rho_C = (\text{pI} - \text{pH})/\text{MW} \tag{1.12}$$

Although useful as a rough indicator for the behavior of macromolecules in, for example, an electric field, this simple relationship is only a rough approximation and does not take into account the compensating effects of counterions on the net charge of a molecule (see Chapter 3 for a detailed treatment of partial and net charges of proteins).

1.5.2 The Unique Protein Sequence

The sequence of a protein is the order of the amino acids that are covalently linked by peptide bonds (Figure 1.21). Considering the diversity of proteins in a cell, it would be interesting to ask how many sequences can be built using the same composition of building blocks. Consider, for example, a simple tripeptide. How many different tripeptides can be built from the 20 common amino acids?

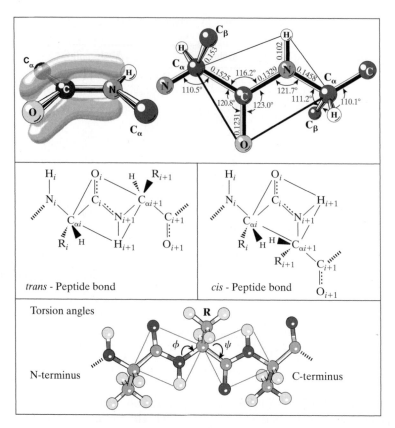

Figure 1.21 The peptide bond. The peptide bond that chemically links two amino acid residues along a polypeptide chain is a C—N bond of an amide linkage. The bond shows partial double-bond character, with the π-electrons distributed between the O—C—N atoms (top panel), and is not freely rotating [bond lengths and angles from Engh and Huber (1991), *Acta Crystallogr. A* **47**: 392–400]. The bond is found predominantly in the *trans*-configuration, but could also adopt the *cis*-form (middle panel). The two freely rotating bonds on either side of the C_α-carbon thus define the torsion angles ϕ (for rotation about the N—C_α bond) and ψ (for rotation about the Ca—C bond).

There are 20 different components, with no constraints on how often each component can occur at any of the three positions. The probability of finding any one of the 20 possible amino acids at position 1 is 1/20. This is the same probability as finding any one amino acid occurring at any one position along a chain. The probability of finding a tripeptide with a specific amino acid sequence can be calculated as the products of the probability of finding a particular residue at each site. For a tripeptide, this would be $1/20 \times 1/20 \times 1/20 = 1/(20)^3 = 1/8000$. There are thus 8000 different possible tripeptide sequences, and the probability of finding any one sequence in a set of random tripeptides is 1 in 8000. For a polypeptide of any length N, the number of possible sequences for that length is 20^N. Thus, for a polypeptide the size of a small protein ($N \approx 100$) there are $20^{100} \approx 10^{130}$ different possible amino acid sequences. This is more than the total number of particles currently thought to be in the universe. This simple analysis assumes that there is an equal probability of placing each of the 20 amino acids at any particular position along the polypeptide chain. The most accurate estimate for the frequency of the occurrence of a polypeptide sequence (or any biopolymer sequence) must take into account the statistical probability for the occurrence of each of the amino acids (Application 1.1). Still, the number of possible sequences of a particular length is astronomically large.

We can now ask: How many different sequences have the identical composition of residues? This question is a simple example of the basic statistical approach to understanding molecular thermodynamics (see Chapters 2 and 4). If we restrict the composition of a peptide to two glycines and one alanine (G_2A composition), there are only three different tripeptide sequences that can be built. This is identical to asking how many different ways can we arrange two particles (the glycine residues) in three boxes (each possible position in the tripeptide). The alanine residue automatically fits into the open position that does not contain a glycine. We define the number of possible arrangments as the variable W.

The strategy for calculating W becomes one of determining the number of boxes available g versus the number of unique particles n that can be placed in the boxes as we sequentially fill them. In placing the first particle, there are three

---

Application 1.1   Musical Sequences

With 20 amino acid building blocks, proteins can have a wide variety of sequences and repeating sequence motifs. One such search for the sequence glutamyl-leucyl-valyl-isoleucyl-serine (ELVIS) found that this motif occurred four times in a database of 25,814 protein sequences (Kaper and Mobley, 1991). This frequency of occurrence (one in about 6500 sequences) is significantly higher than we would expect from the random occurrence of any five amino acids (once in roughly three million random amino acids). As a control, the sequence HAYDN did not occur in this same data set, suggesting a sequence bias against names of classical composers. However, this analysis assumed that all amino acids occur with identical frequencies. If the actual amino acid composition of an average protein is considered, the occurrence of ELVIS is predicted to be 1 in 954,293 sets of random pentapeptides (about 1 in every 4000 polypeptide sequence, assuming 200 amino acids per sequence), which is three times higher than random probability. The sequence HADYN, on the other hand, is expected only once in nearly seven million amino acids (less than half of a purely random occurrence). Thus, proteins apparently treat both noted musicians equally. The survey did find, however, that LIVES was not represented in the data set, indicating that probabilities of specific amino acids are not the only important factors—certain combinations of amino acids are poorly represented. Therefore, the probability of finding ELVIS LIVES would be extremely low.

KAPER, J. B. and H. L. T. MOBLEY (1991), *Science* **253**, 951–952.

available boxes for one of two indistinguishable particles (one of the two glycines), so the ratio of boxes to particles g/n is 3/2. In placing the second particle, $g/n = 2/1$. Finally, for the third particle, there is one box remaining for one particle, defining $g/n = 1$. The number of possible arrangements is thus $W = \Pi g_i/n_i$, for each ith type of particle. In this example $W = (3/2) \times (2/1) \times 1 = 3$ or there are three unique tripeptide sequences that can be built from two glycines and one alanine. The general formula for distributing N total number of particles over g boxes (positions), for n unique types of particles, is a simple binomial expansion (also known as the *Fermi-Dirac permutation*).

$$W = \prod_{i=1}^{N} \frac{g_i!}{n_i!(g_i - n_i)!} \tag{1.13}$$

The probability of finding a single polypeptide having a specific amino acid sequence for a given amino acid composition is simply the inverse of W. Thus, if only a very specific set of compositions of a polymer can be used by the cell, the number of possible sequences for a macromolecule becomes more restricted. Equation 1.13 is the general form for determining the number of ways to arrange a set of particles into identical (*degenerate*) positions. This applies to residues along a sequence or particles in energy levels.

    In general, sequence directs structure, and the three-dimensional structure of two molecules with similar sequences can be assumed to have very similar structures. The degree of sequence homology (identity of amino acids) between two proteins required to specify similar structures has been suggested to be 25% to 30%. It is not true, however, that a unique structure requires a unique sequence (Application 1.2). Molecules with similar structures need not be homologous in sequence. For example, the nucleoprotein H5 and the eukaryotic transcription factor HNF-3/*fork-head* protein are nearly identical in their structures but have only 9 of 72 amino acids in common.

1.5.3 Secondary Structures of Proteins

The regular and repeating structure of a polypeptide is its secondary structure (2°). The term *regular* defines these as symmetric structures. The only symmetric three-dimensional structure that can be constructed from a linear chain of chiral building blocks is a helix. The secondary structure of a protein therefore describes the helices formed by polypeptide chains. These helices are usually held together by hydrogen bonds formed between the amino hydrogens (the hydrogen-bond donors) and the carboxyl oxygens (the hydrogen-bond acceptors). The significance of hydrogen bonds in the secondary structures of proteins was recognized early on by Linus Pauling. The distances between the hydrogen-bond donor and acceptor groups (measured in numbers of atoms and complete turns of the helix) has traditionally been used in the nomenclature of the helices. A 3_{10} helix, therefore, is characterized as having 10 atoms separating the amino hydrogen and carboxyl oxygen atoms that are hydrogen-bonded together to form one complete turn of the helix. Chapter 3

Application 1.2   Engineering a New Fold

One of the problems faced in trying to predict the structure of a protein from its sequence or to design new proteins from scratch is that the rules for protein folding are not well understood. This is further complicated by the observation that two proteins differing by as much as 70% in sequence can have nearly identical three-dimensional structures. With this in mind, protein engineers challenged themselves to take a protein of one structure and, without altering more than 50% of the sequence, design a protein with a completely different fold. The feat was accomplished from studies of two proteins, protein G found at the surface of *Streptococcus* bacteria, and Rop, an RNA-binding protein (Dalal et al. 1997). Protein G is predominantly β-sheet, while Rop is nearly all α-helical. The stability of Rop as an α-helical bundle was dependent on a number of helix-stabilizing amino acids at the interface between the helical coils. Similarly, protein G required certain key amino acids to stabilize the β-sheets. Thus, by replacing 50% of the residues in protein G with the helix stabilizing amino acids in Rop (Figure A1.2), the sheet-like protein was converted to a helix bundle similar to Rop.

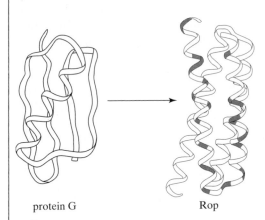

Figure A1.2 Replacement of 50% of the amino acids changes the tertiary structure of the mostly β-sheet protein G to a helical bundle similar to the RNA-binding protein Rop. The shaded regions are the amino acids in protein G that were replaced by helix forming residues from Rop.

protein G   Rop

DALAL, S., S. BALASUBRAMANIAN, and L. REGAN (1997), *Nature Struct. Biol.* **7**, 548–552.

---

shows that the stabilization of helices by hydrogen bonds is dependent on the environment of the polypeptide and, as a consequence, is solvent dependent.

   One of the short-comings in using this convention for naming helices is that extended structures such as the strands of a β-sheet cannot be described in this manner. Each strand of a β-sheet is a regular and repeating two-fold helical structure, but with hydrogen-bonding interactions *between* strands holding the sheet together. Beta-strands are regular and repeating structures and, therefore, are helices. In addition, important structural information is excluded by this nomenclature, including the handedness of the helix. A more accurate method to describe the structural features of helices in a shorthand notation is by its *helix symmetry*. This form of nomenclature has the added benefit that it conforms to the symmetry rules used in

X-ray crystallography and allows the prediction of the diffraction pattern obtained from fiber diffraction studies.

1.5.4 Helical Symmetry

A helix is a structure in which residues rotate and rise in a repeating manner along an axis. This is the same as the definition given above for screw symmetry, except that screw symmetry as applied to a helix is not continuous like the threads of a screw. Each residue is a well-defined point along the helix. Still, we can generate a helix by applying the rotational operator $\hat{c}$ and translational operator T of screw symmetry on a monomer unit. Thus, starting with a residue at position i having coordinates $(x, y, z)_i$, the position of the next residue along a helix is defined as $\hat{c}(x, y, z)_i + T = (x, y, z)_{i+1}$. This form of screw symmetry is known as *helical symmetry* and uses terminology that is analogous to that used to describe the characteristics of threads of a screw (e.g., pitch and handedness, Figure 1.18). The screw axis is the *helix axis*. As with a screw, the pitch P is the transition along the helix axis in one complete turn of the helix.

   A better model for the discrete steps of a helix is the steps of a spiral staircase, which by our definition is also a helix (Figure 1.22). As we climb the spiral staircase, the ascension up each step is the translation part of the helical symmetry operator. This is the *helical rise h* for helices in macromolecules. Each step of the spiral staircase is rotated by some angle θ, which is the *helical angle* or *helical twist*. Thus, the steps rise by a specific vertical distance h and rotate by some angle θ. If the staircase

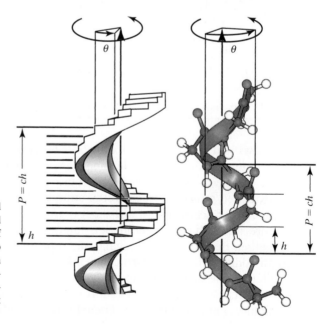

Figure 1.22 Spiral staircase and helix. A spiral staircase is a good model for the discrete residues of a helix. Each step is analogous to the individual residues of a biopolymer helix, and can be described by the same set of parameters: rise (h), pitch (P), repeat (c), and twist (θ).

is sufficiently long, the symmetry of the steps will repeat. The number of steps required to reach this point is the repeat c, and the sum of the vertical rise for one repeat is the pitch P of the helix. If we start at a set of coordinates (x, y, z) such that the z-axis points upward, the coordinates for each step i are related to the next step $i + 1$ by the screw operator

$$
\begin{vmatrix} \cos\theta & -\sin\theta & 0 \\ \sin\theta & \cos\theta & 0 \\ 0 & 0 & 1 \end{vmatrix} \times \begin{vmatrix} x \\ y \\ z \end{vmatrix}_i + \begin{vmatrix} 0 \\ 0 \\ h \end{vmatrix} = \begin{vmatrix} x \\ y \\ z \end{vmatrix}_{i+1} \tag{1.14}
$$

The angle θ can also be described in terms of the helical parameters c and P. The repeat is a rotation about the z-axis by a full $360°$ or 2π radians. The helical angle θ is thus $2\pi/c$. The pitch divided by the repeat is simply the incremental rise at each step $P/c = h$. We can thus rewrite this set of symmetry operators as

$$
\begin{vmatrix} \cos(2\pi/c) & -\sin(2\pi/c) & 0 \\ \sin(2\pi/c) & \cos(2\pi/c) & 0 \\ 0 & 0 & 1 \end{vmatrix} \times \begin{vmatrix} x \\ y \\ z \end{vmatrix}_i + \begin{vmatrix} 0 \\ 0 \\ P/c \end{vmatrix} = \begin{vmatrix} x \\ y \\ z \end{vmatrix}_{i+1} \tag{1.15}
$$

Using this set of symmetry elements, we can mathematically generate the three-dimensional coordinates of a helical structure starting from the structure of an initial building block. This spiral staircase analogy exactly mimics the relationships among the residues of a biopolymer chain.

The symmetry matrix for a helix derived from the pitch and the rise does not, however, yield a unique model. A helix can be either right-handed or left-handed in exactly same manner that a screw is either left- or right-handed. In terms of the helical parameters, a positive rotation of the helical angle $\theta > 0°$ gives a right-handed helix, and a negative rotation $\theta < 0°$ gives a left-handed helix. However, we have defined a convention that all rotations are right-handed. A left-handed helix can be constructed using only right-handed rotations by changing the sign of the translation. Since $\theta/2\pi = 1/c = h/P$ and $-\theta/2\pi = 1/c = -h/P$ either P or h is negative for a left-handed helix.

Now we can consider some specific types of helices. A simple helix is the 3_{10} helix, which has 3 residues per turn (Figure 1.23). The 3_{10} helix is also a right-handed, three-fold screw (Table 1.7). To accurately reflect its helical symmetry, we will use the shorthand notation N_T, where N represents the N-fold rotation operator and T the translation in fractions of a repeat for the symmetry operator. To elaborate on T, in a 3_{10} helix each residue along the chain translates or rises 1/3 of this repeat, or $(1/3)P$ along the helix axis. The parameter T is the numerator of this translation. The helical symmetry of a 3_{10} helix is thus 3_1, or has three-fold screw symmetry. In addition, it tells us that this is right-handed; the handedness is inherent in T. To demonstrate this, consider a helix with 3_2 symmetry. In this case, there are three residues for an exact repeat, and each residue rises 2/3 of a repeat (Figure 1.24). If we simply generate a sequence of residues using this operator, the biopolymer chain is left with gaps. The gaps are filled by moving the adjacent symmetry unit by

Figure 1.23 Structures of the 3_{10} helix and the α-helix. The two types of helices typically observed in proteins are the 3_{10} helix and the α-helix. The α-helix is the most common helix found in globular proteins. Both are stabilized by intramolecular hydrogen bonds between the amino hydrogen and the carbonyl oxygen of the peptide backbone.

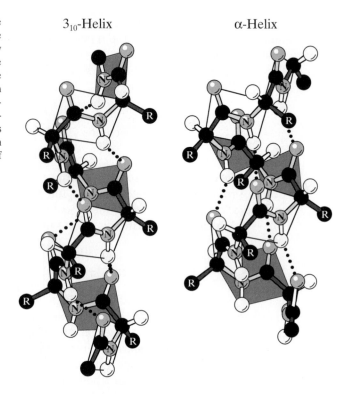

3_{10}-Helix α-Helix

the repeat distance P along the helix axis. Thus, we can change the handedness of the helix without changing the way the helix is generated: All rotations are right-handed with translations in the direction of the thumb of the right hand.

In general, a right-handed helix is designated as N_1 and a left-handed helix as N_{N-1}, for helices with integer numbers of residues per turn. These principles of exact repeats and translations in fractions of repeating units will also be used to

Table 1.7 Helical Symmetry of Macromolecular Helices

Structure Type	Residues per Turn	Rise (nm)	Helical Symmetry
Trans-conformation (polypeptides)	2.0	0.36	2_1
β-sheet	2.0	0.34	2_1
3_{10} helix	3.0	0.20	3_1
α-helix	3.6	0.15	18_5
π-helix	4.4	0.12	22_5
A-DNA	11.0	0.273	11_1
B-DNA	10.0–10.5	0.337	10_1
C-DNA	9.33	0.331	28_3
Z-DNA	−12.0	−0.372	6_5*

*The repeating unit of Z-DNA is two base pairs in a dinucleotide. This gives an average repeat of −12 base pairs per turn of the left-handed double helix.

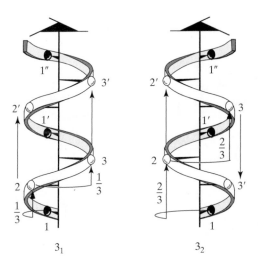

Figure 1.24 Right-handed and left-handed helical symmetry. A helix with 3_1 helical symmetry has each residue rotated $+120°$ and translated 1/3 of the each repeat ($h = P/3$). This generates a right-handed helix. In contrast, 3_2 helical symmetry rotates each residue by the same angle, but translates the residues by 2/3 of a repeat ($h = 2P/3$). When each of the repeating units are filled by the symmetry-related residues, the connections from residue 1′ to 2 to 3′ results in a left-handed helix.

describe the symmetry of crystals, and this shorthand for helical symmetry will be seen in the diffraction of X-rays by molecular fibers of these helices.

The α-helix is the most widely recognized secondary structure in protein biochemistry (Figure 1.23). This helix is right-handed and has a hydrogen bond between the carbonyl oxygen of residue i and the amino group at residue $i + 3$. Unlike the 3_{10}-helix, the hydrogen-bonded residues do not exactly align along the helix axis. Thus, the helical repeat is $c = 3.6$ residues per turn, which gives a helical angle $\theta = 360°/3.6$ residues/turn $= 100°$/residue. The rise is $h = 0.15$ nm per residue and, therefore, the pitch is $P = 0.54$ nm. The symmetry operator that generates such a helix along the z-axis is

$$
\begin{vmatrix} -0.174 & -0.985 & 0 \\ 0.985 & -0.174 & 0 \\ 0 & 0 & 1 \end{vmatrix} \times \begin{vmatrix} x \\ y \\ z \end{vmatrix}_i + \begin{vmatrix} 0 \\ 0 \\ 0.15 \text{ nm} \end{vmatrix} = \begin{vmatrix} x \\ y \\ z \end{vmatrix}_{i+1} \tag{1.16}
$$

The symmetry of the α-helix is 3.6_1. It is more convenient, however, to convert N and T to integers by multiplying their values, in this specific case, by 5. Thus, the helical symmetry for the α-helix is 18_5, which means that there are 18 residues in 5 full turns of the helix. We see in Chapter 6 that this exactly describes the X-ray diffraction pattern of α-helical proteins.

Other standard helices found in macromolecules (and their helical parameters) are listed in Table 1.7. Notice in this table the *trans*-conformation of a peptide and the strands of a β-sheet have two-fold screw symmetry. The two conformations differ in that the *trans*-conformation is fully extended to give a straight backbone, while the backbone of β-strands are pleated like the folds of a curtain (Figure 1.25). In addition, the β-sheets found in proteins are typically twisted and therefore the residues of the strands are not related by exact two-fold screw symmetry. Nonetheless, both the *trans*-conformation and β-strands are regular and therefore

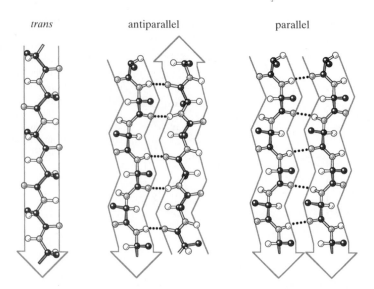

Figure 1.25 Structures of the *trans*-conformation and β-sheets in proteins. The *trans*-conformation is the fully extended form of the polypeptide backbone, with all peptide bonds lying in the same plane. The strands of β-sheets are twisted slightly from this fully extended conformation. Parallel β-sheets are oriented with N- to C-terminus of the strands (arrows) pointing in the same direction, while antiparallel β-sheets have strands oriented in opposite directions.

are helices. Thus, our statement is correct that the only symmetric 3D structure (i.e., secondary structure) that can be built from a linear chain of chiral building blocks is a helix.

1.5.5   Effect of the Peptide Bond on Protein Conformations

The structure of the peptide bond restricts the possible conformations that can be adopted by a protein and restricts the number of freely rotating bonds along the backbone of polypeptides. The C—N bond of the peptide linkage is a partial double bond, and therefore is not freely rotating. The peptide bond can adopt either the *trans*- or the *cis*-configuration (Figure 1.21), much like that of *trans*- and *cis*-1,2-dichloroethylene. The *cis*-configuration is energetically unfavorable because of collisions between the side chains of adjacent residues. Proline is an exception where the two configurations are nearly isoenergetic; the *trans*-configuration is slightly favored (4:1) under biological conditions.

Fixing the peptide bond in one of two configurations leaves only two free bonds along the backbone, one on either side of the C_α carbon. This defines the two torsion angles for the backbone; the ϕ angle at the amino nitrogen-C_α bond and the ψ angle at the C_α-carboxyl carbon bond. The noncovalent interactions between adjacent side chains place energetic constraints on these torsion angles (see Chapter 3). The C_β carbon of alanine represents the minimum interactions that exist between the side chains of the common amino acids. The ϕ and ψ angles of poly-L-alanine, therefore, represent the possible backbone conformations of a typical protein. A graphical plot of the steric energies of interaction as a function of the ϕ and ψ angles was first developed by Ramachandran and Sasisekharan (1968) to represent the sterically allowed conformations of a polypeptide chain (Figure 1.26). The torsion angles of the helical structures are all centered in the allowed regions. Alanine, however, is not a good model for glycine or proline, and is a poor approximation for amino acids with bulky side chains. The side chain of glycine is a hydrogen and thus the torsion angles of this amino acid are not restricted by a C_β carbon. The three-carbon side chain of proline, in contrast, forms a five-member ring with the C_α carbon and amino nitrogen of the backbone; the ϕ angle is therefore constrained at a value of $-60°$, while the ψ angle has two minima at about $-55°$ and about $145°$. Chapter 3 contains a more thorough discussion of the energies of interactions that lead to sequence effects on protein conformations.

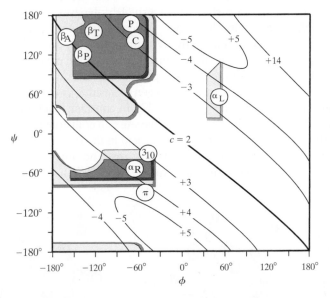

Figure 1.26   Ramachandran plot. The van der Waals interaction energies of an internal Ala residue in a polypeptide chain represents the values of ϕ- and ψ-angles that can be adopted by a typical amino acid residue in the chain. A ϕ,ψ-plot of the energies shows conformations that are sterically allowed (dark-shaded regions), moderately allowed (light-shaded regions), and disallowed (open regions). The curves through the plot represent the angles for helices with a particular repeat c. The secondary structures that are found in proteins have torsion angles that lie within or near allowed and moderately allowed regions of the ϕ,ψ-plot. These include the structures of the right-handed α-helix (α_R), left-handed α-helix (α_L), 3_{10} helix, π-helix, parallel β-sheets (β_P), antiparallel β-sheets (β_A), twisted β-sheets (β_T), polyproline (P), and collagen (C). The helical repeat for protein secondary structures follow the contours through the plot (positive values of c are for right-handed helices and negative values are for left-handed helices).

1.5.6   The Structure of Globular Proteins

The fibrous structural proteins, such as α- and β-keratins and collagen, contain very long helices. Water-soluble and membrane proteins, however, are more globular in shape and therefore must adopt more complex folds. The overall global conformation of a protein is its tertiary structure. Although tertiary structures are highly varied across different proteins, there remains some degree of regularity within them, because local regions adopt regular secondary structures. Groups of helices associate to form supersecondary structures, which are motifs that recur frequently in many different proteins. Finally, the structure of a protein can often be segregated into domains that have distinct structures and functions.

Supersecondary structures.   The structural motifs formed by the association of two or more helices are categorized as supersecondary structures because they are regularly occurring patterns of multiple helices. The simplest example is a β-sheet formed by two parallel or antiparallel β-strands held together by hydrogen bonds (Figure 1.25). Let us consider the antiparallel sheet first. To form a two-stranded antiparallel β-sheet from a single polypeptide, the chain must make a complete 180° change in direction. Typically, the amino acid residues linking the C-terminus of one strand to the N-terminus of the second must make a tight turn. There are two types of β-turns typically observed in protein structures. These are distinguished by the rotation of the peptide bond between the second and third residues. In a type I β-turn, the carbonyl oxygen of residue two is pointed away from the side chains of residues 2 and 3 (Figure 1.27). In contrast, in a type II β-turn this peptide is rotated in the opposite direction. This causes a collision between the carbonyl oxygen and the C_β carbon of the third amino acid in the turn. For this reason, glycine is often observed at this position in type II β-turns. The other common β-turn, the type III β-turn, is simply one turn of a 3_1-helix (Figure 1.23).

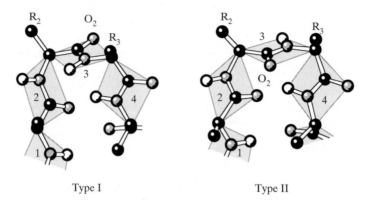

Type I   Type II

Figure 1.27   Type I and type II β-turns. The tight turns formed by four amino acids residues can have the keto oxygen (O_2) of the second residue in the turn pointing away from the flanking side chains R_2 and R_3 (type I β-turn) or in the same direction as the side chains (type II β-turn).

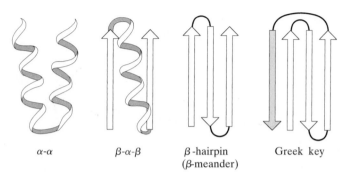

α-α β-α-β β-hairpin   Greek key
                                                    (β-meander)

Figure 1.28   Supersecondary structures in proteins. Shown are examples of recurring motifs composed of α-helices and β-sheets that are observed in globular proteins.

The antiparallel β-sheet can be extended in a simple regular pattern to form a motif known as a *β-hairpin* or, for longer stretches of antiparallel β-sheets, a *β-meander* (Figure 1.28). A more complex arrangement places the first, second, and third strands in a standard β-meander motif, but the fourth strand along the chain folds back to hydrogen bond to the first. The trace of the chain through the strands of the β-sheet and the connecting loop resemble the repeating motif that decorates vases from ancient Greece and is thus called a *Greek key* motif. These extensive sheets allow the hydrogen-bonding atoms in the peptide backbone of β-strands to be efficiently accommodated. Notice that a single β-strand would have two faces of the backbone open to form hydrogen bonds. Bringing two strands together accommodates one hydrogen-bonding face of each strand. Strands can be added without introducing any more additional open hydrogen-bond faces to the sheet. We will see later that intramolecular hydrogen bonding is greatly disfavored in aqueous environments but is favored in nonpolar environments. Thus, these extensive β-sheets are favored in the interior of a globular protein or in the protein regions buried within the bilayer of a membrane. The ultimate extension of a β-sheet is to associate enough strands so that the first strand can hydrogen-bond with the last to form a closed structure. This closed structure is a β-barrel; all the hydrogen-bonding faces of the strands are entirely accommodated by the sheet, leaving none to associate with water.

A two-stranded parallel β-sheet formed by a single polypeptide chain must incorporate a longer stretch of amino acids between the β-strands than found in the β-turns. A long loop must be introduced to span at least the length of one β-strand, and is often observed to adopt an α-helix. This is a β-α-β motif formed by a β-strand followed by an α-helix followed by a β-strand. Notice in this case that even though there is an intervening helix, the two β-strands are hydrogen-bonded to each other. Finally, there are supersecondary structures that involve only α-helices (e.g., the α-helix bundles such as the α-α pair in Figure 1.28).

Domains.   A polypeptide chain can fold into a number of distinct structural and functional regions larger than supersecondary structures called *domains*. A dramatic example of distinct domains is seen in calmodulin (Figure 1.29). This protein

Figure 1.29　Crystal structure of calmodulin. The two calcium binding domains are separated by a long α-helix.

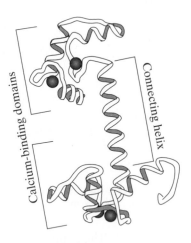

has two calcium-binding domains connected by a single long α-helix. The activation of the protein by calcium binding is associated with a change in the length of this intervening helix, which brings the two domains closer together. Domains are distinguished as much by function as they are by structure. Many enzymes, such as tRNA synthetase, have separate *functional domains* for binding substrates, binding effectors, and interacting with other proteins.

Tertiary structure.　The tertiary structure of a globular protein is the overall three-dimensional conformation of the polypeptide chain. There are a number of ways to represent this structure, each with different degrees of detail. Remember that a structure is really only a model that best fits the physical properties determined for that molecule. The most detailed representation of the structure is a list of atoms in the molecule, specifying the element and the location as (x, y, z) coordinates for each atom (Figure 1.30). However, a list is not a picture of the molecule's 3D shape; does not show where the helices, supersecondary structures, and domains are located; and does not illustrate how these parts are related to each other.

A list of atomic coordinates can be interpreted by molecular graphics programs to allow the visualization of the 3D structure of the molecule with varying degrees of detail. The models developed by Corey and Pauling and later improved upon by Kultun (CPK models) accurately depict the size and shape of molecules by placing each atom at its proper position in 3D space and representing each atom by a sphere at the van der Waals radius. However, the majority of the atoms lie in the interior of a globular protein and thus are hidden by the atoms at the surface of a CPK model. By removing the van der Waals spheres, we can see all the atoms both in the core and the surface. To help visualize the helical structures of the protein, the backbone can be traced by a ribbon, but there may still be too much information to see the secondary structures. This information glut can be further reduced by eliminating the atoms of all but the most critical side chains (for example, those at the active site of an enzyme). Finally, the atoms of the backbone can themselves be replaced by symbols that represent the various types of secondary structures

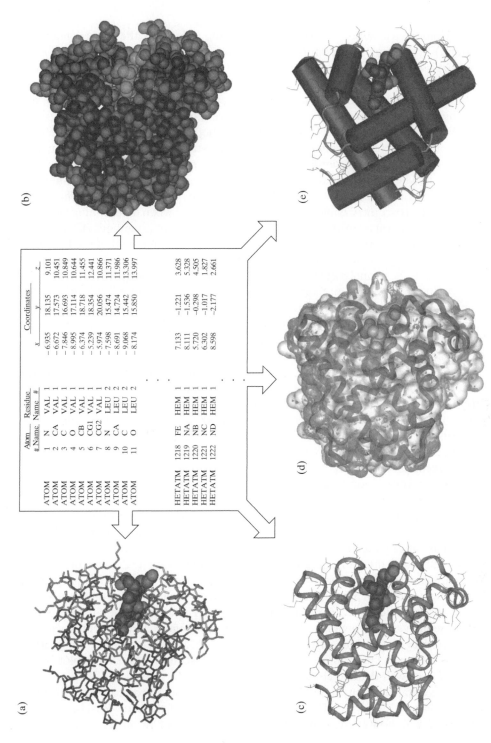

The table shown in the figure:

ATOM	#	Atom Name	Residue Name	#	Coordinates x	y	z
ATOM	1	N	VAL	1	−6.935	18.135	9.101
ATOM	2	CA	VAL	1	−6.672	17.573	10.451
ATOM	3	C	VAL	1	−7.846	16.693	10.849
ATOM	4	O	VAL	1	−8.995	17.114	10.644
ATOM	5	CB	VAL	1	−6.374	18.718	11.455
ATOM	6	CG1	VAL	1	−5.239	18.354	12.441
ATOM	7	CG2	VAL	1	−5.974	20.056	10.866
ATOM	8	N	LEU	2	−7.598	15.474	11.371
ATOM	9	CA	LEU	2	−8.691	14.724	11.986
ATOM	10	C	LEU	2	−9.068	15.442	13.306
ATOM	11	O	LEU	2	−8.174	15.850	13.997
HETATM	1218	FE	HEM	1	7.133	−1.221	3.628
HETATM	1219	NA	HEM	1	8.111	−1.536	5.328
HETATM	1220	NB	HEM	1	5.720	−0.298	4.505
HETATM	1221	NC	HEM	1	6.302	−1.017	1.827
HETATM	1222	ND	HEM	1	8.598	−2.177	2.661

Figure 1.30   Representing the structure of a protein molecule. The structure of a macromolecule is a list of atoms and their (x, y, z) coordinates. This set of *atomic coordinates* can be interpreted to give (a) a stick model, (b) a CPK or van der Waal's surface model, (c) a ribbon model, (d) a solvent accessible surface model, or (e) a simple caricature of the molecule.

(e.g., cylinders for helices; arrows pointing from the N- to the C-terminus of a β-strand) present in a protein. Now by simple inspection of the simplified visual model, it is easy to locate the secondary structures, supersecondary structures, and domains that constitute the tertiary structure of a protein. The most detailed presentation of a molecular model, therefore, may not be the most informative, depending on what features of the structure we are most interested in.

Visual molecular models are not the only methods used to represent the conformation of a protein. An alternative is to define structural features such as helices analytically, using atomic coordinates to assign the structure of each amino acid. This is important when trying to assess the number of amino acids that adopt an α-helix or β-strand conformation in a structure determined from X-ray diffraction (Chapter 6) or NMR spectroscopy (Chapter 12), and is in turn useful when determining how well structure-prediction algorithms (Chapter 4) and spectroscopic methods (Chapter 10) predict the helical structures of a protein. Finally, the simple model of the helices in Figure 1.30 requires first that the amino acids be assigned as being helical or nonhelical. This requires an analytical tool for protein structure analysis.

A simple analytical method to assigning secondary structure is to calculate the ϕ and ψ angles at each C_α carbon along the chain. Although this is an accurate representation of the protein conformation, it may not be definitive in its assignment of helices. The average torsional angles for α-helices, and parallel and antiparallel β-sheets of actual protein structures will generally fall within the expected values for each structure, but are broadly distributed for individual amino acid residues. The regions of the Ramachandran plot for these structures greatly overlap, particularly for the extended structures such as β-sheets, polyproline, and collagen-type conformations. Kabsch and Sander (1983) used the principle that helices and β-sheets are held together by hydrogen bonds, and indeed have traditionally been classified according to their hydrogen-bonding patterns, to derive a method for assigning secondary structures. These analytical methods, however, do not provide information beyond secondary structure.

An alternative method to represent the overall conformation of a globular protein from primary to tertiary structure is through a *contact* or *diagonal plot*. Amino acids along the polypeptide chain that are in close contact in space are plotted to generate a pattern that reflects the regular secondary structures and some tertiary contacts of the protein. The plot is generated by first aligning the protein sequence along the horizontal axis (residues *i*) and along the vertical axis (residues *j*). The distance between any two amino acids (*i* and *j*) is calculated as the distance between, for example, the C_α carbons of the amino acid residues. Within this sequence-versus-sequence grid, a point is placed for each amino acid *i* that falls within a specified range of distance from amino acid *j*. To demonstrate the utility of the method for analyzing protein structures, we will first show how such a plot is constructed from well-defined conformations.

The distance of any amino acid to itself is 0 nm and produces a set of points along a diagonal for *i* = *j*. This is the main diagonal of the plot. Residues *i* and *j*, with *i* ≠ *j*, will show mirror symmetry across this diagonal. In subsequent examples, we will use a distance of 0.4 nm to 0.6 nm between C_α carbons as the range where amino acids are considered to be in close contact. This range excludes residues that directly follow each other along the chain, because their distance is less than 0.4 nm, but, as we will see, it includes amino acids that form α-helices and β-sheets.

For an α-helix, an amino acid at position i is within 0.4 nm to 0.6 nm of residues $j = i + 3$ and $j = i + 4$ (Figure 1.31a). The pitch of the helix is 0.54 nm for a repeat of 3.6 residues per turn. Thus, seven residues that form approximately two turns of an α-helix would generate a set of points that parallel the main diagonal but is displaced by an average of 3.5 amino acids from this diagonal. This is the signature of an α-helix.

Parallel and antiparallel β-sheets each have a unique pattern. The contact plot for two antiparallel β-strands consisting of four amino acids each, with four residues forming the connecting β-turn, is shown in Figure 1.31b. Notice that this pattern of contact points lies perpendicular to the main diagonal. These same strands placed in a parallel arrangement generate a set of points parallel to the main diagonal. In this case, the points are displaced from the diagonal by N residues, where N is the number of amino acids in the loop that connects the C-terminus of the first strand with the N-terminus of the second.

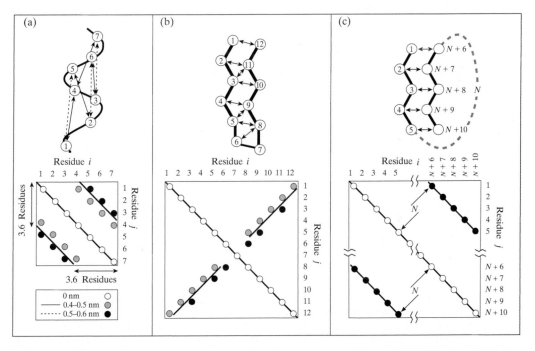

Figure 1.31   Contact plots for α-helix and β-sheets. The points of contact for the residues in (a) an α-helix, (b) antiparallel β-sheet, and (c) parallel β-sheet give characteristic patterns that are the signatures of each form of secondary structure.

A contact plot can also provide some information on the tertiary structure of a protein. For example, if two α-helices are adjacent, there would be additional points to indicate which residues are in close contact. A point or set of points at the far corners of the plot is evidence that the N- and C-terminae of the polypeptide chain are in close proximity in space. This method therefore provides a means to assign secondary, supersecondary, and tertiary structures to amino acids starting with a known conformation of a protein.

Working backward, an experimentally determined contact plot becomes a useful tool for following structural transitions or to assign structural features using data from physical methods that measure distances between amino acids (e.g., by multidimensional high resolution NMR, see Chapter 12) (Figure 1.32). In this latter case, we would first look for signature elements of secondary structures (similar to those seen for α-helices and β-sheets in Figure 1.31). Any additional points that cannot be assigned to secondary elements are therefore associated with the tertiary structure and can be used to construct a complete 3D structure of the molecule.

Protein folds. Each individual polypeptide chain has an associated unique tertiary structure; however, it is now recognized that these unique overall structures may be constructed from a limited number of recurring topological motifs of helices and/or sheets known as *protein folds*. Currently (2003) there are 498 common folds that make up the structures in the protein database (over 21,000 protein structures),

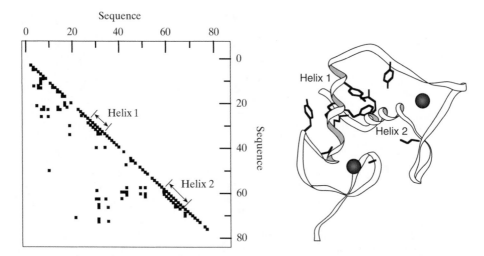

Figure 1.32　Contact plot of a protein from NMR. The relative distances between amino acid residues are obtained from multidimensional high-resolution NMR spectroscopy. When represented as a diagonal plot, these distances provide information to help define the secondary and tertiary structures of the protein. The contact plot for the *zinc-finger* domain of the estrogen receptor shows two α-helical regions, one short β-sheet region near the N-terminus, and two β-turns where the zinc ions bind. A 3D model of the protein is constructed using this distance information and the allowed geometries of amino acids. The structure apparently is stabilized by the zinc ions, as well as the packing of aromatic and alkyl side chains to form a hydrophobic core. [Adapted from Schwabe et al. (1990), *Nature* **348**, 458–461.]

and can be considered the basic building blocks of all tertiary structures. The folds may be structural (as in the TIM barrel, a set of repeating α-β-α motifs originally identified in the structure of triose isomerase, or TIM) or functional (as in the Rossmann fold, which forms a unique topology for binding nucleotides) (Figure 1.33). It is not entirely clear where protein folds fall in the hierarchy of protein structure, since it can be as small as a supersecondary structure, or could be an entire protein tertiary structure. The folds that can be seen within a protein structure, however, can be used to facilitate the classification of proteins into functional or structural classes, and is increasingly used as the modules for predicting tertiary structures of new polypeptide sequences with unknown structures through homology to sequences with known structures.

Quaternary structure.   The noncovalent association of polypeptides to form a multimeric complex defines the quaternary structure of a protein. This level of structure is described first by the types and number of polypeptide subunits. For instance, a complex formed by two identical subunits is a *homodimer,* while that formed by two nonidentical subunits is a *heterodimer.* A classic illustration of the significance of quaternary structure to a protein's function is the comparison of myoglobin and hemoglobin. They both bind oxygen, but they have evolved to perform very specific functions in the body. Myoglobin is responsible for oxygen storage in muscle cells, while hemoglobin serves as a transport protein to carry oxygen from

Figure 1.33   Protein folds. The TIM barrel (*left*) and Rossmann fold (*right*) are commonly occurring topologies seen in many proteins within a structural or functional class. On the left, the one repeating "fold" that is shaded is repeated three times to generate this TIM barrel. The structure of the Rossmann fold is shown with a dinucleotide substrate bound within the binding pocket of the fold.

the lungs to the various tissues, including the muscles. Their primary, secondary, and tertiary structures are nearly identical in all respects, from their amino acid sequences to their helical structures to the global folding of their polypeptide chains to form the unique oxygen-binding pockets to the heme prosthetic group that binds oxygen. They differ in that myoglobin is a monomer, while hemoglobin has the quaternary structure of a heterotetramer composed of two α-chains and two β-chains, or an $\alpha_2\beta_2$ tetramer. Structurally, hemoglobin would seem to be a tetrameric complex of molecules that individually appear to be very similar to myoglobin. This is not how it functions, however. Although hemoglobin has the capacity to bind four oxygen molecules, as would be expected for the tetramer, it has a lower affinity for oxygen than does myoglobin, and the binding of the ligand to the tetramer is cooperative. Thus, the chains of hemoglobin interact and communicate within the tetramer.

The communication among subunits in a protein complex is highly dependent on how they are organized. The complexes of identical or nearly identical subunits are typically symmetric. A complete description of the quaternary structure of a protein such as hemoglobin must therefore include the symmetry relationships among the subunits. Using the point groups described previously, the hemoglobin molecule shows C_2 symmetry (Figure 1.34). This relates the two α-subunits and the two β-subunits by 180° rotations. The α- and β-chains are similar, but not identical. These are related by pseudo-C_2 symmetry axes. The binding of oxygen to one subunit breaks the true C_2 symmetry of the tetramer not only because of the presence of the oxygen, but also because of the conformational changes that are induced. When oxygen binds to one subunit, the iron of the heme is pulled out of the plane of the porphyrin, which tugs on the distal histidine. This places a strain on the overall conformation of the polypeptide chain. In myoglobin, this results in a dramatic change in the tertiary structure of the protein. The subunit binding an oxygen in the hemoglobin tetramer cannot undergo such a dramatic transition in its conformation because of its interactions with the three unbound subunits, which in turn try to maintain the overall symmetry of the quaternary structure. This resistance to the loss of symmetry may account for the low affinity of hemoglobin for the first oxygen molecule. Indeed, after two oxygens are bound, the overall quaternary structure

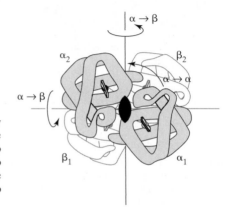

Figure 1.34 Symmetry in hemoglobin quaternary structure. The tetramer of hemoglobin shows true C_2-symmetry, with the two-fold axis perpendicular to the plane of the page. This relates α-subunit to α-subunit, and β-subunit to β-subunit. There are two pseudo two-fold symmetry axes relating α- to β-subunits.

changes in a concerted manner, but the tetramer remains in the C_2 point group. The tetramer therefore appears to have a thermodynamic propensity to remain symmetric (see Chapter 15).

When proteins exist in large clusters, higher order symmetries are possible (Table 1.8). For example, the 10-subunit cluster of hemocyanin is in the D_5 point group, while proteins such as hexokinase have helical symmetry. The polymerized tetramers of sickle-cell hemoglobin that form rigid rods also have helical symmetry.

We can ask the question: Why should protein subunits associate in clusters that are regular and symmetric? If we think about interaction sites in terms of maximizing the number of possible interactions with the minimum number of sites, we can begin to understand why quaternary structures are generally symmetric. Let's start with a simple model in which two identical subunits associate to form a stable dimer. Dimerization requires that two molecules, each having three degrees of rotational freedom, associate to form a single complex. The resulting complex has only three degrees of rotational freedom (the extra three degrees of freedom become vibrations), resulting in a loss of entropy equivalent to 5.5 J/mol K. To overcome the negative entropy of dimerization, we would need to maximize the number of energetically favorable interactions between the subunits. Let us now imagine a subunit with both a site A and its complement A' on its surface. Two such subunits can form a dimer through an A to A' interaction. For this example, we will arbitrarily assign an energy for this interaction to be -5 kJ/mol. In a simple nonsymmetric scheme, the two subunits can dimerize through a single A to A' interaction (Figure 1.35). The interaction energy holding this dimer together would thus be -5 kJ/mol, or -2.5 kJ/mol for each subunit. However, there is an A site of one subunit and an A' site of its partner that are not paired, and thus the potential for interactions is wasted.

In order to maximize the stability of the complex, we can require that each A site pair to an A' site. Thus, the interaction energy would be -10 kJ/mol for the dimer, or -5 kJ/mol per subunit. This automatically defines the relationship of the two subunits as C_2 symmetric. We see that a symmetric dimer effectively maximizes the number of interactions per subunit (Figure 1.35a). If the interaction site A lies 90° relative to A' on each subunit, the complex formed would be a tetramer having four-fold symmetry. Again, this symmetric arrangement maximizes the number of interactions holding the tetramer together such that there is on average one paired interaction site per subunit.

Table 1.8   Examples of Point Group Symmetry in the Quaternary
Structures of Protein Complexes

Protein	Point Group Symmetry
Alcohol dehydrogenase	C_2
Prealbumin	C_2
Hemerythrin from *Phascolopsis gouldii*	D_4
Hemocyanin	D_5
Viral coat proteins	Icosahedral

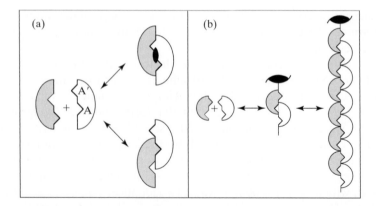

Figure 1.35 Symmetric versus nonsymmetric association of subunits. (a) A possible symmetric and non-symmetric association of two subunits, each having an interaction site A and its complement A'. In the symmetric association, there are two A-A' interactions, while the nonsymmetric complex has only a single A-A' interaction. (b) Subunits related by two-fold screw symmetry would each have two A-A' interactions, except for the subunits at the two ends, which have only a single such interaction.

Forming a dimer such that each subunit is related by screw symmetry as in Figure 1.35b creates only a single *A* to *A'* interaction. In this case, dimerization to a symmetric complex shows no advantage over the asymmetric dimer, and therefore is the same energetically. For screw symmetry, however, each subsequent subunit added to the complex creates an additional *A* to *A'* interaction. When the complex is extended to an infinitely long polymer, the average number of interactions approaches one per subunit. Only the two subunits at the ends will not reach their maximum potential for interactions. As we see in Chapter 4, this type of behavior, in which the initial steps of a process are energetically unfavorable but subsequent steps are favorable, can be easily described and analyzed by a statistical mechanics model. This model is also applicable to the transitions from coils to helices in proteins.

1.6   THE STRUCTURE OF NUCLEIC ACIDS

Nucleic acids are the functional forms of polynucleotides. The structure of the polymers of nucleotides follow many of the same principles described above for polypeptides and proteins, but with some significant differences. First, there are fewer types of nucleic acid building blocks (Figure 1.36). The building blocks are separated into two major groups, as distinguished by the sugar component of the nucleic acid: Ribonucleic acid (RNA) contains a ribose sugar, while 2'-deoxyribonucleic acid (DNA) has a 2'-deoxyribose sugar ring.

The polymers of RNA and DNA are normally constructed from four types of monomers, each distinguished by the *nucleobase* attached to the sugar. The nucleobases are derivatives of either purine or pyrimidine. The purine nucleobases of

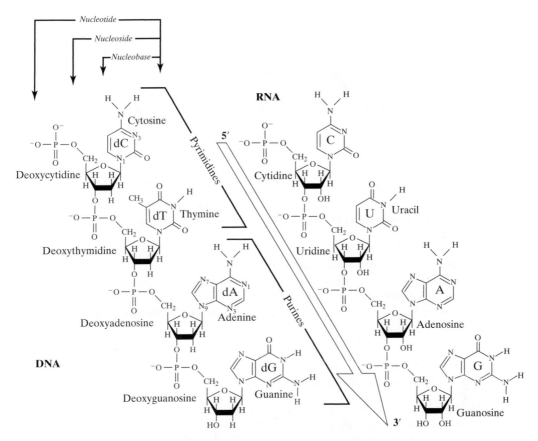

Figure 1.36   Nucleic acids. Ribonucleic acids (RNA) and 2'-deoxyribonucleic acids (DNA) are polymers constructed from nucleotide monomers. The nucleotides are distinguished by the nucleobases, which can be either of the pyrimidine or of the purine type. The polynucleotide sequence is a chain that extends from the 5'-terminus to the 3'-terminus.

both RNA and DNA are guanine and adenine. However, in RNA, the pyrimidine nucleobases are cytidine and uridine, while for DNA they are cytidine and thymine. There are a number of modified nucleotides in the cell. For example, C5-methyl-2'-deoxycytosine (dm^{5}C) accounts for about 3% to 5% of the cytosine in human DNA. The structure of tRNA requires a number of modified bases, including pseudouridine and O6-methylguanine, that affect the hydrogen-bonding interactions important for its tertiary structure.

When linked into a polymer, the resulting polynucleotide chain is highly flexible, with the number of torsion angles for each nucleotide monomer along the polynucleotide backbone being far greater than for the amino acids along a polypeptide backbone (Figure 1.37). The permissible secondary structures are all helical. Double-stranded polynucleotides most frequently involve pairing across strands, rather like the strands of a β-sheet. The tertiary structures of single-stranded polynucleotides,

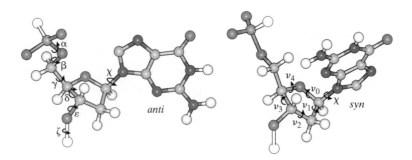

Figure 1.37   Torsion angles in nucleic acids. The structures of nucleic acids are defined by the torsion angles along the phosphoribose backbone (α to ζ), the torsion angles within the sugar ring (v_0 to v_4), and the rotation of the nucleobase relative to the sugar (χ). Rotation about χ places the base either extended from the ribose (*anti*-conformation) or sitting above the ring (*syn*-conformation).

particularly RNA molecules, are very similar to those of polypeptides. Double-stranded DNAs do not show the convoluted folds of single-stranded RNA, but do have well-defined topologies when they are formed into closed circles. We will review the structural properties of nucleic acids here. A more complete treatment of how nucleic acids are stabilized is presented in Chapter 3.

1.6.1   Torsion Angles in the Polynucleotide Chain

The number of possible conformations for RNA and DNA is much greater than that found in proteins, because the number of torsion angles in the phosphoribose backbone is much greater, as shown in Figure 1.37. As with polypeptides, however, there are some restrictions imposed by the chemical bonding. The conformation of the ribose or deoxyribose sugar is characterized by the *sugar pucker,* which defines the direction in which the five atoms of the ring deviate from a plane (Figure 1.38). The sugar pucker is typically defined by the positions of the C2′ and C3′ atoms relative to a plane formed by the C1′, O4′, and C4′ atoms. The face of this plane toward the glycosidic bond is the *endo* face, while the opposite face toward the O3′ oxygen is the *exo* face. A sugar with the C2′ carbon puckered toward the *endo* face is in the C2′-*endo* conformation; one with the C3′ carbon toward the *exo* face is in the C3′-*exo* conformation.

The rotations about the bonds of the sugar ring are not rigidly restricted as in the peptide bond, but are correlated in that any rotation of one bond requires concomitant rotations about the other four bonds to maintain a nonplanar conformation. The torsion angles within the five-member ring can thus be treated as a single set of correlated angles called the *pseudorotation angle* Ψ and is defined as

$$\Psi = \tan^{-1}\left[\frac{(v_4 + v_1) - (v_3 + v_o)}{2v_2(\sin 36° + \sin 72°)}\right] \tag{1.17}$$

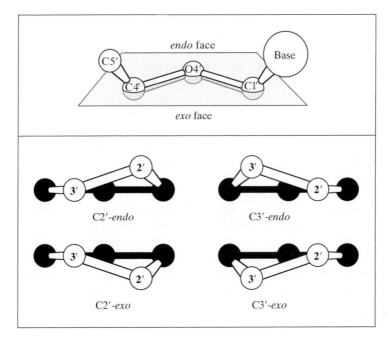

Figure 1.38 Sugar conformations of nucleic acids. The pucker of the sugar ring in RNA and DNA is defined relative to the plane formed by the C1'-carbon, C4'-carbon, and O4'-oxygen of the five-member ring. The *endo* face lies above the plane, toward the nucleobase, while the *exo* face lies below the plane.

The two major sugar conformations are defined as C3'-*endo* for $0° \leq \Psi \leq 36°$ and C2'-*endo* for $144° \leq \Psi \leq 190°$.

The other torsion angle that strongly affects the structure of nucleic acids is the χ-angle around the glycosidic bond (Figure 1.37). The restrictions to rotation around this bond are imposed by the steric clashed between the nucleotide base and the sugar ring. The base can be oriented either *anti* ($-180° \geq \chi \geq -90°$ and $+90° \leq \chi \leq +180°$) or *syn* ($-90° \leq \chi \leq +90°$). In the *anti* conformation, the base extends away from the ribose ring, while in *syn*, the base lies on top of the ring. The base oriented in the *syn* conformation forms a more compact structure, but is sterically hindered, particularly for pyrimidine bases. In a pyrimidine base, the sugar is closest to a six-member aromatic base, while in purines, the smaller five-member ring is closest to the sugar.

Other than these correlated and sterically restricted rotations, the backbones of polynucleotides are generally free to adopt a variety of conformations.

1.6.2   The Helical Structures of Polynucleic Acids

The secondary structures of DNA and RNA are all helical. Unlike polypeptides, however, the regular helices of polynucleotides usually have a minimum of two strands (either from different polynucleotide chains or different regions of the same chain) that are hydrogen-bonded together to form base pairs, base triplets, and even quadruplets. In terms of information storage and transmission, the most important of these are the base pairs found in the genomic DNA of most cells. In cells, the DNA molecule usually exists as two polynucleotide strands held together to form a *duplex*. In a DNA duplex, dG is paired with dC and dA is paired with dT as *Watson-Crick base pairs* (Figure 1.39) to form an antiparallel double-stranded molecule. The sequence of one strand automatically defines the sequence of the opposite strand (the complementary strand). Thus, although double-stranded DNA is composed of two polymers, it behaves as a single molecule; the size of a DNA is therefore characterized by the number of nucleotides or by the number of base pairs in the genome.

  Base pairing is also important in RNA, but because these are typically single-stranded molecules, it is intramolecular, holding various local regions into well-defined secondary and tertiary structures, much like we have already seen with polypeptides. The number and types of base pairing that exist in compactly folded RNA structures, including the structures of tRNA and mRNA, are more varied. They range from the standard Watson-Crick base pairs that hold the helices together in DNA, to mismatched purine-purine, pyrimidine-pyrimidine, and reversed base pairs

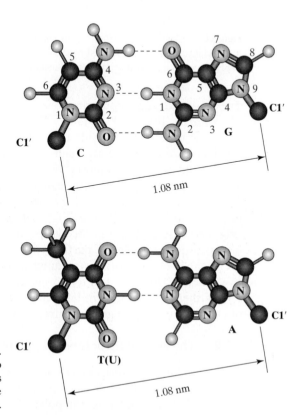

Figure 1.39 Watson-Crick base pairs. Bases interact by hydrogen bonds to form base pairs. The standard base pairs in double-stranded nucleic acids are the C · G and T · A Watson-Crick base pairs.

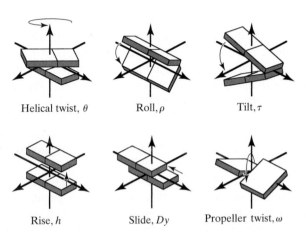

Figure 1.40   Base-pair and base-step parameters of nuclei acid double helices. The structure of double-stranded nucleic acids are defined by the relative conformations of two adjacent base pairs in a base step (e.g., helical twist, roll, tilt, rise, and slide) and the relative conformations of the bases in a base pair (e.g., the propeller twist).

that are not observed in standard duplex DNA, but are important for defining the tertiary structures of RNA molecules.

The backbone conformations are correlated with the conformation of the base pairs, accommodating the twisting, shifting, and sliding of bases and base pairs relative to each other in the helix. The detailed conformation of helical polynucleotides, there-fore, are typically described in terms of the structural relationships between the bases within a base pair, and between base pairs in a stack of base steps. The definitions of these base-pair and base-step structural parameters are summarized in Figure 1.40.

The standard structure of DNA is that of B-DNA and for RNA is A-RNA (Figure 1.41). However, both base-paired double helices can adopt a number of

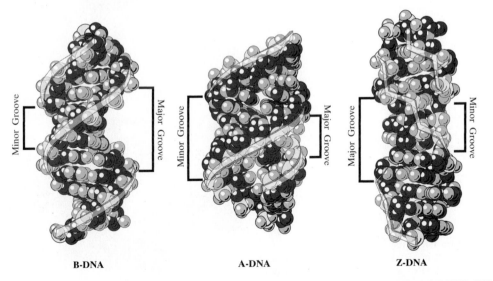

Figure 1.41   Structures of B-DNA and the two alternative double-helical forms of A-DNA and Z-DNA. RNA du-plexes adopt structures similar to A-DNA and rarely to Z-DNA.

different forms—that is, they are *polymorphic*. DNA, for example, can adopt either the standard B-form, the A-form similar to RNA, or even a left-handed form called Z-DNA. From early X-ray diffraction studies on the fibers of naturally occurring and synthetic DNAs, it was clear that the structure of the double helix is strongly dependent on the hydration of the molecule. High humidity (and standard aqueous solutions) favors B-DNA, while low humidity, alcohols, and salts favor the alternative structures of A- and Z-DNA. There is also a strong sequence dependency for the stability of each type of DNA, with Z-DNA favored in alternating sequences of dG·dC base pairs and A-DNA favored in nonalternating dG·dC sequences.

The structures of A-, B-, and Z-DNA have all been studied by X-ray diffraction on fibers, X-ray diffraction on single crystals grown from oligonucleotides, and in solution by NMR, circular dichroism (CD), and linear dichroism. The model of B-DNA was first built by James Watson and Francis Crick (1953) using the fiber-diffraction photographs of Rosalind Franklin (Figure 1.42). B-DNA is characterized as a right-handed antiparallel double helix held together by Watson-Crick base pairs. The structure from fiber diffraction has a rise of $h = 0.34$ nm per base pair and helical repeat of $c = 10$ base pairs per turn. The nucleotides are all in the *anti* conformation and the sugars are primarily C2'-*endo*. The resulting duplex has a wide major groove, which is used by proteins to recognize the DNA sequence, and a narrow minor groove, which is lined by a series of tightly bound water molecules that form a continuous spine. This is the standard form of DNA in aqueous

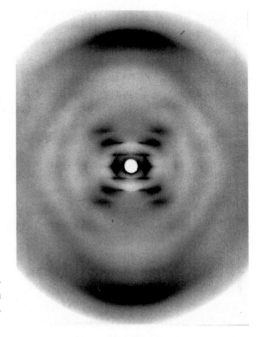

Figure 1.42 Fiber diffraction photograph of B-DNA. X-ray diffraction from a fiber of the lithium salt of B-DNA at 90% humidity. [Courtesy of R. Langridge.]

solution, and for most genetic analyses is sufficient to account for the behavior of genomic sequences.

Single-crystal structures and solution studies support the basic Watson-Crick model for B-DNA but show it to be highly variable. The helical repeat in crystals of B-DNA range from $c = 10.0$ to 10.5 base pairs per turn. In solution, the average repeat is about 10.5 base pairs per turn. In addition, mismatched base pairs occur with variable frequency. A mismatched base pair that is not repaired correctly by the cell will result in mutations along the sequence.

A-DNA is a shorter, broader right-handed cousin to B-DNA. This results from the base pairs being highly inclined relative to the helix axis and a subsequently short rise of only 0.26 nm/base pair. The broadness of this helix comes from the displacement of the base pairs away from the helix axis. When viewed down the axis, A-DNA appears to have a channel through its middle. The bases are again all in the *anti* conformation, and the two strands held together by Watson-Crick base pairing. The sugar pucker, however, is *C3'-endo*. The major groove of the helix is narrower and the minor groove wider than found in B-DNA. Proteins decipher the base-specific information of the double helix through the hydrogen-bond donor and acceptor functions in the major groove; this information is largely inaccessible in the A-conformation.

In contrast, Z-DNA is a narrower and longer left-handed variant of the DNA duplex. Interestingly, the first structure of an oligonucleotide determined from X-ray diffraction of a single crystal was that of Z-DNA. Although the structure formally has a helical repeat of -12 base pairs per turn (this is negative because it is left-handed), it is better described as six repeating dinucleotide units. The nucleotides alternate between the *anti* and the more compact *syn* conformations. The *syn* conformation is sterically restricted in pyrimidine bases and thus Z-DNA is characteristically formed in alternating *anti* pyrimidine/*syn* purine sequences, primarily alternating sequences of d(CpG/CpG) or d(CpA/TpG) dinucleotides. The sugar pucker also alternates between *C2'-endo* for the pyrimidines in *anti*, and *C3'-endo* for the purines in *syn*. All of these alternating structural features give the double helix a zig-zag backbone, which is the root of its name—the Z refers to this zig-zag backbone.

The significance of these alternatives to B-DNA in the cell has been a source of much discussion and speculation. The d(AT)-rich promoter DNA sequence in eukaryotes (the TATA-box) is induced into a highly inclined and displaced structure similar to A-DNA upon binding of the promoter protein. Similarly, the transient DNA/RNA hybrid formed during transcription and the mixed RNA-DNA chain that initiates DNA replication has been shown to be in the A-form by CD, as expected from the strong influence of the RNA nucleotides on these structures.

The biological function of Z-DNA, on the other hand, has been more widely debated, probably because of the radical nature of its structure. Z-DNA differs so much from standard B-DNA that it is not recognized by most DNA-binding proteins. This includes endonucleases, histones that form chromatin structure, and polymerases that transcribe DNA. However, a domain of double-stranded RNA deaminase, an RNA-editing enzyme, has recently been observed to bind specifically to left-handed DNA. This leads to the hypothesis that Z-DNA helps to situate editing enzymes close to the growing mRNA chain and, at the same time, allows the enzyme to recognize which nucleotide to edit by virtue of its distance from the transcribing RNA polymerase.

Although A-, B-, and Z-conformations are currently the best characterized structures of DNA, there is a virtual alphabet soup of alternative structures that have been observed for various DNA sequences. These are not restricted to antiparallel duplex structures. H-DNA is a triple-stranded structure in which a third strand sits in the major groove of a standard right-handed duplex (Figure 1.43). This triple-helical structure can be induced to form in a number of different sequences, primarily sequences that are all purines along one strand and all pyrimidines along the other [e.g., $d(GA)_n \cdot d(CT)_n$]. There is now an entire industry developing around the concept that oligonucleotides can be designed to bind to a standard DNA or RNA duplex in a sequence-specific manner and in this way form a triple-stranded structure that will inhibit transcription and turn off a specific gene. DNA can also adopt four-stranded structures, including G-quartets found at the telomere ends of chromosomal DNAs, and four-stranded Holliday junctions that are involved in the process of exchanging information across DNA duplexes (called *recombination*) (Figure 1.44).

B-DNA itself can adopt a variety of forms, depending on the sequence and the environment. It can be extruded into a cruciform structure in sequences that are related by a true dyad axis (Figure 1.43) or be statically bent along a curved helical axis. Static bending is characteristic of sequences having the repeating motif

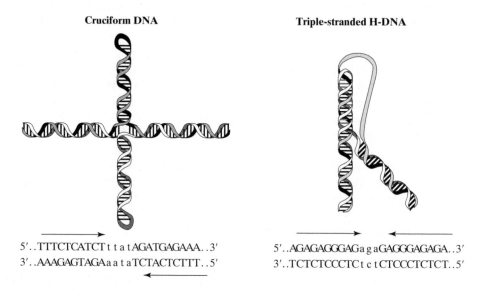

Cruciform DNA

Triple-stranded H-DNA

5′..TTTCTCATCT t t a t AGATGAGAAA..3′
3′..AAAGAGTAGA a a t a TCTACTCTTT..5′

5′..AGAGAGGGAG a g a GAGGGAGAGA..3′
3′..TCTCTCCCTC t c t CTCCCTCTCT..5′

Figure 1.43 Cruciform DNA and triple-stranded H-DNA. Cruciform DNA is formed by inverted repeat sequences, with a dyad axis of symmetry between the two strands of the duplex. Triple-strand H-DNA is stabilized by hydrogen bonding between three nucleobases to form *base triplets*. These are typically formed by sequences that are rich in purines along one strand and pyrimidines in the complementary strand. When drawn as a duplex, the sequence shows mirror symmetry along the stands. [Adapted from Schroth and Ho (1995), *Nucleic Acids Res.* **23**, 1977–1983.]

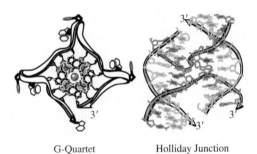

Figure 1.44 Structures of a G-quartet and a Holliday junction.

G-Quartet   Holliday Junction

d(AAAATTTT), known as *A-tracts*. A-tracts cause a local bend of about 20° in linear DNA. DNA bending by proteins has been found to play an important role in a number of biological processes, including the regulation of transcription. Protein-induced bending is more a feature of the flexibility of the DNA and the topology of the DNA-binding surface than any inherent feature of the DNA structure, as in statically bent DNA. A nucleosome wraps 140 base pairs of DNA around nearly two full turns of a circle in essentially a sequence-independent manner.

The helical structures in RNA are more monotonous than that in DNA. The standard structure of RNA is A-RNA, although it can be induced into a left-handed Z-RNA structure that is analogous to Z-DNA. It has never been observed in a B-like structure. In fact, mixtures of ribo- and deoxyribonucleotides appear always to form the A-RNA type structure. This includes duplexes that pair one strand of DNA with a complementary strand of RNA, or even DNA oligonucleotides with a single ribonucleotide incorporated into each strand. The strong preference for this A-helix structure results from the additional 2′-hydroxyl group of the ribose ring. This single hydroxyl group collides with atoms of adjacent nucleotides in all conformations except, apparently, in the A-conformation. In addition, this A-RNA structure allows the hydroxyl group to form a hydrogen bond with the phosphodiester bond of an adjacent nucleotide along the chain. This is not to suggest that RNA structures are less variable than DNA. In fact, since most RNAs fold from a single polynucleotide chain, their tertiary structures can be highly variable. This is further enhanced by unusual bases that are incorporated into the chain and a number of unusual (non–Watson-Crick) base pairs and base triplets involved in the tertiary-structure fold.

1.6.3   Higher-Order Structures in Polynucleotides

Beyond the helical structures of RNA and DNA are the higher-order structures. Single-stranded RNAs fold into tertiary structures that are most analogous to the compact structures of globular proteins. This is not to say that DNA structure stops at the level of helices. The double helix of DNA in the cell is highly supercoiled. This is important for both the compaction of DNA into chromosomes, and as the driving force to induce alternative structures (such as Z-DNA) within local regions of a long DNA sequence. The tertiary structures of RNAs and the correlation between supercoiling and the helical structure of DNA (the topology of DNA) are discussed separately here.

RNA folding. The varied structures of RNA reflects the diversity of functions now recognized for this class of molecule. The classical functions of RNA in the cell have been ascribed to the conversion of the genetic information stored in genomic DNA to the functional sequence of the polypeptide chains in proteins. Messenger RNA (mRNA) is the intermediary message synthesized during transcription of a specific gene, ribosomal RNA (rRNA) is an integral part of the ribosome (the machinery responsible for the synthesis of the polypeptide chain using the information in mRNA), while transfer RNA (tRNA) is the physical link between the amino acid and the triplet code along the mRNA strand that codes for that residue along the peptide chain. More recently, RNA has proven to be a highly talented molecule, capable of catalyzing chemical reactions including the splicing (cutting and rejoining, or ligation) of itself or other RNA molecules, and implicated in the actual aminoacyl-transfer reaction during peptide bond synthesis. More recently, small and micro interference RNA molecules (siRNA and miRNA) have been shown to suppress gene expression by targeting the degradation of specific mRNAs.

The first single-crystal structure determined for a polynucleotide was for tRNA (Figure 1.45). The structure represents many features that are common to other RNA conformations and has served as the model for RNA folding for many years. The standard representation of tRNA as a flat cloverleaf emphasizes the base-paired regions that form standard A-DNA secondary structures, the anticodon loop that complements the coding information in mRNA, and the amino acyl-charging 3′-terminus where the activated amino acids are attached, ready to be translated into the sequence of a polypeptide chain. However, the cloverleaf does not represent the tertiary structure, which includes relationships among the helices and the loops. The actual structure is more compact, with helices stacked against each other and the seemingly unstructured loops folded back into the helical regions to form base triplets that help to stabilize the overall fold.

The dominant feature of the tertiary structure is the L-shape of the conformation, with both the 5′-end and the 3′-amino acyl-charging end terminating one arm and the anticodon loop terminating the perpendicular arm of the L-structure. The structure of tRNA in Figure 1.45 is labeled according to the cloverleaf model to show the complex relationship among the various secondary structures. In fact, there are two distinct domains in the structure, one for each leg of the L-structure. This becomes more evident from a representation of the structure as a topological projection rather than the clover leaf. In this representation, the first domain is clearly formed by the stacked helices of the D stem and anticodon stem, and the second by the stacked helices T stem and amino acyl acceptor stem. The T loop and D loop at the ends of the two domains interact through base triplets to fix the structure at the elbow. Finally, the bases at the anticodon loop, which must remain unpaired in the isolated tRNA molecule, are stacked.

This structure provides a set of basic rules for tRNAs in general, including mitochondrial tRNAs which, when drawn in the standard cloverleaf form do not appear to be related at all to standard tRNAs in prokaryotics or eukaryotes. The number of bases in the anticodon arm is highly variable in mitochondrial tRNAs (ranging from 4 to 9, as compared to the conserved 6 base pairs of other tRNAs), and

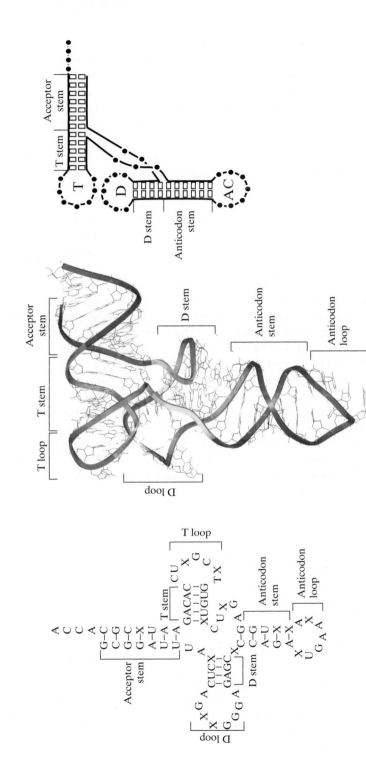

Figure 1.45 Structure of tRNA. The structure is represented as a cloverleaf, as the crystal structure, and as a topological projection.

the size of the D and T loops are highly variable. However, when they are drawn as a topological projection, we can easily see the features that are common for all tRNAs, regardless of the source. The total number of base pairs of the stacked helices in each domain remain relatively constant at 10 to 12 base pairs. If the anticodon stem is shortened or lengthened, the stacked D stem is lengthened or shortened to compensate. Similarly, to maintain the interactions between loops of the two domains, the length of the stacked helices of the T stem and acceptor stem must remain constant. Thus, the overall shape of the tRNA tertiary structure is largely conserved.

The tRNA structure also provides a set of general rules that have been useful for modeling other RNA folds.

- Bases will hydrogen-bond to form base pairs, triplets, etc. whenever possible,
- Base pairs within and between helices will be stacked as much as possible, and
- Base-paired duplex regions adopt the A-conformation.

An important aspect of RNA structure, in terms of its function, is that the sequence-dependent information in the major groove of the A-form helix is not accessible and therefore cannot be read by proteins. Consequently, the current model is that proteins recognize RNA sequences by their unique tertiary structures. One consequence is that base substitutions can be accommodated more readily in RNA as compared to DNA, but requires concomitant substitutions that will not disrupt the overall three-dimensional structure. For example, a guanosine nucleotide that is involved in a $G \cdot C$ base pair in a helical stem region can be replaced by a adenosine, but requires a complementary C to U substitution to maintain a base pair. This covariation in sequence (or *covariance*) can be used to identify regions of base paired nucleotides, but is not restricted to duplex regions. Covariance can also be required to maintain base triplets, etc., and therefore can be used to define tertiary structural elements (points of contact, much like the contact plots for protein tertiary structures), and can be used to predict the complex folds of large and complex RNA molecules (Application 1.3).

Application 1.3   Embracing RNA Differences

Scientists often look for similarities in patterns or motifs to help guide in building molecular models. For RNA structures, however, it may be the differences, specifically the *correlated variations in sequence* for functionally identical molecules, that will help solve the problem of predicting the topologies of their tertiary structures. Sequence variations either can be found to occur naturally, or can be engineered using an *in vitro* selection method. For example, by analyzing the sequences of the P-RNAs (RNA molecules that facilitate the processing of transfer RNAs) from various bacteria, Brown et al., showed that base-paired stem regions of this complex RNA can be reliably identified by their covariations (correlated substitutions across the various sequences). In addition, they observed correlations among three nucleotide positions that are interpreted as evidence for close interactions between distant nucleotides that facilitate formation of the compact tertiary structure of this molecule. For example, nucleotides 94 and 104 of the P8 stem domain show highly correlated occurrence for A and U (indicating formation of a base pair), but this specific covariance depends on position 316 of the P18 domain also being a G. Thus, the analysis predicts a U · A · G base triplet that brings the P18 loop close to the stem of P8. Other covariations in sequences also help to bring the loop of P14 into this region of the tertiary structure. Using both sequence covariations and chemical cross-lining data, Massire et al. constructed a model for the complete tertiary structure of the P-RNA of two bacteria forms of this catalytic molecule.

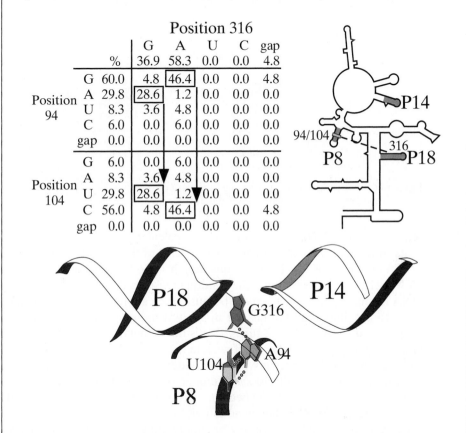

BROWN, J. W., J. M. NOLAN, E. S. HAAS, M. A. T. RUBIO, F. MAJOR, and N. R. PACE (1996), *Proc. Natl. Acad. Sci., USA* **93**, 3001–3006.

MASSIRE, C., L. JAEGER, and E. WESTHOF (1998), *J. Mol. Biol.* **279**, 773–793.

DNA topology. DNA is dominated by the structure of the double helix. However, the gross conformation of DNA extends beyond secondary structure. To describe the higher-order conformational properties of DNA, we start with a linear duplex of B-DNA and ask what happens if its structure is perturbed. Take, for example, a length B-DNA of $N = 147$ base pairs (Figure 1.46). Assuming the average

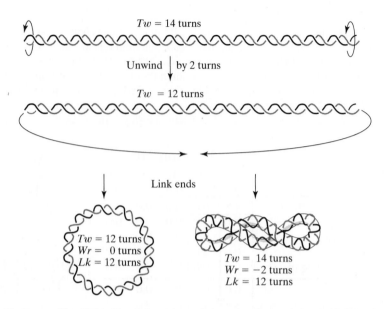

Figure 1.46 Supercoiling DNA. Unwinding 14 turns of covalently closed circular B-DNA by two turns results in a loss in helical twist and generation of two negative supercoils.

repeat $<c>$ of 10.5 base pairs/turn, we can calculate that this length of DNA represents 14 turns of the helix. The number of turns of the double helix defines the *twist* ($Tw = N/<c>$) of the DNA. If this length of B-DNA is now untwisted so that $<c> \approx 12$ base pairs per turn, the resulting twist would be $Tw = 12$ turns. Thus, the DNA is considered to be *unwound* because there are fewer turns in the double helix.

To accommodate the change in the unwinding of the helix, the ends would simply need to rotate by -2 turns. This is not difficult in a linear duplex where the ends are unrestricted. However, if we join the two ends to form a closed circle, then a change in *Tw* will not occur without broader consequences, since there are no free ends to accommodate the loss in number of helical turns. The result is that the closed circular DNA (ccDNA) becomes strained. This is analogous to what happens when a segment of a rubber band is twisted. The strain in the rubber band, as in ccDNA, is absorbed by a writhing or supercoiling of the circle. The *writhe* (*Wr*) of ccDNA, therefore, is correlated to its average twist.

Supercoils can be either positive (*Wr* > 0) or negative (*Wr* < 0). When a supercoil is represented as a duplex coiled around a cylindrical core, a negative supercoil conforms to the standard definition of a left-handed helix (Figure 1.47). This coiled structure is directly analogous to the supercoiled DNA observed in the chromatin structure of eukaryotic nuclei. However, if the core is removed, the free ccDNA now forms crossovers of the duplex, as observed in bacterial plasmid DNAs. There are a number of differences between the two types of supercoiled DNA. First, although both are negatively supercoiled DNA, the crossovers of the free ccDNA are right-handed, while the coiled DNA wraps around the nucleosome proteins in a left-handed direction. In addition, the nearly two turns of coiled DNA wrapped

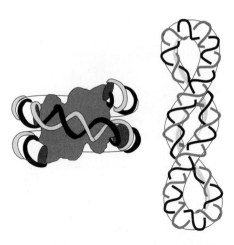

Figure 1.47 Two forms of supercoiled DNA. Negatively supercoiled DNA found in the chromatin structure wraps twice around the nucleosome core proteins in a left-handed direction. Negatively supercoiled DNA in the absence of a core forms right-handed crossovers. [Adapted from Arents and Moudrianakis (1993), *Proc. Natl. Acad. Sci., USA* **90**; 10489.]

around the nucleosome core represents only one negative supercoil. This is a result of the curvature of the protein surface. To wrap a DNA duplex over such a small surface, the entire helix twists, and the number of base pairs in B-DNA goes from about 10.5 to 10.2 base pairs per turn.

The superhelical strain in a ccDNA is reflected in its superhelical density σ, as

$$\sigma = Wr/Tw \tag{1.18}$$

This tells us the number of supercoils for each turn of DNA. The DNAs of both prokaryotic and eukaryotic cells are negatively supercoiled, with $\sigma \approx -0.006$.

The correlation between Tw and Wr is defined by the linking number (Lk) of ccDNA according to the relationship

$$Lk = Tw + Wr \tag{1.19}$$

The linking number, therefore, defines the overall conformation or *topology* of the ccDNA according to the degree to which torsional strain is partitioned between Tw and Wr. The parameter Lk is fixed and can be changed only by breaking the bonds of the phosphodiester backbone of one or both strands of the duplex. In the cell, the change in Lk to generate a new topological state or a *topoisomer* of the ccDNA (Figure 1.48) is the function of *topoisomerases*, enzymes that break one or both strands of the DNA duplex to allow the helical strain to be relieved.

The action of a topoisomerase is always to relieve strain; the lowest energy reference state is thus defined as relaxed, closed circular B-DNA, with $Tw° = N/10.5$ turns, $Wr° = 0$, and $Lk° = Tw°$. Any change in the conformation of ccDNA is relative to this reference state, and brings the molecule to a higher energy state (the thermodynamic relationships and how we get to a higher energy state of ccDNA is treated in detail in Chapter 4). The change in twist is thus $\Delta Tw = Tw - Tw°$, the change in writhe is $\Delta Wr = Wr - Wr°$, and the change in linking number is $\Delta Lk = Lk - Lk°$ (remembering that ΔLk requires a topoisomerase). The relationship between these parameters still holds, with

$$\Delta Lk = \Delta Tw + \Delta Wr \tag{1.20}$$

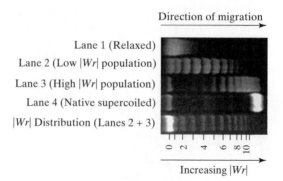

Direction of migration

Lane 1 (Relaxed)
Lane 2 (Low $|Wr|$ population)
Lane 3 (High $|Wr|$ population)
Lane 4 (Native supercoiled)
$|Wr|$ Distribution (Lanes 2 + 3)

0   2   4   6   8   10

Increasing $|Wr|$

Figure 1.48 Topoisomers of a bacterial plasmid. The plasmid pBR322 can exist as a relaxed closed circle ($|Wr| = 0$ turns, Lane 1), as a highly supercoiled closed circle ($|Wr| > 20$ turns, Lane 4), or as a mixed population with the writhe distributed over a broad range of $|Wr|$ values (Lanes 2 and 3, mixing the populations yields a broad distribution of topoisomers). These can be resolved by agarose gel electrophoresis, with the more compactly supercoiled form of the plasmid migrating faster in the electric field than the relaxed closed circular form. [Courtesy of M. N. Ho.]

This relationship tells us that any change in the helical conformation of the DNA duplex is accompanied by a change in the supercoiling of the ccDNA. For example, if 1050 base pairs of B-DNA are converted to A-DNA, ΔTw is $Tw_{A-DNA} - Tw^\circ = 1050/11$ turns $- 1050/10.5$ turns $= 95.5 - 100$ turns $= -4.5$ turns. For a topoisomer with $\Delta Lk = 0$ turns, $\Delta Wr = +4.5$ supercoils according to Eq. 1.20. Alternatively, if $\Delta Lk = -4.5$ turns, then either $\Delta Tw = -4.5$ turns and $\Delta Wr = 0$ supercoils, or $\Delta Wr = -4.5$ supercoils and $\Delta Tw = 0$ turns. The partitioning of superhelical strain can be to unwind the DNA or to supercoil the DNA, or, a case we have not considered yet, a mixture of the two. How structural transitions in ccDNA are induced by negative supercoiling and how strain is thermodynamically partitioned are looked at in Chapter 4.

EXERCISES

1.1 For helices with 8_1, 8_2, and 8_4 symmetry,
    a. What are the rotation angles and the translations (in fractions of pitch) for these helices?
    b. How many objects are in the motifs (monomer, dimer, and so on)?
    c. Might an 8_3 helix exist and, if so, what would be its structural characteristics?

*1.2 A molecule of L-Ala is placed with its chiral center at the origin of a right-handed axis system. The carbonyl carbon has atomic coordinates $[0, 0.15 \cos (35.2°), -0.15 \sin (35.2°)]$, and the C_β carbon (methyl group) at $[0, -0.15 \cos (35.2°), -0.15 \sin (35.2°)]$.
    a. Show how the y and z coordinates were determined for the carbonyl and C_β carbons (the length for this carbon-carbon single bond is 0.15 nm).

*In this and subsequent chapters, exercises that are judged to be more difficult are marked with an asterisk.

b. In this same axis system, determine the (x, y, z) coordinates of the amino nitrogen and the hydrogen. (A C—H bond is 0.11 nm and a C—N bond is 0.13 nm.)

c. Write the operator that generates the coordinates for D-Ala.

d. What are the coordinates for these same atoms in D-Ala?

*1.3 An amino acid is placed with its C_α carbon at $(0, 0.23, 0.05)$ nm of a right-handed axis system.

    a. Calculate the coordinates for all C_α carbons of six amino acids in the extended conformation with the y-axis as the helical axis.

    b. Calculate the coordinates for all C_α carbons of a six amino acid α-helix with its helical axis along the z-axis (radius of an α-helix = 0.226 nm).

1.4 Draw the structure of the tripeptide Ser-Pro-Ala with the Ser-Pro peptide bond in the *cis*-configuration, the other peptide bond in the *trans*-configuration, and all C_α carbons in the *trans*-conformation. Why do you think that the Ser-Pro bond may be more likely to be *cis* than other peptide bonds? What is the structural consequence of placing this peptide bond in *cis*?

*1.5 The following plot shows the amino acid contacts of the bovine pancreatic trypsin inhibitor (BPTI) as determined from the single crystal structure [adapted from T. G. Oas and P. S. Kim (1988), *Nature*, **336**, 42–48].)

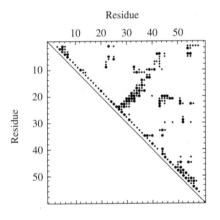

    a. Assign all the secondary structures in the protein.

    b. Sketch the tertiary structure of the protein, giving approximate arrangements of the α-helices and strands of β-sheets.

1.6 Sketch the contact plot for a three-stranded β-meander, with six amino acids in each strand and four residues in the connecting loops.

*1.7 Assume a 5250 base pair, closed circular plasmid with 10 negative supercoils.

    a. Calculate Tw, Wr, Lk, and σ for this plasmid (use 10.5 base pairs/turn for B-DNA).

    b. By adding an intercalator (something that slips between base pairs, such as ethidium bromide), you see that the number of supercoils in the plasmid is now only 8 negative supercoils. Calculate the changes in twist (ΔTw), writhe (ΔWr), linking number (ΔLk), and superhelical density ($\Delta\sigma$) resulting from this intercalator. How was the effect of the intercalator detected?

   c. You have a mixture of a plasmid that is a mixture of topoisomers (with 5 negative supercoils and 5 positive supercoils). From your answers in parts a and b, what is one way to separate these two sets of topoisomers and why?

   d. Your starting plasmid is now A-DNA instead of B-DNA. What is the linking number required to have 10 negative supercoils in this plasmid?

1.8 The covariance data on nucleotides 93, 105, and 214 of P-RNA are shown in the table here. What type of interaction are these nucleotide bases involved in?

		%	Base 214 G	A	U	C	gap
			38.1	54.8	1.2	0.0	5.9
Base 93	G	0.0	0.0	0.0	0.0	0.0	0.0
	A	0.0	0.0	0.0	0.0	0.0	0.0
	U	38.1	32.1	2.4	1.2	0.0	2.4
	C	60.7	5.9	51.2	0.0	0.0	3.6
	gap	1.2	0.0	1.2	0.0	0.0	0.0
Base 105	G	61.9	7.1	51.2	0.0	0.0	3.6
	A	36.9	31.0	2.4	1.2	0.0	2.4
	U	0.0	0.0	0.0	0.0	0.0	0.0
	C	0.0	0.0	0.0	0.0	0.0	0.0
	gap	1.2	0.0	1.2	0.0	0.0	0.0

1.9 You want to develop a contact plot of the C1′ carbons for the tRNA structure shown in Figure 1.45 (typical C1′-C1′ distances are ~1 nm across a base pair, 0.75 nm along an RNA chain).

   a. What distance range(s) would be best for showing the RNA secondary and tertiary structure contacts?

   b. Construct a contact plot for tRNA from the information in Figure 1.45.

REFERENCES

Macromolecular Structure

There are a number of excellent general texts in biochemistry, all of which provide more detail on the basics of macromolecular structure than can be presented in this chapter. Examples are listed below.

MATHEWS, C. K., VAN HOLDE K., and K. G. AHERN (1999) *Biochemistry,* 3rd ed., Benjamin-Cummings, Menlo Park, CA.

HORTON, H. R., MORAN, L. A., OCHS, R. S., RAWN, J. D., and K. G. SCRIMGEAR (2002) *Principles of Biochemistry,* 3rd ed., Prentice Hall, Upper Saddle River, NJ.

VOET, D. and J. G. VOET (2003) *Biochemistry,* 3rd ed., John Wiley & Sons, New York.

Water Structure

TANFORD, C. (1980) *The Hydrophobic Effect: Formation of Micelles and Biological Membranes,* John Wiley & Sons, New York. The classic text on hydration and hydrophobicity.

Protein Structure

KABSCH, W. and C. SANDER (1983) "Dictionary of Protein Secondary Structure: Pattern Recognition of Hydrogen-bonded and Geometric Features," *Biopolymers,* **22**, 2577–2637. This paper describes the hydrogen-bonding and geometric parameters for assigning protein secondary structures in proteins.

RAMACHANDRAN, G. N. and V. SASISEKHARAN (1968) "Conformation of Polypeptides and Proteins," *Adv. Protein Chem.,* **23**, 283–438.

SASISEKHAREN, V. (1962) "Stereochemical Criteria for Polypeptide and Protein Structures," in *Collagen,* ed. N. RAMANATHAN, 39–78, John Wiley & Sons, Madras, India. This landmark paper (as well as Ramachandran and Sasisekharan's work listed here) puts forth the concept that protein conformations can be best described by the main chain torsion angles, ϕ and ψ, of the polypeptide chain.

SCHULZ, G. E. and R. H. SCHIRMER (1979) *Principles of Protein Structure,* Springer-Verlag, New York. A detailed discussion of all aspects of protein structure.

BRANDEN, C.-I. and J. TOOZE (1998) *Introduction to Protein Structure,* 2nd ed., Garland Publishing, New York. A well-written and -illustrated text on the basics of protein structure.

Nucleic Acid Structure

SAENGER, WOLFRAM (1979) *Principles of Nucleic Acid Structure,* Springer-Verlag, New York. A detailed discussion of all aspects of nucleic acid structure.

WATSON, J. D. and F. H. C. CRICK (1953) "A Structure for Deoxyribose Nucleic Acid," *Nature* **171**, 737–738. This landmark paper describes the antiparallel double-helical model for B-DNA.

The single crystal structures of A-DNA, B-DNA, and Z-DNA were first published in the following papers:

SHAKKED, Z., D. RABINOVICH, W. B. T. CRUSE, E. EGERT, O. KENNARD, S. G. DALA, S. A. SALISBURY, and M. A. VISWAMITRA (1981) "Crystalline A-DNA: The X-Ray Analysis of the Fragment d(G-G-T-A-T-A-C-C)," *Proc. Royal Soc. London, Ser. B.* **231**, 479–487.

WANG, A. H.-J., G. J. QUIGLEY, F. J. KOPAK, J. L. CRAWFORD, J. H. VAN BOOM, G. VAN DER MAREL, and A. RICH (1979) "Molecular Structure of a Left-Handed Double-Helical DNA Fragment at Atomic Resolution," *Nature* **282**, 680–686.

WING, R., H. DREW, T. TAKANO, C. BROKA, S. TANAKA, K. ITAKURA, and R. E. DICKERSON (1980) "Crystal Structure Analysis of a Complete Turn of DNA," *Nature,* **287**, 755–758.

CHAPTER

2

Thermodynamics and Biochemistry

Thermodynamics, with its emphasis on heat engines and abstract energy concepts, has often seemed irrelevant to biochemists. Indeed, a conventional introduction to the subject is almost certain to convince the student that much of thermodynamics is sheer sophistry and unrelated to the real business of biochemistry, which is discovering how molecules make organisms work.

But an understanding of some of the ideas of thermodynamics *is* important to biochemistry. In the first place, the very abstractness of the science gives it power in dealing with poorly defined systems. For example, we can use the temperature dependence of the equilibrium constant for protein denaturation to measure the enthalpy change without knowing what the protein molecule looks like or even its exact composition. And the magnitude and the sign of that change tell us something more about unfolding of the protein molecules. Again, modern biochemists continually use techniques that depend on thermodynamic principles. A scientist may measure the molecular weight of a macromolecule or study its self-association by osmotic pressure measurements. All that is observed is a pressure difference, but the observer *knows* that this difference can be quantitatively interpreted to yield an average molecular weight. To use these physical techniques intelligently, we must understand something of their bases—this is what a good deal of this book is about.

In this chapter, we briefly review some of the ideas of thermodynamics that are important to biochemistry and molecular biology. While most readers will have taken an undergraduate course in physical chemistry, it has been our experience that this usually leads, insofar as thermodynamics is concerned, to a fairly clear understanding of the first law and some confusion about the second. Since the aim of this section is the use of thermodynamics rather than contemplation of its abstract beauty, we shall emphasize some molecular interpretations of thermodynamic principles. But it should never be forgotten that thermodynamics does not depend for its rigor on explicit details of molecular behavior. It is, however, sometimes easier to visualize thermodynamics in this way.

2.1   HEAT, WORK, AND ENERGY—FIRST LAW OF THERMODYNAMICS

The intention of biochemistry is ultimately to describe certain macroscopic systems involving multitudes of molecules, in terms of individual molecular properties. The fact is, however, that such systems are so complex that a complete description is beyond the capabilities of present-day physical chemistry. On the other hand, a whole field of study of energy relationships in macroscopic systems has been developed that makes no appeal whatsoever to molecular explanations. This discipline, thermodynamics, allows very powerful and exact conclusions to be drawn about such systems. The laws of thermodynamics are quite exact for systems containing many particles and this gives us a clue as to their origin. They are essentially statistical laws.

The situation is in a sense like that confronting an insurance company; the behavior of the individuals making up its list of insurees is complex, and an attempt to trace out all of their interactions and to predict the fate of any one of them would be a staggering task. But if the number of individuals is very large, the company can rely with great confidence on statistical laws, which say that so many will perish or become ill in any given period. Similarly, physical scientists can draw from their experiences with macroscopic bodies (large populations of molecules) laws that work very well indeed, even though the laws may leave obscure the mechanism whereby the phenomena are produced. Just as the insurance company with only 10 patrons is in a precarious position (it would not be too unlikely for all 10 people to die next month), so is the chemist who attempts to apply thermodynamics to systems of a few molecules. But 1 mole is 6×10^{23} molecules, a large number indeed. However, a bacterial cell may contain only a small number of some kinds of molecules; this means that some care must be taken when we apply thermodynamic ideas to systems of this kind (but see Hill 1963).

For review, let us define a few fundamental qualities.

System.   A part of the universe chosen for study. It will have spatial boundaries but may be *open* or *closed* with respect to the transfer of matter. Similarly, it may or may not be thermally insulated from its surroundings. If insulated, it is said to be an *adiabatic* system.

State of the system.   The thermodynamic state of a system is clearly definable only for systems at equilibrium. In this case, specification of a certain number of variables (two of three variables—temperature, pressure, and volume—plus the masses and identities of all the independent chemical substances in the system) will specify the state of the system. In other words, specification of the state is a recipe that allows us to reproduce the system at any time. It is an observed fact that if the state of a system is specified, its properties are given. The properties of a system are of two kinds. *Extensive* properties, such as volume and energy, require for the definition specification of the thermodynamic state including the amounts of all substances.

Intensive properties, such as density or viscosity, are fixed by giving less information; only the *relative* amounts of different substances are needed. For example, the density of a 1 M NaCl solution is independent of the size of the sample, though it depends on temperature T, pressure P, and the concentration of NaCl.

Thermodynamics is usually concerned with changes between equilibrium states. Such changes may be *reversible* or *irreversible*. If a change is reversible, the path from initial to final state leads through a succession of near-equilibrium states. The system always lies so close to equilibrium that the direction of change can be reversed by an infinitesimal change in the surroundings.

Heat, q. The energy transferred into or out of a system as a consequence of a temperature difference between the system and its surroundings.

Work, w. Any other exchange of energy between a system and its surroundings. It may include such cases as volume change against external pressure, changes in surface area against surface tension, electrical work, and so forth.

Internal energy, E. The energy within the system. In chemistry, we usually consider only those kinds of energy that might be modified by chemical processes. Thus, the energy involved in holding together the atomic nuclei is generally not counted. The internal energy of a system may then be taken to include the following: translational energy of the molecules, vibrational energy of the molecules, rotational energy of the molecules, the energy involved in chemical bonding, and the energy involved in nonbonding interactions between molecules. Some such interactions are listed in Table 2.1. How these interactions affect the structure and stability of proteins and nucleic acids is discussed in Chapter 3.

The internal energy is a function of the state of a system. That is, if the state is specified, the internal energy is fixed at some value regardless of how the system came to be in that state. Since we are usually concerned with energy changes, internal energy is defined with respect to some arbitrarily chosen standard state.

Enthalpy, $H = E + PV$. The internal energy of a system plus the product of its volume and the external pressure exerted on the system. It is also a function of state.

With these definitions, we state the first law of thermodynamics, an expression of the conservation of energy. For a change in state,

$$\Delta E = q - w \tag{2.1}$$

which takes the convention that heat absorbed by a system and work done by a system are positive quantities. For small changes, we write

$$dE = \text{\dj}q - \text{\dj}w \tag{2.2}$$

Table 2.1   Noncovalent Interactions Between Molecules

Type of Interaction	Equation[a]	Order of Magnitude[b] (kJ/mol)
Ion-ion (Charge-Charge)	$E = \dfrac{Z_1 Z_2 e^2}{Dr}$	60
Ion-dipole	$E = \dfrac{Z_1 e \mu_2 \theta}{Dr^2}$	−8 to +8
Dipole-dipole	$E = \dfrac{\mu_1 \mu_2 \theta'}{Dr^3} - \dfrac{3(\mu_1 r\theta'')(\mu_2 r\theta'')}{Dr^5}$	−2 to +2
Ion-induced dipole	$E = \dfrac{Z_1 e^2 \alpha_2}{2D^2 r^4}$	0.2
Dispersion[c]	$E = \dfrac{3h\nu_0 \alpha^2}{4r^6}$	0 to 40

[a] In these equations, e is the charge of a proton (or the magnitude of the charge of an electron), Z is the valence charge, μ is the dipole moment of a dipole, and α is the molecular polarizability. D is the dielectric constant of the medium and r is the distance between the molecules. The factors θ' and θ'' are functions of the orientations of dipoles. (See Moelwyn-Hughes 1961.)

[b] Calculations were made with the following assumptions: (1) molecules are 0.3 nm apart; (2) all charges are 4.8×10^{-10} esu (electron charge); (3) all dipole moments are 2 debye units; and (4) all polarizabilities are 2×10^{-24} cm^3. These are typical values for small molecules and ions. The dielectric constant was taken to be 8, a reasonable value for a *molecular* environment; energies would be lower in aqueous solution, where $D \cong 80\kappa\epsilon_0$. Since the ion-dipole and dipole-dipole interactions depend strongly on dipole orientation, we have given the *extreme* values. For comparison, covalent bond energies range between 120 and 600 kJ/mol.

[c] Dispersion interactions are between mutually polarizable molecules. The charge fluctuations in the molecules with frequency ν_0 interact, producing a net interaction. Since this depends so strongly on distance and becomes important only for very close molecules, a range of values is given.

The slashes through the differential symbols remind us that whereas E is a function of state and dE is independent of the path of the change, q and w do depend on the path. The first law is entirely general and does not depend on assumptions of reversibility and the like.

If the only kind of work done involves change of volume of the system against an external pressure (PdV work),

$$dE = \bar{d}q - PdV \qquad (2.3)$$

Similarly, we write for the change in the enthalpy of a system the general expression

$$dH = d(E + PV) = dE + PdV + VdP$$
$$= \bar{d}q - \bar{d}w + PdV + VdP \qquad (2.4)$$

For systems doing only PdV-type work, $\bar{d}w = PdV$, and

$$dH = \bar{d}q + VdP \qquad (2.5)$$

Equations 2.3 and 2.5 point up the meaning of dE and dH in terms of measurable quantities. For changes at constant volume

$$dE = dq$$
$$\Delta E = q_v \quad \text{for a finite change of state} \tag{2.6}$$

whereas for processes occurring at constant pressure, Eq. 2.5 gives

$$dH = dq$$
$$\Delta H = q_p \quad \text{for a finite change of state} \tag{2.7}$$

That is, the heat absorbed by a process at constant volume measures ΔE, and the heat absorbed by a process at constant pressure measures ΔH. These quantities of heat will in general differ, because in a change at constant pressure some energy exchange will be involved in the work done in the change of volume of the system.

The thermochemistry of biological systems is almost always concerned with ΔH, since most natural biochemical processes occur under conditions more nearly approaching constant pressure than constant volume. However, since most such processes occur in liquids or solids rather than in gases, the volume changes are small. To a good approximation, we can often neglect the difference between ΔH and ΔE in biochemistry and simply talk about the energy change accompanying a given reaction.

Figure 2.1 summarizes the relationships among the quantities q, w, ΔE, and ΔH. Note that we begin with the perfectly general first law and specialize to particular kinds of processes by adding more and more restrictions.

2.2 MOLECULAR INTERPRETATION OF THERMODYNAMIC QUANTITIES

We have seen that from the first law, together with the assertion that the internal energy is a function of the state of a system, powerful and general conclusions can be drawn. Although these have not required molecular models to attest to their validity, the student should keep in mind that quantities such as the internal energy and the energy changes in chemical reactions are ultimately expressible in terms of the behavior of atoms and molecules. It will be worth our while to explore the point in more detail.

Suppose that we ask the following question: If we put energy into a system to give an increase in the internal energy, where has the energy gone? Surely some has appeared as increased kinetic energy; but if the molecules are complex, some must be stored in rotational and vibrational energy and in intermolecular interactions, and perhaps some is accounted for by excited electronic states of a few molecules. Therefore, the question is really one of how the energy is *distributed*.

For thermodynamic properties, we are talking about large numbers of molecules, generally in or near states of equilibrium. The first means that a *statistical*

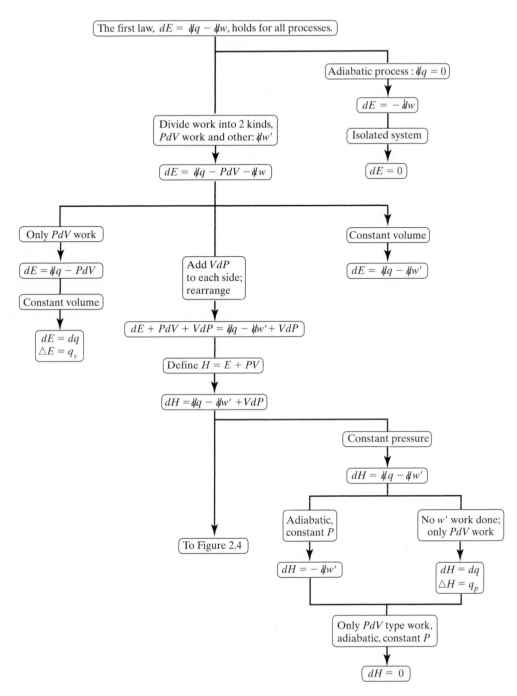

Figure 2.1   Consequences of the first law and the definition of ΔH. The other kind of work $đw'$ may be identified with electrical work, work done in expanding a surface and so forth.

point of view may be taken; we need not follow the behavior of any one molecule. The second implies that we should look for the *most probable* distribution of energy, for we would not expect an equilibrium state to be an improbable one. Although any system might, by momentary fluctuations, occasionally distribute its energy in some improbable way (like having almost all of the energy in a few molecules), the relative occurrence of such extreme fluctuations becomes vanishingly small as the number of molecules becomes very large.

To see the principles involved, let us take a very simple system, a collection of particles that might be thought of, for example, as atoms in a gas or as protein molecules in a solution. Each of these entities is assumed to have a set of energy states available to it, as shown in Figure 2.2. The energy states available to a particle are not to be confused with the *thermodynamic states* of a system of many particles. Rather, they are the quantized states of energy accessible to any particle under the constraints to which the whole collection of particles is subject. Suppose that we have six particles and a total energy of 10ϵ, where ϵ is some unit of energy. Some distributions are shown in Figure 2.2, each of which satisfies the total energy requirement. Now let us say that for any particle, any state is equally probable. This simply means that there is nothing to prejudice a particle to pick a given state. Then the most probable distribution will be the one that corresponds to the largest number of ways of arranging particles over the states. If we label the particles, we see that there are only six ways of making state (a) and many more ways of making either state (b) or (c). The number of ways of arranging N particles, n_1 in one group, n_2 in another, and so forth, is

$$W = \frac{N!}{n_1! n_2! n_3! \cdots n_t! \cdots} \tag{2.8}$$

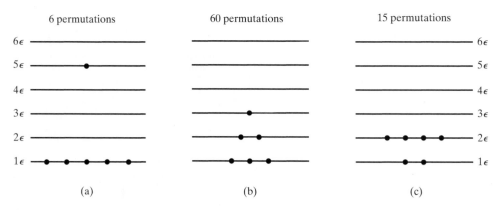

Figure 2.2 Some distributions of particles over energy states, subject to the constraints that $N = 6$ and $E = 10\epsilon$. The numbers are calculated from Eq. 2.8 (remember $0! = 1$). The Boltzmann distribution most closely resembles (b). Of course, there are not enough particles for it to hold accurately in this simple case.

Since the most probable distribution is the one that corresponds to the largest number of arrangements of particles over energy states, the problem of finding that distribution is a problem of maximizing the number W, subject to the restrictions that N and total energy E are constants. Using standard mathematical techniques for handling such problems, the result for a large number of particles is found to be

$$n_i = n_1 e^{-\alpha(\epsilon_i - \epsilon_1)} \tag{2.9}$$

where n_i is the number of particles in state i, n_1 is the number in the lowest state, and ϵ_i and ϵ_1 are the energies of states i and 1, respectively. The constant α turns out to be $1/k_B T$, where k_B is the Boltzmann constant, the gas constant R divided by Avogadro's number, and T is the absolute temperature. A simple derivation is given in Moore (1972). For a more leisurely but very clear discussion see Gurney (1949).

Equation 2.9 is referred to as the *Boltzmann distribution* of energies. It should always be kept in mind that this is not the only possible distribution, and that if we could sample a collection of molecules at any instant we would expect to find deviations from it. It is simply the *most probable* distribution and hence will serve well if the number of particles is large and the system is at or near equilibrium.

One more modification of Eq. 2.9 should be made. We have written a distribution over energy *states,* whereas a distribution over *levels* would often be more useful. The distinction lies in the fact that levels may be degenerate—there may be several atomic or molecular states corresponding to a given energy level. The energy levels of the hydrogen atom will serve as one example and the possible different conformational states corresponding to a given energy for a random-coil polymer as another. (See Chapters 4 and 8.) If each level contains g_i states (that is, if the degeneracy is some integer g_i), levels should be weighted by this factor. Then

$$n_i = \frac{g_i}{g_1} n_1 e^{-(\epsilon_i - \epsilon_1)/k_B T} \tag{2.10}$$

where n_i and n_1 now refer to the number of particles in energy *levels* i and 1, respectively.

Equation 2.10 states that if the degeneracies of all states are equal, the lowest states will be the most populated at any temperature. At $T = 0$, $n_i = 0$ for $i > 1$, which means that all particles will be in the lowest level, while as $T \to \infty$ the distribution tends to become more and more uniform. At high temperatures, no level is favored over any other, except for the factor of degeneracy. Another useful form of Eq. 2.10 involves N, the total number of particles, instead of n_1. If we recognize that $N = \Sigma_i n_i$ (the sum being taken over all levels), then

$$N = \frac{n_1}{g_1} \sum_i g_i e^{-(\epsilon_i - \epsilon_1)/k_B T}$$

or

$$\frac{n_i}{N} = \frac{n_1}{g_1} \frac{g_i e^{-(\epsilon_i - \epsilon_1)/k_B T}}{(n_1/g_1) \sum_i g_i e^{-(\epsilon_i - \epsilon_1)/k_B T}}$$

$$= \frac{g_i e^{-(\epsilon_i - \epsilon_1)/k_B T}}{\sum_i g_i e^{-(\epsilon_i - \epsilon_1)/k_B T}} \tag{2.11}$$

The sum in the denominator of Eq. 2.11 is frequently encountered in statistical mechanics. It is sometimes called (for obvious reasons) the *sum over states* and more often (for less obvious reasons) the *molecular partition function*.

Since Eq. 2.11 gives the fraction of molecules with energy ϵ_i, it is very useful for calculating average quantities. We make use of this idea in subsequent chapters. For example, in Chapter 4 we calculate the average structure of helical molecules in just this way.

2.3   ENTROPY, FREE ENERGY, AND EQUILIBRIUM—SECOND LAW OF THERMODYNAMICS

So far in discussing chemical and physical processes, we have concentrated on the energetics. We have shown that the first law, a restatement of the conservation of energy, leads to exceedingly useful and general conclusions about the energy changes that accompany these processes. But one factor has been pointedly omitted—there has been no attempt to predict the *direction* in which changes will occur. Thus, the first law allows us to discuss the heat transfer and work accompanying a chemical reaction, such as the hydrolysis of adenosine triphosphate (ATP) to yield adenosine disphosphate (ADP),

$$H_2O + ATP \longrightarrow ADP + \text{inorganic phosphate}$$

but gives no indication whether or not ATP will spontaneously hydrolyze in aqueous solution. Intuitively, we would expect that under the given conditions some particular equilibrium will exist between H_2O, ATP, ADP, and inorganic phosphate, but there is no way that the first law can tell us where that equilibrium lies.

As another example, consider the dialysis experiment shown in Figure 2.3. A solution of sucrose has been placed in a dialysis bag immersed in a container of water. We shall assume that the membrane is permeable to sucrose molecules. Either intuition or a simple experiment will tell us that the system is not at equilibrium. Rather, we know and find that sucrose will diffuse through the membrane until the concentrations inside and outside the bag are equal.

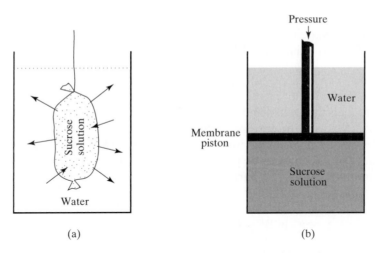

Figure 2.3   (a) A dialysis experiment in which sucrose will diffuse out of a bag and water will diffuse into the bag until equilibrium is attained. This process is irreversible. No work is done. (b) A way of doing the same experiment reversibly. The membrane piston is impermeable to sucrose and permeable to water. If the pressure on the piston is gradually reduced, the same final state (uniform mixing) approached irreversibly in (a) will be approached reversibly. With this arrangement, work will be done.

What is it that determines this position of equilibrium and the spontaneous (irreversible) process that leads to it? A first guess might be made from analogy to mechanical processes. A mechanical system reaches a state of equilibrium when the energy of the system is at minimum. (Think of a ball rolling to the bottom of a hill.) But this will clearly not do here. Dilute sucrose solutions are close to ideal, which means that the energy of the system is practically independent of concentration. In fact, the interaction of sucrose with water is such that the state of lower concentration is actually of higher energy.

A moment's thought shows that the equilibrium state is favored in this case because of its *higher probability*. If we imagine a vast array of such systems, each of which has been left to itself for a considerable period, it is clear that the great majority will have the sucrose distributed quite uniformly throughout. True, there might be a very rare occurrence in which the sucrose concentration was appreciably higher inside or outside. But such systems will be very exceptional, for they would require the remarkable coincidence that a large excess of randomly wandering sucrose molecules happened to be inside or outside of the bag. A closer inspection shows that the main difference between the initial and final states of the system lies in the number of ways W in which sucrose molecules can be distributed over the total volume available to them. There are more ways of putting N sucrose molecules into a large volume than into a small volume. Thus, one of the determinants of the equilibrium state of a system of many particles is the *randomness* of the system or the number of ways W in which the particles of the system may be distributed, whether it be over levels of energy or, as in this instance, over the volume of the system.

For this reason, we define a function of state, which we shall call the *entropy*,

$$S = k_B \ln W \tag{2.12}$$

where k_B is the Boltzmann constant. The logarithm of W is chosen for the following reason: We wish the entropy to be an extensive property; that is, we wish the S of a system that is made up of two parts (1 and 2) to be $S_1 + S_2$. Now if W_1 is the number of ways of distributing the particles in part 1 in the particular state and W_2 the corresponding quantity for part 2, the number of ways for the whole system in this state is $W_1 W_2$,

$$S = k_B \ln W = k_B \ln W_1 W_2 = k_B \ln W_1 + k_B \ln W_2 = S_1 + S_2 \tag{2.13}$$

This description of S, although exceedingly useful for certain problems, is not clearly related to measurable physical quantities. The relationship may be accomplished in the following way. Let us assume that the molecules in a system have available a certain set of energy states and are distributed over these states according to the Boltzmann distribution

$$n_i = n_1 e^{-(\epsilon_i - \epsilon_1)/k_B T} \tag{2.14}$$

Now

$$W = \frac{N!}{n_1! n_2! n_3! \cdots n_n!} \tag{2.15}$$

$$\ln W = \ln (N!) - \ln (n_1!) - \ln (n_2!) - \cdots \tag{2.16}$$

or, using Stirling's approximation ($\ln n! \cong n \ln n - n$)

$$\ln W \cong N \ln N - N - n_1 \ln n_1 + n_1 - n_2 \ln n_2 + n_2 - \cdots \tag{2.17}$$

Since $\sum_i n_i = N$, we obtain

$$\ln W = K - \sum_i n_i \ln n_i \tag{2.18}$$

where K is the constant $N \ln N$.

Let us now consider an infinitesimal change in the state of the system. This is assumed to be a *reversible* change; that is, the system does not depart appreciably from a state of equilibrium but simply changes from the Boltzmann distribution appropriate to the initial state to that appropriate to the infinitesimally different new state,

$$dS = k_B d \ln W \tag{2.19}$$

$$= -k_B d \sum_i n_i \ln n_i = -k_B \sum_i n_i d \ln n_i - k_B \sum_i \ln n_i dn_i \tag{2.20}$$

$$= -k_B \sum_i \ln n_i dn_i \tag{2.21}$$

The first term on the right of Eq. 2.20 vanishes because it equals $\Sigma_i dn_i = d\Sigma_i n_i$ and the total number of particles is constant. The expression for dS can then be evaluated by remembering that $n_i = n_1 e^{-(\epsilon_i - \epsilon_1)/k_B T}$ so that

$$\ln n_i = \ln n_1 - \frac{\Delta \epsilon_i}{k_B T} \tag{2.22}$$

where $\Delta \epsilon_i$ is used to abbreviate $(\epsilon_i - \epsilon_1)$. Again dropping a term that is the sum of dn_i, we obtain

$$dS = k_B \left(\frac{\Sigma \Delta \epsilon_i}{k_B T} \right) dn_i = \frac{1}{T} \sum_i \Delta \epsilon_i dn_i \tag{2.23}$$

Now the sum in Eq. 2.23 is simply the heat absorbed in the process, for if a system has a given set of *fixed* energy levels and a reversible change in the population of the levels occurs, this must involve the absorption or release of heat from the system.

$$dS = \frac{dq_{rev}}{T} \tag{2.24}$$

This is the classic definition of an entropy change. For a finite isothermal process, we say that

$$\Delta S = \frac{q_{rev}}{T} \tag{2.25}$$

The heat q_{rev} is a perfectly defined quantity, since the change is required to be reversible.

Equation 2.25 states that if we wish to calculate the entropy change in the transition between state 1 and state 2 of a system, it is necessary only to consider a reversible path between the two states and calculate the heat absorbed or evolved. As a simple example, we may consider the case of an isothermal reversible expansion of 1 mole of an ideal gas from volume V_1 to volume V_2. For each infinitesimal part of the process, $dE = 0$ (since the gas is ideal and its energy is independent of its volume). Then

$$dq = dw = PdV = \frac{RTdV}{V} \tag{2.26}$$

and

$$\frac{q_{rev}}{T} = \frac{RT}{T} \int_{V_2}^{V_1} \frac{dV}{V} = R \ln \left(\frac{V_2}{V_1} \right) \tag{2.27}$$

so

$$\Delta S = R \ln \left(\frac{V_2}{V_1} \right) \tag{2.28}$$

The entropy of the gas increases in such an expansion.

We can arrive at exactly the same result from the statistical definition of entropy. Suppose that we divide the volume V_1 into n_1 cells, each of volume V; therefore, $V_1 = n_1 V$. The larger volume V_2 is divided into n_2 cells of the same size; $V_2 = n_2 V$. Putting one molecule into the initial system of volume V_1, there are n_1 ways in which this can be done. For two molecules there will be n_1^2 ways. For an Avogadro's number ($\mathfrak{N}$) of molecules, the number of ways to put them into V_1 is $W_1 = n_1^{\mathcal{N}}$. If the larger volume V_2, containing V_2/V cells, is occupied by the same $\mathcal{N}$ molecules, $W_2 = n_2^{\mathcal{N}}$. From Eq. 2.12,

$$\Delta S = k_B \ln n_2^{\mathcal{N}} - k_B \ln n_1^{\mathcal{N}}$$
$$= k_B \ln \left(\frac{n_2}{n_1} \right)^{\mathcal{N}} = R \ln \left(\frac{n_2}{n_1} \right) = R \ln \left(\frac{V_2/V}{V_1/V} \right) \tag{2.29}$$

where we have made use of the fact that $\mathcal{N} k_B = R$. Then

$$\Delta S = R \ln \left(\frac{V_2}{V_1} \right) \tag{2.30}$$

Note that the cell volume V has canceled out of the final result; it is just a device to allow us to compare the number of ways of making the systems 1 and 2. Note also that it was necessary for the constant in Eq. 2.12 to have the value k_B (the Boltzmann constant) in order that the two methods of approach would lead to the same numerical value of ΔS.

A very similar calculation could be carried out for our examples of sucrose diffusing out of a dialysis bag. In this case, however, the calculation from q_{rev}/T would be more difficult, for we would have to imagine some reversible way in which to carry out the process. One such method is depicted in Figure 2.3b. The dialysis bag is replaced by a membrane piston, permeable to water but not to sucrose. To the piston is applied a pressure equal to the osmotic pressure of the solution (see Chapter 13). If this pressure is gradually reduced, the piston will rise, doing work on the surroundings and diluting the sucrose solution. Since once again the ideality of the system requires $\Delta E = 0$, heat must be absorbed to keep the system isothermal.

Both the classic and statistical calculation will again lead to the same result, which is formally identical to that for the gas expansion. Per mole, we have

$$\Delta S = R \ln \left(\frac{V_2}{V_1}\right) = R \ln \left(\frac{C_1}{C_2}\right) \tag{2.31}$$

where C_1 and C_2 are concentrations.

A very similar calculation allows us to determine the entropy of mixing for an ideal solution made up of N components. Assume that there are N_i molecules in the whole solution; $N_0 = \sum\limits_{i=1}^{N} N_i$. We assume that the molecules are about the same size, so that each can occupy a cell of the same volume. Since the mixture is assumed to be ideal, there will be no volume change in mixing, and the total number of cells in the solution will be N_0, the total number of molecules. The entropy of mixing is defined as

$$\Delta S = S(\text{solution}) - S(\text{pure components}) \tag{2.32}$$

We are concerned only with mixing entropy, so we may say that each of these quantities is given by $k_B \ln W$, where W is the number of distinguishable ways of putting molecules in cells. Then

$$\Delta S_m = k_B \ln \frac{W(\text{solution})}{W(\text{pure components})} \tag{2.33}$$

But all arrangements of molecules in the pure components are the same, so W (pure components) $= 1$, and $\Delta S_m = k_B \ln W$ (solution). But W (solution) is just the number of ways of arranging N_0 cells into groups with N_1 of one type, N_2 of another, and so forth. Therefore, we have the familiar expression

$$\Delta S_m = k_B \ln \frac{N_0!}{N_1! N_2! \cdots N_N!} \tag{2.34}$$

Since all of the Ns are large, we may use Stirling's approximation,

$$\Delta S_m \cong k_B (N_0 \ln N_0 - N_0 - N_1 \ln N_1 + N_1 - N_2 \ln N_2 + N_2 - \cdots) \tag{2.35}$$

or, since $\sum\limits_{i} N_i = N_0$,

$$\Delta S_m = k_B \left(N_0 \ln N_0 - \sum_{i} N_i \ln N_i \right) \tag{2.36}$$

or

$$\Delta S_m = k_B \left(\sum_{i} N_i \ln N_0 - \sum_{i} N_i \ln N_i \right) \tag{2.37}$$

$$= -k_B \sum_{i} N_i \ln \frac{N_i}{N_0} = -k_B \sum_{i} N_i \ln X_i \tag{2.38}$$

where X_i is the mole fraction of i. Multiplying and dividing by Avogadro's number, we obtain the final result

$$\Delta S_m = -R \sum_i n_i \ln X_i \qquad (2.39)$$

where n_i is the number of moles of component i. This expression says that even if there is no interaction between the molecules, the entropy of the mixture is always greater than that of the pure components, since all X_is are less than unity, and their logarithms will be negative quantities.

It is evident from the above examples and from Eq. 2.12 that the entropy can be considered as a measure of the randomness of a system. A crystal is a very regular and nonrandom structure; the liquid to which it may be melted is much more random and has a higher entropy. The entropy change in melting can be calculated as

$$\Delta S = \frac{q_{rev}}{T} - \frac{\Delta H_{\text{melting}}}{T_{\text{melting}}} \qquad (2.40)$$

Since the melting of a crystalline solid will always be an endothermic process and since T is always positive, the entropy will always increase when a solid is melted. Similarly, a *native* protein molecule is a highly organized, regular structure. When it is *denatured*, an unfolding and unraveling takes place; this corresponds to an increase in entropy. Of course, this is not the whole story; upon denaturation the interaction of solvent with the protein may change in such a way either to add or subtract from this entropy change.

Whenever a substance is heated, the entropy will increase. We can see this as follows: At low temperatures, only the few lowest of the energy levels available to the molecules are occupied; there are not too many ways to do this. As T increases, more levels become available and the randomness of the system increases. The entropy change can, of course, be calculated by assuring that the heating is done in a reversible fashion. Then, if the process is at constant pressure, $dq_{rev} = C_p dT$, where C_p is the constant pressure heat capacity;

$$dS = \frac{C_p dT}{T} \qquad (2.41)$$

or, in heating a substance from temperature T_1 to T_2,

$$\Delta S = \int_{T_1}^{T_2} \frac{C_p}{T} dT \qquad (2.42)$$

If C_p is constant,

$$\Delta S = C_p \ln \left(\frac{T_2}{T_1} \right) \qquad (2.43)$$

The definition of the entropy change in an infinitesimal, reversible process as dq_{rev}/T allows us a reformulation of the first law for reversible processes, since

$$dE = dq - dw = TdS - PdV \qquad (2.44)$$

This leads to one precise answer to a question posed at the beginning of this section: What are the criteria for a system to be at equilibrium? If and only if a system is at equilibrium, an infinitesimal change will be reversible. We may write from Eq. 2.44 that if the volume of the system and its energy are held constant,

$$dS = 0 \qquad (2.45)$$

for a reversible change. This means that S must be either at a maximum or minimum for a system at constant E and V to be at equilibrium. Closer analysis reveals that it is the former. If we isolate a system (keep E and V constant), then it will be at equilibrium only when the entropy reaches a maximum. In any nonequilibrium condition, the entropy will be spontaneously increasing toward this maximum. This is equivalent to the statement that an isolated system will approach a state of maximum randomness. This is the most common way of stating the *second law of thermodynamics*.

Isolated systems kept at constant volume are not of great interest to biochemists. Most of our experiments are carried out under conditions of constant temperature and pressure. To elucidate the requirements for equilibrium in such circumstances, a new energy function must be defined that is an explicit function of T and P. This is the *Gibbs free energy, G*:

$$G = H - TS \qquad (2.46)$$

We obtain the general expression for a change dG by differentiating Eq. 2.46:

$$dG = dH - TdS - SdT$$
$$= dE + VdP + PdV - SdT - TdS \qquad (2.47)$$

This collection of terms simplifies if we note that $dE = TdS - PdV$ if the process causing dG is reversible. Then

$$dG = VdP - SdT \quad \text{reversible process} \qquad (2.48)$$

And now, if the system is also constrained to the normal laboratory conditions of constant T and P,

$$dG = 0 \quad \text{reversible process} \qquad (2.49)$$

This may be interpreted in the same way as Eq. 2.45. The value of G is at an extremum when a system constrained to constant T and P reaches equilibrium. In this case it will be a minimum.

A similar quantity, the *Helmholtz free energy*, is defined as $A = E - TS$. It is of less use in biochemistry, because $dA = 0$ for the less frequently encountered conditions of constant T and V. *Free energy* will mean the Gibbs free energy here. Some of these consequences of the second law are summarized in Figure 2.4.

The Gibbs free energy is of enormous importance in deciding the direction of processes and positions of equilibrium in biochemical systems. If the free-energy change calculated for a process under particular conditions is found to be negative, that process is spontaneous, for it leads in the direction of equilibrium. Furthermore, ΔG as a combination of ΔH and ΔS emphasizes the fact that both energy minimization and entropy (randomness) maximization play a part in determining the position of equilibrium. As an example, consider the denaturation of a protein or polypeptide;

$$\Delta G_{den} = \Delta H_{den} - T\Delta S_{den} \qquad (2.50)$$

Now we expect ΔS_{den} to be a positive quantity for the following reason: By definition, we have $\Delta S_{den} = R \ln (W_{den}/W_{native})$; since the denatured protein has many more conformations available to it than does the native, it follows that $W_{den}/W_{native} \gg 1$. Furthermore, the breaking of the favorable interactions (hydrogen bonding and the like) that hold the native conformation together will surely require the input of energy, so ΔH will also be positive. If we examine the behavior of Eq. 2.50 with positive ΔS and ΔH, we see that ΔG could be either positive or negative, depending on the values of ΔH and ΔS. However, at some low T, $\Delta G > 0$, whereas at sufficiently high T, $\Delta G < 0$. Thus, the native state should be stable at low T and the denatured state stable at high T. This is in fact found, and ΔH and ΔS estimated from simple considerations are of the order of magnitude of those observed (see Exercise 2.5).

The same problem can be considered from a slightly different point of view, which emphasizes the close relation of the Boltzmann distribution to problems of order and energy. Consider the energy of a polypeptide to be represented schematically by Figure 2.5. There is, we assume, one energy minimum corresponding to the ordered conformation and a host of higher energy conformations corresponding to the random-coil form. Of course, this is a gross oversimplification, and it must be understood that Figure 2.5 is a schematic representation of a multidimensional energy surface. We may write the Boltzmann equation for the ratio of numbers of molecules in two forms as

$$\frac{n_{den}}{n_{native}} = \frac{g_d}{g_n} e^{-(\epsilon_d - \epsilon_n)/k_B T} \qquad (2.51)$$

We can then identify the degeneracies of these levels with the number of conformations corresponding to each:

$$g_n = W_n = 1 \qquad (2.52)$$

$$g_d = W_d \gg 1 \qquad (2.53)$$

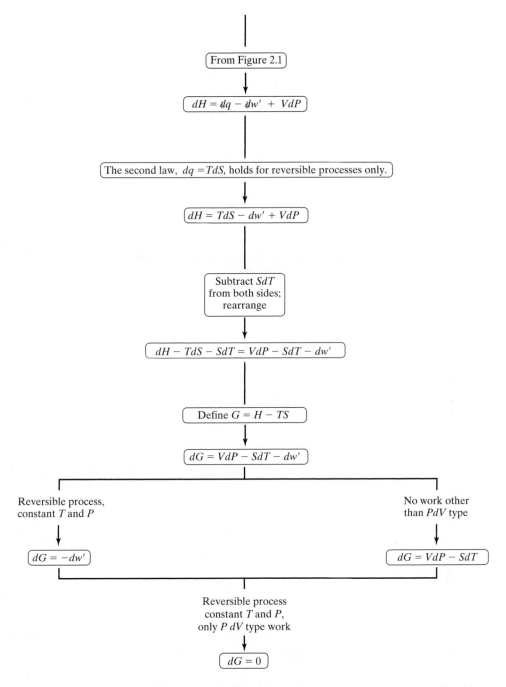

Figure 2.4 A continuation of Figure 2.1, summarizing some relations derived from the second law. Note that here q, w, and w' are those for reversible processes only.

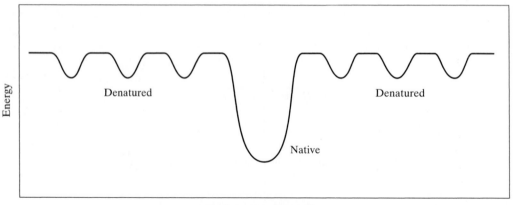

Figure 2.5 A *highly* schematic representation of the energy states of a protein. We assume that the molecule has only a single native state and a large number of denatured conformations of equal energy. These assumptions are certainly far too severe, but they allow a simple calculation.

Now, we can rewrite Eq. 2.51 as

$$\frac{n_d}{n_n} = e^{\ln (W_d/W_n)} e^{-(\epsilon_d - \epsilon_n)/k_B T}$$

$$= e^{-[(\epsilon_d - \epsilon_n) - k_B T \ln (W_d/W_n)]/k_B T} \qquad (2.54)$$

or, putting quantities on a molar basis,

$$\frac{n_d}{n_n} = e^{-[(E_d - E_n) - RT \ln (W_d/W_n)]/RT} \qquad (2.55)$$

Both sides of this equation are recognizable. In the exponent, if we neglect the difference between ΔE and ΔH and note that the quantity $R \ln (W_d/W_n)$ is an entropy change, we have

$$\frac{n_d}{n_n} = e^{-(\Delta H - T\Delta S)/RT}$$

$$= e^{-\Delta G/RT}$$

$$\ln \left(\frac{n_d}{n_n}\right) = -\frac{\Delta G}{RT} \qquad (2.56)$$

For the special case where n_d/n_n is the ratio of molecules at equilibrium, $\Delta G = 0$. However, most readers are familiar with the equation

$$\ln \left(\frac{n_d}{n_n}\right) = \ln K = -\frac{\Delta G^0}{RT} \qquad (2.57)$$

where K is the equilibrium constant and ΔG^0 is the *standard* free energy. The term ΔG^0 is a measure of the *stability* of the products relative to the reactants at equilibrium (we discuss the standard state in greater detail in Section 2.4 and derive Eq. 2.57 more rigorously in Chapter 13). Here it points out again the role of randomness: The denatured state may be favored in some circumstances, just because it corresponds to more states of the molecules. Conversely, it should be noted that only because of the energetically favorable residue interactions does the native form described above exhibit stability. Were these not present, the entropy change would drive the reactions toward the random coil at *any* temperature above absolute zero.

It should not be supposed that the stability of real macromolecular conformations is as simple as implied here. For one thing, the solvent cannot, in general, be neglected, and solute-solvent interactions may play a very important role. To take a particular example, it has been argued that the stability of some macromolecules may derive in part from *hydrophobic bonding.* Such bonding may lead to a very different temperature dependence of stability. In the breaking of a hydrophobic bond, nonpolar groups are separated from one another and put in contact with the solvent. In aqueous solution, such groups are expected to become surrounded by shells of "ice-like" water (Figure 1.9). The immobilization of a large number of water molecules should correspond to an entropy decrease—perhaps so much so to override the entropy increase accompanying the macromolecule's gain of conformational freedom. If this is the case, the overall entropy change for the transition from the ordered conformation of the macromolecule to the random one would involve a *decrease* in entropy. Then the temperature dependence should be a reverse of the above example; *low* temperatures should favor disorganization. Such behavior may be observed when we see that some multisubunit proteins dissociate into their individual subunits at temperatures around 0°C.

2.4   THE STANDARD STATE

In chemistry, a *standard state* is defined so that all chemical reactions can be compared under a defined set of reference conditions. For example, the standard state for liquids and solids is the most stable form at 298 K under 1 atm pressure. The standard state for gases has been defined for a temperature of 298 K and partial pressures (not total pressures) of each component at 1 atm. Thus, any change in a system can be related to a point of reference, which is the standard state, and the corresponding standard thermodynamic quantities of H^0, S^0, and G^0. We can readily see from Eq. 2.46 that

$$G^0 = H^0 - TS^0 \tag{2.58}$$

and that

$$\Delta G^0 = \Delta H^0 - T\Delta S^0 \tag{2.59}$$

For a mixed population of components, for example, a solution with multiple solutes, the standard state is defined in an analogous fashion to that of mixtures of gases but, instead of partial pressures of 1 atm, the concentrations for each solute are defined at 1 M. The free energy of a system, then, can be perturbed from this standard state by changing any of these parameters, for example, deviations in the concentrations from 1 M. In the case where we have a solution of some solute A in a pure solvent, we can see that

$$G = G^0 + RT \ln a \qquad (2.60)$$

where a is the activity (or effective concentration) of A. In a dilute solution, a is approximately equal to the molar concentration $[A]$ and therefore

$$G = G^0 + RT \ln[A] \qquad (2.60)$$

The obvious problem with this expression is that logarithms are applied to unit-less quantities. This is where the standard state becomes useful in another way. In fact, the $\ln[A]$ quantity in Eq. 2.60 should be written as $\ln([A]/1\text{ M})$, for $[A]$ in units of moles/l. This way, G becomes explicitly a measure of the Gibbs free energy for a system relative to the standard state where the concentrations of solutes are indeed 1 M.

For a solution containing multiple components of a reaction (reactants R and products P), the change in free energy is

$$\Delta G = \Delta G^0 + RT \ln \left(\frac{\Pi[P_i]}{\Pi[R_i]} \right) \qquad (2.61)$$

Notice here that if the concentrations of all reactants and all products are 1 M, we reiterate the standard state and $\Delta G = \Delta G^0$. Thus, the standard state is defined as the thermodynamic state where the free energy of the system is the standard free energy.

However, the standard state of a system may not be the most stable state (it may not be possible to hold all reactants and products of a reaction at 1 M concentration). A closed system (in which no additional components or external energy is introduced) will eventually reach a state where concentrations of reactants and products do not change (that is, they are at equilibrium) and, therefore, there is no thermodynamic driving force to change the state of the system ($\Delta G = 0$). At this point, the concentrations of materials are likely in fact not 1 M, but will reach their equilibrium concentrations ($[R]_{eq}$ for reactants and $[P]_{eq}$ for products). From Eq. 2.61, we can see then that

$$0 = \Delta G^0 + RT \ln \left(\frac{\Pi[P_i]_{eq}}{\Pi[R_i]_{eq}} \right) \qquad (2.62)$$

which leads to the familiar form

$$\Delta G^0 = -RT \ln K_{eq} \qquad (2.62)$$

Thus, the standard free energy can be directly calculated from the equilibrium constant.

Finally, the standard or reference biochemical state is not necessarily the same as that of the chemical definition. For example, the standard solvent system in biochemistry is an aqueous buffer. The standard state, therefore, is the stable form of the solvent as well as the solutes that, for water, is a neutral solution at pH = 7.0, where the proton and hydroxide ion concentrations are at 10^{-7} M, rather than at 1 M as defined under the standard chemical state. We must, therefore, recognize that ΔG^0 determined for a standard biochemical system is defined with not all components at 1 M, but is still determined at equilibrium.

2.5   EXPERIMENTAL THERMOCHEMISTRY

We now have the definitions for the thermodynamic parameters of enthalpy, entropy, free energy, and their relationship to equilibrium, but how are these parameters actually determined for a biochemical system? The K_{eq} can be readily determined from the concentrations of all components at equilibrium (the details for how this is done are discussed in later chapters), which then allows us to calculate ΔG^0. The next sections discuss some of the experimental approaches for delineating the ΔH^0 and ΔS^0 components of ΔG^0.

2.5.1   The van't Hoff Relationship

Although we have not stated this explicitly, it should be clear from Eqs. 2.59 and 2.62 that both K_{eq} and ΔG^0 are temperature dependent quantities. Rearranging Eq. 2.62 to

$$\ln K_{eq} = -\frac{\Delta G^0}{RT} = -\left(\frac{\Delta H^0 - T\Delta S^0}{RT}\right) \tag{2.64}$$

or

$$\ln K_{eq} = \frac{1}{R}\left(\Delta S^0 - \frac{\Delta H^0}{T}\right) \tag{2.65}$$

results in an expression known as the van't Hoff relationship, indicating that $\ln K_{eq}$ is a linear function of $1/T$. Differentiating Eq. 2.63 versus $1/T$ gives

$$\frac{d \ln K_{eq}}{d(1/T)} = \frac{-\Delta H^0}{RT} = \frac{-\Delta H^0}{R}\left(\frac{1}{T}\right) \tag{2.66}$$

which tells us that the enthalpy of a reaction can be determined by measuring the equilibrium constant for a system at multiple temperatures. The enthalpy of the system can then be calculated from the slope $(-\Delta H^0/R)$ of the resulting linear van't Hoff plot of $\ln K_{eq}$ versus $1/T$ (Figure 2.6). This is known as the van't Hoff enthalpy

Figure 2.6 Van't Hoff plot for the thermal unfolding of the coiled coil domain of the GCN4 transcription factor. [Adapted from Holtzer et al. (2001), *Biophys. J.* **80**, 939–951.]

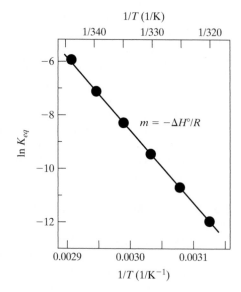

(ΔH_{vH}^0). In addition, we would expect from Eq. 2.63 that the y-intercept of the van't Hoff plot should give $\Delta S^0/R$, assuming that ΔH_{vH}^0 is temperature independent, which, for narrow temperature ranges, may be reasonable. However, over large ranges of T, this assumption is not true, and certainly not when extrapolated to $1/T = 0$ (or T to infinity). ΔS^0 is thus more reliably calculated directly from the values of $\ln K_{eq}$ and the ΔH_{vH}^0 determined from the van't Hoff plot.

A different form of the van't Hoff relationship can be derived by differentiating $\ln K_{eq}$ in Eq. 2.65 relative to dT, rather than to $d(1/T)$, to give

$$\frac{d \ln K_{eq}}{dT} = \frac{\Delta H_{vH}^0}{RT^2} \tag{2.67}$$

Although not as convenient for interpretation as the linear van't Hoff plot, this form is mathematically easier to manipulate as we try to relate the van't Hoff enthalpies to the enthalpic values determined by other methods.

2.5.2 Calorimetry

Calor in Latin means "heat"; *calorimetry,* therefore, is literally "the measurement of heat" (q). Most chemistry students are familiar with a "bomb" calorimeter, which is used to measure the heat exchanged between a system and its surroundings under constant volume (q_v). In such an experiment, for example, when studying the combustion of an organic compound in a sealed container, the resulting heat released is a direct measure of the change in internal energy (ΔE) of the system, according to Eq. 2.6. In biochemical reactions, unlike the bomb calorimetry experiments, the system is typically under constant pressure rather than constant volume and therefore Eq. 2.7 tells us that q_p under these conditions is a measure of ΔH.

There are several methods for measuring heat and the changes in heat in a biochemical system. The current *microcalorimeters* are designed to accurately study thermochemical processes in very small volumes (as low as 100 μl). The two most common techniques are differential scanning calorimetry, which is typically used to monitor the changes associated with phase transitions, and isothermal titration calorimetry, which monitors the energies of interaction between molecules upon mixing. A third, emerging method, which is not discussed here, is photoacoustic calorimetry, where one "listens" for the changes in thermal energy resulting from a dynamic change in the system.

Differential scanning calorimetry. Differential scanning calorimetry (DSC) has generally been applied to studying the simplest of biochemical reactions, which is a unimolecular transition of a molecule from one conformation to another. In the basic DSC experiment (Figure 2.7), energy is introduced simultaneously into a sample cell (containing a solution with the molecule of interest) and a reference cell (containing only the solvent), raising the temperatures of both identically over time. The difference in the input energy required to match the temperature of the sample to that of the reference at each step of the scan is the amount of excess heat either absorbed (in an endothermic process) or released (in an exothermic process) by the molecule in the sample.

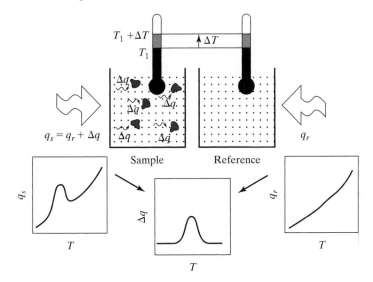

Figure 2.7 Experimental setup for a differential scanning calorimetry experiment. The amount of heat required to increase the temperature by the same increment (ΔT) of a sample cell (q_s) is higher than that required for the reference cell (q_s) by the excess heat absorbed by the molecules in the sample (Δq). The resulting DSC scans with the reference subtracted from the sample shows how this excess heat changes as a function of temperature.

Consider a very simple DSC experiment in which the sample contains, for example, an aqueous buffered solution of a protein, while the reference contains just the aqueous buffer without the protein. During the scan, more energy is required to bring the sample to the same temperature as the reference because of the excess heat absorbed by the protein (Figure 2.7). When corrected for the mass of the protein in the sample cell, the excess heat (dq) absorbed for the incremental change in temperature (dT) will be a measure of the heat capacity at constant pressure (C_p) of just the protein, as given by

$$C_p = dq/dT \qquad (2.68)$$

or, similarly under constant pressure,

$$C_p = dH/dT \qquad (2.69)$$

If we start at a temperature where the protein is in its folded state, that heat capacity would be for the protein in its native folded form (C_N). At some point, the protein will undergo thermal unfolding, and the subsequent slope will change and represent the heat capacity of the denatured protein (C_D). At the point of transition, there will be an associated phase change and a heat of transition that, for protein denaturation, is endothermic. The average calorimetric enthalpy ($<\Delta H^0>$) associated with denaturation can be determined by integrating the area under the curve between T_1 and T_2, where T_1 is the temperature at which all of the protein is in its native form and T_2 where it is denatured (Figure 2.8).

$$\langle \Delta H^0 \rangle \propto \int_{T_1}^{T_2} C_p dT \qquad (2.70)$$

Notice that, unlike the assumptions made in the van't Hoff relationship, the heat capacities and the associated enthalpies are temperature dependent throughout the temperature range of the scan. In addition, although the temperature is constantly

Figure 2.8 DSC for protein denaturation shows the heat capacity (C_P) changing from that of the native form [$C_N(T)$] to that of denatured form [$C_D(T)$]. At the point of transition is the excess heat capacity associated with denaturation. The integrated areas under the curve from T_1 (fully native) to T_2 (fully denatured) is the enthalpy for denaturation. At the midpoint temperature (T_m), the protein is half native and half denatured; thus the associated enthalpies for the folded native [$\Delta H_N(T)$] and denatured [$\Delta H_D(T)$] protein are equal and of opposite signs. [Adapted from S.-I. Kidokoro and A. Wada (1987), *Biopolymers* **26**, 213–229.]

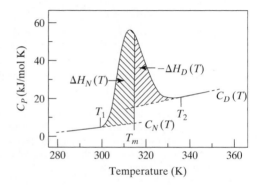

rising, each point along the DSC scan (if the scan rate is slower than the rate of transition) can be considered to be at or near equilibrium (and therefore, ratios of the ending and starting forms of the protein can be treated as equilibrium constants).

The temperature at which half the protein is denatured is the melting temperature (T_m) for the protein (or the midpoint temperature for denaturation). This is obviously the point where [denatured] = [native] protein $(K_{eq} = 1$ at the T_m for the simple two-state reversible denaturation process, where there are no significant populations of intermediate folded forms of the protein). At this point, the integrated areas under the transition curve are identical and therefore, the contributions of the enthalpies for the native and denatured forms are equal (Figure 2.8). In addition, the overall enthalpy at the T_m (ΔH_m^0) can be equated with the calorimetric enthalpy $(<\Delta H^0>)$ without significant error. Thus, the associated $\Delta G_m^0 = 0$, and ΔS_m^0 can be calculated by

$$\Delta S_m^0 = \Delta H_m^0 / T_m \tag{2.71}$$

In this case, ΔH_m^0 and ΔS_m^0 are correct at the T_m. At any temperature $(T \neq T_m)$, the enthalpy $[\Delta H^0(T)]$, entropy $[\Delta S^0(T)]$, and free energy $[\Delta G^0(T)]$ difference between the native and denatured states can be shown to be

$$\Delta H^0(T) = \Delta H_m^0 + \int_{T_m}^{T} \Delta C_p \partial T \tag{2.72}$$

$$\Delta S^0(T) = \Delta H_m^0 / T_m + \int_{T_m}^{T} \frac{\Delta C_p}{T} \partial T \tag{2.73}$$

where ΔC_p is the difference in heat capacity between the native and denatured states, and

$$\Delta G^0(T) = \Delta H^0(T) - T \Delta S^0(T) \tag{2.74}$$

These relationships are explicitly correct only for transitions that involve only two states that are fully reversible. For transitions where intermediates are sufficiently long lived to contribute to the population, or if the transition is not entirely reversible, determinations of the parameters are not as straightforward.

To relate this calorimetric enthalpy to the van't Hoff enthalpy (ΔH_{vH}^0), we start by defining the calorimetric enthalpy at a particular temperature $[\Delta H^0(T)]$ as the difference between the enthalpies of native (H_n^0) and denatured (H_d^0) forms of the protein.

$$\Delta H^0(T) = H_n^0(T) - H_d^0(T) \tag{2.75}$$

We can show that the ratio of denatured and native forms of the protein at some temperature $[K(T)]$ is also proportional to the ratios of the enthalpies of the two forms of the protein.

$$K(T) = \frac{[denatured]}{[native]} = -\frac{\Delta H_d^0(T)}{\Delta H_n^0(T)} \tag{2.76}$$

If $K(T)$ is treated as an equilibrium constant, then the van't Hoff relationship in Eqs. 2.67 and 2.76 can be applied to derive the following function for ΔH_{vH}^0

$$\Delta H_{vH}^0(T) = -RT^2 \frac{d \ln K(T)}{dT} \tag{2.77}$$

Recalling that $d \ln x/dy = (1/x)\, dx/dy$, we can use Eqs. 2.69 and 2.76 to show that

$$\Delta H_{vH}^0(T) = -RT^2 \left(\frac{\Delta C_n(T)}{\Delta H_n^0(T)} - \frac{\Delta C_d(T)}{\Delta H_d^0(T)} \right) \tag{2.78}$$

This tells us that ΔH_{vH}^0 is dependent on the fractions of the native and denatured forms of the proteins, and the associated temperature dependence on the excess heat capacities [$\Delta C_n(T)$ and $\Delta C_d(T)$] and enthalpies [$\Delta H_n^0(T)$ and $\Delta H_d^0(T)$] in this simple two-state transition. Experimentally, ΔH_{vH}^0 is determined from the temperature dependence of K_{eq}. Equation 2.78 shows that if this is not a simple two-state denaturation process, then ΔH_{vH}^0 will not match the calorimetric $<\Delta H^0>$. For example, if denaturation involves formation of a relatively long-lived intermediate structure, the resulting DSC curve will be a convolution of the heat capacities and enthalpies of all relevant forms of the protein. Compared to predictions from the two-state model, there might be a broadening or skewing of the DSC curve, disparities between the van't Hoff and experimentally determined T_m, or deviations of the calculated fractions of the denatured and native proteins from the expected values of 0.5 at the T_m. The contribution of each component to the overall shape of the DSC curve can be deconvoluted by applying a statistical mechanics model to describe the transition (Chapter 4). A ratio of $\Delta H_{vH}^0/<\Delta H^0> = 1$, however, is evidence that the phase transition indeed is a simple two-state process.

Differential scanning calorimetry can also be used to study the thermochemistry of two or more interacting molecules, as in the binding of a ligand to a protein, or the interaction between two proteins.

$$A + B \longrightarrow C$$

In such cases, there would be a contribution of the heat capacities and enthalpies from each component (A and B) and the complex between them (C) (Figure 2.9) to the thermodynamic system. Once again, the contributions of each can be delineated by applying statistical mechanics models for binding to the system (as discussed in Chapters 4 and 15). The complications in this approach are that the fractions of bound and free proteins may be temperature dependent, and each component can undergo thermal denaturation during the DSC experiment.

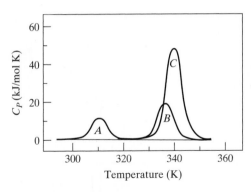

Figure 2.9 The DSC scan for the association of two molecules (A and B) to form a complex (C) will have components from A, B, and C, in proportions that depend on the equilibrium concentrations at each temperature along the scan. [Adapted from I. Jelesarov and H. R. Bosshard (1999), *J. Mol. Recognit.* **12**, 3–18.]

Isothermal titration calorimetry.   Although DSC can be used to determine the energetics of a binding reaction, the more typical method to study the thermochemistry of interacting molecules is by isothermal titration calorimetry (ITC). At first glance, ITC appears conceptually to be simpler than DSC, because it measures directly the amount of heat absorbed or released by a reaction that results from mixing together two or more components. In the simplest case, a ligand (L) is introduced into a solution of protein (P) that can bind to form a simple binary complex (PL). A more detailed discussion of ligand binding is presented in Chapter 13. Here, we consider only this very simple bimolecular reaction. The resulting release or absorption of heat from the reaction is dependent on the enthalpy at a particular temperature $[\Delta H^0(T)]$ and the number of moles of complex formed (n_{PL}, which is dependent on the volume V of the sample and the concentration of the complex $[PL]$).

$$q = \Delta H^0(T)n_{PL} = \Delta H^0(T)V[PL] \tag{2.79}$$

Modern instruments actually measure the amount of energy required to maintain a constant temperature between the sample cell and a reference over time (dq/dt, where t is time and not temperature). Consequently, similar to DSC, q is the integrated area of the resulting peak as you monitor the compensating energy, but in ITC, q is integrated as a function of time (Figure 2.10).

For the following equilibrium (with the association constant K_a)

$$P + L \overset{K_a}{\rightleftharpoons} PL$$

we can easily show that

$$[PL] = [P_T]\left(\frac{K_a[L]}{1 + K_a[L]}\right) \tag{2.80}$$

where $[L]$ is the concentration of unbound ligand, and $[P_T]$ is the total concentration of protein ($[P_T] = [P] + [PL]$, where $[P]$ is the concentration of unbound protein). Substituting Eq. 2.80 into Eq. 2.79, we get

$$q = \Delta H^0(T)V[P_T]\left(\frac{K_a[L]}{1 + K_a[L]}\right) \tag{2.81}$$

Figure 2.10 Isothermal titration calorimetry (ITC) involves incremental addition over time of a ligand (circles) into a solution of a protein. At each titration point, formation of a complex results in the release or absorption of heat (q), which is the heat of complex formation. At the initial time intervals (t_1 and t_2), the amount of free protein is sufficient to bind ligand according to the concentrations of free ligand and protein in the solution. In later intervals, however, the amount of free protein becomes limiting until, at the end of the titration (t_4), there is no protein available to form a complex and, therefore, no excess heat is absorbed or released, even with the added ligand.

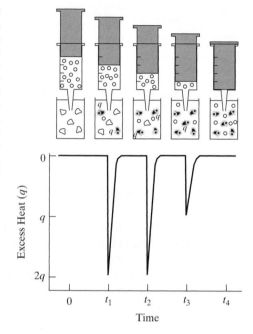

From Eq. 2.81, it is obvious that if q is measured, the enthalpy of the binding reaction can be determined if K_a is known or K_a can be determined if $\Delta H^0(T)$ is known. However, neither can be determined from a single experiment. Therefore, in an actual experiment, a fixed quantity of protein is titrated incrementally with smaller amounts of ligand at defined time intervals, and the area under each peak is the heat absorbed or released at time interval i (q_i) for the amount of ligand added at that interval (Figure 2.10). Thus, for any step i along the titration, the area under the ith peak along the titration is

$$q_i = \Delta H^0(T)V[P_T]\left(\frac{K_a[L]_i}{1 + K_a[L]_i} - \frac{K_a[L]_{i-1}}{1 + K_a[L]_{i-1}}\right) \tag{2.82}$$

Without knowing K_a to start, it is impossible to know $[L]$. What is known is the total amount of ligand added ($[L_T]$), or the ratio of total ligand added relative to the total protein ($R = [L_T]/[P_T]$). The resulting ITC data can thus be presented as a plot of q_i as a function of R (Figure 2.11). Values for K_a and $\Delta H^0(T)$ can thus be determined by fitting the resulting data with Eq. 2.82, but with the following expression:

$$[L]_i = \frac{[L_T] - [P_T] - 1/K_a \pm \sqrt{([L_T] - [P_T] - 1/K_a)^2 - 4[L_T]}}{2} \tag{2.83}$$

substituted for $[L]_i$ (this is simply the concentration of the ligand at each increment i expressed in terms of the known quantities of $[L_T]$ and $[P_T]$).

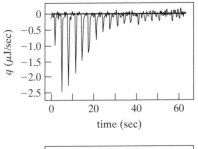

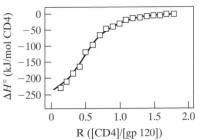

Figure 2.11 The thermochemistry of the binding reaction between core binding domain of the exterior glycoprotein (gp120) from the HIV-1 virus and the CD4 receptor of the targeted host cell. In this ITC study, the heat change over time (top graph) is integrated and normalized to the amount of the CD4 receptor injected. This yields the enthalpy change ($\Delta H^0 = -263$ kJ/mol) and an association constant of $K_a = 5 \times 10^6$ M^{-1} when the resulting titration curve (bottom graph) is fit using Eq. 2.82 and 2.83. [Data from Myszka et al. (2000), *Proc. Natl., Acad. Sci., USA* **97**, 9026–9031.]

Once K_a and $\Delta H^0(T)$ are determined, $\Delta G^0(T)$ can be calculated as $\Delta G^0(T) = -RT \ln K_a$ and $\Delta S^0(T)$ from the standard equation $\Delta G^0(T) = \Delta H^0(T) - T\Delta S^0(T)$. The heat capacity associated with the binding reaction can be determined by repeating the ITC experiment at different temperatures. Equation 2.69 tells us that ΔC_p should be the slope of the resulting plot of $\Delta H^0(T)$ versus T (Figure 2.12). The effect of T on $\ln K_a$ allows us to calculate ΔH_{vH}^0. Once again, ΔH_{vH}^0 will be identical to the calorimetric enthalpy only if the binding process is truly a two-state process between the free and bound forms of the protein, and if K_a measures the entire population of free and bound protein and ligand.

One limitation of ITC is that it cannot be directly applied to studying the thermodynamics of a binding reaction in which K_a is very large—the practical upper limit is estimated to be $K_a \approx 10^9$ M^{-1}. In such tight binding reactions, each incremental addition of ligand results in complete binding of the ligand ($[PL] \approx [L_T]$).

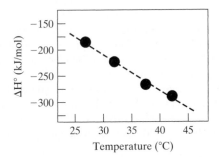

Figure 2.12 The heat capacity of the HIV-1 gp120 and CD4 receptor binding reaction in Figure 2.11 is determined by measuring the change in enthalpy over a range of temperatures. A $C_p = -7.5$ kJ/mol° is calculated from the slope when ΔH^0 is plotted relative to the temperature. [Data from Myszka et al. (2000) *Proc. Natl., Acad. Sci., USA* **97**, 9026–9031.]

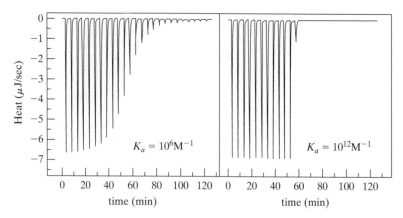

Figure 2.13 The ITC trace of a binding reaction having a measurable association constant $(K_a = 10^6 \text{ M}^{-1})$ results in a smooth titration curve that can be fit to yield the ΔH^0 and K_a of the reaction. A binding reaction with a very strong association $(K_a = 10^{12} \text{ M}^{-1})$, however, shows heats of complex formation that are identical for all intervals where the ligand added is less than the protein in solution, but becomes zero when the concentration of ligand added becomes greater than that of the protein. This results in a very sharp titration curve that cannot be fit by Eqs. 2.82 and 2.83.

Consequently, since $[P] = [P_T] - [P_L]$, when $[L_T] \geq [P_T]$, $[P] \approx 0$. In other words, q_i measured at each titration increment where $[L_T] < [P_T]$ will all be essentially the same, reflecting the complete formation of the complex PL, and equal to the amount of ligand added. For each increment where $[L_T] > [P_T]$, q_i will be zero, since there will be no free protein left to bind ligand (Figure 2.13). The resulting titration will show a sharp transition from complete binding of added ligand to no binding. The resulting binary ITC trace cannot be accurately fit using Eqs. 2.82 and 2.83 and, therefore, K_a and ΔH^0 cannot be determined for the binding reaction. One strategy to circumvent this problem is through a competitive ITC experiment, where a weak ligand initially bound to a protein is displaced by one that binds the protein more tightly (Application 2.1).

Application 2.1   Competition Is a Good Thing

In the area of drug discovery and design, one of the more important properties to optimize is the affinity of an inhibitor for a protein, with tighter binding generally regarded as indicating a more effective inhibitor. Trying to study the thermochemistry of very tightly binding inhibitors, however, can be difficult. For example, the energetics for binding of a second-generation inhibitor (KNI-764) to HIV-1 protease cannot be studied directly by ITC, because of the very high association constant of this compound. But, all is not lost—rather than trying a direct measurement, Velazquez, Kiso, and Freire (2001) designed a competitive assay for this determination. The assay starts by detailing the thermochemistry of a weaker binding inhibitor (Ac-pepstatin) to the viral protease. With the association constant of Ac-pepstatin in hand $(K_{Ac} = 2.4 \times 10^6 \text{ M}^{-1})$, this weak inhibitor-protease complex is titrated and displaced by the stronger binding KNI-764 compound. The resulting ITC titration reflects the thermochemistry of the dissociation and competitive reassociation steps, and yields an apparent association constant (K_{app}). The energetics for direct binding of KNI-764 can thus be deduced, and the binding constant for this tight binding inhibitor (K_{KNI}) calculated from the expression

$$K_{app} = \frac{K_{KNI}}{(1 + K_{Ac}[\text{Ac-pepstatin}])} \tag{A2.1}$$

This demonstrates the power of a classic thermodynamic cycle, which allows the determination of an immeasurable quantity indirectly by determining the parameters for the steps that leads to the same end point, but through a circuitous route. In this case, K_{KNI} was estimated to be $3.1 \times 10^{10}\ M^{-1}$, which is well beyond the practical limits of a direct ITC measurement.

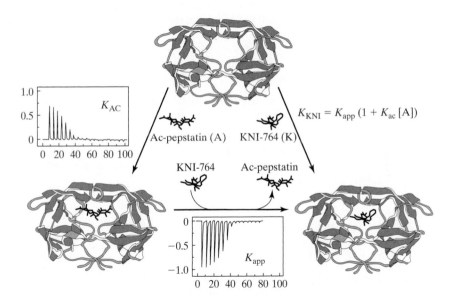

Figure A2.1 The association constant for the tight binding KNI-764 HIV-1 protease inhibitor (K_{KNI}) was determined by first measuring the binding constant for the weaker inhibitor Ac-pepstatin (K_{Ac}), followed by a competitive study to determine an apparent association constant (K_{app}) for the displacement of Ac-pepstatin by KNI-764.

VELAZQUEZ-CAMPOY, A., Y. KISO, and E. FREIRE (2001) "The Binding Energetics of First and Second Generation HIV-1 Protease Inhibitors: Implications for Drug Design," *Arch Biochim Biophys.* **390**, 169–175.

EXERCISES

[Note: The first six exercises in this chapter are intended mainly for the review of the student's previous experience with physical chemistry. They may draw on details that have not been mentioned in this chapter.]

2.1 Calculate q, w, ΔE, and ΔH for the isothermal expansion of 2 moles of an ideal gas from 16 liters to 95 liters at 0°C.

2.2 The standard-state enthalpy change for the oxidation of palmitic acid

$$CH_3(CH_2)_{14}COOH(s) + 24O_2(g) \longrightarrow 17CO_2(g) + 16H_2O(l)$$

is −9805 kJ.
  a. Calculate ΔE.
  b. Calculate the work done when 1 mole is oxidized at 1 atm pressure.

2.3 The heat of melting of ice at 1 atm and 0°C is +5.9176 kJ/mole. The density of ice under these conditions is 0.917 g/cm³ and the density of water is 0.9998 g/cm³.
  a. Calculate $\Delta H - \Delta E$.
  b. Calculate ΔS.
  *c. What will the melting point be at 10 atm? Derive necessary equations, make necessary assumptions.

2.4 Assuming that the osmotic pressure is given by $\pi = RTC/M$, work out the ΔS for the process depicted in Figure 2.3. Assume that you start with 1 g of sucrose in 100 ml and increase the volume to 1 liter.

*2.5 Assuming that a polypeptide chain has only one α-helical conformation and that there are three possible orientations for each amino acid residue in the random-coil state, calculate ΔS for the conformational change

$$\alpha\text{-helix} \longrightarrow \text{random coil}$$

for a polypeptide of 100 residues. What value of ΔH per residue would be required to make the melting point (the temperature at which the equilibrium constant equals 1) be 50°C? Compare this with the hydrogen-bond energy, estimated to be 0 to 12 kJ/mol.

2.6 Two energy levels of a molecule are separated by 1×10^{-15} erg. The degeneracy of the higher level is twice that of the lower. Calculate
  a. the relative populations of these levels at 0°C
  b. the temperature at which they will be equally populated

*2.7 Closed circular DNA molecules can be supercoiled. The energy required to put i supercoil twists into a mole of DNA molecules is given approximately by $\Delta E = Ki^2$, where $K \cong 0.4$ kJ/mol. Consider a circular DNA molecule that has been broken and the ends rejoined. Show that the number of molecules with i supercoil twists after rejoining is given by a Gaussian distribution about $i = 0$, and calculate the relative numbers of molecules with 0, 1, 2, and 10 supercoil twists at 37°C.

2.8 The equilibrium constant (K_{eq}) for a binding reaction was seen to increase monotonically as the temperature (T) increased from 290 K to 315 K (see accompanying table). Construct a van't Hoff plot for the data to determine ΔH^0, ΔS^0, and ΔG^0 for this reaction at room temperature (298 K). Of these parameters, which ones require that we specify the temperature?

T (°K)	K_{eq} (M^{-1})
290	1.2×10^5
295	1.3×10^5
300	1.4×10^5
305	1.5×10^5
310	1.6×10^5
315	1.7×10^5

2.9 The van't Hoff enthalpy for the binding study in Exercise 2.8, when compared to the calorimetric enthalpies, was found to match very well the values from ITC, but not from DSC. What may be the possible explanation for this discrepancy?

2.10 Using the thermodynamic cycle in Figure A2.1, derive Eq. A2.1 for the apparent affinity constant (K_{app}) for the competition assay between the weak and tight inhibitor binding to HIV-1 protease.

REFERENCES

Physical Chemistry

Thermodynamics and elementary statistical mechanics are well treated in any number of the better physical chemistry texts. We recommend especially:

ATKINS, P. W. (2002) *Physical Chemistry,* 7th ed., W. H. Freeman, New York.

MOORE, W. J. (1972) *Physical Chemistry,* 4th ed., Prentice Hall, Upper Saddle River, NJ. More detailed and rigorous than most of the available texts at this level.

Physical Chemistry for Biological Sciences

There are also several texts directed toward students of the biological sciences. We have found the following to be especially good:

EISENBERG, D. and D. CROTHERS (1979) *Physical Chemistry with Applications to the Life Sciences,* Benjamin-Cummings, Menlo Park, CA.

TINOCO, I., JR., K. SAUER, J. C. WANG and J. D. PUGLISI (2002) *Physical Chemistry: Principles and Applications in Biological Sciences,* 4th ed., Prentice Hall, Upper Saddle River, NJ.

Basic Thermodynamics

In the realm of more specialized books, the following have been especially useful:

DENBIGH, K. G. (1955) *The Principles of Chemical Equilibrium,* Cambridge University Press, Cambridge, UK. A very different book—discursive, somewhat long, but beautifully clear.

GIBBS, J. W. (1948) *The Collected Works of J. W. Gibbs,* Yale University Press, New Haven, CT. Not a useful text, but worth examining for the elegance of the analysis.

GURNEY, R. W. (1949) *Introduction to Statistical Mechanics,* McGraw-Hill Book Company, New York. An excellent book for the beginner in statistical thermodynamics.

HILL, T. L. (1963) *Thermodynamics of Small Systems,* W. A. Benjamin, New York. The application of statistical methods to a number of kinds of problems of interest to the biochemist.

KIRKWOOD, J. G. and I. OPPENHEIM (1961) *Chemical Thermodynamics,* McGraw-Hill Book Company, New York. A high-level, authoritative treatment; terse.

KLOTZ, I. (1957) *Energetics in Biochemical Reactions,* Academic Press, New York. A very brief, nonmathematical introduction to thermodynamics for biochemists. Some nice examples and explanations.

MOELWYN-HUGHES, E. A. (1961) *Physical Chemistry,* Pergamon Press, Ltd., Oxford, UK.

MOROWITZ, H. J. (1970) *Entropy for Biologists,* Academic Press, New York. A brief (and good) thermodynamics text.

Calorimetry

The theory and methods of DSC and ITC techniques as they apply to biochemical systems are discussed in greater detail in the following references.

INDYK, LAWRENCE and HARVEY F. FISHER (1998) "Theoretical Aspects of Isothermal Titration Calorimetry," *Meth. Enzymology* **295**, 350–364.

JELESAROV, ILIAN and HANS RUDOLF BOSSHARD (1999) "Isothermal Titration Calorimetry and Differential Scanning Calorimetry as Complementary Tools to Investigate the Energetics of Biomolecular Recognition," *J. Mol. Recognition* **12**, 3–18.

KIDOKORO, SHUN-ICHI and AKIYOSHI WADA (1987) "Determination of Thermodynamic Functions from Scanning Calorimetry Data," *Biopolymers* **26**, 213–229.

LEVITT, STEPHANIE and ERNESTO FREIRE (2001) "Direct Measurement of Protein Binding Energetics by Isothermal Titration Calorimetry," *Curr. Opinion Struct. Biol.* **11**, 560–566.

SACHEX-RUIZ, JOSE M. (1995) "Differential Scanning Calorimetry of Proteins," in *Subcellular Biochemistry*, vol. 24, ed. B. B. Biswas and S. Roy, 133–176. Plenum Press, New York.

3

Molecular Thermodynamics

3.1 COMPLEXITIES IN MODELING MACROMOLECULAR STRUCTURE

One of the basic principles in biochemistry is that the information needed to fold a macromolecule into its native three-dimensional (3D) structure is contained within its sequence. This principle was first demonstrated by Anfinsen (1961, 1973), who showed that unfolded (*denatured*) ribonuclease spontaneously refolded (*renatured*) to an enzymatically active form. A long-sought goal in physical biochemistry is to accurately predict the 3D structure of a macromolecule starting with its sequence—this is the *folding problem* for macromolecules. In this chapter, we describe methods to model the conformations of macromolecules by using the basic principles of thermodynamics at the level of individual molecules. This includes a description of the interactions that facilitate the proper folding of macromolecules and how these interactions are formulated into energy functions that are useful for modeling macromolecular structures and behavior through *molecular simulation*. However, we stress that the folding problem has not been solved and that the principles described here represent only steps toward a general solution to the problem.

In theory, all the chemical properties of a macromolecule, including its 3D structure, can be predicted from an accurate description of the total thermodynamic state of the system at the atomic level. Although atoms are most accurately described by *quantum mechanics* (Chapter 8), most descriptions to date are only approximate. In this chapter, we review the methods of classical Newtonian physics to describe the thermodynamic properties of macromolecules.

Even with this classical approach, modeling macromolecular structures is complicated by the large number of atoms in the system. Insulin, a small protein of 51 amino acid residues, is composed of over 760 atoms, nearly 400 of which are non-hydrogen atoms. If we include ions and water molecules that are associated with each molecule, the system becomes even more unwieldy. For example, an insulin molecule in a 1 mM solution would, on average, interact with over 10^8 water molecules.

Finally, the conformations of the molecules are highly variable, with both the macromolecule and its solvent environment assuming any of a large number of different structures at any time. Each macromolecule has many degrees of freedom, not only in rotation and translation in solution, but also internal degrees of freedom at each freely rotating bond. Thus, the problem in trying to model macromolecular structures de novo stems from the large size and complexity of the system. The strategy, therefore, is to simplify the system. We will start by discussing some of these simplifications and the assumptions required to make them work.

3.1.1   Simplifying Assumptions

One of the first simplifications in trying to model a system of biological molecules in solution is to assume that the average behavior of the system can be represented by a single molecule (Figure 3.1). In this model, a single macromolecule and its associated solvent is isolated in a box. If the single molecule in this box is truly dynamic, it will eventually sample all the possible conformations accessible to the system. The properties of a population of molecules is thus represented by the time-averaged behavior of a single macromolecule in an isolated box. By making this a periodic (repeating) box, the contents of one box are identical to that of all the other boxes, and anything that leaves the box from one direction must simultaneously enter it from the opposite direction so that the concentration of material remains constant. Molecules are not allowed to move freely in or out of the box except in this periodic manner. This limitation may appear trivial, but in fact it is important if we consider that many biological molecules are dramatically affected by self-association, association with salts, and so on. One of the logistical problems in this type of model is to define a box that is large enough to accurately simulate the system, including the behavior

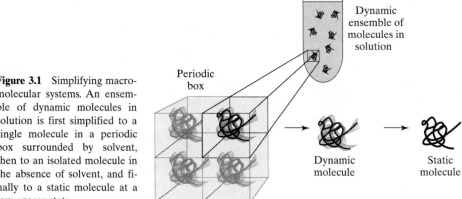

Figure 3.1 Simplifying macromolecular systems. An ensemble of dynamic molecules in solution is first simplified to a single molecule in a periodic box surrounded by solvent, then to an isolated molecule in the absence of solvent, and finally to a static molecule at a low-energy state.

of the bulk solvent. From an experimental point of view, there are now a number of very exciting methods to detect and study the structure and behavior of single molecules (Chapter 15). However, one needs to be cautioned that the characteristics of each individual molecule is unique and, therefore, many single-molecule measurements must be made in order to understand the average properties of a population.

Many macromolecular systems are still too unwieldy to simulate even when isolated in a periodic box. The next step in reducing the size and complexity of the system is to remove much or all the individual solvent and ions in the box. Obviously, if a molecule is treated in vacuo, in the absence of a solvent environment, there is no need for a physical box to contain the system. The problems with this simplification are obvious. Nearly all macromolecules make some contact with water, with the degree of solvent interaction defining the properties of the molecule. Thus, both hydrophilic and hydrophobic effects are largely ignored. However, a simulation in vacuo does reduce the number of atoms in the system by at least an order of magnitude. A number of approximations have been incorporated into the various molecular simulation methods as attempts to include the effects of the solvent without explicitly including solvent molecules as part of the system. We will discuss some of these approximations for treating the solvent and how they contribute to our understanding of macromolecular folding.

Finally, we can make the assumption that the native conformation of a molecule is the one with the lowest overall potential energy. The dynamic properties of the system are ignored in this case. Nonetheless, this general principle lays a foundation for methods that try to study and predict macromolecular structure, and does help to simplify the overall system.

We will start at the lowest level of molecular structure (the atom), and work through methods that attempt to rebuild the original complex system in a series of manageable steps.

3.2   MOLECULAR MECHANICS

3.2.1   Basic Principles

The best predictions of the structure and physical properties for a molecule come from an exact quantum mechanical treatment of every atom within a molecular system (Chapter 8). However, this is only analytically possible for the hydrogen atom. Using approximations to the wavefunctions for larger atoms introduces errors that are compounded as the molecule increases in size and complexity. The alternative is to apply a classical rather than quantum mechanical treatment to describe the interactions between atoms.

According to classical mechanics, the total energy E within a system includes both the kinetic K and the potential energies V, as discussed in Chapter 2 and as summarized in Eq. 3.1.

$$E = K + V \tag{3.1}$$

The kinetic energy used here includes all the motions of the atoms in the system, and thus is the sum of their kinetic energies. The potential energy of a macromolecule in its various conformations can be represented by a multidimensional surface with hills and valleys (Figure 3.2); any particular conformation corresponds to one point on this surface.

The description of molecular interactions is based on the principles of Newtonian physics. This is *molecular mechanics*, where molecular motions are determined by the masses of and the forces acting on atoms. The nuclei contribute the mass while the electrons provide the force of interaction between atoms. Thus, in classical molecular mechanics, the electrons and nuclei are treated together.

The basic relationship in molecular mechanics is Newton's second law of motion, which relates the force F along a *molecular trajectory* (the distance vector **r**) to the acceleration a of a mass m along that trajectory.

$$F = ma \qquad (3.2)$$

To simplify this discussion, we will only consider one component (for example x) of **r**. The potential energy surface, therefore, becomes an *energy profile* (Figure 3.2), but the relationships that we derive in one dimension apply to all three directions for the molecular trajectory.

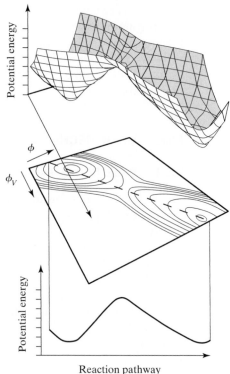

Figure 3.2 The potential energy surface for rotating a tyosine side chain in the protein pancreatic trypsin inhibitor. The potential energy is plotted as a function of the ring dihedral angle ϕ defined by the atoms C_α-C_β-C_γ-$C_{\delta 1}$ of the tyrosyl residue and a virtual dihedral angle ϕ_V, which measures the angle of the ring relative to the plane of the peptide bond (defined by the atoms C_β-C_γ-$C_{\delta 2}$ of the tyrosyl residue and the amino nitrogen of the next amino acid). The three-dimensional surface is projected onto a flat topographical map of the surface, where each contour represents an isoenergy level. The cross-section through the two-dimensional map (dotted line through) shows the profile of the potential energy along the reaction pathway. [Data from J. A. McCammon et al. (1983), *J. Am. Chem. Soc.* **105**, 2232–2237.]

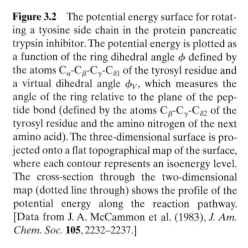

Reaction pathway

The force exerted in the direction of r is related to the potential energy by

$$F = -\frac{\partial V}{\partial r} \tag{3.3}$$

The force applied on an atom thus depends on how the potential energy changes as the distance between interacting atoms changes. The local gradient in the potential energy defines the *force field* in molecular mechanics. There are a number of different force fields used for modeling the structures of macromolecules, each having its own distinctive advantages and disadvantages. Our discussion will focus on energy functions that are common features of macromolecular force fields.

Newton's first law of motion is the special case for a system at equilibrium, where the net force is defined as $F = 0$. A system at equilibrium thus has $-\partial V/\partial r = 0$, which means that the molecule sits at a potential energy minimum. This is the basic principle behind *energy minimization* methods, which attempt to find the lowest energy conformation of a macromolecule, providing a static picture of the system at equilibrium.

The kinetic energy K of an atom is related to its velocity v or, equivalently, its momentum p.

$$K = \frac{1}{2}mv^2 = \frac{1}{2}\frac{p^2}{m} \tag{3.4}$$

The parameter K describes the dynamic change in the atomic positions at any time t. Thus, the methods of *molecular dynamics* are used to simulate the time-dependent changes in a system.

We discuss the application of molecular force fields to the simulation of molecular properties using energy minimization and molecular dynamics in greater detail later in this chapter. First, we must define the potential energy functions describing atomic and molecular interactions that are common to macromolecular force fields. This is followed by a discussion of how protein and nucleic acid structures are stabilized through these interactions.

3.2.2   Molecular Potentials

A description of the total potential energy in a macromolecular system must include the intermolecular interactions among molecules and the intramolecular interactions among atoms within the molecule. The potential energy of a single, isolated molecule depends only on the intramolecular interactions. The total intramolecular potential energy V_{total} is thus the sum of two types of interactions, the bonding V_{bonding} and the nonbonding interactions $V_{\text{nonbonding}}$. For N number of atoms in the molecule, we can write

$$V_{\text{total}} = \sum_{i=1}^{N} (V_{\text{bonding}} + V_{\text{nonbonding}})_i \tag{3.5}$$

Every conceivable conformation of a system has a corresponding value of V_{total}. If we plot V_{total} as a multidimensional surface where the coordinates are the positions of all atoms in the system, this is the energy surface for the system. To calculate the total potential, it is often sufficient to add together the interactions between pairs or small groups of adjacent atoms. These may be either *bonded* or *nonbonded* pairs or groups.

The bonding interactions are the covalent bonds that hold the atoms together, while the nonbonding interactions include electrostatic, dipolar, and steric interactions. In molecular mechanics, the potential energy functions are derived empirically, based on how molecules behave, as opposed to *ab initio* derivations from quantum mechanics. In many instances, the empirically derived functions are more accurate because they are based on the macroscopic properties that are actually observed for macromolecules, while quantum mechanical functions are often very approximate. We should note, however, that many parameters for these properties (for example, partial charges of atoms in peptide bonds) are derived from quantum mechanical calculations.

3.2.3 Bonding Potentials

The chemical bond that holds two atoms together is conceptually the easiest interaction in the total potential energy surface to understand. It also dominates the surface because of its magnitude ($\sim$150 to >1000 kJ/mol). The bond energy is the energy absorbed in breaking a bond or released in forming a bond (Table 3.1). Bond energies are therefore enthalpic energies.

The bonding energy is modeled in a force field as a distance-dependent function. An equilibrium distance r_0 is defined for the standard length of the chemical bond where, according to Eq. 3.3, no force is exerted on either of the bonded atoms. For any distance $r \neq r_0$ the atoms are forced toward r_0 by the potential energy surface. Therefore, how this force is applied depends on how V is defined.

The potential energy profile for a chemical bond is an anharmonic function with a minimum value at r_0, a steep ascending curve for $r < r_0$, and a more gradual ascending curve that approaches $V = 0$ for $r > r_0$ (Figure 3.3). The simplest form of the potential treats the chemical bond as a spring. Like a bond, a spring has an

Table 3.1 Average Dissociation Energies of Chemical Bonds in Organic Molecules

Bond Type	Bond Energy (kJ/mol)
C—H	408
C—C	342
C=C	602
C=N	606
C=O	732
O—H	458

Source: From A. Streitwieser, Jr., and C. H. Heathcock (1976), *Introduction to Organic Chemistry*, Macmillan, New York.

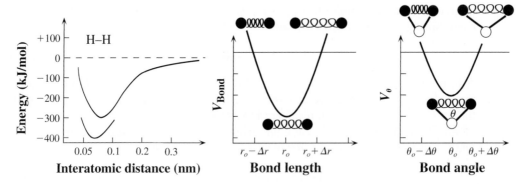

Figure 3.3   The potential energies for a hydrogen-hydrogen bond, and for deforming a covalent bond V_{bond} and the bond angle between three bonded atoms V_θ as treated by simple spring models. The potential energies calculated from quantum mechanics (upper curve) is compared to the experimental values (lower curve) for a hydrogen-hydrogen covalent bond (H—H). The deformations associated with any bond length are modeled, in the simplest case, as harmonic springs, with a spring constant for stretching $+\Delta r$ or compressing $-\Delta r$ a bond from the equilibrium bond length r_o. The spring model greatly overestimates the potential energy for stretching to large Δr. Deformations to the bond angle θ_o can be similarly treated as a spring between the 1-3 atoms of the three bonded atoms.

equilibrium length, and stretching or compressing the spring from r_0 requires an applied force. The potential energy of a chemical bond V_{bond} is thus dependent on the equilibrium potential (V_{bond}^0) and a function that describes the deformation of the spring from r_0, and the spring constant k_{bond}.

$$V_{bond} = V_{bond}^0 + k_{bond}(r - r_0)^2 \qquad (3.6)$$

The treatment of the chemical bond as a spring is approximate. V_{bond} as described by Eq. 3.6 is symmetric about r_0 and is thus a simple *harmonic* function. Consequently, the spring model matches the steep ascent of the potential energy profile for compressing a chemical bond, as described by the quantum mechanical model. However, it also defines a steep potential for bond stretching and therefore does not allow the extension to long distances that ultimately leads to dissociation. Thus, this approximation depicts molecules more tightly bonded than they really are. The harmonic spring model obviously could not be applied in simulating a true chemical reaction, such as those catalyzed by enzymes. However, for macromolecules at or near equilibrium, where the atomic fluctuations are small, this is a good approximation.

In addition to being stretched and compressed, a bond can be bent and twisted. The lateral bending of a bond is not explicitly treated in most molecular mechanics force fields. For a *three-atom center* held by two bonds, bending falls into the category of deformations to the bond angle. The bond angle θ is defined as the angle between three bonded atoms A—B—C (Figure 3.3). We can think about a deformation to the bond angle as a compression or extension of a spring that connects atom A to atom C. Thus, the potential energy function V_θ for the bond angle can be treated in a manner similar to V_{bond} with θ_0 defining the equilibrium bond angle and k_θ the spring constant for deforming this angle.

$$V_\theta = V_\theta^0 + k_\theta(\theta - \theta_0)^2 \tag{3.7}$$

Alternatively, since A and C are the number 1 and 3 atoms of the three-atom center, the distance between the two atoms is called the 1–3 distance (r_{1-3}), and the potential for the bond angle is often referred to as the 1–3 potential (V_{1-3}). V_{1-3} is defined identically as V_θ, where r_θ is the equilibrium distance for the standard bond angle, and k_{1-3} is the spring constant for bending the bond angle.

$$V_{1-3} = V_{1-3}^0 + k_{1-3}(r - r_0)^2 \tag{3.8}$$

The twisting of a bond defines the dihedral angle ϕ around the central bond between atoms B and C of the four-atom center A—B—C—D (Chapter 1). The potential energy function for ϕ, V_ϕ, depends on the type of the bond connecting B to C. Single bonds, for example, are relatively free to rotate, while double bonds have very distinct energy minima at $\phi = 0°$ and $180°$. V_ϕ is not treated as a simple harmonic spring function, but takes the form of the periodic function.

$$V_\phi = V_\phi^0 + V_n \cos\left(\gamma + \frac{\phi}{n}\right) \tag{3.9}$$

In Eq. 3.9, V_n is the torsion force constant (equivalent to a spring constant), n is the period of the function, and γ is the phase angle that defines the position of the minima. For a single bond, the function defines three minima, at $\phi = 60°$, $180°$, and $300°$ (Figure 3.4), associated with the *staggered* conformations around the bond. The height of each potential barrier is V_n. For a double bond, there are two minima, at $\phi = 0°$ and $180°$. Thus, $n = 2$ and $\gamma = -180°$ in Eq. 3.9. In analogy to the bond

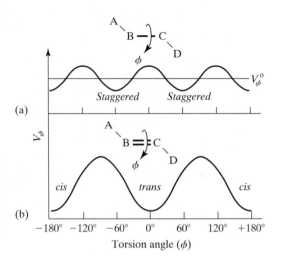

Figure 3.4 Potential energies for rotations about the dihedral angle ϕ for a single bond (a) and a double bond (b). Curves were calculated from Eq. 3.9, with $n = 3$ and $\gamma = 0°$ for a single bond and $n = 2$, and $\gamma = \pm180°$ for a double bond.

angle, the potential for the dihedral angle is often referred to as the 1-4 potential. However, it is more difficult to think of the 1-4 potential in terms of a spring connecting atoms A to D.

These are the minimum definitions of the bonding potentials. Certain force fields include explicit functions to define the planarity of aromatic groups, such as the bases of nucleic acids and the aromatic amino acid side chains. However, much of this can be handled by stringent definitions of the bond angles and dihedral angles. In general, most force fields reduce the bonding interactions between atoms in a molecule to that of simple harmonic springs. The potential energies for these interactions are very large, but they do not drive the folding of macromolecules because they are approximately the same for all conformations of a molecule. The conformations of macromolecules are defined by the weaker interactions between nonbonded atoms.

3.2.4   Nonbonding Potentials

The nonbonding potentials define all of the interactions that are not directly involved in covalent bonds. We describe these briefly in Chapter 1. Here, we provide a more detailed discussion of the potential energy functions for each interaction. The two broad categories of noncovalent interactions are the intermolecular interactions (those between molecules, and between a molecule and the solvent) and the intramolecular interactions (those between the atoms or groups of atoms within a single molecule). Both types of interactions include charge-charge, dipolar, dispersion, and steric interactions. The potential energy functions for nonbonding interactions (Table 2.1) have two common features. First, they are distance dependent and, second, the long and medium range interactions (electrostatic and dipolar) are strongly dependent on the polarizability of the intervening medium, as measured by the dielectric constant ($D = 4\pi\varepsilon_0$).

The potential energy functions for the nonbonding interactions are inversely related to some power n of the distance (r) between atoms (as in $1/r^n$). The range at which a particular interaction becomes dominant depends on n. For large r, $1/r^n$ approaches zero more rapidly for higher values of n. Conversely, for small r, $1/r^n$ approaches ∞ more rapidly for higher values of n (Figure 3.5). Thus, functions that depend on high powers of r (where n is large) are short-range interactions, while those with low powers of r (n is small) are long-range interactions.

3.2.5   Electrostatic Interactions

The treatment of the electrostatic potential V_e between two unit charges Z_1 and Z_2 is given by Coulomb's law

$$V_e = \frac{Z_1 Z_2 e^2}{Dr} \tag{3.10}$$

The interaction is directly proportional to the product of the two charges (the charge of a proton, $e = 1.602 \times 10^{-19}$ C), and inversely proportional to the dielectric

Figure 3.5 Potential energy functions inversely related to r^n. As the distance r increases, functions that are dependent on higher powers (n) of r approach $E = 0$ more rapidly and are therefore shorter-range interactions. Electrostatic interactions have an $n = 1$ relationship, while dipole-dipole interactions have $n = 3$, London dispersion forces have $n = 6$, and steric repulsion forces have $n = 12$.

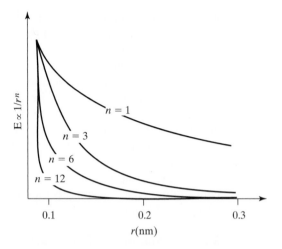

constant of the medium D and the distance separating the two charged species r. V_e is thus defined for charges that are paired—it is a *pairwise* interaction. The potential for single isolated charges in a dielectric medium is modeled by the self-energy (see Chapter 1 for a discussion of the self-energy). It should be noted that Coulomb's law as expressed in Eq. 3.10 does not explicitly account for shielding of the charges from counterions in solution. Methods to treat the more complex electrostatic potentials of macromolecules in solution are discussed later in this chapter. However, the simple potential function in Eq. 3.10 is the form typically incorporated into molecular mechanics force fields.

The effect of the dielectric constant D on electrostatic interactions is discussed in some detail in Chapter 1. An accurate treatment of the dielectric constant allows molecular mechanics force fields to account for the effects of a solvent on molecular structure without explicitly incorporating solvent atoms into the model. There are many strategies for assigning the dielectric constant to a macromolecule such as a globular protein. One approach is to define a boundary that distinguishes the interior from the exterior of the protein. The dielectric constant for exposed atoms can then be set to a value similar to that of the solvent. If the solvent is water, $D = 78.5\kappa\varepsilon_0$ at the exterior of the protein. The interior of the protein is then treated as a low-dielectric cavity. Typical values for the dielectric constant for the interior of a protein range from $1\kappa\varepsilon_0$ to $20\kappa\varepsilon_0$, with a good approximation being $3.5\kappa\varepsilon_0$. This is, of course, a rough estimate, since the true dielectric character must vary continuously throughout the molecule. Therefore, other, more sophisticated models have been sought.

One approach is to treat the dielectric constant as a distance-dependent variable. This strategy is based on the assumption that two interacting atoms are likely to be separated by a polarizable medium in the intervening space at long distances, while two closely spaced atoms will have fewer intervening polarizable atoms. A simple function to describe a distance-dependent dielectric is

$$D = f(r)\kappa\varepsilon_0 \tag{3.11}$$

From this definition, the dielectric constant approaches that for a vacuum at close distances (or zero, if the distances between atoms are allowed to approach $r = 0$), while at long distances it approaches that of water. At intermediate distances, the dielectric constant would be estimated to be somewhere between the two extremes.

Alternatively, the dielectric constant can be described as a local function of the protein density ρ at each point and the dielectric constants of water D_W and the protein D_P.

$$D = (1 - \rho)D_W + \rho D_P \tag{3.12}$$

The assumption here is that the highly compact core of a folded molecule excludes water and is thus more dense, while less dense regions will be mixtures of protein and solvent. The problem is that such a function is difficult to incorporate into a standard force field for atomic interactions, since density is a gross measure of molecular structure. However, this relationship has been used to simulate the gross topology of proteins from simple models (see Chapter 4).

3.2.6   Dipole-Dipole Interactions

A separation of the centers of positive and negative charges (δ_+ and δ_-, indicating full and partial charges) in a group give rise to a dipole. This is characterized by a dipole moment, μ, a vector quantity whose magnitude is given by the product δr, where r is the charge-charge distance. The direction of the vector is conventionally taken from δ_- to δ_+. The Coulombic interaction between two dipoles can be approximated by considering only the distance between the dipoles and the dielectric constant of the medium separating two dipoles. The significance of direction is illustrated by considering the interaction between two dipoles oriented in different directions (Figure 3.6). In this analysis, we consider two dipole moments separated by a distance vector $\mathbf{r}$. A simple potential energy function for dipole-dipole interaction is given in Table 2.1.

If the two dipoles lie side by side, their moments can be oriented either parallel or antiparallel to each other. In the antiparallel orientation, where the positive ends interact with the negative ends, we would expect an attractive force (or a negative potential). In contrast, we would expect a repulsive force (or a positive potential) for the parallel orientation of the dipoles. The dipole-dipole interaction is

$$V_{dd} = \frac{\mu_1 \cdot \mu_2}{D|\mathbf{r}|^3} - \frac{3(\mu_1 \cdot \mathbf{r})(\mu_2 \cdot \mathbf{r})}{D|\mathbf{r}|^5} \tag{3.13}$$

which depends on the orientation of the two dipole moments relative to the distance vector. If the two dipoles are oriented either parallel or antiparallel to each other, but are arranged side by side, then both μ_1 and μ_2 will be perpendicular to $\mathbf{r}$. Thus, $\mu_1 \cdot \mathbf{r} = 0$ and $\mu_2 \cdot \mathbf{r} = 0$, and Eq. 3.13 reduces to

$$V_{dd} = \frac{\mu_1 \cdot \mu_2}{D|\mathbf{r}|^3} \tag{3.14}$$

Figure 3.6 Potential energy functions for dipole-dipole interactions. Dipoles that are arranged side by side in parallel and antiparallel directions have dipole moments that are perpendicular to the distance vector **r**. The potential energy is calculated by Eq. 3.14. The potential energies of the head-to-tail and head-to-head alignments of dipoles, however, must be evaluated using the more general relationship in Eq. 3.13.

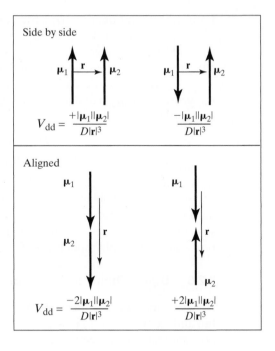

V_{dd} for dipoles that are aligned parallel and antiparallel are $-2|\mu_1||\mu_2|/D|\mathbf{r}|^3$ and $+2|\mu_1||\mu_2|/D|\mathbf{r}|^3$, respectively, from Eq. 3.13.

Equation 3.13 gives us an intuitive understanding of dipole-dipole interactions. However, in many force fields, V_{dd} is incorporated into the potential function for electrostatic interactions by treating each atom as a monopole having a defined partial valence (Table 3.2). Then Coloumb's law in Eq. 3.10 can be used directly to evaluate the interaction between the individual atoms that constitute the dipole. This approach also treats charge-dipole interactions without the need for a separate function in the force field. The interaction between charges or permanent dipoles with induced dipoles are not normally treated separately in molecular mechanics force fields. Induced dipole-induced dipole interactions are included in the van der Waals interactions, which are discussed in Section 3.2.7.

3.2.7   van der Waals Interactions

Two noble gas atoms will attract each other, although neither has a permanent charge or dipole moment. The attractive force derives from an instantaneous and short-lived imbalance in the electron distribution of an atom that generates a temporary dipole. This temporary dipole induces the electron distribution of a neighboring atom to polarize in order to minimize electron-electron repulsion between the atoms. The resulting synchronous interaction is thus an induced dipole-induced

Table 3.2 Examples of Partial Charges of Atoms in
Proteins Calculated from Quantum Mechanics

Amino Acid	Atom Type	Charge
Backbone	N	−0.36
	H_N	+0.18
	C_α	+0.06
	H_α	+0.02
	C	+0.45
	O	−0.38
Ser	C_β	+0.13
	H_β	+0.02
	O_γ	−0.31
	H_γ	+0.17
Tyr	O_η	−0.33
	H_η	+0.17
Cys	S_γ	+0.01
	H_γ	+0.01

Source: From Momany et al. (1975), *J. Phys. Chem.* **79**,
2361–2381.

dipole interaction. London (1937) showed that this attraction, known as *London dispersion forces*, is a natural consequence of quantum mechanics.

The magnitude of the attractive potential is dependent on the volume and the number of polarizable electrons in each interacting group. The London dispersion potential V_L between two uncharged atoms is

$$V_L = -\frac{3I\alpha_1\alpha_2}{4r^6} \tag{3.15}$$

where I is the ionizing energy, and α_1 and α_2 are the polarizability of each atom. The inverse relationship at r^6 makes this a very short-range interaction, with the attraction between atoms dropping off dramatically for even a small increase in distance. All atoms are polarizable to some extent and therefore show this short-range attractive interaction. However, for groups that have a permanent charge or dipole moment, this interatomic attraction is dwarfed by the larger electrostatic interactions at longer distances.

With all atoms attracted to each other, we would expect r to approach zero to minimize the potential energy. We know, however, that this cannot be the case. Counteracting this attraction is a repulsive force, acting at extremely short distances, that keep atoms at respectable distances. At the atomic level, this is associated with the repulsion of electrons clouds and, to a lesser extent, from nucleus-nucleus repulsion. These two repulsions dominate the potential energy function for two atoms at closest approach.

The simplest model to accommodate the repulsive force is the *hard sphere approximation*, which treats each atom as an impenetrable spherical volume. However, this model is too stringent to accurately represent the behavior of atoms. A more accurate model is to define a repulsive potential V_R that acts at extremely short ranges. A typical function is

$$V_R = \frac{k}{r^m} \tag{3.16}$$

where m is some value between 5 and 12. The function approaches that of a hard sphere for the typical value of $m = 12$.

Together, the attractive London dispersion and the repulsive potentials produce an equilibrium distance at which the two opposing energies become equal. The van der Waals radii r_{vdw} for neutral atoms are defined so they sum to give the equilibrium distance (Table 3.3). The two interactions are treated as a single van der Waals potential V_{vdw}. The function that best describes this balance between attraction and repulsion is thus

$$V_{vdw} = \frac{A}{r^m} - \frac{B}{r^6} \tag{3.17}$$

where A and B are constants that describe the magnitude of the repulsive and attractive terms, respectively, and m is the power of the repulsive term (usually between 5 and 12). In a *6-12 potential* or *Lennard-Jones potential, m* = 12. This approaches the hard sphere model for steric interactions (Figure 3.7).

At r_{vdw}, the net force on the two atoms is zero. Thus, A and B can be estimated from Eqs. 3.3 and 3.17 for any pair of atoms (Table 3.4). The steep increase in V_{vdw} at distances that are significantly shorter than the sum of the r_{vdw} for two atoms is overcome only by formation of a chemical bond.

3.2.8   Hydrogen Bonds

Often included in the list of weak interactions of macromolecules are the hydrogen bonds formed between a hydrogen-bond donor and acceptor pair (Figure 1.6). Linus Pauling was the first to emphasize the significance of hydrogen bonds in the folding of macromolecules. This is now a well-accepted principle, although how the potential energy for the interaction should be properly treated is still debated.

Table 3.3   van der Waals Radii r_{vdw} of Atoms in Biological Molecules

Atom Type	r_{vdw} (nm)
H (aromatic)	0.10
H (aliphatic)	0.12
O	0.15
N	0.16
C	0.17
S	0.18

Source: From A. Bondi (1964), *J. Phys. Chem.* **68**, 441–451.

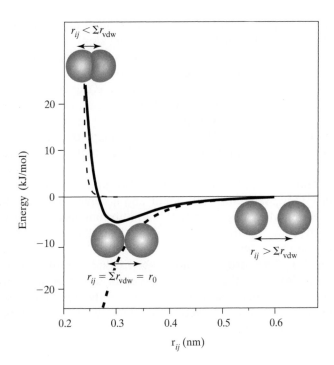

Figure 3.7 The van der Waals potential is a sum of the very short-range attraction between atoms (London dispersion forces) and the extremely short-range steric repulsion between atoms. Together, the two functions define an optimal distance r_0, which is sum of the van der Waals radii r_{vdw} of the two atoms.

Table 3.4 Coefficients for the Repulsive (A) and Attractive (B) Terms of the 6-12 van der Waals Potential in Eq. 3.17

Atomic Interaction	A (kJ-nm^{12}/mol)	B (kJ-nm^6/mol)	r_0 (nm)
H$\cdots$H	1.84×10^{-8}	1.92×10^{-4}	0.240
H$\cdots$C	1.57×10^{-7}	5.27×10^{-4}	0.290
H$\cdots$N	1.11×10^{-7}	5.15×10^{-4}	0.275
H$\cdots$O	1.03×10^{-7}	5.11×10^{-4}	0.272
H$\cdots$P	6.35×10^{-7}	1.43×10^{-3}	0.310
C$\cdots$C	1.18×10^{-6}	1.52×10^{-3}	0.340
C$\cdots$N	8.90×10^{-7}	1.51×10^{-3}	0.325
C$\cdots$O	8.49×10^{-7}	1.51×10^{-3}	0.322
C$\cdots$P	4.49×10^{-6}	4.12×10^{-3}	0.360
N$\cdots$N	6.63×10^{-7}	1.50×10^{-3}	0.310
N$\cdots$O	6.30×10^{-7}	1.50×10^{-3}	0.307
N$\cdots$P	3.44×10^{-6}	4.08×10^{-3}	0.345
O$\cdots$O	5.97×10^{-7}	1.51×10^{-3}	0.304
O$\cdots$P	3.28×10^{-6}	4.10×10^{-3}	0.342
P$\cdots$P	1.68×10^{-5}	1.12×10^{-2}	0.380

The equilibrium distance r_0 is the sum of the van der Waals radii of the two interacting atoms.

Source: From F. Jordan (1973), *J. Theor. Biol.* **30**, 621–630.

As we recall from Chapter 1, the hydrogen bond is an interaction between a polarized D—H bond (where D is the hydrogen-donating atom) and the polarized nonbonding orbitals of an acceptor (A). Thus, the bond is primarily a dipole-dipole interaction; many of the properties of a hydrogen bond can be accounted for by treating it as a simple dipole-dipole interaction. For example, the energy attributed to a hydrogen bond (about 4 to 48 kJ/mol) is significantly weaker than a typical covalent or coordinate bond, and is closer to the magnitude of the potentials for dipole-dipole interactions (Figure 1.6). The optimal alignment of the three atoms (D—H···A) to give a head-to-tail arrangement of the dipole moments is predicted by the potential function in Eq. 3.13 for dipole interactions.

However, in certain ways the hydrogen bond is also similar to covalent bonds. There is an optimal distance separating the donor and acceptor that ranges from 0.26 to 0.30 nm (Table 1.3). More importantly, this equilibrium distance is less than the sum of the respective r_{vdw} for the atoms involved. Thus, the hydrogen bond is truly distinct from nonbonded interactions.

The potential energy of a hydrogen bond can be treated in a force field as the interaction between the dipoles of the donor and acceptor atoms, along with a potential that accommodates the special structural features of the bond V_{HB}. The optimum length of the hydrogen bond is accommodated by including a positive repulsive term and a negative attractive term in V_{HB}.

$$V_{HB} = \frac{C}{r^{12}} - \frac{D}{r^6} \qquad (3.18)$$

In this definition of V_{HB}, C and D are proportionality constants that are specific for each donor-acceptor pair. Thus, the potential allows for shorter-range interactions than does Eq. 3.13 for the standard dipole-dipole interaction. However, unlike the harmonic model for covalent bonds, the hydrogen bond is anharmonic and, therefore, can readily dissociate. The V_{HB} potential typically contributes only about 2 to 8 kJ/mol to the dipole-dipole interaction of the hydrogen bond.

Equation 3.18 describes the potential energy for the hydrogen-bond interaction between a donor and an acceptor, but it does not answer the question of whether hydrogen-bonding interactions stabilize the structures of macromolecules in aqueous solution. The significance of this problem can be seen from an analysis of the energy of hydrogen bonds formed between molecules of N-methyl acetamide (NMA) in various solvents. NMA dimerizes in solution by forming hydrogen bonds between the amino N—H group and the carboxyl oxygen (Figure 3.8). Thus, it is a good model for the type of hydrogen bonds found in proteins and nucleic acids.

Klotz and Franzen (1962) followed the degree of dimerization of NMA as a function of concentration in various solvents (Table 3.5). Their results show that dimerization is favored in carbon tetrachloride (CCl_4), but unfavorable in water. The interpretation is that water competes for the hydrogen-bonding potential of the donors and acceptor groups of NMA, thereby reducing the probability for dimerization. For each hydrogen bond formed between two molecules of NMA, there is a net loss of one hydrogen bond to water. If ΔH^0 is of the same order of magnitude

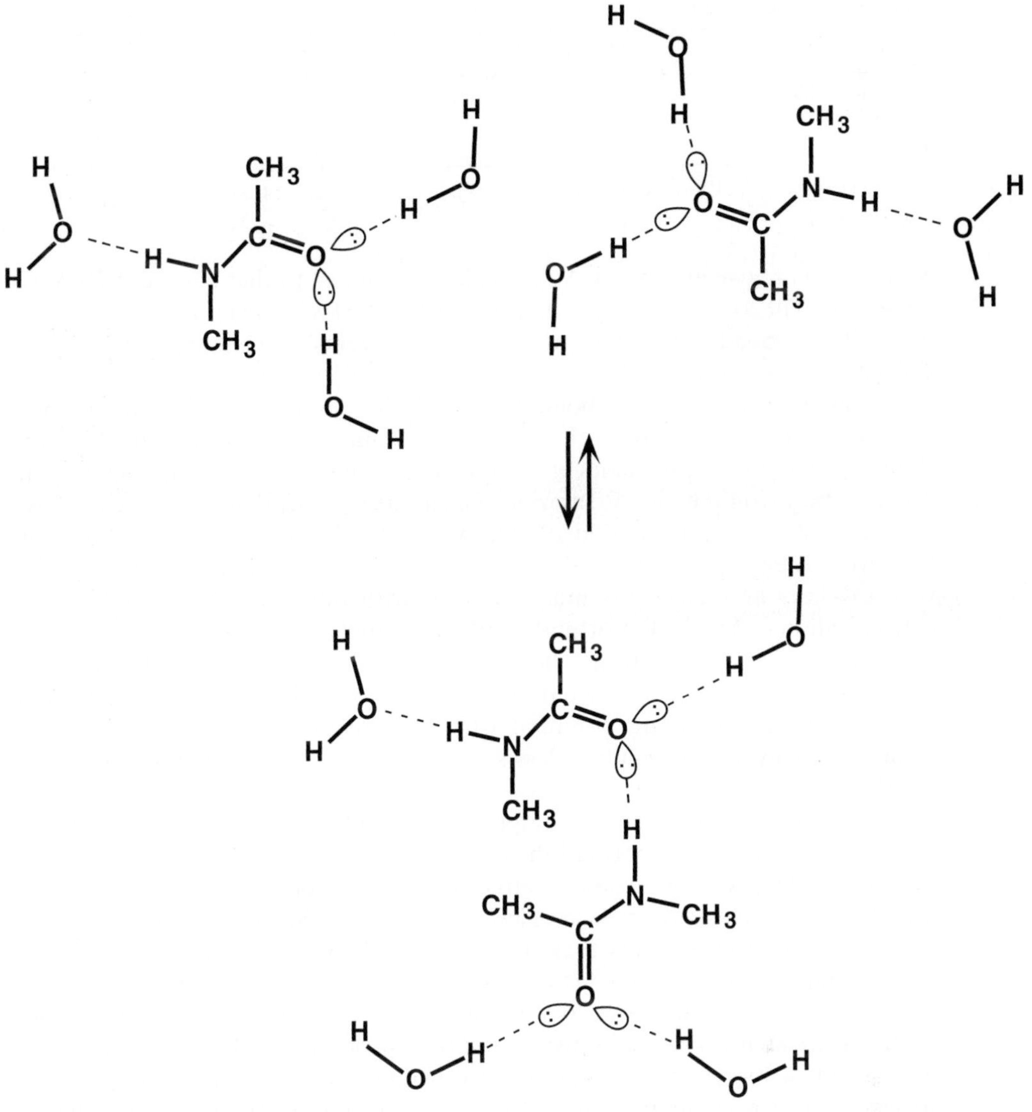

Figure 3.8 Hydrogen-bonding interactions of N-methylacetamide (NMA) in water. The competing interactions are between NMA and the solvent, and between two NMA molecules to form a dimer. The higher concentration of water favors the fully solvated monomers.

for the interaction of NMA to NMA and NMA to water, the higher concentration of bulk solvent greatly favors hydrogen bonding to water.

This is borne out from the temperature dependence of dimerization. The negative enthalpy in CCl_4 indicates that a hydrogen bond is inherently stable. However, in water $\Delta H^0 = 0$ kJ/mol, indicating that the enthalpy change for the hydrogen

Table 3.5   Energies for Dimerization of n-Methylacetamide (NMA) at 25°C

Solvent	ΔH^0 (kJ/mol)	ΔS^0 (J/mol K)	ΔG^0 (kJ/mol)
Carbon tetrachloride	−17	−45	−3.8
Dioxane	−3.3	−16	1.6
Water	0.0	−41	12.8

Source: Data from I. M. Klotz and J. S. Franzen (1962), *J. Am. Chem. Soc.* **84**, 3461.

bond formed between two NMA molecules is identical to that between NMA and the solvent. The loss in entropy from bringing two NMA molecules together to form the hydrogen bond is nearly identical in both solvents, and in water this negative ΔS^0 makes ΔG^0 positive.

Thus, whether hydrogen bonding is a stabilizing or destabilizing interaction in macromolecules, according to this analysis, depends on the solvent environment. When the donor-acceptor pairs are sequestered from competing interactions with water, as they would be in the interior of a globular protein or in the stacked bases of nucleic acids, the formation of intramolecular hydrogen bonds is favorable. However, when these groups are accessible to the solvent, hydrogen bonding between donors and acceptors must compete with the water, which is in higher concentration (~55 M). Unfortunately, the potentials for the dipole and the short-range interactions of a hydrogen bond, as defined by Eqs. 3.13 and 3.18, do not account for these competing solvent effects. Therefore, force fields that include an explicit hydrogen-bonding potential will often overestimate the effects of intramolecular hydrogen bonds unless solvent molecules are explicitly included in the system.

It should be noted that this simple approach of counting hydrogen bonds between donors and acceptors, and those that are made or broken with the solvent may be too simplistic. For example, although each hydrogen bond formed between a donor and acceptor does require breaking an equivalent number of hydrogen bonds to water, those waters that are released will be incorporated into the bulk solvent and will experience nearly the same number of interactions there as they did with the macromolecule (in other words, as far as the water is concerned, there is effectively no loss or gain of hydrogen-bonding potential). Thus, it has been suggested that a more accurate method to evaluate numbers of hydrogen bonds formed and broken should consider the donor and acceptor groups coming into and out of a vacuum relative to an aqueous environment. Thus, we see that the debate concerning the contribution of hydrogen bonds to protein stability is far from being resolved.

3.3   STABILIZING INTERACTIONS IN MACROMOLECULES

We will now discuss how the various interactions described above contribute to the stability of protein and nucleic acid structures. This is not an attempt to describe folding pathways but rather to illustrate how each interaction affects the native structures of macromolecules, and how the energy functions can be used to model the potential energy surfaces for the molecules.

3.3.1   Protein Structure

To understand how the structures of proteins are stabilized by the different interactions described above, we will consider the interactions of a single amino acid residue in the middle of a polypeptide chain. This residue has two freely rotating bonds (associated with the torsion angles ϕ and ψ) along the peptide backbone (Figure 3.9). Using the energy functions described above, we can develop a potential energy surface for this residue as it rotates about ϕ and ψ. For simplicity, we will consider the potential surface for one residue within a polyalanine chain. This model represents the interactions involving the backbone and the β-carbon of the side chains and is therefore relevant for all amino acids except Gly and Pro.

The starting conformation places the Ala residue in the extended chain conformation ($\phi = \psi = -180°$). We will first fix ϕ to $180°$ and analyze the potential energies of the Ala at position i relative to the immediately adjacent residue $i + 1$ as it rotates about ψ. This generates a potential energy profile for the conformations sampled along a trajectory for ψ (Figure 3.10). The total potential energy at each torsion angle $V_{\phi,\psi}$ of a residue is

$$V_{\phi,\psi} = \sum (V_{\text{bonding}} - V_{\text{nonbonding}}) \tag{3.19}$$

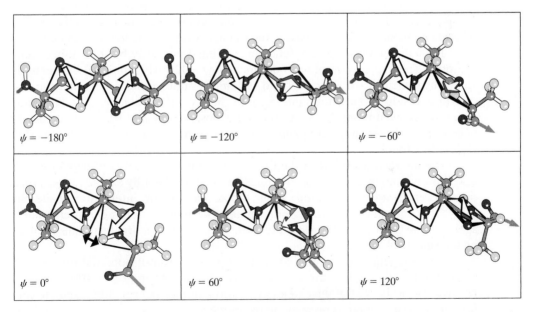

Figure 3.9 Intramolecular interactions between internal alanine residues in a polyalanine chain. The rotation along ϕ is fixed at $-180°$, and ψ rotated from $-180°$ to $+120°$. The dipole moment of the peptide bonds are shown as arrows in the plane of the bond. The close steric interaction between the hydrogen atoms of the amino groups at $\psi = 0°$ is highlighted by a double arrow.

Figure 3.10 Potential energy profiles for the alanine residue in Figure 3.9 as ψ is rotated. The total potential energy V_{total} is the sum of the van der Waals potential V_{vdw}, as calculated by Eq. 3.17; the dipole-dipole potential V_{dd}, as calculated by Eq. 3.13; and the dihedral potential V_{ψ}, as calculated by Eq. 3.9. Notice that V_{dd} and V_{ψ} are plotted on greatly expanded vertical scales, and contribute less to the overall potential.

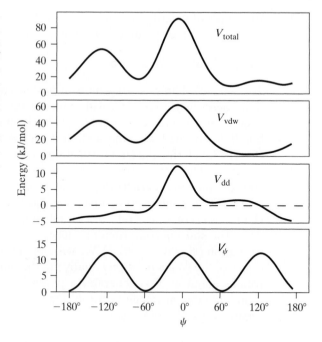

In this example, $V_{bonding}$ is defined only by the torsional energy of ψ. For this single C—C bond, molecular mechanics uses the periodic function in Eq. 3.9 and adds to this the specific van der Waals and dipole-dipole interactions. The remainder of the interactions are nonbonding. We can see that the van der Waals interactions dominate $V_{\phi,\psi}$.

The van der Waals energy function is defined primarily by short-range steric collisions. The first major collision occurs at $\psi = -120°$ (Figure 3.10) between the amino hydrogen in the peptide bond (linking residues i and $i + 1$) and the C_{β}-carbon of residue i. This steric interaction is highly unfavorable, $V_{vdw} \approx 40$ kJ/mol. However, an even more dramatic collision is observed at $\psi = 0°$. Here, the amino hydrogens of the two adjacent peptide bonds are separated by less than 0.1 nm, with $V_{vdw} \approx 60$ kJ/mol.

The interactions between the dipole moments of the adjacent peptide bonds also show minimum and maximum values. In this case, the fully extended conformation has the lowest associated dipole-dipole energy for interaction, while the $\psi = 120°$ conformation is a local maximum. The most unfavorable interaction is again at $\psi = 0°$. It should be noted that the dipole and dihedral terms are dwarfed by the van der Waals potential.

Taken together, the potential energies give an overall potential profile with three distinct minima and three maxima. The conformation at $\psi = 0°$ is the most unfavorable, while that of the extended conformation (ψ near $\pm 180°$) and the conformations at $\psi = \pm60°$ represent energy minima.

We can repeat this calculation for all values for ϕ to produce a map of the energies for all ϕ and ψ torsion angles (Figure 3.11). The resulting ϕ, ψ plot maps the

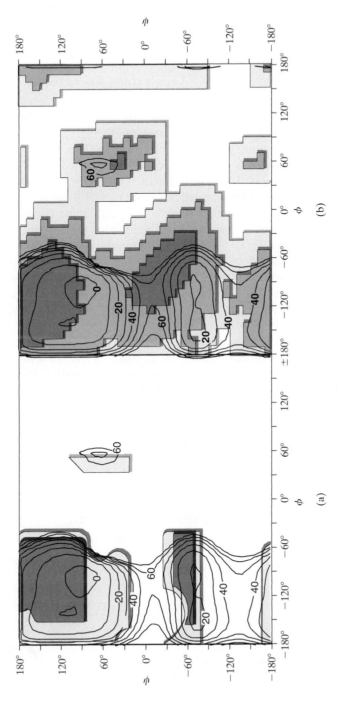

Figure 3.11 ϕ, ψ-plot for the potential energy of an internal alanine. The standard Ramachandran plot (a) compared to a plot for the occurrence of ϕ- and ψ-angles observed in the single-crystal structures of proteins (b). Contours of the ϕ, ψ-plot represent 10 kJ/mol potentials. The sterically allowed regions of the Ramachandran plot are highlighted as dark-shaded regions, while moderately allowed regions are shown as light-shaded regions. Unshaded regions are not allowed. The plot of ϕ- and ψ-angles from protein crystal structures are for all amino acids excluding glycine and proline, and residues with *cis*-peptide bonds. Dark-shaded regions represent highly populated regions, while unshaded regions are essentially unpopulated. [Data from A. L. Morris et al. (1992), *Proteins Struct. Funct. Genet.* **12**, 345–364.]

potential energy surface that represents the *conformation space* of possible structures along a polypeptide chain. The two obvious exceptions are glycine, which does not have a C_β-carbon and is therefore not as constrained as Ala, and proline, which has more restricted torsion angles ($\phi \approx -60°$ and $-60° < \psi < +140°$).

In addition, bulky side chains impose even more steric restrictions to the potential energy. However, for this discussion, we will use the potentials for poly(Ala) as being general for most amino acids in a general polypeptide chain. There are three distinct minima observed for this potential surface. The broad region at $180° > \psi > 60°$, for $180° > \phi > -60°$ represents the extended conformations of the polypeptide chain, including β-sheet conformations (Figure 1.26). The other primary energy minimum at $\psi \approx -60°$ is associated with right-handed helical conformations (the α-helix and 3_{10} helix). Finally, the narrow local minimum at $\phi \approx \psi \approx 60°$ represents the left-handed helix and type II β-turns.

The resulting potential energy surface for polyalanine correlates very well with the standard Ramachandran plot calculated using a hard sphere approximation for steric collisions (Figure 3.11). Thus, the potential energies of the polypeptide chain are dominated by steric interactions along the poly(Ala) chain. However, when the actual ϕ- and ψ-angles from protein-crystal structures are mapped onto this surface, the polypeptide chain samples a larger area of conformational space than is predicted by the potential energy functions, including areas that are considered to be disallowed. Studies that measure the energies for macromolecular packing in crystals show that the models used for atomic collisions may exaggerate the effect of steric collisions by as much as five- to ten-fold. Softening the repulsive term of the Lennard-Jones function would reduce the magnitude of the energy barriers that straddle the allowed regions of the ϕ, ψ-plots, thereby bringing the predicted distribution of protein conformations more in line with the observed distributions. This discrepancy is particularly acute in the region centered at $\phi \approx -90°$ and $\psi \approx 0°$, which is calculated to be a significant potential energy barrier because of the close approach of N_i to $NH_{(i+1)}$ (Figure 3.12). To accommodate this discrepancy, it has been suggested that the van der Waals potential be softened for this particular nitrogen. Alternatively, the definitions of the force field may need to be redefined to treat the interaction between the amino hydrogen and the π-electrons of N_i as being favorable.

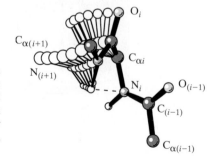

Figure 3.12   Close approach between the amino nitrogen N_i of residue i and the hydrogen of the amino nitrogen $NH_{(i+1)}$ of an adjacent residue $i + 1$ as the $C_{\alpha i}$—N_i bond is rotated. The dashed line represents a distance between the atoms of 0.236 nm, which is less than the sum of their van der Waals radii. [Adapted from P. A. Karplus (1996), *Protein Sci.* **5**, 1406–1420.]

3.3.2   Dipole Interactions

At what point do we start to see a significant contribution from dipolar interactions? This weaker interaction in polypeptide structures is most prevalent when the moments align and sum vectorially, as has been proposed for α-helices (Figure 3.13). The dipole moments of the peptide bonds in an α-helix align approximately along the helical axis to form a *macroscopic dipole* or *helix dipole*,

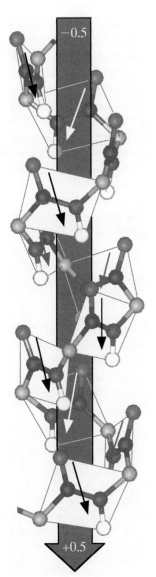

Figure 3.13 Macroscopic dipole of an α-helix in polypeptides and proteins. The dipole moments of individual peptide bonds (arrows in the peptide plane) sum to form an overall enhanced dipole moment aligned along the helical axis. The magnitude of the moments is estimated to be $Z = -0.5$ at the C-terminus of the chain (*top*) and $Z = +0.5$ at the N-terminus.

with the positive end ($Z \approx +0.5$) at the N-terminus and the negative end ($Z \approx -0.5$) at the C-terminus.

The observed cooperativity in the formation and elongation of helical structures, and the stabilization of charges at the ends of α-helices have been attributed to the summation and reinforcement of the peptide dipole moments in the helix dipole. For example, $Glu_{20}Ala_{20}$ was observed to be more stable as an α-helix than $Ala_{20}Glu_{20}$. In 0.1 M salt, the melting temperature T_M of $Glu_{20}Ala_{20}$ is 42° lower than the T_M of $Ala_{20}Glu_{20}$ (Figure 3.14). In both sequences, the 20 Ala residues form the helix. The sequences were studied at pH = 10, so that the Glu residues on either end are fully ionized and therefore adopt a more extended structure because of electrostatic repulsion between the acidic side chains. The negatively charged Glu residues placed at the N-terminus stabilizes the α-helical structure of the Ala_{20} block because of the favorable interaction between the negatively charged Glu side chains and the positive end of the helix dipole. Similarly, the string of Glu residues at the C-terminus destabilizes the helix through unfavorable interactions with the negative end of the dipole. The difference in T_M between the sequences was reduced to 17°C in 0.41 M salt, reflecting the screening of the charged Glu side chains by the ions in solution.

The magnitude of the helix dipole has been estimated by measuring its effect on the pK_a of a titratable acid (such as succinic acid) covalently bonded to the N-terminus of a short α-helix. This is compared to the pK_a of the same acid attached to a nonhelical peptide. We would expect that the positive end of the helix dipole would enhance the acidity of the succinate group and thus lower its pK_a in the helix-forming peptide as compared to the nonhelical peptide. The difference in the pK_a reflects the magnitude of the dipole moment. This is calculated from ΔpK_a by treating the dipole stabilization as a perturbation ΔG_P^0 to the intrinsic free energy for dissociation of the protonated of the acid. The perturbation energy can be calculated by

$$\Delta G_P^0 = 2.303 \, RT \, \Delta pK_a \qquad (3.20)$$

The titration curves for succinic acid in the helical *versus* nonhelical peptides show a pK_a shift of -0.51 pK_a units. This is equivalent, from Eq. 3.20, to about 2.9 kJ/mol of stabilization from the helix dipole. However, the observed changes in pK_a were independent of the length of the peptide, indicating that there is no observed reinforcement of the dipoles. In addition, the magnitude of the helix dipole estimated

Figure 3.14 Melting profiles of the peptides $Glu_{20}Ala_{20}Phe$ (solid circles) and $Ala_{20}Glu_{20}Phe$ (open circles). The percent α-helix for each peptide was determined by circular dichroism spectroscopy (Chapter 10). The melting temperature T_M of each peptide is defined at 50% α-helix. [Data from Takahashi et al. (1989), *Biopolymers* **28**, 995–1009.]

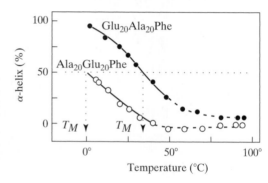

from ΔG_P^0 is 3.5 debye. This is nearly identical to the dipole of an isolated peptide bond (3.7 debye). Thus, the ionizable group sees the helix dipole as a local effect that does not extend beyond one to two turns of the helix. This may simply reflect the short-range nature of ion-dipole interactions, suggesting that there is no macromolecular dipole moment.

3.3.3   Side Chain Interactions

It is much more difficult to assess the contribution of side chain–side chain interactions to the stability of proteins because they are very flexible and can adopt a large number of different conformations. Side chain interactions differ between amino acids that are exposed to a solvent at the surface of a globular protein or packed into the interior of a protein.

Amino acid side chains in the interior of a folded protein are less flexible and, therefore, have greatly reduced entropy. The loss in entropy for each CH_2 methylene group of a hydrocarbon chain has been estimated to correspond to a free energy of about 2 kJ/mol at standard temperatures. In addition to this loss in conformational entropy, placing amino acids with charged side chains or side chains with hydrogen-bonding potential into the protein interior destabilizes the folded state of the protein. Recall from Chapter 1 that burying a charged Lys residue requires up to 40 kJ/mol of self-energy. Similarly, an uncompensated hydrogen-bond donor or acceptor group that is isolated in the interior of the protein without its hydrogen-bonding partner has a self-energy associated with each of the partially charged atoms of the side chain. Thus, charged and hydrogen-bonding side chains are typically exposed or, when found in the interior of a globular protein, are compensated for by a counter charge or a hydrogen-bonding partner.

Counteracting these destabilizing effects are the favorable van der Waals interactions and hydrophobic interactions associated with the more compact and less polar environment of the protein interior. When the increased van der Waals interactions are taken into account, the free energy of packing a methylene group into a solid core, such as the protein interior, is estimated to be 0.6 kJ/mol favoring the folded state. We discuss the contribution of hydrophobic interactions in a later section of this chapter.

Side chains at the surface of a globular protein are not as restricted as those buried in the interior and therefore do not show as great a loss in conformational entropy. However, they are also not as tightly packed. A typical amino acid side chain that is exposed to a solvent, therefore, contributes very little to the overall stability of a protein. However, charged amino acid side chains at the surface can interact electrostatically to affect the stability of folded proteins and the stability of protein complexes.

3.3.4   Electrostatic Interactions

The effect of electrostatic interactions on the stability of proteins is dependent on the Coulombic interactions between the charges. For example, residues His 31 and Asp 70 in phage T4 lysozyme form an ion pair. If either is mutated to the uncharged Asn amino acid, the stability of the protein is reduced by 12 to 20 kJ/mol. This is mirrored by changes to the pK_a of the remaining acid or base group (Anderson et al. 1990).

The treatment of the electrostatic potential by Coulomb's law in Eq. 3.10 is overly simplistic for a protein in solution. In a buffered solution, formal and partial charges are partially compensated by interactions with counterions, particularly at the solvent-exposed surface of a protein. For example, removing a negative charge near the surface of a protein, such as subtilisin or ribonuclease T1, decreases the pK_a of a neighboring His residue by 0.4 units. The effective dielectric constant between the charges was thus estimated to be $40\kappa\varepsilon_0$, which is significantly higher than expected for amino acid residues buried in a protein yet only half that in water. This suggests that the charges at the residues are effectively compensated by *screening* from counterions at the protein surfaces.

Counterion screening is treated by the more general Poisson-Boltzmann (PB) relationship for electrostatic interactions.

$$\nabla D(\mathbf{r})\nabla V_e(\mathbf{r}) - D(\mathbf{r})K^2 \sinh[V_e(\mathbf{r})] + \frac{4\pi Ze\rho(\mathbf{r})}{k_B T} = 0 \qquad (3.21)$$

In this relationship, ∇ is the gradient (a vectorial derivative relative to the positional vector $\mathbf{r}$). For the electrostatic potential $V_e(\mathbf{r})$, the distance-dependent dielectric constant is $D(\mathbf{r})$, the charge density is $\rho(\mathbf{r})$, and $K = 8\pi(Ze)^2 I/D(\mathbf{r})k_B T$. An accurate electrostatic potential can be derived by integrating Eq. 3.21 for a given ionic strength I of the solution.

$$I = \sum_{i=1}^{N} \frac{Z_i^2 c_i}{2} \qquad (3.22)$$

where Z_i and c_i are the charge and concentration of each ionic species i in solution, summed over all ions N.

There are a number of numerical methods that can be used to solve the PB relationship, nearly all of which make some type of simplifying assumption concerning the definition of $\mathbf{r}$. For example, the Born model for transferring an ion from a medium of dielectric D_1 to a another of dielectric D_2 uses a *cavity* radius of r to define the volume of solvent excluded by the ion. The electrostatic free energy ΔG^{Born} according to this model is

$$\Delta G^{\text{Born}} = \left[\frac{(Ze)^2}{2r}\right]\left[\frac{1}{D_2} - \frac{1}{D_1}\right] \qquad (3.23)$$

We notice that this is similar in form to the definition of the self-energy in Eq. 1.4, with the cavity radius being equivalent to the Stokes's radius of the ion.

One method that requires no assumption for $\mathbf{r}$ is the finite difference solution to the PB relationship (FDPB). In this method, the macromolecule and any solvent are mapped onto a three-dimensional grid. Each grid point is a well-defined positional vector in space and can be assigned a partial charge, ionic strength, and dielectric constant. The FDPB is numerically evaluated for each grid point. This

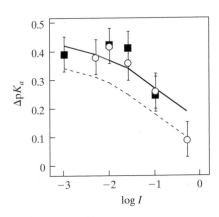

Figure 3.15   Effect on the pK_a of a histidine residue in subtilisin from mutating an acidic amino acid to a neutral amino acid. Shifts in the pK_a of His 64 observed at different ionic strengths I resulting from mutating Glu 156 to Ser (solid squares) and Asp 99 to Ser (open circles). [Data from A. J. Russell et al. (1987), *J. Mol. Biol.* **193**, 803–813.] The shifts predicted from the free differential Poisson-Boltzmann model are shown for Glu 156 to Ser (solid line) and Asp 99 to Ser (dashed line). [Calculated values from M. K. Gilson and B. H. Honig (1987), *Nature* **330**, 84–86.]

method has been used to model a number of electrostatic processes, including the effect of point mutations on protein structure and function (Figure 3.15), the pH dependence of protein-folding and -unfolding, and formation of macromolecular complexes.

We see that the specific conformation adopted by a particular peptide sequence is dependent on the interactions between side chains, and the effect of the side chains on the backbone interactions. The propensities for amino acids to form secondary structures and to affect tertiary structure are discussed in greater detail in Chapter 4.

3.3.5   Nucleic Acid Structure

The basic principles that determine polynucleic acid structures are very similar to those for proteins. In the case of DNA, there are some very well-defined conformations and transitions between conformations that allow us to study specifically how the potential energy functions affect structural stability. The A, B, and Z forms of duplex DNA (Figure 1.40) are all antiparallel helix pairs, with standard Watson-Crick type base-pairing, and are interconvertible. We will focus our discussion primarily on these three forms of duplex DNA, with the understanding that many of these same principles apply also to RNA and nonduplex forms.

We can imagine that potential energy contour maps similar to those for polypeptides can be constructed for nucleic acids by analyzing the torsions about all freely rotating bonds. However, although nucleic acids are less complex than polypeptides in that they are constructed from 4 rather than 20 basic building blocks, the number of different local conformations is much greater because there are more torsion angles about which rotation is possible (10 for each nucleotide as compared to 2 per residue along the backbone of peptides) (Figure 1.36). Indeed, the structure of DNA is *polymorphic*, capable of adopting a large variety of conformations, including those that are not double stranded. A potential energy surface for nucleic acids that would be analogous to the ϕ, ψ plots for proteins must therefore be 11-dimensional. However, this can be greatly simplified in nucleic acids if we consider only those torsion angles that directly relate to the structure of the duplex.

The pucker of the sugar ring from planarity is strongly correlated to the double-helical structure of nucleic acids (Figure 1.37). For example, the sugar pucker in right-handed B-DNA is typically C2'-endo, while that of A-DNA and A-RNA is C3'-endo. The energetic barrier to the transition between A- and B-DNA has been attributed to the inversion of the sugar conformation. X-ray diffraction studies of deoxyoligonucleotide single crystals show this simple rule to be generally, but not universally true. The sugars of nearly all nucleotides in the A-form (whether DNA or RNA) are in the C3'-endo conformations. The deoxyribose of B-DNA however is highly variable (Table 3.6). Thus the sugar pucker is not an absolute indicator of the duplex structure.

The χ-angle for the rotation around the glycosidic bond also affects the structure of the duplex (Figure 1.36). Placing a nucleotide base in the *syn* conformation is considered to be unfavorable because of the steric collisions between the base and the sugar ring. It is generally accepted that purines adopt the *syn* conformation more readily than do pyrimidines. Although the pyrimidine bases are smaller overall, the ring closest to the sugar is larger. Thus, left-handed Z-DNA, which is characterized by a pattern of *anti-syn* conformations of nucleotides along each strand, is favored by alternating pyrimidine-purine sequences, with the purine base adopting the *syn* conformation.

The sugar pucker and glycosidic angles are correlated in a manner analogous to the ϕ, ψ-angles of the polypeptide chain. For example, the energy profile shows that steric contacts of a purine or pyrimidine base with the sugar depends on the rotation about χ but is specific for particular sugar puckers. The χ-angles for all three standard forms of duplex DNA lie at potential energy minima along profiles of the respective sugar puckers (Figure 3.16). It is also clear that for left-handed

Table 3.6   Sugar Conformations in the Single-Crystal Structure of the Self-Complementary B-DNA Duplex of d(CGCGAATTCGCG)

Nucleotide	Sugar Pucker	Nucleotide	Sugar Pucker
C1	C2'-endo	G24	C3'-endo
G2	C1'-exo	C23	C1'-exo
C3	O4'-exo	G22	C2'-endo
G4	C2'-endo	C21	C1'-exo
A5	C1'-exo	T20	C1'-exo
A6	C1'-exo	T19	C1'-exo
T7	O4'-endo	A18	C2'-endo
T8	C1'-exo	A17	C2'-endo
C9	C1'-exo	G16	C2'-endo
G10	C2'-endo	C15	O4'-endo
C11	C2'-endo	G14	C1'-exo
C12	C1'-exo	G13	C2'-endo

The nucleotides are numbered 1 to 12 from the 5'-terminus to the 3'-terminus of one chain, and 13 to 24 from 5' to 3' of the second chain, so that the nucleotide C1 is paired with G24. The C1'-exo and C3'-exo sugar conformations are in the same family as C2'-endo, while C2'-exo and C4'-exo are in the C3'-endo family.

Source: From R. E. Dickerson and H. R. Drew (1981), *J. Mol. Biol.* **149**, 761–786.

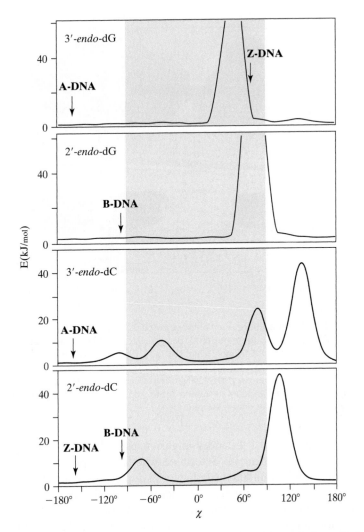

Figure 3.16   Potential energy profiles of nucleotide conformations. The nucleobases 2'-deoxyguanine and 2'-deoxycytosine are rotated about the χ-torsion angle. The shaded region represents the *syn* conformation of the nucleotides. The deoxyribose sugar of the nucleic acids are fixed in either the 2'-endo or 3'-endo conformations. Average conformations of A-DNA, B-DNA, and Z-DNA are indicated.

Z-DNA the χ-angle defines the sugar pucker as 3'-*endo* for the guanine bases that are *syn*.

The sugar pucker of a nucleotide can be characterized by the pseudorotation angle Ψ defined in Eq. 1.17. A plot of the potential energy as a function of Ψ and χ maps the potential energy surface for purines and pyrimidines (Figure 3.17). Not surprisingly, these profiles are different for purines and pyrimidines, but are not strongly dependent on whether the sugar is a ribose or 2'-deoxyribose. The potential energy surfaces are sequence specific but independent of the backbone.

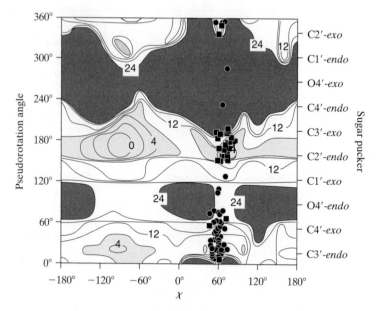

Figure 3.17   Potential energy map of the sugar pucker (pseudorotation angle ψ of the ribose ring) versus the χ-torsion angle for a purine 2′-deoxynucleotide. Light-shaded areas represent regions with total potential energy <8 kJ/mol, while the dark shaded areas have energies >48 kJ/mol. [Adapted from A. Saran et al. (1973), *Theor. Chim. Acta (Berlin)* **30**, 31–44.] The sugar conformations and χ-angles of the guanine bases found in Z-DNA single crystals are plotted. The guanines at the 3′-terminus of each strand (squares) all fall in allowed regions of the map, while all other guanines (circles) are constrained by the structure to higher energy regions.

We can map the Ψ- and χ-angles measured from DNA single-crystal structures onto an energy surface to determine whether the potential functions for the force fields accurately simulate the behavior of duplex DNA structures. We analyze Z-DNA structures here to compare the energetics for the *syn* and *anti*-conformations of double-stranded DNA structures. From these Ψ, χ-plots we see, as with the ϕ, ψ-plots for proteins, that the structures generally fall in the energy wells of the potential energy surface. However, there are a large number of purines in the *syn* conformation that fall at a saddle point of this energy surface. Again, since the potential energies are dominated by steric interactions, this suggests that the hard sphere approximation to the Lennard-Jones potential is too restrictive.

Other interactions that are important for all duplex structures of nucleic acids are hydrogen bonding between bases, base-stacking interactions, and charge repulsion at the negatively charged phosphoribose backbone. The torsion angles of the nucleic acid backbone, as in proteins, define a range of conformations available for the polymer structure. However, the interactions between the nucleotide bases determine which conformations are most likely to be observed for any particular nucleotide sequence. These are primarily hydrogen-bonding interactions and the vertical-stacking interactions between bases of the nucleotides.

3.3.6   Base-Pairing

The B-DNA model derived by Watson and Crick suggested the proper pairing of bases that holds the two strands together (Figure 1.38). They proposed that only certain tautomers of the bases provided the matching hydrogen-bonding donor and acceptor groups that allow cytosines to pair with guanine, and thymine (uracil) to pair with adenine bases (Figure 1.38). However, other hydrogen-bonding patterns have been observed, including Hoogsteen and mismatched base pairs, base triplets, base quartets, and reverse base pairs (Figure 3.18). Unusual base pairs are particularly prevalent in the structures of RNAs, and nearly all possibilities can be found in the structure of tRNAs.

The stability of these various types of base pairs depends on the nature of the hydrogen bonds. It is experimentally difficult to determine the contribution of hydrogen-bonding to the stability of polynucleotide structures because it is difficult to

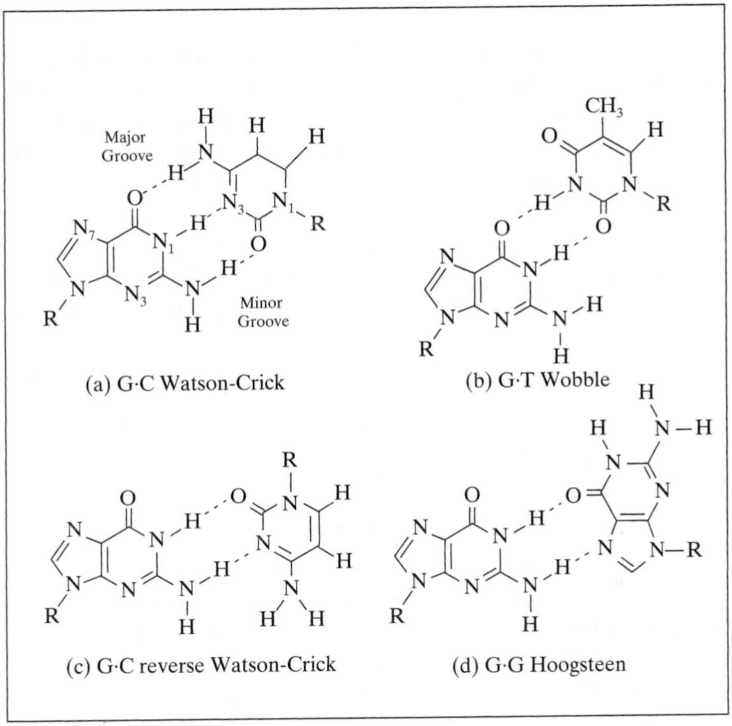

(a) G·C Watson-Crick

(b) G·T Wobble

(c) G·C reverse Watson-Crick

(d) G·G Hoogsteen

Figure 3.18   Examples of base pairs found in DNA and RNA. The standard G·C base pair matches the purine and pyrimidine bases to form three hydrogen bonds. In this Watson-Crick base pair, the sugars (R) are oriented toward the same face of the base pair, defining a major groove face and a minor groove face. A mismatched G·T wobble base pair has only two hydrogen bonds between the purine and pyrimidine nucleotides, while a G·G Hoogsteen base pair holds together two purine bases with two hydrogen bonds. Reverse base pairs have their sugars in an antiparallel orientation and, thus, do not have major and minor groove faces.

separate base-pairing from base-stacking interactions. Most estimates of hydrogen-bonding strengths between bases therefore come from studies on isolated nucleotides or isolated nucleotide bases. However, in these studies with isolated nucleotides, we cannot assume that the bases always form standard Watson-Crick type base pairs, but must consider the competing Hoogsteen, reverse Watson-Crick, and reverse Hoogsteen base pairs as well—all, potentially, in equilibrium.

The pairing of isolated nucleotides in water is weak for the same reason that N-methylacetimide dimerization is weak. The hydrogen bonds between bases must compete with hydrogen bonds to water. However, NMR studies of nucleotide monophosphates do provide a qualitative estimate for base pairing potentials, with

$$G \cdot C > G \cdot U > G \cdot A$$
$$A \cdot U \approx A \cdot C \gg A \cdot G$$
$$G \cdot G \approx C \cdot C > C \cdot U > C \cdot A$$

Thus, in general, purine-pyrimidine base pairs are more stable than purine-purine base pairs.

The base-pairing potentials determined in organic solvents are more quantitative and better mimic the hydrogen bonds formed between stacked aromatic bases of polynucleotides (Table 3.7). There is a strong solvent dependency in the strength of the base pairs. Therefore, to estimate the contribution of hydrogen-bonding to a polynucleotide structure, we must carefully choose which solvent best mimics the stacked bases of the polynucleotide chain. Base-pairing is favored in chloroform over dimethyl sulfoxide, as reflected in the greater loss in entropy in the more polar solvent. The enthalpic change (about 24 kJ/mol) for the standard purine-pyrimidine base pairs is comparable in both solvents, and corresponds to a release of 8 to 12 kJ/mol of

Table 3.7 Thermodynamic Parameters for Dimerization of Nucleobases Through Hydrogen Bonds

Dimer	Solvent	ΔG^0 (kJ/mol)	ΔH^0 (kJ/mol)	ΔS^0 (J/mol K)
G·G	DMSO[a]	4.2	-4 ± 4	
C·C	DMSO[a]	5.7	-1.7 ± 1.5	
G·C	DMSO[a]	-3.2 ± 0.4	-24	-66
A·A	CHCl$_3$[b]	-2.8 ± 0.2	-16 ± 3	47 ± 8
U·U	CHCl$_3$[b]	-4.5 ± 0.2	-18 ± 2	44 ± 4
A·U	CHCl$_3$[b]	-11.5 ± 0.6	-26 ± 2	49 ± 5
A·U	Vacuum[c]		-60	
G·C	Vacuum[c]		-87	
U·U	Vacuum[c]		-39	
C·C	Vacuum[c]		-66	

[a] Data were collected by NMR in dimethyl sulfoxide (DMSO) at 37°C [from R. A. Newmark and C. R. Cantor (1968), *J. Am. Chem. Soc.* **90**, 5010–5017].

[b] Data were collected by infrared spectroscopy at 25°C [from Y. Kyogoku et al. (1967), *J. Am. Chem. Soc.* **89**, 496–504].

[c] Data were collected by mass spectrometry [from I. K. Yanson et al. (1979), *Biopolymers* **18**, 1149–1170].

Source: From W. Saenger (1984) *Principles of Nucleic Acid Structure*, ed. C. Cantor, Springer-Verlag, New York.

energy for each hydrogen bond formed. If the loss in entropy is assumed to be comparable for all Watson-Crick base pairs, ΔG^0 can be estimated to be about -10 kJ/mol for the formation of a G·C base pair and about -1 kJ/mol for an A·T base pair.

The energies of the hydrogen bonds in the base pairs are influenced by the substituent groups of each base. For example, the A·T base pair is more stable than the analogous A·U base pair, but the only difference is the methyl group at the 5-position of the pyrimidine ring (Figure 1.35). The association of 9-ethyladenine with various 1-cyclohexyluracil derivatives shows that the electronegativity at the hydrogen-bond donor and acceptor atoms affects the stability of the resulting base pair (Table 3.8). The order of base-pair stabilities in Table 3.8 mirrors the inductive effects of the substituent groups on the electronegativity of the O4 of the uracil base, indicating that the base pairs in these studies are not of the Watson-Crick type. A methyl group at the 5 position donates electrons to the O4, making it a better hydrogen-bond acceptor. Alternatively, the substituent of 5-fluorouracil is electron withdrawing, making the O4 a worse hydrogen-bond acceptor.

3.3.7   Base-Stacking

The other significant interaction between the aromatic bases of nucleic acids is base-stacking. In nearly all regular structures of nucleic acids, the bases lie with their flat faces parallel, like coins in a vertical stack (Figure 3.19). The energetics of stacking interactions have been determined for nonbase-paired and base-paired dinucleotides (Table 3.9). For the unpaired bases, a purine-purine stacking is more stable than pyrimidine-purine, which is more stable than pyrimidine-pyrimidine. For base-paired dinucleotides, the trend is less well defined. It is stronger for sequences rich in d(C·G) base pairs and weaker for d(T·A) base pairs. Interestingly, the strength of base-stacking correlates very well with the melting of DNA duplexes. The mechanism for melting nucleic acid duplexes is discussed in greater detail in Chapter 4.

Studies that measure the temperature-dependence of base-stacking show a negative enthalpy and entropy associated with this interaction (Table 3.10). This has led to the conclusion that base-stacking is defined by enthalpic potential terms, including dipole interactions and van der Waals interactions, as opposed to solvent rearrangement. However, this interpretation is inconsistent with the observation

Table 3.8   Relative Association Constants K_A for Adenine Pairing with Various Uracil Bases (U) Having Substituent Groupsat the C5-Position

Substituent Group	K_A (M^{-1})
H-(Uracil)	100
CH_3-(Thymine)	130
F-(5-Fluorouracil)	240
I-(5-Iodouracil)	220

Source: From Y. Kyogoku, R. C. Lord, and A. Rich (1967), *Proc. Natl. Acad. Sci., USA* **57**, 250–257.

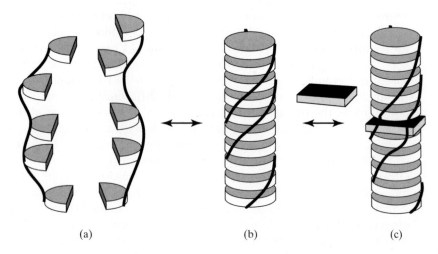

Figure 3.19 Bases of nucleic acids form vertical stacks upon forming helical structures (b). This is interrupted by disruption of the helical structure by melting (a) or by intercalation of a planar hydrophobic molecule (c).

Table 3.9 Energies for Base-Stacking between and within Nucleobases

Compound	ΔH^0 (kJ/mol)	ΔS^0 (J/mol K)	ΔG^0 (kJ/mol)
Nucleobases and nucleosides			
Purine	-17.3 ± 0.8	-54	-1.8
Ribosylpurine	-10.3 ± 0.4	-29	-1.6
2′-Deoxyadenosine	-27 ± 4	-74	-4.1
Cytidine	-11.5 ± 0.4	-41	0.3
Uridine	-11.1 ± 0.4	-41	1.2
Thymidine	-10 ± 1	-37	0.24
Dinucleotides			
ApA	-35 ± 2	-118 ± 6	0
CpC	-35 ± 2	-124 ± 6	1.8
ApU	-30 ± 2	-102 ± 7	0.0

Source: Values for base-stacking between nucleobases from P. O. P. Ts'o (ed.), *Basic Principles in Nucleic Acid Chemistry,* 1; 453–584. Academic Press, New York (1974); and for within dinucleotides from J. T. Powell et al. (1972), *Biopolymers* **11**, 235–340.

that nucleic acid structures are disrupted by organic solvents and that large planar hydrophobic groups *intercalate* or insert themselves between the base pairs.

This apparent contradiction is resolved when we assign the general phenomenon of parallel-stacked bases to solvent interactions, while the enthalpic dipolar and van der Waals potentials account for specificity in the stacking interactions. The large negative entropy can be attributed to the loss of conformational entropy required to order the bases into discrete stacks. Thus, parallel stacking of the bases

Table 3.10   Base-Pair-Stacking Energies for Paired Dinucleotides in B-DNA Calculated from Quantum Mechanics

Dinucleotide Pairs	Stacking Energy (kJ/mol dinucleotide)
↓C·G↑ / ↓G·C↑	−60.15
↓C·G↑ ↓T·A↑ / ↓A·T↑ ↓G·C↑	−43.33
↓C·G↑ ↓A·T↑ / ↓T·A↑ ↓G·C↑	−40.45
↓G·C↑ / ↓C·G↑	−39.95
↓G·C↑ ↓C·G↑ / ↓G·C↑ ↓C·G↑	−34.06
↓T·A↑ / ↓A·T↑	−27.09
↓G·C↑ ↓A·T↑ / ↓T·A↑ ↓C·G↑	−27.09
↓G·C↑ ↓T·A↑ / ↓A·T↑ ↓C·G↑	−27.95
↓A·T↑ ↓T·A↑ / ↓A·T↑ ↓T·A↑	−22.14
↓A·T↑ / ↓T·A↑	−15.75

Arrows indicate the direction of the two chains from the 5′-end to the 3′-end.

Source: From R. L. Ornstein et al. (1978), *Biopolymers* **17**, 2341–2360.

effectively excludes water from the interior of a nucleic acid duplex. One consequence is that the hydrogen-bonding groups are largely sequestered from the competing interactions with water and thus hydrogen bonding becomes favorable.

The parallel stacking of base pairs also maximizes the van der Waals interactions between bases. All known structures of duplex DNA, for example, have the bases separated by 0.34 to 0.37 nm, the average sum of van der Waals radii of the base atoms. In addition to the transient dipoles associated with the van der Waals interactions, there are permanent dipoles that help to define the relative orientations of the bases in the plane. The magnitude and direction of the dipole moments in the bases can be calculated from quantum mechanics (Chapter 8). Thus, both the van der Waals and dipole interactions contribute to the enthalpy and the sequence specificity of stacking.

The stability of a particular base pair within a stem of a DNA or RNA duplex depends on the hydrogen bonds formed between the bases and the stacking interactions with base pairs flanking either side. This is reflected in the free energies for locally melting or opening of a base pair (Table 3.11). The analysis of local opening shows that G·C base pairs are about 10 kJ/mol more stable than A·U base pairs, which corresponds to nearly a 60-fold higher probability for opening an A·U base pair.

3.3.8   Electrostatic Interactions

It is ironic that the model for DNA structure proposed by Linus Pauling, just prior to Watson and Crick's structure of B-DNA, placed the negatively charged phosphates at the center of a triple helix. Pauling, who first recognized the significance of

Table 3.11   Free Energy for Opening the Central G·C or A·U Base Pair in a Three-Base-Pair Triplet of RNA

Triplet			ΔG^0 (at 25°C, kJ/mol)
G	G	G	30.9
C	C	C	
G	G	A	27.8
C	C	U	
A	G	G	27.8
U	C	C	
A	G	A	25
U	C	U	
A	A	A	16
U	U	U	
A	A	G	17.1
U	U	C	
G	A	A	17.1
C	U	U	
G	A	G	18
C	U	C	

Source: From J. Gralla and D. M. Crothers (1973), *J. Mol. Biol.* **78**, 301–319.

hydrogen bonds in the structure of proteins, believed that the primary interaction holding the strands of DNA together was electrostatic. In his model, the DNA was a triple helix held together by sodium ions that form salt bridges to the negatively charged phosphates of the nucleic acid backbone. In hindsight, this makes very little sense. We now know that the negatively charged phosphates are responsible for keeping the backbones of DNA and RNA strands well separated. Single-stranded DNA, where the distances between negatively charged phosphates of the backbone are maximized, is favored in low-salt solutions. Duplexes are favored at high salt concentrations, where the charges of the phosphates are well screened. Even in a buffered solution containing relatively high concentrations of counterions, the phosphates remain at least partially negatively charged.

The effective charge of the polynucleotide backbone depends on the extent to which cations accumulate at the negatively charged phosphates. To study these screening effects, the polynucleotide backbone is treated as a linear array of regularly spaced charges, rather than as a detailed atomic model. A duplex of B-DNA, for example, is modeled as a cylinder with negative point charges spaced according to the helical rise of $h = 0.34$ nm for each base pair. Since each base pair has two phosphate groups, the average spacing of charges along the helical axis is $b = h/Z = (0.34$ nm$)/2 = 0.17$ nm. Cations from the solution accumulate along the nucleic acid backbone to screen the negative charges of the phosphates. We can thus ask the question: What is the effective charge of DNA in a buffered solution? To address this question, we will start by deriving a relationship that approximates the screening effects on the free energy of the polyelectrolyte and then consider the direct binding of cations to the backbone.

The simplest approach to this problem is to start with Coulomb's law in Eq. 3.10, and expand it to include all charges of the nucleic acid backbone. If there are N unitary charges in the linear array, the electrostatic potential is the sum of the interactions of each charge j with all other charges k along the backbone,

$$V_e = \sum_{j=1}^{N} \sum_{k \neq j}^{N} \frac{e^2}{Dr_{jk}}$$ (3.24)

for $j \neq k$, and where r_{jk} is the distance between charges j and k along the chain, for an array of N charges.

The effect of cation screening is included into this potential by introducing the Debye-Hückel screening parameter κ into the charge term e of Eq. 3.24, where κ is

$$\kappa = \left(\frac{8\pi e^2}{100Dk_BT} \right)^{1/2} I^{1/2}$$ (3.25)

We see that the screening factor is dependent on I, the ionic strength of the solution. This comes from the integration of the Poisson-Boltzmann equation for point charges in a uniform dielectric medium.

Equation 3.25 describes the diffuse counterion atmosphere along the DNA backbone. It does not tell us how direct cation-binding through *condensation* defines this atmosphere. Gerald Manning showed that the direct condensation of small cations such as Na^+ and Mg^{2+} to a polyelectrolyte such as DNA depends only on Z (the valence of the ion) and b, defined above. Whether a counterion condenses onto the DNA depends on its charge density, as reflected in the dimensionless parameter ξ.

$$\xi = \frac{e^2}{(Dbk_BT)}$$ (3.26)

If $\xi > 1$, condensation will occur. For a polyelectrolyte in water at 25°C, $\xi = 0.71/b$ (if b is given in nm). Thus, for B-DNA ($b = 0.17$ nm), ξ is estimated to be 4.2, and therefore cations will condense onto this polyelectrolyte.

The fraction of the charge on the phosphate that remains uncompensated is $1/Z\xi$. This is 0.24 for DNA in an aqueous Na^+ environment, which means that 76% of the phosphate charge will be compensated, according to the Manning theory. The

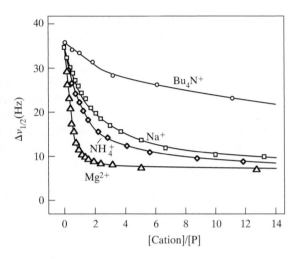

Figure 3.20 Displacement of sodium ions from DNA, as monitored by ^{23}NA-NMR. The linewidth of the sodium ion at half its height ($\Delta \nu_{1/2}$) sharpens as the ion is released. In the competition assay, both Mg^{2+} and NH_4^+ displace sodium at low concentrations of cations relative to phosphates of the DNA and are thus effective competitors for binding. The linear region of the titration curve is used to determine the average number of sodium ions bound to each phosphate of the DNA chain. [Data from M. L. Bleam et al. (1980), *Proc. Natl. Acad. Sci. USA* **77**, 3085–3089.]

degree of Na^+ condensation onto a DNA backbone has been experimentally estimated by ^{23}Na-NMR spectroscopy from the linewidths of free and bound ions (Figure 3.20). By titrating the sodium form of DNA with competing cations, the number of ions along the DNA backbone has been determined to be 0.75 Na^+ per phosphate at $I \leq 0.5$ M which is in excellent agreement with the prediction. The fractional charge was observed to be independent of the ionic strength of the solution, as expected from Eq. 3.25.

Similarly, in Mg^{2+}, the charges of DNA are predicted from Eq. 3.26 to be 88% compensated. The affinity of ions for the negative phosphates increase as the charge of the ion increases, as we would expect from Coulomb's law. Thus, the divalent Mg^{2+} cation binds more tightly to the DNA backbone than does Na^+. Mg^{2+} ions have been observed to coordinate directly to the oxygens of the phosphoribose backbone in single-crystal structures of nucleic acids. Na^+ are seldom observed directly in crystal structures. From dialysis titrations, the binding of the magnesium ion to DNA is maximally 0.85 to 0.86 per phosphate, which is again in excellent agreement with the predictions. Thus, even at high concentrations of divalent cations, the charges of the phosphates are not entirely neutralized.

Electrostatic interactions are primarily responsible for the association of peptides and proteins to nucleic acids. Protein binding to a polynucleotide chain results in the release of cations. The number of monovalent cations released upon binding an oligolysine peptide to single-stranded polynucleotides is summarized in Table 3.12. The 0.74 to 0.93 K^+ released per positive charge of the peptide suggests that the lysine side chain interacts with more than one single phosphate group of the nucleotides.

The conformations of DNA and RNA sequences result from a balance among hydrogen-bonding and stacking of aromatic bases, the electrostatic repulsion between negatively charged phosphates, and interactions with the solvent. Therefore, the general principles that dictate the folding of proteins into a native conformation

Table 3.12   Number of Waters and K^+ Released from Polynucleotides upon Binding of Oligolysine with Charge Z

Polynucleotide	Waters Released	K^+ Released
poly(U)	$(2 \pm 1)Z$	$(0.74 \pm 0.04)Z$
poly(A)	$(15 \pm 5)Z$	$(0.90 \pm 0.08)Z$
poly(C)	$(25 \pm 5)Z$	$(0.93 \pm 0.07)Z$
poly(dT)	$(15 \pm 5)Z$	$(0.76 \pm 0.07)Z$

Source: From D. P. Mascotti and T. M. Lohman (1993), *Biochemistry* **32**, 10568–10579.

are applicable to nucleic acids, and vice versa. The folding problem in biochemistry is thus common to all macromolecules in the cell.

3.4   SIMULATING MACROMOLECULAR STRUCTURE

With the potential energy functions and their effects on the structures of proteins and nucleic acids defined, we can apply these relationships to simulate the structural properties of macromolecules. Molecular simulations include the application of a force field to model the lowest potential energy of a conformation, the dynamic properties of macromolecular structure, and variations on these. The method used depends on the question being asked. Some appropriate applications will be discussed for each technique, but these should not be considered to be an exclusive list.

It is critical in a molecular simulation to start with an initial model that makes chemical and biochemical sense. In all of these techniques, the positions of atoms are perturbed in small increments, and thus it is difficult to sample all the possible arrangements of atoms in conformational space. The measure for the success of a simulation is that the results reproduce the properties observed from experimental studies on the molecule. Therefore, it is not surprising that some of the most successful applications of these methods incorporate as much empirical information as possible into their starting models.

In addition, it is important that the overall system be accurately defined. We can ask, for example, whether it is essential that all the solvent molecules be explicitly included in the model, or whether it is sufficient to model the system in vacuo and rely on the approximations in the force fields to adequately model solvent interactions. This is often a balance between keeping the overall system simple and keeping it reasonably accurate. The degree to which assumptions are applied to simplify a system also depends on what question is asked.

Finally, it is important to understand the limitations of force fields. Certain force fields were derived explicitly for proteins and are not applicable to other macromolecules. In addition, the parameters are not interchangeable between force fields. Most recently, molecular simulation methods (particularly molecular dynamics and energy minimization) have become important tools in facilitating the process of determining macromolecular structures by X-ray diffraction from single crystals (Chapter 6) and high-resolution nuclear magnetic resonance (NMR) spectroscopy (Chapter 12). These experimental data are incorporated into the force field as an additional potential energy function. This allows these experimental methods to include the thermodynamic properties of the molecular structure and thus allows some structures to be determined with a minimal amount of data. However, we should note that the resulting structures may be greatly influenced by the peculiarities of the force fields applied in solving the structure if there are not sufficient data to overcome the inaccuracies in the potential functions.

3.4.1   Energy Minimization

The goal in energy minimization is to find the conformation of a macromolecule that is at the lowest potential energy. The process is actually an exercise in finding a single optimal solution to a set of multivariable equations. The equations are defined by the potential energy functions of the force field as described in Eq. 3.5, and the solution is that set of (x, y, z) coordinates for all atoms that lies in the lowest energy well of the potential energy surface. However, there is not an analytical solution to the problem. Therefore, the solution requires following the contours of the energy surface by sampling different conformations across this surface (Figure 3.21) and in this way finding the lowest energy well along this surface. This process is possible because Newton's laws of motion are deterministic. For example, the force applied on any atom is given by Eq. 3.3. If F can be calculated for an atom at some position r and again after a small incremental change in position Δr, F can be predicted at any new position $r + \Delta r$. This is the force that drives an atom toward a potential well.

  Determined in this manner, F is a straight-line approximation of a curved surface, representing the steepest slope for the descent toward the energy well. By definition, the trajectory for the *steepest descent* will overshoot the bottom of the well unless it is dampened. F calculated at the new position will push the atom back in the opposite direction and, eventually, the simulation will converge at the bottom of the well. This is not a very efficient way to approach the potential energy minimum,

Figure 3.21 Energy minimization brings the total energy of a molecular conformation to a low energy well of the potential energy surface. The force F_i applied at each step i is calculated as the slope of the energy along the molecular trajectory r. The minimized energy falls into a local well and may not represent the global minimum of the system.

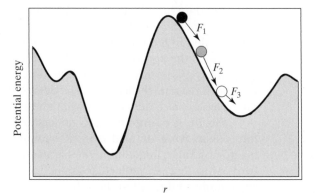

but it illustrates the general problem. There are more efficient and accurate methods for finding the minimum of the potential energy surface, but detailed descriptions of these methods are beyond the scope of this text.

A more fundamental problem in energy minimization, regardless of how efficiently the method converges, is that F will always push the atom to the bottom of the energy well in which it already sits. Thus, the molecule will find the *local minimum* of the surface. It can not locate the absolute lowest energy conformation along the potential surface (the *global minimum*) unless it already sits is that well. This and other problems can be addressed by molecular dynamics.

3.4.2   Molecular Dynamics

Energy minimization treats only the potential energy term of the total energy, and thus it treats the macromolecule as a static entity. The kinetic energy term of the relationship describes the dynamic properties of the atoms in a molecule or the momentum of the atoms in space as described by Eq. 3.4. The momentum is related to the force exerted on the atom F and the potential energy by

$$F = \frac{\partial p}{\partial t} = -\frac{\partial V}{\partial r} \tag{3.27}$$

Thus, the kinetic energy is related to the potential energy functions of the force field. To see how they are related, consider an atom at some time t with an initial coordinate $x(t)$. Its coordinate after some short time interval Δt is given by the Taylor series,

$$x(t + \Delta t) = x(t) + \left[\frac{dx(t)}{dt}\right]\Delta t + \left[\frac{d^2x(t)}{dt^2}\right]\frac{\Delta t^2}{2} + \cdots \tag{3.28}$$

The term $[dx(t)/dt]$ is the velocity $v = (2K/m)^{1/2}$, while $[d^2x(t)/dt^2]/2$ is the acceleration $a = F/m$. Although the series is exact when expanded infinitely, it is typically truncated to the second derivative term when it is used in a molecular dynamics simulation. Thus, for a given mass, the position of an atom is defined as a function of the kinetic and potential energies. In other words, the position of an atom after a time Δt is dependent on how the total energy of the atom partitions between its kinetic and potential energies.

The kinetic energy of an atom is related to the state of the system, in particular the temperature T. At any time, we can define an instantaneous temperature corresponding to an atom's kinetic energy by

$$K = 3k_B T \tag{3.29}$$

where k_B is the Boltzmann constant, and the factor 3 accounts for the three directional components (x, y, z) for the velocity. We recall from Eq. 3.1 that the total energy of a system is the sum of the potential energies and kinetic energies. Since V is temperature independent, T affects only the kinetic energy term. If we set T and the total energy of the system is conserved, we will know K and V. At low temperatures, E is primarily in the potential term, while at high T, it is primarily kinetic.

For an ensemble of atoms in a macromolecule, the velocity of the system (v) is a probability function of the temperature, as given by the Maxwell-Boltzmann relationship:

$$P(v)\partial v = \left[\frac{m}{2\pi k_B T}\right]^{3/2} e^{\{-mv^2/2k_B T\}} 4\pi\, v^2\, \partial v \tag{3.30}$$

In this relationship, $P(v)$ is the probability that a mass will have a velocity v at a temperature T (Figure 3.22). The average temperature of the distribution increases with the velocity, and thus the kinetic energy increases. The probability g that any atom will have a particular velocity along the x-, y-, or z-axis is defined by a Gaussian distribution. Along the x-axis, for example, $g(v_x)$ is

$$g(v_x)\partial v_x = \left[\frac{m}{2\pi k_B T}\right]^{1/2} e^{\{-mv_x^2/2k_B T\}} \partial v_x \tag{3.31}$$

Thus, in a molecular dynamics simulation of a macromolecule, the time interval Δt and the average temperature $\langle T \rangle$ must be defined. From $\langle T \rangle$, the initial velocities along the x-, y-, and z-axes for the simulation are randomly set to give a Gaussian distribution for the ensemble of atoms. This defines the kinetic energy of the system that, along with the potential energy force field, defines the total energy. At $T = 0$ K, $E = V$. It should be noted, however, that T for many simulations may appear to be unreasonably high. This results from the approximations in the force fields. For example, the extremely steep potential of the covalent bond (Figure 3.3) suggests that a bond does not dissociate unless K and T are extremely large ($T > 1000$ K for a typical C—C bond). This also affects more typical simulations, in that bonds will also be prevented from stretching. Thus, atomic positions may seem unreasonably rigid, even at $T = 300$ K.

The time interval allows the Taylor series in Eq. 3.28 to be numerically integrated to define new atomic positions. However, Δt must be small for the simulation to be accurate, typically on the order of femtoseconds to picoseconds. For a process that requires, for example, 1 msec (Table 3.13), the simulation at the upper range of $\Delta t = 1$ psec requires 10^6 calculations per atom per coordinate. This limits most simulations to relatively small systems. One simplification to the system is to

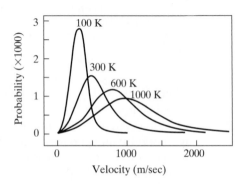

Figure 3.22   Boltzmann distribution of velocities for particles at different temperatures. The probability that a population of particles will have a particular velocity is dependent on the temperature of the system, according to Eq. 3.31. The average velocity of particles and the width of the distribution both increase with increasing temperature.

Table 3.13   Time Scale of Molecular Processes in Proteins and Nucleic Acids

Process	Time Range
All macromolecules	
Vibrations of bonded atoms	10–100 fsec
Elastic vibrations of globular regions	1–10 psec
Torsional libration of buried groups	0.01–1 nsec
Relative motions of hinged globular domains	0.01–100 nsec
Allosteric transitions	10 μsec–sec
Local denaturation	10 μsec–sec
Nucleic acids	
Longitudinal motions of bases in duplexes	10–100 fsec
Lateral motions of bases in duplexes	0.1–1 psec
Global stretching	0.1–10 psec
Global twisting	0.1–10 psec
Change in sugar pucker	1–100 psec
Global bending	0.1–100 nsec
Proteins	
Rotations of surface side chains	10–100 psec
Rotations of medium-sized internal side chains	0.1 msec–sec

Source: From J. A. McCammon and S. C. Harvey (1987), *Dynamics of Proteins and Nucleic Acids*, Cambridge University Press, Cambridge, UK.

reduce the number of atoms by not treating hydrogen atoms explicitly but uniting their properties with those of the attached heavy atom. This *united atom* approach allows, for example C—H to be treated as a single mass.

Despite these limitations of molecular dynamics, the method has found a large variety of applications. A low-temperature simulation (i.e., at or near ambient temperatures) facilitates the process of finding the lowest energy conformation of a system. A conformation, for example, that has been energy minimized to a local potential energy well can be forced out of that well to sample other points in search of the global minimum by increasing the temperature and thus the kinetic energy of the system (Figure 3.23). At any time interval, the state of the system along the dynamics trajectory can be slowly and incrementally reduced to $T = 0$ K to sample a new energy well. Thus, the molecule is allowed to *melt* during the molecular dynamics simulation, and *reanneal* during energy minimization. The entire process is known as *simulated annealing*. This combination is a powerful tool for searching conformational space. However, unless the molecular dynamics calculation is run for an infinitely long time, we can never be certain that the global minimum has been reached. Alternatively, a high-temperature simulation allows us to study a number of truly dynamic processes, including the stability of structures, activated complexes, and Brownian motion.

3.4.3   Entropy

Up to this point, we have discussed a number of potential energy functions and how these relate to the structures of polypeptides and polynucleotides. We should stress that they are not free energies but the enthalpic contribution to the free energy.

Figure 3.23 Simulated annealing. A molecular conformation that is trapped in a local energy minimum (solid circle) can traverse over a potential energy barrier during simulated annealing. In this process, the molecule is heated to a higher average temperature. This temperature defines the kinetic energy of the system and the velocity distributed according to Eq. 3.31. Over a specified time interval Δt, the temperature is reduced during energy minimization to sample a different well along the potential energy surface.

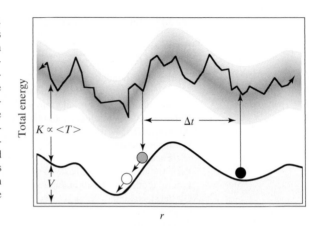

What we have not dealt with is the contribution of entropy to macromolecular structure. There are two forms of entropy in a macromolecular system: the entropy inherent in the macromolecule and the solvent entropy. We will first consider the inherent entropy of a macromolecular structure, or the *conformational entropy*. Solvent entropy is treated in Section 3.4.4.

Conformational entropy. The degree of conformational freedom in a macromolecule is a measure of its conformational entropy. This is related to the degree of degeneracy at each energy level of the molecule and can be roughly estimated for a biopolymer from the degrees of rotational freedom about the freely rotating bonds of the chain. Consider a polypeptide of n amino acids. This chain is characterized by $n - 1$ peptide bonds, and $(n - 2)$ possible rotation angles about ϕ and ψ. We have seen above that the two rotation angles exhibit three distinct minima each. Thus, there are three degrees of conformation for each torsion angle, or there are nine possible conformations at each C_α carbon. Using this approximation, we can estimate the number of different conformations W that are possible for a polypeptide containing n number of residues. This can be calculated by considering the number of possible conformations g for each amino acid and number of amino acids in the chain.

$$W = g^{(n-2)} \tag{3.32}$$

Thus, for a 100 amino acid polypeptide with $g = 9$ for ϕ and ψ, there are $9^{98} \approx 3 \times 10^{93}$ different possible conformations. This represents, at one extreme, the number of isomers of random structures that can be formed. This method of estimating W is thus called *isomer counting*. If the native structure of a protein represents only one conformation (a very simplistic assumption), then the entropy of this native form is calculated as $S_N = k_B \ln 1 = 0$. The difference between the entropy of the native structure and that of all possible random structures (S_R) is

$$\Delta S = S_N - S_R = k_B[\ln(1) - \ln(W)] = k_B(n - 2) \ln g \tag{3.33}$$

In this calculation, ΔS would be about -2 kJ/mol K for a 100 amino acid polypeptide. The large number of possible conformations in a polypeptide backbone leads to a problem known as *Levinthal's paradox*. The paradox states that if in the course of folding, a protein is required to sample all possible conformations, the protein will never find its native structure. In the above example, sampling one conformation every picosecond (10^{12} sec) would require 10^{74} years to cover all 3×10^{93} possible conformations. Either the number of possible conformations is grossly exaggerated or proteins fold by some mechanism that eliminates many conformations, or both.

From the ϕ, ψ-plots similar to that of Figure 3.11, the actual number of allowed conformations at each C_α-carbon has been estimated to be 3.5 per amino acid residue. This greatly reduces the number of possible conformations, but it is still ~2×10^{53} for a 100 residue polypeptide. If a conformation is sampled every picosecond, it would still require over a billion years to sample just 1% of the possible structures. Apparently, reducing the degrees of freedom along the backbone does not resolve Levinthal's paradox.

The paradox would be resolved if protein folding proceeds through a limited set of intermediate conformations rather than through a random process. These intermediates have been referred to as *molten globules*, compact metastable states of proteins with pronounced native-like secondary structure and a less-ordered tertiary structure. In the cell, certain protein-folding pathways are facilitated by *molecular chaperones*, a set of accessory-folding proteins that, it has been suggested, stabilize the molten globule state. This suggests that the conformational entropy of a macromolecule must be redefined to account for the nonrandom process of folding.

An approach to estimating conformational entropy that takes into account intermediate states is to compute the density of the folded (ρ_F) versus that of the unfolded states (ρ_U). Since each biopolymer has a well-defined mass, the density reflects the effective volume occupied by each chain. For a gas, the entropy is dependent on its volume, with a compression to a smaller volume V_1 relative to the initial volume V_0 resulting in a change in entropy.

$$\Delta S = k_B \ln\left(\frac{V_1}{V_0}\right) \tag{3.34}$$

The analogy for folding a biopolymer is that a chain in a large volume can sample more different conformations than one restricted to a smaller volume. Thus, folding a biopolymer to a compact structure reduces the volume and lowers the entropy. During each step of folding, the chain samples successively more compact (higher-density) states. The problem is to estimate the density of each state. The effective volume for any state along the folding pathway can be estimated as a probability for observing a chain with a radius of gyration r. The density of conformations between r and $r + \partial r$ for a chain of n monomers is the probability of observing a residue at some distance r from the molecular center $W(r)$. This is the Gaussian distribution

$$W(r) = g^{(n-2)}\left(\frac{3}{2\pi\langle r^2\rangle}\right)^{3/2} e^{-3r^2/2\langle r^2\rangle} \tag{3.35}$$

$W(r)$ for a random chain is related to the number of conformations for each monomer g and the effective radius of the random chain (the mean square radius $\langle r^2 \rangle$; see Chapter 4 for a more detailed discussion of random chain conformations). This can be expressed as the probability of observing a particular density state $W(\rho)$ relative to the density of some reference state ρ_0, by assuming the volume occupied by the chain to be a sphere. Thus, from Eq. 3.35 and the equation for the density of a linear chain of n monomers confined to a sphere is

$$\rho = \frac{n}{V} = \frac{3n}{4\pi r^3} \tag{3.36}$$

$W(\rho)$ is estimated as

$$W(\rho) = g^{(n-2)}\left(\frac{\rho_0}{n}\right)\left(\frac{6}{n}\right)^{1/2} e^{-(3/2)(\rho_0/\rho)^{2/3}} \tag{3.37}$$

It can be readily shown that ΔS is

$$\Delta S = \frac{-3k_B}{2}\left[\left(\frac{\rho_0}{\rho_2}\right)^{2/3} - \left(\frac{\rho_0}{\rho_1}\right)^{2/3}\right] \tag{3.38}$$

where ρ_1 and ρ_2 are the densities of any two states relative to the density of the native state ρ_0. When we reach the native state, the entropy approaches zero. Notice that in this definition of conformational entropy, the isomers and the degeneracy are not counted. ΔS is dependent only on the relative densities of each state along the folding pathway. If protein folding proceeds from a less compact form to more compact form, then Eq. 3.38 indicates that there is a successive decrease in the entropy of the system. However, the number of possible conformations decreases as the structure becomes more compact. The volume of an unfolded protein has been estimated experimentally to be 3.3 ± 0.3 times larger than that of the native protein. The corresponding loss in entropy during folding, according to Eq. 3.38, is -28 J/mol K, which is significantly lower than the -2 kJ/mol K estimated from isomer counting. The molten globule state has been estimated to be only $50\% \pm 8\%$ larger in volume than the native state, or 16 J/mol K higher in entropy relative to the folded state. The properties of random chains and structures that can be used to model compact intermediate states will be described in greater detail in Chapter 4.

Normal mode analysis. An alternative computational measure for the entropy of a conformation is to ask how rigid the structure is. This can be computed by considering the vibrational modes of a molecule as it becomes excited at a frequency ω. The calculation is called a *normal mode analysis* of the molecule. These modes are analogous to the vibrational modes observed in infrared spectroscopy (see Chapter 9). This introduces a dynamic term (the kinetic energy) to the system and is primarily a correction to the potential energy functions of the force field. Recall that the total energy is the sum of V and K. K is defined in Eq. 3.29 as a linear function of temperature. The total energy can thus be defined by as a temperature-dependent function

$$E = V + Tf \tag{3.39}$$

where f is the temperature-independent component of K. Dividing by T, we obtain

$$\frac{E}{T} = \frac{V}{T} + f \tag{3.40}$$

We remember that for this relationship, the temperature-independent component (in this case f) is the entropy. ΔS, therefore, is reflected in K as a molecular vibrational excitation from an equilibrium state. For a flexible molecule, K is large for the vibrational modes and thus ΔS is large.

In a normal mode analysis, the chemical bond is described as a spring, as we discuss in Section 3.2.3. What happens to the kinetic energy when the spring oscillates at a frequency ω? We will start by restricting this discussion to displacement of atom i with mass m_i along the axis x for a single atom. The restoring force for a one-dimensional oscillator is $F = -k\partial x$. Equation 3.3 defines $F = m\partial^2 x(t)/\partial t^2$. Thus, the motion along the x-axis is

$$-k\partial x = \frac{m_i \partial^2 x(t)}{\partial t^2} = \frac{\partial^2 q_i(t)}{\partial t^2} \tag{3.41}$$

where we can define a mass-weighted coordinate along the x-axis as q_i (where $q_i = m_i^{1/2} x$). The solution to Eq. 3.41 is the harmonic function

$$q_i(t) = A_i \cos(\omega t + \delta) \tag{3.42}$$

where $\omega = (k/m_i)^{1/2}$ and δ is the phase of the oscillator. We can verify that this is the solution by substituting back into Eq. 3.41. The kinetic energy associated with this oscillation is simply

$$K = \frac{1}{2}\left(\frac{\partial q_i}{\partial t}\right)^2 \tag{3.43}$$

This analysis has been applied, for example, to show that left-handed Z-DNA is more rigid than B-DNA, with ΔS estimated to be about 15 J/mol K between the two conformations.

3.4.4   Hydration and the Hydrophobic Effect

Macromolecules all interact with water to some extent. Indeed, the sequestering of nonpolar groups away from water, which is the original definition of hydrophobicity, is the primary driving force that folds macromolecules into compact tertiary structures. Although this is an easy concept to understand—we all know that water and oil do not mix—the effect is much more difficult to define at the atomic level and is thus difficult to model. There are many basic concepts in hydration and hydrophobicity that remain widely debated in the literature. For example, a well-defined set of continuously hydrogen-bonded waters has been found to sit in the minor grooves of duplex DNAs (Figure 3.24). However, it is not clear whether this spine of hydration should have a stabilizing or destabilizing effect on a DNA duplex. Although the hydrogen-bonding interactions may be considered stabilizing, water molecules in such a well-defined network are no longer freely exchanging with the bulk of the solvent and thus decrease the entropy of the system.

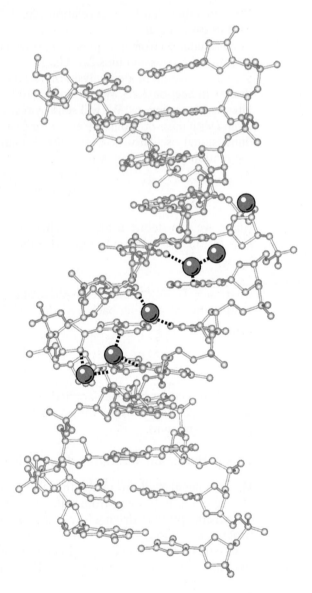

Figure 3.24 The single-crystal structure of DNA duplexes shows regular patterns of waters along the grooves. In B-DNA, this spine of waters (dark-shaded circles) lies in the minor groove and forms a well-defined set of waters hydrogen bonded to the DNA and to each other.

The difficulty in modeling the hydrophobic effect in macromolecular structures is compounded by the debate over the thermodynamic basis for the effect. The hydrophobic effect is believed to be an entropic effect. Indeed, a process that has a positive ΔH^0 and ΔS^0 is often considered to be driven by hydrophobicity. We recall from Chapter 1 that the hydrogen-bonded network of water is highly dynamic. Thus, bulk water is highly entropic. The origin of the hydrophobic effect has been interpreted as coming from waters at the first hydration shell, forming a highly static, cage-like clathrate water structure (Figure 1.9) that surrounds the exposed surfaces of nonpolar atoms. The van der Waals interactions between the solvent and the hydrated atoms are favorable, but the loss in entropy of these immobilized waters is unfavorable. The current estimate is that the number of degrees of freedom for each water molecule locked into a clathrate-like structure is reduced by a factor of 2. This is consistent with the observation that the transfer of alkanes from an organic solvent to water at 25°C is primarily entropy driven. However, at 110°C, this transfer is enthalpically driven, although the overall free energy of transfer remains relatively independent of the temperature. Since most biological macromolecules operate at about 25°C, the entropy-driven model is often invoked. For this discussion, we will treat hydrophobicity in terms of the free energy of hydration ΔG_H^0, but we will use the more widely accepted entropic explanation to interpret some of the models.

A simple relationship to describe the free energy for hydrating a molecule, both polar and nonpolar, is

$$\Delta G_H^0 = \Delta G_{\text{vdw}}^0 + \Delta G_{\text{cav}}^0 + \Delta G_e^0 \qquad (3.44)$$

where ΔG_{vdw}^0 is the favorable van der Waals interactions of the first-shell waters with the atoms of the macromolecule, ΔG_{cav}^0 is the unfavorable entropic process of forming a clathrate cavity in the solvent to fit the molecule, and ΔG_e^0 are the other electrostatic (and dipolar) interactions between the molecule and the solvent dipole. Notice that ΔG_e^0 is not relevant for nonpolar molecules. ΔG_{vdw}^0 and ΔG_{cav}^0 are both associated with the first-shell waters and therefore are related to the surface tension of the water at the molecular interface. We can think of this as shrink-wrapping the molecule with a taut surface. The surface is approximated by the *solvent-accessible surface* (SAS), the effective surface of the molecule that is hydrated.

There are two common methods to calculate the SAS of molecule (Figure 3.25), both of which can conceptually be thought of as rolling a probe with the radius of a water molecule ($r_W \sim 0.14$ nm) across the van der Waals surface of the molecule to determine which surfaces are exposed to and which are buried from solvent. In one calculation, the SAS is defined by the path traveled by the center of the probe. The resulting surface is analogous to one in which the effective radius of a water molecule r_W is added to the van der Waals radius of each exposed atom. Thus, an atom that is exposed to a solvent has a larger effective radius in order to account for the waters at the first hydration shell. An alternative calculation also uses a rolling ball, but only the points of contact between the water and molecular surfaces are counted toward the SAS. Both methods are valid, and each has its advantages and disadvantages, which we will not discuss in detail here. The resulting SAS from

Figure 3.25 The solvent-accessible surface (solid line) of a molecule is calculated by rolling a ball of radius r_W over the molecular surface. The solvent-accessible surface is defined (a) by the path traveled by the center of the ball, or (b) by the contacts made between the ball and the van der Waals surface.

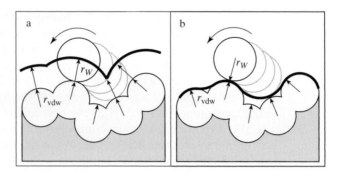

both calculations are smoothed versions of a molecule's van der Waals surface (Figure 1.30).

A semiempirical approach to converting the area into free energies is to use an *atomic solvation parameter* (ASP) that defines the energy for transferring an atomic surface to water. These are derived from the partition coefficients of small molecules for an organic solvent relative to water. The partition coefficient P is treated as an equilibrium constant between the molecule in water and the organic solvent. A typical organic solvent for these studies is octanol, which is primarily nonpolar. From the partition coefficient, ΔG_H^0 can be determined for the molecule by

$$\Delta G_H^0 = -RT \ln P \tag{3.45}$$

From the calculated solvent-accessible surface, the ASP for each atomic surface can be derived by determining how each surface type contributes to ΔG_H^0. This deconvolution of the hydration free energies requires that ΔG_H^0 be determined for a number of molecules. For the atomic surfaces of proteins, ASP values have been determined from the ΔG_H^0 of each amino acid (Table 3.14). Atomic solvation parameter values for nucleic acids were derived from the ΔG_H^0 of small molecules (Table 3.15). With the ASP values, we can estimate ΔG_H^0 for an unknown molecular surface by

$$\Delta G_H^0 = \sum (\text{ASP}_i \times \text{SAS}_i) \tag{3.46}$$

Here, ASP_i and SAS_i represent the atomic solvation parameter and solvent-accessible surface for each surface type exposed on a molecule. Using such an analysis, ΔG_H^0 for all 20 common amino acids has been derived relative to Gly (Table 3.14). These free energies are comparable to the experimental free energies for transferring amino acids from octanol to water.

The resulting ΔG_H^0 for nucleic acids has been useful for predicting the effect of sequence on the relative stabilities of A-DNA and Z-DNA, as compared to the standard B-DNA conformation of the duplex. This is possible only because the structures of both the starting and the altered conformations are well defined. The ΔG_H^0

Table 3.14 Comparison of ΔG_H^0 for Hydrating Amino Acids from Experiment and Various Theoretical Calculations

Amino Acid Type	ΔG_H^0 (kJ/mol)			
	Experimental (octanol)[a]	Calculated (solvent-accessible surfaces)[b]	Nonpolar Surface (Å^2)[c]	"Hydrophobic Effect" for Side Chain Burial[d]
Ala	1.7	2.8	86	4.2
Arg	−5.64	−8.7	89	4.6
Asn	−3.4	−2.5	42	−0.4
Asp	−4.33	−4.9	45	−0.4
Cys	5.52	1.6	48	0.0
Gln	−1.2	−0.91	66	2.1
Glu	−3.6	−3.1	69	2.1
Gly	0	0	47	0.0
His	0.7	2.6	129	5.4
Ile	10.1	7.8	155	11
Leu	9.56	7.8	164	12
Lys	−5.56	−2.6	122	8.0
Met	6.92	9.9	137	9.6
Phe	10.1	9.5	194	9.6
Pro	4.0	4.9	124	8.0
Ser	−0.2	0.04	56	0.8
Thr	1.44	2.1	90	4.6
Trp	12.7	11	236	12
Tyr	5.40	6.6	154	6.7
Val	6.84	6.2	135	9.2

All values are given relative to Gly and therefore represent the hydration of the side chain.

[a] Determined from the partition of amino acids from octanol to water [from J.-L. Fauchere and V. Pliska (1983), *Eur. J. Med.-Chim. Ther.* **18**, 369–375].

[b] Calculated from solvent-accessible surfaces [from D. Eisenberg and A. D. McLachlan (1986), *Nature* **319**, 199–203].

[c] Summed from values of non-polar surface areas [from G. J. Lesser and G. D. Rose (1990), *Proteins Struct. Funt. Genet.* **8**, 6–13].

[d] Calculated using a consensus $\Delta G_{transfer}^0$ of 104 J/Å^2 for buried aliphatic surfaces and 67 J/Å^2 for aromatic surfaces (1995) [from P. A. Karplus (1997), *Protein Science* **6**, 1302–1307].

Table 3.15 Atomic Solvation Parameters (ASP) for Atoms in Nucleic Acids

Group	Surface Type	ASP (kJ/mol · nm^2)
Nucleobase	Aromatic carbon	14
	Methyl carbon	18
	Aromatic nitrogen and oxygen	−28
Ribose	Alkyl carbon	18
	Alkyl oxygen	−16
Phosphate	Oxygen and phosphorous	−41

Source: From T. F. Kagawa et al. (1993), *Nucleic Acids Res.* **21**, 5978–5986.

for a particular sequence as B-DNA and as Z-DNA reflects the relative stability of the sequence in the two conformations (Figure 3.26).

One problem with this approach to modeling solvent interactions is that the ΔG_H^0 calculated in this manner underestimates the hydrophobic effect (Table 3.14). This discrepancy may arise from neglecting the contribution of excluded volume and underrepresenting the surface tension of water. By calculating only the exposed surface, the isolation of nonpolar surfaces from the solvent is not included in ΔG_H^0. This would be particularly problematic for compactly folded structures such as globular proteins. The correlation between the predicted and experimental results, however, indicates that this is a useful strategy for comparing the solvent interactions between defined structures.

The overall ΔG_H^0 of a macromolecule must also take into account the free energy for hydrating both the nonpolar and polar atoms of a molecule. The ΔG_{vdw}^0 and ΔG_{cav}^0 terms of Eq. 3.44 should also apply to the exposed surfaces of polar atoms. The factor that makes the hydration of polar surfaces favorable is the ΔG_e^0 term. The approximation works well for analyzing DNA structures because, to a first approximation, the electrostatic interactions for any conformation of a DNA duplex is assumed to be sequence independent. The amount of exposed surface from the charged phosphate groups is essentially unchanged for different base sequences in a given conformation. All B-DNA sequences will have approximately the same electrostatic interaction with the solvent. In addition, according to the Manning treatment in Eq. 3.26, cation screening of DNA duplexes are largely independent of the solvent-ion environment.

These approximations do not hold for protein structures, since the electrostatic interactions within the protein and with the solvent differ for different conformations

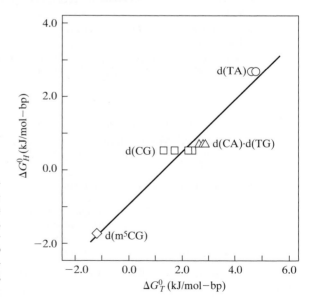

Figure 3.26 Comparison of the difference in hydration free energy ΔG_H^0 calculated from solvent-accessible surfaces of Z-DNA versus B-DNA, and the free energy required to induce a transition from B-DNA to Z-DNA (ΔG_T^0). [Data from T. F. Kagawa et al. (1989), *Biochemistry* **28**, 6642–6651.]

of the polypeptide chain. Therefore, ΔG_e^0 must be evaluated using an electrostatic potential function from Eq. 3.21. One of the primary problems, however, is that this electrostatic term (as with all electrostatic interactions) is highly solvent dependent. Indeed, when the free energy for the transfer of amino acids from one solvent type to another ($\Delta G_{transfer}^0$) was compared for several solvent types, one sees that only the nonpolar amino acids have a consistent set of $\Delta G_{transfer}^0$. This has lead to the suggestion that only the burying of the nonpolar components of a molecule should be considered to contribute toward the "hydrophobic effect" in folding and that this part can be properly treated by SAS calculations (Table 3.14). Certainly burying polar groups into the interior of a globular protein or into a membrane will be associated with an energetic cost, but this should be treated separately from hydrophobicity. Effectively, this tells us that ΔG_e^0 should be removed from Eq. 3.44, and be evaluated separately, employing one or more of the many methods described in this chapter for treating electrostatic interactions.

3.4.5   Free Energy Methods

Up to this point, we have discussed the enthalpy of a system as defined by the potential energy functions for molecular interactions, the entropy inherent in a macromolecule, and the energy of the solvent interactions. If the solvent is explicitly included as water molecules in a system (e.g., a periodic box as in Figure 3.1), then it need not be treated separately. The problem now is to put all of this together to determine the free energy of the entire system, since this is the fundamental measure of macromolecular stability. The stability of, for example, the native conformation of a protein is a comparison of its free energy G_N to the free energy G_U of the unfolded state of the system. Thus, the free energy for a stable system is actually reflected by $\Delta G = G_N - G_U$. This is the relative free energy that is typically determined experimentally. Furthermore, if the system is at equilibrium, then this is the standard free energy ΔG^0 for folding.

We can approach the problem of simulating free energies of a system in a number of ways. Perhaps the simplest is to recall that ΔG is a function of the enthalpy and entropy. Thus, G can be determined by calculating the potential energy for a sufficiently large sampling of conformations. For this discussion, we will define Γ as one particular set of coordinates for the atoms in a system (i.e., one point in conformation space) and $V(\Gamma)$ as the potential for that molecular conformation. The probability of observing this conformation depends on the overall potential of the conformation, expressed as the integral $\int \partial \Gamma e^{-V(\Gamma)/k_B T}$ relative to the possible states of the molecule. The Gibbs free energy for N number of atoms in a system is defined as

$$G = -k_B T \ln\left[\frac{\int \partial \Gamma e^{-V(\Gamma)/k_B T}}{N! \Lambda^{3N}} \right] \tag{3.47}$$

where Λ is a temperature-dependent function. If we define two states with free energies of G and G^*, with potentials $V(\Gamma)$ and $V^*(\Gamma)$, ΔG can be shown to be

$$G^* - G = \Delta G = -k_B T \ln\left[\frac{\int \partial\Gamma e^{-\beta V^*(\Gamma)}}{\int \partial\Gamma e^{\beta V(\Gamma)}}\right]$$

$$= -k_B T \ln\left[\frac{\int \partial\Gamma e^{-\beta\{V^*(\Gamma)-V(\Gamma)\}}e^{-\beta V(\Gamma)}}{\int \partial\Gamma e^{-\beta V(\Gamma)}}\right]$$

$$= -k_B T \ln\left\langle e^{-\beta\{V^*(\Gamma)-V(\Gamma)\}}\right\rangle \tag{3.48}$$

where $\left\langle e^{-\beta\{V^*(\Gamma)-V(\Gamma)\}}\right\rangle$ represents an average over a Boltzmann population of conformations sampled in the system. This average is simply the sum of $e^{-\{V^*(\Gamma)-V(\Gamma)\}/k_B T}$ for all conformations divided by the total number of conformations sampled.

The free energy can be evaluated using a number of different methods that sample conformation space. Thus, if the various conformations are simulated using normal mode analysis, the system would be averaged over the conformations at the excitation frequency ω of the simulation. More typically, the conformations are generated and evaluated by a molecular-dynamics simulation to give a time-averaged ΔG.

Yet another alternative is a *Monte Carlo simulation*. As the name suggests, there is an element of chance introduced into the simulation. In this method, a starting conformation Γ is generated as an arbitrary arrangement of atoms in the system. This is slightly perturbed to a new conformation Γ^*, and the energy recalculated. If $V(\Gamma^*) < V(\Gamma)$, the new conformation is kept. However, if $V(\Gamma^*) > V(\Gamma)$, the new conformation is either kept or discarded randomly. This is accomplished by comparing $e^{-\Delta V(\Gamma)/k_B T}$, which varies between 0 and 1, to a random number i_R between 0 and 1. If $e^{-\Delta V(\Gamma)/k_B T} > i_R$, the new conformation is kept. A Monte Carlo simulation repeated many times will lead toward a minimum free energy and not simply to the lowest energy of the potential surface (as in energy minimization). Keeping high-energy conformations during the simulation reduces the chances of being trapped in a local free energy minimum.

These free energy calculations can be applied to as simple or as complex a system as computational resources allow. A normal mode analysis for the free energy, for example, takes into account the conformational flexibility of the macromolecule, but not the solvent entropy. The most accurate description of the system would include all atoms for waters, ions, and so on that define the solvent environment, along with the macromolecule. These systems can be treated by molecular dynamics or Monte Carlo simulations. However, both must allow the molecule to sample sufficiently large areas of conformation space to accurately describe the average state of the system. It would be difficult, for example, to start with a completely random arrangement of atoms in a polypeptide chain, and expect either method to generate a native conformation with the lowest free energy. This brings us back to Levinthal's paradox.

One very successful application of free energy calculations is *free energy perturbation*. In this case, we make no attempt to solve the folding problem generally, but use free energy calculations to calculate the consequences of a small change in a system. For example, how does a point mutation affect the structure of a protein, or how does a change in sequence affect the stability of a DNA duplex? In this method, we attempt to simulate the effects of small changes in composition on the larger conformation of a macromolecule. The free energy of a system is calculated as the molecule is gradually perturbed from the starting state to a new state. The method works well because the new conformation at each step is expected to be close to the well-defined conformation of the previous step.

A simple example of this method is the application of free energy perturbation to study the sequence-dependent stability of Z-DNA. In this simulation, a dC·dG base pair was converted to a dT·dA base pair in both the B-DNA and Z-DNA conformations. The hydrogen atom at the C5 position of the cytosine bases was gradually mutated to a methyl group and the amino group at the N4 position mutated to a keto oxygen by incrementally changing the mass, radius, and partial charge of the atoms at each position, until the cytosine was replaced by thymine. Similarly, the guanine was mutated to adenine. The free energies were evaluated by molecular dynamics over a 10 ps simulation. The difference in ΔG^0 for d(C·G) versus d(T·A) base pairs in Z-DNA was determined to be about 8 kJ/mol for a simulation in vacuo and 12 kJ/mol with solvent atoms explicitly included in the simulation, as compared to the experimentally determined value of 2 kJ/mol difference. Although the values for ΔG^0 from the simulations are significantly higher than the experimental values, they point in the right direction. This illustrates the amount of error inherent in molecular mechanics calculations.

Thus, molecular simulations can be used to describe the thermodynamic properties of macromolecules if the system is adequately defined. This requires a description not only of the macromolecular structure, but also the surrounding solvent. Molecular simulation methods do not currently solve the folding problem, but can give significant insight into how structures are stabilized. These same methods can be applied to problems other than macromolecular stability, including ligand binding and enzyme kinetics. The level of accuracy of the simulation depends on the level of detail to which the system is defined, and also on the question being asked.

EXERCISES

3.1   The amino acids His 31 and Asp 70 are separated by ~0.35 nm in T4 lysozyme. Calculate the Coloumbic potential between the two charged side chains (make the assumption that these are point charges in a dielectric medium of $40\kappa_0$). Estimate the pK_a of the His residue in this ion pair.

3.2 Starting with the general potential energy function for dipole-dipole interactions, write a reduced function for each of the following arrangements of dipoles.

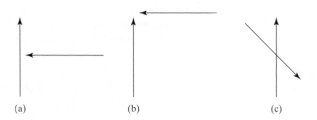

(a)   (b)   (c)

*3.3 The optimal distance for the van der Waals interaction between two carbonyl oxygen atoms is $r_o = 0.353$ nm. The energy for this interaction is -20.78 kJ/mol.
   a. Estimate the repulsive parameter A and dispersion parameter B for the Lennard-Jones 6-12 potential.
   b. Calculate the energy at $r = 0.33$ nm and at $r = 0.4$ nm, using the parameters in (**a**).

3.4 Derive the force function for a Coulombic interaction, a dipole-dipole interaction, and a van der Waals interaction.

*3.5 Calculate the entropy loss as a polypeptide chain folds from the completely denatured state to the compact intermediate to the folded state. Assuming that the entropy at each state represents the number of possible conformations for that state and that the native folded state represents a single conformation, estimate the number of conformations for a polypeptide chain in the fully unfolded state and the compact intermediate state.

3.6 Double-helical A-DNA has a helical rise of 0.26 nm. Apply the Manning theory for polyelectrolytes to estimate a value for ξ for the A-form of DNA. Will sodium ions condense on this polypeptide? If so, estimate the net charge per phosphate.

3.7 The potential barrier for changing the glycosidic angle of a pyrimidine nucleotide from $\chi = -140°$ to $\chi = -2°$ (with a 2'-*endo* sugar pucker) is approximately 120 kJ/mol (Figure 3.16). Calculate the kinetic energy and the temperature required to overcome this potential energy barrier.

3.8 The solvent-accessible surfaces for each atom type of the central dC·dG base pair of d(Gm^{5}CG)·d(m^{5}CGm^{5}C) (where m^{5}C is 5-methyl cytosine) as A-DNA, B-DNA, and Z-DNA are listed in the accompanying table. Which conformation would be favored by this sequence?

	Base			**Ribose Backbone**			
	C	**CH$_3$**	**O**	**N**	**C'**	**O'**	**P**
A-DNA	0.142	0.251	0.160	0.376	0.880	0.193	0.655
B-DNA	0.135	0.242	0.156	0.385	0.979	0.212	0.635
Z-DNA	0.157	0.253	0.189	0.320	0.904	0.203	0.710

3.9 The dipole moment of a peptide bond is 3.7 Debye in water. Assuming that a hydrogen bond is essentially a dipole-dipole interaction, estimate the energy of a hydrogen bond between two peptides in water and in the interior of a protein (neglect the competing interactions with the solvent).

REFERENCES

Molecular Mechanics

McCAMMON, J. A. and S. C. HARVEY (1987) *Dynamics of Protein and Nucleic Acids*, Cambridge University Press, Cambridge, UK. Although slightly outdated, this text discusses in detail the various methods used in molecular simulations.

LONDON, F. (1937) "The General Theory of Molecular Forces." *Trans. Faraday Soc.* **33**, 8. In this paper, London describes the origin of the attractive force between atoms in terms of induced dipole-dipole, induced dipole-quadrupole, and induced quadrupole-quadrupole interactions resulting from fluctuating charge distributions.

The two most widely used force fields (CHARMM and AMBER) are described in detail by

BROOKS, B. R., R. E. BRUCCOLERI, B. D. OLAFSON, D. J. STATES, S. SWAMINATHAN, and M. KARPLUS (1983) "CHARMm: A Program for Macromolecular Energy, Minimization, and Dynamics Calculations," *J. Comput. Chem.* **4**, 187–217.

WEINER, P. K. and P. A. KOLLMAN (1984) "AMBER: Assisted Model Building with Energy Refinement: A General Program for Modeling Molecules and Their Interactions," *J. Comput. Chem.* **2**, 287–309.

Stabilizing Interactions in Macromolecules—Protein Structure

Comprehensive compendia of chapters on the protein-folding problem include

CREIGHTON, T. E. (ed.) (1992) *Protein Folding*, W. H. Freeman and Company, New York.

EISENBERG, D. S., and F. M. RICHARDS (eds.) (1995) "Protein Stability," in *Advances in Protein Chemistry*, vol. 46, Academic Press, San Diego.

There are a number of articles that point out the complexities in modeling protein stability

KARPLUS, P. A. (1996) "Experimentally Observed Conformation-Dependent Geometry and Hidden Strain in Proteins," *Protein Sci.* **5**, 1406–1420.

SHARP, K. A. and S. W. ENGLANDER (1994) "How Much Is a Stabilizing Bond Worth?" *Trends Biochem Sci.* **19**, 526–529.

Articles of particular interest that describe specific issues in protein structure include

ANDERSON, D. E., W. J. BECKTEL, and F. W. DAHLQUIST (1990) "pH Dependent Denaturization of Proteins: A Single Salt Bridge Contributes 3–5 kcal/mol to the Free Energy of Folding of T4 Lysozyme," *Biochemistry* **29**, 2403–2408.

ANFINSEN, C. B. (1973) "Principles That Govern the Folding of Protein Chains," *Science* **181**, 223–230. The classic paper describing the spontaneous folding of proteins to their native form.

ANFINSON, C. B., E. HABER, M. SELA, and F. H. WHITE (1961) "The Kinetics of Formation of Native Ribonuclease During Oxidation of the Reduced Polypeptide Chain," *Proc. Natl. Acad. Sci. USA* **47**, 1309–1314.

BURLEY, S. K. and G. A. PETSKO (1988) "Weakly Polar Interactions in Proteins," *Adv. Protein Chemistry* **39**, 125–189.

Stabilizing Interactions in Macromolecules—Nucleic Acid Structure

JAYARAM, B. and D. L. BEVERIDGE (1996) "Modeling DNA in Aqueous Solutions," *Annu. Rev. Biophys. Biomol. Struct.* **25**, 367–394. Reviews computational methods for treating the ion atmosphere of DNA.

LOUISE-MAY, S., P. AUFFINGER, and E. WESTHOF (1996) "Calculations of Nucleic Acid Conformations," *Curr. Opin. Struct. Biol.* **6**, 289–298. This is a review of recent quantum mechanical, Monte Carlo, and molecular dynamics calculations on nucleic acid structures.

MANNING, G. S. (1978) "The Molecular Theory of Polyelectrolyte Solutions with Applications to the Electrostatic Properties of Polynucleotides," *Quart. Rev. Biophys.* **11**, 179–246.

SAENGER, W. (1983) *Principles of Nucleic Acid Structure,* ed. C. R. Cantor, Springer-Verlag, New York. Chapters 3 and 6 are introductory discussions of the forces that stabilize nucleic acid structures.

Simulating Macromolecular Structure

Two very good general texts in this area include

BURKERT, U. and N. L. ALLINGER (1982) *Molecular Mechanics*, American Chemical Society, Washington, DC.

MCCAMMON, J. A. and S. C. HARVEY (1987) *Dynamics of Proteins and Nucleic Acids,* Cambridge University Press, Cambridge, UK.

Hydrogen Bonding

KLOTZ, I. M. and J. S. FRANZEN (1962) "Hydrogen Bonds Between Model Peptides Groups in Solution," *J. Am. Chem. Soc.* **84**, 3461–3466.

PAULING, L. (1960) *The Nature of the Chemical Bond,* Cornell University Press, 3rd ed. Chapter 12 of this classic text outlines the history of the hydrogen bond, from interactions between simple inorganic compounds to those in proteins and nucleic acids.

Hydration and Hydrophobicity

EISENBERG, D. and A. D. MCLACHLAN (1986) "Solvation Energy in Protein Folding and Binding," *Nature* **319**, 199–203. First to apply the concept of hydration free energy calculations from solvent accessible surfaces to study protein structure.

MEYER, E. (1992) "Internal Water Molecules and H-Bonding in Biological Macromolecules: A Review of Structural Features with Functional Implications," *Protein Sci.* **1**, 1543–1562. This is a more recent review of hydrogen bonding, focusing particularly on the internal hydrogen bonds observed in high-resolution protein crystal structures.

P. A. KARPLUS (1997) "Hydrophobicity Regained," *Protein Sci.* **6**, 1302–1307. Discusses the pitfalls of treating hydration free energies using solvent-accessible surface calculations for proteins, and how going back to the original definition of hydrophobicity may rescue the method.

CHAPTER 4

Statistical Thermodynamics

4.1 GENERAL PRINCIPLES

Most of our observations of macromolecular properties in solution and in cells reflect the average behavior of a population. Even the native structure of a protein represents an ensemble of different conformations, all with the common property of showing biological activity. Statistical mechanics allows us to describe the distribution of molecular conformations that contribute to this population under a given thermodynamic state. The average properties of a macromolecule can be represented as either the time-average behavior of a single molecule or the behavior of an ensemble of many molecules at any instant in time. One of the basic postulates in a statistical-mechanical treatment of thermodynamic properties, or *statistical thermodynamics,* is that the time-average and ensemble average behaviors are identical. A time-average conformation of a structure can be modeled by carrying out a molecular dynamics simulation over a very long period of time (Chapter 3). In this chapter, we discuss the behavior of molecular ensembles as described by statistical-mechanical methods.

Statistical mechanics is most often applied to the study of macromolecular systems that involve multiple thermodynamic states. This includes the transitions between different conformations in biopolymers, the assembly of multiple subunits into multicomponent complexes, and the binding of multiple ligands to macromolecules. In this chapter, we focus on the application of statistical mechanics to structural transitions in well-defined biopolymers and see how this leads to approaches for predicting the conformations of proteins and nucleic acids. The application of these same statistical concepts to model the mechanisms of ligand binding are treated in Chapter 15.

4.1.1   Statistical Weights and the Partition Function

The simplest multistate system is one in which a molecule can exist in one of two forms, A or B. If the two forms are in equilibrium, with the reaction $A \leftrightarrow B$, their relative concentrations are expressed as the equilibrium constant $K_{eq} = [B]/[A]$. A and B are the two possible states of the molecule in this system (in this discussion, we will use capital letters to designate the macroscopic states of a system, for example, the overall conformation, as opposed to microscopic states that we specify as, for example, the conformation of individual residues along a biopolymer chain). From this, we can ask the question: What is the probability of observing the molecule as B relative to the probability of observing it as A? This is by definition K_{eq}. However, for multistate systems, the more general problem is to determine the probability of finding B versus all possible states ($P_B = [B]/\Sigma[\text{all forms}]$). In the simple example above, this is specifically $P_B = [B]/([A] + [B])$, which is redefined as a function of K_{eq}.

$$P_B = \frac{[B]}{[A] + [B]} = \frac{\dfrac{[B]}{[A]}}{\dfrac{[A]}{[A]} + \dfrac{[B]}{[A]}} = \frac{K_{eq}}{1 + K_{eq}} = f_B \tag{4.1}$$

In this simple two-state system, A serves as the reference state to which B is compared. As we expand upon this, it will generally be useful to compare all states of a system to a reference state. We therefore define the *statistical weight* ω as the concentration of all species at some state relative to that of a reference state (in this case, B as compared to the reference state A). The two states of this particular example can be represented by the statistical weights $\omega_A = [A]/[A] = 1$ and $\omega_B = [B]/[A]$, so that Eq. 4.1 can be rewritten as

$$P_B = \frac{\omega_B}{\omega_A + \omega_B} = \frac{\omega_B}{1 + \omega_B} \tag{4.2}$$

This may appear to be a trivial redefinition of the equilibrium constant. However, for systems with multiple states, ω_J can be treated as a microequilibrium constant, referring all species in any macroscopic thermodynamic state J of a system to those of the reference state (which we will now designate as ω_0). In contrast, K_{eq} in such cases is a measure of the macroscopic behavior of the overall system, relating the concentrations of all products to all reactants.

When treated as an equilibrium constant, ω is related to the difference in the free energy between any state $J(G_J^0)$ and the reference state (G_0^0). If each state of the system is unique, with a particular energy level, then

$$G_J^0 - G_0^0 = \Delta G_J^0 = -RT \ln \omega_J \tag{4.3}$$

or

$$\omega_J = e^{-\Delta G_J^0/RT} \tag{4.4}$$

For the specific case of $A \leftrightarrow B$, P_B is given by Eq. 4.5.

$$P_B = \frac{e^{-\Delta G_B^0 / RT}}{1 + e^{-\Delta G_B^0 / RT}} \tag{4.5}$$

It is clear from this relationship that the state having the lowest energy is the most probable state of a system. Thus, if $\Delta G_B^0 < 0$, then $\omega_B > 1$ and $P_B > 0.5$ or a majority of the molecules will be in form B. This is in accord with chemical intuition.

For a system having N possible states, the sum over all states (the denominator of this relationship) is simply the sum of the statistical weights for all states J. How the energy of each individual (microscopic) thermodynamic state relates to the overall (macroscopic) thermodynamic behavior of the system is defined by the *partition function Q*.

In our simple example, each state represents a single form of the molecule, A or B. However, if a particular state represents more than one isoenergetic form of the molecule, it is *degenerate,* and there is a higher statistical probability of observing that state. Consider, for example, two tossed coins that can land with either their head side (H) or tail side (T) up. The three possible states for the two coins are HH, TT, and HT. Both the HH and TT states are unique and thus the number of different combinations that result in either state (the degeneracy of each state, g_{HH} and g_{TT}) is 1. The HT state, however, represents two possible combinations with one coin landing heads up and the other tails up ($g_{HT} = 2$). The HT combination is thus twice as probable as either HH or TT. This is consistent with the second law of thermodynamics, which favors more degenerate states. Therefore, the degeneracy of each state J is reflected in the intrinsic entropy of that state ($S_J = R \ln g_J$). The degeneracy of each state is incorporated into the partition function as

$$Q = \sum_{J=0}^{N} g_J \omega_J \tag{4.6}$$

The probability P_J of observing a particular state J of a system therefore depends on the degeneracy and the statistical weight of that state relative to all possible states, where Q is now defined as in Eq. 4.6.

$$P_J = \frac{g_J \omega_J}{Q} \tag{4.7}$$

This result is identical to the Boltzmann distribution, as given in Chapter 2 by Eq. 2.9.

These are the basic relationships for the statistical-mechanics treatment of thermodynamic processes. We will describe applications of statistical mechanics to a number of structural transitions in polypeptides and polynucleotides in order to illustrate how ω and Q are derived and utilized. Before discussing the specifics of each system, in each case we need to define a model that describes the transitions between structures of the particular biopolymers.

4.1.2   Models for Structural Transitions in Biopolymers

The question of the probability of observing one of two possible states is the simplest case of a broader statistical-mechanics treatment of transitions between two or more defined forms of a molecule. A complete description must consider all the possible states of a system. Consider the often-asked question concerning the structure of a biological macromolecule: What is its secondary structure? This limits the problem to transitions between a set of well-defined possible states, such as between a random coil and an α-helix or between a β-strand and an α-helix. For polynucleic acids, the two states can be, for example, the duplex and single-stranded forms of DNA or RNA (i.e., the melting and annealing of the double helix), or the standard right-handed (B-form) and the left-handed (Z-form) duplexes of DNA. We will therefore present a set of simple *two-state models* for structural transitions in biopolymers.

  For these two-state transitions, we will again consider the transition from the starting or reference state A to a new state B. Now we consider how the state of each individual residue of the biopolymer contributes to the macroscopic forms A and B. For a biopolymer, the reference state A will have all residues of the chain in the a conformation. This can be described as

$$A = \ldots aaaaaa \ldots \tag{4.8}$$

It should be noted that Eq. 4.8 is not a mathematical expression, but an indication of two equivalent ways to describe a state. The lowercase designation is used to distinguish the conformations of the residue as opposed to the entire chain. The new state B will have all residues in the b conformation, as described by

$$B = \ldots bbbbbb \ldots \tag{4.9}$$

From this point on, we will refer to each state by the overall statistical weight for that state (e.g., ω_0 for the reference state and ω_j for each state j, where j can be used to count the number of residues in the b form). We will also include the degeneracy of each state within ω_j rather than treating it separately.

  If n number of residues simultaneously convert from a to b, then there truly are only two possible states of the molecule, with statistical weights ω_0 and ω_n, in the latter case where all $j = n$ number of residues are in the b conformation (Figure 4.1). This is an *all-or-none* transition, since all the residues are either a (for $j = 0$ for ω_0) or b (for $j = n$ for ω_n). The transition of each residue within the biopolymer is described at the residue level by defining a statistical weight for each residue ($s = [b]/[a]$). Thus, s represents the probability that a residue converts from a to b (for simplicity, we will consider here only a homopolymer so that s is identical for each residue along the chain). The statistical weight of the overall state ω_n is thus the product of s for each residue in a biopolymer of n residues in length. In this case, all s_j are identical, and equal s, so

$$\omega_n = \prod_{j=1}^{n} s_j = s^n \tag{4.10}$$

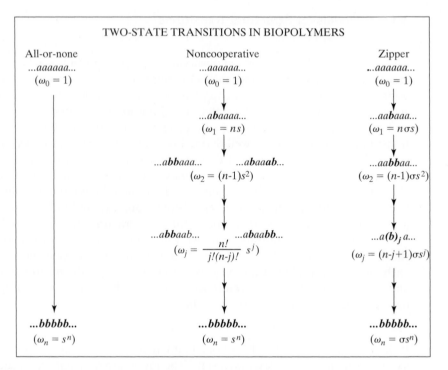

Figure 4.1 Models and statistical weights for the thermodynamic states during structural transitions in biopolymers. The highly cooperative all-or-none model, noncooperative model, and the zipper model are compared to show the transition from the reference form *a* to form *b* for each residue along a chain.

An all-or-none transition is highly cooperative, with *n* representing the *cooperative length* of a transition. This represents one extreme example of how biopolymers convert from one form to another.

An alternative model allows any residue to behave separately and independently when converting from *a* to *b* (Figure 4.1). In this *noncooperative* model, the reference state ω_0 again starts with all residues as *a*. In this case, the biopolymer converts stepwise from *a* to *b*, with each transition at the residue level contributing to an increase in the fraction of *b* observed. Each step *j* represents a different state, with ω_j being the statistical weight for each state *j*, where *j* also represents the total number of residues in the form *b*. Again, *s* describes the probability of each residue converting from *a* to *b*,

$$s = \frac{[b]}{[a]} = e^{-\Delta G^0_{b-a}/RT} \tag{4.11}$$

Thus, the transition can be thought of as a series of microequilibria, with *s* being the microequilibrium constant for extending the form *b* through the polymer by one residue. If *s* is identical for all residues, as it is in a homopolymer, $\omega_j = s^j$. We notice that since biopolymers have a direction, each residue is unique even in a homopolymer but, again for simplicity, we will ignore the end effects for now. Thus, *aabaaa* and *aaaabaa* represent two unique states. For a chain of n residues, there are n unique combinations with a single residue b. This is the degeneracy of the state where $j = 1$; therefore $g_1 = n$ and $\omega_1 = ns$. The degeneracy of each state is included in ω_j as in Eq. 4.12.

$$\omega_j = g_j s^j \tag{4.12}$$

The form of the partition function for this model becomes apparent by considering a simple example where $n = 4$. In this case, it is easy to show that $\omega_0 = 1$ (with $g_0 = 1$), $\omega_1 = 4s$ ($g_1 = 4$), $\omega_2 = 6s^2$ ($g_2 = 6$), $\omega_3 = 4s^3$ ($g_3 = 4$), and $\omega_4 = s^4$ ($g_4 = 1$). The sum of these states is the partition function

$$Q = 1 + 4s + 6s^2 + 4s^3 + s^4 = (1 + s)^4 \tag{4.13}$$

The general form of the partition function for n number of residues is thus the simple polynomial expression

$$Q = (1 + s)^n \tag{4.14}$$

When Eq. 4.14 is expanded, the coefficients of the power of s (the degeneracy factors) are in this case the binomial coefficients

$$g_j = \frac{n!}{j!(n - j)!} \tag{4.15}$$

which is just the number of ways of dividing n residues into j in state b and $n - j$ in state a. So,

$$\omega_j = \frac{n! s^j}{j!(n - j)!} \tag{4.16}$$

The probability of observing any state j, now averaged across all states $\langle P_j \rangle$ is

$$\langle P_j \rangle = \frac{\omega_j}{Q} = \frac{\dfrac{n! s^j}{j!(n - j)!}}{1 + \displaystyle\sum_{j=1}^{n} \frac{n! s^j}{j!(n - j)!}} \tag{4.17}$$

In the case where $s \approx 1$, which corresponds to $\Delta G^0 \approx 0$, the residues are energetically indifferent as to which form each adopts. In other words, each residue can randomly be observed as either the a or b form. The most probable state of the system will then be the one that is most degenerate. For a chain of n residues, this is $\omega_{n/2}$. Thus, for $n = 4$, the most probable state is ω_2 where $g_j = 6$.

We can now ask the question: What is the fraction of residues in the system that are in the b form? The fraction in b at each state is simply the number of b residues, j, relative to the total number of residues, $f_j = j/n$. The total fraction of residues as b in the whole system is the average probability of observing b, $\langle P_b \rangle$. This probability is the sum of all the states having at least one residue as b, with each state weighted by f_j, relative to all the possible states of the system. This is given by

$$\langle P_b \rangle = \frac{\sum_{j=1}^{n} \left(\frac{j}{n} \right) \omega_j}{Q} = \frac{1}{1 + s} \tag{4.18}$$

(See Chapter 15 for a demonstration of this equality.) The probability of observing conformation a is simply $\langle P_a \rangle = 1 - \langle P_b \rangle$, which is equivalent to

$$\langle P_a \rangle = \frac{\sum_{j=1}^{n} \left(\frac{n - j}{n} \right) \omega_j}{Q} \tag{4.19}$$

Finally, the ratio $\langle P_b \rangle / \langle P_a \rangle$ is obviously s, as we expect.

This noncooperative model, in which each step is separate and independent, is used in Section 4.4.1 to describe a random walk and in Chapter 15 to describe noncooperative ligand binding to macromolecules with multiple binding sites. However, most transitions in the secondary structure of biopolymers fall somewhere between this noncooperative model and the fully cooperative all-or-none model. Many of these transitions are best described by the more complex *zipper model*.

The zipper model dissects the structural transition into a number of discrete steps, as in the noncooperative transition above. However, in the zipper model, starting the transition is more difficult than extending it. Therefore, the first step is of high energy and consequently low probability. This *initiation step* provides a *nucleation* point for the transition. Once initiated, the extension or *propagation* of the transition through the polymer occurs via a series of lower energy and therefore higher probability steps; the transition "zips" along the chain relatively freely. The zipper model therefore describes structural transitions in biopolymers that are cooperative in this particular way.

The partition function for the zipper model is derived from the basic relationships of the noncooperative model. The only difference is that the statistical weight for the first step of the transition ω_1 must now include a factor that accounts for the extra energy required to nucleate the transition. In the ω_1 state, in which a single residue adopts form b within the chain, the conformation is

$$\dots aabaaa \dots$$

In the zipper model, this is the high-energy nucleation step that initiates the transition from a to b. Thus, the statistical weight of state 1 is not simply $\omega_1 = s$, but must include a *nucleation parameter* σ to represent the probability for initiating the transition. For the zipper model, $\sigma \ll 1$. The physical meaning of σ is discussed in greater detail for specific examples of conformation transitions in polypeptides and polynucleotides. Thus, the statistical weight for the state $j = 1$ is $\omega_1 = \sigma s$.

For the next step of the transition, we can imagine two possibilities. In one case, the next transition simply extends b to an immediately adjacent residue to give the conformation

$$\ldots aabbaa \ldots$$

Alternatively, the next transition may occur at a nonadjacent residue to give, for example, the conformation

$$\ldots abaaba \ldots$$

These two possibilities are distinguished by the zipper model. In the first case, $\omega_2 = \sigma s^2$, while for the second case, $\omega_2 = \sigma^2 s^2$. The probability of observing two adjacent residues as opposed to two isolated residues as b is thus $\sigma s^2/\sigma^2 s^2 = 1/\sigma$. If $\sigma \ll 1$ (σ is on the order of 10^{-4} for the coil-to-helix transition in a polypeptide), the probability of having two isolated residues as b (which is essentially the probability of two nucleating events) is extremely low and the existence of such states is neglected entirely in the simplest form of the zipper model. Thus, the zipper model confines the transition within contiguous residues along the chain. Each subsequent addition of residues along this contiguous region simply introduces one additional s in ω_j for each state j. The term s is called the *propagation parameter* since it represents the statistical weight for extending or propagating the transition by one residue. This can be shown to apply to all states for $j \geq 1$. Thus, the expression for ω_j can be rewritten as

$$\omega_j = \sigma s^j \tag{4.20}$$

If we consider only the contiguous states as being in either the a or b form, the degeneracy for any state j, for $j = 1$ to n, is the number of ways that a segment of j adjacent residues can fit into n possible positions.

$$g_j = (n - j + 1) \tag{4.21}$$

The reference state $j = 0$ is treated as an exception, with $g_0 = 1$. The partition function for the zipper model is thus

$$Q = 1 + \sum_{j=1}^{n} \omega_j = 1 + \sum_{j=1}^{n} (n - j + 1)\sigma s^j$$

$$= 1 + \sigma \sum_{j=1}^{n} (n - j + 1)s^j \tag{4.22}$$

The average probability of observing b in the distribution of molecules $\langle P_b \rangle$ is dependent on the probability of observing all states that include residues as b (ω_j for $j \geq 1$), and the fraction of residues at each state as b ($f_j = j/n$).

$$\langle P_b \rangle = \frac{\sigma \sum_{j=1}^{n} [f_j \omega_j]}{Q} = \frac{\sigma \sum_{j=1}^{n} \left[\left(\frac{j}{n} \right)(n - j + 1)s^j \right]}{Q} \tag{4.23}$$

The cooperativity of the transition as defined by the zipper model is dependent on the magnitude of σ relative to s. If $\sigma = 1$, the transition is noncooperative since we no longer discriminate between the first and the subsequent transitions to b. This case is identical with the noncooperative model described above. Alternatively, if $\sigma \ll s$, then the transition is highly cooperative and approaches the all-or-none model. In this case, those molecules that have nucleated will have all residues transformed entirely to the b form, while molecules that have not nucleated will all remain as a. Because the cooperativity of the transition is dependent on σ, it is often called the *cooperativity coefficient*. These properties of the zipper model will become more apparent as we present specific examples of structural transitions in polypeptides and polynucleotides.

We notice that the summations in both Eqs. 4.22 and 4.23 include the term s^j but not σ. Thus, the length of a polymer affects the contribution of the propagation parameter but not of the nucleation parameter to $\langle P_b \rangle$. One way to experimentally resolve the two terms is to study the effect of different lengths of a biopolymer on $\langle P_b \rangle$. According to Eq. 4.23, when $n = 0$, $\langle P_b \rangle = \sigma$. The nucleation term, therefore, can be determined by extrapolating $\langle P_b \rangle$ to a biopolymer of zero length.

The zipper model is applicable in describing a large variety of processes involving macromolecules. This includes the assembly of subunits to form multimeric complexes and the formation of regular structures along a biopolymer chain. Since the zipper model is a defined pathway for converting one form of a molecule to another, it can apply not only to the equilibrium properties but also to the kinetic behavior of such transitions by defining σ and s in terms of kinetic constants. However, this is not discussed in detail here.

In applying the statistical-mechanical treatment of the zipper model to the study of structural transitions in biopolymers, we need to give physical meanings to the nucleation and propagation parameters. This will allow us to define each state and thus to define the partition function for the system. The nucleation parameter is typically intrinsic to the type of transition being described. The nucleation parameter determined for the coil-to-helix transition in a polypeptide of a specific length, for example, should be applicable to the same transition in a chain of any length. However, it may not be applicable to the transition from a β-strand to a helix. Similarly, the propagation parameter is dependent on the type of biopolymer and the type of transition observed in the system. The parameter s is dependent on the interactions that help to stabilize or destabilize each residue as b versus a along a chain and thus reflects the intrinsic stability of b versus a, as well as perturbations induced by changes in the external environment. We will see, for example, that s is dependent on the intrinsic ability of the hydrogen bond to hold a residue in a helix in a polypeptide, but that the stability of the hydrogen bond itself is dependent on the external factors of temperature and solvent composition. Thus, s is described by Eq. 4.24, where ΔG_I^0 is the intrinsic energy difference between b and a, and δ is a perturbation to the system induced by changes in the external state of the system.

$$s = e^{-(\Delta G_I^0 + \delta)/RT} \tag{4.24}$$

This dependence of s on external factors provides the impetus for the system to partition the molecular population from the reference state to some new state.

It is beyond the scope of this chapter to derive detailed partition functions for all possible transitions in biopolymers. We will focus on a limited set of examples to illustrate how statistical mechanics predicts the behavior of well-defined polypeptides and polynucleotides, and how these can be extended to predicting the structures of proteins and double-stranded DNAs.

4.2   STRUCTURAL TRANSITIONS IN POLYPEPTIDES AND PROTEINS

4.2.1   Coil-Helix Transitions

The transition between a random coil and a helix structure is an important component in the folding pathway for polypeptides and proteins. The term *random coil* refers to a set of structures that are unfolded or denatured relative to the regular structures of helices in polypeptides and the compact tertiary structures of proteins. A more detailed description of random structures is given in Section 4.4. To understand the properties of these transitions, we will start by discussing what the nucleation and propagation terms mean when the zipper model is applied to the coil-to-helix transition in a polypeptide of identical amino acids (a *homopolymer*), and then use a statistical-mechanical treatment of the zipper model to describe the thermodynamics of this transition.

This discussion must begin with a review of some of the energetic terms that help to stabilize helices, with the α-helix as an example. As we recall from Chapter 3, the hydrogen bond formed between the carbonyl oxygen of residue j and the amide hydrogen of residue $j + 3$ helps to stabilize the structure of the α-helix (Figure 4.2). However, this intramolecular hydrogen bond is energetically favorable only when isolated from competing interactions with the solvent (see Table 3.5, Chapter 3). Thus, many shorter oligopeptides adopt an α-helical conformation only in an organic solvent such as 2,2,2-trifluoroethanol (TFE). Still, certain long polypeptides can be induced to form α-helices in solution. For example, poly-[γ-benzyl-L-glutamate] undergoes a coil-to-helix transition upon heating. In this case, δ of Eq. 4.24 is the amount of heat absorbed ΔH^0, when forming the helix. The positive sign of ΔH^0 indicates that the transition is entropy driven, consistent with the desorption of water from the helix. The transition is highly cooperative for long polypeptides and thus can be modeled by the zipper model.

In applying the zipper model to this transition, we need to define the nucleation parameter σ and propagation parameter s for the coil-to-helix transition. The high cooperativity of the transition suggests that σ is very small relative to s. To understand this at the molecular level, we can start by considering the difference in entropy between the two possible conformations (the *conformational entropy*). We recall that a random coil is best described as an ensemble of conformations that are irregular (i.e., not symmetric), as a helical structure is. The transition from a random

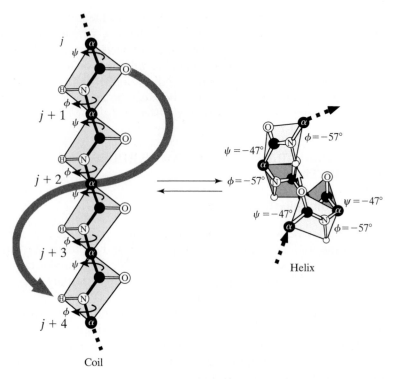

Figure 4.2 Nucleation step in the formation of an α-helix in a polypeptide chain. The torsion angles ϕ and ψ around each C_α-carbon must be fixed to average values of $\phi \approx -57°$ and $\psi \approx -47°$ to form a single intramolecular hydrogen bond between residue j and $j + 3$.

coil to an α-helix starts with a set of energetically degenerate structures and results in a single highly symmetric structure. Thus, the coil is entropically favored by conformational entropy over the helix.

The difference in conformational entropy between the coil and the helix can be estimated by considering the degrees of freedom for the two states. There are two freely rotating bonds (the ϕ- and ψ-angles; see Chapter 3) for each amino acid (except for Pro) along a polypeptide chain. As we recall from Chapter 2, the entropy of a system is defined as $S = R \ln W$, where W is the number of ways of arranging the particles of a system. In this case, W can be described in terms of the distribution of torsion angles ϕ and ψ around the C_α-carbon of each amino acid. For an α-helix, $\phi = -57°$ and $\psi = -47°$. These represent the mean for a population of structures called an α-helix but, for this discussion, are treated as discrete angles. In an α-helix, there is only one possible set of ϕ, ψ angles at each residue; therefore, $W = 1$ and $S = 0$.

In a random coil, there are three distinct minima for ϕ and ψ for each amino acid residue (except for proline). We can therefore estimate that there are nine possible discrete conformations for each residue in a random coil. Thus, the conformational entropy of the coil is $S = R \ln 9 = 18$ J/mol K. For the transition of one residue from a coil and to a helix, therefore, $\Delta S = -18$ J/mol K for each residue. The value of s has been experimentally determined to be ~1 for poly-[γ-benzyl-L-glutamate] in aqueous buffers, equivalent to a free energy of 0. This suggests that the loss in conformational entropy is effectively balanced in the propagation term by stabilizing effects, including the formation of the intramolecular hydrogen bond and, for this particular polypeptide, hydrophobic interactions between the side chains. This is further supported by the observation that $s \approx 5$ for the transition in dioxane, a solvent that is more favorable to the formation of the intramolecular hydrogen bond than is water.

The nucleation parameter σ has been estimated to be 2×10^{-4} for poly-[γ-benzyl-L-glutamate]. To interpret σ in terms of the coil-to-helix transition, we must first make the assumption that the nucleation event requires the formation of at least one full turn of an α-helix. Thus, the ϕ- and ψ-angles of four amino acids must be set in order to initiate formation of the helix (Figure 4.2). This is equivalent to $\Delta S = -67$ J/mol K (or -20 kJ/mol for $T\Delta S$) of entropic energy. However, there is only a single stabilizing hydrogen bond that can be formed within these four residues. For the first step of the transition, $\omega_1 = \sigma s$. If s accounts for the balance between the stabilizing and destabilizing energies for one residue, σ must account for the additional conformational entropy of the remaining three residues that are also fixed in forming the first turn of the helix. The overall energy for this nucleation event can be estimated from the additional loss in entropy as ~15 kJ/mol, which corresponds to $\sigma \approx 10^{-3}$. This is very similar to the experimentally determined σ.

For the heat-induced coil-to-helix transition in poly-[γ-benzyl-L-glutamate] (Figure 4.3), s is a temperature-dependent variable. If s is treated as an equilibrium constant between the coil and helix forms of each residue, then, starting with the standard van't Hoff relationship $\partial \Delta G^0/\partial T = \Delta H^0/T$, it can be readily shown that

$$\ln s = \frac{-\Delta H^0}{RT} + c \qquad (4.25)$$

The constant c is evaluated at $\ln s = 0$. For long polypeptides, this is approximately at the midpoint temperature of the transition T_M, so that $c = \Delta H^0/RT_M$. Thus, at any temperature, s is related to temperature by

$$\ln s = \frac{-\Delta H^0}{R}\left(\frac{1}{T} - \frac{1}{T_m}\right) \qquad (4.26)$$

With these parameters, the temperature-dependent transition for poly-[γ-benzyl-L-glutamate] can be accurately modeled by the statistical-mechanics treatment of the zipper model (Figure 4.3). This treatment also allows us to understand some of behavior of the coil-to-helix transition. The cooperativity of the transition increases with increasing n. In short polymers, the nucleation term is dominant during all steps of the transition. As n increases, the propagation parameters become dominant toward the end of the transition. This helps to distribute the higher free energy associated with the nucleation term over a larger number of residues, making the overall transition more probable at the higher temperatures.

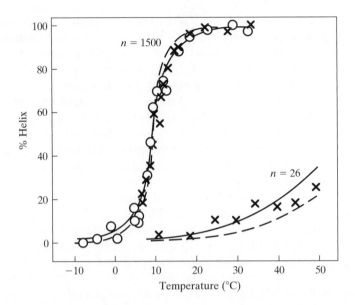

Figure 4.3 The temperature-dependent coil-to-helix transition in poly-[γ-benzyl-L-glutamate]. The transition from a coil to an α-helix structure are compared for two lengths of the polymer ($n = 1500$ and $n = 26$ amino acid residues). Data are from Doty and Yang (circles) and Doty and Iso (X), measured in a solution containing dichloroacetic acid and 1,2-dichloroethane, in a 7:3 ratio. The curves are simulated using Eq. 4.22 for the zipper model. The propagation parameter s is temperature-dependent, as defined by Eq. 4.25. The solid curves are calculated using $\sigma = 2 \times 10^{-4}$, while the dashed curves were calculated with $\sigma = 1 \times 10^{-4}$. The smaller value for σ results in a more cooperative transition than observed from the data. Both the cooperativity and the extent of the transition are dependent on the length of the polypeptide. [Adapted from B. H. Zimm and J. K. Bragg (1959), *J. Chem. Phys.* **31**, 526.]

It is clear that σ and s are dependent on the nature of transition and the sequence of biopolymer. In the case of a polypeptide, the nature of the helix and the properties of the amino acid side chains strongly affect both parameters. For example, sequences of the type $(AAAAK)_n$ A converts from 3_{10} helices to α-helices as n increases from 3 to 4. Since the sequences are essentially identical, side-chain interactions cannot account for the length dependence of the transition. This suggest that the α-helix is more difficult to initiate but is more readily propagated, compared to the 3_{10} helix. The difference in nucleation can be attributed again to the difference in the conformational entropy between the two types of helices. In an α-helix, hydrogen bonds are formed between residues j and $j + 3$, while for the 3_{10} helix, the hydrogen bonds are between j and $j + 2$. Thus, there is one additional residue that must be fixed to form the first turn of an α-helix compared to a 3_{10} helix. We can estimate the difference in conformational entropy to be equivalent to fixing one additional set of ϕ, ψ-angles in α-helix as compared to the 3_{10} helix.

For the α-helix, both σ and s are sequence dependent. For example, in lysine-containing polymers $\sigma = 1 \times 10^{-3}$, which is 10-fold more favorable for initiating helix formation than in poly-[γ-benzyl-L-glutamate]. Recently, a series of oligopeptides that are stable as α-helices have been used to determine the ability of each amino acid to affect the stability of the helix. The sequences of these oligopeptides have the general form

$$\ldots hhh \ X \ hhh \ldots$$

where h represents amino acids that favor formation of the helix and X represents any amino acid. These peptides are also referred to as *host-guest* peptides, with the amino acid X residing as a guest in the host α-helical sequence. The effect of X on the stability of the helix reflects the intrinsic ability of this amino acid to adopt the helical conformation. When compared to $X = $ Gly, this can be expressed as the effect of the side chain of the guest amino acid on the free energy for stabilizing the helix, ΔG_H^0 (Table 4.1). Alternatively, if we assume that σ is not dependent on X, then ΔG^0 is related to the propagation parameter of the zipper model s.

Table 4.1   Parameters for α-Helix Formation of Amino Acid X in Host-Guest Peptides

Amino Acid	ΔG_α^0 [a]	ΔG_α^0 [b]	s [c]
Ala	−3.17	−3.25	1.07
Arg	−2.80	—	1.03
Lys	−2.68	—	0.94
Leu	−2.55	−2.55	1.14
Met	−2.06	−2.35	1.20
Trp	−1.85	—	1.11
Phe	−1.69	—	1.09
Ser	−1.44	−1.15	0.76
Gln	−1.36	−1.98	0.98
Glu	−1.11	—	0.97
Cys	−0.95	—	0.99
Ile	−0.95	−1.61	1.14
Tyr	−0.70	−0.95	1.02
Asp	−0.62	—	0.68
Val	−0.58	−1.40	0.95
Thr	−0.45	—	0.82
Asn	−0.29	−0.74	0.78
His	−0.29	—	0.69
Gly	0.00	0.00	0.59
Pro	~12	—	0.19

Differences in free energy are given in kJ/mol.

[a] Determined from urea denaturation of the host peptide acetyl-EWEALEKKLAALE-X-KLQALEKKLEALEHG. ΔG^0 for amino acid X is given relative to $X = $ Gly. ΔG^0 for the charged amino acids are dependent on the ionic strength of the solution. Values are given for 1 M NaCl [data from O'Neil and DeGrado (1990), *Science* **250**, 646–651].

[b] Determined for the tripeptide sequence EXK within the host peptide succinyl-YSEEEEKKKKE-X-KEEEEKKKK-NH$_2$ at 4°C and neutral pH conditions [data from Lyu et al. (1990), *Science* **250**, 669–673].

[c] Data from Wojcik et al. (1990), *Biopolymers* **30**, 12–24.

The reverse transition from helix to coil can be described using the same set of parameters and definitions for the coil-to-helix transition. Therefore, we will not explicitly derive the partition function for this transition. We will instead discuss the probability of initiating the helix-to-coil transition at the middle versus the ends of the helix, to illustrate the differences in the mechanism of the forward and reverse transitions.

Consider in this case a polypeptide with n number of residues, again with a representing the coil form and b the helix form of each residue. We can use the same nucleation and propagation parameters defined previously for the coil-to-helix transition, and treat the states of the system as the exact reverse of the coil-to-helix transition. The initial state will have n residues in a helical form and thus the statistical weight for this state is $\omega_n = \sigma s^n$. The chain can be represented as

$$bbbbb \ldots bbbbb$$

If a single residue at either end converts from b to a, the statistical weight for the new state will be $\omega_{n-1} = 2\sigma s^{n-1}$. The degeneracy of this state is 2 because the transition can initiate from either end.

$$abbbb \ldots bbbbb \quad \text{or} \quad bbbbb \ldots bbbbba$$

Alternatively, if the transition initiates in the middle of the polypeptide at some residue i, the chain is separated into two isolated helical regions of $i - 1$ and $n - i$ in length.

$$\ldots bbbbbabbbbb \ldots$$

Each isolated segment of helix has its own set of statistical weights, σs^{i-1} for the segment at the N-terminus and σs^{n-i} at the C-terminus. The overall statistical weight for this state is $\omega_{n-1} = (n - 2)\sigma^2 s^{n-1}$ where $n - 2$ represents the possible ways of arranging a single residue as a in a chain, but excluding the two end residues. The σ^2 factor is required since each of the two continuous stretches of helix must now have its own nucleation parameter. The probability of melting a single residue at one end of the helix, P_e, as opposed to the middle P_m is thus given by

$$\frac{P_e}{P_m} = \frac{2\sigma s^{n-1}}{(n - s)\sigma^2 s^{n-1}} = \frac{2}{(n - s)\sigma} \tag{4.27}$$

If $\sigma = 2 \times 10^{-4}$ then $P_e/P_m = 1$ when $n = 2 + 2/\sigma \approx 10^4$. Thus, for short helices with $n < 10^4$ residues, the transition initiates at the ends and migrates toward the middle of the chain. This minimizes the number of nucleation parameters required in each state. In contrast, for long helices with $n > 10^4$ residues the large number of possible nucleation points in the middle of the chain overcomes the additional σ required for this pathway. The denaturation of globular proteins follows these types of helix-to-coil mechanisms. Both the coil-to-helix and helix-to-coil transitions are cooperative and can be described by the zipper model.

4.2.2   Statistical Methods for Predicting Protein Secondary Structures

We can imagine that once s and σ are determined for all amino acids for all possible secondary structures, a statistical model can be developed to accurately predict the *propensity* for (probability of) sequences within polypeptides to adopt helical structures. However, for globular proteins, the environment experienced by amino acid residues at the surface differs from those buried in the interior of the protein. Since the propagation parameters are dependent not only on sequence but also on these external factors, it becomes difficult to specifically assign values for s and σ unless we already know whether a particular residue is buried or surface accessible. Still, attempts have been made to define the average propensities for amino acids to adopt specific secondary structures in average proteins.

Although statistical mechanics have not yet been routinely applied to this problem, many features of the zipper model have been incorporated into statistical methods to predict the secondary structures of globular proteins. The familiar method developed by Chou and Fasman (1974) starts with a set of parameters that reflects the average propensity for amino acids to adopt α-helices, $\langle P_\alpha \rangle$; β-sheet, $\langle P_\beta \rangle$; or turns, $\langle P_T \rangle$ (Table 4.2).

Table 4.2   Propensities of Amino Acids to Form α-Helices $\langle P_\alpha \rangle$ and β-Sheets $\langle P_\beta \rangle$

α-Residues	$\langle P_\alpha \rangle$	α-Assignment	β-Residues	$\langle P_\beta \rangle$	β-Assignment
Glu	1.44 ± 0.06	H_α	Val	1.64 ± 0.07	H_β
Ala	1.39 ± 0.05	H_α	Ile	1.57 ± 0.08	H_β
Met	1.32 ± 0.11	H_α	Thr	1.33 ± 0.07	h_β
Leu	1.30 ± 0.05	H_α	Tyr	1.31 ± 0.09	h_β
Lys	1.21 ± 0.05	h_α	Trp	1.24 ± 0.14	h_β
His	1.12 ± 0.08	h_α	Phe	1.23 ± 0.09	h_β
Gln	1.12 ± 0.07	h_α	Leu	1.17 ± 0.06	h_β
Phe	1.11 ± 0.07	h_α	Cys	1.07 ± 0.12	h_β
Asp	1.06 ± 0.06	h_α	Met	1.01 ± 0.13	I_β
Trp	1.03 ± 0.10	I_α	Gln	1.00 ± 0.09	I_β
Arg	1.00 ± 0.07	I_α	Ser	0.94 ± 0.06	i_β
Ile	0.99 ± 0.06	i_α	Arg	0.94 ± 0.09	i_β
Val	0.97 ± 0.05	i_α	Gly	0.87 ± 0.05	i_β
Cys	0.95 ± 0.09	i_α	His	0.83 ± 0.09	i_β
Thr	0.78 ± 0.05	i_α	Ala	0.79 ± 0.05	i_β
Asn	0.78 ± 0.06	i_α	Lys	0.73 ± 0.06	b_β
Tyr	0.73 ± 0.06	b_α	Asp	0.66 ± 0.06	b_β
Ser	0.72 ± 0.04	b_α	Asn	0.66 ± 0.06	b_β
Gly	0.63 ± 0.04	B_α	Pro	0.62 ± 0.07	B_β
Pro	0.55 ± 0.05	B_α	Glu	0.51 ± 0.06	B_β

Listed are values compiled from the crystal structures of 64 proteins, and the assignments as former (H and h), indifferent (I and i) and breakers (b and B) for each type of structure.

From P. Y. Chou (1989), in *Prediction of Protein Structure and the Principles of Protein Conformation*, ed. G. D. Fasman, 549–586, Plenum Press, New York.

The $\langle P_\alpha \rangle$ for each amino acid is defined by first determining the mole fraction of that amino acid actually observed to form the α-helix structure in a data set of protein crystal structures ($\chi_{\alpha i}$ = occurrence of amino acid type i in an α-helix/total occurrence of amino acid type i in the data set). The average amino acid, represented by $\langle \chi_\alpha \rangle$, is the average of $\chi_{\alpha i}$ for all 20 common amino acids. For each amino acid type, $\langle P_\alpha \rangle$ is the fraction of that amino acid in a helix relative to the average amino acid, defined as $\chi_{\alpha i}/\langle \chi_\alpha \rangle$. Values for $\langle P_\beta \rangle$ and $\langle P_T \rangle$ were similarly determined for each type of amino acid (Table 4.2). Not surprisingly, Gly has a low $\langle P_\alpha \rangle$, but a high $\langle P_T \rangle$. The third residue in a type II β-turn is typically Gly. The $\langle P_\alpha \rangle$ and $\langle P_\beta \rangle$ for Pro are both low. Pro is therefore both an α-helix and β-sheet breaker because of the restricted ϕ- and ψ-angles around the C_α-carbon of this amino acid.

The values for $\langle P_\alpha \rangle$ and $\langle P_\beta \rangle$ are analogous to an average propagation term $\langle s \rangle$ in the zipper model for the coil-to-α-helix or the coil-to-β-sheet transitions. The low-probability nucleation event is accounted for by the method in assigning regions of α-helix or β-sheets. For example, one method for using $\langle P_\alpha \rangle$ is to categorize amino acids as strong helix formers (H), average helix formers (h), weak helix formers (I), indifferent (i), weak helix breakers (b), or strong helix breakers (B) (Table 4.2). By inspection, helices are considered to be initiated in regions where there is a contiguous set of strong helix formers (H and h, with I counted as half a helix former), and to propagate through amino acids categorized as H, h, I, i, and in isolated cases as b (Figure 4.4). The helix is terminated, however, when two or more helix breakers are encountered. This is repeated for each type of secondary structure. One feature of this approach is that a group of amino acids with strong preferences to form a particular type of structure is required as a point of nucleation for that structure. Once initiated, the conformation will extend as far as possible because of the cooperativity in the transition. The conformation is terminated either at points where the system no longer supports the structure or where the formation of an alternative structure can be initiated. These are the essential features of the zipper model as we described it above.

There are several variations on the basic Chou and Fasman method; each differs by how the structural propensities are applied or how they are defined. However, these methods are all found to be limited to about 70% reliability in predicting secondary structures in globular proteins (Figure 4.5). One reason for this limitation may be that the $\langle P_\alpha \rangle$, $\langle P_\beta \rangle$, and $\langle P_T \rangle$ determined from single-crystal structures represent the propensities in an average solvent environment and do not account for the contribution of tertiary structure to the stability of helices and coils. For example, turns typically are found at the surfaces of globular proteins and thus amino acids at the turns should be hydrophilic. The method developed by Rose (1978) to predict turns is based on the hydropathy of amino acids (Table 1.6). In this method, regions of a sequence that are not predicted to be helical and that have negative hydropathies can be assigned to turns.

A more fundamental problem in not taking tertiary structure into account is that the prediction methods may be ignoring the actual mechanism of secondary-structure formation during protein folding. The paradigm of a hierarchical sequence of folding from primary to secondary to tertiary structure may not be correct. One recent model for folding involves a collapse of the polypeptide chain into a relatively compact structure, and the environment of the collapsed structure

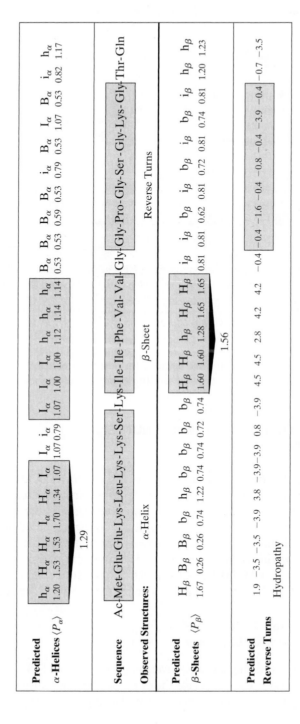

Figure 4.4 Application of the Chou and Fasman method for prediction of α-helices and β-sheets and the Rose method for prediction of turns in proteins. The ⟨P_α⟩ and ⟨P_β⟩ and the associated $H, h, I, i; b, B$ designations from Table 4.2 and the hydropathies from Table 1.6 are assigned to each amino acid of adenylate kinase. The values of ⟨P_α⟩ and ⟨P_β⟩ listed below residues predicted as helix and coil, respectively; are averages across those residues. The regions predicted by the prediction methods to be α-helices, β-sheets, and turns are compared to the observed in the single-crystal structure of the protein. [Adapted from G. E. Schultz and R. H. Schirmer (1979), *Principles of Protein Structure*, 121, Springer-Verlag, New York.]

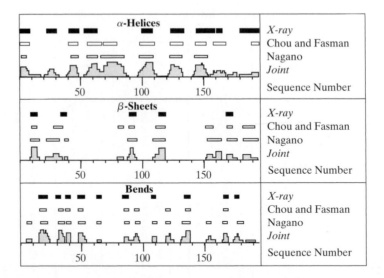

Figure 4.5 Comparison of methods for predicting protein secondary structure. The two widely used methods of Chou and Fasman and of Nagano to predict regions of α-helices, β-sheets, and turns in adenylate kinase are compared to the structures assigned by X-ray diffraction and by a joint prediction that combines the methods of Chou and Fasman, Nagano, Barry and Friedman, and Kabat and Wu. [Adapted from Schulz et al. (1974), *Nature* **250**, 140.]

defines the ability of an amino acid to adopt a helical structure. We discuss the statistical methods to model this type of folding pathway in Section 4.4.3.

4.3   STRUCTURAL TRANSITIONS IN POLYNUCLEIC ACIDS AND DNA

The structural transitions in polynucleic acids can be analyzed and modeled using the same basic statistical-mechanics treatment of the zipper model already described. This includes the annealing of single-stranded polynucleotides to form double-stranded DNA or RNA, or the reverse transition with the double-stranded helices melting to their respective single strands. The transitions are analogous to the coil-helix and helix-coil transitions in polypeptides. The transitions between different helical conformations (e.g., from standard B-DNA to A-DNA or to left-handed Z-DNA) within linear or circular DNAs have also been successfully treated by statistical mechanics.

4.3.1   Melting and Annealing of Polynucleotide Duplexes

The processes of annealing (or renaturing) and melting of DNA or RNA double-helices show *hysteresis*. The curves for the two transitions are distinct (Figure 4.6), indicating that they follow different mechanisms. The midpoint temperature for melting

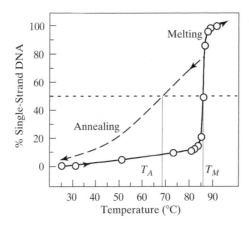

Figure 4.6 Curves for melting and annealing (renaturation) of double-stranded DNA. The solid line shows the highly cooperative melting curve for fully duplex DNA, while the dashed curve represents the less cooperative renaturation of the duplex. The midpoint temperatures, defined at 50% double-stranded DNA, for melting T_M is higher than that for annealing T_A. [Adapted from G. L. Zubay, W. W. Parson, and D. E. Vance (1995), *Principles of Biochemistry*, 638. Wm. C. Brown Publishers, Dubuque, IA.]

double-stranded DNA, T_M, is higher than the annealing temperature T_A, reflecting differences in the structure of the half-melted and half-renatured DNAs. We first discuss the mechanism for melting double-stranded polynucleotides because of its similarity to the helix-to-coil transition in polypeptides. This is followed by a discussion of the annealing mechanism, emphasizing the specific aspects that differ from melting and from the coil-to-helix transition in polypeptides. As with the helix-to-coil transition, we start by describing the melting behavior of simple homopolymers.

The melting of *homoduplexes* of identical base pairs follows a mechanism that is analogous to the helix-to-coil transition already described, although the molecular details differ. In this discussion, the melting of the duplex will refer to the separation of the two strands by breaking the stabilizing interactions that hold the base pairs together. This is typically attributed to the hydrogen bonds between the nucleotide bases but, as we will see, must also include base-stacking interactions. The two are related in that unstacking in a duplex of the bases typically requires that the hydrogen bonds break. The reverse, however, is not true.

The structural meanings of the nucleation and propagation parameters differ from those for polypeptides, but the properties of the transition in the two types of biopolymers are nearly identical. The melting of double-stranded DNA or RNA is again treated as a transition from state *a* to *b*, where the fully base-paired duplex represents the reference state. Thus, to describe this mechanism by the zipper model (Figure 4.1), we need to define each state in terms of the energetic contributions to the nucleation and propagation parameters. At the base-pair level, a polynucleotide duplex is stabilized by hydrogen bonding and stacking between nucleotides, while the single-stranded form is favored by an increase in entropy as well as repulsion between the negatively charged phosphates of the backbone. The propagation parameter *s*, therefore, is dependent on temperature and salt concentrations as reflected in the dependence of T_M on the amount of salt in the solution (Figure 4.7).

Although melting of a DNA or RNA duplex is typically thought of as the breaking of hydrogen bonds that hold the two strands together, it is loss in base-stacking that is normally observed. Typical melting studies use absorption spectroscopy

Figure 4.7 Dependence of the midpoint temperatures for melting T_M DNA duplexes on the G + C content of a sequence and the salt concentration in solution. [Data from J. Marmur and P. Doty (1962), *J. Mol. Biol.* **5**, 109.]

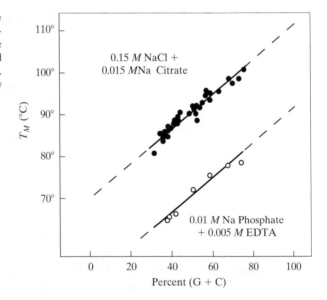

(see Chapter 11), which monitors the unstacking of bases, but not directly the breaking of hydrogen bonds; therefore, T_M measured in this way reflects the unstacking of base pairs rather than the breaking of hydrogen bonds. Base-stacking is a very strong interaction, with many single-stranded polynucleotides, particularly polypurine sequences, being completely stacked at room temperature. However, for most random sequences, the bases are mostly unstacked upon melting. Thus, s for extending the melted duplex by one base pair is primarily associated with the loss of one stacking interaction.

Unstacking also destabilizes base pairs that have not been melted. A base pair that is immediately adjacent to a melted base pair also loses one stacking interaction, even if the bases remain hydrogen bonded. This is reflected in the nucleation parameter σ. The nucleation parameter is thus attributed to the loss of an additional stabilizing stacking interaction ΔG^0_n that is not accounted for by s.

$$\sigma = e^{-\Delta G^0_n/RT} \tag{4.28}$$

As with the helix-to-coil transition in polypeptides, the melting of the duplex is favored at the ends of the polymer chain. The statistical weight for melting a single base pair at either end of the duplex is $\omega_1 = g_1\sigma s = 2\sigma s$. In contrast, melting a base pair in a place other than the two ends would result in the loss of two stacking interactions, one with each adjacent unmelted base pair. The statistical weight for this state is $\omega_1 = (n - 2)\sigma^2 s$. Thus, the probability of melting a base pair in the middle as opposed to at one of the ends of the duplex is $(n - 2)\sigma/2$. As with the helix-to-coil transition in polypeptides, the ends will melt first in duplexes for $n < 2/\sigma$, but there would be a higher statistical probability for melting base pairs in the middle when $n > 2/\sigma$.

For duplexes of mixed sequences, s is dependent on the stability of each base pair along the sequence. This is reflected in the dependence of T_M as a function of G + C content in a sequence (Figure 4.7). Although this is typically attributed to the additional hydrogen bond in a G·C base pair, we should remember that base-pair stacking also contributes. Indeed, T_M appears to be linearly related to stacking energies, as calculated from quantum mechanics (Figure 4.8). Finally, melting is initiated around long stretches of T·A base pairs, since ω_1 is a function of both σ and s (Figure 4.9).

The reverse transition that anneals or renatures single strands to form the duplex is analogous to the coil-to-helix transition, but is so significantly more complicated that we will not attempt a derivation of the partition function for the transition. However, we will discuss the properties of the nucleation and propagation parameters as they are relevant to how this transition differs from the melting of the duplex, causing hysteresis. The nucleation event for annealing is bimolecular, requiring two strands to come together to form a short base-paired region (Figure 4.9). The nucleation parameter therefore must be dependent on the concentration of the complementary single strands. This can be accommodated by defining the term κ, which is dependent on the overall concentration of single strands. The statistical weight for the formation of the first base pair is thus

$$\omega_1 = \kappa \sigma s \tag{4.29}$$

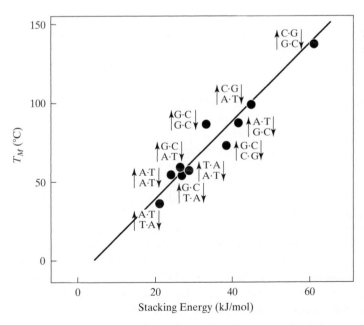

Figure 4.8 Correlation between DNA melting temperatures T_M and nearest neighbor stacking energies. Energies are shown for two base pairs calculated in the B-DNA conformations. Arrows indicate 5′ to 3′ directions of the strands. [From O. Gotoh and Y. Takashira (1981), *Biopolymers* **20**, 1033–1042.]

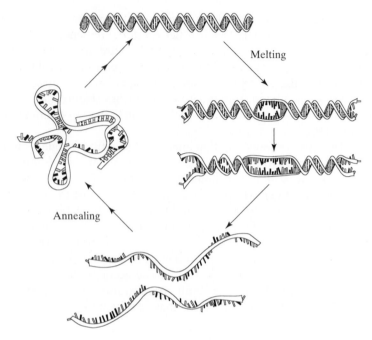

Figure 4.9   Comparison of models for melting and annealing DNA duplexes. The melting transition initiates at the ends and at regions rich in A·T base pairs in the middle of the duplex. As the temperature increases, the melted regions extend and other nucleation points appear. Finally, the duplex is completely melted to the two complementary single strands. The initial step for annealing the two single strands is bimolecular. During this process, the duplex may sample a number of intermediate states having both interstrand and intrastrand base pairs in trying to find the most stable combination of base pairs.

An added complication is that the nucleation event is also sequence dependent. Since C·G base pairs are more stable than T·A base pairs, we expect the nucleation event to be more probable at cytosine or guanine nucleotides. Thus, s must be dependent on the nucleotides being paired.

We can begin to appreciate the complexity of the system by imagining that the nucleation point for annealing brings together two regions along the chain that may not be ultimately paired in the fully duplex DNA. Thus, the two strands must sample a number of different possible base-pairing combinations prior to reaching the most stable duplex form. If strands are annealed too rapidly, they can be kinetically trapped in higher-energy states where not all the bases are optimally paired and the duplex incorporates mispaired or unpaired bases. In addition, because annealing is concentration dependent, annealing a dilute DNA solution could result in the individual single strands forming intramolecular duplex structures that may not lead to formation of the full duplex without remelting the structure. This is particularly acute with short sequences, where the concentration-dependent factor κ becomes significant.

In contrast to melting, which starts as a well-defined structure and proceeds along a specific pathway, the annealing transition must sample many possible thermodynamically degenerate forms at each state. This greatly increases the entropy of each annealed state. Thus, a duplex that is half-melted represents a very narrow distribution of structures, while a half-renatured duplex is much more heterogeneous in composition. As a consequence, the T_M and the cooperativity for melting will always be greater than the T_A for renaturation (Figure 4.6).

4.3.2   Helical Transitions in Double-Stranded DNA

The annealing and melting of DNA and RNA are only two of many structural transitions available to polynucleotides. DNA is highly polymorphic (see Chapter 1), and can adopt a large number of different helical structures. The reference state for DNA under physiological conditions is right-handed double-helical B-DNA. However, DNA can be induced to form (among other structures) A-DNA, left-handed Z-DNA, or triple-stranded H-DNA; it can also be extruded as a cruciform or melted to single-stranded DNAs. The transitions between the various double-helical forms have been observed in linear polymers and within isolated regions of closed circular DNAs.

The transition from B-DNA to A-DNA in linear duplexes can be treated directly by Eq. 4.22 for the zipper model. In this case, s reflects the difference in energy between base pairs as A-DNA versus B-DNA, and σ is dependent on the extra energy ΔG_J^0 required to form a junction between the two structures. The transition from B- to A-DNA of a homopolymer of poly(dG)·poly(dC) can be induced by adding ethanol to the solution (Figure 4.10). A-DNA is stabilized by nonaqueous solvents, including alcohols, because it is dehydrated relative to B-DNA. The model for this transition fits the data with $\Delta G_J^0 = 5$ to 8 kJ/mol or $\sigma = 0.14$ to 0.04. This relatively small energy for the nucleation term is likely to be associated with the small difference in the base-pair stacking between A-DNA and B-DNA.

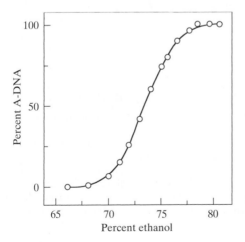

Figure 4.10 Ethanol-induced B-DNA to A-DNA transition in poly-[dG·dC]. The curve is simulated using a statistical-mechanics treatment of the zipper model, with the nucleation energy set at 5 kJ/mol. [Data from V. I. Ivanov, and D. Y. Krylov (1992), *Methods in Enzymology* **211**, 111–127.]

The transition from B-DNA to left-handed Z-DNA in linear duplexes can be treated in the same manner, with σ defined by the junction between the two opposite-handed helices, and s dependent on factors such as salt, alcohol, or temperature, all of which help to stabilize Z-DNA. However, this transition can also be induced by negative superhelical stress. We discuss this transition in some detail in Section 4.3.3.

4.3.3   Supercoil-Dependent DNA Transitions

The more interesting transitions in DNA that can be accurately modeled by statistical thermodynamics are the helical transitions in double-stranded, closed circular DNA (ccDNA). These transitions are induced by negative supercoiling and thus do not require the addition of dehydrating agents such as alcohols or salts. The DNA molecules in most organisms are negatively supercoiled. In eukaryotes, the DNA is coiled around histone proteins to form nucleosomes. The prokaryotic nucleoid and plasmid DNAs are found in a negatively supercoiled state. In addition, Lui and Wang (1987) have proposed that both positive and negative supercoils are induced by the passage of RNA polymerase through the DNA double strand (Figure 4.11) during transcription in eukaryotes and prokaryotes. Thus, these supercoil-induced transitions are considered to be more physiologically relevant in that the conditions necessary for them have been found to exist in all

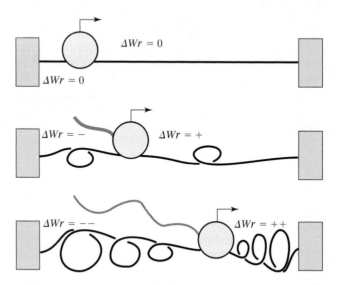

Figure 4.11 Model for RNA polymerase-induced supercoils in DNA duplexes that are topologically fixed at the ends (boxes). The polymerase initially binds to the relaxed duplex, with the no apparent superhelical strain ($Wr = 0$) in the DNA. As the polymerase transcribes along the DNA to form a strand of mRNA, the strands of the DNA duplex must separate, inducing positive superhelical strain ($Wr = +$) in front of the transcription complex, and negative superhelical strain ($Wr = -$) behind the transcription complex. The strain both in front of and behind the polymerase becomes more exaggerated as the complex transcribes further along the DNA.

cells. The partition function for these types of transitions can be treated with the standard zipper model, except that both the nucleation and propagation parameters must now include the effect of the transition on the overall topological state of the ccDNA. To understand how this property is incorporated into the model, we will discuss the transition from B-DNA to Z-DNA in ccDNA. We must first review the topological states of ccDNA.

The transitions between various helical conformations in ccDNA are dependent on the topological state of the DNA. As we recall from Chapter 1, the topology of ccDNA is described by the linking number Lk, which links the supercoiling or writhe Wr to the helical twist Tw. The effect of supercoiling on a DNA transition is determined by studying the behavior of the *topoisomers* of ccDNA at discrete values of Lk. A transition from standard B-DNA to any of the alternative forms involves a change in Tw. For a particular topoisomer, any change in Tw must be compensated for by an equivalent change in Wr such that Lk remains unaltered $(Lk = Wr + Tw)$. The structural transitions within ccDNA can be accurately modeled using statistical thermodynamics methods by accurately describing the energies of the various topological states.

As with all other systems, we must start by defining the reference or starting state ω_0. The reference topological state is relaxed (no supercoils), closed circular B-DNA, in which $Wr_0 = 0$ and $Tw_0 = (1 \text{ turn}/10.5 \text{ bp})n$, for n base pairs in a ccDNA. Thus, Lk_0 is equal to $Tw_0 = (1 \text{ turn}/10.5 \text{ bp})n$. Any transition to a new state ω_j is accompanied by a change in Tw ($\Delta Tw = Tw_j - Tw_0$), and a concomitant change in Wr ($\Delta Wr = Wr_j - Wr_0$). Therefore, ΔLk represents the topological state of a ccDNA topoisomer relative to this reference state.

To derive the partition function for this type of transition, we must first describe the energetics of each state in relation to the nucleation and propagation parameters of the zipper model. Rather than trying to present an abstract derivation for a general transition, we will consider the specific case of a localized transition within ccDNA from standard right-handed B-DNA to left-handed Z-DNA (the *B-Z transition*). This is a good working example because the resulting partition function can be used to very accurately model the behavior of simple repeating sequences in ccDNA, and also of more complex sequences found in genomes. In addition, the approach yields reliable predictions for the formation of Z-DNA in genomes and therefore illustrates the extent to which statistical thermodynamics methods can be successfully applied to structure prediction. Finally, the method used to derive a partition function for this transition is directly applicable to any supercoil-dependent structural transition in ccDNA.

We will consider in this discussion a B-Z transition localized within a region of n alternating d(C·G) base pairs in negatively supercoiled ccDNA (Figure 4.12). The conformational changes in a particular topoisomer include unwinding of the DNA helix ($\Delta Tw < 0$) and a concomitant relaxation of negative supercoils (ΔWr becomes less negative). It is this reduction in the superhelicity that favors the formation of the left-handed structure (indeed any underwound structure) in negatively supercoiled DNA. Although we will discuss the transition of base pairs from B- to Z-DNA,

Figure 4.12 Possible states for a closed circular DNA having a 12 base pair alternating d(C · G) sequence. In the reference state, all the DNA is B-form ($\Delta Tw = 0$) and there are no supercoils ($\Delta Wr = 0$); therefore, $\Delta Lk = 0$. In a different topoisomer state, ΔLk is reduced to -2, for example. At $\Delta Lk = -2$, the topological strain can partition between negative supercoiling ΔWr or unwinding of the helical twist of the duplex ΔTw. If ΔTw is localized within the d(C·G) insert, the base pairs can be induced to convert from right-handed B-DNA to left-handed Z-DNA.

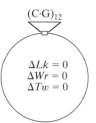

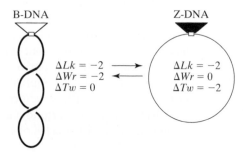

we must remember that Z-DNA is characterized by an alternating *anti-syn* conformation of nucleotides (Chapter 1) and that pyrimidines are sterically inhibited from adopting the *syn* conformation. Thus, Z-DNA sequences are typically alternating pyrimidines and purines, characterized by a dinucleotide (two-base pair) repeat unit.

Each base pair j that converts to Z-DNA defines a new state of the system with a statistical weight ω_j. The statistical weights for the propagation terms in ω_j are dependent on the energy, $\Delta G^0_{\Delta Lk}$, of the overall topological state of the system. We can think of this energy as the sum of the energies for ΔTw and ΔWr.

$$\Delta G^0_{\Delta Lk} = \Delta G^0_{\Delta Tw} + \Delta G^0_{\Delta Wr} \qquad (4.30)$$

The term $\Delta G^0_{\Delta Tw}$ is simply the difference in free energy between Z-DNA versus B-DNA for each base pair that adopts the Z-form, ΔG^0_{Z-B} (see Table 4.3). The term $\Delta G^0_{\Delta Wr}$ is the free energy of the superhelicity of the ccDNA of n base pairs in size, and is defined as

$$\Delta G^0_{\Delta Wr} = K\Delta Wr^2 \qquad (4.31)$$

This relationship has the same form as the energy for deforming a spring, with the proportionality constant ($K = 1100\ RT/N$ for a ccDNA of N base pairs) being analogous to the spring constant for ccDNA. One supercoil in a ccDNA of 1100 bp in length has a supercoiling free energy of 4 kJ/mol. The dependence of $\Delta G^0_{\Delta Wr}$ on ΔWr^2 means that the lowest energy state has $\Delta Wr = 0$, and the introduction of any degree of supercoiling (whether positive or negative) results in a higher energy state.

The free energy of any topoisomer with a defined ΔLk is given by

$$\Delta G^0_{\Delta Lk} = \Delta G^0_{\Delta Tw} + K\Delta Wr^2 \qquad (4.32)$$

Table 4.3 Propagation Free Energies ΔG_P^0 (kJ/mol Per Base Pair) Required to Extend Z-DNA after the Initial Step to Overcome the Nucleation Free Energy for the B-Z Transition

Dinucleotide	ΔG_P^0
CG/CG	1.2
CA/TG	2.9
TA/TA	4.9
CC/GG	7.0
TC/AG	7.0
AA/TT	9.1

$\Delta G_N^0 = 21$ kJ/mol per B-Z junction. Values are listed for the base pairs of dinucleotides in the alternating *anti-syn* repeat of Z-DNA.

From P. S. Ho (1994), *Proc. Natl. Acad. Sci. USA* **91**, 9549–9553.

The energy of the system can thus partition between the unwinding of the DNA helix, ΔTw, and supercoiling, ΔWr. If $\Delta Tw = 0$, the DNA duplex remains in the B-form and $\Delta G_{\Delta Lk}^0 = K\Delta Wr^2$. Alternatively, for each base pair that adopts the Z-conformation, the helix will unwind by $\Delta Tw = Tw_Z - Tw_B$, where Tw_Z is the helical twist of Z-DNA (-1 turn/12 bp) and Tw_B is the helical twist of B-DNA ($+1$ turn/10.5 bp). Since $\Delta Wr = \Delta Lk - \Delta Tw$, Eq. 4.32 becomes Eq. 4.33.

$$\Delta G_{\Delta Lk}^0 = \Delta G_{Z-B}^0 + K(\Delta Lk - \Delta Tw)^2$$
$$= \Delta G_{Z-B}^0 + K(\Delta Lk - [Tw_Z - Tw_B])^2$$
$$= \Delta G_{Z-B}^0 + K(\Delta Lk - j[-0.179 \text{ turns/bp}])^2 \tag{4.33}$$

Like the coil-to-helix transition, the nucleation parameter for the B-Z transition, σ, is dependent on fixing a number of residues in an unstable conformation. In this case, however, the nucleating event is not the formation of one turn of Z-DNA, but a structure that splices a left-handed helix to a right-handed helix. This is a *B-Z junction*. There are no free ends in circular DNA; thus, a segment of Z-DNA within a B-DNA duplex must be flanked at both ends by B-Z junctions. The energy required to form a B-Z junction ($+21$ kJ/mol/junction) was determined by studying the effect of length on ΔG_{Z-B}^0, and extrapolating to a Z-DNA insert of 0 length. The B-Z junction is best described as the melting of four base pairs of double-stranded DNA. In addition to this melting energy, σ must also take into account the effect that each junction has on ΔWr of the ccDNA. Since each B-Z junction acts as a transition point between the opposite-handedness of the two DNA forms, the average twist ΔG_{Z-B}^0 is by definition 0 turns/junction or $\langle \Delta Tw \rangle = -0.4$ turns/junction relative to ω_0. At a given ΔLk, $\Delta G_{\Delta Wr}^0 = K[\Delta Lk - (-0.4)]^2$. Two junctions are absolutely required for a B-Z transition of any length, even for only a single base pair as Z-DNA. Thus, the nucleation parameter for a single B-Z junction is

$$\sigma = e^{-(21 \text{ kJ/mol} + K[\Delta Lk - (-0.4)]^2)/RT} \tag{4.34}$$

or, for both junctions of an isolated segment of Z-DNA,

$$\sigma = e^{-(42\,\text{kJ/mol}+K[\Delta Lk+0.8]^2)/RT} \tag{4.35}$$

The mechanism for the B-Z transition can thus be described in terms of the zipper model (Figure 4.13). In the initial step, the first base pair that adopts the Z form automatically induces the formation of two B-Z junctions. Z-DNA is propagated by extending the left-handed conformation as the two junctions migrate in divergent directions along the sequence. Each new state of the system ω_j has j base pairs as Z-DNA. This must be confined within n base pairs of the ccDNA, so ω_j is

$$\omega_j = (n - j + 1)\sigma s^j$$
$$= (n - j + 1)e^{-(42\,\text{kJ/mol}+K[\Delta Lk+0.8]^2)/RT}e^{-(\Delta G^0_{Z-B}+K[\Delta Lk-j\{-0.179\,\text{turns/bp}\}]^2)/RT}$$
$$\tag{4.36}$$

The partition function for the B-Z transition can therefore be derived by incorporating ω_j into the general function for the statistical-mechanics treatment of the zipper model to give the probability for any given number of base pairs (j) as

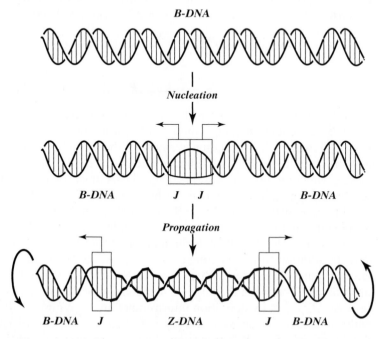

Figure 4.13 Zipper model for the supercoil-induced B-Z transition in closed circular DNA. The high-energy nucleation step requires melting ~8 base pairs to form two B-Z junctions (J). The junctions migrate in opposing directions as the intervening base pairs wind in the left-handed direction to form Z-DNA. [Adapted from P. S. Ho (1994), *Proc. Natl. Acad. Sci., USA* **91**, 9549–9553.]

Z-DNA $\langle P_j \rangle$ as

$$\langle P_j \rangle = \frac{\omega_j}{1 + \sum_{j=1}^{n} \omega_j} = \frac{(n - j + 1)\sigma s^j}{1 + \sum_{j=1}^{n} (n - j + 1)\sigma s^j} \tag{4.37}$$

Since the fraction of Z-DNA formed is directly reflected in the topological properties of the ccDNA, the probability of having the left-handed conformation is defined by the probability of each state that includes at least one base pair as Z-DNA, and the fraction of Z-DNA, f_Z, in those states. This is expressed as the average ΔTw, $\langle \Delta Tw \rangle$ for a topoisomer having a particular ΔLk. Therefore, f_Z can be expressed as the degree of unwinding of the supercoiled ccDNA for the B-Z junctions and as converting j base pairs to Z-DNA, $f_Z = -0.8 + j(-0.179)$ turns, so that

$$\langle \Delta Tw \rangle = \frac{\sum_{j=1}^{n} f_Z \omega_j}{1 + \sum_{j=1}^{n} \omega_j} \tag{4.38}$$

The values of $\langle \Delta Tw \rangle$ can be directly determined from ΔWr for each ΔLk of a topoisomer in a distribution. The topoisomers in a distribution are resolved by two-dimensional gel electrophoresis (Figure 4.14). The first (vertical) dimension of the gel resolves topoisomers according to ΔWr. Recall that supercoiled DNA, whether positive or negative, migrates faster through a gel than does the relaxed closed circular form (Figure 1.47). Thus, the slowest band has $\Delta Wr = 0$ (the band at the top of the gel), with each successively faster migrating band representing one additional supercoil (either positive or negative).

In the second dimension, the DNA in the gel is exposed to an intercalating agent that inserts between the stacked base pairs and thus unwinds the DNA (e.g., induces a negative ΔTw) which, for each topoisomer, must be compensated for by making ΔWr less negative. Since this change in ΔWr in the second dimension is useful in resolving the positively and negatively supercoiled DNAs, ΔLk can be determined from the bands that migrated coincidently in the first dimension. For example, in Figure 4.14, the amount of intercalator added prior to electrophoresis in the second dimension was sufficient to unwind the DNA by $\Delta Tw = -22$ turns; consequently, the ΔWr of each topoisomer will become more positive by $+22$ turns to maintain a fixed ΔLk. The topoisomer with $\Delta Wr = -22$ turns in the first dimension is now characterized as having $\Delta Wr = 0$ (and is therefore the slowest and left-most band) in the second dimension, while all toposiomers with $\Delta Wr < -22$ turns will now be positively supercoiled and thus migrate faster in the second (horizontal) dimension. This second dimension, therefore, can be thought of as resolving the topoisomers so that they can be indexed according to their actual ΔLk. Each resolved band corresponds to a single topoisomer, with the ΔLk counted, as in Figure 4.14.

The resolution of positive and negative supercoils in itself is not very interesting. However, the addition of intercalator in the second dimension also removes the

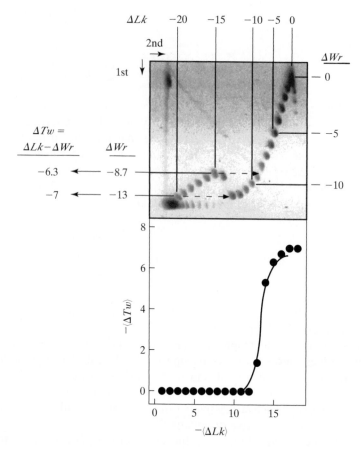

Figure 4.14   Two-dimensional gel electrophoresis analysis of a B-Z transition in a 40 base pair sequence inserted into the closed circular plasmid pBR322. In the vertical (first) dimension, the topoisomers are resolved according to their writhe ΔWr. In the horizontal (second) dimension, the intercalator chloroquine has been added to relax approximately 22 negative supercoils. This allows the topoisomers to be resolved according to their ΔLk. The change in twist ΔTw is estimated from the ΔLk determined from the second dimension and the ΔWr from the first dimension. These are plotted as $-\langle Tw \rangle$ versus $-\langle Lk \rangle$ to show the high cooperativity of the B-Z transition. The curve was calculated using values for $\Delta G^0_{(Z-B)}$ and ΔG^0_J from Table 4.3. [Data from Howell et al. (1996), *Biochemistry* **35**, 15373–15382.]

negative superhelical energy that holds the d(C·G) insert as left-handed Z-DNA. The result is that topoisomers with different values of ΔLk but identical ΔWr (and therefore comigrate in the first dimension) can be resolved in the second dimension because the ΔTw stored in the form of left-handed DNA is repartitioned into ΔWr. In short, the discontinuity seen in the gel represents those topoisomers that were involved in the B-Z transition, and the difference between ΔLk and ΔWr is $\langle \Delta Tw \rangle$ for each of those topoisomers. At the $\Delta Lk = -15$ turns (counted along the second, horizontal direction), we see that this topoisomer comigrated with a band associated with $\Delta Wr = -8.7$ turns, indicating that, in the absence of the intercalator, this topoismer had a $\Delta Tw = \Delta Lk - \Delta Wr = -6.3$ turns, which is equivalent to 35 base pairs of left-handed Z-DNA. From Eq. 4.31, $\Delta G^0_{\Delta Wr}$ for this topoisomer is 46.8 kJ/mol for a ccDNA of 4401 base pairs, while $\Delta G^0_{\Delta Lk}$ is estimated to be 139.3 kJ/mol if all of this ΔLk were in writhe. The difference between this $\Delta G^0_{\Delta Lk}$ and $\Delta G^0_{\Delta Wr}$ is the free energy associated with the B-Z transition for this topoismer ($\Delta G^0_{\Delta Tw}$), which is 71.4 kJ/mol. We can readily see, therefore, that with 21 kJ/mol required for each of the two B-Z junctions, we estimate a ΔG^0_{Z-B} of 0.8 kJ/mol per base pair, or 1.6 kJ/mol per CG dinucleotide, which compares favorably with the value in Table 4.3.

The $\langle \Delta Tw \rangle$ observed from the gels and calculated by the statistical mechanics treatment of the zipper model shows that the zipper model accurately describes the supercoil-induced formation of Z-DNA, and the application of the model to fit the data from the gel yields more accurate values for ΔG^0_{Z-B}. The B-Z transition is highly cooperative for sequences with a high propensity for forming Z-DNA [e.g., alternating $d(CG)_n$]. However, as the propensity to become left-handed becomes reduced [e.g., alternating $d(CA)_n \cdot d(TG)_n$ dinucleotides], the transition becomes less cooperative because the propagation terms start to approach the magnitude of the nucleation term.

The relationships derived here can be directly applied to analyzing and modeling supercoil-induced transitions from B-DNA to any of the various underwound forms of DNA including cruciform DNA and triple-stranded DNA (Figure 1.42). All that is required is that the unwinding and the energetic difference between the altered structure and B-DNA, and for the junctions between each structure and B-DNA, be incorporated into the definition of the nucleation and propagation parameters of the partition function.

4.3.4   Predicting Helical Structures in Genomic DNA

The statistical methods described for analyzing and modeling DNA secondary structures can be directly adapted for predicting the DNA sequences that have the highest probability for adopting that structure. Again, the best example is the prediction of left-handed Z-DNA in a genomic sequence. The statistical thermodynamic model derived in Section 4.3.3 can be applied to analyzing any sequence, not just homopolymers, of Z-DNA. In this case, a unique propagation parameter is assigned to each type of base pair (Table 4.3). The only modification needed to account for this is for each region of a sequence to be treated as unique. Thus, the various possible states are not degenerate. The partition function thus treats the propagation parameter of each base pair individually.

A genomic sequence can be analyzed for Z-DNA as overlapping sections of the DNA. The propensity for regions of the DNA sequence to adopt the left-handed conformation can be compared to that of an average random sequence to determine its statistical significance. One of the great advantages of studying Z-DNA is that its presence in genomic DNAs has been studied by many experimental methods. For example, regions predicted by the statistical methods to form Z-DNA correlate very well with the probability for the binding of anti-Z-DNA antibodies to DNAs,

Figure 4.15 Comparison of the sequences predicted and experimentally observed to form Z-DNA in the double-stranded form of the ϕX-174 viral genome. The propensity of sequences to form Z-DNA is interpreted as the number of kilobase pairs (1000 base pairs) that needs to be searched in a random sequence to find a sequence with a comparable probability of forming Z-DNA. The predictions are compared to results from tabulating the location of anti-Z-DNA antibodies bound at sites along the genome. The fraction of antibodies bound at each site reflects the ability of those sequences to adopt the Z-conformation. The four best sequences correlate well between the predictions and the experimental results. [Adapted from P. S. Ho et al. (1986), *EMBO J.* **5**, 2727–2744.]

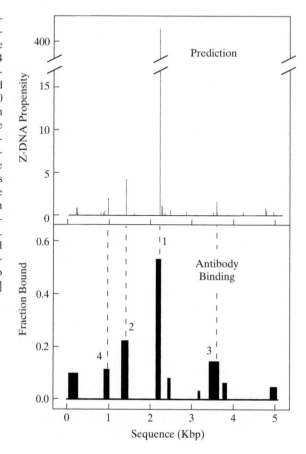

such as the double-stranded, closed circular form of the ϕX174 viral genome (Figure 4.15).

4.4   NONREGULAR STRUCTURES

Up to this point, we have considered only the regular helical structures of biopolymers. However, statistical methods can also be used to model and study nonhelical structures, including the tertiary structures of compact proteins and polynucleotides, and truly random structures such as coils. Both of these problems are related. To model or predict the tertiary structure of a protein, for example, we can search for that set of conformations from all possible conformations that are most probable for a given state of the system. In a brute force approach, this becomes an exercise in trying to describe all the possible conformations to find the most probable state by statistical mechanics. Thus, the two types of problems are related in that we need methods to model and analyze random structures. We will start by looking at the properties of random linear systems, and then apply this to the more complicated three-dimensional tertiary structures of macromolecules.

4.4.1   Random Walk

A random structure is one in which the conformation of each monomer unit along a chain is entirely independent of all other units. This is analogous to allowing a blind-folded disoriented person to wander aimlessly across a field (Figure 4.16). The path-way of this *random walk* traces what would be the completely random chain conformation in two dimensions. The random walk, as with a biopolymer chain, has an origin (O) and an endpoint (E). The length of each step l in its simplest form is equivalent to the distance between each pair of monomers. The analogy is that N steps along the path is equivalent to N bonds linking residues together along the chain. The random walk can be described in two dimensions (as in a trip across a field), or in one or three dimensions. The most accurate description of the random conformation for a biopolymer is in three dimensions. However, many of the prop-erties for the simple one-dimensional random walk also apply to two and three di-mensions, so we will begin with this as an example.

In a one-dimensional random walk, we start at the origin O and allow each step to move either forward ($+$) or backward ($-$) along a line. The statistical weight or probability of a forward step is f and of a backward step is b. Since this is random, the probability of stepping forward is equal to the probability of stepping backward ($f = b$, and $f + b = 1$). Thus we can represent a random set of steps in one dimen-sion as

$$+ + - - - - + - + + - + + - - - +$$

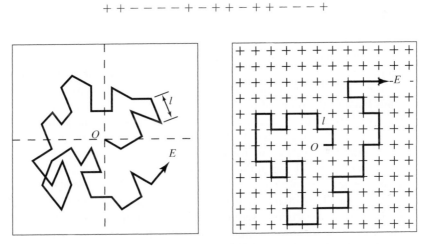

Figure 4.16   Random walks in a two-dimensional field. The walk across a field initiates at a point of origin O and ends at a point E to trace a random chain. In this walk, steps can be of different lengths, and turns can be at any angle. The length of each step l is analogous to the distance between each residue along the chain. A walk across a two-dimensional grid generates a set of well-defined struc-tures for a chain that are simple-but-exact minimal models for the tertiary structures of polypeptides and polynucleotides.

The probability of this particular sequence occurring is

$$ffbbbbfbffbffbbbf$$

For N total steps, there are j steps forward and $N - j$ steps backward. The overall probability for a sequence of steps occurring is thus $f^j b^{N-j}$. The number of different combinations of N steps is exactly analogous to the model for a noncooperative transition between two states of N residues. In fact, the random walk in one dimension is conceptually and mathematically equivalent to a noncooperative random two-state transition, with a forward step being one state and the reverse step being the second state for each residue. The sum of all the possible combinations of random steps is $(f + b)^N$. This is similar to the partition function in Eq. 4.13 and can be interpreted in the same manner as Q was in that example.

The number of different combinations of steps, with the number of steps forward equal to j (W_j), is equivalent to the degeneracy of each state in the noncooperative two-state transition model,

$$g_j = W_j = \frac{N!}{j!(N - j)!} \tag{4.39}$$

The probability of having j steps forward and thus $N - j$ steps backward, P_j, is analogous to Eq. 4.17 for the probability of observing any state j in the two-state transition model,

$$P_j = \left[\frac{N!}{j!(N - j)!} \right] f^j b^{(N-j)} \tag{4.40}$$

We have already shown that for a completely random process ($f = b$) the most probable sequence of steps will be the most degenerate. This occurs for $j = N/2$. Thus, the most probable set of steps will have exactly half the steps forward and half backward, or the most probable sequence of steps relative to any other sequence leads back to the point of origin. This does not mean, however, that all sequences of steps will always lead back to the origin. In fact, for any randomly chosen sequence of steps, it is more likely that $E \neq O$. Consider, for example, the case of $N = 4$. The most probable number of forward steps is $j = 2$ because $W_2 = 6$. These six sequences of steps all end at the origin. However, the number of sequences leading to $E \neq O$ is the sum of all combinations where $j \neq N/2$. This is, $1 + 4 + 4 + 1 = 10$, or there is a 10 to 6 chance that any particular sequence will not end at the origin. The probability of a particular sequence ending somewhere other than O increases as N increases, first because the number of the sequences of steps increases and, second, because the degeneracy of the sequences of steps approaches that of the $j = N/2$. Thus, for an ensemble of random macromolecules, individual molecules can have distinct conformations. We can therefore ask the question: What is the average behavior of the ensemble?

4.4.2   Average Linear Dimension of a Biopolymer

One important measurable quantity of a biopolymer's structure is its average linear dimension. This is described by the *root-mean-square end-to-end distance* $\langle L^2\rangle^{1/2}$ or, alternatively, the *root-mean-square radius* $\langle R^2\rangle^{1/2}$. These quantities are related to the *root-mean-square displacement* from the origin $\langle d^2\rangle^{1/2}$, which is the distance we expect to be from O, on average, after N steps of a random walk. The parameter $\langle d^2\rangle^{1/2}$ is in turn related to but different from the *mean-displacement* $\langle d\rangle$ which is a measure of the average net displacement after N steps of the random walk.

We would expect for an ensemble of completely random walks that $\langle d\rangle = 0$. However, the root-mean-square displacement $\langle d^2\rangle^{1/2}$, a measure of the average absolute displacement from the origin at the end of the random walk, will not be zero. The difference between $\langle d^2\rangle^{1/2}$ and $\langle d\rangle$ is evident if we remember that it is more probable that any sequence of steps will not end at the origin. There will be, for example, a number of sequences ending three steps in front of ($d = +3$) and an equal number at three steps behind the origin ($d = -3$). The average net displacement for these two sets of sequences is $\langle d\rangle = 0$. However, $\langle d^2\rangle$ would be the average of $(+3)^2$ and $(-3)^2$ which is nine, and $\langle d^2\rangle^{1/2} = 3$. Thus, on average, this set of random steps all end three steps from the origin, with no indication as to whether a particular walk ends in front of or behind the origin.

We can show that $\langle d\rangle$ and $\langle d^2\rangle^{1/2}$ is a function of the number of steps N in the random walk. For a given path, we can draw a vector from the origin to the end of the random walk; this vector $\mathbf{d}$ will be the vector sum of the vectors corresponding to the individual steps $\mathbf{l}_j$, which have magnitude l_j,

$$\mathbf{d} = \sum_{j=1}^{N}\mathbf{l}_j \tag{4.41}$$

The magnitude of the square of the vector is $\langle d^2\rangle$. To obtain this, we simply multiply it by itself; that is, we take the *dot product* of the vector with itself,

$$\langle d^2\rangle = \mathbf{d}\cdot\mathbf{d} = \left(\sum_{j=1}^{N}\mathbf{l}_j\right)\left(\sum_{k=1}^{N}\mathbf{l}_k\right)$$

$$= \sum_{j}\sum_{k}\mathbf{l}_j\cdot\mathbf{l}_k \tag{4.42}$$

The summation in Eq. 4.42 represents all the possible products between individual step vectors. These products will be of two kinds: when $j = k$, we are multiplying a vector by itself and will simply get l_j^2, if we multiply two different steps, we obtain $l_j l_k \cos\theta$ where θ is the angle between the two sters. For a one-dimensional walk, $\cos\theta = \pm 1$. If we consider a large number of random paths, the term $l_j l_k$ will equally likely be positive or negative, so they drop out. We find, then,

$$\langle d^2\rangle = \sum_{j=1}^{N}l_j^2 = Nl^2 \tag{4.43}$$

and

$$\langle d^2 \rangle^{1/2} = \left(\sum_{j=1}^{N} l_j^2 \right)^{1/2} = N^{1/2} l \tag{4.44}$$

Thus, $\langle d^2 \rangle^{1/2}$ is dependent on $N^{1/2}$ or, in terms of a biopolymer, the square root of the number of steps along a chain. These relationships derived in one dimension are also applicable to random walks in two and three dimensions, and thus are applicable to the description of the properties of the tertiary structures of macromolecules.

The root-mean-square displacement from the origin for a random walk also tells us the average absolute distance between the first and the last residues of a biopolymer. This is defined as the root-mean-square end-to-end distance of the chain ($\langle L^2 \rangle^{1/2} = \langle d^2 \rangle^{1/2} = N^{1/2} l$) and reflects the flexibility of the chain. Thus, the flexibility of the biopolymer $\langle L^2 \rangle^{1/2}$ is a function of the effective distance l between joints in the chain and the number of units separated by l. The parameters N and l need not explicitly be the number of residues and the distance between residues, respectively, since the pathway between any two residues is not entirely random. For simple organic polymers, l is simply a function of the bond length as constrained by the bond angle. In polypeptides, l is additionally constrained at each amino acid residue by the geometry of the peptide bond and the interactions between residues. N therefore is the number of functional units or segments of a polymer that appears to behave randomly, and l is the effective length of these random segments.

For a biopolymer, constraints are such that the chain behaves as if it were not flexibly jointed at each backbone atom, but as if it were a freely jointed chain of N-effective segments. The length of each of these segments defines the *persistence length*. The persistence length of double-stranded DNA, for example, has been estimated from hydrodynamic measurements to be ~45 nm. Thus, the flexibility of long duplexes of DNA can be described in terms of ~130 base-pair segments connected by universal joints.

Application 4.1   LINUS: A Hierarchic Procedure to Predict the Fold of a Protein

The recent successful methods for predicting various levels of protein structure are more simplistic than the detailed molecular mechanics calculations discussed in Chapter 3. The simple exact models discussed in Chapter 4 treat amino acids as simple points along a regular grid. These reproduce the overall topological properties of compact proteins structure rather nicely, but the structural information in the models is very low. Intermediate between detailed molecular mechanics calculations and simple exact models is LINUS (an acronym for *Local Independently Nucleated Units of Structure*). This algorithm uses *hierarchic condensation* to fold the chain locally to a low-energy structure, then progresses along the chain, evaluating each local conformation in the context of its interactions with the remainder of the protein. The result is that local structure is treated with some detail, but the distant interactions between amino acids along the chain are also considered.

LINUS treats a polypeptide chain as autonomous 50-residue segments, with the conformations of increasingly large blocks (3 to 6 to 48) of amino acids generated and evaluated within each segment. To simplify the calculations, the amino acids are reduced to the basic atoms of the peptide backbone, the C_β-carbon (except for Gly), and one or two atoms at the γ-position (except for Ala), which effectively reduces the number of torsion angles to a maximum of 3 for each residue (ϕ, ψ, plus one for the side chain). The interactions between residues are calculated using a simplified energy function that considers only steric repulsion, hydrogen bonding, hydrophobicity, and a function to chase the conformation toward $\phi > 0$ (except for Gly).

Despite the simplicity of the algorithm, the method has shown some success in predicting the secondary and supersecondary structures for five of seven test proteins. Within these five successful cases, the helices, sheets, and turns were predicted with an accuracy at the residue level ranging from 72% to 95%, with an average accuracy of about 84%. An example of one such prediction is that of Eglin C (Figure A4.1). It is unclear at this point whether the strategy implemented in LINUS will be general, given the fundamental problems with hierarchic folding models.

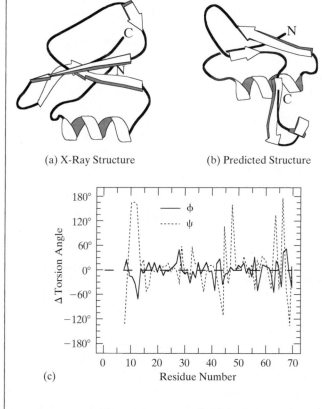

(a) X-Ray Structure   (b) Predicted Structure

(c)

Figure A4.1 The predicted structure of Eglin C predicted by LINUS is compared with the X-ray structure. The predicted tertiary structure (b) resembles that of the X-ray structure, (a) with some errors in the orientation of the two long β-strands. The plot (c) shows the deviations in ϕ, ψ-torsion angles (Δ torsion angle) for each amino acid residue from 8–70.

SRINIVASAN, R. and G. D. ROSE (1995), *Proteins* **22**; 81–99.

The effective size of a randomly organized but compactly folded polymer is not well represented by $\langle L^2 \rangle^{1/2}$. A ccDNA, for example, has no end-to-end distance because there are no ends to a circle. A more appropriate measure of the effective size for such molecules, including the compact folded structure of a polypeptide, is the root-mean-square radius $\langle R^2 \rangle^{1/2}$. This reflects the average displacement of each step or residue from the center of mass of a macromolecule and is thus a function of the mass of each residue m^j and the distance that residue is from the center r^j. These relationships are described by Eq. 4.45 for linear chains folded into a compact structure (e.g., the structure of a protein) and by Eq. 4.46 for a closed circular molecule (e.g., ccDNA).

$$\langle R^2 \rangle^{1/2} = \left[\frac{\sum m_j r_j^2}{\sum m_j} \right]^{1/2} \tag{4.45}$$

As we might expect, the root-mean-square radius is, like the root-mean-square end-to-end distance, proportional to $N^{1/2}$, assuming all masses are approximately equal

$$\langle R^2 \rangle^{1/2} = \left(\frac{Nl^2}{6} \right)^{1/2} \tag{4.46}$$

From Eqs. 4.44 and 4.46, we can show that $\langle R^2 \rangle^{1/2} = \langle L^2 \rangle^{1/2}/6^{1/2}$ for a linear molecule.

4.4.3 Simple Exact Models for Compact Structures

The random walk provides a set of relationships for understanding the general behavior of biopolymers. However, statistical methods can be used to derive more detailed models for the properties of compact macromolecules, including the mechanism of folding proteins and nucleic acids (e.g., tRNA; Figure 1.43) into their respective tertiary structures. These models start with the one-dimensional random walk and simply extend it to higher dimensions. The premise is that if a set of all the possible conformations of a macromolecule can be generated, then statistical mechanics can distinguish from this set the most probable conformations. Unfortunately, the number of possible random structures that need to be constructed and analyzed for the complete three-dimensional structure of a macromolecule can be astronomical. We had already shown that even with all the constraints placed on a polypeptide, there are still about 10^{53} possible backbone conformations for a chain of 100 amino acid residues (see Chapter 3). It would be impossible to generate this number of structures from a random walk.

An alternative approach is to sample the possible conformations of a single macromolecule by molecular simulation (e.g., using the Monte Carlo method; Chapter 3). Remember that in statistical mechanics, the time-average behavior of a single molecule is equivalent to the average behavior of an ensemble. The advantage to the molecular simulation approach is that the energies of interactions between residues can help to direct each step of the walk (Application 4.1). However, these molecular simulation methods allow only small incremental changes in the structure at each step. Thus, it is not possible to ensure that all the available conformational space will be sampled. A compromise is to use the energetics of the interactions to define a restricted walk.

A number of different physical and energetic restrictions can be imposed on a random walk by defining very explicitly the rules for each step in the walk. The structures generated in this way are simple in their geometry and very exact, and are called *simple exact models*. In one type of simple exact model, the walk is confined to the grid points of a two- or three-dimensional lattice (Figure 4.16). Starting at the origin, each successive step j falls on an immediately adjacent grid point. This walk is defined as self-avoiding so that no two residues can occupy the same grid point. The number of unique structures in this system is greatly reduced because the degrees of freedom for each successive step is low. For a two-dimensional square grid, each step (except for the first) has a maximum of three possibilities—in a self-avoiding walk, a step back to the previous grid point is not allowed. Similarly, the number of available grid points decreases as the chain extends from 1 to N residues. Thus, each step is very well defined and exact, and the model defines a restricted number of conformations.

The interaction energies responsible for protein folding and unfolding (at the secondary and tertiary structural levels) can be incorporated into a random walk by defining favorable and/or unfavorable interactions between residues. To study the effect of hydrophobicity on protein folding, for example, a peptide sequence can be constructed with only two types of residues, H for hydrophobic residues and P for polar or hydrophilic residues. If an energy ϵ is assigned for breaking each hydrophobic-hydrophobic (HH) interaction, then the most stable form of the structure will be the one with the maximum number of intact HH contacts. This defines the native conformation or the reference state of the protein. The energy for each conformation generated by this self-avoiding walk through a grid is thus explicitly defined. This set of simple exact models and the thermodynamic states of these models can provide insight into the mechanisms of macromolecular folding.

As a simple example, we will consider the conformations of the simple hexapeptide sequence H-P-H-P-P-H that are generated by a walk along a two-dimensional grid. Again, what we learn from this simplified walk in two dimensions can be extrapolated to the more realistic case in three dimensions. The resulting structures resemble simple ball-and-stick toys (Figure 4.17); this type of model for compact structures is thus aptly called the *hexamer toy model* (HTM). The number of HH contacts h can simply be counted to determine the energy for each conformation of the HTM. Each state includes all the conformations having the identical number of HH contacts and thus the same energy $h\epsilon$. Assuming that ϵ is temperature independent and that pressure-volume effects can be neglected, $h\epsilon$ would represent the enthalpy for folding, $\Delta H^0_{\text{folding}}$ for each state. The number of conformations with the identical h HH contacts represents the degeneracy for each state g_h and therefore $\Delta S_{\text{folding}} = R(\ln g_{h(\text{folded})} - \ln g_{h(\text{unfolded})})$. Therefore, if only the HH contacts contribute to the enthalpy of folding, then $\Delta G^0_{\text{folding}} = h\epsilon - RT(\ln g_{h(\text{folded})} - \ln g_{h(\text{unfolded})})$.

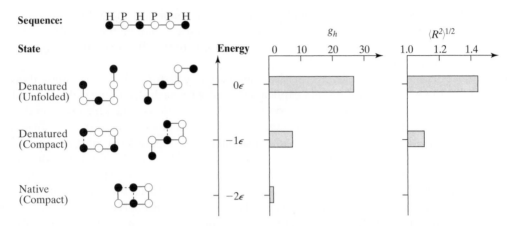

Figure 4.17   Comparison of thermodynamic states for a simple hexamer toy model (HTM) for compact structures folded by hydrophobic (HH) contacts. Residues in the hexamer sequence are defined as H for hydrophobic and P for polar (hydrophilic). The lowest energy native state has the largest number of HH contacts (h), as shown by the dashed lines. There are two different denatured states, one with $h = 1$ to give a compact denatured state, and the more extended forms at $h = 0$. Only two examples of each are shown. The energies $-h\epsilon$, degeneracies g_h (which are counted as the number of unique structures in each state), and average root-mean-square radii $\langle R^2 \rangle^{1/2}$ are compared for each thermodynamic state. [Data from Dill et al. (1993), *Protein Science* **4**, 561–602.]

There are three distinct states for the HTM (Figure 4.17). The native state has two HH contacts, an intermediate state has one HH contact, and the highest energy denatured state has no HH contacts. However, the fully denatured state has the highest entropy, representing the largest population of degenerate structures. An interesting property of the HTM is that the most compact state is the native conformation, but the average $\langle R^2 \rangle^{1/2}$ of the intermediate state ($h = 1$) is also relatively compact when compared to the extended structures of the fully denatured structures. These *compact denatured* forms of the polypeptide have been proposed as the intermediate states in the folding pathway for proteins. They represent conformations in which a large proportion of the hydrophobic core of the protein has formed (in this case, 50% of the HH interactions) in a relatively compact structure.

A statistical-mechanics analysis of the HTM shows a sigmoidal curve for the heat denaturation of the compact native conformation (Figure 4.18). The partition function Q is derived by defining the degeneracy of each state h as g_h and the statistical weight $\omega_h = g_h e^{-h\epsilon/RT}$. The distribution of states undergoes a cooperative and temperature-dependent transition from the native conformation at $T = 0$, to the compact intermediate, and finally to the fully denatured states as T approaches ∞ (Figure 4.18). The associated $\Delta G^{\circ}_{\text{folding}}$ increases with temperature as the entropy of the system increases from the low degenerate native state to the higher degenerate denatured states. To account for the cold denaturation of proteins, ϵ can be defined as being temperature dependent (Figure 4.18). This complicates the thermodynamic interpretation only slightly. Thus, the minimal and simple HTM for a small peptide, having only hydrophobic interactions to distinguish between the

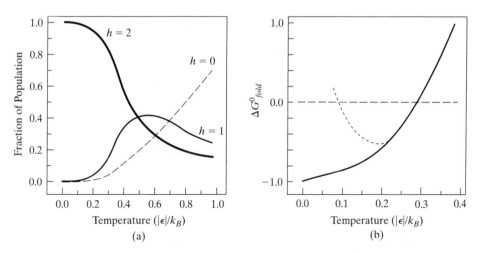

Figure 4.18   Thermal denaturation and the effect of ϵ on the free energy of folding $\Delta G_{\text{folding}}^0$ of HTMs. (a) The compact native state (heavy solid curve) with $h = 2$ represents the entire population at temperatures of $T = 0|\epsilon|/k_B$ (i.e., in the lowest thermal state of the system), where k_B is the Boltzmann constant. As T increases, the population first shifts to the compact denatured intermediates (thin solid curve, $h = 1$), then to the fully extended forms (dashed curve, $h = 0$) of the denatured states. (b) Heat denaturation of the compact structures result from the increase in $\Delta G_{\text{folding}}^0$ as the temperature increases (solid curve). If the energy of the hydrophobic contacts are also temperature-dependent, the structures can also be shown to undergo cold denaturation at low temperatures (dashed curve). [Data from Dill et al. (1993), *Protein Science* **4**, 561–602.]

various thermodynamic states, mimics the behavior of the denaturation and folding of proteins. This has also been applied to the study of the folding pathways for tertiary structures in RNA. The most significant disadvantage of the simple exact models, however, is that they do not yield complete conformations of molecules, with all atoms accurately defined. The topologies of the native structures from these models, however, do resemble those of actual proteins (Figure 4.19).

The various examples presented in this chapter illustrate the diverse application of statistical mechanics to the study of the thermodynamic properties of macromolecular structures. The prerequisites for applying the methods include an accurate model for each state of the system, and accurate definitions for the statistical weights and the degeneracy for each state. Once the intrinsic and extrinsic terms of the nucleation and propagation parameters are derived, the statistical-mechanics approach can yield very reliable predictions for a macromolecular structure from simple thermodynamic principles. We should note that many of the principles described here for thermodynamic behavior of molecules can be extended to kinetic models by first defining a pathway for a process, then replacing the equilibrium constants (micro and macro) with kinetic constants and energy terms with activation energies in the partition function. Such an approach can result in significant insights into, for example, how a protein can fold over a reasonable time period by overcoming Levinthal's paradox (Application 4.2).

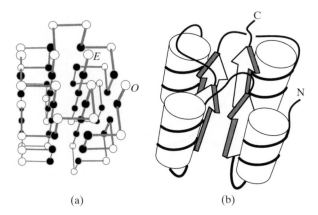

(a)   (b)

Figure 4.19 The topology of compact structures constructed from simple exact models in three-dimensions. (a) A chain of 67 hydrophobic (solid spheres) and polar (open spheres) amino acids is modeled on a three-dimensional grid, with the packing defined by hydrophobic-hydrophobic (HH) contacts. Helices are assigned from the model as regular and repeating segments, while strands are assigned as straight segments along the chain. (b) These are represented by cylinders for the helices and arrows for the strands to show the type of topology represented by this compact structure. The resulting topology is a four-helix, four-strand bundle that is similar to those observed in a number of proteins. [Adapted from Dill et al. (1993), *Protein Science* **4**, 561–602.]

Application 4.2   Folding Funnels: Focusing Down to the Essentials

Statistical thermodynamic lattice models tell us something about the properties of the folded, unfolded, and intermediate states, but not the folding process itself. If we were to take Levinthal's paradox and extend it to ask how long it would take to fold a protein into its native form through purely random walks, then the answer is that a protein would never fold within the lifetime of a cell if a thorough search through all conformation space is required to find the lowest energy form. By applying a kinetic model to these simple lattice models, and making some simplifying assumptions, we can derive a general mechanism for folding (called a *folding funnel*) that simplifies the folding landscape and thus overcomes the Levinthal problem. The discussion here focuses on the first construction of a kinetic folding funnel, but we should recognize that there has been significant work in refining and extending this model to make it more realistic.

The basic principle in constructing the folding funnel is that a random coil will fold into a set of random condensed states. For simplicity, these states will be modeled as all possible maximally compact conformers constructed around the simple exact lattice models (some of which are native or native-like, while others are nonnative type conformations). The interconversion between individual states is diffusive and local (that is, once there is a collapse, one cannot change the global conformation except through a series of incremental local changes). In the lattice model, these local reconfigurations are ones in which a single residue is randomly moved to a nearby vacant lattice point, and pairs of residues can rotate 90°, while maintaining the backbone chain.

The thermodynamic assumption is still the same, that the native conformation is the lowest energy (which again can be assigned to the one having the highest number of nonbonding interactions).

What was observed from this simulation is that the starting random structure quickly collapses down to compact states with a majority of the possible nonbonding interactions established. The actual folding to the native states occurs from this set of compact, nonnative states through the local reconfigurations of the chain. When the results of this type of simulation are presented as an energy versus conformation plot, it is easy to see the steep funnel that leads from the compact states to the lowest energy native state (Figure A4.2). This idea of a folding funnel allows us to overcome Levinthal's paradox without evoking a specific folding pathway for a particular protein, but makes the profound statement that "folding may well proceed as a convergence of multiple pathways among families of interconverable dense conformations. . . . "

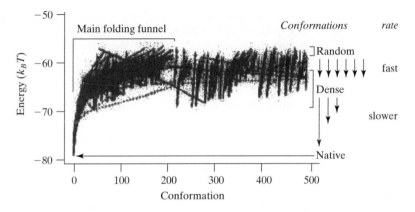

Figure A4.2 The folding funnel resulting from a kinetic model for folding simple lattice models of proteins. A "foldable" protein sequence is subjected to a simluation where random states will quickly collapse to establish lower energy dense states, but that local reconfiguration to the lowest energy native state is slower. Points show individual conformations along the conformational axis (x), and lines connect conformations that occur relatively rapidly through local reconfigurations.

LEOPOLD, P. E, M. MONTAL, and J. N. ONUCHIC (1992), *Proc. Natl. Acad. Sci., USA* **89**; 8721–8725.

EXERCISES

4.1 For a 10 amino acid peptide, compare the degeneracy of the system if the coil-to-helix transition follows the all-or-none model, the noncooperative model, and zipper model. Evaluate the partition function for all three models (using $\sigma = 2 \times 10^{-4}$ and $s = 1$).

*4.2 We start with an equilibrium distribution of particles at energy levels.

Energy	Initial Number of Particles
$\dfrac{2h\nu}{k_BT}$	c
$\dfrac{h\nu}{k_BT}$	b
0	a

Estimate the probability that one in 10^6 particles redistributes, with the overall energy level remaining the same. In this case, we will use the vibrational energy of HCL, which is $h\nu/k_BT = 14$ at room temperature. This exercise will demonstrate that the most probable distribution is significantly more likely than a slightly perturbed distribution.

4.3 Calculate the degeneracies and the entropies for tossing 10 coins as all heads, and as half heads and half tails.

4.4 Using the Chou and Fasman $\langle P_\alpha \rangle$ values in Table 4.2, estimate the helical propensity of the guest-host peptide YSEEEEKKKKE-X-KEEEEKKKK, for X = Gly, Asn, Ile, Met, Leu, and Ala. How do these values compare to the thermodynamic values in Table 4.1?

4.5 Estimate σ for the transition from a 3_{10} helix to an α-helix, assuming that σ for the α-helix is $\sim 10^{-4}$ and depends only on the difference in conformational entropy.

4.6 The ribonucleotide-binding domain of the transacting transcriptional activator (*tat*) protein from HIV type I has the sequence

FITKALGISYGRKKRRQRRRPPQ

Using Table 4.2, predict the secondary structure(s) of this sequence.

4.7 The nucleation term for melting of double-stranded RNA is estimated to be 10^{-7}. What length of duplex will have an equal probability of melting at the ends, as opposed to the middle?

4.8 Two commonly used plasmids for gene cloning are the closed circular plasmids pUC19 (2686 base pairs in size) and pBR322 (4361 base pairs). During the cloning procedure, a plasmid is cut with an endonuclease to form a linear duplex, the gene is ligated, and two ends of the DNA are brought back together in order to reform the closed circular plasmid. From the persistence length, estimate the root-mean-square end-to-end distance and the root-mean-square radius of linear and closed circular form of the two plasmids.

4.9 Alternating sequences of n d(C·G) base pairs can form Z-DNA under negative superhelical stress in closed circular DNAs (size = $4269 + n$). The midpoints of the transition were observed at topoisomers with the following ΔLk:

N	ΔLk
12	-12.5
24	-15.5
48	-22.0

Calculate the nucleation and propagation parameters for this transition.

4.10 In the hexamer toy model (HTM), the assumption is made that interactions between polar amino acids, and between polar and hydrophobic amino acids, do not contribute to the stability of the molecule. Is this a good assumption for molecules in aqueous solution? How would this assumption change if the HTM were for an integral membrane protein? Estimate the degeneracy and ΔS for the lowest energy level of the sequence P-H-H-P-H-P (where H's are hydrophobic and P's are polar) relative to that of the sequence H-P-H-P-P-H in aqueous solution.

4.11 A simplified version of the HTM is the tetramer toy model (TTM), which involves only four residues. For the tetramer sequence H-P-P-H (where H's are hydrophobic and P's are polar residues),
   a. Draw all possible conformations of the tetramer.
   b. Assuming that an H-H interaction is stabilizing, how many energy levels are there in the TTM?
   c. Calculate ΔS for each energy level relative to the lowest energy conformation.
   d. What are the free energy relationships for each energy level?

REFERENCES

General Principles

CHANDLER, D. (1987) *Introduction to Modern Statistical Mechanics,* Oxford University Press, New York.

DICKERSON, R. E. (1969) *Molecular Thermodynamics,* Benjamin-Cummings, Menlo Park, CA. Chapter 2 is a very readable introduction to the fundamentals of statistical mechanics.

Polypeptide and Protein Structure

CHOU, P. Y. and G. D. FASMAN (1974) "Prediction of Protein Conformation," *Biochemistry* **13**, 222–245. This seminal paper describes the now well adopted method for predicting the secondary structures in proteins from their sequence.

FASMAN, G. D. (ed.) (1989) *Prediction of Protein Structure and the Principles of Protein Conformation,* Plenum Press, New York. An older but still relevant compilation of review chapters on protein structure and prediction.

ROSE, G. D. (1978) "Prediction of Chain Turns in Globular Proteins on a Hydrophobicity Basis," *Nature* **272**, 586–590.

SCHULTZ, G. E. and R. H. SCHIRMER (1979) *Principles of Protein Structure,* Springer-Verlag, New York.

ZIMM, B. H. and J. K. BRAGG (1959) "Theory of the Phase Transition Between Helix and Random Coil in Polypeptide Chains," *J. Chem. Phys.* **31**, 526–535. The classic treatment of the coil-to-helix transition in polypeptides by statistical mechanics.

Nucleic Acid Structure

COZZARELLI, N. R. and J. C. WANG (1990) *DNA Topology and its Biological Effects,* Cold Spring Harbor Laboratory Press, Cold Spring Harbor, NY.

Ho, P. S. (1994) "The Non-B-DNA Structure of d(CA/TG)$_n$ Does Not Differ from That of Z-DNA," *Proc. Natl. Acad. Sci., USA* **91**, 9549–9553. A recent review of statistical methods to study and predict Z-DNA formation in negatively supercoiled DNA.

Ivanov, V. I. and D. Y. Krylov (1992) "A-DNA in Solution as Studied by Diverse Approaches," *Methods in Enzymology* **211**, 111–127. A good review of the experimental and statistical approaches to studying A-DNA in solution.

Liu, L. and J. C. Wang (1987) "Supercoiling of the DNA Template During Transcription," *Proc. Natl. Acad. Sci. USA* **84**, 7024–7027.

Peck, L. J. and J. C. Wang (1983) "Energetics of B- to Z Transition in DNA," *Proc. Natl. Acad. Sci., USA* **80**, 6206–6210. The initial derivation of the partition function for the B-DNA to Z-DNA transition in supercoiled DNA.

Saenger, W. (1984) *Principles of Nucleic Acid Structure,* Springer-Verlag, New York.

Random Structures

Tinoco, I., Jr., K. Sauer, J. C. Wang and J. D. Puglisi (2002) *Physical Chemistry,* 4th Ed. Prentice Hall, Upper Saddle River, NJ. Chapter 11 is a clear presentation of the random walk and of the derivations for the linear and radial measures of random structures.

Simple Exact Models

Dill, K. A., S. Bromberg, K. Yue, K. M. Fiebig, D. P. Yee, P. D. Thomas, and H. S. Chan (1995) "Principles of Protein Folding—A Perspective from Simple Exact Models," *Protein Science* **4**, 561–602. Recent review of simple exact models and their application to protein-folding.

5

Methods for the Separation and Characterization of Macromolecules

5.1 GENERAL PRINCIPLES

In the preceding chapters, we have discussed the diversity, complexity, and dynamics of macromolecular structure. Now we turn to experimental methods for determining these structures and examining their interactions and transformations.

At present, we possess two powerful techniques, X-ray diffraction and NMR, that allow us to determine the structures of many macromolecules in astonishing detail; we will describe these in later chapters. But before we can apply such physical methods meaningfully, we must first purify the macromolecule of interest, separating it from the multitude of molecules, both large and small, found in most biochemical preparations. When it is purified, we usually begin the analysis by obtaining certain preliminary information about each biopolymer. Before we set out to determine the exact structure of a protein, for example, there are certain simple questions that we need to answer:

1. Does the protein exist, under physiological conditions, as a simple polypeptide chain or as an aggregate of several chains?
2. If the latter is the case, are the chains of one or several kinds? What are their approximate molecular weights?
3. What is the *approximate* size and shape of the native protein molecule: Is it a globular or fibrous protein?
4. What is its ionic character: Is it most rich in acidic or basic residues?

This chapter describes a group of techniques that are commonly employed to answer questions like these. These are experimental methods in which molecules are moved by forces acting on them; they utilize, therefore, what we call *transport* processes. These various techniques will separate macromolecules on the basis of

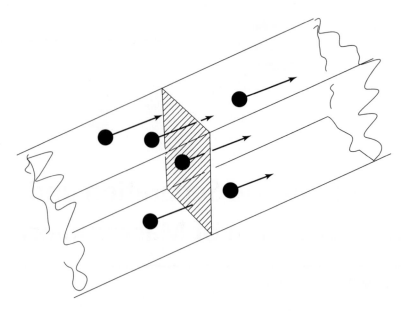

Figure 5.1   The concept of flow. Here we consider a case in which flow is occurring in one dimension, as will occur in sedimentation or electrophoresis experiments, or in diffusion from a planar boundary. We consider a 1 cm^2 area perpendicular to the direction of molecular transport, and define, J_i, the flow of a particular substance (i) as the net number of moles or grams of i that cross this surface in 1 second.

mass, size, shape, charge, or some combination of these parameters. Thus, they can also be employed in preparative separation and purification. Equally important, these methods can be used for the *analysis* of mixtures. Indeed, without techniques like sedimentation and gel electrophoresis for such analyses, it is difficult to see how modern molecular biology could have evolved.

All transport processes can be described in terms of a quantity called *flow* (J), which is defined as the number of mass units (moles or grams) crossing 1 cm^2 of surface in 1 sec (see Figure 5.1). This is really a familiar concept, for it is quite analogous to the idea of electric current—the flow of electricity in a wire. The different transport processes that we shall consider in this chapter all have in common the general concepts of material flow described above. They differ in the forces that *drive* that flow. We shall begin with the most elementary—the diffusional flow driven by the thermal energy common to all molecules.

5.2   DIFFUSION

Diffusion is most often associated with the random *Brownian motion* of molecules in solution. It is indeed possible to measure diffusion directly from this kind of motion and this is described in Chapter 7. However, here we consider situations in which diffusion contributes to observable mass transport. A typical example is

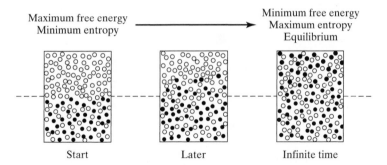

Figure 5.2 Diffusion from an initially sharp boundary between solution (*below*) and solvent (*above*).

shown in Figure 5.2, in which solvent has been carefully layered over a solution to create a *boundary* at which the solute concentration changes abruptly from a value C_0 to zero. Common sense tells us that this situation cannot persist; the random motion of solute molecules will ultimately make the concentrations uniform throughout the system. From a thermodynamic viewpoint, the entropy of the system must increase, or the free energy decrease, accompanied by an evening of the concentration distribution of the solute.

5.2.1 Description of Diffusion

The transport process carrying out this redistribution is the diffusion of solute, and net transport of matter by diffusion will occur whenever there is a concentration gradient. Once that concentration gradient is abolished, and concentration is everywhere uniform, no further *net* flow of matter will occur (although molecules are still moving to and fro). For one-dimensional diffusion the relationship between diffusional flow and concentration gradient is expressed by *Fick's first law*:

$$J = -D\left(\frac{\partial C}{\partial x}\right) \tag{5.1}$$

where D is the *diffusion coefficient* of the diffusing substance. Equation 5.1 corresponds to the common-sense notion that the rate of flow will depend on the steepness of the gradient. When the gradient is gone, net flow stops. The diffusion coefficient, D, measures the rapidity with which a substance diffuses. We shall leave the molecular interpretation of this important quantity until later, and consider first various situations in which the phenomenon of diffusion can be of importance. In addition to diffusion across a planar concentration boundary, as in Figure 5.2, other examples important in biology and biochemistry are shown in Figure 5.3. Some, like diffusion of substances into or out of a cell (Figure 5.3a), are complicated to analyze because the problem is three-dimensional. On the other hand, a simple but illuminating example is one-dimensional diffusion from a very thin layer, as shown in Figure 5.3b. This situation is encountered, for example, in

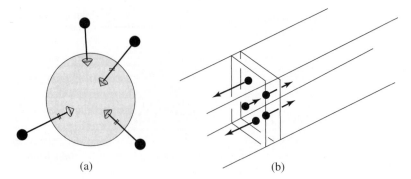

(a)   (b)

Figure 5.3 Other important examples of diffusion. (a) Diffusion of nutrient into an idealized spherical cell. This is a three-dimensional problem, but mathematical analysis is possible. (b) Diffusion from an initially thin layer. This situation is encountered in all zonal centrifugation and electrophoresis experiments, as described later in this chapter.

gel electrophoresis or zonal ultracentrifugation (see later) and understanding the diffusional behavior is important to maximizing resolution in such techniques. As we shall see, the example in Figure 5.3b can also reveal some fundamental facts about diffusion.

If we are to understand any of the processes depicted in Figure 5.2 or 5.3, Eq. 5.1 will not be enough. What we need is an equation that describes the concentration as a funtion of both spatial position and time, following some initial concentration distribution at $t = 0$. This suggests that we have to seek the solution to some differential equation, given initial and boundary conditions. What additional general information do we have? The answer is the conservation of mass. This can be expressed in the very general way shown in Figure 5.4. We consider a thin slab, of thickness Δx, and cross section A perpendicular to the direction of flow. The mass flowing in during time Δt is $J(x)A\Delta t$, whereas that flowing out is $J(x + \Delta x)A\Delta t$, so the change in mass, Δw, is

$$\Delta w = J(x)A\Delta t - J(x + \Delta x)A\Delta t \tag{5.2}$$

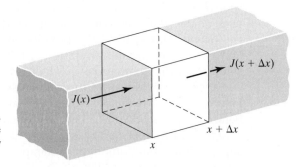

Figure 5.4 The continuity equation. The change of concentration within the slab is determined by differences in flow in and flow out.

If we divide by the volume of the slab, $A\Delta x$, we obtain the change in concentration

$$\frac{\Delta w}{A\Delta x} = \Delta C = \frac{(J(x) - J(x + \Delta x))}{\Delta x}\Delta t \tag{5.3a}$$

or

$$\frac{\Delta C}{\Delta t} = \frac{-\Delta J}{\Delta x} \tag{5.3b}$$

In the limit of infinitesimal increments, we get

$$\left(\frac{\partial C}{\partial t}\right) = -\left(\frac{\partial J}{\partial x}\right) \tag{5.4}$$

This result, sometimes called the *continuity equation,* is just a statement of the conservation of mass—what goes in must come out, or the concentration within will change. If we put this together with Fick's first law (Eq. 5.1) we get

$$\left(\frac{\partial C}{\partial t}\right) = -\frac{\partial}{\partial x}\left(-D\frac{\partial C}{\partial x}\right) \tag{5.5a}$$

or

$$\left(\frac{\partial C}{\partial t}\right) = D\left(\frac{\partial^2 C}{\partial x^2}\right) \tag{5.5b}$$

which is *Fick's second law.* This differential equation, together with appropriate initial and boundary conditions, governs all one-dimensional diffusion behavior. Not surprisingly, this equation is very similar to the differential equation describing heat conduction. For two- and three-dimensional problems, more complex but analogous equations pertain. To take a specific one-dimensional example, consider diffusion in the $\pm x$ direction from a thin slab of thickness δ, centered at $x = 0$ as shown in Figure 5.5b. The concentration in the slab, at $t = 0$ is C_0. We assume the boundaries for diffusion are $\pm\infty$, so that $C = 0$ at $x = \pm\infty$ at all t. Given these initial and boundary conditions, the unique solution for Eq. 5.5b is

$$C(x, t) = \frac{C_0\delta}{2(\pi Dt)^{1/2}}e^{-x^2/4Dt} \tag{5.6}$$

at $x \gg \delta, t \gg 0$. This is diagrammed in Figure 5.5; it is shown as a Gaussian distribution about $x = 0$, which would be an exact representation were the initial distribution infinitely thin ($\delta = 0$). It is of interest to calculate the average displacements of molecules from a starting point at $x = 0$. Obviously, since $+$ and $-$ displacements

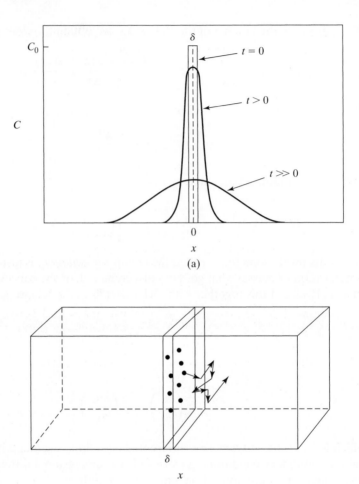

Figure 5.5 (a) Diffusion from a narrow zone. The concentration distribution approximates a Gaussian error function, which broadens with time. (b) The same process as viewed from the point of view of molecular random walks.

are equally likely $\langle x \rangle = 0$. But what is $\langle x^2 \rangle$? This can be found by evaluating the ratio of the integrals:

$$\langle x^2 \rangle = \frac{\displaystyle\int_{-\infty}^{\infty} x^2 C(x, t)\, dx}{\displaystyle\int_{-\infty}^{\infty} C(x, t)\, dx} = \frac{\displaystyle\int_{0}^{\infty} x^2 C(x, t)\, dx}{\displaystyle\int_{0}^{\infty} C(x, t)\, dx} \qquad (5.7)$$

where the change in limits is justified because both integrands are even functions. Inserting Eq. 5.6, and canceling common terms, we find

$$\langle x^2 \rangle = \frac{\displaystyle\int_0^\infty x^2 e^{-x^2/4Dt}\, dx}{\displaystyle\int_0^\infty e^{-x^2/4Dt}\, dx} \tag{5.8}$$

Consulting any good table of definite integrals, we find that

$$\langle x^2 \rangle = 2Dt \tag{5.9}$$

Compare this result to Eq. 4.43. If we consider that the diffusing molecule is making random steps in solution, and that the number of these steps is proportional to the time of diffusion, then it is clear that the two equations are the same! This is saying that diffusion from a thin layer is nothing different from a random walk from an origin. Equation 5.9 also makes clear the dimensions of the diffusion coefficient: [distance]^2/time. Traditionally, diffusion coefficients of macromolecules are usually expressed in the cgs units of cm^2/sec. Returning now to Figure 5.5, we note that as the initially narrow band of solute spreads with time, the width of the band increases with the square root of time, consistent with Eq. 5.9 (see Exercise 5.1).

Solutions to the differential equation for diffusion (Eq. 5.5b) have been found for many other conditions. For example, diffusion from a sharp boundary between solvent and solution proceeds as shown in Figure 5.6. Exact solutions are known for this case, and many others. The interested reader is referred to references at the end of this chapter.

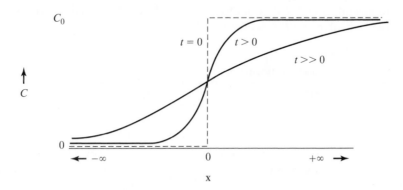

Figure 5.6 The spreading of an initially sharp boundary (as in Figure 5.2) by diffusion. The concentration distribution at time t is given by the equation

$$C(x, t) = \frac{C_0}{2}\left\{ 1 - \frac{2}{\sqrt{\pi}} \int_0^{x/(2Dt)^{1/2}} e^{-y^2}\, dy \right\}$$

5.2.2   The Diffusion Coefficient and the Frictional Coefficient

The molecular interpretation of the diffusion coefficient was first deduced in early research by Albert Einstein. He found the simple relationship:

$$D = \frac{kT}{f} = \frac{RT}{\mathcal{N}f} \qquad (5.10)$$

where k is the Boltzmann constant, $\mathcal{N}$ is Avogadro's number, and f is a quantity called the *frictional coefficient*. The frictional coefficient measures the resistance encountered by the molecule in moving through the solvent. We shall encounter this quantity, which is determined by the size and shape of the molecule, as well as the viscosity of the solvent, in all methods that involve transport through solution. For the simplest case, a sphere of radius a, the frictional coefficient is given by Stokes's law:

$$f_0 = 6\pi\eta a \qquad (5.11)$$

where η is the viscosity of the solvent. Thus, for spherical molecules D is simply related to the radius as

$$D_0 = \frac{RT}{6\pi\mathcal{N}\eta a} \qquad (5.12)$$

For certain special forms of nonspherical particles, it has also been possible to calculate exact values for the frictional coefficient (see Table 5.1 and Figure 5.7) and approximate values can be obtained for more complex shapes (see below).

The equations in Table 5.1 and the curves in Figure 5.7 are given as ratios f/f_0, where f_0 is the Stokes's law value for a spherical molecule of the same volume. The nonspherical molecule will always have a larger frictional coefficient than its spherical equivalent, so f/f_0 is always ≥ 1.

Table 5.1   Frictional Coefficient Ratios

Shape	f/f_0	a_e
Prolate ellipsoid	$\dfrac{P^{-1/3}(P^2 - 1)^{1/2}}{\ln\left[P + (P^2 - 1)^{1/2}\right]}$	$(\alpha\beta^2)^{1/3}$
Oblate ellipsoid	$\dfrac{(P^2 - 1)^{1/2}}{P^{2/3}\tan^{-1}\left[(P^2 - 1)^{1/2}\right]}$	$(\alpha^2\beta)^{1/3}$
Long rod	$\dfrac{(2/3)^{1/3}P^{2/3}}{\ln 2P - 0.30}$	$\left(\dfrac{3\beta^2\alpha}{2}\right)^{1/3}$

In these equations, $P = \alpha/\beta$, where α is the semimajor axis (or the half-length for a rod) and β is the minor axis (or radius of a rod). The quantity a_e is the radius of a sphere equal in volume to the ellipsoid or rod, so $f_0 = 6\pi\eta a_e$.

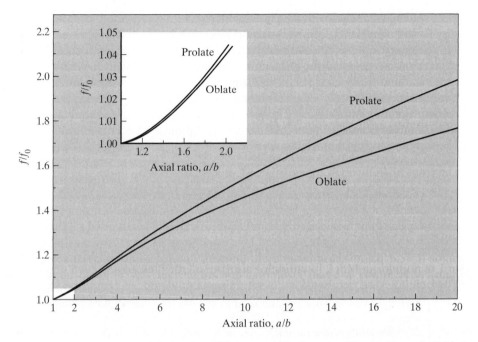

Figure 5.7   The dependence of frictional coefficient on particle shape for ellipsoids. The ratio f/f_0 is the frictional coefficient of a particle of the given axial ratio divided by the frictional coefficient of a sphere of the same volume. Note that for values of f/f_0 close to 1, it is almost impossible to distinguish between prolate and oblate ellipsoids.

There are a number of ways in which diffusion coefficients can be determined experimentally. One can utilize, for example, the rate of transfer of solute across a porous diaphragm, in effect using Stokes's first law to calculate D. Quite exact measurements of diffusion coefficients can be made from the analysis of the spreading of a boundary between solution and solvent (as in Figure 5.6), but this method is laborious and requires large amounts of material; it is rarely used today.

Most measurements of D are now made by *dynamic light-scattering* methods, which are described in Chapter 7. This technique has the advantage over earlier methods because only small amounts of solution are required and rapid measurement is possible.

5.2.3   Diffusion Within Cells

A question of major interest to biologists is how diffusion proceeds within the cytoplasm or nucleus of a cell. Such information is vital to understanding metabolic and regulatory processes—for example, finding how fast "second messengers" can travel from the cell membrane to the DNA in the nucleus. It is now possible to make such measurements, by a number of techniques. One of the most widely employed techniques employs the *photobleaching* of dye labels. The molecule to be studied is labeled with a fluorescent dye, and then injected into the cell. An intense laser beam is focussed on a small region, to bleach the dye. This leaves a spot within a microscopic image of the cell that does not exhibit fluorescence. However, as labeled molecules diffuse into this volume, the fluorescence recovers, at a rate depending on D. In fact, the half time for recovery is proportional to $1/D$, and the proportionality constant can be measured by experiments with the same substance in aqueous solution. An example of the use of this technique to measure diffusion of DNA molecules with cells is given in Application 5.1.

Application 5.1   Measuring Diffusion of Small DNA Molecules in Cells

To know how fast small nucleic acid molecules can move once they are inserted into cells is important for applications such as gene delivery and antisense therapy. Therefore, Lukacs et al. have studied the diffusion of a graded series of double-strand DNA molecules in HeLa cells. The DNAs were labeled with a fluorescent dye, and ranged in size from 21–6000 bp. They were microinjected into either cytoplasm or nucleus. Photobleaching was via an argon-ion laser at 488 nm. The kind of data obtained are shown in Figure A5.1. In all cases, the diffusion coefficients in cytoplasm were much less than in aqueous solution, the ratio D_{cyto}/D_w decreasing sharply with DNA length. In the nucleus,

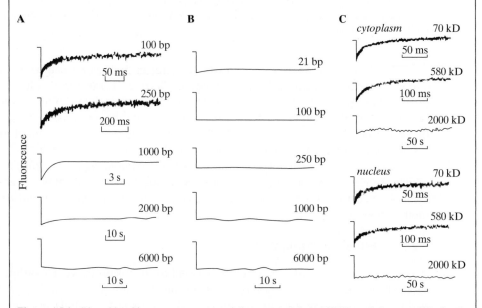

Figure A5.1   Photobleaching measurements of fluorescein-labeled DNA and dextran diffusion in HeLa cells. Recovery curves are for DNA in cytoplasm (A) and nucleus (B), and for dextran in cytoplasm and nucleus (C). Note that *none* of the DNA samples can diffuse at significant rates in the nucleus, whereas dextran molecules can. Thus, the hindrance in nucleus is specific for nucleic acid.

the DNA did not diffuse at all. The results indicate that effects other than simple viscous drag must be involved in hindering the diffusion—most probably entanglement or binding to cellular constituents is involved.

LUKACS, G. L., HAGGIE, P., SEKSEK, O., LECHARDUER, D., FREEDMAN, N., and
    A. S. VERKMAN (2000), *J. Biol. Chem.* **275**, 1625–1629.

Measurement of D really gives us only one piece of useful molecular information—the frictional coefficient. What this in itself tells about a macromolecule is limited. If we knew the molecule was exactly spherical, we could use Stokes's law to calculate its radius according to Eq. 5.11. In fact, you will find that values of "Stokes's radii" measured in this fashion are frequently quoted in the literature. But these must be treated with caution, for they can be completely meaningless numbers if the molecule deviates significantly from being spherical (consider Exercise 5.8). If something is known from other sources concerning the shape of the molecule (i.e., if it is known to be a rod, for example) then a value of f can be useful in estimating dimensions.

We turn now to another hydrodynamic technique, *sedimentation,* that can provide considerably more information concerning macromolecular size and shape.

5.3   SEDIMENTATION

We shall introduce the concept of sedimentation using a simple but adequate mechanical analysis, noting that a more elegant thermodynamic theory exists [see Chapter 13]. Imagine a solute molecule in a solution that is held in a rapidly spinning rotor (Figure 5.8). Forgetting for the moment the random pushes and pulls that the molecule receives from its neighbors and that account for its diffusion, we may say that there are three forces acting on it. If the rotor turns with an angular velocity ω (radians per second), the molecule will experience a *centrifugal* force proportional to the product of its mass (m) and the distance (r) from the center of rotation, $F_c = \omega^2 rm$. At the same time, the molecule displaces some solution and is buoyed up by it; this will give rise to a counterforce equal to that which would be exerted on the mass (m_0) of solution displaced, $F_b = -\omega^2 rm_0$. Finally, if as a result of these forces the molecule acquires a velocity v through the solution, there will be a resisting *frictional force* $F_d = -fv$, where f is the frictional coefficient. Under these circumstances, a molecule initially at rest will acquire a velocity just great enough to make the total force acting upon it zero.

$$F_c + F_b + F_d = 0 \tag{5.13a}$$

$$\omega^2 rm - \omega^2 rm_0 - fv = 0 \tag{5.13b}$$

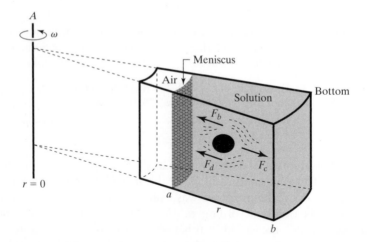

Figure 5.8   Diagram of a sedimentation experiment (not to scale). The sector-shaped cell is in a rotor spinning about the axis A at an angular velocity ω. The molecule is acted on by centrifugal, buoyant, and frictional drag forces. The cell has been given a sector shape because sedimentation proceeds along radial lines; any other shape would lead to concentration accumulation near the edges, with accompanying convection. The angular velocity (ω) is given by $(2\pi/60)$ times (revolutions per minute).

For the mass of solution displaced, m_0, we may substitute the product of the particle's mass times its partial specific volume ($\bar{v}$) (see Chapter 13 for exact definition of $\bar{v}$), times the solution density (ρ): $m_0 = m\bar{v}\rho$. Therefore,

$$\omega^2 rm(1 - \bar{v}\rho) - fv = 0 \tag{5.14}$$

Next, we multiply Eq. 5.14 by Avogadro's number, to put things on a molar basis, and rearrange, placing the molecular parameters on one side of the equation and the experimentally measured ones on the other.

$$\frac{M(1 - \bar{v}\rho)}{\mathcal{N}f} = \frac{v}{\omega^2 r} = s \tag{5.15}$$

The velocity divided by the centrifugal field strength ($\omega^2 r$) is called the *sedimentation coefficient*, s. According to Eq. 5.15, it is proportional to the molecular weight multiplied by the *buoyancy factor* $(1 - \bar{v}\rho)$ and inversely proportional to the frictional coefficient. It is experimentally measurable as the ratio of velocity to field strength. The units of s are seconds. Since values of about 10^{-13} sec are commonly encountered, the quantity 1×10^{-13} sec is called 1 Svedberg, named after T. Svedberg, a pioneer in sedimentation analysis. Svedberg units are conventionally denoted by the symbol S.

The quantity $\bar{v}$ (pronounced "vee-bar") in Eq. 5.15 can be thought of as the inverse of the effective density of the macromolecule. If this density is equal to that of the solution, $(1 - \bar{v}\rho) = 0$ and $s = 0$. If the molecule is less dense than the solution, then $(1 - \bar{v}\rho)$ is negative and we have *flotation* instead of sedimentation. This can happen, for example, for some lipoproteins in aqueous solution.

5.3.1   Moving Boundary Sedimentation

Determining the sedimentation coefficient.   What happens when a centrifugal field is applied to a solution of large molecules? Figure 5.9 shows the concentration changes that occur in an ultracentrifuge cell in which the molecules were initially uniformly distributed throughout the solution. When the field is applied, all begin to move, and a region near the meniscus becomes entirely cleared of solute. Thus, a *moving boundary* is formed between solvent and solution; this travels down the cell with a velocity determined by the sedimentation velocity of the macromolecules. We can, therefore, measure the sedimentation coefficient by following the rate of this boundary motion; this is what an *analytical ultracentrifuge* is designed to do. A schematic diagram of a modern analytical centrifuge is shown in Figure 5.10. The rotor spins (up to 70,000 rpm) in an evacuated chamber (to reduce friction heating) and its speed and temperature are closely controlled. In one mode of operation, the progression of sedimentation is observed by a *scanning absorption optical system.* The ultracentrifuge is equipped with a light source and monochromator, so that a wavelength can be selected at which the solute absorbs light maximally. The ultracentrifuge cell is illuminated by this light each time it revolves through the optical path. Focusing mirrors are arranged to form a cell image; this image is then scanned by a photomultiplier as it is displaced in the r direction. A double-sector cell is employed so that the photomultiplier alternately observes solution and solvent; thus, the whole system acts as a double-beam spectrophotometer. At each point r, the absorbance difference between the solution sector and the reference sector is recorded. The kind of output obtained is shown in Figure 5.9; it is a series of graphs of absorbance versus r at a succession of times in the sedimentation experiment. The boundary position is usually defined as the midpoint of the absorbance step, and can

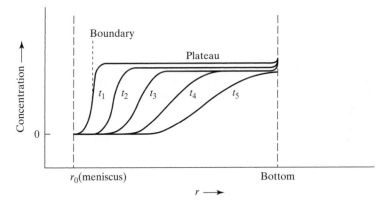

Figure 5.9   A moving boundary sedimentation experiment. The successive graphs, obtained at regular intervals after the beginning of sedimentation, show the concentration of the solute as a function of distance. Cell bottom and meniscus are marked. There is a solute-free region, a boundary region, and a plateau region. The sedimentation coefficient is measured from the midpoint of the boundary. The boundary broadens with time as a consequence of diffusion.

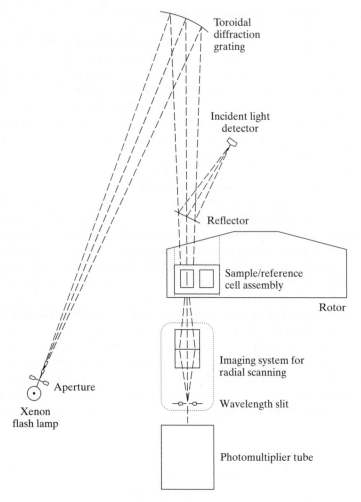

Figure 5.10 A schematic drawing of the absorption optical system for a modern analytical ultracentrifuge. The entire system is contained within the vacuum chamber housing the rotor. The diffraction grating allows choice of monochromatic light from about 200–800 nm. [Courtesy of Beckman-Coulter, Inc.]

be recorded at successive times during the experiment. Since the velocity of sedimentation v can be set equal to dr_b/dt, we have, from Eq. 5.15,

$$\frac{dr_b}{dt} = r_b\omega^2 s \tag{5.16}$$

Integrating gives

$$\ln[r_b(t)/r_b(t_0)] = \omega^2 s(t - t_0) \tag{5.17}$$

where $r_b(t)$ and $r_b(t_0)$ are positions at t and t_0, respectively. A graph of $\ln r_b(t)$ versus t will be a straight line with slope $\omega^2 s$, which allows us to calculate s.

The boundary does not remain sharp as it moves down the cell. Rather, it broadens because of the phenomenon of diffusion. Since D is inversely proportional to f, which in turn depends on molecular size, diffusion will be very slow for large molecules and fast for small ones. This has important effects on sedimenting boundaries. If the diffusion coefficient were zero, the boundary would remain infinitely sharp as it traversed the cell. Such a situation is *approximated* by very large molecules at high rotor speeds. Figure 5.11 contrasts the behavior of chambered nautilus hemocyanin (a) with $M = 3,500,000$ and its subunit (b) with $M = 350,000$. In a given length of time, the subunit has sedimented only about 1/5 as far, but the boundary has spread much more.

Below the boundary there is always a *plateau* region, which in most cases will extend nearly to the bottom of the cell (see also Figure 5.9). Here the concentration is independent of r and almost (but not quite) the same as the initial concentration. There is a slight *radial dilution effect* that occurs as sedimentation proceeds so that at any point in the plateau region $C_p/C_0 = (r_m/r_b)^2$, where C_0 is the initial concentration, r_m is the meniscus position, and r_b is the boundary position (see Figure 5.12). At the bottom of the cell, of course, the solute "piles up" in a dense layer.

The sedimentation coefficient is a characteristic property of each biological macromolecule. However, because it depends on $\bar{v}$, ρ, and f, all of which depend on temperature and the buffer solution in which the molecule is dissolved, it is necessary for comparative purposes to correct values obtained under various experimental conditions to *standard conditions*. These have been chosen to correspond to a hypothetical sedimentation in pure water at 20°C. The variation of f with T and the buffer solutions results entirely from its dependence on viscosity η. Hence, we may write

$$s_{20,w} = s_{T,b} \frac{(1 - \bar{v}\rho)_{20,w}\eta_{T,b}}{(1 - \bar{v}\rho)_{T,b}\eta_{20,w}} \tag{5.18}$$

Here $s_{T,b}$ is the value measured under experimental conditions, and $s_{20,w}$ corresponds to the value expected at the hypothetical standard condition. In practice, we

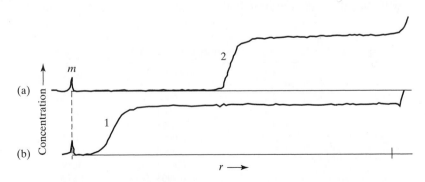

Figure 5.11 Data on (a) sedimentation of chambered nautilus hemocyanin ($M = 3.5 \times 10^6$) and (b) hemocyanin subunits at about the same time of sedimentation. Both are homogeneous, but the boundary for the latter has spread much more due to more rapid diffusion.

Figure 5.12 The radial dilution effect. A thin laminum containing a group of solute molecules, within the "plateau" region of the concentration gradient, is followed with time. The volume of the laminum increases as it moves with the particles, because (1) the cell is sector shaped with sides converging toward the axis of the rotor, and (2) the particles at the front edge see a slightly greater centrifugal field than those at the back. Because the volume increases, but the number of particles remains unchanged, the concentration of solute decreases with time.

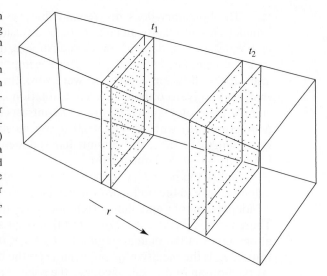

usually do not know the variation of $\bar{v}$ with T and the buffer composition; fortunately, such variation is small in most cases. Furthermore, it is practical to factor the viscosity term into a *relative buffer viscosity* (almost independent of T) and a temperature factor. Then

$$s_{20,w} \cong s_{T,b} \frac{(1 - \bar{v}\rho_{20,w})}{(1 - \bar{v}\rho_{T,b})} \left(\frac{\eta_{T,b}}{\eta_{T,w}} \right) \left(\frac{\eta_{T,w}}{\eta_{20,w}} \right) \qquad (5.19)$$

Equation 5.19 allows correction of an observed sedimentation coefficient to standard conditions. Values of ρ and $\eta_{T,b}/\eta_{T,w}$ are available for many buffer solutions, and the variation of water viscosity with temperature is given in Table 5.2.

It is also often important to take into account the fact that s may depend on macromolecule concentration. Any interactions among the sedimenting molecules will alter the sedimentation behavior. In the most usual case the molecules may be

Table 5.2 Density and Viscosity of Water as a Function of Temperature

T (°C)	ρ (gm/ml)	η (cp)[a]
0.0	1.0004	1.780
5.0	1.0004	1.516
10.0	1.0002	1.308
15.0	0.9996	1.140
20.0	0.9988	1.004
25.0	0.9977	0.891
30.0	0.9963	0.798

[a]Poise are the cgs units of viscosity. Thus, the viscosity of water at 20.0° is 1.004×10^{-2} poise = 1.004 centipoise (cp).

thought of as interfering with one another so as to make the frictional coefficient increase with concentration. If

$$f = f^0(1 + kC + \cdots) \qquad (5.20)$$

where f^0 is the value of f at $C = 0$, then

$$s = \frac{s^0}{1 + kC} \qquad (5.21)$$

and since k is usually positive, s will decrease with increasing C. As might be expected, this effect is most pronounced with highly extended macromolecules. Figure 5.13 shows representative data for serum albumin (a compact globular protein) and a small RNA. While the two substances have nearly the same s^0, the concentration dependence for the RNA is much greater; this is a common feature of extended molecules like nucleic acids. Fortunately, DNA and RNA absorb ultraviolet light very strongly, allowing us to use very dilute solutions and thereby to largely escape from these concentration-dependent effects.

Also shown in Figure 5.13 are data from a rather different kind of system, in which s *increases* with increasing C. In all cases where such behavior has been observed, it appears to result from a rapid monomer-polymer equilibrium. In some cases (a monomer-dimer equilibrium, for example), such systems will yield only a single boundary, with the sedimentation coefficient dependent on the proportions of monomer and polymer. Since high concentrations favor polymer formation, the s value increases with C. In any event, it is the sedimentation coefficient extrapolated to zero concentration ($s^0_{20,w}$) that is characteristic of the macromolecule itself. Many values of sedimentation coefficients have been measured; a few are listed in Table 5.3. What can we learn from them?

Interpreting the sedimentation coefficient.   It is clear from Eq. 5.15 that the sedimentation coefficient will, in general, increase with increasing molecular

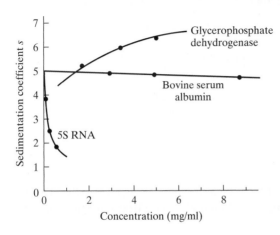

Figure 5.13 Concentration dependence of the sedimentation coefficient. Bovine serum albumin (BSA) and the ribosomal nucleic acid (5S RNA) exhibit behavior typical of compact and extended macromolecules, respectively. The behavior of honeybee glycerophosphate dehydrogenase is that to be expected for a reversibly associating substance. [Data from R. L. Baldwin (1957), *Biochem. J.* **65**, 503; D. G. Comb and T. Zehavi-Willmer (1967), *J. Mol. Biol.,* **23**, 441; and R. R. Marquardt and R. W. Brosemer (1966), *Biochem. Biophys. Acta,* **128**, 454.]

Table 5.3   Some Sedimentation and Diffusion Data

Substance	$s^0_{20,w} \times 10^{13}$ (sec)	$D^0_{20,w} \times 10^7$ (cm²/sec)	$\bar{v}_{20}$ (cm³/g)	$M_{s,D}$
Lipase	1.14	14.48	0.732	6,667
Lysozyme	1.91	11.20	0.703	14,400
Serum albumin	4.31	5.94	0.734	66,000
Catalase	11.3	4.10	0.730	250,000
Fibrinogen	7.9	2.02	0.706	330,000
Urease	18.6	3.46	0.730	483,000
Hemocyanin (snail)	105.8	1.04	0.727	8,950,000
Bushy stunt virus	132	1.15	0.740	10,700,000

weight. However, the relationship is not a simple linear one, for s also depends on f, which in turn depends on the size, shape, and hydration of the macromolecule. Thus, a highly asymmetric molecule might have a lower sedimentation coefficient than a globular protein of lower molecular weight. For example, compare the values for catalase and fibrinogen in Table 5.3.

If macromolecules were both spherical and unhydrated, we could derive a direct relationship between $s^0_{20,w}$ and M. This is because the frictional coefficient for an unhydrated sphere of radius a_0 is given by

$$f_0 = 6\pi\eta a_0 \tag{5.22}$$

and a_0 is in turn related to the molecular weight via the anhydrous molecular volume, V_0.

$$\frac{4}{3}\pi a_0^3 = V_0 = \frac{M\bar{v}}{\mathcal{N}} \tag{5.23}$$

If we solve Eq. 5.23 for a_0 and insert the result in Eq. 5.22, we obtain the following expression for the frictional coefficient of an unhydrated sphere:

$$f_0 = 6\pi\eta(3M\bar{v}/4\pi\mathcal{N})^{1/3} \tag{5.24}$$

Inserting this result into Eq. 5.15 gives Eq. 5.25.

$$s^0_{20,w} = \frac{M(1 - \bar{v}\rho)}{\mathcal{N}6\pi\eta(3M\bar{v}/4\pi\mathcal{N})^{1/3}} \tag{5.25a}$$

$$= \frac{M^{2/3}(1 - \bar{v}\rho)}{6\pi\eta\mathcal{N}^{2/3}(3/4\pi)^{1/3}\bar{v}^{1/3}} \tag{5.25b}$$

Here $\eta, \bar{v}$, and ρ correspond to values in water at 20°C, usually expressed in cgs units. According to Eq. 5.25b, the sedimentation coefficient for unhydrated spherical molecules should be proportional to the 2/3 power of the molecular weight. Indeed, we can rearrange Eq. 5.25b to yield

$$s^* = \frac{s_{20,w}^0 \bar{v}^{1/3}}{(1 - \bar{v}\rho)} = \frac{M^{2/3}}{6\pi\eta N^{2/3}(3/4\pi)^{1/3}} \tag{5.26}$$

We then see that the combination of quantities on the left (which we shall call, for convenience, s^*) should be, for any anhydrous and spherical macromolecule, determined uniquely by its molecular weight. In the log-log plot shown in Figure 5.14, a straight line with a slope of 2/3 is predicted.

It is instructive to plot data for a collection of real proteins on this graph. Note that many of them fall on or close to a line parallel to that predicted by Eq. 5.26. These are the *globular* proteins, and their behavior is explained by the fact that each of them does not depart too far from sphericity, and all are hydrated to about the same limited degree. Other proteins, like fibrinogen and myosin, fall far below the line; these are highly asymmetric proteins, which have high frictional coefficients.

While it does not allow us to determine molecular weight unambiguously, Figure 5.14 has certain uses. No real macromolecule could correspond to a point *above or to the left of* the upper line, for that would mean that it had a frictional coefficient less than that predicted for an unhydrated sphere of the same molecular weight. Conversely, this tells us that the upper line marks the *minimum* molecular weight possible for any molecule with a given value of s^*. If we are willing to assume that an unknown molecule is a typical globular protein, we can use the lower line in Figure 5.14 to estimate M from s^*. But such a procedure can be hazardous; note that if we attempted to use it for the fibrous protein myosin, we would predict a value of $M \cong 100,000$, much lower than the correct value of 540,000!

In order to use sedimentation data in this or other ways, it is necessary to know a value for $\bar{v}$. There are two ways to obtain this. The first, which is applicable to all substances, is to measure $\bar{v}$ from solution density measurements. The variation in density for a dilute solution of a molecule of partial specific volume $\bar{v}$ is given to a good approximation by

$$\rho = \rho_0 + (1 - \bar{v}\rho_0)C \tag{5.27}$$

where ρ_0 is the solvent density, and C is the concentration in gm/cm^3. Unfortunately, density measurements of very high precision (on the order of one part in 10^5 or 10^6) are needed to get $\bar{v}$ values good to a few percent at practical biopolymer concentrations. Such measurements are very tedious and difficult. Fortunately, there is a way to circumvent this difficulty, at least for proteins. It has been shown that the $\bar{v}$ of a protein can be quite accurately calculated from its amino acid composition, on the assumption of additivity of *residue volumes*. Tables of such volumes have been developed that yield good results when compared with direct measurement (see Zamyatnin in the reference section in this chapter).

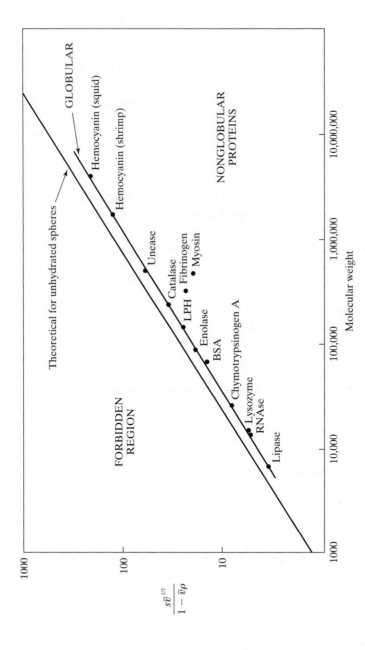

Figure 5.14 Graphs of $s\bar{v}^{1/3}/(1 - \bar{v}\rho)$ vs. molecular weight on a log-log scale. The upper line corresponds to Eq. 5.25b. Data for various proteins are also plotted on this graph; those for *globular* proteins fall on or close to an empirical line of correct slope (lower line on graph). Note that some *fibrous* proteins like myosin or fibrogen deviate markedly from this relationship.

If the molecular weight of a macromolecule is known, the sedimentation coefficient can be used to obtain semiquantitative information about its size and shape. As a first approximation, we might calculate the frictional coefficient, f, using a rearranged Eq. 5.15. This then allows us to calculate the Stokes's radius, as described for diffusion measurements (see p. 220). However, as mentioned earlier, the significance of this number should not be exaggerated; if the molecule is nonspherical, it has no exact physical meaning.

For a slightly more sophisticated analysis, Eq. 5.15 can be used to calculate f, and Eq. 5.24 to calculate f_0. The ratio, f/f_0, will inevitably be somewhat greater than unity. The problem in interpretation comes from the fact that this departure from unity can be accounted for by many possible combinations of different degrees of hydration and different shapes. If we know the hydration from some other kind of measurement, the analysis can be taken a bit further. We can rewrite f/f_0 in terms of f_{sp}, the hypothetical frictional coefficient for a spherical molecule of a given hydration.

$$f/f_0 = (f/f_{sp})(f_{sp}/f_0) \tag{5.28}$$

Thus, f_{sp}/f_0 is a factor depending *only* on hydration; it compares a hydrated sphere to an unhydrated one. If we assume that the total volume of a hydrated molecule is simply the sum of the anhydrous volume and the volume of hydrating water, it is easy to show that

$$f_{sp}/f_0 = (1 + \delta)^{1/3} \tag{5.29}$$

where δ is the volume of water of hydration per unit volume of anhydrous macromolecule. The factor f/f_{sp} is a pure *shape* factor, measuring how much the molecule differs from sphericity. This can be interpreted quantitatively if the shape of the molecule is known. For example, if we know that a given molecule is rodlike, the equation given in Table 5.1 could be used to estimate the length/diameter ratio.

Unfortunately, it has not been possible to deduce *exact* expressions for frictional coefficients for any but the simplest (and often unrealistic) models: spheres, ellipsoids, rods, and the like (see Figure 5.7 and Table 5.1). There exists, however, a theory that allows us to approximate rather accurately the frictional coefficients for molecules of quite complicated shapes. In 1954, J. Kirkwood analyzed the frictional behavior of a molecule made up of a number of subunits, and showed how the hydrodynamic interaction between those subunits could be taken into account. For a particle of N identical subunits, each with frictional coefficient f_1, the frictional coefficient of the assembly can be written as

$$f_N = Nf_1\left(1 + \frac{f_1}{6\pi\eta N}\sum_{i=1}^{N}\sum_{j\neq i}^{N}\frac{1}{R_{ij}}\right)^{-1} \tag{5.30}$$

Here R_{ij} is the distance between centers of subunits i and j, and the summation is taken over all pairs (except $i = j$). This is an exceedingly powerful equation, for in principle it allows us to calculate the frictional coefficient of *any* object by modeling it in terms of a large number of small subunit beads (see Bloomfield et al. 1967). Such calculations are inexact for very asymmetric shapes because of certain approximations in the theory. However, the Kirkwood method is quite useful and practical for the study of multisubunit proteins. Suppose we have a protein made up of N identical subunits, each with sedimentation coefficient s_1, where

$$s_1 = \frac{M_1(1 - \bar{v}\rho)}{\mathcal{N}f_1} \tag{5.31}$$

If we combine Eqs. 5.15, 5.30 and 5.31, and note that $M_N = NM_1$, we obtain a simple expression for the ratio of the sedimentation coefficient of the N-mer to that of the subunit.

$$\frac{s_N}{s_1} = 1 + \frac{f_1}{6\pi\eta N} \sum_{i=i}^{N} \sum_{j\neq i}^{N} \frac{1}{R_{ij}} \tag{5.32}$$

The ratio s_N/s_1 will depend on the geometry of the molecule, through the subunit-subunit distances R_{ij}. If we write f_1 in Eq. 5.32 as $f_1 = 6\pi\eta a_1$ (where a_1 is the Stokes's radius of the subunit) the result becomes even simpler.

$$\frac{s_N}{s_1} = 1 + \frac{a_1}{N} \sum_{i=i}^{N} \sum_{j\neq i}^{N} \frac{1}{R_{ij}} \tag{5.33}$$

The Stokes's radius, can, of course, be calculated from s_1 and M_1. Equation 5.33 allows us to test various postulated geometries of subunit arrangement, provided that s_N, s_1, M_1, and N are known. Table 5.4 gives examples of predicted values for s_N/s_1 for various possible quaternary arrangements of a tetramer. Computer programs have now been devised that allow the calculation of f_N or s_N for very complex shapes (see de La Torre et al. 1994).

Calculation of *M* from *s* and *D*.   We have emphasized that the sedimentation coefficient, by itself, cannot yield an unambiguous value for the molecular weight. This is because of the presence in Eq. 5.15 of the frictional coefficient, which depends in complicated ways on size, shape, and hydration. However, it is possible to

Table 5.4   Predicted Sedimentation Coefficient Ratios for Different Tetramer Structures

Structure	s_4/s_1
Linear	2.083
Square-planar	2.353
Tetrahedral	2.500

Calculated assuming spherical subunits in contact, Eq. 5.33.

eliminate f by making use of measurements of the diffusion coefficient. If we divide Eq. 5.15 by Eq. 5.10, we obtain

$$\frac{s}{D} = \frac{M(1 - \bar{v}\rho)/\mathcal{N}f}{RT/\mathcal{N}f} = \frac{M(1 - \bar{v}\rho)}{RT} \tag{5.34}$$

in which s and D must be measured under (or corrected to) the same conditions. Equation 5.34, which has been named the *Svedberg equation,* was one of the first equations used to show that biopolymers are giant molecules. It has been used less frequently recently because of the inconvenience of classical methods for measuring D. However, the development of the fast, accurate laser-scattering method (Chapter 7) may make it once again an attractive approach. (Some results from s/D calculations are shown in Table 5.3).

Analysis of mixtures in moving boundary sedimentation. So far, we have dealt only with sedimentation of a single macromolecular component. If there are multiple components present (as in Figure 5.15), it is often possible to determine both the sedimentation coefficient and amount of each, provided that their boundaries can be resolved. The enemy of resolution is diffusion, which blurs boundaries and makes them overlap. Since boundary spreading due to diffusion proceeds at a rate determined only by the diffusion coefficient of the component concerned, whereas the rate of boundary *separation* depends on the actual velocities of boundary motion, higher rotor speeds will in general lead to greater resolution.

It is possible to compensate for the smearing effect of diffusion by taking advantage of a peculiar difference between sedimentation transport and diffusive transport. The distance traveled by sedimenting molecules is very nearly proportional to the time of sedimentation. However, molecules wandering about in diffusion will, on average, travel distances proportional to the *square root* of time (see Eq. 5.9). This means that as sedimentation proceeds for longer and longer times, the *separation* between different components becomes more and more dominant over

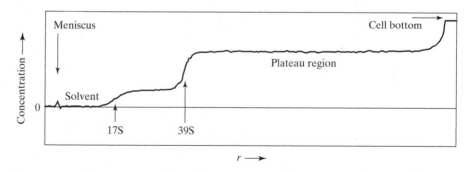

Figure 5.15 The concentration profile expected for sedimentation analysis of a two-component solute that does not show reversible association-dissociation reactions. The amounts of the two components can be approximately calculated from the plateau heights, when corrected for radial dilution. The data shown are for shrimp hemocyanin.

the spreading of any one boundary. Researchers van Holde and Weischet (1975) have utilized this fact to extrapolate out the diffusion smearing. Figure 5.16 shows how the diffusion-smeared boundary produced by a single, homogeneous substance can be corrected to demonstrate this homogeneity.

The development of extremely rapid computing power has allowed new kinds of analysis that would not have been conceivable a short time ago. An extreme case is the fitting of data, by interactive methods, to the differential equation governing sedimentation and diffusion, the *Lamm equation*.

$$\left(\frac{\partial C}{\partial t}\right) = -\frac{1}{r}\left\{\frac{\partial}{\partial r}\left[s\omega^2 r^2 C - Dr\left(\frac{\partial C}{\partial r}\right)\right]\right\} \tag{5.35}$$

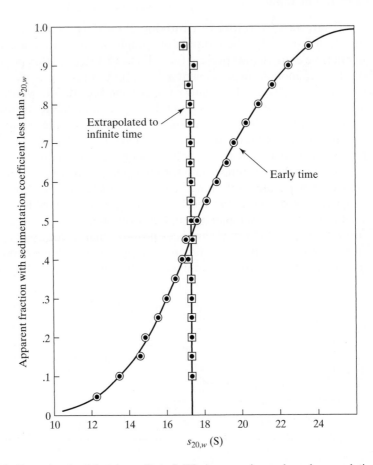

Figure 5.16 Removing the deleterious effect of diffusion smearing on boundary resolution. Data are from a study of squid hemocyanin. The curve with the circles depicts the apparent distribution of sedimentation coefficients at an early time during an experiment. The line through the squares shows the apparent distribution when extrapolated to infinite time. The fact that it is a vertical "step" function shows that the sample is homogeneous, with $s_{20,w} = 17.2\ S$.

This partial differential equation contains both sedimentation and diffusion terms, and s and D are parameters that may be derived by fitting experimental data to it. Furthermore, the equation can be generalized to include multiple components, which may or may not be in chemical equilibrium. An example of the power of the technique is shown in Figure 5.17, which depicts the "mass spectrum" of a commercial sample of bovine serum albumin. In addition to the main (monomeric) component with a mass of about 64,000 Da, small amounts of dimer and trimer are also revealed. The development of methods like this promises to enormously extend the power of sedimentation methods.

5.3.2   Zonal Sedimentation

The moving boundary technique, powerful as it can be, suffers from a number of disadvantages that have limited its use. First, an expensive and specialized instrument, the analytical ultracentrifuge, is required. Second, the recording of concentrations depends entirely on optical methods, which often require fairly high concentrations. Finally, and perhaps most important, the moving boundary technique has difficulty in resolving complex mixtures, since only the *slowest* component is wholly resolved from others. If a narrow *zone* of material could be placed

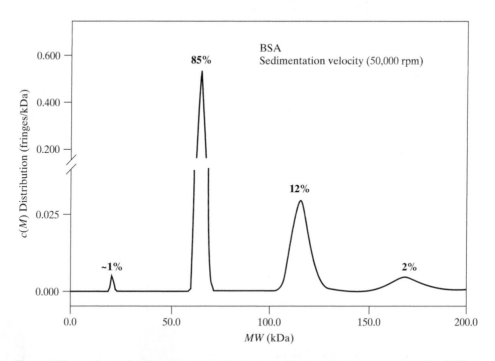

Figure 5.17   Analysis of sedimentation velocity data via fitting to the Lamm equation (Eq. 5.35). A serum albumin sample is shown to consist mostly of monomer, but small amounts of dimer and trimer were also present. [Courtesy of Drs. S. Anderson and D. Malencik.] The analysis is according to methods devised by P. Shuck (see, for example, Lebowitz, et al, 2002).

at the meniscus and sedimented through the cell, resolution would be enhanced, because each component would form its own sedimenting zone. However, a moment's reflection shows that this is not so easy to achieve. Placing a zone containing macromolecules on top of a solvent will lead to immediate convection, because the solute-containing zone will be more dense than the solvent. However, stable zonal sedimentation can be accomplished if the solution is layered onto a *density gradient* produced by some inert substance (Figure 15.18). Such a density gradient can be produced using a substance such as sucrose or glycerol by carefully layering solutions of decreasing density, one on the other with a mixing device of the kind shown in Figure 5.19. In this way, a smooth gradient of any steepness desired can be made.

The gradient produced in this way is, of course, not indefinitely stable; it will eventually disappear by diffusion. But this process is slow, and experiments with a duration of several hours can be performed without too much change occurring in the gradient. The macromolecular solution must be layered onto the gradient with care, and it must be less dense than the top sucrose solution. It is important that the solution layered onto the gradient not be too concentrated and dense. Otherwise the band will be unstable as it moves into the gradient, and resolution will be lost because of localized convection.

The advantage in the resolving power of the gradient technique over the conventional sedimentation transport experiment should be obvious. In the conventional method, only the most slowly sedimenting material can be recovered in pure form, whereas the density gradient method completely resolves components from one another. The crude-appearing sampling method actually works well and little remixing is produced. More sophisticated and reproducible techniques have been developed for emptying the tubes when high precision is needed. From the tube number in which a given component appears, the distance traveled can be estimated and *s* can be calculated. However, the variation in both density and viscosity along the sucrose gradient makes the velocity vary during sedimentation and complicates the calculations. We note that the *velocity* of sedimentation at point *r* will be given by

$$v = s\omega^2 r = \frac{M(1 - \bar{v}\rho)\omega^2 r}{\mathcal{N}f}$$
(5.36)

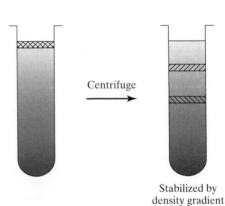

Centrifuge

Figure 5.18 The principle of zonal sedimentation. A thin layer containing two different components will separate into two zones after transport. The zone is stabilized against convection by a density gradient.

Stabilized by density gradient

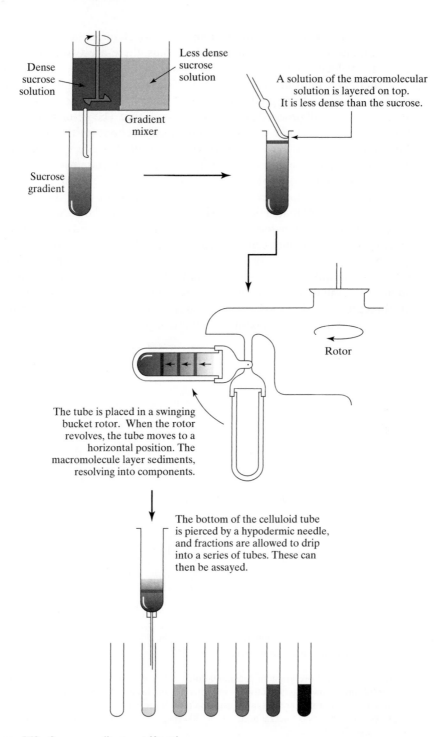

Figure 5.19   Sucrose gradient centrifugation.

The frictional coefficient is proportional to η, which increases with r because of the gradient, and $(1 - \bar{v}\rho)$ decreases with r. Thus, there are two factors that tend to make the velocity become slower and slower as the band moves down the tube. On the other hand, the value of r in Eq. 5.36 is increasing, tending to accelerate the motion. If the form of the gradient is chosen correctly, these factors will balance and sedimentation will proceed at a nearly constant rate. The gradient that will produce this effect is called an *isokinetic* gradient, and directions for the construction of isokinetic gradients have been worked out for a number of experimental situations. Even with an isokinetic gradient, it is advisable to include with the sample two or more marker substances of known sedimentation coefficient to provide an internal calibration scale. However, even with the greatest of care the density gradient method still cannot match the precision of moving boundary sedimentation. Density gradient centrifugation does, however, have the enormous advantage that a wide variety of assay techniques can be used to detect the various components. For example, materials can be specifically labeled with radioactive isotopes or discriminating enzymatic or immunological assays can be used. In this way, the sedimentation of a minor component in a crude mixture can be followed, a feat impossible with the optical methods used in most analytical ultracentrifuges. Figure 5.20 shows the sedimentation of an enzyme mixture in a sucrose gradient.

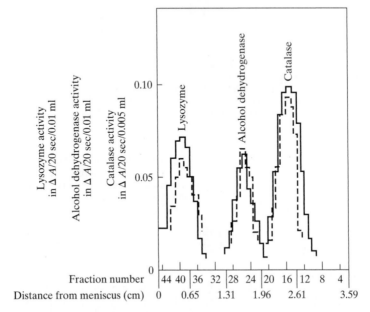

Figure 5.20 A sucrose gradient sedimentation velocity experiment with mixtures of three enzymes. The solid lines show enzyme activities in a mixture of the three enzymes in buffer. The dashed lines show activities in a mixture prepared using a crude bacterial extract as the solution medium. In each case, sedimentation was for 12.8 hr at 3°C, 37,700 rpm. [From R. B. Martin and B. Ames (1961), *J. Biol. Chem.*, **236**, 1372.]

As you may have already realized from Figure 5.19, density gradient sedimentation has another enormous advantage over sedimentation in the analytical ultracentrifuge—it can be used as a *preparative* technique. Indeed, it is one of the most powerful biophysical methods for separating and isolating macromolecules from complex mixtures. Last, but by no means least, is the fact that density gradient centrifugation, unlike moving boundary sedimentation, is available to almost every biochemistry laboratory. No special instrument is needed—only a swinging bucket rotor of the type shown in Figure 5.19 and a standard preparative ultracentrifuge in which to spin it.

5.3.3   Sedimentation Equilibrium

In the moving boundary sedimentation experiment, the ultracentrifuge is operated at a rotor speed sufficient to pull a boundary away from the meniscus, leaving pure solvent behind. But suppose we ran at a much lower speed, so the centrifugal force was much smaller. What would happen? The answer is depicted in Figure 5.21: Instead of forming a solution-solvent boundary, the flow of solute caused by sedimentation simply creates a concentration gradient in the cell. With time of sedimentation, this gradient becomes steeper and steeper, but as it does so, backflow due to *diffusion* becomes more and more pronounced. We know this by Fick's first law (Eq. 5.1):

$$J_D = -D\left(\frac{\partial C}{\partial r}\right) \tag{5.37}$$

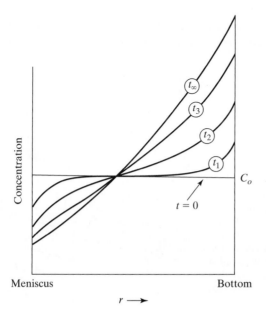

Figure 5.21 The approach to sedimentation equilibrium. As the concentration gradient created by sedimentation becomes more pronounced, backflow by diffusion increases until a final gradient is approached, which represents an equilibrium between these opposed processes.

Eventually, the gradient will become steep enough so that the backflow due to diffusion exactly balances the outward flow produced by sedimentation *at every point in the cell*. This is called the state of *sedimentation equilibrium*. The sedimentation flow J_s is just the product of the velocity with which molecules sediment and their local concentration (see Figure 5.22). Therefore, from Eq. 5.15,

$$J_s = v_s C = \frac{M(1 - \bar{v}\rho)\omega^2 r}{\mathcal{N}f} C \tag{5.38}$$

When sedimentation equilibrium is attained, the net flow at every point in the cell $(J = J_s + J_D)$ will be zero.

$$0 = \frac{M(1 - \bar{v}\rho)\omega^2 r C}{\mathcal{N}f} - D\left(\frac{dC}{dr}\right) \tag{5.39}$$

Rearranging, and noting that $D = RT/\mathcal{N}f$ (Eq. 5.10),

$$\frac{RT}{\mathcal{N}f}\left(\frac{dC}{dr}\right) = \frac{M(1 - \bar{v}\rho)\omega^2 r C}{\mathcal{N}f} \tag{5.40}$$

We now write the derivative as dC/dr, because at equilibrium, C is a function only of r, not also of t. We can factor out the $\mathcal{N}f$ on both sides of Eq. 5.40 and again rearrange to

$$\frac{d\ln(C)}{d(r^2)} = \frac{M(1 - \bar{v}\rho)\omega^2}{2RT} \tag{5.41}$$

This equation tells us that at sedimentation equilibrium, $\ln C$ will be a linear function of r^2, with slope $M(1 - \bar{v}\rho)\omega^2/2RT$. Alternatively, we can integrate Eq. 5.41 to obtain

$$\ln[C(r)/C(r_0)] = \frac{M(1 - \bar{v}\rho)\omega^2}{2RT}(r^2 - r_0^2) \tag{5.42a}$$

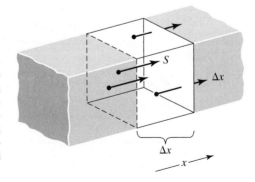

Figure 5.22 The relation between molecular velocity and flow. In time Δt, a molecule with velocity v moves a distance $\Delta x = v\Delta t$. Thus, all molecules in a slab extending back this distance from the surface S will pass through S in Δt seconds. The amount of solute passing will be the concentration times the slab volume: $\Delta w = CS\Delta x = CSv\Delta t$. Because the flow is defined as $J = \Delta w/S\Delta t$, we find $J = Cv$.

or

$$C(r)/C(r_0) = e^{\frac{M(1-\bar{v}\rho)w^2}{2RT}(r^2-r_0^2)} \tag{5.42b}$$

where r_0 is a reference point (the meniscus, for example). Thus, $C(r)$ is an exponential function of r^2. A graph of $C(r)$ (expressed as absorbance) as a function of r is shown for a real experiment in Figure 5.23a, and the corresponding linear graph of $\ln(C)$ versus r^2 is given in Figure 5.23b.

From such data, the molecular weight may be determined, provided that $\bar{v}$ has been measured or estimated. The method is rigorous and involves no assumptions whatsoever about particle shape or hydration. In fact, Eq. 5.42 can be derived on very rigorous thermodynamic grounds, as is shown in Chapter 13. Since no denaturants (such as SDS) are involved, the molecular weight obtained will be that of the actual molecular structure existing in solution under the conditions studied. Thus, if a protein consists of a number of subunits, the molecular weight of the multisubunit

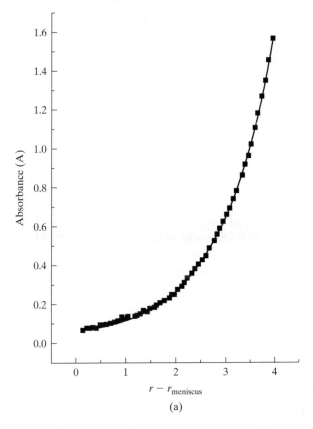

(a)

Figure 5.23   Data on the sedimentation equilibrium of monomeric squid hemocyanin. (a) C versus r curve, (b) $\ln C$ versus r^2 curve. The data fit the curve predicted for an ideal, homogeneous component of 382,000 Da.

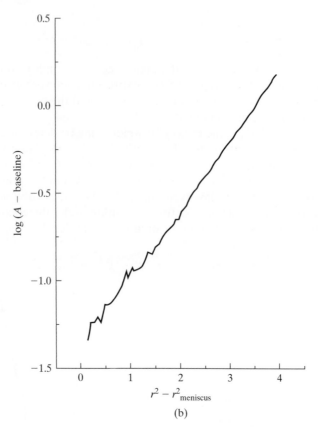

Figure 5.23   *(Continued)*

structure will be obtained. Protein–nucleic acid complexes can be studied equally well. An example that uses both sedimentation equilibrium together with a number of other techniques is shown in Application 7.1.

The range of molecular weight over which the sedimentation equilibrium method can be applied is extraordinarily broad (see Table 5.5). Substances with molecular weights below 100 and above 10,000,000 have been investigated with

Table 5.5   Some Sedimentation Equilibrium Results

Substance	Molecular Weight from Chemical Formula	Molecular Weight from Sedimentation Equilibrium
Sucrose	342.3	341.5
Ribonuclease	13,683	13,740
Lysozyme	14,305	14,500
Chymotrypsinogen *A*	25,767	25,670
Whelk hemocyanin	–	13,200,000

comparable accuracy. This is because the rotor speed can be adjusted to give comparable results with large or small molecules.

If the macromolecular solute is not homogeneous but instead consists of several components (for example, different aggregation states of a given protein subunit), the analysis becomes more complicated, but it can still provide valuable information. There are essentially two methods of analysis that can be employed; the choice will depend on the circumstances.

Multiexponential curve fitting.   Suppose we have a few components (say, two to four). Each will be distributed in the cell at sedimentation equilibrium in an exponential gradient, as in Eq. 5.42b. The steepness of each concentration gradient will depend on the molecular weight (see Figure 5.24a). The sum of the component concentrations will *not* be a simple exponential function and will exhibit curvature on an $\ln C_t$ versus r^2 plot (Figure 5.24b). One way to analyze such a situation is to find a combination of simple exponential functions that closely fits the observed concentration curve. This will then give the amount and molecular weight of each component. There are computer algorithms that can

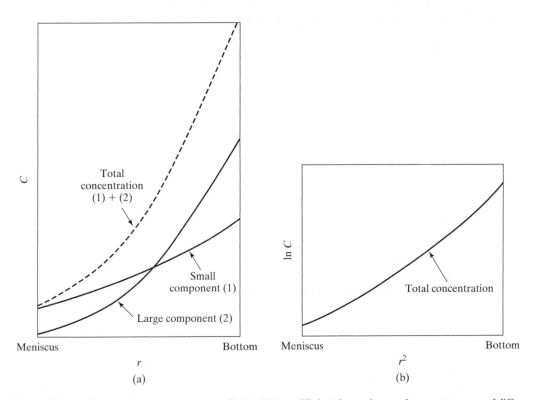

Figure 5.24   (a) The concentration curves at sedimentation equilibrium for a mixture of two components of different molecular weight (curves 1 and 2). Each is an exponential in r^2. The total concentration ($C_1 + C_2$) is *not* a simple exponential. (b) The $\ln C_t$ versus r^2 plot corresponding to the curve in (a).

do this efficiently, at least for a small number of components. See, for example, Schuster and Laue (1994); Cole and Hansen (1999).

Use of average molecular weights.   It is easy to show that for sedimentation equilibrium of a mixture of n solute components of molecular weights M_i

$$\frac{d\ln C_t}{d(r^2)} = \frac{\omega^2(1 - \bar{v}\rho)}{2RT} M_{wr} \tag{5.43}$$

where M_{wr}, the *weight average molecular weight* of the material at point r in the cell is defined as

$$M_{wr} = \sum_1^n C_i M_i / \sum_1^n C_i = \sum_1^n C_i M_i / C_t \tag{5.44}$$

the concentrations C_i being those at point r in the cell. Note that Eq. 5.43 is just like Eq. 5.41 with M_{wr} substituting for M and the total concentration C_t replacing the concentration of an individual component, C. At each point in the cell, the slope of the $\ln C_t$ versus r^2 plot gives the weight average molecular weight of the mixture at that point. In Chapter 14, we see how this kind of analysis can be used in the study of associating macromolecules.

Experimental methods.   Obviously, sedimentation equilibrium requires an analytical ultracentrifuge, as the concentration must be measured along the cell while the rotor is spinning. Most modern instruments make this measurement using the kind of scanning optical system described in Section 5.3.1 to measure solute absorbance. However, even more sensitive interferometric and fluorescence techniques now make possible, measurement with greater precision and at lower concentrations.

In order for equilibrium to be attained in a reasonable time, it is desirable to use a short solution column—only about 1 to 3 mm in height. This means that a relatively small volume of solution—on the order of 30 to 100 μl—is required. Under these conditions, most macromolecules will reach equilibrium in less than 24 hours, but the time becomes greater if very large macromolecules are studied.

With advances in ultracentrifuge and computer technology, it has become possible to analyze associating systems with rigor. For details see Schuster and Laue (1994). An example is presented in Chapter 14, Section 14.2.

5.3.4   Sedimentation Equilibrium in a Density Gradient

There is a quite different kind of sedimentation equilibrium experiment that has proved especially useful in the study of nucleic acids. Suppose, as a specific example, that a solution containing a small amount of high-molecular-weight nucleic acid and a high concentration of a dense salt such as cesium chloride is placed in an ultracentrifuge cell and spun for a long time at high speed. The salt will eventually reach sedimentation equilibrium, giving a concentration gradient like that described by Eq. 5.42. This will yield a density gradient in the cell. The experiment becomes interesting if the salt concentration has been so chosen that at some point r_0 in the cell, $\rho(r_0) = 1/\bar{v}$, where $\bar{v}$ is the specific volume of the macromolecule. The quantity $(1 - \bar{v}\rho)$ will then be positive above this point and negative below it, so we expect the sedimentation of the macromolecule to converge toward the point r_0. Eventually, equilibrium will be established between the tendency for the macromolecules to diffuse away from r_0 and the effect of sedimentation that drives them back. The macromolecular solute will come to equilibrium in a band centered on r_0 (see Figure 5.25). If the molecular weight is large, the band will be narrow, because s will be large and D small.

It can be shown that the distribution of solute concentration is approximately Gaussian, following the equation

$$C(r) = C(r_0)e^{-(r-r_0)^2/2\sigma^2} \tag{5.45a}$$

where the standard deviation σ is given by

$$\sigma = \left(\frac{RT}{\omega^2 r_0 M \bar{v}(d\rho/dr)}\right)^{1/2} \tag{5.45b}$$

Here $d\rho/dr$ is the density gradient established by the salt.

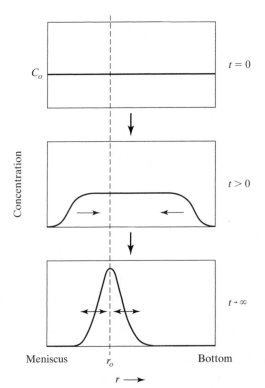

Figure 5.25 The principle of density gradient sedimentation equilibrium. The point r_0 is the isodensity point.

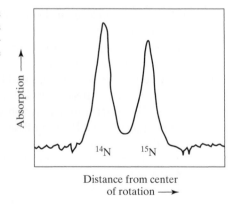

Figure 5.26 Density gradient sedimentation equilibrium of a mixture of DNA from bacteria grown on ^{15}N and ^{14}N media. Note that separation of the bands is excellent, even though the densities differ only by 0.014 g/μm^3.

Although it is theoretically possible to measure M of nucleic acids in this manner, it is difficult, and little used today, where simpler gel methods (see Section 5.4) or sequencing are available. Rather, the greatest utility of the density gradient sedimentation equilibrium experiment lies in the exquisite sensitivity of the band *position* to solute density. The classic example is the Meselson-Stahl experiment (1958), which demonstrated semiconservative replication of DNA. In this case, the technique allowed the resolution of nucleic acids differing only in their content of ^{14}N and ^{15}N in their bases. See Figure 5.26. It was this same density-gradient sedimentation equilibrium technique that led to the discovery of *satellite* DNAs. The density at which a DNA molecule bands in such a gradient is a function of the $(G + C)/(A + T)$ ratio. Certain repetitive DNA sequences that are especially rich in either G/C or A/T base pairs will form satellite bands to the main-band DNA. Their discovery gave the first solid indication of the complexity of the eukaryotic genome. The density-gradient sedimentation experiment is still used as a preparative method for such DNA, to purify DNA, and to separate DNA from RNA and DNA/RNA hybrids. A variety of high-density salts (for example, chlorides or sulfates of rubidium or cesium) are used to prepare gradients that are best suited for different isolations or analyses.

5.4   ELECTROPHORESIS AND ISOELECTRIC FOCUSING

The great majority of the polymers of biological interest are electrically charged. Like low-molecular-weight electrolytes, polyelectrolytes are somewhat arbitrarily classified as "strong" or "weak," depending on the ionization constants of the acidic or basic groups. They may be strong polyacids, such as the nucleic acids; weak polybases, such as poly-*l*-lysine; or polyampholytes, such as the proteins. This polyelectrolyte character has a considerable influence on the solution behavior of these substances. It is not surprising, then, that there has been considerable interest in methods for estimating the charge carried by biopolymers.

Even more important are the manifold ways in which electric charge differences can be used to separate and analyze mixtures of biopolymers. At the present time, electrophoretic methods are used in every area of biochemistry and molecular biology; they represent, in toto, the most important physical technique available to scientists working in these fields. Indeed, most of the remarkable advances in molecular biology over the past few decades would have been impossible without electrophoretic methods.

5.4.1 Electrophoresis: General Principles

The transport of particles by an electrical field is termed *electrophoresis*. Formally, the process is very like sedimentation, for in both cases a field, which is the gradient of a potential, produces the transport of matter. However, since electrophoresis will depend on the charge on a macromolecule rather than its mass, it provides a different handle for the analysis and separation of mixtures.

An exact theory of electrophoresis has been very difficult to develop. Serious problems arise because electrophoresis must almost always be carried out in aqueous solutions containing buffer ions and salts in addition to the macromolecule of interest; at the very least, the counterions that have dissociated from the macromolecule to provide its charge must be present. This means that we study the molecule not alone, but in the presence of many other charged particles, and these will both influence the local field and interact with the macromolecule, making analysis difficult. To begin, then, let us consider an idealized, simplified situation, an isolated charged particle in a nonconducting medium (Figure 5.27a). We may write the basic equation for electrophoresis in exactly the same fashion as for sedimentation; we assume that the force on the particle, when suddenly applied, causes an immediate acceleration to a velocity at which the viscous resistance of the medium just balances the driving force. The force experienced by a particle in an electrical field is given by Coulomb's law,

$$F = ZeE \tag{5.46}$$

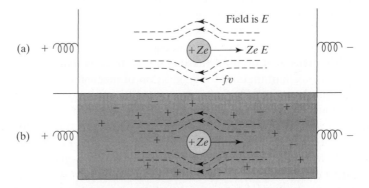

Figure 5.27 (a) An idealized model for electrophoresis in which a particle of charge Ze is placed in an electric field in a nonconducting solvent. (b) A more realistic model, in which a charged macromolecule is subjected to an electric field in an aqueous salt solution. The small ions form an ion atmosphere around the macromolecule in which ions of charge opposite to that of the macromolecule predominate. This ion atmosphere is distorted by the field and by the motion of the particle.

where Z is a positive or negative integer corresponding to the number of charges, e is the magnitude of the proton charge, and E is the electrical field in units of potential per centimeter. Since the resistance to motion is given by $-fv$, where f is the frictional factor, and v the velocity, the condition that the net force on the particle be zero in steady motion tells us that

$$fv = ZeE \qquad (5.47)$$

Equation 5.47 can be rearranged to define an *electrophoretic mobility U*, a quantity very analogous to the sedimentation coefficient.

$$U = \frac{v}{E} = \frac{Ze}{f} \qquad (5.48)$$

As with the sedimentation coefficient, U is the ratio of velocity to the strength of the driving field (compare Eq. 5.15). If the particle happens to be spherical, Stokes's law (Eq. 5.11) applies and we can write

$$U = \frac{Ze}{6\pi\eta a} \qquad (5.49)$$

where a is the particle radius and η the solvent viscosity. In principle, this equation could be generalized to take into account hydration and deviations from spherical shape, as in the cases of sedimentation and diffusion. However, in dealing with real macroions in aqueous solution, there are more formidable complications to consider first. In such an environment, the macromolecule is surrounded by a *counterion atmosphere,* a region in which there will be a statistical preference for ions of opposite sign (see Figure 5.27b). Since the field pulls in opposite directions on the macroion and its counterions, the atmosphere becomes asymmetric, and hence produces an electrostatic force on the macroion opposite to that produced by the applied external field. To account for this, and other complications, many attempts have been made to introduce corrections into Eq. 5.49. Only recently (see Chapter 13, Section 13.2.4) have advances been made that allow quite accurate measurements of protein charge under some circumstances.

The net result of all of these complications is that electrophoresis, while providing an excellent method for separation of macromolecules, has until now proved less useful than methods such as sedimentation for obtaining quantitative information about macromolecular structure. Even under circumstances where Eq. 5.49 is not accurate, it still predicts one point of interest; at the pH for which a protein or other polyampholyte has zero charge (the isoelectric point), the mobility should be zero. At lower pH values a protein will be positively charged and move toward the negative electrode, and at higher pH values transport will be in the opposite direction. As we shall see in Section 5.4.6, this behavior has been utilized to develop the method of *isoelectric focusing.*

The absolute values of mobilities in solution can be measured by *moving boundary electrophoresis* or *free electrophoresis,* which is analogous to moving boundary sedimentation. However, this technique is almost never used at the present time, for it requires large amounts of material to yield a quantity, the absolute mobility, that we have seen to be of limited utility. The interested reader will find a description of free electrophoresis in the older texts.

The tremendous resurgence in the use of electrophoresis in recent years has been due almost entirely to the development of what may be generally termed *zonal* techniques. In these methods, a thin layer or zone of the macromolecule solution is electrophoresed through some kind of matrix. The matrix provides stability against convection, as does a sucrose gradient in zonal sedimentation. In addition, in many cases the matrix acts as a molecular sieve to aid in the separation of molecules on the basis of size.

The kind of supporting matrix used depends on the type of molecules to be separated and on the desired basis for separation: charge, molecular weight, or both. A list of some commonly used materials and their applications is given in Table 5.6. Almost all electrophoresis of biological macromolecules is at present carried out on either polyacrylamide or agarose gels. For this reason, the term *gel electrophoresis* is more commonly employed than *zonal electrophoresis*. These gels can be prepared in a wide range of concentrations and, in the case of polyacrylamide gels, the density of cross-linking can also be readily controlled. For some purposes, composite agarose-polyacrylamide gels are employed, or gels in which there is a gradient in matrix concentration.

The principle of the technique is illustrated in Figure 5.28. The gel has been cast as a slab between glass plates; it contains the buffer to be used. Indentations ("wells") are cast in the top of the slab to provide multiple lanes for electrophoresis. Each macromolecular solution is applied in a thin layer in one well. When a voltage is applied, the zone of macromolecules migrates down each lane. If several components of different mobility are present, they will separate during electrophoresis, just as the zones of molecules of different sedimentation rate separate in zonal centrifugation. Usually, a dye of high mobility is added to the macromolecular solution; its migration serves to mark the progress of the experiment. The dye also serves as a convenient measure of mobility; the *relative mobility* U_{ri} of each component i is defined by

$$U_{ri} = \frac{U_i}{U_d} = \frac{d_i}{d_d} \qquad (5.50)$$

Table 5.6   Some Media for Zonal Electrophoresis

Medium	Conditions	Principal Uses
Paper	Filter paper moistened with buffer, placed between electrodes	Small molecules: amino acids or nucleotides
Polyacrylamide gel	Cast slabs; cross-linked	Proteins and nucleic acids
Agarose gel	As polyacrylamide, but no cross-linking	Very large proteins, nucleic acids, nucleoproteins, etc.

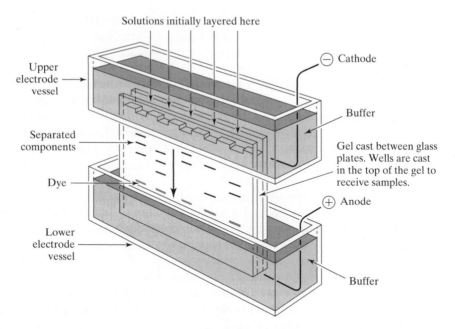

Figure 5.28 A typical slab gel electrophoresis apparatus. This apparatus is designed for small (8.3 cm × 10.3 cm) gel slabs, which can be 0.5 mm or 1 mm thick. Multiple samples are run on each gel; each is initially placed in a small "well" cast in the top of the gel.

where U_d is the dye mobility and d_i and d_d are the distances that component i and dye, respectively, have moved by the conclusion of the experiment. Because all lanes in a slab gel are subjected to the same electrical field, comparison of mobilities can be easy and accurate.

It must be emphasized that the mobility exhibited by a macromolecule in gel electrophoresis is generally different from the mobility that would be found for the same substance in the same buffer in free electrophoresis. This is because a gel will almost always act as a molecular sieve. A macromolecule cannot move as easily through a network as it can when free in solution. In fact, a very simple relationship has been discovered between relative mobility and gel concentration, as shown in Figure 5.29. Such graphs are called *Ferguson plots,* after their developer, H. A. Ferguson. They all obey the general equation

$$\log U_{ri} = \log U_{ri}^0 - k_i C \tag{5.51}$$

where C is the gel concentration and U_{ri}^0 is the relative mobility of component i when $C = 0$, that is, the relative mobility in free electrophoresis. The constant k_i depends on the molecular size of component i; large molecules will have large values of k, whereas very small molecules will have small values, and hence will behave almost the same way in a gel as they do in free electrophoresis. This is readily understandable, because to a small molecule the highly swollen gel matrix presents a resistance little different from that of the buffer solution.

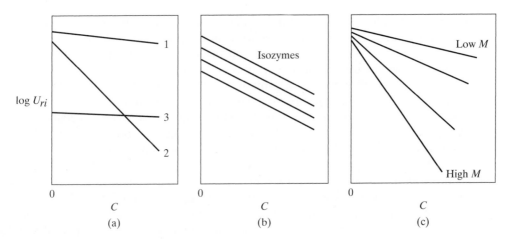

Figure 5.29   Ferguson plots for a number of commonly encountered situations. See the text.

With these few general concepts, we can explain much of what is observed in gel electrophoresis of proteins and nucleic acids. First, consider a miscellaneous group of proteins that are of various sizes and have various charges; their behavior will be as shown in Figure 5.29a. Some are small but highly charged. These will have a large U_{ri}^0 but a small k_i as shown by line 1. Others may be large, and also highly charged (line 2), or small, and have a small charge (line 3). Obviously, the mobilities of these proteins, relative to one another, will depend greatly on the gel concentration. In fact, even the order in which they migrate on the gel can be changed by changing C, because the Ferguson plots in Figure 5.29a cross one another. Interpretation is obviously difficult.

There are, however, some special cases in which interpretation is simplified. Consider, for example, the set of Ferguson plots shown in Figure 5.29b. This is what we would expect for a set of *isozymes,* protein molecules that have essentially the same mass and size but that differ in net charge. The Ferguson plots will be parallel lines. Such behavior is a good test for isozymes.

5.4.2   Electrophoresis of Nucleic Acids

Separation of nucleic acids by molecular weight.   An even more important special case of gel mobility behavior is that depicted in Figure 5.29c, which clearly corresponds to a set of molecules that have approximately the same mobility in free electrophoresis but differ in size. How is this possible? One type of molecule that could approximate this kind of behavior is a highly extended coil like DNA, with a charge proportional to its length. For such a *freely draining coil* the frictional coefficient is also nearly proportional to its length. It is a direct consequence of Eq. 5.48 that the mobility in free electrophoresis will be essentially independent of molecular weight for a set of large molecules of this kind. If the mobility in free electrophoresis U_{ri}^0 (in Eq. 5.51) is the same for all molecules in a mixture, separation on the gel will depend *entirely* on the molecular sieving; therefore, the molecules will be separated primarily on the basis of their molecular sizes. This principle is widely used in the electrophoretic analysis of nucleic acids. Figure 5.30 illustrates the electrophoresis of a mixture of DNA fragments produced by the restriction endonuclease digestion of a bacterial plasmid. Note the remarkably high resolution that is obtained. A graph of the logarithm of the DNA molecular weight (expressed here in base pairs) versus relative mobility is shown in Figure 5.31. For a given gel concentration, usually there will be some molecular weight range in which $\log M_i$ and U_{ri} are approximately linearly related. A series of fragments, such as those shown in Figure 5.31 for which the molecular weights are exactly known from DNA sequencing, serves as an excellent calibration set for the estimation of the molecular weights of unknown DNA molecules. An advantage of slab gels for such an analysis is that one or more calibration lanes can be run in parallel with lanes containing the samples to be analyzed.

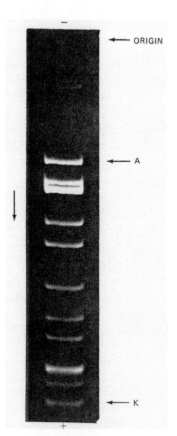

Figure 5.30 Gel electrophoresis of defined fragments of DNA produced by the action of the restriction endonuclease CfoI on the bacterial plasmid pBR 322.

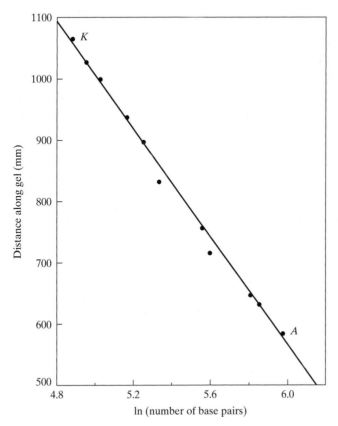

Figure 5.31 A graph of $\log M_i$ versus mobility for the fragments shown in Figure 5.30. The molecular weight, M, is expressed in base pairs; it is precisely known for each fragment. It will be noted that two fragments migrate anomalously. This is not unusual; there *are* sequence effects on mobilities, often associated with bending of specific sequences.

Pulsed field electrophoresis. Although the range of DNA molecular sizes that can be analyzed by the methods described above can be extended up to about 100 Kbp by using very low-concentration agarose gels, separation of still larger molecules requires other methods. This is because very large DNA molecules tend to become hung up in the gel network and are unable to migrate. Several ingenious methods have been devised to circumvent this problem; most operate on the principle that a periodic, brief reversal or change in direction of the field will allow the DNA molecules to unhook from their entanglement and resume migration (see Figure 5.32). Using methods of this kind, it has been possible to separate DNA molecules up to 10,000 Kbp in length; for example, the DNA molecules from whole yeast chromosomes can be resolved by this method. Pulsed-field electrophoresis has been immensely valuable in analyses of entire genomes of a number of organisms.

Effects of DNA tertiary structure on electrophoretic mobility. Since the migration of DNA in a gel depends primarily on molecular dimensions, it is possible

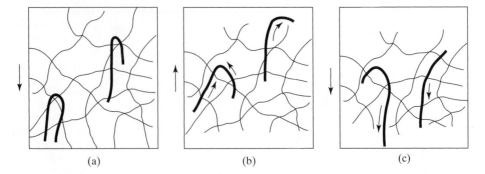

(a)   (b)   (c)

Figure 5.32 The principle of pulsed field electrophoresis. (a) The molecules are electrophoresing but becoming entangled in the gel; (b) the field is briefly reversed, allowing untangling; (c) the normal direction of electrophoresis is resumed. This cycle is repeated over and over.

to use electrophoresis to measure DNA bending and supercoiling. It has been found, for example, that a bent, short piece of DNA will migrate more slowly than a linear, rodlike DNA of the same overall length, and protocols exist to determine both the position and angle of the bend from electrophoresis experiments (see Application 5.2). In a similar way, circular DNA molecules migrate differently from linear isomers, and supercoiling, which compacts the overall structure, leads to more rapid migration. As shown in Figure 5.33, it is possible to separate individual topoisomers by this technique.

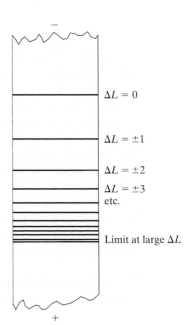

$\Delta L = 0$

$\Delta L = \pm 1$

$\Delta L = \pm 2$

$\Delta L = \pm 3$
etc.

Limit at large ΔL

Figure 5.33 Separation of topoisomers by one-dimensional electrophoresis. Positive and negative supercoils containing the same absolute number of superhelical turns will move at the same rate.

Application 5.2   Locating Bends in DNA by Gel Electrophoresis

Gel electrophoresis of sets of DNA restriction fragments, like that shown in Figure 5.30, often contain particular sequences that migrate anomalously slowly. In the early 1980s, a number of lines of evidence indicated that these might correspond to sequences containing intrinsic bends (Marini et al. 1982). But if such bends occurred at specific points, how could they be localized?

An ingenious technique devised by Drs. H.-M. Wu and D. Crothers (1984) allows this to be done with precision through simple gel electrophoresis measurements. The principle of the method is indicated in Figure A5.2a; the DNA region suspected of containing a bend is cloned in two (or more) tandem copies into some convenient vector of known sequence. Cleavage with different restriction nucleases (each of which must cut at only one site in the repeat) will give a series of fragments that are of identical length but which bear the bend site in different positions. Bending does produce retardation in mobility, and those fragments bent near the middle will be retarded the most—those with the bend near an end are "almost like" straight DNA. Gel electrophoresis of the set of fragments produces a pattern like that shown in Figure A5.2b. Plotting the mobility versus the position of restriction cutting, as shown, allows prediction of the point of greatest mobility. Cutting at this point must correspond to cutting at the bend itself.

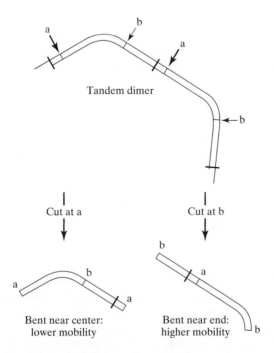

Figure A5.2a   Principle of the method. If the tandem repeat is cut at site a, a curved DNA will result, whereas cutting at site b puts the curve near an end. The fragment obtained from cut a will migrate more slowly.

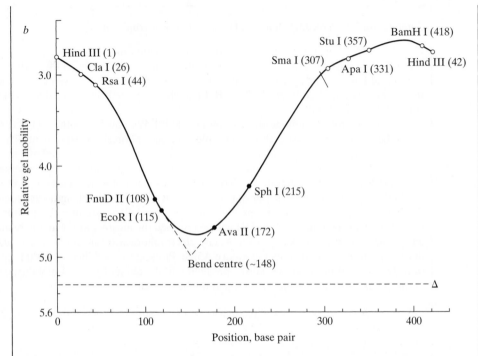

b

Figure A5.2b   Application of the technique to kinetoplast DNA. A series of restriction sites is used to produce differently permuted fragments, all of the same length. Plotting their relative mobilities versus cutting position locates the site of bend. [From Wu and Crothers (1984), *Nature*, **308**, 509–513, with permission from *Nature*.]

In the example shown, the bend is located in a particular region of trypanosome kinetoplast DNA. This sequence contains a striking series of equally spaced $(A/T)_5$ tracts, which account for the bending of the DNA.

The technique can also be used to detect bending that is not inherent in the DNA sequence but is produced by binding of proteins to particular sites (Wu and Crothers 1984).

MARINI, J., S. LEVENE, D. M. CROTHERS, and P. T. ENGLUND (1982), *Proc. Natl. Acad. Sci. USA* **79**, 7664–7668.

WU, H.-M. and CROTHERS, D. M. (1984), *Nature* **308**, 509–513.

However, in one-dimensional gels like that shown in Figure 5.33, positive and negative topoisomers with the same degree of supercoiling are not separated, for they are equally compact. To accomplish their separation, *two-dimensional gels* of the kind shown in Figure 5.34 are employed. Here, the topoisomer mixture has been electrophoresed in one direction in the same manner as in Figure 5.33. After electrophoresis, the gel is soaked in an intercalating dye such as chloroquine, and reelectrophoresed in a direction perpendicular to the first. The dye unwinds DNA as it intercalates, which results in the generation of positive supercoils in compensation.

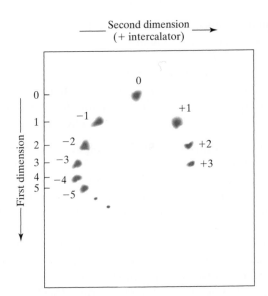

Second dimension
(+ intercalator) →

First dimension →

Figure 5.34   Resolution of positive and negative superhelical DNA by two-dimensional electrophoresis. See text.

Thus, topoisomers that were originally negatively coiled become *less* supercoiled, while those that were positively coiled become *more* supercoiled, and pairs of topoisomers of the same absolute level of supercoiling are separated. Even more subtle features of supercoiled DNA can be revealed using two-dimensional gels, as shown in Chapter 1.

5.4.3   SDS-Gel Electrophoresis of Proteins

General principles.   The behavior shown in Figure 5.29c also serves as the basis for *SDS-gel electrophoresis,* a very widely used technique for the estimation of the molecular weights of polypeptide chains. In this method, the protein to be studied is first heated in a dilute solution of a detergent such as sodium dodecyl sulfate (SDS). This breaks down all native quaternary, tertiary, and secondary structures in the protein. Usually, a reducing agent such as β-mercaptoethanol is also added to reduce any disulfide bonds. The protein is then electrophoresed in the presence of SDS. It is found that the separation proceeds on the basis of polypeptide-chain weight, and is nearly independent of the charge on the polypeptide.

Although this remarkable behavior is still not fully understood, a key to it can be found in two observations. First, SDS at a given solution concentration binds to many different proteins at a constant *weight-weight* ratio. This means that there is a defined number of bound SDS charges per amino acid residue; therefore, the charge contributed by SDS (which will generally greatly outweigh the intrinsic protein charge) is proportional to protein molecular weight. Furthermore, it has been demonstrated by hydrodynamic and neutron-scattering studies that the complexes between SDS and proteins are extended structures. Thus, we once again have conditions comparable to those in DNA electrophoresis—elongated molecules with both charge and friction proportional to molecular weight. As in that case, this predicts that the free mobility will be essentially independent of molecular weight, and that separation will be by the effect of sieving.

As in the case of DNA, there exists, for each particular gel type and concentration, an approximately linear relationship between the logarithm of the protein molecular weight and mobility (see Figure 5.35). Such a calibration curve must be established in every experiment using a series of molecules of known size.

Analysis of multisubunit structures by SDS gel electrophoresis.   It is important to note that SDS not only unfolds polypeptide chains, but also dissociates multichain proteins into their individual subunits. Thus, it can never tell us the "native" molecular weight of such a protein. Indeed, if the protein is composed of *n* identical subunits, its multisubunit nature will not even be detected by SDS gel electrophoresis alone. However, this technique can reveal the different kinds of subunits that make up the quaternary structure of a complex protein, and indicate their relative amounts. To obtain the exact stoichiometry, however, an independent measure of the molecular weight of the intact, multisubunit protein is required. This can be obtained from other techniques, such as sedimentation equilibrium (Section 5.3.3) or light scattering (Chapter 7). To take a simple example: If we were to find a protein to have a native molecular weight of 100 KDa, and SDS-gel electrophoresis reveals subunits of 20 KDa and 60 KDa, we should feel confident in asserting that the native protein contains one 60 KDa subunit and two 20 KDa subunits. It would help, in such an example, to measure the relative intensities of the bands, remembering that each protein will take up stain in proportion to its weight. Thus, in the case above, we should expect the intensities of the 60 KDa and 20 KDa bands to be in about the ratio of 6 : 4 (since there are *two* of the smaller subunit). Such simple analysis has had at times a

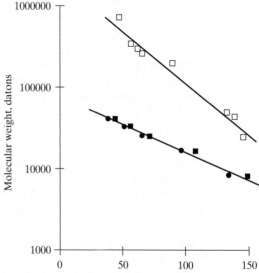

Figure 5.35   Demonstrating that both DNA and proteins give approximately linear graphs of log (molecular weight) versus mobility. The samples shown were run under identical conditions. DNA samples (open squares) are restriction fragments from a bacterial plasmid. Proteins are polymers of the globular domain of histone H5, crosslinked with either glutaraldehyde (squares) or dithiobis (succinimidyl propionate) (circles). [Courtesy of G. Carter.]

major impact on biochemistry. Consider, for example, the histone octamer that forms the core of the nucleosome. When nucleosomes were first discovered, it was found that each contained four small histone proteins (H2A, H2B, H3, and H4) in approximately equal concentrations (see Figure 5.36). A measurement of the total mass of protein in the nucleosome indicated that there must be *two* of each—an octamer of histones.

There is one circumstance under which SDS gel electrophoresis will not lead to complete dissociation into subunits. This occurs if some of the subunits are held together by disulfide bonds. To allow complete dissociation, SDS gel electrophoresis is often carried out in the presence of a reducing agent such as β-mercaptoethanol or dithiothreitol that cleaves disulfide bonds. Indeed it is wise, in working with an unknown protein, to conduct parallel electrophoresis experiments with and without a reducing agent. To use our hypothetical example again, suppose we found in

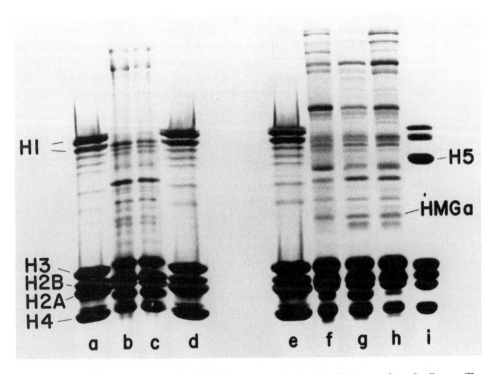

Figure 5.36 SDS-gel electrophoresis of acid-soluble nuclear proteins from a number of cell types. The major proteins present are the nucleosomal core histones, H3, H2B, H2A, and H4, plus the larger lysine-rich histones, H1 and H5. There is also a large number of minor nonhistone proteins. The gel has been somewhat overloaded with the histones in order to reveal these minor proteins. Lanes *a*, *d*, and *e* are from calf thymus; lanes *b*, *c*, *f*, *g*, and *h* are from various yeast strains. Lane *i* is from chicken erythrocytes. The samples were run on a 15% polyacrylamide slab gel, on an apparatus similar to that shown in Fig. 5.28. Migration is from the top toward the anode (*bottom*). The gel was stained with the dye Coomassie blue and photographed. [Courtesy of Dr. J. Davie.]

the presence of β-mercaptoethanol subunits of 30 KDa and 20 KDa, instead of the 60 KDa and 20 KDa found before. We could infer from this that the 60 KDa unit was in fact a disulfide-linked dimer. A great deal of information can be gleaned from a simple technique.

While SDS-gel electrophoresis is a quick and inexpensive way to estimate chain weight, it should be used with caution, for a number of potential pitfalls can lead to serious error. Some proteins, for example, bind more or less SDS than the average amount, which will make them run anomolously on the gel. If the protein itself has a very large positive or negative charge, this charge may not be negligible compared to the charge produced by the bound SDS. In extreme cases, this can lead to errors as large as 50% in the estimate of molecular weight. To take a specific example, histone H1 (Figure 5.36) has a molecular weight of about 21,000, but it behaves in gel electrophoresis like a molecule of 30,000 Da, because of its strong positive charge.

Stacking gels and discontinuous buffer systems.   The high resolution seen in gels such as the one illustrated in Figure 5.36 is obtained by the use of a *stacking gel* and a discontinuous buffer system. To understand what these terms mean, consider the arrangement shown in Figure 5.37. On top of the *running gel* in the tube or slab there is placed a stacking gel of lower concentration and thus of higher porosity. The sample is layered on top of this. Note that quite different buffers

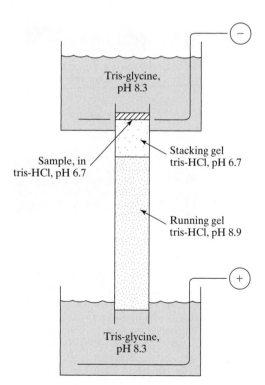

Figure 5.37 The principle of stacking in gel electrophoresis. (See the text.)

are used in the upper reservoir, in the stacking gel, and in the running gel. A critical feature for the operation of the system shown in this figure is that the amino proton of glycine is only feebly dissociated below pH 9; the equilibrium

$$^+H_3NCH_2COO^- \Leftrightarrow H_2NCH_2COO^- + H^+$$

lies well to the left, toward the electrically neutral zwitterion form. If a fraction α of the zwitterion has dissociated, glycine molecules can be thought of as each spending a fraction of time α in the anionic form and the observed mobility will be αU_{G^-}, where U_{G^-} is the mobility of anionic glycine. At pH values below 9, the mobility therefore becomes very low.

The efficiency of the system in producing sharp bands depends on the fact that the mobility of glycine is less than the mobility of chloride ion. When the voltage is applied, and ions begin to move toward the anode, the Cl^- ions will begin to move faster than the glycine. But any tendency for a gap to appear in the ion concentration will lead to a drop in conductivity in the boundary region. However, the current must be the same at every point in the gel column; therefore, the voltage gradient will increase across this boundary, and the slower glycine ions will be accelerated and catch up with the boundary. If only chloride and glycine are present, a sharp glycine-chloride boundary will move down the gel. The sharpness of this boundary is maintained, since as glycine falls back it is accelerated by the increased potential gradient and if any glycine anion were to penetrate the chloride-containing solution, it would be decelerated by the much weaker potential gradient existing in that medium of higher conductivity.

In the system as depicted, negatively charged protein molecules will have mobilities in the stacking buffer intermediate between those of chloride and glycine. This is the kind of system that would be appropriate for SDS-protein complexes. Consequently, the proteins will be concentrated into a series of very thin *stacks* between the glycine and chloride. Each will be placed according to its relative mobility. Since the stacking gel is dilute, very little molecular sieving occurs here.

When the boundary passes from the stacking gel into the running gel, two important changes occur. First, the higher pH in the running gel favors the formation of more gly$^-$ ion; therefore, the effective mobility of glycine (relative to proteins) increases. Second, the higher gel concentration in the running gel begins to impede the protein molecules. Consequently, glycine overpasses the proteins, and a glycine-chloride boundary moves ahead. The proteins are now electrophoresing in a uniform glycine buffer, and are being separated by the molecular sieving effect. Each protein stack enters the running gel as a very thin, concentrated layer, even though the proteins may have been present in a more diffuse band when first applied to the gel.

The advantages of such a system are obvious. Rather sizable protein loads can be applied, and yet high resolution is possible because of the stacking effect. We have examined one particular system for the sake of clarity, but there are many buffer systems that can be used for different requirements. The discontinuous-buffer stacking systems can be used with electrophoresis of proteins under nondenaturing conditions or in the presence of SDS for molecular size analysis.

In fact, the present range of gel electrophoresis techniques for protein studies is truly enormous, with many special variants (use of urea, nonionic detergents, and so forth) for special applications. The reader is referred to the excellent manual by Hames and Rickwood (1981) or the compact handbook by Patel (1994) for details. One major advance should be noted in particular. It has become possible to improve resolution in many cases by the use of *two-dimensional gel* techniques. In such methods, one lane of a slab gel is run under one set of conditions. The excised lane is then placed at the top of a gel slab and electrophoresis in the second dimension takes place under different conditions. The number of possible combinations is virtually limitless.

5.4.4   Methods for Detecting and Analyzing Components on Gels

So far, we have said little about the problem of locating and quantifying bands on gels. There are *many* possible techniques, a feature that adds to the power and versatility of gel electrophoresis. Most techniques can be classed in one of four categories:

Staining.   The gel is soaked in a solution of a dye or fluorescent material that binds to the substances to be measured. For nucleic acids, ethidium bromide is most often used; it yields a brilliant fluorescence on binding. Proteins are most commonly stained with the dye Coomassie blue. The stained gels are washed of excess staining reagent, and usually are photographed. The amounts of various components can then be measured (approximately) by densitometer scanning of the negative. To be reliable, this requires that the stain react equally with all components and exhibit a linear concentration-intensity relationship. Greater sensitivity is gained with the use of silver salts, which deposit metallic silver at the protein bands. However, silver staining is prone to nonlinearity.

Autoradiography.   If the material being analyzed is radioactive, the gel can simply be placed in contact with a photographic film, which will be blackened in those areas corresponding to bands. Again, if care is taken to ensure the linearity of the film response, quantification is possible by scanning the film.

Direct radioactivity scanning.   Devices like *phosphoimagers* are coming into more and more use; they allow direct recording of radioactivity from the gel.

Blotting.   The use of radioactivity has permitted the development of some very elegant methods for the detection of specific substances in bands on slab gels. The Southern blotting technique, named for its inventor, E. Southern, is an example (see Figure 5.38). Suppose that we wish to determine whether a particular DNA sequence is present in any of the bands that have been resolved on an agarose gel of a DNA mixture. The DNA is denatured in the gel with NaOH; the gel is then laid on a nitrocellulose sheet that is underlain with several sheets of filter paper. The single-stranded DNA molecules are then either squeezed (by pressure) or electrophoresed onto the nitrocellulose. Since lateral diffusion of the DNA is very slow, a replica of the gel pattern is created on

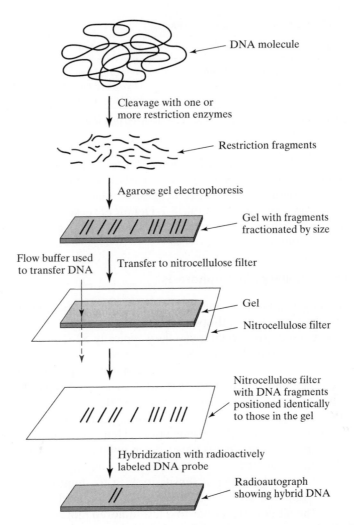

DNA molecule

Cleavage with one or
more restriction enzymes

Restriction fragments

Agarose gel electrophoresis

Gel with fragments
fractionated by size

Flow buffer used
to transfer DNA

Transfer to nitrocellulose filter

Gel

Nitrocellulose filter

Nitrocellulose filter
with DNA fragments
positioned identically
to those in the gel

Hybridization with radioactively
labeled DNA probe

Radioautograph
showing hybrid DNA

Figure 5.38   The principle of Southern blotting of DNA fragments. [From Watson, *Molecular Biology of the Gene,* 4th ed. Copyright 1987 by J. D. Watson, reprinted by permission of Benjamin-Cummings Publishers.]

the nitrocellulose. The sheet is then baked to fix the single-stranded DNA molecules in place. It is then treated with a solution containing a radioactive *probe,* single-stranded DNA molecules that are the complement of the sequence to be sought. Conditions are such that these can hybridize to any partner strands that may be affixed to the nitrocellulose. The sheet is then washed to remove unhybridized probe DNA, and subjected to autoradiography or scanning. Only those bands on the blot containing DNA fragments complementary to the probe DNA will be radioactive and expose the film. This technique has proved enormously powerful in locating particular sequences in enzymatic digests of DNA.

Table 5.7   Electrophoretic Transfer Techniques

Method	Material Transferred	Transferred to	Probe
Southern	Denatured DNA	Nitrocellulose	Radioactive DNA or RNA
Northern	RNA	Diazotized paper[a]	Radioactive DNA or RNA
Western	Proteins	Diazotized paper[a]	Radioactive antibody

[a]Paper to which a diazo compound has been coupled that will covalently bind to the RNA or protein.

Two other techniques, facetiously termed *Northern* and *Western* blotting, have been developed to probe for particular RNA sequences or proteins, respectively, in electrophorograms. The basic principle is similar to that used in Southern blots, and some features are given in Table 5.7. Details of these methods and references to the literature are given in Rickwood and Hames (1982), Hames and Rickwood (1981), and Patel (1994).

5.4.5   Capillary Electrophoresis

In recent years, a new technique has emerged that possesses certain advantages over gel electrophoresis in both speed and sensitivity. In *capillary electrophoresis,* separation occurs in a fused quartz capillary, typically $50–100 \, \mu m$ in diameter and 20–60 cm long. Since the electrical resistance in such a small cross section will be large, quite high voltage gradients can be applied without passing a large current with consequent heating. This high voltage allows for rapid separation—often in minutes, rather than in the hours required with gels. Because the fine capillary inhibits convective mixing, it is not necessary to use a gel matrix, although such is sometimes added to obtain molecular sieving effects. Detection is usually by UV absorbance or fluorescence, as components pass a given point in the capillary. Because of the very small sample volumes used (in the nanoliter range), *extremely* small amounts of material can be analyzed. Mass sensitivity is in the femtomole (10^{-15} mol) scale with UV detection and down to the zeptomole (10^{-21} mol) scale with fluorescence detection. The latter corresponds to a few hundred molecules! The sharpness and rapidity of separation that can be obtained is indicated by Figure 5.39. Of course, like all techniques, capillary electrophoresis has its limitations. The presence of charged groups on the walls of the capillary produces an electroosmotic flow of solvent at most pH values. More seriously, some proteins may be absorbed or even denatured by interaction with the surface (the surface area/volume ratio is very high). Such effects can be alleviated by coating the capillary with a polymeric material; linear polyacrylamide is often used.

5.4.6   Isoelectric Focusing

We have noted that every polyampholyte, such as a protein, will have one pH at which its electrophoretic mobility is exactly zero, the isoelectric point (pI). This fact is used to separate macromolecules in the technique called *isoelectric focusing.*

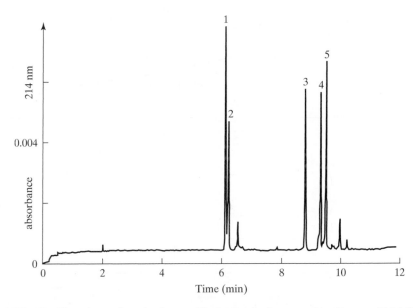

Figure 5.39 Capillary electrophoresis of a set of basic proteins in a coated capillary at pH 4.5. Proteins are 1—lysozyme, 2—cytochrome c, 3—myoglobin, 4—trypsinogen, 5—α-chymotrypsinogen. [From Schmalzing et al. (1993), *J. Chromatog. A.,* **652**, 149–159, with kind permission of Elsevier Science, NL Burgerhartsraat 25, 1055 Amsterdam, The Netherlands.]

Suppose we were able to create, in a gel or in a capillary, a stable pH gradient between the anode and cathode. Molecules would migrate until each reached the point at which the pH equaled its isoelectric point (see Figure 5.40). Bands would form at different points in the pH gradient and would remain fairly sharp, for any macromolecule that diffused away from its band position would no longer be isoelectric, and would experience a force pulling it back.

How can such a pH gradient be established and maintained? There is always a tendency for the cathode compartment to become basic, through the reaction

$$2H^+ + 2\,e^- \longrightarrow H_2$$

Similarly, the anode compartment tends to become acidic, through reactions like

$$2H_2O \longrightarrow 4H^+ + 4e^- + O_2$$

Normally, the pH changes resulting from these reactions are confined to regions near the electrodes, even in weakly buffered solutions. However, if the gel or capillary contains a mixture of low-molecular-weight ampholytes [molecules such as $^+H_3N(CH_2)_nCOO^-$, for example], these will tend to partially sort themselves out in the gradient; those with lower pI concentrating near the anode, those with higher pI near the cathode. Their buffering capacity will then stabilize the pH gradient, allowing isoelectric focusing of macroions. Of course, this is not a true equilibrium, for electrolysis is continually proceeding, but the gradient and the bands can be stable

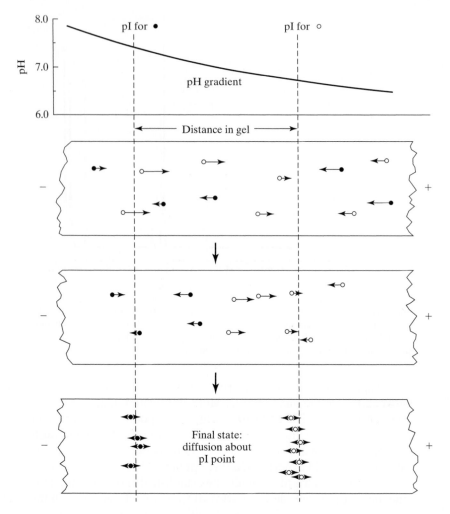

Figure 5.40   The principle of isoelectric focusing. If a stable pH gradient exists in a gel or capillary, each protein molecule will migrate toward the position where it is isoelectric. Here, two kinds of protein molecules (open and filled circles), initially uniformly distributed in the system, migrate toward their respective pI values.

for rather long times. The technique can show remarkable resolution of proteins with closely spaced isoelectric points, as shown in Figure 5.41. The method can be used for either analytical or preparative purposes. Recently, the commercial availability of preprepared gels with immobilized ampholytes has made the technique even simpler and faster.

   A particularly discriminating use of isoelectric focusing is in the *O'Farrell* technique, which utilizes two-dimensional slab gels. The proteins are separated by

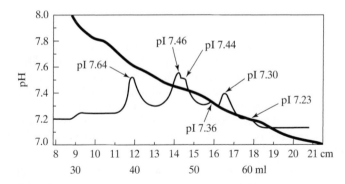

Figure 5.41   Separation of hemoglobin components by isoelectric focusing. Note that the differences in pI that can be resolved are extremely small.

isoelectric focusing in the first dimension, and are then subjected to SDS-gel electrophoresis in the second dimension. Thus, the x- and y-coordinates of a spot on the two-dimensional gel slab depend on the charge and molecular weight, respectively. An example of such a gel is shown in Figure 5.42.

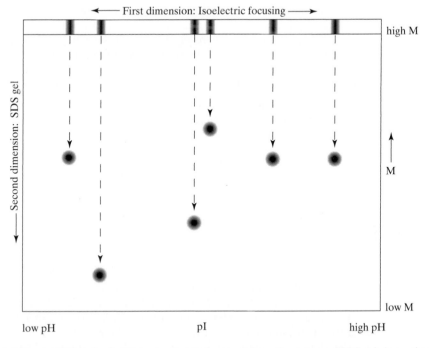

Figure 5.42   An O'Farrell gel, utilizing isoelectric focusing in one dimension and SDS-gel electrophoresis in the second dimension.

EXERCISES

Note: Because sedimentation, diffusion, and electrophoresis are techniques developed decades ago, calculations involving them have been traditionally carried out in cgs units rather than the more modern SI system. It would only cause confusion in interpreting the literature to change at this point, so we shall continue to use cgs. In some of the following problems, you will use the *viscosity* (η) of solutions. The cgs units of viscosity are *poise*, which have units of gm·cm/sec. Viscosity data useful in several problems are found in Table 5.2, expressed in *centipoise* (1 cp $= 10^{-2}$ poise). The gas constant, R, is 8.314×10^7 erg/deg.mol in cgs units.

5.1. The diffusion spreading of a narrow band or zone increases with the *square root* of time. Demonstrate this from Eq. 5.6. [Hint: First obtain an expression for C/C_{max}, where C_{max} is the concentration at $x = 0$. Then solve for $x_{1/2}$, the *half width* of the band.]

5.2. From the result in Exercise 5.1, calculate the half-width for initially narrow bands of sucrose ($D_{20,w} = 4.59 \times 10^{-6}$ cm^2/sec), serum albumin ($D_{20,w} = 5.94 \times 10^{-7}$ cm^2/sec) and Bushy stunt virus ($D_{20,w} = 1.15 \times 10^{-7}$ cm^2/sec) after one hour of diffusion in water at 20°C.

5.3. What is the largest possible value for the diffusion coefficient that a molecule of $M = 50{,}000$ Da can have at 20°C? (You may assume $\eta = 1 \times 10^{-2}$ poise, and $\bar{v} = 0.73$ cm^3/gm.)

5.4. Consider a small spherical particle of mass m, initially at rest, which is acted on by a constant force F turned on at $t = 0$. Solve the equation of motion for this particle, which must be $F - fv = m\, dv/dt$, where v is the velocity. Show that it approaches a constant maximum velocity $v_{max} = F/f$. If the particle is spherical, of radius 100 Å, and of density 1.5 gm/cm^3, calculate the time required to attain 99% of the final velocity. You may assume the solvent medium to be water, with $\rho = 1.00$ gm/cm^3, $\eta = 1 \times 10^{-2}$ poise.

5.5. A sedimentation velocity experiment was performed with subunits of aldolase at 3°C, with a rotor speed of 51,970 rpm. Data for the boundary position versus time are given below. The time of the first scan has been taken as zero.

r_b (cm)	$t - t_0$ (sec)
6.3039	0
6.4061	3,840
6.4920	7,680
6.5898	11,520
6.6777	15,360

  a. Calculate the sedimentation coefficient. Correct to $s_{20,w}$, assuming that $\bar{v} = 0.73$ at 3°C and 0.74 at 20°C. The viscosity and density of the buffer, relative to water are 1.016 and 1.003, independent of temperature. Other data are in Table 5.2.
  b. Estimate the minimum molecular weight for this protein.
  c. Calculate f/f_0 if the protein has an actual $M = 52{,}000$ Da. What does the result suggest about aldolase subunits under these conditions?

*5.6. A spherical protein of diameter R dimerizes and rearranges to form a prolate ellipsoid with long axis equal to $2R$ and volume equal to the sum of the original monomer volumes. What will be the percent increase or decrease in (a) the diffusion coefficient and (b) the sedimentation coefficient?

*5.7. Using the Kirkwood theory, predict the percent increase or decrease in sedimentation coefficient when a spherical protein dimerizes by point contact between the two spheres. Compare it with the result in Exercise 5.6. Which do you think is the more realistic model?

5.8. Data are given below for two proteins.

Protein	$s^{\circ}_{20,w} \times 10^{13}$	$D^{\circ}_{20,w} \times 10^{7}$	$\bar{v}_{20}$
Concanavalin (jack bean)	6.40	5.10	0.730
Myosin (cod)	6.43	1.10	0.730

a. Calculate M for each.
b. Calculate a Stokes's radius for each.
c. Calculate f/f_0 for each.
d. Assuming that each is a sphere but hydrated enough to account for f/f_0, calculate the required hydration.
e. Assuming that each is a prolate ellipsoid, hydrated to an extent of 0.2 cc H_2O/cc protein, estimate α/β for each.

5.9. The data given below describe the variation of $s^{\circ}_{20,w}$ and $D^{\circ}_{20,w}$ for a protein as a function of pH. Explain what happens to the protein at low and high pH, assuming that it is in the native state between pH 5 and pH 6.

pH	$s^{\circ}_{20,w} \times 10^{13}$	$D^{\circ}_{20,w} \times 10^{7}$
2	2.93	7.91
3	3.02	8.00
4	3.89	8.00
5	4.41	5.90
6	4.40	5.92
7	4.15	5.61
8	3.60	4.86
9	2.25	3.08
10	2.20	2.97

5.10. a. A protein with $\bar{v} = 0.72$ cm³/g is studied by sucrose gradient centrifugation at 5°C. If the gradient runs from 10 to 30% sucrose, by what percent will the sedimentation coefficient decrease as the protein proceeds from the meniscus to the bottom of the tube? The following data for sucrose solution at 5°C are necessary.

Percent Sucrose	ρ (g/cm^3)	η (centipoise)
10	1.0406	2.073
30	1.1315	4.422

b. If the meniscus is 8.0 cm, and the bottom of the tube 16.0 cm from the center of rotation, by what percent will the *velocity* of sedimentation change in traversing the tube?

5.11. Show how you might use sucrose gradient centrifugation to demonstrate that protein synthesis takes place on polyribosomes (rather than on monoribosomes) and that these polyribosomes contain mRNA. You may assume you have access to any radiolabeled compounds that are needed. Describe the experiments in some detail.

5.12. A virus is studied by sedimentation equilibrium. The data are

$$\text{rotor speed} = 800 \text{ rpm (remember, } \omega \text{ must be in rad/sec)}$$
$$T = 4°C$$
$$\bar{v} = 0.65 \text{ cm}^3/\text{gm}$$
$$\rho = 1.002 \text{ gm/cm}^3$$
$$r_b = 7.30 \text{ cm (cell bottom)}$$
$$r_a = 7.00 \text{ cm (meniscus)}$$

At equilibrium, a graph of ln C versus r^2 is found to be linear, and $C_b/C_a = 8.53$. What is the molecular weight of the virus?

5.13. This problem pertains to the proteins described in Exercise 5.8. You will need some of the results from that exercise.
   a. Suppose each protein were studied by sedimentation equilibrium. Calculate the rotor speed (rpm) required to give a 3:1 ratio in concentration between the ends of the solution column. You may assume that the column is 3 mm high, with $r_a = 6.70$ cm, $r_b = 7.00$ cm, and $T = 20°C$.
   b. The time to attain sedimentation equilibrium (within 0.1%) is given approximately by

$$t_e = \frac{0.7(r_b - r_a)^2}{D}$$

where D is the diffusion constant and all quantities are in cgs units. Calculate the time in hours for each protein.

*5.14. To get an idea of the effect of heterogeneity on sedimentation equilibrium, calculate the theoretical ln C_t versus r^2 curve for a mixture of equal weights of the virus discussed in Exercise 5.12, and another virus half as large, with the same $\bar{v}$. You may assume, as a good approximation, that each kind of particle will be at concentration C_0 at the point $[(r_b^2 + r_a^2)/2]^{1/2}$ where r_b and r_a are the bottom and meniscus, respectively. Estimate the apparent molecular weight at the meniscus and bottom.

5.15. The first sedimentation equilibrium experiment ever performed was by Svedberg and Fåheus in 1926. In their study of horse hemoglobin, they obtained the following data.

r (cm)	C (arbitrary units)
4.16	0.388
4.21	0.457
4.26	0.486
4.31	0.564
4.36	0.639
4.41	0.732
4.46	0.832
4.51	0.930
4.56	1.061
4.61	1.220

Other data: $\bar{v} = 0.749$, $\rho = 1.00$ gm/cm^3, $T = 293.3°$K, $\omega = 9.12 \times 10^2$ rad/sec.

a. Calculate M; compare with sequence value 64,715 gm/mol.

b. Is hemoglobin homogeneous in molecular weight? (In 1926, whether proteins were homogeneous or not was a major question.)

*5.16. The bouyant densities of *E. coli* DNA and salmon sperm DNA in CsCl are 1.712 and 1.705 g/cm^3, respectively. Suppose that a mixture of these two DNAs, each of $M = 2 \times 10^6$ g/mol, is studied by density-gradient sedimentation equilibrium. The rotor speed is 45,000 rpm, $T = 25°$C, and the density gradient in the region of banding (7 cm from the center of rotation) is 0.1 g/cm^4. Compare the peak separation to the band width (as measured by the standard deviation σ) under these conditions. Is resolution possible? Would the resolution be improved by dropping the rotor speed to 30,000 rpm? You may assume that the density gradient is proportional to ω^2.

5.17. An invertebrate hemoglobin is found, under native conditions, to have a sedimentation coefficient of about 4.4 S and a diffusion coefficient of about 6×10^{-7} cm^2/sec (both at 20°C, H$_2$O). The parameter $\bar{v}$ is estimated to be 0.73 cm^3/gm. The following data are found from SDS-gel electrophoresis:

- After treatment with β-mercaptoethanol, the protein migrates as a doublet. The bands have traveled 10.0 and 10.6 cm.

- In the absence of β-mercaptoethanol treatment, only the 10.0 cm band of the doublet is seen, but there is a new band at 5.6 cm.

- A series of standard proteins on the same gel migrates as shown below.

Protein	M	d (cm)
Phosphorylase b	94,000	0.5
Bovine albumin	67,000	1.1
Ovalbumin	43,000	3.9
Carbonic anhydrase	30,000	6.6
Trypsin inhibitor	20,100	9.3
α-Lactalbumin	14,400	11.7

Describe the subunit structure of this protein. (Don't expect it to behave like human hemoglobin; invertebrate hemoglobins are often quite different from the mammalian types.)

5.18. Consider two DNA molecules: one 400 base pairs long, the other 800 base pairs long. Assuming that each is a rigid rod of diameter 25 Å, with a length of 3.4 Å per base pair, calculate the ratio of the electrophoretic mobilities of these two molecules using the very approximate Eq. 5.48. Note that the charge will be proportional to the length.

5.19. The digestion of chromatin by a nuclease, such as micrococcal nuclease, tends to cleave the DNA at the junctures between nucleosomes to make a series of oligonucleosomes. To determine the DNA "repeat size," the DNA from such a digestion is electrophoresed on a polyacrylamide slab gel parallel to a lane of restriction fragments of a bacteriophage DNA of known sizes. The following data give the distances migrated on such a gel. Deduce the repeat size of the calf chromatin.

Phage DNA Fragments		Fragments of Calf Nuclear DNA	
Size (base pairs)	***d* (cm)**	**Number**	***d* (cm)**
794	11.5	I	30.5
642	13.6	II	19.2
592	14.6	III	14.4
498	16.7	IV	11.5
322	22.5		
288	24.1		
263	25.4		
160	33.0		
145	34.2		
117	37.2		
94	40.0		

REFERENCES

General

CANTOR, C. R. and P. R. SCHIMMEL (1980) *Biophysical Chemistry,* W. H. Freeman, San Francisco. Chapters 10 and 11 cover particle hydrodynamics in somewhat greater depth and detail than herein.

Diffusion

CRANK, J. (1979) *The Mathematics of Diffusion,* 2nd ed., Oxford University Press, Oxford, U.K. The ultimate source for theory of diffusion.

EDELSTEIN-KESHET, L. (1988) *Mathematical Models in Biology,* Random House, New York. Chapter 10 contains useful applications in biology, including diffusion.

Sedimentation

BLOOMFIELD, V., W. O. DALTON, and K. E. VAN HOLDE (1967) "Frictional Coefficients of Multisubunit Structures. I. Theory," *Biopolymers* **5**, 135–148. Application of the Kirkwood theory to the calculation of frictional coefficients for molecules of various shapes.

Cole, J. L. and J. Hansen (1999) "Analytical Ultracentrifugation as a Contemporary Biomolecular Research Tool," *J. Bio. Mol. Tech.* **10**, 163–176. A good current review.

de La Torre, G. S. Navarro, M. Lopez-Martinez, F. Diaz, and J. Lopez-Cascales (1994) "HYDRA: a computer program for the prediction of hydrodynamic properties of macromolecules," *Biophys. J.* **67**, 530–531.

Fujita, H. (1975) *Foundations of Ultracentrifugal Analysis,* John Wiley & Sons, New York. A most detailed treatise on sedimentation theory.

Lebowitz, J., M. S. Lewis, and P. Shuck, (2002) "Moden Analytical Ultracentrifugation in Protein Science—A Tutorial Review," *Protein Science* **11**, 2067–2079.

Schuster, T. M. and T. M. Laue (eds.) (1994) *Modern Analytical Ultracentrifugation,* Birkhäuser, Boston. A collection of useful papers on techniques and applications.

Sheller, P. (1981) *Centrifugation in Biology and Medical Science,* John Wiley & Sons, New York. Entirely devoted to density gradient sedimentation; many practical hints.

van Holde, K. E. and W. O. Weischet (1975) "Boundary Analysis of Sedimentation—Velocity Experiments with Monodisperse and Paucidisperse Solutes," *Biopolymers* **17**, 1387–1403. A technique to remove diffusion smearing from sedimentation data.

Zamyatnin, A. A. (1984) "Amino Acid, Peptide, and Protein Volume in Solution," *Ann. Rev. Biophys. Bioeng.* **13**, 145–165. A good discussion of $\bar{v}$, its measurement or calculation.

Electrophoresis

Hames, B. D. and D. Rickwood (1981) *Gel Electrophoresis of Proteins,* IRL Press Ltd., London.

Karger, B. L., Y.-H. Chu, and F. Foret (1995) "Capillary Electrophoresis of Proteins and Nucleic Acids," *Ann. Rev. Biophys. Biomol. Struct.* **24**, 579–610. A thorough discussion of techniques and applications of capillary electrophoresis.

Osterman, L. A. (1984) *Methods of Protein and Nucleic Acid Research,* vol. 1, parts I and II, Springer Verlag, Berlin and New York. A summary of electrophoresis and isoelectric focusing.

Patel, D. (1994) *Gel Electrophoresis: Essential Data,* Wiley, Chichester, U.K. A compact little compendium of recipes and useful data for practical applications.

Rickwood, D. and B. D. Hames (1982) *Gel Electrophoresis of Nucleic Acids,* IRL Press Ltd., London.

Schwartz, D. C. and C. R. Cantor (1984) "Separation of Yeast Chromosome-Sized DNAs by Pulsed Field Gradient Gel Electrophoresis," *Cell* **37**, 67–75. An early application of pulsed-field electrophoresis.

6

X-Ray Diffraction

If a picture is worth a thousand words, then a detailed three-dimensional macromolecular structure is priceless to a physical biochemist. There is no better way to understand how macromolecules function in a cell than to have a visual image of their parts and how they interact. Our first encounter with the components of a living cell usually comes from peering through a microscope, where light is either reflected from a surface or transmitted through an object. In both cases, the scattered light is focused by a series of lenses to form an image of the object (Figure 6.1). In this chapter, we discuss methods for seeing the atoms of a macromolecule by X-ray diffraction. This method is analogous in many ways to light microscopy in that X-rays are scattered by the atoms of the macromolecule. However, X-rays cannot be focused by lenses to form an image of a molecule. Instead, the X-rays are scattered from a regular repeating array of molecules (that is, from a single crystal) to give a pattern that represents the macromolecular order and structure. The structure must be reconstructed using mathematics as the lens to *transform* the pattern back into the original structure. Since the crystals are imperfect and information is lost during the transformation, the structure that we get is not a true image, as from microscopy, but a model of the structure (see Chapter 1). Building a model that best fits the experimental data is similar to solving for the solutions to a set of multiple equations. We first discuss the requirements necessary for determining the structure of a macromolecule at the atomic level (to *atomic resolution*). This is followed by a description of what crystals are and how they are grown, and how X-ray diffraction from crystals is used to solve structures of macromolecules to atomic resolution. Finally, we discuss fiber diffraction, a traditional method that uses the repeating symmetry of fibrous biopolymers (Chapter 1) to construct models that represent the average conformation of these biopolymers without the need to crystallize them.

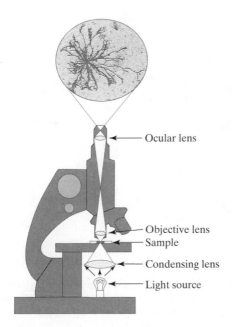

Figure 6.1 The light microscope. A microscope forms an image by focusing the light scattered from a sample, such as a melanophore on a slide, through a series of lenses. The image can be magnified infinitely, but the resolution is limited by the wavelength of visible light to about 0.2 microns. [Courtesy of P. McFadden.]

Ocular lens

Objective lens
Sample
Condensing lens
Light source

6.1   STRUCTURES AT ATOMIC RESOLUTION

Although a number of new, higher-resolution microscopes have been developed (Chapter 16), the only method that currently yields reliable structures of *macromolecules* at atomic resolution is X-ray diffraction from single crystals. This technique requires three distinct steps: (1) growing a crystal, (2) collecting the X-ray diffraction pattern from the crystal, and (3) constructing and refining a structural model to fit the X-ray diffraction pattern. No one step is any more important than the others, and all three steps must be completed to determine the structure of a macromolecule.

A molecular structure *solved to atomic resolution* means that the positions of each atom can be distinguished from those of all other atoms in three-dimensional space, without the need to apply additional assumptions concerning the structure of the molecule. The closest distance between two atoms in space is the length of a covalent bond. Since the typical covalent bond is approximately 0.12 nm, we need to "see" two atoms separated by this distance as distinct particles (Figure 6.2). There are theoretical and practical limitations to resolving a structure to this fine a level. First, the system must have the atoms of its molecules held rigidly and, second, each molecule or group of molecules in the system must have identical conformations. Any fluctuation in the positions of the atoms in the molecule or any significant deviations of molecules from a single conformation would result in an averaging of the structure, which would blur our vision and thereby reduce the resolution of the technique. As we will see later, only a single crystal of a molecule has the potential to satisfy both of these requirements.

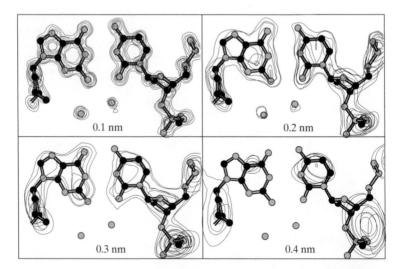

Figure 6.2 Resolving molecules to the atomic level. The information content increases as structures are determined to higher resolution (in this case, lower numbers are better). The 0.1 nm resolution structure of a dG·dC base pair in a crystal of a DNA fragment shows details of each atom in the molecule, as well as the solvent structure surrounding the molecule. At 0.2 nm and 0.3 nm resolution, the structure of the nucleotides are still discernable, but by 0.5 nm resolution, only the presence of the strongly diffracting phosphates of the backbone can be unambiguously distinguished.

We must now find a radiation source that allows us to see two atoms separated by only 0.12 nm. The limit of resolution LR of any optical method is defined by the wavelength λ of the incident radiation.

$$LR \cong \frac{\lambda}{2} \qquad (6.1)$$

This is a consequence of the wave properties of light. An extension of the Heisenberg uncertainty principle that results from treating light as a wave states that the position of a particle cannot be fixed to better than about half the wavelength of the radiation used to examine that particle. The resolution of a light microscope, for example, is restricted by the wavelength of visible light (λ = 400 nm to 800 nm) to about 200 nm, or 0.2 μm (about the size of the organelles in a cell). If we require the resolution of the technique to be about 0.12 nm to resolve the atoms of a macromolecule, the wavelength of light required for our atomic microscope would necessarily be <0.24 nm. This falls into the X-ray range of the electromagnetic spectrum (Figure 6.3). However, as we discussed above, X-rays cannot be focused and thus cannot form an image of an object in the same manner as a light microscope. We rely on the constructive and destructive interference caused by scattering radiation from the regular and repeating lattice of a single crystal to determine the structure of macromolecules. We must therefore first obtain a crystal of the molecule.

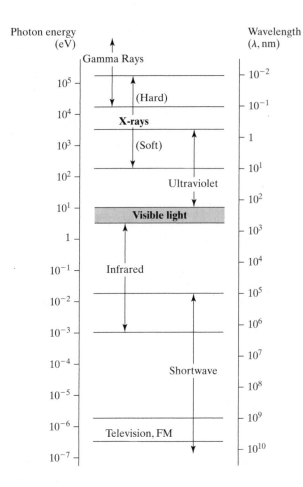

Figure 6.3 Electromagnetic spectrum. Visible light falls in the wavelength range of 400 to 800 nm, with corresponding energies of about 5 to 10 eV. X-rays are shorter wavelength (0.1 to 10 nm) and consequently higher energy (10^2 to 10^5 eV). [Adapted from J. A. Richards, Jr., F. W. Sears, M. R. Wher, and M. W. Zemansky (1960) *Modern University Physics*, 600. Addison-Wesley, Reading, MA.]

6.2   CRYSTALS

6.2.1   What Is a Crystal?

Crystals are solids that are exact repeats of a symmetric motif. What makes crystalline quartz different from noncrystalline glass? Both have essentially the same chemical composition. The most telling difference is that the molecules in a crystal are arranged in an orderly fashion, while the molecules of an amorphous solid, like a glass, are disordered. By *ordered, we mean* regular, symmetric, and repeating (see Chapter 1). This is particularly evident when we compare broken pieces of glass with pieces of a crystal. A glass will shatter to form random shards, with no relationship between the shape of the intact and broken pieces. A crystal, however, can be cleaved very specifically to give fragments that are smaller versions of the original crystal. This can be done almost indefinitely, until we reach the basic unit that describes a crystal, called the *unit cell*. Thus, a crystal can be generated from a molecule by repeating a set of translational or rotational symmetry operators indefinitely. This has great utility; to determine the structure of a crystal, we need only determine the structure of the least symmetric component of the unit cell. Thus, in describing a crystal, we need only describe the least symmetric unit and the symmetry that gives us a crystal (Figure 6.4).

Any symmetric system can be reduced to a level that is not symmetric. For example, the $\alpha_2\beta_2$-tetramer of a hemoglobin molecule with identical α-subunits and β-subunits can be constructed by applying a two-fold rotation to an $\alpha\beta$-dimer (Figure 1.33). There is no true symmetry relationship between the α- and β-subunits of this dimer. In a crystal, the level at which there is no symmetry is aptly called the *asymmetric unit*. By analogy, the $\alpha\beta$-dimer can be considered to be the asymmetric unit of the hemoglobin tetramer in solution.

Starting with the asymmetric unit, we can now apply rotational or screw operators to construct the *lattice motif*. In the simplest hemoglobin example, a C_2 rotational symmetry operator applied to the $\alpha\beta$-dimer would generate the hemoglobin tetramer, which would be the lattice motif of a hemoglobin crystal.

The lattice motif is translated in three dimensions to form a regular and repeating array, called the *crystal lattice,* with each repeated motif forming a point within the lattice (Figure 6.4). The *lattice points* can be connected to form the corners of three-dimensional boxes. These boxes are the *unit cells;* each unit cell contains all the atoms of the lattice motifs and the asymmetric unit. We see that the edges of this box form the axes of the crystal system. The edges of the unit cell define a set of unit vector axes **a, b**, and **c**, with the unit cell dimensions as their respective lengths a, b, and c. These vectors need not be at right angles, and the angles between the axes are denoted as α between the **bc**-axes, β between the **ac**-axes, and γ between the **ab**-axes. The relationship between the lengths a, b, and c and angles α, β, and γ of the unit cell axes defines the unique shape and size of the unit cell. However, there are constraints placed on the shape of the unit cell.

A *crystal* is a stacking of unit cells repeated in three dimensions to build a lattice, leaving no space between the unit cells. This automatically constrains the shape

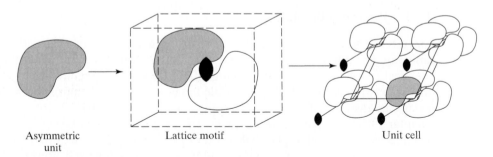

Asymmetric   Lattice motif   Unit cell
unit

Figure 6.4 Components of a crystal. The asymmetric unit is that part of the crystal that shows no symmetry. A symmetry operator (for example, a C_2 rotational axis) generates the lattice motif. Repeating this motif by translation generates the corners of the unit cell, which is the basic repeating unit of the crystal lattice.

of a unit cell to a parallelepiped, with four edges to a face and six faces to the unit cell. A crystal is constructed by translating the unit cells in three dimensions to fill a volume, the unit cell is constructed by translating the repeating motif (the lattice points), and the lattice motif is generated by applying symmetry operators to the asymmetric unit.

Since all unit cells within a crystal are identical, the *morphology* of a crystal is defined by the size and shape of a single unit cell (the lengths of a, b, and c, and the angles α, β, and γ) and the symmetry of the motif. Each level of the crystal, with the exception of the asymmetric unit, can be generated using mathematical operators; solving a crystal structure requires only that we determine the conformation of the atoms in the asymmetric unit. In the case of our hemoglobin example, the structure of the entire hemoglobin tetramer in a crystal can be solved by determining only the structure of a single $\alpha\beta$-dimer in the asymmetric unit.

Within the constraints placed on the shape of the unit cell, there are 14 unique crystal lattices that can be constructed. These are the *Bravais lattices* (Figure 6.5). The combination of the 32 symmetry types (the point groups, Chapter 1) along with the 14 Bravais lattices describes all the possible morphologies of crystals, yielding the shapes and symmetry of 230 different *space groups*. The space group assigned to a crystal uniquely defines the number of asymmetric units that assemble to form the unit cell of the crystal. Each space group specifies the lattice type and the symmetry of the unit cell.

All symmetry operators described by the various point groups can potentially be incorporated into a crystal lattice. However, there are constraints imposed by the definition of a unit cell and by the asymmetry of most biological polymers. Symmetry operators that invert the configuration of a chiral center are not allowed in crystals of biological macromolecules; mirror symmetry, which relates L- and D-stereoisomers, will not be found in crystals of naturally occurring biological macromolecules (for an exceptional case, see Figure 1.20). Thus, the two allowed types of symmetry operators that are observed in crystals of biological molecules are the rotational and screw symmetry operators. This reduces the 230 potential space groups to only 65 that are relevant to naturally occurring biological macromolecules. The designation for symmetry elements in the crystal are identical to those described in Chapter 1 (see Table 1.4).

The rotational components of crystal symmetry are restricted by the angles α, β, and γ. Since C_2 (or simply designated as 2 for a two-fold rotation) or 2_1 symmetry relates two objects by rotation through 180°, these symmetry elements require at least two sets of rectangular faces in a unit cell. Four-fold rotational symmetry requires that at least one set of the faces also be a square. We notice that five-fold rotation is not allowed in standard crystallography. A five-fold rotation or screw axis defines a pentagonal face and, since regular pentagons cannot be packed in three-dimensional space without leaving gaps, we cannot define a unit cell with one face having five edges (this rule has recently been challenged, but so far the controversy has not been resolved). Three-fold rotations require the three unit cell lengths and angles to be identical. This can be accommodated by placing

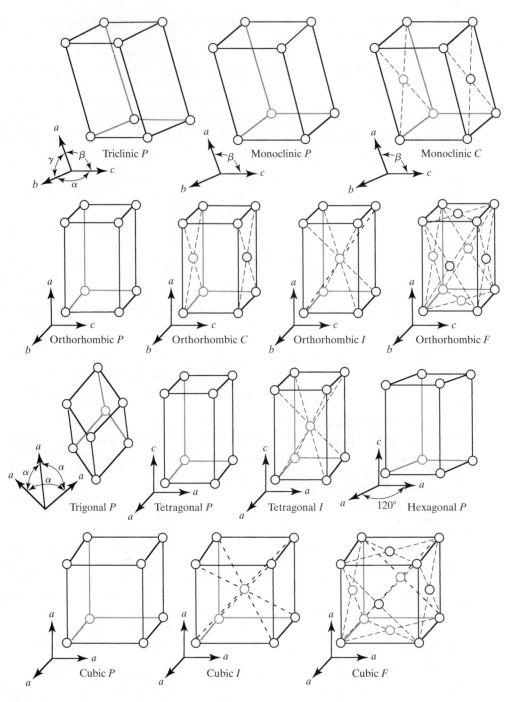

Figure 6.5 The 14 Bravais lattices in crystallography. [Adapted from G. H. Stout and L. H. Jensen (1989), *X-Ray Structure Determination, a Practical Guide,* 2nd ed., 50. John Wiley & Sons, New York.]

the three-fold axis at the corner of a unit cell or having two adjacent three-fold axes on a face. A six-fold rotational or screw symmetry occurs on a hexagonal face and relates two edges of the unit cell by rotation through $60°$. These rotations and their associated screw operators constitute the extent of the symmetry types that are possible for the unit cell of macromolecular crystals. In addition, the presence of two orthogonal symmetry axes automatically defines a third orthogonal symmetry axis. For example, a unit cell having $2_1 2_1$ symmetry (two perpendicular two-fold screw axes) must actually have $2_1 2_1 2$ or $2_1 2_1 2_1$ or higher symmetry. The symmetry axes in a unit cell need not all intersect in the center. However, if two axes do intersect, the third axis must also intersect. Conversely, for two nonintersecting axes, the third axis must be nonintersecting.

With these rules in mind, we can now discuss some of the 14 Bravais lattices and their space groups in more detail. To start, we must add one qualifier to our definition of a unit cell. Although lattice points are necessarily found at the corners of the unit cell, they are not restricted to the corners. Additional lattice points may be found at the center of the faces or the center of mass of the unit cell. If lattice points are found only at the corners, the unit cell is *primitive* and the lattice is given the designation P. If lattice points are found at each face of an opposing pair, we have a centered or C lattice. Lattice points at all six faces define a face-centered or F lattice. Finally, a unit cell containing a lattice point at the center of mass is body-centered and designated as an I lattice. There is a correspondence between the placement of lattice points and symmetry axes. For example, a unit cell that is an orthorhombic lattice ($a \neq b \neq c, \alpha = \beta = \gamma = 90°$) can have lattice points at the corners, the centers of the faces or the centers of the unit cell, defining P, C, I, or F as possible crystal lattices.

Together, the lattice type and symmetry of the unit cell defines the *space group* of the crystal. The shorthand abbreviations for space groups incorporate the lattice type (L) and the symmetry of the unit cell (R_T, where R is the rotation and T designates the translational element of the symmetry operator) in the form $LR_T R_T R_T$. Thus, the space group $P2_1 2_1 2_1$ is a primitive unit cell having two-fold screw axes parallel to each of the three crystallographic axes.

We will continue our discussion of Bravais lattices with the unit cell having the lowest potential for symmetry. This is the triclinic lattice, where $a \neq b \neq c$, and $\alpha \neq \beta \neq \gamma \neq 90°$ or $120°$. All edges of the unit cell are of different lengths and all faces are inclined. Since none of the faces are orthogonal or hexagonal, there can be no symmetry axes through any face. Similarly, the lattice can only be primitive. Thus, a triclinic unit cell automatically defines a $P1$ space group. If one angle is set at $90°$, then a second angle must also be $90°$, forming a monoclinic unit cell. This requires a space group having a two-fold rotation or screw axis. We notice that even though there are two orthogonal faces, there can only be one symmetry axis. The introduction of a second symmetry axis automatically requires the definition of a third axis of symmetry, which requires a third orthogonal face. That is not allowed in a monoclinic space group. Thus, a monoclinic unit cell can be either $P2$ or $P2_1$. Finally, an additional lattice point can be added to the nonorthogonal face of the monoclinic unit cell, generating a C lattice. Again, an I or F lattice requires that all three unique faces be orthogonal, which is contrary to the definition of a monoclinic unit cell. These restrictions to symmetry and placement of lattice points in the unit cell define the 14 Bravais lattices and the associated space groups (Table 6.1).

We see that the lattice type along with the symmetry of the unit cell defines the space group of the unit cell. The lengths and angles of the unit cell define the unit cell parameters, and the space group along with the unit cell parameters define the crystal morphology. We will see later that it is the crystal morphology that dictates the general characteristics of the X-ray diffraction pattern. Different crystals that have identical unit cell lengths and angles and are in the same space group are said to be *isomorphous*. Their X-ray diffraction pattern should also appear to be very similar.

We now understand that all the molecules within the crystal are generated by applying symmetry operators to a single asymmetric unit. Thus, a crystal is nothing more than a single asymmetric unit mathematically replicated in three-dimensional space. Alternatively, we can see that all the molecular properties of a crystal can be attributed to those of the asymmetric unit. We can assert, with few exceptions, that all the molecules or groups of molecules that constitute the asymmetric unit in a single crystal suitable for X-ray diffraction studies have essentially identical conformations. Therefore, to solve the structure of a crystal, we need only solve the structure of the asymmetric unit.

Table 6.1 Sixty-Five Possible Space Groups in Macromolecular Crystals

Lattice Type	Possible Bravais Lattices	Crystal Shape	Possible Space Groups
Triclinic	P	$a \neq b \neq c$ $\alpha \neq \beta \neq \gamma \neq 90°$	$P1$
Monoclinic	P, C	$a \neq b \neq c$ $\alpha = \gamma = 90°, \beta \neq 90°$ or $120°$	$P2, P2_1, C2$
Orthorhombic	P, C, I, F	$a \neq b \neq c$ $\alpha = \beta = \gamma = 90°$	$P222, P2_12_12_1, P2_12_12, P222_1,$ $C222, C222_1, F222, I222, I2_12_12_1$
Tetragonal	P, I	$a = b \neq c$ $\alpha = \beta = \gamma = 90°$	$P4, P4_1, P4_2, P4_3, I4, I4_1, P422,$ $P42_12_1, P4_122, P4_12_12, P4_222,$ $P4_22_12, P4_32_12, P4_322, I422, I4_122$
Trigonal	P	$a = b \neq c$ $\alpha = \beta = 90°, \gamma = 120°$	$P3, P3_1, P3_2, P321, P312,$ $P3_112, P3_121, P3_212, P3_221,$
	R (Rhombohedral)	$a = b = c$ $\alpha = \beta = \gamma < 120° (\neq 90°)$	$R3, R32$
Hexagonal	P	$a = c \neq b$ $\alpha = \gamma = 90°, \beta = 120°$	$P6, P6_1, P6_2, P6_3, P6_4, P6_5, P622,$ $P6_122, P6_322, P6_522, P6_422, P6_522$
Cubic	P, I, F	$a = b = c$ $\alpha = \beta = \gamma = 90°$	$P432, P4_132, P4_232, P4_332, F432,$ $F4_132, I432, I4_132$

6.2.2   Growing Crystals

Now that we know what a crystal is, how can we go about growing a crystal of a macromolecule? There is no straightforward answer to this question because the field of crystallization is more an art than a science. Nonetheless, we can describe the general theory concerning crystallization. Remember that a crystal is a solid. However, macromolecules typically are in aqueous solution. Thus, we must somehow bring the molecule out of solution or *precipitate* the molecule in an orderly fashion to form a crystal.

A molecule comes out of solution when its concentration exceeds its intrinsic solubility S^0 (S^0 is not actually an intrinsic property, but is dependent on external factors, including temperature, pressure, and the solvent). Growing a crystal requires bringing the concentration of the material in solution to supersaturation (Figure 6.6). The need for supersaturating the solution will become evident when we discuss the mechanism of crystal growth later. The concentration of the molecule can be increased by removing the solvent to decrease the overall volume of the solution. For example, a macromolecule with an intrinsic solubility of 2 mM in solution at 1 mM, can be precipitated by simply decreasing the overall volume to less than half the original volume. For small organic and inorganic compounds, crystals can be grown by simply evaporating solvent from solution.

A second strategy for precipitating a molecule is to decrease S^0 for a fixed concentration of material. For example, given the same 1 mM solution above, changing

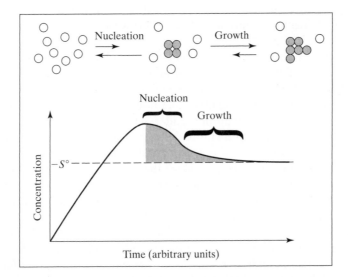

Figure 6.6   Mechanism of crystallization. The initial step in crystallization is the nucleation of a minimum crystal lattice. This is a low probability step that occurs in a supersaturated solution. The crystal grows by adding molecules to the surface of the seed, and occurs at concentrations close to the intrinsic solubility S^0 of the molecule.

the temperature from $27°$ to $4°C$ may be sufficient to reduce the intrinsic solubility of the molecule by one-half. Alternatively, we can affect S^0 by modifying the solvent (for example, by increasing or decreasing the ionic strength of the solution). The optimal solubility of polyelectrolytes, such as polypeptides or polynucleotides, generally falls within a specific range of salt concentrations. If the ionic strength of the solution is increased above this range by addition of salt or by evaporation, or decreased below this optimal range by dialysis, S^0 will decrease and the molecule will precipitate. These are the standard *salting in* and *salting out* effects exploited in many purification techniques.

6.2.3   Conditions for Macromolecular Crystallization

How can we selectively grow a crystal out of solution as opposed to producing an amorphous precipitate? There are a number of factors that are important in crystallizing a molecule. Perhaps the most important is the purity of the sample. We can view purity in terms of biochemical purity and structural purity. A sample is considered to be *biochemically pure* when each macromolecule in the sample has the same molecular formula. This is clearly the ideal situation and it is not generally achieved. In most cases, a macromolecule must be better than 95% pure to produce a crystal.

The term *structural purity* refers to the conformation of the molecules in a particular sample. As we see in Chapters 2 and 3, the energetic difference between various conformations of a macromolecule can be very small. Even a chemically pure sample may represent a highly heterogeneous population of conformations. Thus, the crystallization solutions and methods are chosen to favor the native conformation of the macromolecule.

We see that crystallization of a macromolecule requires finding a solution that balances solubility with structural homogeneity. These are likely to be under very specific conditions that constitute a minor subset of the conditions that fulfill either criterion. Thus, even though the conditions under which a molecule has a stable conformation and the conditions under which the macromolecule is insoluble are known, it may still be difficult to define the specific conditions under which a molecule will crystallize. For this reason, finding the conditions for crystallization is essentially a shotgun method, in which the molecule is placed under a large number of different solution conditions in the hope that one condition can be found that satisfies both the solubility and conformational requirements for crystal growth.

There is an increasingly large number of macromolecules (both proteins and nucleic acids) that have been crystallized from a large variety of different solutions. However, crystals of nearly all proteins and nucleic acids can be grown from a small subset of these conditions. There are now over 50 different buffer and salt solutions (Table 6.2) that are widely used as starting points to search for crystallization conditions for macromolecules. A more scientific approach to crystallizing macromolecules is still desirable.

Table 6.2   Screening Solutions in Sparse Matrix Methods for Crystallizing Proteins and Nucleic Acids

Crystallization Solutions			
Salt	**Buffer**	**Precipitant**	**Molecules Crystallized**
Proteins			
None	0.1 M Tris	2 M Ammonium sulfate	Tropomysin
			*Eco*R1-DNA complex
			Monellin
0.2 M Na citrate	0.1 M Tris	30% Polyethylene glycol (PEG)	Lysozyme
			Myoglobin
			Ribonuclease A
			Insulin
0.2 M Na citrate	0.1 M Cacodylate	30% PEG	Lysozyme
			Pepsin
Nucleic acids			
12 mM Spermine,	40 mM	10% 2-methyl-2,4-dimethylpentane diol (MPD)	d(CG)$_3$ Z-DNA
20 mM Mg^{2+},	Cacodylate		G-quartet DNAs
80 mM Na$^+$	pH 7.0		DNA-adriamycin
0.5 mM Spermine, 15 mM Mg^{2+}, 2 mM BaCl$_2$	pH 6.5	7% 2-Propanol	Phe-tRNA
2 mM CaCl$_2$, 10 mM Mg^{2+}	pH 7.0	15% MPD	Group I intron (from *Azoarcus*)
			12-Base pair RNA

To understand the techniques for crystallizing a macromolecule, we need to compare the mechanisms for crystal growth to those for amorphous precipitation. In comparison with the molecule in solution, the highly ordered molecules in a crystal lattice have significantly lower entropy. The external degrees of rotational and translational freedom in a crystal lattice are much reduced from those in solution. For example, two molecules in solution each have complete translational and rotational freedom in three dimensions. If two molecules associate to *nucleate* the formation of a crystal lattice, their movements are exactly correlated and the external degrees of freedom are reduced by a factor of 9. Thus, the difference in entropy between the dimer and monomer states of the two molecules is $\Delta S^0 = -R \ln (2) = -18$ J/mol K. Two molecules, however, do not form a stable nucleation complex. It has been estimated that the contents of at least four unit cells must come together in a highly cooperative manner to form a stable and unique nucleation lattice. This is equivalent to the assembly of 16 asymmetric units for a relatively highly symmetric space group such as $P2_12_12_1$. The minimum ΔS^0 for the formation of this nucleation lattice would be $-R \ln (144) = -42$ J/mol K. In solution, the molecules will have a distribution of conformations, but in the crystal there is only a single conformation. Therefore, there is an additional loss in

conformational entropy during crystallization. This latter value is more difficult
to estimate.

The growth step of the crystal is envisioned to be single molecules adding to
the surfaces of the nucleating lattice. The loss in entropy at these subsequent steps is
expected to be less than for the nucleation step. Crystal growth, therefore, occurs in
two distinguishable steps: (1) a low-probability nucleation step to form a seed, and
(2) a higher-probability growth (or propagation) step to increase its size (Figure 6.6).
The two steps in crystal growth are analogous to the steps in the zipper model for
structural transitions in macromolecules (see Chapter 4).

We see that nucleation of crystal growth requires a very large driving force.
The necessary driving force for nucleation comes from bringing the concentration
of macromolecules well above their intrinsic solubility (*supersaturation*). The subsequent addition of single molecules at the crystal surfaces occurs at concentrations below supersaturation, near the intrinsic solubility. Thus, growth is a series of
microequilibrium steps for macromolecules in solution at the surface of the crystal lattice.

Nucleation is the more important of the two steps. If the solution does not
reach supersaturation, a nucleus is not formed and a crystal cannot grow. On the
other hand, a solution that becomes supersaturated very rapidly will form multiple
nuclei that will rapidly deplete the molecules from the crystallization solution. This
results in a shower of tiny crystals. There must be a balance between the two extremes in order to grow single crystals that are sufficiently large to provide high-
resolution X-ray diffraction data. Thus, the macromolecules in solution must be
concentrated in a very well-controlled manner.

A number of methods have been developed to facilitate the process of crystallizing macromolecules (Application 6.1). Two widely used strategies are vapor
diffusion and microdialysis, both of which are designed to vary the solution environment, reducing the solubility of a macromolecule in a controlled manner. In
both cases, the molecule is initially dissolved in a buffered solution, usually containing a mixture of solvents and salts. This sample solution is physically separated
from a larger volume of solvent called a reservoir. The sample and reservoir solutions are placed in a closed system, and solvent or salts are removed from (or
sometimes added to) the sample solution to render the macromolecules less soluble. In vapor diffusion methods (Figure 6.7), solvents are transferred according to
the vapor pressure of the sample versus the reservoir. In this typical system, water
is drawn from the sample so that the volume of the sample systematically decreases until the vapor pressure of the water in the sample and that in the reservoir have equilibrated. The two most common vapor diffusion techniques are the
hanging drop method, where the sample literally hangs above the reservoir, and
the sitting drop method, where the sample sits in a well surrounded by the reservoir. In the microdialysis method, solvent is transferred by equilibrating the osmotic pressures of the sample and reservoir across a semipermeable membrane
(see Chapter 13).

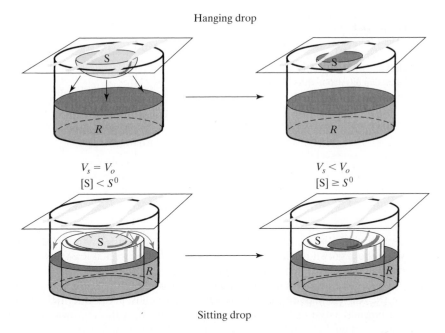

Hanging drop

$$V_s = V_o$$
$$[S] < S^0$$

$$V_s < V_o$$
$$[S] \geq S^0$$

Sitting drop

Figure 6.7   Vapor diffusion methods of crystallization. In the hanging drop method of vapor diffusion, a sample in solution is suspended above a reservoir, R, that contains a high concentration of a precipitant. The lower vapor pressure of the reservoir draws water from the sample solution, S, to reduce the volume of the sample, V_S, below its initial volume, V_o. Consequently, the concentration of molecules in the sample solution, $[S]$, increases to above the intrinsic solubility S^0 of the molecule, resulting in precipitation or crystallization. In the sitting drop method, the sample solution sits in a well rather than hanging suspended, but otherwise the two methods are the same.

Application 6.1   Crystals in Space!

Crystallographers will go to great lengths to grow larger and better crystals, including going into space. In one case, Lawrence DeLucas became a crystallographer in space as a member of the Columbia space shuttle crew in 1992. The general idea for growing protein crystals under microgravity conditions is to provide more uniform solutions to the growing crystal by reducing convective flow. In addition, crystals grown under Earth's gravitational pull will settle either to the bottom of a drop in a hanging drop experiment or to the bottom of a glass well in a sitting drop experiment. Consequently, at least one surface will be excluded from solution and cannot grow with the remainder of the crystal.

   One of the first demonstrations of how microgravity can improve the size and quality of protein crystals was seen in a set of experiments on the space shuttle STS-26 flight in 1988. On this shuttle mission, a special vapor diffusion apparatus, based on the hanging drop method, was used to crystallize the proteins γ-interferon D, porcine elastase and isocitrate lyase. The most dramatic effects were seen in isocitrate lyase (Figure A6.1)—the

Figure A6.1 Comparison of the diffraction intensity from crystals of isocitrate lyase grown in space (on U.S. space shuttle STS-26 flight, triangles) and on Earth (circles). Resolution limits are labeled in angstrom units (0.1 nm).

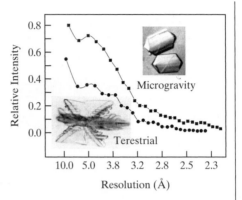

crystals grown on Earth were highly dendritic and not suitable for diffraction studies, while those grown on flight STS-26 were clean prisms with at least a two-fold increase in diffraction intensity across all resolution ranges.

  Crystals grown in the microgravity environments of the U.S. Skylab and space shuttle, the Russian MIR space station, and, more recently, in the International Space Station are typically more ordered and up to 50% larger than those grown on Earth. Unfortunately, it is unlikely that extraterrestrial crystallography will become a routine method of growing macromolecular crystals in the near future.

DeLucas, L. J. et al. (1989) "Protein Crystal Growth in Microgravity," *Science* **246**, 651–654.

  With any luck, we obtain a single crystal of a macromolecule that will diffract X-rays. The process of crystallization is but one leg in the triad necessary to solve a structure. With a crystal in hand, we can now collect X-ray data to solve the structure of the crystal.

6.3 THEORY OF X-RAY DIFFRACTION

As we discussed above, the wavelength of X-ray radiation is well suited for resolving atoms separated by the distance of a covalent bond. The energy of a quantum of this radiation is approximately 8000 eV, which is approximately the energy of electrons in their orbitals. This equivalence of energy leads to interactions so that the electrons of an atom will primarily be responsible for the scattering of X-rays. The number of electrons in a given volume of space (the *electron density*) determines how strongly an atom scatters X-rays. The interference of the scattered X-rays leads to the general phenomenon of *diffraction*. In this section, we briefly review the general theory of diffraction, extend this to X-ray diffraction, and apply X-ray diffraction to single crystals. We can then consider what can be learned about the crystal morphology and how we can transform the diffraction data into the structure of the molecules in the crystal.

In X-ray diffraction, we treat all electromagnetic radiation as waves. The general theories and consequences of diffraction are applicable to any energy of radiation; the diffraction of visible light is often used to illustrate the principles involved in X-ray diffraction from a crystal. We start our discussion by considering the components of diffraction—scattering and interference.

Scattering simply refers to the ability of objects to change the direction of a wave. With visible light, the simplest example is the reflection from a mirror; the mirror simply changes the direction of the light waves. Nonreflecting objects will also scatter light waves. An object placed in the path of a light from a point source cannot cast a sharp shadow because of scattering from its edges. The origin of scattering can be best developed if we start with Huygen's principle, which states that every point along a wavefront can be considered the origin of a new wavefront, as we see in Figure 6.8a. The velocity of this new wavefront is equal to that of the original. For an unimpeded wavefront, the secondary wavefront can be constructed by drawing circles with radius $r = vt$ (where v is velocity and t is time) at points along the starting wavefront and connecting the tangents to each of these circles (Figure 6.8b). Objects placed in the path of a wavefront act as points of propagation for new wavefronts. The entirely new wavefront is called a *scattered wave*.

If we now place two point objects (A and B) in the path of the wavefront, each of the two points will propagate a new wavefront having identical wavelengths and velocities (Figure 6.9). The offset in the maximum amplitudes of the two waves (their relative phase) depends on the positions of A and B relative to the origin of the initial wavefront. At some position in space, the wave propagating from A will reinforce the scattered wave from B through constructive interference if the two scattered waves are in phase. Alternatively, the wave from A will reduce the amplitude of the wave from B through destructive interference if the amplitudes are out of phase. This is called *diffraction;* the sum of the two waves propagated from A and B result in an amplitude that is dependent on the relative positions of A and B and is also dependent on where the new wavefronts are being observed. Concomitantly, if we make several observations of the amplitude of the new wavefront at different positions, we can extrapolate the information to determine the positions of the diffracting objects A and B relative to the origin of the

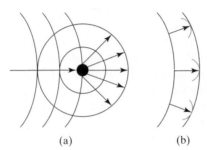

(a)　　　　(b)

Figure 6.8 Huygen's principle of diffraction. (a) Each point in front of a wavefront acts as a point of propagation for a new wavelet, which sums to form a new wavefront. Each point in front of the incident wavefront generates a wavelet having the same velocity as the wavefront, represented as a set of concentric circles emitted from the point. (b) The new wavefront is formed by connecting the tangents of the wavelets from all points of propagation.

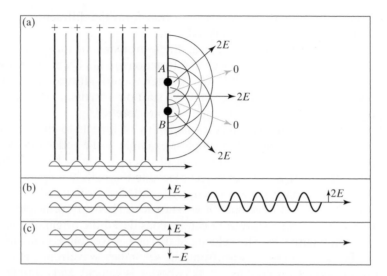

Figure 6.9 Constructive and destructive interference of scattered waves. (a) Two scattering points placed in front of an incident wavefront act as points of propagation. (b) The amplitudes E of the resulting wavelets from the scattering points can sum to form a new wave with twice the amplitude along the vectors direction for $2E$ in (a). (c) Waves that are 180° out of phase annihilate each other to give a net amplitude of zero along the vectors 0 in (a).

initial wavefront. This is how X-ray diffraction is used to solve the structure of molecules in single crystals. In the remainder of this chapter, we develop the conceptual tools necessary for understanding how the positions of atoms are determined by the diffraction of X-rays.

6.3.1 Bragg's Law

In 1912, W. L. Bragg developed a simple relationship to explain how diffraction relates to the relative positions of point objects in space. To derive Bragg's law of diffraction, we will first simplify our conceptual model of diffraction and think of lattice points in the crystal as parallel planes. Stacking a set of reflecting planes at regularly spaced intervals d creates a simple model of a one-dimensional crystal (Figure 6.10). In this model, a wave of X-rays (with wavelength λ) is incident on the reflecting planes at an angle θ. The wave is scattered by reflection from the planes at an identical angle θ. We can ask at this point which values of θ will result in constructive versus destructive interference. We assume that the distance to our point of observation is very large compared to d, so the individual paths of scattered light are essentially parallel. Because we have a large number of planes, we observe constructive interference only when the reflected waves are perfectly in phase (peaks aligned with peaks, node with nodes, and valleys with valleys). This only occurs when the difference in the length of the path of the incident and the reflected

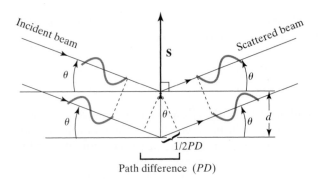

Figure 6.10 Bragg's law of diffraction. An incident beam of X-rays hits a regular array of reflecting planes at an angle θ and is reflected at the same angle. Constructive interference of the reflected or scattered X-rays occurs when the path difference between adjacent planes (spaced by a distance d) is equal to some integer number of wavelengths. The scattering vector ($\mathbf{S}$) is perpendicular to the reflecting planes and its length equal to the inverse spacing between these planes.

waves of each plane, PD, is equal to some integer n of the wavelength of the incident X-rays.

$$PD = n\lambda \tag{6.2}$$

This path difference is related to the distance separating the reflecting planes by the simple trigonometric relationship,

$$\frac{1}{2}PD = d\sin\theta \tag{6.3}$$

Substituting Eq. 6.3 into Eq. 6.2, we derive Bragg's law of diffraction

$$2\,d\sin\theta = n\lambda$$

or

$$2\sin\theta = \frac{n\lambda}{d} \tag{6.4}$$

Rearranging Eq. 6.4 gives

$$\frac{2\sin\theta}{\lambda} = \frac{n}{d} \tag{6.5}$$

This simple relationship makes two profound statements concerning the properties of diffraction. First, Eq. 6.4 tells us that the angles at which diffraction is observed is quantized for integer values (0, 1, 2 ...) of n. For an infinite row of scatterers, the only condition at which a reflection will be observed is when the Bragg law condition is met. Under all other conditions, the scattered radiation is annihilated.

Second, there is a reciprocal relationship between the Bragg angle θ and the spacing between the reflecting planes (d). This means that larger spacing of repeating units in a crystal results in smaller diffraction angles. Alternatively, we can transform the relationship into a direct one by defining a new term, the *scattering vector* ($\mathbf{S}$), which has a magnitude $|\mathbf{S}| = n/d$, and a direction perpendicular to the reflecting plane. Equation 6.5 can thus be rewritten as the direct relationship

$$\frac{2 \sin \theta}{\lambda} = \frac{n}{d} = |\mathbf{S}| \tag{6.6}$$

where $\mathbf{S}$ has units of 1/length and, consequently, allows us to introduce a concept known as reciprocal space. This concept will be developed further in this chapter but, for now, we can consider $\mathbf{S}$ as a means of directly relating the geometry of the scattering planes to the distance separating the observed reflections (Figure 6.10).

Some immediate questions become obvious concerning Bragg's law. Is reflection an accurate way of depicting a diffraction event? What is reflecting the X-rays? What sits between the reflecting planes (a void)? What happens in three dimensions?

6.3.2   von Laue Conditions for Diffraction

Clearly, treating X-ray diffraction as simple reflection from parallel planes is not an entirely accurate representation of the actual process. First, diffraction is a consequence of scattering from atoms. Atoms scatter X-rays, not because they are shiny spheres but because they are oscillating dipoles. Thus, the scattered waves do not necessarily travel along the single path of a reflected wave; atoms scatter X-rays in all three dimensions. Second, reflection from a plane can only occur when the incident angle θ is not equal to zero. Thus, if our one-dimensional crystal of parallel planes is exactly perpendicular to the incident X-ray beam, we predict no reflection. Scattering from the atoms of a crystal has no such constraint. Scattering is a general phenomenon, and reflection from a plane is merely a specific example of it. Still, the data observed during an X-ray diffraction experiment are often called *reflections* because, as we will see, Bragg's law correctly describes the conditions of constructive interference in X-ray diffraction. Let us now develop a slightly different model that more accurately describes X-ray diffraction from atoms.

We will start with the simplest case by building a one-dimensional crystal in which single atoms are spaced along the crystallographic **c**-axis (Figure 6.11). If we treat these atoms as scattering points, then the incident X-rays can be scattered in any direction. We will now orient this crystal with the **c**-axis perpendicular to the incident radiation. Now imagine scattering from the points in all directions. Only certain directions of the scattered radiation will reinforce; these will be such that the path difference PD between the rays scattered by adjacent atoms corresponds to an integral number of wavelengths. Notice that this condition is no different from that for deriving Bragg's law. If γ is the angle between the direction of the scattered radiation and the row of scatterers, this condition is given by

$$l\lambda = c \cos \gamma \tag{6.7}$$

where l is an integer and c is the spacing between points (or the length of the unit cell along **c**). If the row of points is very long, complete annihilation of the scattered

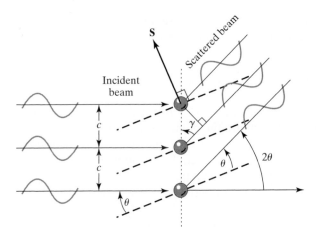

Figure 6.11 von Laue conditions for diffraction. A set of scattering atoms arranged in a regular array are spaced by a distance c along the vertical axis. Constructive interference occurs when the angle γ of the scattered beam relative to the crystal axis conforms to the von Laue conditions for scattering. These conditions are analogous to Bragg's law when reflecting planes (dashed lines) at each scattering atom form an angle θ relative to both the incident and the scattered beams. The diffraction angle relative to the incident beam is 2θ.

waves will occur at all other angles. Then, from such a row, radiation will be scattered only along the surfaces of cones, with the conical axis lying coincident with the crystallographic **c**-axis. This is illustrated in Figure 6.12, where the distance to the point of observation is large compared to the size of the crystal and the crystal serves as the point of origin for the scattered X-rays.

If the incident radiation makes an angle γ_0 other than 90° with the row of scatterers, Eq. 6.7 must be modified to be

$$l\lambda = c(\cos \gamma - \cos \gamma_0) \tag{6.8}$$

If we expand this to a two-dimensional array, with spacing a and c along the **a** and **c** axes, Eqs. 6.8 and 6.9 must be satisfied simultaneously,

$$h\lambda = a(\cos \alpha - \cos \alpha_0) \tag{6.9}$$

Equation 6.9 describes a second set of cones that is now coincident with the crystallographic **a**-axis. We see that reinforcement will occur only where the two sets

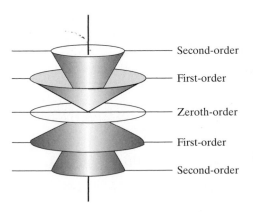

Second-order

First-order

Zeroth-order

First-order

Second-order

Figure 6.12 An incident beam of X-rays causes a set of scattering cones from a one-dimensional crystal aligned along the vertical axis. Each cone makes an angle 2θ relative to the incident beam to conform to the von Laue conditions for diffraction. The intersection of each cone with a piece of flat photographic film is an arc. Each arc is a layer line representing the order of the reflection, the integer index l in Eq. 6.6. In a three-dimensional crystal, each axis of the unit cell generates a set of concentric cones, with the conical axes aligned parallel with the crystallographic axes.

of cones intersect, which is a line. For a three-dimensional array, such as an orthorhombic crystal with spacings a, b, and c, a third condition must be simultaneously satisfied.

$$k\lambda = b(\cos \beta - \cos \beta_0) \tag{6.10}$$

Equations 6.8 to 6.10 are called the von Laue conditions for diffraction. The three resulting cones will intersect at points that are the points of reflection.

These conditions are satisfied simultaneously only for certain specific orientations with respect to the vector describing the incident beam. Suppose that the crystal is arbitrarily oriented with respect to the incident beam. We can ask the following question: If we consider an angle 2θ between the incident beam and the reflected beam, as illustrated in Figure 6.11 (we will relate this to the diffraction angle later), is there reinforcement of the scattered X-rays in this direction? If so, we have a reflection and the von Laue conditions must be satisfied. To simplify the nomenclature at this point, we write $\underline{\alpha} = \cos \alpha$, $\underline{\beta}_0 = \cos \beta_0$, and so forth. We see that these are the direction cosines for the diffracted and the incident X-ray beams as vectors relative to the crystal axes. The following relationships must apply to the direction cosines of the two vectors in an orthorhombic space group

$$\underline{\alpha}^2 + \underline{\beta}^2 + \underline{\gamma}^2 = 1 \tag{6.11}$$

$$\underline{\alpha}_0^2 + \underline{\beta}_0^2 + \underline{\gamma}_0^2 = 1 \tag{6.12}$$

and if the angle between the two vectors is 2θ, we have

$$\cos 2\theta = \underline{\alpha}\underline{\alpha}_0 + \underline{\beta}\underline{\beta}_0 + \underline{\gamma}\underline{\gamma}_0 \tag{6.13}$$

These are standard geometric relationships. Finally, if we square the von Laue conditions, we find

$$\frac{h^2\lambda^2}{a^2} = \underline{\alpha}^2 - 2\underline{\alpha}\underline{\alpha}_0 + \underline{\alpha}_0^2 \tag{6.14}$$

$$\frac{k^2\lambda^2}{b^2} = \underline{\beta}^2 - 2\underline{\beta}\underline{\beta}_0 + \underline{\beta}_0^2 \tag{6.15}$$

$$\frac{l^2\lambda^2}{c^2} = \underline{\gamma}^2 - 2\underline{\gamma}\underline{\gamma}_0 + \underline{\gamma}_0^2 \tag{6.16}$$

or summing,

$$\left(\frac{h^2}{a^2} + \frac{k^2}{b^2} + \frac{l^2}{c^2}\right)\lambda^2 = 1 - 2(\underline{\alpha}\underline{\alpha}_0 + \underline{\beta}\underline{\beta}_0 + \underline{\gamma}\underline{\gamma}_0) + 1$$
$$= 2(1 - \cos^2 \theta) \tag{6.17}$$

By a standard identity, $1 - \cos 2\theta = 2\sin^2\theta$, therefore

$$\left(\frac{h^2}{a^2} + \frac{k^2}{b^2} + \frac{l^2}{c^2}\right)\lambda^2 = 4\sin^2\theta$$

or

$$\lambda\left(\frac{h^2}{a^2} + \frac{k^2}{b^2} + \frac{l^2}{c^2}\right)^{1/2} = 2\sin\theta \tag{6.18}$$

This is the von Laue equation. It uses the defined axes of the unit cell and holds in three dimensions for *all* planes of atoms in a crystal in which the **a**-, **b**-, and **c**-axes are orthogonal.

   How are Bragg's law and the von Laue conditions related? To find the relationship, we need only ask what the reflecting planes are and what the Bragg angle θ represents. Figure 6.11 compares the von Laue row of scatterers for a one-dimensional crystal to the Bragg reflecting planes, which by definition form equal angles with both the incident and the scattered beams. The diffraction angle can now be defined relative to the direction of the incident beam. Since the incident beam is fixed, this is a more convenient reference than the scattering plane, which is variable. We readily see that the angle of the scattered beam relative to the incident beam is twice the Bragg's angle, 2θ. In practice, this is the angle we measure, called the diffraction or scattering angle. Thus, for each value l in Eq. 6.6, equivalent to n in Eq. 6.5, there is a series of planes that generate a continuous set of scattered beams. For each value of l, the scattered beams form a set of cones in which the angle of the cone relative to the incident beam is the diffraction angle 2θ, and the axis of the cone lies parallel to the direction of the one-dimensional crystal. We notice that $l = 0$ conforms to the conditions for diffraction, and yields a plane of scattered X-rays, with $2\theta = 0$. Under the von Laue equation, h, k, and l, which are called the *Miller indices,* define the integer number of wavelengths that result in an observed reflection from a three-dimensional crystal. Thus, for a given set of Miller indices h, k, and l, Bragg's law and the von Laue equation are equal.

$$\left(\frac{h^2}{a^2} + \frac{k^2}{b^2} + \frac{l^2}{c^2}\right)^{1/2} = \frac{2\sin\theta}{\lambda} = \frac{n}{d} \tag{6.19}$$

We notice from this expression that the von Laue conditions is an expansion of Bragg's law to three dimensions, with the distance between Bragg reflecting planes (d_{hkl}) now defined for specific lattice indices h, k, and l. In addition, from Eq. 6.6 and 6.19, the scattering vector **S** can also be defined in three dimensions such that

$$|\mathbf{S}| = \left(\frac{h^2}{a^2} + \frac{k^2}{b^2} + \frac{l^2}{c^2}\right)^{1/2} \tag{6.20}$$

or that the vector has coordinates (h/**a**, k/**b**, l/**c**) where, again, **a**, **b**, and **c** define the edges of the crystal unit cell in real space. Again, each **S** is a vector perpendicular to the virtual reflecting plane consistent with the von Laue conditions.

If we were to record the cones of diffraction from a one-dimensional crystal using a planar sheet of photographic film, the diffraction pattern that we observe would be the intersection of each cone with a plane, giving an arc (Figure 6.13). Each line is called a layer line and is numbered according to the value l. As the crystal is expanded in three dimensions, each additional dimension yields a set of cones whose diffraction angle satisfies the von Laue conditions. The observed diffraction from a two- or three-dimensional crystal is the intersection of each series of cones as described above. The resulting points of reflection can be seen by comparing the intersection of a film plane with each set of cones from a two-dimensional crystal (Figure 6.13).

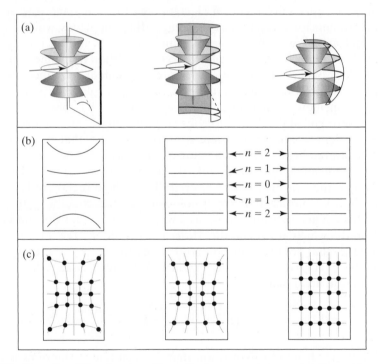

Figure 6.13 Recording diffraction data. The reflecting cones from a crystallographic axis can be recorded using a piece of photographic film that is flat, cylindrical and wrapped around the cones, or spherical (a). The resulting layer lines form one-dimensional arcs on the flat film, and straight lines on both the cylindrical and spherical films (b). The spacing between layer lines increases on the cylindrical film as the order of the layer lines n increases from zero, but remains constant for all layer lines on spherical film. The diffraction from a three-dimensional crystal are points at the intersection of the cones from the three crystallographic axes (c). These points are at the intersection of arcs on flat film, at the intersection between arcs and lines on cylindrical film, and the intersection of straight layer lines on spherical film. A theoretically spherical film or a curved detector would result in an undistorted diffraction pattern.

Although the von Laue conditions describe diffraction at the angle 2θ relative to the incident beam, it should also hold true for -2θ, which occurs in the opposite direction. In the one-dimensional crystal in Figure 6.11, this results in two sets of related cones oriented in opposite directions for each layer line n. We will see later that not only do we expect to see reflections at 2θ and -2θ, but that the intensities of these related reflections are identical. This is known as *Friedel's law,* and the reflections related by Friedel's law are called *Friedel pairs.* Friedel's law simply states that a reflection with Miller indices (h, k, l) should be identical for one at $(-h, -k, -l)$ (indices for these reflections are abbreviated as *hkl* and *-h-k-l*).

6.3.3   Reciprocal Space and Diffraction Patterns

We can now expand on our discussion of reciprocal space. You should recognize that although the unit cell exists in a real lattice, the reciprocal lattice is imaginary and its construction is useful primarily to describe conditions under which diffraction should occur according to Bragg's law and the von Laue conditions. We will start the discussion with orthorhombic and higher-symmetry crystal lattices, then extend these to lower symmetry lattices. The von Laue equation, Eq. 6.17, says that we will observe reflections at scattering angle 2θ for certain values of $h/\mathbf{a}$, $k/\mathbf{b}$, and $l/\mathbf{c}$. In the notation of reciprocal space, these certain values are written as $h\mathbf{a}^*$, $k\mathbf{b}^*$, and $l\mathbf{c}^*$, where for orthorhombic and higher symmetries $a^* = 1/a$, $b^* = 1/b$, and $c^* = 1/c$. That is, in reciprocal space the unit cell vectors are $\mathbf{a}^*$, $\mathbf{b}^*$, and $\mathbf{c}^*$ with lengths a^*, b^*, and c^* in nm^{-1} (as opposed to nm for the direct unit cell in real space). The directions of the principal axes $\mathbf{a}^*$, $\mathbf{b}^*$, and $\mathbf{c}^*$ in an orthorhombic unit cell correspond to the directions of $\mathbf{a}$, $\mathbf{b}$, and $\mathbf{c}$ in the real space lattice.

For nonorthorhombic space groups, however, these angular relationships $(\alpha^*, \beta^*, \text{ and } \gamma^*)$, are not as readily defined. For example, consider the angle β, which relates $\mathbf{a}$ and $\mathbf{c}$ in the direct lattice. Another way to think about this is that $\mathbf{a}$ is a vector that extends at an angle β from the $\mathbf{b}$-$\mathbf{c}$ plane and $\mathbf{c}$ is the vector that extends at an angle β from the $\mathbf{a}$-$\mathbf{b}$ plane. The reciprocal lattice is constructed using the scattering vector $\mathbf{S}$, which, as you recall, is perpendicular to the reflecting plane. Now consider the $\mathbf{a}$-$\mathbf{b}$ plane not as a plane of the unit cell lattice, but as a reflecting plane. The scattering vector that is perpendicular to this reflecting plane with length $1/b$ is now $\mathbf{b}^*$. Similarly, $\mathbf{a}^*$ is the vector with length $1/a$ that is perpendicular to the $\mathbf{b}$-$\mathbf{c}$ reflecting plane, and the angle β^* relates $\mathbf{a}^*$ to $\mathbf{c}^*$ (Figure 6.14). From this construction, we can readily show that β is the complement of β^*—consequently, the opposite angle of the parallelipid in the reciprocal lattice must also be β. The general relationships between the reciprocal crystal lattice and the crystal lattice in real space for monoclinic and triclinic lattices are described in Table 6.3.

The beauty of the reciprocal space concept is that, with the general relationships, the von Laue equation (Eq. 6.18) can be rewritten as

$$\lambda(h^2\mathbf{a}^{*2} + k^2\mathbf{b}^{*2} + l^2\mathbf{c}^{*2} + 2hk\mathbf{a}^*\mathbf{b}^*\cos\gamma^* + 2hl\mathbf{a}^*\mathbf{c}^*\cos\beta^*$$

$$+ 2kl\mathbf{b}^*\mathbf{c}^*\cos\alpha^*)^{1/2} = 2\sin\theta = \frac{\lambda}{d_{hkl}} \tag{6.21}$$

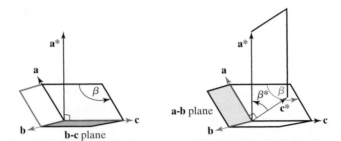

Figure 6.14 Construction of a reciprocal unit cell from a unit cell in real space. The left panel describes the reciprocal axis **a*** as the scattering vector that is perpendicular to the **b-c** reflecting plane in the real space unit cell (where the **b** axis points out of the plane of the page). Similarly, the **c*** axis is perpendicular to the real space **a-b** plane. The angle relating the **a*** and **c*** axes is β^* (which is complementary to the real unit cell angle β).

which is valid for *all* crystal symmetries. In this expression, d_{hkl} represents the spacing between a reflecting plane at the unit cell origin and at a reciprocal lattice point with coordinates (h, k, l). Furthermore, it is easy to directly visualize Bragg's law. For this discussion, the reciprocal lattice is assumed to represent the geometry (distances and angular relationships) of the observed X-ray diffraction pattern. We start with the reciprocal lattice plane **a*-c***, and place a crystal at some point A along the path of an incident X-ray in this plane (Figure 6.15). Since A is the position of the crystal, X-rays can scatter in any direction from this point, which is represented by a circle inscribed with a radius length of $1/\lambda$. The incident beam enters this circle at some point B and exits at some point O. We notice that O is the origin of the reciprocal lattice. In addition, we can simply state that any lattice point L that intersects the circle defines the conditions for diffraction; therefore, AL represents the scattering expected from the crystal that corresponds exactly to the lattice point at L. The three reference points $O, B,$ and L allow us to define the lengths and angles for scattering in reciprocal space. The distance OL between the reciprocal lattice points causing scattering is $1/d_{hkl}$ in the Bragg notation. Since OLB is inscribed in a semicircle, it must be a right angle. Then, by definition, $\sin(OBL) = OL/OB = (1/d_{hkl})/(2/\lambda)$ where θ is defined in Figure 6.15 as the angle OBL. But after rearranging, this is simply Bragg's law $\lambda = 2d_{hkl} \sin \theta$ for $n = 1$. Thus, θ is the Bragg angle and BL a Bragg reflection plane.

Vectorially, we can see the conditions for scattering by representing the incident beam (AO) by $\mathbf{s_0}$ and the scattered beam (AL) by $\mathbf{s}$, both of which have lengths in reciprocal space of $1/\lambda$. We also see that OL is the vector distance between Bragg planes (BL and the plane parallel to BL, but intersecting O), which is the *scattering vector* $\mathbf{S}$. In Figure 6.15, $\mathbf{S}$ has components $-a^*$ and $5c^*$, so that $\mathbf{S} = -a^* + 5c^*$. In three dimensions, for the general Bragg plane $(h\ k\ l)$,

$$\mathbf{S} = h\mathbf{a}^* + k\mathbf{b}^* + l\mathbf{c}^* = (h\mathbf{a}^*, k\mathbf{b}^*, l\mathbf{c}^*) \tag{6.22}$$

Thus, the components of $\mathbf{S}$ are simply any value of $(h\mathbf{a}^*, k\mathbf{b}^*, l\mathbf{c}^*)$ that describes a reflection.

Table 6.3   Relationship Between Unit Cell Parameters in Real Space and Reciprocal Space

Lattice Type	Real Space	Reciprocal Space
Orthorhombic and higher symmetry	a	$a^* = \dfrac{1}{a}$
	b	$b^* = \dfrac{1}{b}$
	c	$c^* = \dfrac{1}{c}$
	$\alpha = 90°$	$\alpha^* = 90°$
	$\beta = 90°$	$\beta^* = 90°$
	$\gamma = 90°$	$\gamma^* = 90°$
	V	$V^* = \dfrac{1}{V} = a^*b^*c^*$
Monoclinic		
	a	$a^* = \dfrac{1}{a \sin \beta}$
	b	$b^* = \dfrac{1}{b}$
	c	$c^* = \dfrac{1}{c \sin \beta}$
	$\alpha = 90°$	$\alpha^* = 90°$
	$\beta \neq 90°$	$\beta^* = 180° - \beta$
	$\gamma = 90°$	$\gamma^* = 90°$
	V	$V^* = \dfrac{1}{V} = a^*b^*c^* \sin \beta^*$
Triclinic		
	a	$a^* = \dfrac{bc \sin \alpha}{V}$
	b	$b^* = \dfrac{ac \sin \beta}{V}$
	c	$c^* = \dfrac{ab \sin \gamma}{V}$
	$\alpha \neq 90°$	$\cos \alpha^* = \dfrac{\cos \beta \cos \gamma - \cos \alpha}{\sin \beta \sin \gamma}$
	$\beta \neq 90°$	$\cos \beta = \dfrac{\cos \alpha \cos \gamma - \cos \beta}{\sin \alpha \sin \gamma}$
	$\gamma \neq 90°$	$\cos \gamma = \dfrac{\cos \alpha \cos \beta - \cos \gamma}{\sin \alpha \sin \beta}$
	V	$V^* = a^*b^*c^* \sqrt{1 - \cos^2 \alpha^* - \cos^2 \beta^* - \cos^2 \gamma^* + 2 \cos \alpha^* \cos \beta^* \cos \gamma^*}$

From G. H. Stout and L. H. Jensen (1989) *X-Ray Structure Determination, a Practical Guide*, 2nd ed., 37. John Wiley & Sons, New York.

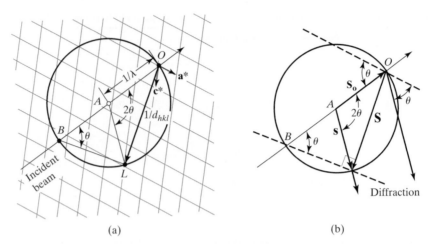

(a)   (b)

Figure 6.15   Conditions for diffraction in reciprocal space. (a) A point of origin O for the scattered X-ray beam is defined at the origin of a unit cell of the reciprocal lattice. A point A is the crystal placed along the incident beam at a distance $1/\lambda$ from O. A circle with a radius of $1/\lambda$ is drawn with A at the center. The point where the circle intersects the incident beam is labeled point B. Any other lattice point L of the reciprocal lattice that intersects the circle represents a reflection in reciprocal space. (b) Bragg's law is derived by defining the diffraction angle θ as the angle OBL, and the trigonometric relationship between the scattering vector **S** and the diameter of the circle. The vector AL is the direction of scattered beam from the crystal in real space. This is shown as the bold arrow extending from the origin O and at an angle 2θ relative to the incident beam.

There is a unique scattering vector for each reflection. Also, the length of **S** is the distance from the origin to the reciprocal space lattice point hkl ($|\mathbf{S}| = 1/d_{hkl}$). Finally, AL in Figure 6.15 must be the direction of the scattered X-rays in real space. We can see again that **S** is perpendicular to the reflecting plane. In addition, we can now define additional vectors in reciprocal space that describe the incident and reflected beams ($\mathbf{s_0}$ and **s**, respectively, both with lengths $1/\lambda$). The path difference between the incident and reflected beams is now seen to be the scattering vector **S** ($\mathbf{s_0} - \mathbf{s} = \mathbf{S}$).

We notice that the angle OAL is 2θ. Using this angle, we could derive the equations for the von Laue conditions for diffraction (which is left as an exercise for the student at the end of this chapter). When extended to three dimensions, the circle describing all possible directions for diffraction becomes a sphere (called the *Ewald sphere*) and the scattering vector **S** is the vector from the origin of the reciprocal lattice to the point of intersection between the vertices of the reciprocal lattice and the surface of this sphere (Figure 6.16). We see that the scattering vectors are discrete, each specified by a unique set of Miller indices h, k, and l. We can think of the Miller indices h, k, and l as integral counters for the reciprocal lattice planes along $\mathbf{a^*}$, $\mathbf{b^*}$, and $\mathbf{c^*}$ that intersect the Ewald sphere to result in an observed reflection.

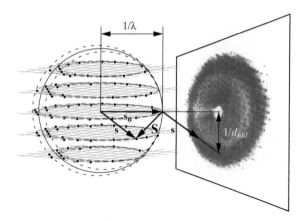

Figure 6.16 The reflection sphere in reciprocal space. The extension of the analysis in Figure 6.15 to a three-dimensional crystal is to draw a sphere with radius $1/\lambda$. Each reciprocal lattice point that intersects the surface of the sphere (filled points) is a reflection in reciprocal space (we should note that since the X-ray is not entirely monochromatic, i.e., there is some spread $\Delta\lambda$, the surface of the sphere has some depth, as represented by the dotted surface, which allows more of the lattice points to intersect and thus to provide for more reflections to come under diffraction conditions). The points included in the volume of the sphere of reflections can come into diffraction condition as the crystal is rotated.

Thus, another way of defining the von Laue conditions is that we will only observe diffraction under the conditions where

$$\mathbf{a} \cdot \mathbf{S} = h$$
$$\mathbf{b} \cdot \mathbf{S} = k$$
$$\mathbf{c} \cdot \mathbf{S} = l \tag{6.23}$$

Rotating the crystal in real space rotates the reciprocal space lattice about O, allowing different sets of reciprocal lattice points to intersect with the surface of the sphere to cause scattering. Reflections will come into the sphere and exit the sphere, allowing us to measure reflections from various parts of reciprocal space.

What does an actual diffraction pattern look like? First, let us consider a simple experiment to record the X-ray diffraction pattern on photographic film. Although film is seldom used today to actually collect data to solve molecular structures, the methods help to illustrate the basic principles of X-ray diffraction. Indeed, many modern X-ray detectors used in crystallography are nothing more than electronic equivalents of photographic film. A single recording of a crystal at some fixed orientation allows us to sample a very limited number of reciprocal lattice points. Such a *still photograph* records only those reflections resulting from those lattice planes that happen to intersect with the Ewald sphere (Figure 6.17). A larger area of reciprocal space can be sampled by oscillating the crystal about an axis. In this oscillation kind of photograph, larger sets of reflections will intersect the surface of the Ewald sphere, resulting in broad arcs of reflections (lunes) being recorded.

Figure 6.17   Still and rotation diffraction patterns. Diffraction resulting from the intersection of reciprocal lattice points with the Ewald sphere. (a) If the crystal is held *still,* the chance intersection is small, resulting in a relatively few observed reflections. (b) When the crystal is rotated, the associated lattice is also rotated and to intersect the Ewald sphere, allowing even more reflections to be observed (shaded regions).

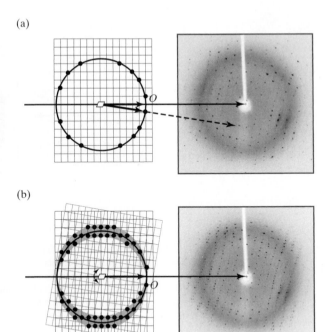

Finally, as the crystal is systematically rotated, the entire reciprocal lattice can be sampled, providing a picture of all reflections to come into view. By properly aligning the crystal and filtering the reflected beams, we can record one specific plane of reflections (usually representing slices through the three-dimensional reciprocal space lattice) that will be useful for determining the geometry and symmetry (morphology) of the crystal unit cell. This concerted rotation of the crystal (and consequently the reciprocal lattice) and filters to selectively record specified planes of the lattice is called *precession photography* (Figure 6.18). You will notice that although the reflections are arranged in a well-ordered array, as one would expect since this represents the reciprocal lattice, the intensities of the reflections vary dramatically: It is these differences in intensities that relate to the types and positions of atoms within the unit cell and therefore is the information that will be used to solve a single crystal structure.

6.4   DETERMINING THE CRYSTAL MORPHOLOGY

Before we discuss the complete analysis of each reflection and how this is used to solve the structure of a molecule, we will discuss the general features of the diffraction pattern and how this relates to the morphology of the unit cell. Bragg's law in Eq. 6.5 tells us that the spacing of the indexed reflections are inversely proportional to the lengths of the crystal unit cell. In addition, these spacings are not

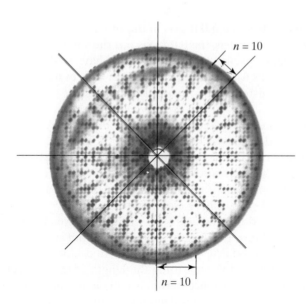

Figure 6.18 Precession photograph of the tetragonal crystal of lysozyme. The photograph was recorded along the four-fold symmetry axis. The photograph is indexed using the vertical and horizontal primary axes shown. An alternative set of primary axes for indexing is indicated along the diagonals. In this latter case, the crystal unit cell will be defined to be larger than the set chosen. The distance between 10 diagonal layer lines is smaller than that of 10 vertical layer lines, which corresponds to a larger unit cell along the diagonal. [Courtesy of P. A. Karplus.]

affected by the number, types, or positions of the atoms in the unit cell. We can therefore use these spacings to directly determine the lengths of the unit cell in the crystal, regardless of the molecules in the crystal. In addition, the orientation of the reflections in the reflection sphere mirrors the orientation of the primary axes in the unit cell (i.e., the angles between the edges of the unit cell). Finally, the symmetry within the unit cell explicitly defines the symmetry of the diffraction pattern. Thus, by observing the spacing and pattern of reflections on the diffraction pattern, we can determine the lengths and angles between each side of the unit cell, as well as the symmetry or space group in the unit cell; together these define the morphology of the crystal.

The regular pattern recorded on a precession photograph (for example, in Figure 6.18) can be used to define the unit cell parameters of a crystal. Each reflection in the diffraction pattern can be assigned to a unique set of Miller indices from the von Laue conditions for diffraction. For us to learn about the morphology of the crystal unit cell, we must first index the precession photograph in terms of the three Miller indices h, k, and l.

We can see by inspection where the three principal axes are in the diffraction pattern (these are the axes that will intersect at the center of the diffraction pattern, and thus, defines the origin at $h = 0$, $k = 0$, and $l = 0$), but what are their identities? The $h\,0\,0$ axis corresponds to the **a**-axis of the unit cell, the $0\,k\,0$ axis to the **b**-axis, and the $0\,0\,l$ axis to the **c**-axis. For the low symmetry triclinic unit cell, the **a**-axis of the crystal unit cell is defined as the shortest of the three axes. The reciprocal relationship for diffraction automatically defines the $h\,0\,0$ axis as having the largest spacing between diffraction layers. Similarly, the **c**-axis is the longest edge of a triclinic unit cell, making the $0\,0\,l$ axis the one with the smallest spacing between layer lines. Thus, the $h\,0\,0$ axis in the precession photograph has the largest spacing between reflections, the $0\,0\,l$ axis has the smallest spacing, and the $0\,k\,0$ axis has intermediate spacing. Other lattices will have well-defined rules for assigning $\mathbf{a}, \mathbf{b}$, and $\mathbf{c}$, and thus for identifying the Miller indices according to the symmetry of the unit cells, as listed in the *International Tables of Crystallography* (Hahn 1989).

Having indexed the reflections, we can now define the geometry of the unit cell, which is the lengths of and the angles between each principal axis. The angle between the $0\,0\,l$ and the $0\,k\,0$ axes defines the angle α of the unit cell, that between $h\,0\,0$ and $0\,0\,l$ defines β, and that between $h\,0\,0$ and $0\,k\,0$ defines γ. The spacing between observed reflections along each principal axis is related to the number of unit cell vectors that intersect the Ewald sphere (see Figure 6.17). Indeed, to calculate the diffraction angle 2θ, we need only measure the distance between n number of reflections on the film D_n, and know the distance between the film and the crystal R. Consider a system where the crystal is aligned and rotated along the crystallographic $\mathbf{c}$-axis, and we are interested in measuring the length of the unit cell along this axis. To do this, we measure the distance between some n number of reflections along the $0\,0\,l$ principal axis, or any line of reflections parallel to $0\,0\,l$. We see that the ratio

$$\frac{D_n}{R} = \tan 2\theta \tag{6.24}$$

This same angle relates the radius of the Ewald sphere $(1/\lambda)$ and the repeat distance $(1/d_{00l})$ of the reciprocal lattice that intersects the sphere as

$$\frac{n/d_{00l}}{1/\lambda} = \sin 2\theta \tag{6.25}$$

Therefore, we can see that the length of the crystallographic $\mathbf{c}$-axis is given by

$$d_{00l} = \frac{n\lambda}{\sin[\tan^{-1}(D_n/R)]} \tag{6.26}$$

In addition to these geometric parameters, the diffraction pattern can also give the symmetry of the crystal and thus allow the space group to be assigned to the crystal. There are numerous different combinations of conditions to define space groups. We will not attempt to go through all of these, but instead we will give a simple example of how the space group can affect the intensity of the diffraction pattern.

Let us consider the simple one-dimensional crystal with regularly spaced points along the $\mathbf{c}$-axis, just as we did to define the von Laue conditions for diffraction (Figure 6.11). The diffraction pattern will show mirror symmetry according to Friedel's law, but no additional special conditions are placed on the intensities. In the one-dimensional case, this means that the intensities at $l = n$ will be identical to those at $l = -n$.

We can now introduce a simple symmetry element to the crystal, a 2_1 (two-fold screw) axis parallel to the **c**-axis of the unit cell (Figure 6.19). For each atom A in the unit cell, there will be a symmetry-related atom A' at one-half the unit cell along the **c**-axis, rotated $180°$ relative to A. Diffraction from A and A' will be a single resultant reflection. For the $0\,0\,l$ axis, the reflections result from a projection of A and A' onto the crystal **c**-axis (this will become evident later). We notice that this symmetry relationship defines the phase and intensities of the resulting reflection from atom A and its symmetry mate A' as being identical, and the unit cell appears to be exactly one-half the length of the actual unit cell. The reciprocal relationship of the diffraction pattern to the spacing of lattice points in the crystal defines the diffraction angle, and thus the spacing of the diffracted spots is twice that expected for the actual unit cell when the crystal is aligned perpendicular to the incident beam. In terms of the Miller indices, there is a special condition: Only the even integer reflections ($l = 2n$) are observed, while all odd reflections ($l = 2n + 1$) cancel and are unobserved. When extrapolated to a three-dimensional crystal having a 2_1 axis parallel to the **c**-axis, the symmetry imposes this special condition on the intensities of

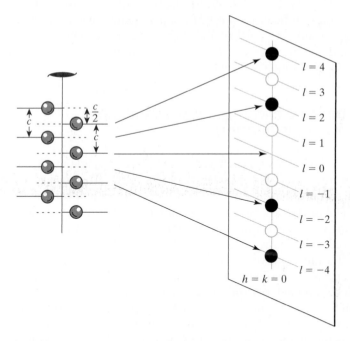

Figure 6.19   Systematic absences caused by a 2_1 axis along the crystallographic **c**-axis. A set of atoms A is spaced by a distance c along the vertical axis. A two-fold screw axis generates a set of symmetry-related atoms A' that are rotated $180°$ and translated by $c/2$ relative to the atoms A. The resulting scattered beam appears to come from a unit cell that is half the length of the actual unit cell. The corresponding diffraction angle will be twice that expected from the unit cell and, therefore, the reflections along the principal axis $0\,0\,l$ will be spaced twice the distance expected. This appears as the absence of reflections at all odd values of l along this axis in the precession photograph.

the reflections only on the 0 0 *l* principal axis. All reflections off this primary axis will show intensities at all values of *n*.

We see that this is an exclusionary condition. That is, the special condition tells us that no reflections should be observed for the odd reflections along the 0 0 *l* axis. It does not, however, guarantee that a reflection will be observed for all even *l* indices along this axis; thus, the observation that all odd *l*-indexed reflections along 0 0 *l* are missing defines a 2_1 symmetry axis that is parallel to **c**-axis of the unit cell. A crystal in the $P2_1$ space group will therefore have all the odd-indexed reflections along one of the principal axes missing. If the odd-indexed spots along an orthogonal axis are also missing, then there is an additional orthogonal 2_1 axis. Notice that this additional 2_1 symmetry axis automatically defines a third 2 or 2_1 or higher symmetry axis that is orthogonal to the first two. Thus, a monoclinic $P2_1$ space group is extended to an orthorhombic $P2_12_12$ or $P2_12_12_1$ lattice, or higher.

Each different space group specifies its own unique set of special conditions for observed and unobserved reflections along the principal and diagonal axes. For the space group $P2_12_12_1$, reflections are permitted at $h = 2n$, $k = 2n$, and $l = 2n$ (i.e., all even h, k, and l reflections) along the principal axes $h\,0\,0$, $0\,k\,0$, and $0\,0\,l$, respectively. At the odd indices along these axes, the intensities of the reflections are zero; these reflections are *systematic absences* in the diffraction pattern. Thus, the space group for a particular crystal can be determined by testing the pattern of observed and unobserved diffraction reflections in the diffraction pattern against the special conditions, as specified in Hahn (1989).

6.5   SOLVING MACROMOLECULAR STRUCTURES BY X-RAY DIFFRACTION

The model that we developed above places a single atom, and thus a single Bragg reflecting plane, within the unit cell of a crystal. The crystals that we are interested in will certainly have more than a single atom in a unit cell (more than 10,000 for a hemoglobin crystal). How does this affect the treatment of diffraction by Bragg's law or the von Laue conditions? What can we say about the crystal morphology? Let us consider the more complicated situation of two atoms in a unit cell. In the Bragg's law formulation, we simply space pairs of reflecting planes in a regular array to represent two atoms in the unit cell of a crystal. The reflections from each pair of planes can be summed to form a new resultant wave that can be considered a single reflection from a single plane. Since each pair of planes is regularly spaced within the crystal, the relative phasing of the summed reflection is the same for each pair of reflecting planes. Thus the diffraction angle is the same as before but related to the regular spacing of the *reflecting pairs*. Stated differently, the diffraction angle of a crystal is inversely proportional to the length of the unit cells in a crystal, regardless of the number of atoms that are in the unit cell.

However, the phase of the resultant wave reflected from the pair of reflecting planes may not be the same as that of the incident radiation, and the phase from each individual plane that makes the pair may not be that of the resultant wave. In addition, the amplitude of the resultant wave may not be identical to that of the incident radiation. Similarly, the atoms in a macromolecule affect the amplitude and phase of the unit cell reflections. This information gives us the structure of the molecule. However, in order to determine the positions of the atoms in the molecule (that is, to solve the structure of the crystal), we must deconvolute each reflection into the phase and amplitude contributions from each of the individual reflections from each atom in the molecule. This (as we will see) is the primary problem in trying to solve the structure of molecules in a single crystal by X-ray diffraction.

To solve the structure of a molecule within the crystal, we must determine the elemental type and position of each atom of that molecule. For a crystal of a biological molecule, this generally includes solving the structure of the ordered solvent molecules as well as the macromolecule. Defining the atom type means that we determine the number of electrons for that atom, while defining the *atomic position* means that we determine the (x, y, z) coordinates of each atom in the unit cell. We will see that in reality, this involves determining the electron density at a position in the unit cell. The most straightforward definition of atomic position is to use a set of x, y, and z values in the standard orthogonal Cartesian coordinate system. However, this becomes cumbersome when the unit cell is not orthogonal. A more convenient method is to define (x, y, z) coordinates in terms of fractions of the unit cell lengths. *Fractional cell coordinates* are the fractional distances of atoms between the origin and the next unit cell, and therefore have values between 0 and 1. The position of an atom in the standard coordinate system (with nm as the unit of length) can be defined as $(x\mathbf{a}, y\mathbf{b}, z\mathbf{c})$.

To understand how both the position and type of atoms can be determined, we must return to the beginning of our discussion on X-ray diffraction and relate these two properties to the amplitudes and the positions of the reflections in the X-ray diffraction pattern. In an X-ray diffraction experiment, the intensity of each reflection is given by the intensity of a single scattering vector $I(\mathbf{S})$. The structure of the molecule defines $I(\mathbf{S})$, and it is the decomposition of this information that is essential for solving the structure of the crystal. To understand the process of and the problems inherent in solving the structure of a macromolecule from single-crystal X-ray diffraction, we must understand how the measured quantity $I(\mathbf{S})$ is derived from the molecular structure. This comes from understanding how waves from different points of propagation combine to give the overall observed amplitude of a resultant summed wave.

6.5.1   The Structure Factor

Thus far we have determined the dimensions and the shape of the unit cell by examining the X-ray diffraction pattern. If the molecule consists of a single atom at the origin (i.e., at each lattice point) of the unit cell, we have solved the structure. However, a macromolecule has many atoms located far from the origin. To treat the scattering from a macromolecule, we will start by describing diffraction from a discrete scatterer at the origin, see how the diffracted X-ray is affected when the atom is displaced from the origin, and finally treat the scattering from multiple atoms.

As we recall, a point placed in the path of a wavefront acts as the origin of a new wavefront. Thus, if we treat light as a wave, an atom that scatters X-rays can be thought of as a point of origin for the scattered wave (Figure 6.20). If the atom is at the origin of a unit cell, then we can describe the observed amplitude of the scattered wave $E = |\mathbf{E}|$ at some point along an axis x and some time t distant from

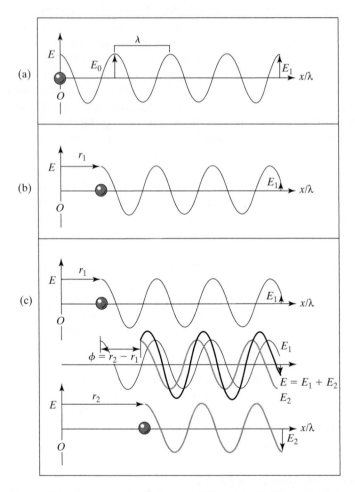

Figure 6.20 Propagation of waves. (a) A point placed at the origin O of the unit cell propagates a wave with a maximum amplitude E_0. At some point x/λ, the instantaneous amplitude is observed as E_1. (b) If the atom is displaced from the origin by a distance r_1, the amplitude of the wave is observed to be different from that propagated from the origin. (c) The wave propagated from a second scatterer at a distance r_2 from the origin will have an observed amplitude E_2. The wave resulting from both scatterers has an amplitude that is the sum of the two waves ($E = E_1 + E_2$), which is dependent on the phase difference ϕ between the two scatterers.

the origin by the standard equation for the propagation of a wave as a cosine function of the maximum amplitude, $|\mathbf{E_0}|$ (see Chapter 8 for a more detailed treatment of light).

$$E(x, t) = |\mathbf{E_0}|\cos 2\pi\left(\nu t - \frac{x}{\lambda} \right) \tag{6.27}$$

The equation for a wave that is shifted in phase by some fraction of a wave ϕ is

$$E(x, t) = |\mathbf{E_0}|\cos 2\pi\left(\nu t - \frac{x}{\lambda} + \phi \right) \tag{6.28}$$

To simplify this discussion, we will define $2\pi(\nu t - x/\lambda)$ as some phase angle ω and $2\pi\phi$ as the shift in the phase angle α, or that

$$E(\omega) = |\mathbf{E_0}|\cos(\omega + \alpha) \tag{6.29}$$

The sum of the two angles becomes

$$|\mathbf{E_0}|\cos(\omega + \alpha) = |\mathbf{E_0}|\cos \alpha \cos \omega - |\mathbf{E_0}|\sin \alpha \sin \omega \tag{6.30}$$

and since $\sin x = -\cos(x + \pi/2)$

$$= |\mathbf{E_0}|\cos \alpha \cos \omega + |\mathbf{E_0}|\sin \alpha \cos(\omega + \pi/2) \tag{6.31}$$

We can express the orthogonality between $\cos(\omega)$ and $\cos(\omega + \pi/2)$ by using "imaginary" numbers, $i = \sqrt{-1}$, since imaginary numbers and real numbers are orthogonal. Then,

$$E(\omega) = |\mathbf{E_0}|\cos \alpha \cos \omega + i|\mathbf{E_0}|\sin \alpha \cos \omega$$
$$= |\mathbf{E_0}|\cos \omega(\cos \alpha + i \sin \alpha) \tag{6.32}$$

This notation has the properties we need to keep track of the phases. Thus, any wave with a phase shift α is composed of a real and an imaginary component, and the amplitude E can be written in terms of $|\mathbf{E}| = |\mathbf{E_0}| \cos \omega$ as

$$E = |\mathbf{E}|\cos \alpha + i|\mathbf{E}|\sin \alpha \tag{6.33}$$

This form of the wave expression is convenient, because the real components can be summed separately from the imaginary components. From Eq. 6.33, we can see that a wave can be represented vectorially in a system with one axis defined as the real component $\cos \alpha$ and the orthogonal axis as the imaginary component $\sin \alpha$. The vector $\mathbf{E}$ in this axis system has a magnitude of $|\mathbf{E}| = |\mathbf{E_0}| \cos \omega$, and a direction defined by the phase angle α. This is known as an Argand diagram (Figure 6.21), and will help us to visualize the vectorial sum of multiple scattered waves in terms of their amplitudes and phase angles.

Figure 6.21 Argand diagram for a wave vector (**E**) with real (cos α) and imaginary (sin α) components of the phase angle. In this system the real component (Re) of a vector is Re(a + ib) = a while the imaginary component is Im(a + ib) = b

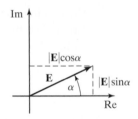

For molecules consisting of many scattering atoms, we would need to sum the amplitudes of all waves. Although the expression in Eq. 6.33 and the associated Argand diagram are convenient for visualizing the various components of the scattered waves, the addition of multiple waves becomes mathematically easier if we express the cosine and sine functions in their exponential forms

$$\cos x = \frac{e^{ix} + e^{-ix}}{2} \tag{6.34}$$

and

$$i \sin x = \frac{(e^{ix} - e^{-ix})}{2} \tag{6.35}$$

This allows us to reformulate Eq. 6.33 in terms of a complex function as follows:

$$|\mathbf{E}|\cos \alpha + i|\mathbf{E}|\sin \alpha = |\mathbf{E}|e^{i\alpha} \tag{6.36}$$

Equation 6.27 can then be rewritten as

$$E(x, t) = |\mathbf{E_0}|e^{i\omega} \tag{6.37}$$

where it is understood that we will always take the real (physically meaningful) part of the expression. The convenience of Eq. 6.37 for summing waves can be seen when we consider the resultant wave generated by two scattering atoms that are displaced from the origin by the distances r_1 and r_2. These distances are defined in unitless fractions of wavelengths, which allows us to define the amplitudes of the scattered wave from the first atoms as

$$E_1 = |\mathbf{E_0}|e^{2\pi i(vt - x/\lambda + r_1)} = |\mathbf{E}|e^{2\pi i r_1} \tag{6.38}$$

and from the second wave as

$$E_2 = |\mathbf{E_0}|e^{2\pi i(vt - x/\lambda + r_2)} = |\mathbf{E}|e^{2\pi i r_2} \tag{6.39}$$

The relative positions of the two atoms in space can be defined as a single variable $\phi = |r_2 - r_1|$; thus Eq. 6.39 can be rewritten as

$$E_2 = |\mathbf{E_0}|e^{2\pi i(r_1 + \phi)} = E_1 e^{2\pi i\phi} \tag{6.40}$$

The observed amplitude for the scattering from the two atoms is simply the sum of the two waves,

$$E = E_1 + E_2 = E_1(1 + e^{2\pi i\phi}) \tag{6.41}$$

We see that the amplitude of the summed waves is increased or decreased by a factor related to the difference in positions of the two atoms, which results in a phase shift between the two summed waves. For two waves that are exactly in phase, $\phi = 0$ cycle and $E = E_1(1 + e^0) = 2E_1$. The two waves add and show constructive interference. For two waves that are out of phase by $\pi, \phi = 1/2$ cycle and $E = E_1(1 + e^{i\pi}) = E_1(1 + \cos \pi + i \sin \pi) = 0$, taking the real part. These waves exactly cancel from destructive interference. The observed amplitudes for reflections from crystals will generally fall somewhere in between these two extremes.

If the two atoms are different types of elements, each atom will have a different number of electrons occupying a given volume in space (this is the *electron density* at that point in space). The intrinsic amplitude of the scattered X-ray from each type of atom, $|\mathbf{E}|$, is dependent on the electron density of the scatterer. The higher the electron density at any point in space, the higher the amplitude of the scattered light. We can include this into the equation for wave propagation by including an *atomic scattering factor f* for each atom. For each atom in a molecule, we can thus define an atomic scattering vector $\mathbf{f}$ with real and imaginary components as

$$\mathbf{f} = fe^{2\pi i\delta} \tag{6.42}$$

where δ is the phase angle for the single atom. This angle is defined by the position of the atom in real space [by the distance vector $\mathbf{r} = (x\mathbf{a} + y\mathbf{b} + z\mathbf{c})$] and the scattering vector ($\mathbf{S}$), such that $\delta = \mathbf{S} \cdot \mathbf{r}$, so that Eq. 6.42 can be rewritten as

$$\mathbf{f} = fe^{2\pi i\mathbf{S} \cdot \mathbf{r}} \tag{6.43}$$

where $\mathbf{S} \cdot \mathbf{r}$ for any atom j is

$$\mathbf{S} \cdot \mathbf{r}_j = (ha^*, kb^*, lc^*)(x_j\mathbf{a} + y_j\mathbf{b} + z_j\mathbf{c}) = hx_j + ky_j + lz_j \tag{6.44}$$

For multiple atoms in the molecule of a unit cell, we simply add each of the atomic scattering vectors to give a summed vector called the *molecular scattering factor* $\mathbf{F}$. The scattering factor is specific for each reflection with Miller indices ($h\ k\ l$) (and the associated scattering vector $\mathbf{S}$) and includes all atoms in the unit cell,

$$\mathbf{F}(hkl) = \mathbf{F}(\mathbf{S}) = \sum_{j=1}^{N}\mathbf{f}_j = \sum_{j=1}^{N}f_je^{2\pi i\mathbf{S} \cdot \mathbf{r}_j} \tag{6.45}$$

where N is the number of atoms in the unit cell.

The vectorial sum of all $\mathbf{f}_j$ to give $\mathbf{F}(\mathbf{S})$ can be seen in an Argand diagram (Figure 6.22) which, by analogy to E, has a magnitude $|\mathbf{F}(\mathbf{S})|$ and an overall phase angle α_{hkl} for that reflection. The amplitude of $\mathbf{F}(\mathbf{S})$ is dependent on the types and number of atoms that act as scatterers in the unit cell. It does not matter in which

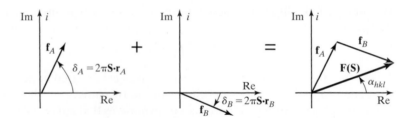

Figure 6.22 Adding the atomic scattering components of two atoms ($\mathbf{f}_A$ and $\mathbf{f}_B$) to give the summed molecular structure factor, $\mathbf{F}(\mathbf{S})$. Both $\mathbf{f}_A$ and $\mathbf{f}_B$ have their own phase angles defined as δ_A and δ_B, and equal to $2\pi\mathbf{S}\cdot\mathbf{r}_A$ and $2\pi\mathbf{S}\cdot\mathbf{r}_B$. The resulting $\mathbf{F}(\mathbf{S})$ has an overall phase angle α_{hkl} that corresponds to the sum of the individual phases of $\mathbf{f}_j$ according to Eq. 6.48.

order the $\mathbf{f}_j$ vectors are summed, the resulting magnitude and phase angles of $\mathbf{F}(\mathbf{S})$ are the same. Thus we make use of Eq. 6.36 and write Eq. 6.45 in terms of the real and imaginary components of each $\mathbf{f}_j$ to give

$$\mathbf{F}(hkl) = \mathbf{F}(\mathbf{S}) = \sum_{j=1}^{N} f_j e^{2\pi i \mathbf{S}\cdot\mathbf{r}_j} = \sum_{j=1}^{N} f_j(\cos 2\pi\mathbf{S}\cdot\mathbf{r}_j + i\sin 2\pi\mathbf{S}\cdot\mathbf{r}_j) \quad (6.46)$$

and

$$\mathbf{F}(\mathbf{S}) = |\mathbf{F}(\mathbf{S})| e^{i\alpha_{hkl}}$$

Finally, from the Argand diagram, we see that

$$|\mathbf{F}(\mathbf{S})| = \left[\left(\sum_{j=1}^{N} f_j \cos 2\pi\mathbf{S}\cdot\mathbf{r}_j \right)^2 + \left(\sum_{j=1}^{N} f_j \sin 2\pi\mathbf{S}\cdot\mathbf{r}_j \right)^2 \right]^{1/2} \quad (6.47)$$

and

$$\alpha_{hkl} = \tan^{-1} \frac{\displaystyle\sum_{j=1}^{N} f_j \sin 2\pi\,\mathbf{S}\cdot\mathbf{r}_j}{\displaystyle\sum_{j=1}^{N} f_j \cos 2\pi\,\mathbf{S}\cdot\mathbf{r}_j} \quad (6.48)$$

This tells us that the real and imaginary components of individual $\mathbf{f}_j$'s can be summed separately to give the corresponding real and imaginary components of the overall $\mathbf{F}(\mathbf{S})$.

So far, we have treated the diffracting elements as fixed and discrete points in space. As discussed above, X-rays are scattered by electrons, and we know from quantum mechanics that electrons should be treated as a probability distribution in space. This means that X-ray scattering is dependent on the electron density ρ, the number of electrons per unit volume. At any point in the unit cell, $\mathbf{r}$, we can replace f_j with an electron density, $\rho(x, y, z) = \rho(\mathbf{r})$. If more electrons occupy a given volume of space, that volume will show stronger diffraction. We can rewrite the scattering from each atom j that contributes to the electron density as

$$\mathbf{f}_j = \int_x \int_y \int_z \rho(\mathbf{r})e^{2\pi i \mathbf{S} \cdot \mathbf{r}} dxdydz \qquad (6.49)$$

If we consider the space in a crystal as having a continuous probability of finding electrons from all of the atoms, then the molecular structure factor is described by integrating over the volume (V) of the unit cell

$$\mathbf{F}(\mathbf{S}) = \int \int \int V\rho(\mathbf{r})e^{2\pi i \mathbf{S} \cdot \mathbf{r}} dxdydz \qquad (6.50)$$

where $dxdydz$ has been replaced by $Vdxdydz$ (using the fractional cell coordinates) and the integrals are over the fraction 0 to 1. This is essentially equivalent to Eq. 6.45 for discrete atoms, because positions where atoms lie are positions that will show the highest probabilities of finding electrons. Similarly, heavier atoms have higher numbers of electrons for a given volume in space. The $e^{2\pi i \mathbf{S} \cdot \mathbf{r}}$ factor depends on the positions of those scatterers in space. Both Eqs. 6.45 and 6.50 tell us that if we know the type and position of each scatterer relative to the origin of the crystal (that is, if we know the structure of the molecule), we can calculate the molecular structure factor for that crystal, which we call $\mathbf{F}(\mathbf{S})_{\text{calc}}$. Later we will use $\mathbf{F}(\mathbf{S})_{\text{calc}}$ to determine how well the final solved structure fits the observed X-ray diffraction data. Returning to the original question, how do we determine the structure of the molecules from the structure factor?

We recognize Eq. 6.50 as a Fourier series. In this case, the structure factor, which is a function of the scattering in reciprocal space, is written in terms of the electron densities in real space. A diffraction pattern of any kind (whether from X-ray or electron diffraction) is a representation of reciprocal space and looks nothing like the objects scattering the radiation. The Fourier transform as given in Eq. 6.51 is a function that gives the electron density at any particular point in real space in terms of the scattering vector in reciprocal space

$$\rho(\mathbf{r}) = \frac{1}{V} \int dV^* e^{-2\pi i \mathbf{S} \cdot \mathbf{r}} \mathbf{F}(\mathbf{S}) \qquad (6.51)$$

where V^* is the volume element in reciprocal space, and V is the real space volume of the unit cell.

Since diffraction occurs in a regular pattern and, according to the von Laue conditions, at discrete points, the electron densities can be calculated from a sum of the $\mathbf{F}(\mathbf{S})$ for all Miller indices ($h\ k\ l$). This is a Fourier series for discrete reflections as opposed to the continuous Fourier transform of Eq. 6.51.

$$\rho(\mathbf{r}) = \frac{1}{NV} \sum_h \sum_k \sum_l \mathbf{F}(\mathbf{S})e^{-2\pi i \mathbf{S} \cdot \mathbf{r}} \qquad (6.52)$$

or, since $\mathbf{F}(\mathbf{S}) = |\mathbf{F}(\mathbf{S})|e^{i\alpha_{hkl}}$,

$$\rho(\mathbf{r}) = \frac{1}{NV}\sum_h\sum_k\sum_l|\mathbf{F}(\mathbf{S})|e^{-2\pi i\mathbf{S}\cdot\mathbf{r}+i\alpha_{hkl}} \qquad (6.53)$$

Thus, if we know the magnitudes $|\mathbf{F}(\mathbf{S})|$ and phase angles α of all possible reflections h, k, and l, we can calculate the electron density at any point $\mathbf{r}$ from the origin in the unit cell. When the electron densities calculated from Eq. 6.53 are displayed for the (x, y, z) positions for all values of $\mathbf{r}$ in the crystal unit cell, we generate an *electron density map* for the unit cell of the crystal (Figure 6.23). This map shows the positions and the types (the number of electrons for a given volume) of each atom in the asymmetric unit of the crystal.

A convenient method for interpreting structural information from an electron density map is to plot the map as a set of contours, as in a geographical topography map (Figure 6.23a). Each set of concentric contours represent peaks of electron density, normally centered at an atom or at groups of atoms, depending on

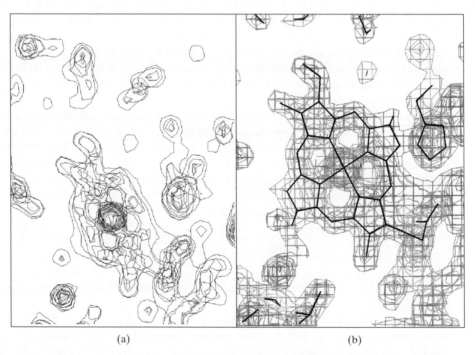

(a)   (b)

Figure 6.23 Electron density maps. The electron density calculated from the Fourier transform in Eq. 6.52 can be represented (a) by a contour map or (b) as a set of chicken wires. In this figure, the electron density of the heme-binding pocket of myoglobin is shown. In (a), four sections of the contour map are overlapped to show the electron density at the heme and the surrounding amino acid residues. An enlarged view of this same set of electron densities is shown in (b) as surrounding the model of the heme (solid lines).

the resolution. The map can be constructed to present the electron density of the unit cell in three-dimensional space by first drawing contours of electron density in one plane of the unit cell. This plane represents one section of the map. The three-dimensional unit cell is sectioned along the axis perpendicular to the contour plane. When the sections are stacked, we generate a three-dimensional view of the electron density in the unit cell. A molecular model can thus be derived from the electron density map by connecting the centers of each peak with the chemical bonds in the molecule. With the advent of three-dimensional computer graphics, such contouring techniques are being supplemented by a simpler "chicken wire" representation of the electron density map (Figure 6.23b). Both types of representations have their utility in solving a macromolecular structure.

Equations 6.45 and 6.50 tell us that all atoms and their associated electrons contribute to the structure factor of each observed reflection, and all observed reflections must therefore be used to completely determine the types and positions of each atom in space. Because we will never in practice be able to measure *all* the possible reflections, we must actually truncate the Fourier series in Eq. 6.53 at some finite values of (h k l). This means we can never have a completely accurate description of the contents of the unit cell, and that we get less information (lower resolution) concerning the structure when there are fewer reflections included in the series. If we can measure the structure factors for a large number of scattering vectors (**S**), then we can readily calculate the electron density at any point in space and thus solve the structure of the crystal. There is a problem, however. Although **F(S)** can be calculated from a known set of atomic positions, it cannot be directly measured in a diffraction experiment. We can measure the intensity [$I(\mathbf{S})$] of the diffracted X-ray for each reflection, and the structure factor is related to the intensity; but, as we see in the next section, the phase information is not given by the measured intensities, and that precludes directly solving the atomic structure of a crystal from the observed X-ray diffraction data.

6.5.2   The Phase Problem

We showed above that the structure factor can be derived starting with a wave description of light propagating from two or more scattering points. Knowing the structure factor, we can calculate the electron density at any position in the crystal unit cell and thus solve the structure of the molecules within the crystal. However, there is a problem in measuring the structure factor. We have devices that detect both the amplitude and phase for light of long wavelengths. In Chapter 12, we discuss nuclear magnetic resonance (NMR) spectroscopy and show how the Fourier series measured in the microwave region can be transformed directly into the NMR spectrum. Similarly, we see in Chapter 9 that infrared absorption can be detected as a Fourier series that can be transformed directly into the infrared spectrum. Unfortunately, the devices that we have available to detect short-wavelength light measure total energy. This will depend on the time period of energy collection and the energy per unit time, which is the intensity I of light. The total energy collected will be proportional to the intensity, and it can be shown (for instance, see Halliday et al.

2001) that the intensity of a light wave is proportional to its amplitude E, squared. In terms of the complex notation of Eq. 6.45, the intensity of a reflection $h\,k\,l$ is

$$I(h\,k\,l) = I(\mathbf{S}) = |\mathbf{F}(\mathbf{S})|^2 = \mathbf{F}(\mathbf{S})\mathbf{F}^*(\mathbf{S}) \tag{6.54}$$

where $\mathbf{F}^*(\mathbf{S})$ is the *complex conjugate* of $\mathbf{F}(\mathbf{S})$, which is

$$\mathbf{F}^*(\mathbf{S}) = \sum_{i=1}^{N} f_j e^{-2\pi i \mathbf{S}\cdot\mathbf{r}} \tag{6.55}$$

One outcome from this set of expressions is that $I(\mathbf{S}) = I(-\mathbf{S})$, or that a reflection at hkl has the same intensity as a reflection at $-h$-k-l. As you recall, Friedel's law tells us that the von Laue conditions for scattering at an angle of 2θ (and its associated hkl indices) will also define a reflection at -2θ (for $-h$-k-l). We now see that the intensities of the Friedel pairs (reflections at hkl and $-h$-k-l or at $\mathbf{S}$ and $-\mathbf{S}$, respectively) should also be identical.

A more bothersome outcome is that in determining $\mathbf{F}(\mathbf{S})$ from $I(\mathbf{S})$, we lose critical information for solving the structure of the molecule. From Eq. 6.54, we can show that $|\mathbf{F}(\mathbf{S})| = I(\mathbf{S})^{1/2}$. Thus, we can determine the amplitude of the structure factor directly from the observed intensities, but we do not have the phase information. A simple analogy in mathematics is to take $\sqrt{4} = |x|$. In this case, $x = \pm 2$. We know that the magnitude of $x = 2$, but do not know its sign. This is known as the *phase problem* in crystallography. Referring to the Argand diagram in Figure 6.22, we know the magnitude of the structure factor $|\mathbf{F}(\mathbf{S})|$, but not its direction. How critical is this loss of information? Let us go back to a very simple situation where we have two atoms (A and B), with atom A at the origin of the unit cell and B displaced by $\mathbf{r}$ from the origin (Figure 6.24). The phase angle for A is $\delta_A = e^{2\pi i \mathbf{S}\cdot\mathbf{r}_A} = e^0 = 1$ (that is, this atomic scattering vector lies along the real axis). For B, $\delta_B = e^{2\pi i \mathbf{S}\cdot\mathbf{r}_B}$. The resulting molecular structure factor $\mathbf{F}(\mathbf{S})$ will have

Figure 6.24 Effect of shifting the origin of the unit cell on the overall phase angle of two atoms. The left panels represent two atoms, with atom A at the origin of the unit cell and B displaced by $\mathbf{r}$ from A (*top left*), and the resulting atomic scattering vectors $\mathbf{f}_A$ and $\mathbf{f}_B$ to give a phase angle δ_B for atom B and α for the sum of the two atoms (*lower left*). When the origin of the unit cell is shifted, this is equivalent to shifting the position of the two atoms by some distance $\mathbf{R}$ (*upper right*). The result is that the phases of both atoms and the overall phase angle α are rotated by an additional angle $2\pi\mathbf{S}\cdot\mathbf{R}$ (*lower right*).

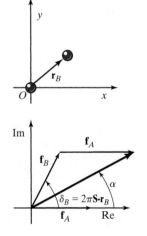

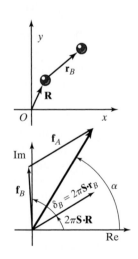

an amplitude and direction that is the sum of these vectors. Now, by shifting the origin of the unit cell by some distance vector $\mathbf{R}$, we add $\mathbf{R}$ to both $\mathbf{r}_A$ and $\mathbf{r}_B$ to give $\delta_A = e^{2\pi i \mathbf{S}\cdot(\mathbf{r}_A+\mathbf{R})} = e^{2\pi i \mathbf{S}\cdot\mathbf{r}_A}e^{2\pi i \mathbf{S}\cdot\mathbf{R}}$ and $\delta_B = e^{2\pi i \mathbf{S}\cdot(\mathbf{r}_B+\mathbf{R})} = e^{2\pi i \mathbf{S}\cdot\mathbf{r}_B}e^{2\pi i \mathbf{S}\cdot\mathbf{R}}$. In other words, by shifting the origin of the unit cell, both $\mathbf{f}_A$ and $\mathbf{f}_B$ and consequently $\mathbf{F(S)}$ are rotated by the angle $2\pi\mathbf{S}\cdot\mathbf{R}$. What happens if we were to allow $\mathbf{R}$ to be any value? In this case, $2\pi\mathbf{S}\cdot\mathbf{R}$ can be any angle added to $\mathbf{F(S)}$ and, consequently, the resultant structure factor sweeps out a circle with a radius $= |\mathbf{F(S)}|$ (Figure 6.25). This is exactly analogous to not knowing the phase angle α_{hkl} when $|\mathbf{F(S)}|$ is calculated as $I(\mathbf{S})^{1/2}$, which is that the molecular structure factor sweeps out a circle in the Argand diagram with a length $= I(\mathbf{S})^{1/2}$. Thus, without knowledge of the phase angle α_{hkl} for the molecular structure factor, we do not know where the origin of the unit cell is and, without the origin, it is impossible to determine $\mathbf{r}$ (or the absolute $x, y,$ and z coordinates) of the individual atoms. The consequence of this loss in information can be seen in Figure 6.26, which compares the electron density calculated only from the amplitudes of $\mathbf{F(S)}$ and not the phases to that calculated using only the phases.

Why bother with X-ray diffraction if we apparently cannot directly solve structures using this method? Obviously, the technique would not be useful and would not have become such a widely used and powerful technique if there were not ways around the phase problem. In the following sections, we will discuss some general methods for solving the phase problem, and thus for solving crystal structures. These include molecular replacement, direct methods, and isomorphous replacement. In molecular replacement, a known structure of a related molecule is used as the initial model for the unknown structure. This method has been generally useful for solving the structures of various nucleic acids and of structural variants of known protein structures. This method, however, will not work for a completely unknown structure.

Direct methods use various tricks to solve the phase problem, either by trying all possible phase combinations for each $\mathbf{S}$ and simply finding that combination that best fits the overall data to solve the structure, or by using the phase information for each atom inherent in the intensity data to retrieve some information concerning the relative positions of atoms in the crystal (this is known as the *Patterson method*). These methods are currently useful only for directly solving the structures of small

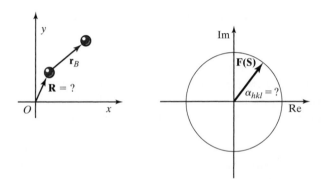

Figure 6.25 When the location of the unit cell is not known. If the origin of the unit cell is unknown, this is equivalent to shifting all of the atoms in the unit cell by some unknown distance $\mathbf{R}$ (left panel). The result is that all of the atomic scattering vectors are related by any angle $2\pi\mathbf{S}\cdot\mathbf{R}$ and, consequently, the overall phase angle α_{hkl} is unknown (right panel). The corollary is that if α cannot be determined, then the location of the unit cell origin is unknown.

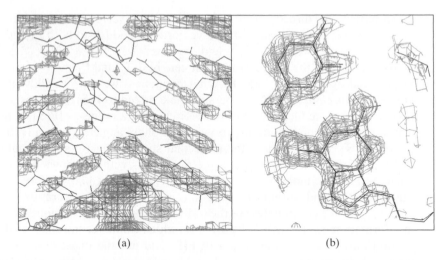

(a)   (b)

Figure 6.26 Electron density calculated from the two components of $\mathbf{F}(h\,k\,l)$. In (a), the electron density of a DNA crystal was calculated using only $|\mathbf{F}(h\,k\,l)|$ from the X-ray diffraction data. The map does not fit the model of the DNA structure, but resembles the pattern expected for the Patterson function. In (b), the same map was calculated using only the phase information for $\mathbf{F}(h\,k\,l)$ with $|\mathbf{F}(h\,k\,l)|$ set at 1.0 for all reflections. The resulting map very closely resembles the dC·dG base pair in the structure. This demonstrates the importance of the phasing information over the magnitude of the structure factor.

molecules (up to 100 to 300 nonhydrogen atoms) because of the exponential growth in the phase problem as the size of the molecule increases. We will discuss the Patterson method in some detail, since it is also a useful first step for the more general method of isomorphous replacement.

In isomorphous replacement, specific atoms in the crystal are tagged with heavy atoms, atoms that have very high electron densities and therefore can strongly perturb the X-ray diffraction pattern. Once the positions of these heavy atoms are located within the crystal (using direct methods or the Patterson method), the overall phase of the original molecule can be estimated. This is a general technique for solving the structures of completely unknown structures of macromolecules including nucleic acids, proteins, and complexes of these.

At the end of this section, we discuss some of the newer methods for solving the phase problem, including anomalous dispersion with synchrotron sources.

Molecular replacement.   As we have shown above, once we have a rough model for a structure, we can estimate $\mathbf{F}(\mathbf{S})$ for all values of $(h\,k\,l)$ for that structure. If the structure that we are trying to solve is very similar to one that has been previously determined, we can simply use those elements within the known structure that are common to both to generally "phase the data." This is perhaps the simplest method for solving the structure of a macromolecule, and it has been applied to solve structures as small as oligonucleotides and as large as virus particles. There are a number of specific criteria that must be met before molecular replacement will work. The first is that the previously determined structure must be very similar to the one we are trying to solve. If they are dissimilar to any great degree, the phasing information will not be sufficiently accurate to give good electron density maps. This limits the types of crystal problems that are amenable to this method. One class of problems for which molecular replacement is an appropriate method is the determination of the structure of a mutant protein from the structure of the native protein, or the structure of homologous proteins from different species. Another class of problems for which molecular replacement has been successfully used is in solving the structures of double-helical oligonucleotides. In this latter case, the chemical and physical properties of the molecules are very well understood from other experimental methods. In addition, the regularity of the structures allows reasonable models to be built entirely from the symmetry of the structures (see Chapter 1).

In these situations, the number of atoms that differ between the known and unknown structures is small, while the number of those that are identical both chemically and in spatial location is large. The common regions supply both the atomic scattering factors (from the chemical formula) and the phase information (from the atomic coordinates of the atoms) for a large number of scatterers. In the case of a mutant protein structure solved from a native structure, the two proteins may differ by less than one amino acid out of a hundred. Thus, the calculated structure factors can be very accurate. We can simply exclude the atoms that are not shared between the known and the unknown structures and calculate an electron density map for the regions that differ. This is known as a *difference electron density map* (or an *omit map*), since atoms that are dissimilar are omitted from the **F(S)** calculation (see Figure 6.27). The quality of the difference map depends on

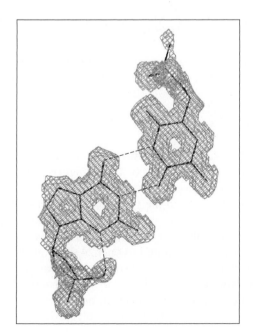

Figure 6.27 The omit map calculated for the overhanging bases of the DNA fragment d(GCGCGCG)·d(TCGCGCG). The electron density was calculated using only the six dC·dG base pairs of the duplex region (underlined) to phase the data. The residual electron density that is not accounted for by the six base pairs is shown to be that of a dG·dT reverse-wobble base pair (the model is shown as solid lines, and the hydrogen bonds in the base pair as dashed lines). [Adapted from Mooers et al. (1997), *J. Mol. Biol.* **269**, 796–810.]

the accuracy of the phases calculated from the starting model. With a good difference map, however, we can use the residual electron density to build those parts of the unknown structure to fit the observed densities.

The other constraint to molecular replacement is that the model must sit in the correct orientation and absolute position within the unit cell. If both of these criteria are not met, even a perfect model will not properly phase the diffraction data. This is particularly critical when using a model derived from other methods (e.g., from structures determined by NMR), or from a structure having very different unit cell parameters. Obviously, the simplest case is one where the model that will be used to phase the new data comes from a structure that sits in exactly the same orientation and position within the crystal, and the crystals are isomorphous.

When the simplest case does not hold, then the starting model must somehow be placed in the new unit cell and properly positioned. This is normally accomplished by using a series of rotation and translation functions to "fit" the model to the electron density. Again, the success of this method depends on how closely the initial structure matches the unknown structure.

The Patterson function. Contained within the observed intensities of the X-ray diffraction data are the *relative* phases of the individual atomic scattering vectors, and thus information on the relative positions of the atoms in the unit cell. In trying to solve the crystal structure, why do we not simply use the observed intensities to construct a Fourier series that will be some function of the atomic positions? This is the basis for the Patterson function P. Using Eq. 6.38, the Patterson function can be defined as

$$P(x, y, z) = \frac{1}{V}\sum_h\sum_k\sum_l I(\mathbf{S})e^{-2\pi i \mathbf{S}\cdot\mathbf{r}} = \frac{1}{V}\sum_h\sum_k\sum_l \mathbf{F}(\mathbf{S})\mathbf{F}^*(\mathbf{S})e^{-2\pi i \mathbf{S}\cdot\mathbf{r}} \quad (6.56)$$

If the transform of $\mathbf{F}(\mathbf{S})$ is $\rho(\mathbf{r})$, it is not surprising that the transform of $\mathbf{F}^*(\mathbf{S})$ is $\rho(-\mathbf{r})$. The resulting Patterson function has the form,

$$P(x, y, z) = \sum_{j=1}^{N}\sum_{k=1}^{N}\rho_j(\mathbf{r}_j)\rho_k(-\mathbf{r}_k) \quad (6.57)$$

The density map calculated from the Patterson function reveals peaks that correspond to the vector differences between the atomic positions (Figure 6.28).

We can see from this that there is a loss of information as we go from atoms in real space to the vector differences derived from the Patterson function. A very real indicator of this lost information is found in the symmetry of a Patterson map. The symmetry of a crystal unit cell can be described by one of 230 different space groups. The symmetry of a Patterson map, however, is limited to only 24 space groups. These 24 Patterson space groups can be generated by first removing all the translational elements of the symmetry operators from the original crystal space group. Thus, two-fold screws become two-fold rotational axes, and so on. In addition, all Patterson maps are centrosymmetric; therefore, there is always a symmetry axis at the origin.

What does this mean in terms of the positional information available from the Patterson function? The physical description of this function is that each peak of a Patterson map corresponds to a distance vector separating two atoms. These are not absolute distances from the origin of the unit cell, but relative distances between individual atoms. Consider the simple case where two atoms A and B are located at positions (x_A, y_A, z_A) and (x_B, y_B, z_B) in the unit cell of a crystal (Figure 6.24). For $i = A$ and $j = B$ in the Patterson function, the difference vector $(\mathbf{r}_i - \mathbf{r}_j) = (\mathbf{r}_A - \mathbf{r}_B) = (x_A - x_B, y_A - y_B, z_A - z_B)$. Similarly, for $i = B$ and $j = A$, $(x_B - x_A, y_B - y_A, z_B - z_A)$. Finally, for $i = A$ and $j = A$, and for $i = B$ and $j = B$, the distance vectors are zero, placing two vectors at the origin of the Patterson map. This means that if we plot the Patterson function in a manner similar to an electron density map, there will be three peaks, two corresponding to a distance vector separating A from B and B from A, called the *cross vectors,* and the third corresponding to the *self-vectors* at the origin. The magnitudes of the peaks are related to the product of the electron density of the two atoms separated by the distance vector. If two different vectors are coincident, the Patterson peaks are additive. The relative magnitudes would thus be $1:1:2$ for the peaks at $(x_A - x_B, y_A - y_B, z_A - z_B), (x_B - x_A, y_B - y_A, z_B - z_A)$ and $(0, 0, 0)$, respectively. We see that for a molecule having a large number of atoms, the self-vectors at the origin very quickly become the dominant feature of a Patterson map, and the number of cross vectors grows as the square of the number of atoms.

There are a number of important concepts that must be stressed in understanding the utility and the limitations of the Patterson function. First and foremost, the function defines relative distances and not absolute positions. However, this does not prevent us from using the Patterson method to solve structures. Consider an atom A that sits at the end of a distance vector $\mathbf{r}$ from a two-fold symmetry axis. There would be a symmetry-related atom A' at some distance $-\mathbf{r}$ from the axis. The distance separating A from A' is $\mathbf{r} - (-\mathbf{r}) = 2\mathbf{r}$. Thus, in the Patterson map, there would be a single vector projecting from the origin with a length of $2\mathbf{r}$. To use this infomation in determining the absolute position of an atom in the unit cell, we must find a section of the Patterson map that reflects the symmetry of the unit cell. The Patterson map itself does not necessarily assume the symmetry of the unit cell, but certain sections (called the *Harker sections*) reflect some of the symmetry elements of the original space group of the crystal (Figure 6.28). For space groups containing rotational or screw-symmetry axes, these occur at the origin of $\mathbf{a}, \mathbf{b}$, and $\mathbf{c}$, or some fraction of the unit cell along $\mathbf{a}, \mathbf{b}, \mathbf{c}$.

The usefulness of the Harker section is that distance vectors for symmetry-related atoms lie in this plane. For example, if we have a single atom A at some position (x_A, y_A, z_A) and there is a two-fold screw parallel to the crystallographic $\mathbf{a}$-axis, then a symmetry-related atom A' must sit at a position $(x_A + 1/2, -y_A, -z_A)$. The difference vector relating A and A' is $(1/2, 2y_A, 2z_A)$. Thus, at the Harker section in the $\mathbf{b}$-$\mathbf{c}$ plane at $1/2$ along the $\mathbf{a}$-axis of the unit cell, there will be a Patterson peak corresponding to the coordinates $(2y_A, 2z_A)$. The absolute coordinates y_A and z_A of atom A can thus be determined directly from the Patterson peak in the Harker plane by taking half the x- and y-coordinates of the observed Patterson peaks.

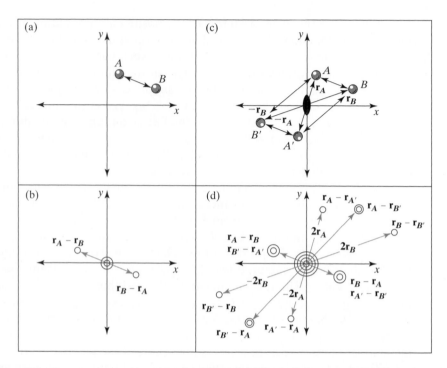

Figure 6.28 Patterson maps of two atoms in a unit cell. (a) Two unique atoms A and B in a unit cell are displaced from the origin by the distance vectors $\mathbf{r}_A$ and $\mathbf{r}_B$. (b) The Patterson map for the two atoms shows two cross vectors, one for the vector from A to B and the other from B to A. The two self-vectors (A to A and B to B) result in two contours at the origin of the map. (c) Two additional atoms, A' and B', are generated in a crystal with two-fold rotational symmetry. Although there are still only two unique atoms, there are now four additional cross vectors. If this is a Harker section in the Patterson map, the additional cross vectors are $2\mathbf{r}_A$ and $2\mathbf{r}_B$. This allows us to determine $\mathbf{r}_A$ and $\mathbf{r}_B$ (or the atomic coordinates of A and B) directly from the Patterson map.

Notice, however, that we cannot determine the coordinate x_A from this Harker section. This requires additional information, which could come from the Patterson peaks in a perpendicular Harker section, if this is allowed by the symmetry of the unit cell.

Why can we not simply use the Harker planes to determine the coordinates in three-dimensional space of all the atoms in a molecule? For a single atom in the asymmetric unit, this would be a trivial problem, and for a small number of atoms it is a tractable problem. However, as the number of atoms increases, so do the number of symmetry-related atoms and correspondingly the number of Patterson peaks in the Harker sections. At some point it becomes impossible to resolve the individual peaks in the Patterson map, even from three perpendicular Harker sections. Thus, the Patterson method is really useful only for locating a small number of atoms within the unit cell. How can this be a useful method in solving structures of macromolecules? We see in the next section that by determining the positions,

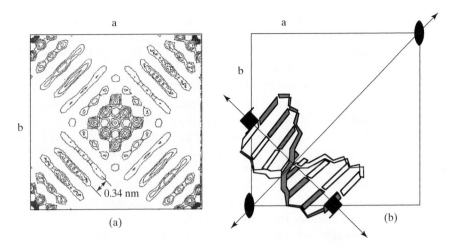

a

a

b

b

0.34 nm

(a)

(b)

Figure 6.29   Patterson map of B-DNA. (a) The Patterson map of an eight base pair duplex DNA in a tetragonal crystal shows regular densities spaced by 0.34 nm. The duplex must therefore be B-DNA (with a rise of 0.34 nm), with the helical axis lying in the plane and aligned diagonally to the crystallographic axes **a** and **b**. (b) The asymmetric unit of this fragment is one strand of the duplex as indicated by the volume of the asymmetric unit. The second strand is generated by two-fold rotation. This automatically places the asymmetric unit on the two-fold axis of the crystal and allows the structure to be solved entirely from the Patterson map and the symmetry of the crystal lattice.

and thus the phases, of a few particular atoms, the overall phase of the protein can be estimated.

For crystals of nucleic acids, there are special situations where the Patterson map can go a long way in solving the structure of the molecule. Because the base pairs of the DNA duplex are stacked along an axis, the large number of redundant atom-atom distances between base pairs will tend to dominate the Patterson map. In this case, we can specifically determine the orientation of the base pairs relative to the crystal axes, determine the helical rise of the structure (and thus the general conformation of the molecule), and with some luck the position and orientation of the molecule within the unit cell (Figure 6.29). This can provide a starting model that can readily be used to solve the structure of all the atoms in the molecule by molecular replacement.

Multiple isomorphous replacement.   The final method that we discuss in detail for solving molecular structures (particularly those of macromolecules) is isomorphous replacement. The *isomorphous* part of this method indicates that we will be using crystals whose unit cells are nearly identical in shape and size and are in identical space groups. The *replacement* part of the method indicates that we will be substituting, or in many cases adding heavy atoms to the molecules in the unit cell. The general method is as follows.

1. The X-ray diffraction data are collected for a crystal of a protein or other biological molecule. This is called the *native* data set, since it comes from the unmodified crystal. The structure factors from the native crystal will be called F_P, from the protein crystallographer's nomenclature, although it can apply to crystals of any biological molecule.

2. Isomorphous crystals are obtained of the identical molecule in which a strongly diffracting atom (a heavy atom, such as a metal or a halogen) is attached at a specific location. The X-ray diffraction data for these crystals are the *heavy atom derivative* data sets. The structure factor for the modified crystals will be called F_{PH}, since it reflects the structure of the starting native molecule and the heavy atom(s) in the crystal.

3. A derivative data set is appropriately scaled to the native set, and a difference data set representing the scattering from only the heavy atom is obtained by subtraction. The structure factors of the difference data F_H reflect only the scattering of X-rays by the heavy atoms.

4. The F_H are used to determine the positions and the phases of the heavy atoms in the unit cell (for example, by the Patterson method described above).

5. This process is repeated for at least one additional heavy atom derivative.

6. The phases of at least two heavy atom derivatives are used to estimate the phase for the native data set to solve the structure of the macromolecule in the native crystal.

We start this discussion with the isomorphous heavy atom crystals. There are three criteria for obtaining a successful heavy atom derivative. First, the procedure should not significantly affect the structure of the crystal or the molecules within the crystal (that is, the crystals must remain isomorphous with the native crystal). Second, the modifications must be specific. This means that we know what amino acid or nucleic acid residue is being modified. Finally, the number of heavy atoms within the asymmetric unit should be small. This makes it easier to locate the positions of the heavy atoms in the unit cell.

There are a number of methods for specifically introducing one or more heavy atoms into a macromolecular structure and maintaining the characteristics of the native crystal. The classical method is to start with the native crystal of the unmodified macromolecule and soak metals or other atoms into the crystal (Application 6.2). A list of common derivatives that are specific for certain residue types in proteins and polynucleotides is listed in Table 6.4. Any dramatic changes in the morphology of the crystal becomes obvious if the crystal cracks during the soaking procedure. More subtle modifications either to the crystal morphology or the macromolecular structure are indicated when the unit cell parameters and space group are determined from the X-ray diffraction pattern of the modified crystal.

Table 6.4 Heavy Atom Derivatives for Macromolecular Crystals

Heavy Atom	Specificity
Proteins	
$AgNO_3$	His, Cys (minor)
$K_2Pd(Br$ or $Cl)_4$	Arg, His
Hg acetate	His, Cys
p-chloromercuric benzene sulphonate (PCMBS)	His
Se	Selenomethionine (incorporated during synthesis)
Nucleic Acids	
Cu	Guanine bases
Pt	Guanine bases
I	Iodouridine (incorporated during synthesis)
Br	Bromouridine (incorporated during synthesis)

Application 6.2   The Crystal Structure of an Old and Distinguished Enzyme

The enzyme urease, which catalyzes the hydrolysis of urea, was first isolated in crystalline form in 1926 by James B. Sumner. Urease was in fact the first enzyme to be crystallized, demonstrating that enzymes were chemical compounds with distinct identities. It was also found to be the first enzyme to utilize nickel in catalysis. The crystal structure of urease, however, was not solved until 1995 in the laboratory of P. Andrew Karplus (Jabri et al. 1995), some 70 years after the first crystals were obtained. The enzyme for structural analysis was isolated from the bacterium *Klebsiella aerogenes,* rather than jack bean. However, the two forms of the enzyme are homologous, with more than 50% of their sequences being identical.

The structure was solved by multiple isomorphous replacement, with 40 heavy atom compounds screened to find five workable derivatives of the crystal (Table A6.2). The phasing of the diffraction data was aided by anomalous scattering from the heavy atoms and incorporation of selenomethionine into the protein by growing bacteria in media containing the modified amino acid. The resulting structure showed the enzyme quaternary structure to be a trimer of heterotrimers. Three complexes of the α, β, and γ subunits associate to form a triangular unit (Figure A6.2). In addition, an unusual, although not unique, carbamylated lysine was observed to bridge the two nickels in the active site. The structure raises some interesting questions, including how this active site evolved. The structure of the catalytic site shows great similarity to that of adenosine deaminase, a zinc-containing enzyme. The two enzymes also share some similarities in their catalytic mechanisms, including a metal-coordinated water that attacks the amide carbon of the substrates, and the tetrahedral intermediate that releases an ammonia to form the product. This is a clear case of divergent evolution of the catalytic sites dictated by the requirements of the substrate rather than convergent evolution driven by the requirements of the catalytic mechanism.

Table A6.2   Crystallographic Data and Results for Urease

Crystal	Resolution of Data	Number of Unique Reflections	Final R Factor	Nonhydrogen Protein Atoms	Solvent Molecules
Native	2 Å	58,334	18.5%	6002	215
Apoenzyme	2.8 Å	20,532	18.4%	5944	157
$HOHgC_6H_4CO_2Na$	3.3 Å	11,027			
$EuCl_2$	3.3 Å	12,210			
$Hg_2(CH_3COO)_2$	2.5 Å	28,709			
$C(HgOOCCH_3)_4$	2.4 Å	29,672			
$(CH_3)_3Pb(CH_3COO)$	2.4 Å	23,486			
Se-Met	3.0 Å	20,332			

Data from Jabri et al. (1995).

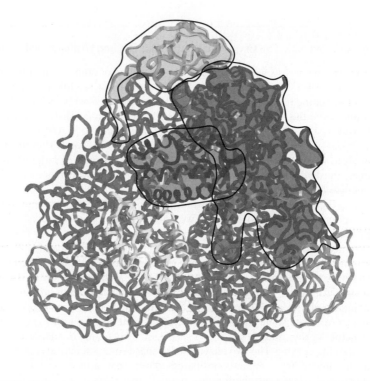

Figure A6.2   Quaternary structure of urease from the bacterium *Klebsiella aerogenes*. The structure of the protein consists of a heterodimer of α, β, and γ subunits to form a unit with a molecular weight of 83,000 mg/mol. Three of these units then associate to form a larger trimer.

JABRI, E., M. B. CARR, R. P. HAUSINGER, and P. A. KARPLUS (1995), *Science* 2: **68**, 998–1004.

In this case, only the magnitudes of the observed and calculated structure factors are compared. Generally, R factors range between 70% (0.7) for a random fit to a value approaching 0% for an ideal perfect fit (Figure 6.31). An ideal fit is not achievable, primarily because of the limitations inherent in the quality of the data and imperfections in the crystal. For macromolecules, a general rule is that a model refined to an R factor of 20% (0.2) or better indicates a good fit, and thus can be presented as being a solution to the structure. To achieve this level of correlation between the model and the data, the structure of the macromolecule along with a large portion of the solvent (water, ions, etc.) must be well defined. Unfortunately, an artificially low R factor can be achieved for a poor model by simply adding more solvent to account for the electron density in the unit cell. We see that solving the structure of a macromolecule from X-ray diffraction data is highly dependent on the construction of molecular models and on the quality of these models. There are a number of steps where model bias can creep in, but several objective criteria are available to evaluate the quality of the resulting models.

Recently, crystallographers have recognized the need for a measure for how well the model fits that is not biased by the model itself. The approach that is now widely used is to first sequester a fraction (about 5% to 10%) of the experimental data from the total data set to use as a test set. The model is refined only against the remaining data (the working set). During refinement of the model, and R factor is calculated for both the working set and the test set according to Eq. 6.58. This latter R factor is called R_{free}; recognizing that it should be free of any bias from the model. If during the course of refinement the model is truly improving, then R_{free} should continue to be reduced. At some point, however, R_{free} will reach a plateau and, even if further refinement or additions to the model reduces the working R factor, the model is considered to be as good as the data allow. Upon further additions to the

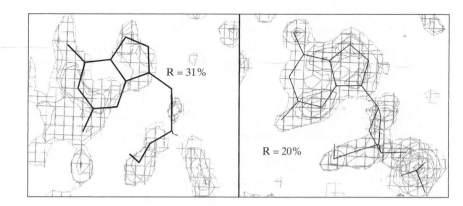

Figure 6.31   Effect of refinement on structure. The guanine nucleotide of a DNA fragment is shown with its electron density map prior to refinement and after refinement. Prior to refinement, the R factor is 31%. The structure is refined against the data to an R factor of 20%, which is one criterion of a good fit of the model to the data.

model, R_{free} may actually start to increase even as R for the working set continues to decrease—this indicates that the data are being overfit (more variables are being introduced than the data allow).

Other methods for phasing X-ray diffraction data. Of the three methods that we have discussed, multiple isomorphous replacement by heavy atoms is the most common technique for solving a previously undetermined macromolecular structure. The usual method for determining the positions of the heavy atoms is from difference Patterson maps. However, there are alternative methods to locating heavy atoms, and the structures of very small proteins and polynucleic acids can be solved by the increasingly more sophisticated direct methods, particularly as more powerful computational resources become available.

An alternative approach to determining the positions of heavy atoms is to take advantage of the properties inherent to these atoms. Atoms with high electron densities not only scatter X-rays, they also absorb X-rays. When the energy of the radiation used in the diffraction experiment can be absorbed by (that is, near the *absorption edge* of) the scattering atom, the assumption that the atomic scattering factor f is independent of the phase is no longer valid. This leads to a phenomenon known as *anomalous dispersion*, which we observe as a breakdown of Friedel's law—then the difference in intensities between Friedel pairs can be used to determine the phase of the heavy atoms. We will discuss anomalous scattering in some detail, because an extension of this method can be used to estimate the phase of the macromolecule using only a single heavy atom derivative.

Our assumption that f is phase independent comes from treating the scatterers as unperturbed electrons. In reality, the interaction of light with any particle requires some change in the properties of that particle. Classically, an electron that scatters X-rays will be induced to vibrate or oscillate. When the wavelength of the radiation approaches the absorption edge of an atom, the electron will resonate at its absorption frequency v_0. This is not significant for the light atoms that constitute the majority of scatterers in macromolecules, but it becomes important for heavy atoms, such as those used in isomorphous replacement. For these heavy atoms, f can be treated as the sum of three terms: f_o (the atomic scattering factor for a spherically distributed electron), and the perturbations f' and if'' to the real and imaginary components of the atomic scattering factor. Thus, for these atoms even the atomic scattering factor includes a component that will affect the phase of the scattered radiation.

The corrections to the isotropic f_o term affect each member of the Friedel pair differently, with

$$f_+ = f_o + f' + if'' \tag{6.59}$$

and

$$f_- = f_o - f' - if'' \tag{6.60}$$

a $2\theta = 180°$, which is the incident beam reflected back on itself. This represents the theoretical limit to θ. The practical limit in terms of instrument design is $2\theta \approx 110°$, which corresponds to $\sin \theta = 0.82$, or $d = 0.094$ nm resolution for CuK_α radiation. This is the wavelength generally used in solving the structures of biological macromolecules.

The two wavelengths typically used in X-ray crystallography are the K_α bands emitted when electrons bombard a copper or a molybdenum anode (Table 6.5). Since $\lambda = 0.154$ nm for X-rays from copper radiation, the best resolution that can be theoretically obtained from this source is $d = 0.077$ nm, while for molybdenum radiation $d = 0.036$ nm. Why do we not use the shortest possible wavelength of radiation to achieve the highest resolution data? The problem lies in a different resolution problem, that is, in the ability to resolve closely spaced reflections in the X-ray diffraction pattern. Although we have treated the reflections from X-ray diffraction as single points, in reality they are normal distributions representing the quantum mechanical probability that a photon will be observed at a particular point in space. This means that each reflection has a width in real space, usually at least $1°$ across θ at half-height. The separation between two reflections must therefore be greater than $1°$ for us to be able to distinguish them as discrete points. This places a limitation on the longest edge of the unit cell that can be studied using either Cu or Mo radiation. For example, a unit cell with a maximum length of 5 nm (a small protein), will have reflections separated by about $2°$ (as estimated from Bragg's law for CuK_α radiation). That same crystal using Mo radiation shows reflections separated by only about $0.8°$, which would not allow the reflections to be resolved. Thus, crystals of biological molecules, with their large unit cell dimensions, require the use of the longer wavelength of X-rays, which is typically CuK_α radiation.

We have shown above that atomic resolution depends on the diffraction angle of the reflections used to determine the parameters for each atom in the crystal unit cell. Now we ask the question: How much data are required to obtain atomic resolution? The number of reflections N is equal to the number of reciprocal lattice points contained in the volume (V^{-1}) of the sphere of reflections (recall that the sphere of reflections is defined in reciprocal space and therefore its volume is in units of $1/nm^3$).

$$N = \frac{4}{3}\pi|S|^3/V^{-1}$$

Table 6.5   X-Ray Radiation and Resolution

Radiation	CuK_α	MoK_α
λ	0.15418 nm	0.07107 nm
$(d_{hkl})_{min} = \lambda/2$	0.07709 nm	0.03554 nm

From G. H. Stout and L. H. Jensen (1989) *X-Ray Structure Determination, a Practical Guide*, 2nd ed., 37. John Wiley & Sons, New York.

or

$$N = \frac{4}{3}\pi V/d^3 \qquad (6.61)$$

in real space.

Finally, let us ask the question: At what point do we have enough data to solve a structure to atomic resolution? To solve a structure, we must assign values to four parameters for each atom in the molecule. To solve the structure to atomic resolution, these parameters must be determined independently. This means that at least one reflection must be collected for each atomic parameter.

The atomic parameters determined from X-ray diffraction are the (x, y, z) coordinates of the atom, and some indication of the type of electron distribution at this position. This fourth parameter is often the *temperature factor,* or *B factor.* The *B* factor was originally introduced as a measure of the thermal motion of the atom. For a given atom of n electrons, a higher *B* factor indicates that these electrons occupy a larger volume. Assigning the properties of atoms to a single parameter assumes that the thermal motions of atoms are isotropic. In fact, the vibrations of atoms in a molecule are anisotropic, and thus the best model for a molecule should include the thermal motion in all three directions in space $(B_x, B_y,$ and $B_z)$. This is observed only in the highest resolution structures, because it dramatically increases the number of parameters required to define an atom. For macromolecules, the *B* factor is usually treated as a single isotropic value except where very high resolution data is available.

In addition to thermal motion, the *B* factor also reflects how often an atom is positioned at a particular point in space. This is the *occupancy* of an atom at a specific position. For example, an atom may be very rigidly held in one place and thus occupy a very well-defined volume in space. However, if it sits in two different positions in different molecules in the crystal, the occupancy at any one position represents only half the electrons of the atom. This results in a lower occupancy and a higher *B* factor for that atom. In practice, the *B* factor is a measure of both thermal motion and occupancy, which reflects the overall *disorder* of the atom.

Thus, we see that at least four unique reflections are required to define the parameters of each atom in the asymmetric unit. For 350 atoms, which occupy a volume of about 6 nm^3, this translates to a minimum of $N = 1400$ reflections. From Eq. 6.45, we see that this number of reflections can be collected at $d = 0.26$ nm or at $2\theta = 34°$. This diffraction angle is significantly lower than we would expect for atomic resolution from Bragg's law. Our calculation, however, greatly overestimates the number of reflections that can be used to solve a molecular structure because not all the reflections in the sphere of reflections are unique.

The number of *unique* data points that can be collected for a specific resolution limit is restricted by the symmetry of the diffraction pattern. For example, Friedel's law tells us that $I(h, k, l) = I(-h -k -l)$, indicating that even for a crystal with the lowest symmetry ($P1$), only half the data (or one hemisphere of the sphere of reflections) is unique. Additional symmetry may reduce this even more. For example, for a crystal with $V = 25$ nm^3 (as in the case of the 350 atom molecule), we would expect 104,720 reflections at 0.1 nm resolution, of which 52,360 would be unique for a $P1$ lattice. For the higher symmetry lattice $P2_12_12_1$, only 13,090 would be unique. This larger number of data points available from a low symmetry space group of a given unit-cell volume corresponds to the fewer asymmetric units contained in this lattice and, thus, the larger number of atoms within the asymmetric unit. Fortunately, for the same volume, molecules in higher symmetry lattices require fewer reflections to solve their structure because of the symmetry relationships among the molecules. The two effects (the number of unique reflections and the number of atoms in the asymmetric unit) just balance. For example, a crystal with $V = 25$ nm^3 in the $P2_12_12_1$ lattice would have four asymmetric units in the unit cell, with each typically containing about 350 atoms. In the $P1$ lattice, there would be a single asymmetric unit containing more than 1400 atoms. Thus, for a $P2_12_12_1$ lattice containing 350 atoms, there are potentially 13,090 unique reflections or 9 per parameter at $d = 0.1$ nm, if all the reflections are reliably collected to this level.

What prevents us from solving all single crystal structures to atomic resolution? We are first limited by the quality of the crystal. Thus far, we have treated crystals as exact repeats of unit cells. In these unit cells, the molecules are rigid and have identical conformations. We showed very early on in this chapter that any deviation from this ideal case would necessarily reduce our ability to resolve the atoms of the structure. In fact, the crystals of macromolecules are not perfect. The molecules are not entirely static, and are not exactly identical across all unit cells. This often results in parts of molecules being unresolved, even if most of the structure is well defined and of high quality. Thus, the structure that is solved has been averaged over the unit cells of the crystal, and the resolution to which a structure can be solved depends on just how close the unit cells are to being identical. This is the primary limitation to the resolution of the data that can be collected. We can work the logic backward and show that this loss in resolution due to time or spatial averaging of the structure will result in fewer reflections observed at high diffraction angles. Indeed, any amount of disorder will tend to broaden and reduce the intensity of all the reflections observed from the crystal. To minimize thermal disorder, crystallographic data can be collected at low temperatures (to liquid-nitrogen or even liquid-helium temperatures). This, however, does not eliminate any structural disorder that is inherent in the crystal unit cell.

The resolution of the reflections is also limited by the size of the crystal. We can simply state that the larger a crystal is, the more repeating scatterers there are in the path of the X-ray beam. This translates into sharpening of all reflections, and thus more data that can be reliably measured. This brings us back to the initial problem in X-ray crystallography, that is, to grow better crystals.

6.6   FIBER DIFFRACTION

A single crystal has all its molecules and the atoms in the molecules arranged in well-ordered and repeating arrays. Can X-ray diffraction provide structural information on molecules that do not form ordered three-dimensional crystalline arrays? The answer is yes, if there is inherent symmetry within the molecule itself to cause constructive and destructive interference of the diffracted X-rays. We observe this in long helical biopolymer fibers. We discuss fiber diffraction here, first for historical reasons. The structural features of the helices of biopolymers (the α-helix from Linus Pauling's work and the DNA duplex from Franklin and Wilkins as interpreted by Watson and Crick) were first determined from fiber diffraction studies. Second, the method still is applicable for providing low-resolution structures of macromolecular complexes that are not amenable to single-crystal X-ray diffraction, including prions and sickle-cell hemoglobin fibers.

We can think of the molecules in oriented fibers as being aligned and therefore ordered in one dimension. If the molecules have a regular secondary structure, then the inherent symmetry of the helices will generate repeating units that are aligned along the fiber axis. We can therefore treat an exact repeat of the helix (see Chapter 1) as a crystalline unit cell with well-defined lattice points. The molecules within the fiber, however, are randomly rotated relative to each other. Thus, the unit cells are rotationally disordered and averaged across the width of the fiber. When treating fiber diffraction in the same manner that we do single-crystal diffraction, we first need to investigate the unit cell, the general properties of the diffraction pattern from this unit cell, and the effect of the individual residues on the diffraction pattern. We can then see how the diffraction pattern can provide us with information on the structural parameters of the helix, including the helical symmetry, pitch, and radius.

In fiber diffraction, the repeat of the helix produces layer lines along the long axis of the fiber (the z-axis), like that from a one-dimensional crystal (Figure 6.12). The diffraction along this long axis are the meridinal reflections (Figure 6.34). The molecules in a fiber may also pack in a semicrystalline array around the diameter to produce a set of layer lines that lie perpendicular to the z-axis. These are the equatorial reflections. In the remainder of this section, we see how the meridinal reflections provide information on the pitch and symmetry of a helix, while the equatorial reflections give us the radius of the helix.

6.6.1   The Fiber Unit Cell

The regular and repeating unit in a helical fiber of a biopolymer is not a box defined by a parallelepiped but a series of stacked repeating cylinders. It is more convenient,

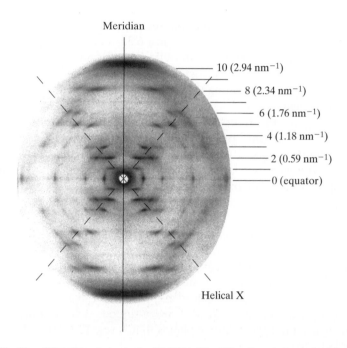

Meridian

10 (2.94 nm^{-1})

8 (2.34 nm^{-1})

6 (1.76 nm^{-1})

4 (1.18 nm^{-1})

2 (0.59 nm^{-1})

0 (equator)

Helical X

Figure 6.34   The fiber diffraction photograph of B-DNA. The diffraction photograph of the lithium form of a DNA fiber (recorded at 66% humidity) shows the helical X expected for helical structures and 10 layer lines spaced according to n/P in nm^{-1} between the origin and the exact repeat of the pattern. This indicates that the fiber is B-DNA. [Courtesy of R. Langridge].

therefore, to define the unit cell in cylindrical coordinates (Figure 6.35). In this case, the unit cells are all aligned relative to a single axis (the helix axis, which is defined as the z-axis). Any point within this unit cell has coordinates r, ϕ, and z, representing the distance from the helix axis, the rotation about the helix axis, and the vertical distance along the helix axis. Since diffraction from fibers must conform to Bragg's law for constructive and destructive interference, these coordinates can be treated

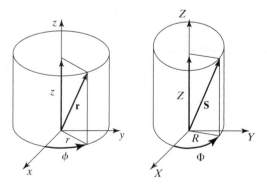

Figure 6.35   Cylindrical unit cell in real space and in reciprocal space. A point in a cylindrical unit cell is defined by the parameters z for the vertical length, r for the radius, and ϕ for the angle of the point from the x-axis. This defines a vector **r** for the position of a point relative to the origin of the unit cell. In reciprocal space, the analogous parameters are Z, R, and Φ to define the scattering vectors **S** relative to the origin.

in reciprocal space by defining the reciprocal lattice parameters R, Φ, and Z to describe the scattering vector, $\mathbf{S}$ (Figure 6.35). Since the helices exactly repeat along the z-axis, the parameter Z will define the spacing of layer lines in the diffraction pattern in reciprocal space, just as the length of the c-axis of a unit cell defines the spacing of layer lines along the $\mathbf{c}^*$-axis in reciprocal space for the one-dimensional crystal in Figure 6.11.

Because the fibers are not crystalline, there are very few reflections in a fiber diffraction pattern. Although there is far from enough information to determine the positions of atoms in the helix, we can show how detailed information on the fundamental structure can be derived for a helical biopolymer by discussing the type of patterns that we expect for different types of helices, and comparing these to patterns observed for actual helices of polypeptides and polynucleotides.

We will start by deriving the diffraction pattern for a continuous helix based on the diffraction from a line that is wrapped around the stacked cylindrical unit cells of the fiber (Figure 6.36). However, the helices of biopolymers are not continuous. They contain discrete residues (amino acid, nucleotide, or saccharide building blocks) that are regularly spaced along this continuous helix. A more representative fiber diffraction pattern for this discontinuous helix will then be derived by introducing a set of spaced lattice planes that is regularly spaced like the residues along the helix axis.

6.6.2   Fiber Diffraction of Continuous Helices

To derive the fiber diffraction pattern for a continuous helix, we start by defining the structure factor for the helix. The most general expression for the structure

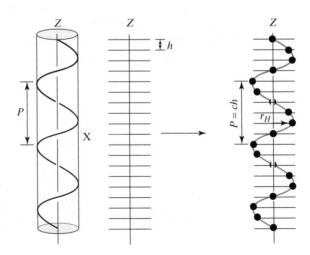

Figure 6.36 Continuous and discontinuous helices. A continuous helix is a line wrapped around the surface of a cylinder. The intersection of this line with a set of lattice lines defines the positions of the residues of a discontinuous helix.

factor in this polar coordinate system $F(R, \Phi, Z)$ is related to the electron density of an infinite solid cylinder by the Fourier series.

$$F(R, \Phi, Z) = \int_0^\infty r\,dr \int_0^{2\pi} d\phi \int_{-\infty}^{+\infty} dz \rho(r, \phi, z) e^{2\pi i [rR(\cos\phi\sin\Phi - \sin\phi\cos\Phi) + zZ]} \tag{6.62}$$

where the exponential term is analogous to the $e^{2\pi i \mathbf{S} \cdot \mathbf{r}}$ term in Eq. 6.43 for single crystals.

This is greatly simplified if we assume that the helix is a line wrapped on the surface of the cylinder: Then lattice points occur only at the radius of the helix, $r = r_H$. At all other values, $r \neq r_H$ and $\rho(r, \phi, z) = 0$. This means that $F(R, \Phi, Z)$ is defined only for $r = r_H$. Using a trignometric identity for ϕ and Φ,

$$F(R, \Phi, Z) = \int_0^{2\pi} d\phi \int_{-\infty}^{+\infty} dz\, e^{2\pi i [r_H R\cos(\phi - \Phi) + zZ]} \tag{6.63}$$

The unit cell has a length equal to the pitch of the helix, P. Therefore, any point along z can be described in terms of the fractional unit cell relative to the pitch $(z = P_\phi/2\pi)$. If we index the reflections along Z relative to the reciprocal lattice, $Z = n/P$ (where n is the index for the Bragg reflection or observed layer line), then the integral simplifies to a form that is dependent only on ϕ.

$$F\left(R, \Phi, \frac{n}{P}\right) = \int_0^{2\pi} e^{2\pi r_H R\cos(\phi - \Phi) + in\phi}\, d\phi \tag{6.64}$$

This is similar in form to the Bessel functions, $J_n(x)$, which describe the amplitudes of dampened sine and cosine functions for the variable x.

$$J_n(x) = \left(\frac{1}{2\pi i^n}\right) \int_0^{2\pi} e^{ix\cos y + iny}\, dy \tag{6.65}$$

For $x = 2\pi r_H R$, the structure factor has the form

$$F\left(R, \Phi, \frac{n}{P}\right) = J_n(2\pi r_H R) e^{i\pi(\phi + \pi/2)} \tag{6.66}$$

The order of each Bessel function, n, is equivalent to the index of each layer line of the diffraction pattern (the zeroth-order Bessel function describes the intensity distribution for the $n = 0$ layer line; see Figure 6.37). As the index n increases to

Figure 6.37 Bessel functions, $J_n(x)$. Bessel functions are shifted to higher values of x and are reduced in intensity $J_n^2(x)$ as the order of the function n is increased. The intensity $I(\mathbf{S})$ along the layer lines in a fiber diffraction pattern is proportional to J_n^2.

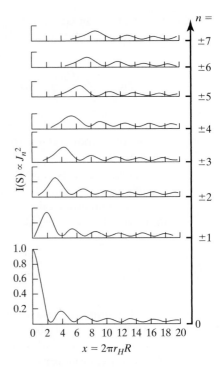

higher orders, the maximum amplitude shifts to higher values of $2\pi r_H R$ (equivalent to the scattering vector $\mathbf{S}$). For each value of n, the amplitude decreases as $2\pi r_H R$ increases.

To construct the fiber diffraction pattern for a continuous helix, we start by defining the meridian axis Z and the equatorial axis R, remembering that an X-ray diffraction pattern reflects the symmetry of the unit cell in reciprocal space (Figure 6.38a). The intensity at each layer line n is calculated as $I = F(R, \Phi, n/P)$ $F^*(-R, \Phi, n/P)$ with the spacing between layer lines equal to $1/P$ along the meridian axis. For each layer line n, $J_n(x) = J_n(-x)$, so the pattern shows mirror symmetry across the meridian axis. Similarly, $J_n(x) = J_{-n}(x)$ so the pattern also has mirror symmetry across the equatorial axis. The maximum intensities at all layer lines form the familiar X pattern associated with helical structures (the *helical X*). A line that connects the maximum intensity of the layer lines forms an angle δ relative to the meridinal axis. For a continuous helix, this δ is related to the inclination angle of the helix relative to the z-axis. Indeed, the diffraction pattern of the continuous helix is drawn as a helical X through the origin of the reciprocal lattice, with each arm of the X placed at an angle equal to δ. The intersection of this X with each layer line indicates the position at which we would expect to observe a reflection from the Bessel function.

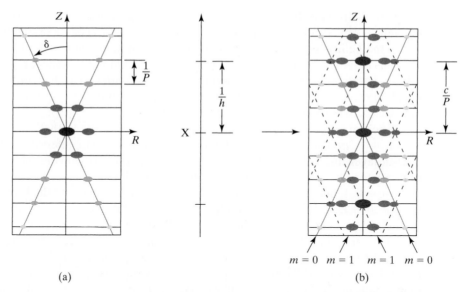

$$m = 0 \quad m = 1 \quad m = 1 \quad m = 0$$

(a)   (b)

Figure 6.38   The diffraction pattern of continous and discontinuous helices. The fiber diffraction pattern of a continuous helix (a) has layer lines spaced by $1/P$ and a helical X at an angle δ relative to the meridian axis. The pattern for a discontinuous helix having an integral number of residues in an exact repeat is derived by the intersection of the fiber diffraction from the continuous helix and the reciprocal lattice of the unit cell. The spacing of the lattice lines indicates the position of the repeating pattern. The pattern for the discontinuous helix repeats the helical X at the repeat c of the helix. In this example, $c = 3$ residues per turn for a 3_1 helix. The order of the Bessel function at each layer line is defined by $n = l - cm$, where l is the index of the layer line and m is an integer indexing the repeat. The original repeat of the helical X pattern has $m = 0$, the first repeat has $m = 1$, and so on. [Adapted from C. R. Cantor and P. R. Schimmel (1980) *Biophysical Chemistry* Part II, 799. W. H. Freeman, New York.]

6.6.3   Fiber Diffraction of Discontinuous Helices

Although the diffraction pattern is regular for a continuous helix, it does not repeat as we would expect for a repeating structure such as a helix of a biopolymer. This arises because continuous helices are dependent only on the pitch and the radius of the helix. The pattern will not repeat until we define distinct unit cells and repeating lattice motifs within the unit cells that can act as reflecting planes. This is analogous to the repeated lattice motifs of the single-crystal unit cell that act as reflecting planes in Bragg's law.

We incorporate the exact helical repeat c, as defined for a helix in Chapter 1, into the function that describes the spacing of the layer lines, Z. For the continuous helix, $Z = n/P$. For the discontinuous helix, we define a new index for each layer line l, such that $Z = l/P$, with l being a function of n and c. This means that the X pattern of the continuous helix will exactly repeat itself along the meridinal axis.

There are two distinct situations that must be considered in accounting for the helical repeat. In the simplest case, the helix exactly repeats in one turn (e.g., where the helical symmetry is explicitly c_1, as defined in Chapter 1). The value of l for this helix is $l = (cm + n)$, where we use m to index the helical repeat. Thus, we can determine which order Bessel function contributes to each layer line by trying all the possible values of m that satisfy the equation $n = l - cm$. Take the simple case of the 3_{10} helix for a polypeptide chain, which, as we recall from Chapter 1, has 3_1 helical symmetry. The Bessel functions that contribute to the $l = 0$ layer line are $|n| = -3m$, which are 0, 3, and 6 for $m = 0, 1$, and 2. The intensity distribution at each layer line will be dominated by the Bessel function having the lowest $|n|$, or $n = 0$ for $l = 0$ in this case. Similarly, for $l = 1$, $|n| = 1 - 3m$, which are 1, 2, and 5 for $m = 0, 1$, and 2. The dominant Bessel function is $n = 1$, and the layer line will be spaced at $Z = 1/P$. This is no different from what is expected from the continuous helix. However, for $l = 2$, the dominant Bessel function is $|n| = 1, 2$, and 5 (for $m = 1, 0$, and -1). Thus, the intensity distribution at $l = 2$ is predicted to be exactly identical to $l = 1$. We can similarly show that the pattern at $l = 3$ is identical to $l = 0$. Thus, the overall diffraction pattern exactly repeats at $l = c$. This is presented diagramatically in Figure 6.38b. If we start with the X pattern of the continuous helix at $l = 0$, we can generate the pattern for the discontinuous helix by simply repeating the helical X at $l = c$ and $l = -c$. These represent $l = mc$ for $m = \pm 1$. We can therefore interpret the index m as representing the scattering vector for one unit cell repeated along the helical axis both above and below the original unit cell, in reciprocal space. This is analogous to the Miller indices (hkl), which count the number of reciprocal lattices along the **a**-, **b**-, and **c**-axes of the unit cell in a single crystal. If each layer line is spaced as $Z = 1/P$, then the pattern is repeated at $c/P = 1/h$ (where h is the helical rise).

In the more complex situation, the helix does not exactly repeat in a single turn. The α-helix, for example, has a helical symmetry of 18_5, or 18 residues in 5 turns of the helix of the exact repeat. If we represent the helical symmetry more generally as c_T where c is the number of residues per exact repeat and T the number of turns in the exact repeat, it can be shown that the layer lines of the diffraction pattern will be regularly spaced by $Z = 1/TP$. Each layer line l will have a pattern of intensities defined by the Bessel functions with order $n = (l - mc)/T$. Since the order of the Bessel functions must be an integer, the trick is to find values of m that yields integer values for n from this equation. In these cases, the diffraction pattern is still a repeat of the helical X from the continuous helix with the repeat pattern centered at the layer line $l = c$ for $m = \pm 1$ and spaced at $Z = 1/h$. Each successively higher order of the Bessel function (from 0 to 1 to 2, etc.) occurs at $l = nV$ for the helical X at $m = 0$. In other words, the actual repeating pattern reflects the integer number of turns of the helix required to exactly repeat the helix (Figure 6.39). In the case of the α-helix, this would be at layer line $n = 18$.

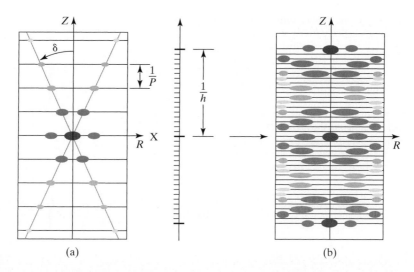

Figure 6.39 Fiber diffraction pattern for a helix with a nonintegral number of residues per turn. An α-helix, which has 3.6 residues per turn, has a helical symmetry of 18_5. The resulting fiber diffraction is predicted to be the intersection of the pattern from the continuous helix (a) with a reciprocal lattice representing the number of residues in the exact repeat of the helix (18 in this case). The resulting pattern therefore has 18 layer lines, with the spacing between lines equal to $1/cP$. In the α-helix, this spacing is $1/5P = 0.37 \text{ nm}^{-1}$. [Adapted from C. R. Cantor and P. R. Schimmel (1980) *Biophysical Chemistry*, Part II, 799. W. H. Freeman, New York.]

The practical application of these formulae is that the X-ray fiber diffraction pattern can be used to determine the structural properties of a helical fiber, including the helical symmetry and pitch. As with single-crystal diffraction, a reflection will be observed when the reciprocal lattice intersects the sphere of reflections. However, for a fiber, the lattice points lie only along the helix axis. In a typical fiber diffraction study, the helix axis must be tipped in order to observe the repeating pattern along the meridinal axis (Figure 6.40). The angle at which the fiber is tipped allows the reciprocal lattice for one repeat to intersect the sphere of reflection to cause scattering. Otherwise, the meridinal reflections will not repeat in the diffraction pattern, and we cannot determine the repeating unit of the helix.

The radius of the helix r can be determined from the packing of the cylindrical unit cells. Since this is perpendicular to the helix axis, the packing of the helices affects the reflections along the equatorial axis. The most efficient packing of cylinders is a hexagonal array (Figure 6.41a). There are two distinct repeating patterns in this array: one related to r, the other to $\sqrt{3}r$. Again, since the diffraction pattern reflects the helical parameters in reciprocal space, we would expect to observe reflections along the equatorial axis spaced at $R = 1/\sqrt{3}r, 1/r, 2/\sqrt{3}r, 2/r$, and

Figure 6.40 Recording a fiber diffraction pattern. The fiber diffraction pattern of a helix represents the intersection of the fiber lattice with the sphere of reflection in reciprocal space. If the fiber is aligned exactly perpendicular to the incident X-ray beam, the exact repeat of the lattice may not intersect the sphere, and therefore the diffraction pattern may not show the layer line for the exact repeat along the meridinal axis. The fiber must be tipped to bring the repeat of the lattice onto the sphere to record the repeating pattern.

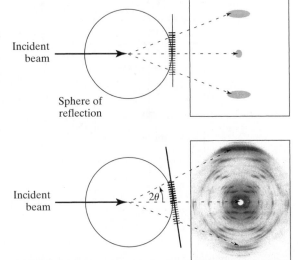

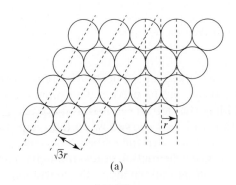

Figure 6.41 Equatorial reflections. The equatorial reflections of a fiber diffraction pattern result from the semicrystalline packing of the helices along the diameter. (a) The helices are regularly spaced by their radius r and by $\sqrt{3}r$. (b) The reflections along the equator of the fiber diffraction pattern, therefore, are spaced by $1/\sqrt{3}r$, $1/r$, $2/\sqrt{3}r$, $2/r$ and so forth.

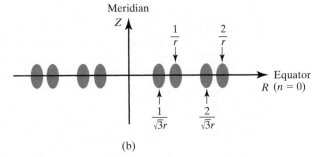

so on (Figure 6.41b). Thus, the width of a helix can be determined from the spacing between equatorial reflections.

Fiber diffraction has a venerable history. Before crystallography, it was the only way to use X-ray diffraction to study the structures of biopolymers such as polypeptides and polynucleotides. Although the data were severely limited, models for biopolymers were made that were consistent with the data. Of course, this did not prove that the model was correct, but many models proved to be very good, and the method contributed significantly to our understanding of biopolymer structures. For example, the structure of the B-DNA duplex derived from fiber diffraction very accurately describes the mechanisms of semiconservative replication. Still, the only reliable information concerning the helical structure are the pitch, the exact repeat, and the radius. We cannot, for example, determine the handedness of the helix, let alone the position of every atom in space. Indeed, the early models for DNA and RNA, including the details of the Watson and Crick type base pairs, were not confirmed as being correct until single crystals of these structure were solved nearly 20 years later.

EXERCISES

6.1 Cryoelectron microscopy can resolve structures to 0.7 nm.
  a. What is the approximate wavelength λ (in nm) and energy (in eV) of the radiation used in the method?
  b. What type of structure(s) might you be able to resolve using this method (such as helices, domains, subunits of hemoglobin, and so on)?

*6.2 Starting with the equation for the molecular structure factor $\mathbf{F(S)}$,
  a. Show that the intensity of reflections along the $(h\,0\,0)$ axis will be zero for all odd values of h when a two-fold screw axis is aligned along the crystallographic **a**-axis. [Hint: start by placing an atom at a specific set of x, y, and z coordinates (in fractional unit cells), and generate its symmetry related atom in the unit cell.]
  b. Show that Friedel's law is correct.

6.3 For an atom with atomic coordinates $x = 1.51$ nm, $y = 0.35$ nm, and $z = 0.50$ nm in a tetragonal unit cell with lengths $a = b = 20.0$ nm, $c = 30.0$ nm,
  a. Calculate the x, y, z coordinates in fractional unit cell lengths.
  b. If this is the only unique atom in the unit cell, calculate the phase angle for the $h = 2, k = 3, l = 4$ reflection.

*6.4 Derive the von Laue condition for diffraction for the two-dimensional reciprocal crystal lattice in Figure 6.19.

6.5 The following is the Patterson map of a molecule composed of identical atoms in a plane. Each contour represents the intensity of the peak at each position.

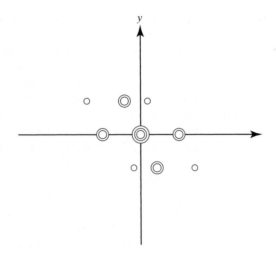

a. How many atoms are in the molecule?
b. Draw the structure of the molecule, and show how this structure would give the Patterson map shown here.

6.6 The precession photograph in Figure 6.14 was recorded with a crystal to film distance of 7.65 cm (the spacing between reflections was measured to be 0.15 cm). The third dimension for this crystal is 3.777 nm.
a. Calculate the unit cell dimensions of this crystal.
b. Propose one possible space group for this crystal.
c. If the third axis has 4_1 symmetry, what systematic absences would you expect to see?
d. Estimate the number of unique reflections one would expect for a data set at 0.133 nm resolution.

6.7 You collected X-ray diffraction data from a protein crystal and from an isomorphous crystal of a derivative having a single selenomethionine derivative. Contours for the selenomethionine heavy atom were observed at the Harker sections of the difference Patterson map at $x = 0.43$, $y = 0.22$, $z = 0.5$ and at $x = 0.43$, $y = 0.5$, $z = 0.35$ (in fractional unit cell coordinates).
a. What are the atomic coordinates of the heavy atom?
b. Calculate F_H for the reflection $h = k = l = 2$.

6.8 For a π-helix (4.4 residues per turn):
a. Write the helical symmetry of this helix for full repeats.
b. Predict the lowest-order Bessel function that would contribute to each layer line starting from 0 to l = full repeat.
c. Equatorial reflections were observed at 0.020, 0.023, 0.035, and 0.040 nm^{-1}. What is the radius of the helix?

6.9 For A-DNA (11 bp/turn, 0.26 nm rise):
a. Predict the two lowest-order Bessel functions that will contribute to layer lines $l = 0$ and $l = 5$.

b. Calculate the spacing between layer lines in reciprocal space and in real space (assuming a fiber to film distance of 12 cm).

c. The diameter of A-DNA is approximately 2.2 nm. Estimate the spacing of the equatorial reflections in nm^{-1}.

d. Sketch the fiber diffraction pattern expected for Z-DNA (2 nm diameter).

REFERENCES

Crystallization of Macromolecules

DUCRUIX, A. and R. GIEGÉ (1992) *Crystallization of Nucleic Acids and Proteins, a Practical Approach,* Oxford University Press, Oxford, UK. Gives a thorough description of how to crystallize macromolecules and what is known about the theory of macromolecular crystallization.

There are a number of *factorial* or *sparse matrix* methods to screen initial crystallization solutions for proteins and nucleic acids. Some are described by:

BERGER, I., C.-H. KANG, N. SINHA, M. WOLTHERS, and A. RICH (1996) "A Highly Efficient Matrix for the Crystallization of Nucleic Acid Fragments," *Acta Cryst.* **D52**, 465–468. Conditions for the crystallization of DNA fragments.

DOUDNA, J. A., C. GROSSHANS, A. GOODING, and C. E. KUNDROT (1993) "Crystallization of Ribozymes and Small RNA Motifs by a Sparse Matrix Approach," *Proc. Natl. Acad. Sci., USA* **90**, 7829–7833. Solutions for the crystallization of RNA fragments.

JANCARIK, J. and S.-H. KIM (1991) "Sparse Matrix Sampling: A Screening Method for Crystallization of Proteins," *J. Appl. Cryst.* **24**, 409–411. Some conditions used for protein crystallizations.

X-Ray Diffraction

The texts with the most thorough description of the theories and methods of X-ray crystallography:

BLUNDELL, T. L. and L. N. JOHNSON (1976) *Protein Crystallography,* Academic Press, London. The general methods of crystallography as applied to proteins.

DRENTH, J. (1994) *Principles of Protein X-Ray Crystallography,* Springer-Verlag, New York.

HAHN, T. (ed.) (1988) *International Tables of Crystallography, Brief Teaching Edition of Vol. A,* 2nd ed., Kluwer Academic Press, New York. A teaching guide that shows how the International Tables of Crystallography are used.

STOUT, G. H. and L. H. JENSEN (1989) *X-Ray Structure Determination, a Practical Guide,* 2nd ed., John Wiley & Sons, New York. A complete description of crystallography in general.

HALLIDAY, D., R. RESNICK, and J. WALKER (2001) *Fundamentals of Physics,* 6th ed., John Wiley & Sons, New York.

Fiber Diffraction

CANTOR, C. and P. R. SCHIMMEL (1980) *Biophysical Chemistry,* Part II, W. H. Freeman, New York, Chapter 14.

WILSON, H. R. (1966) *Diffraction of X-Rays by Proteins, Nucleic Acids and Viruses,* St. Martin's Press, New York.

Some papers that describe recent applications of fiber diffraction to study protein and nucleic acid structures include:

INOUYE, H. and D. A. KIRSCHNER (1997) "X-Ray Diffraction Analysis of Scrapie Prion: Intermediate and Folded Structure in a Peptide Containing Two Putative Alpha-Helices," *J. Mol. Biol.* **268**, 375–389.

LUI, K., V. SASISEKHARAN, H. T. MILES, and G. RAGHUNATHAN (1996) "Structure of Py·Pu·Py DNA Triple Helices. Fourier Transforms of Fiber-Type X-Ray Diffraction of Single Crystals," *Biopolymers* **39**, 573–589.

MU, X. Q. and B. M. FAIRCHILD (1992) "Computer Models of a New Deoxy-Sickle Cell Hemoglobin Fiber Based on X-Ray Diffraction Data," *Biophys. J.* **61**, 1638–1646.

7

Scattering from Solutions of Macromolecules

In the preceding chapter, we described the scattering of X-radiation from a crystal lattice. We saw that from such data it is often possible to extract detailed information concerning macromolecular structure. Much of what we now know concerning the static structures of biopolymers comes from X-ray diffraction studies.

Unfortunately, not all macromolecules or macromolecular complexes can be crystallized. Most are soluble, however, and considerable information can be obtained from studying the scattering of radiation from their solutions. Experiments of this kind can often reveal conformational changes that take place in macromolecules in response to changes in the solvent environment or interactions with other molecules. Three kinds of radiation are widely employed for such experiments: visible light, X-rays, and neutrons. Each has particular advantages and can be used to obtain certain kinds of information. We shall discuss each in turn, but begin with a general treatment of the scattering phenomenon.

7.1   LIGHT SCATTERING

7.1.1   Fundamental Concepts

We begin by examining a simple example, the scattering of radiation by a single molecule. We shall make the light monochromatic (by using filters, for example, or by using a laser) and linearly polarized, although we shall remove the latter restriction later. The light is directed along the x-axis in Figure 7.1 and polarized with its electric vector in the z direction. The molecule is at $x = 0$, $y = 0$, and $z = 0$. We assume λ, the wavelength, to be so long that we can put the molecule *at* the origin, without having to worry about its size. Later, we shall consider what happens if there are many molecules in the scattering sample or if the molecule is very large or if the wavelength of the radiation is very small. In this chapter, we shall treat radiation according

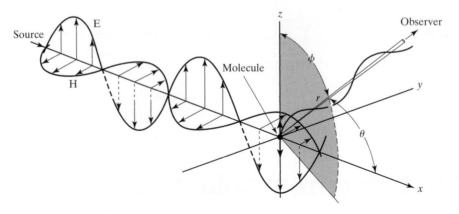

Figure 7.1 Scattering of linearly polarized radiation by a particle. Both electric and magnetic field vectors are shown for the incident radiation. Only the electric vector of the radiation scattered in the particular direction given by θ and ϕ is shown. Radiation scattered in this direction is polarized in the plane defined by the z-axis and r.

to the classical electromagnetic theory. This is appropriate for scattering studies, where transitions between quantized energy states are not involved. In later chapters dealing with the quantized absorption or emission of radiation, a quantum-mechanical approach will be more appropriate.

We can write the equation for the electric vector at point x and time t as in Eq. 6.27.

$$\mathbf{E}(x, t) = \mathbf{E}_0 \cos 2\pi\left(\nu t - \frac{x}{\lambda}\right) \qquad (7.1)$$

where λ is the wavelength and ν the frequency of the light. At the point $x = 0$ where the molecule resides, the electric field oscillates with time.

$$\mathbf{E}(0, t) = \mathbf{E}_0 \cos 2\pi\nu t \qquad (7.2)$$

According to classical electromagnetic theory, this oscillation of the external field should produce corresponding oscillation of the electrons within the molecule. *Such oscillating charges will cause the molecule to have an oscillating dipole, which will in turn act as a minute antenna, dispersing some of the energy in directions other than the direction of the incident radiation.* The scattered radiation arises from the oscillating dipole moment, a vector determined by the molecular polarizability, α of the molecule.

$$\boldsymbol{\mu} = \alpha\mathbf{E} = \alpha\mathbf{E}_0 \cos 2\pi\nu t \qquad (7.3)$$

If the molecule is isotropic, the dipole moment will be in the direction of the electric vector, that is, along the z-axis. This oscillating moment will act as a source of radiation; we may define the radiated field in terms of the coordinates shown in Figure 7.1.

The instantaneous amplitude of the electric field produced by an oscillating dipole, observed at a distance r from the dipole and at an angle ϕ with respect to the direction of polarization (the z-axis), is given by electromagnetic theory.[1]

$$E_r = \left[\frac{\alpha E_0 \, 4\pi^2 \sin \phi}{r\lambda^2} \right] \cos 2\pi \left(\nu t - \frac{r}{\lambda} \right) \tag{7.4}$$

The term in the square brackets in Eq. 7.4 is the part of interest to us. It represents the amplitude of the scattered wave. The intensity of the radiation (the energy flow per square centimeter) depends on the square of the amplitude. We wish to compare the intensity i of the scattered radiation to the intensity I_0 of the incident radiation, which is proportional to the square of *its* amplitude, E_0. Therefore,

$$\frac{i}{I_0} = \frac{(\alpha E_0 \, 4\pi^2 \sin \phi)^2 / (r\lambda^2)^2}{E_0^2} = \frac{16\pi^4 \alpha^2 \sin^2 \phi}{r^2 \lambda^4} \tag{7.5}$$

This equation tells a great deal about the scattered light. Its intensity falls off with r^2, as radiation from a point source must. The intensity of scattering increases rapidly with decreasing wavelength.[2] The intensity depends on the angle ϕ; there is no radiation along the direction ($\phi = 0$) in which the dipole vibrates. A graph in polar coordinates of the radiation intensity looks like the doughnut-like surface in Figure 7.2a.

In some cases, unpolarized radiation is used in light-scattering experiments. Since this may be regarded as a superposition of many independent waves, polarized in random directions in the yz plane, we may imagine that the resulting scattering surface would correspond to the addition of surfaces such as that shown in Figure 7.2a, rotated at random with respect to one another about the x-axis. The resulting surface is seen in Figure 7.2b; it looks somewhat like a dumbbell, with the narrowest part in the yz plane. The equation for the scattering of unpolarized radiation can be easily obtained by averaging over all directions in the yz plane (see Tanford 1961). The result differs from Eq. 7.5 only in that $(1 + \cos^2 \theta)/2$ is substituted for $\sin^2 \phi$.

$$\frac{i}{I_0} = \frac{8\pi^4 \alpha^2}{r^2 \lambda^4} (1 + \cos^2 \theta) \tag{7.6}$$

Here, θ is the angle between the incident beam and the direction of observation. The distribution of scattering intensity in the xy plane is shown in Figure 7.2c; evidently, the scattering is symmetrical in forward and backward directions.

---

[1] The electromagnetic field surrounding a moving accelerating charge [the displaced electron(s)] will consist of three parts: the static field, depending only on the magnitude of the charge; the induction field, depending on its velocity; and the radiation field, depending on the acceleration of the charge. Only the last concerns us here, for the first two fall off rapidly with distance from the charge (see Feynman et al., 1963).
[2] The strong dependence of scattering on λ accounts for the blue of the sky, since we observe the sunlight scattered by the air and its contaminants, and the blue light is scattered more than red.

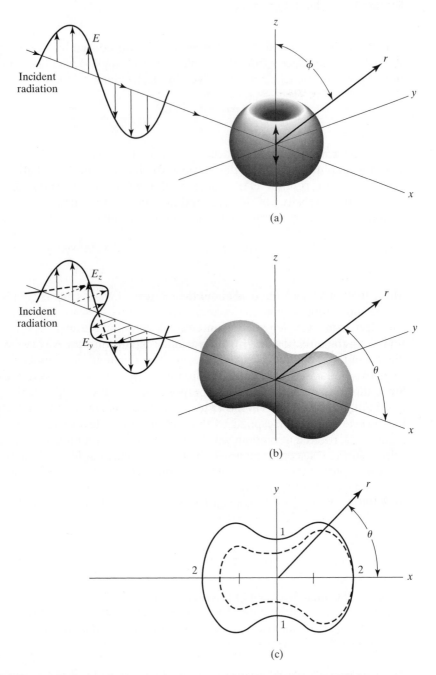

Figure 7.2 (a) Distribution of intensity of scattering of incident light polarized in the z direction. In this figure, and in (b), the distance from the origin to the surface along a direction r shows the intensity in that direction. (b) Distribution of scattering intensity for unpolarized incident light. (c) A section in the xy plane of the surface of (b). The solid line shows the intensity of scattering as a function of θ for Rayleigh scattering (small particle). The dashed line would be observed if the wavelength of the radiation were comparable to particle dimensions. (See Section 7.1.3.)

Equation 7.6 provides a completely adequate description of the scattering of light by a single small, isotropic particle. However, as it stands, this equation is not of much use to the biophysicist, who usually wishes to study solutions containing many macromolecules. We must therefore consider the situation that arises when the scattering is by a collection of particles.

7.1.2   Scattering from a Number of Small Particles: Rayleigh Scattering

We assume a solution contains N identical molecules per cm^3. The molecules are presumed to be much smaller than the wavelength of the radiation, and the solution is so dilute that interparticle interference effects are negligible. Under these conditions, the intensity of the light scattered by these particles (in excess of that scattered by solvent) will just be the sum of the intensities for all the particles. Hence, we can obtain the result by multiplying Eq. 7.6 by N.

$$\frac{i_\theta}{I_0} = \frac{N8\pi^4\alpha^2}{r^2\lambda^4}(1 + \cos^2\theta) \tag{7.7}$$

To put this in a useful form, it is necessary to express the polarizability, α, in terms of some more easily measurable quantity. Because we are considering light in the visible region of the spectrum, the refractive index provides such a quantity. Equation 7.7 measures the *excess* scattering over solvent scattering, so the appropriate expression is for the *excess* polarizability, due to the presence of the macromolecular solute. A convenient measure of polarizability in the optical frequency range is the square of the refractive index. It can be shown that

$$n^2 - n_0^2 = 4\pi N\alpha \tag{7.8}$$

where n is the refractive index of the solution and n_0 is the refractive index of the solvent. We can rearrange the equation to give

$$(n - n_0)(n + n_0) = 4\pi N\alpha \tag{7.9a}$$

or

$$\alpha = \frac{(n + n_0)}{4\pi}\frac{(n - n_0)}{C}\frac{C}{N} \tag{7.9b}$$

where we have multiplied and divided by the weight concentration C (gm/ml). At low concentrations $n + n_0 \cong 2n_0$, and $(n - n_0)/C \cong dn/dC$, where dn/dC is an easily measured change in n with a change in C, called the *specific refractive index increment*. So,

$$\alpha \cong \frac{n_0}{2\pi}\left(\frac{dn}{dC}\right)\frac{C}{N} \tag{7.10}$$

Inserting this result into Eq. 7.7 we find

$$\frac{i_\theta}{I_0} = \frac{N8\pi^4}{r^2\lambda^4}\frac{n_0^2}{4\pi^2}\left(\frac{dn}{dC}\right)^2\frac{C^2}{N^2}(1 + \cos^2\theta) \tag{7.11a}$$

or

$$\frac{i_\theta}{I_0} = \frac{2\pi^2 n_0^2}{r^2\lambda^4}\left(\frac{dn}{dC}\right)^2\frac{C^2}{N}(1 + \cos^2\theta) \tag{7.11b}$$

We note that $N = C\mathcal{N}/M$, where $\mathcal{N}$ is Avogadro's number and M is the molecular weight of the macromolecule. This gives

$$\frac{i_\theta}{I_0} = \frac{2\pi^2 n_0^2 (dn/dC)^2}{r^2\lambda^4\mathcal{N}}CM(1 + \cos^2\theta) \tag{7.12}$$

Equation 7.12 tells us that the excess scattering produced by a solution containing a weight concentration C of particles of molecular weight M depends on the product CM. It also depends on the angle with respect to the incident beam, θ, but it is symmetrical regarding forward and backward scattering, if the scattering particles are small compared to the wavelength of the light. For such scattering, which is called *Rayleigh scattering*, we can define a quantity, the *Rayleigh ratio*, which corrects for the $1 + \cos^2\theta$ term in Eq. 7.7, and is therefore independent of angle.

$$R_\theta = \frac{i_\theta}{I_0}\frac{r^2}{(1 + \cos^2\theta)} \tag{7.13}$$

Then, from Eq. 7.12, we have

$$R_\theta = \frac{2\pi^2 n_0^2 (dn/dC)^2}{\mathcal{N}\lambda^4}CM = KCM \tag{7.14}$$

where

$$K = \frac{2\pi^2 n_0^2 (dn/dC)^2}{\mathcal{N}\lambda^4} \tag{7.15}$$

These equations show that light-scattering measurements can be used for molecular-weight determination. With real solutions, the equations must be modified to take into account the nonideality of the solution. Ideality was implied when we said that the scattering of N particles was N times the scattering from a single particle; this will be true only if the particles are entirely independent of one another. If there are correlations between their positions (if even, for example, the center of one particle cannot penetrate into the space occupied by another), the scattering will be modified by this *solution nonideality* (see Chapter 13). A precise calculation can be carried out on the basis of the thermodynamic theory of concentration fluctuations (see Tanford 1961). The calculation will not be repeated here, for it is rather lengthy and leads to a minor modification in the equation. Specifically, if we write Eq. 7.14 in the following form for an ideal solution

$$\frac{KC}{R_\theta} = \frac{1}{M} \tag{7.16}$$

then for a nonideal solution we can expand in a power series about C.

$$\frac{KC}{R_\theta} = \frac{1}{M} + 2BC + \cdots \tag{7.17}$$

where B is a measure of nonideality, the *second virial coefficient* (see Chapter 13). For nonideal solutions, the scattering must be measured at several concentrations, and the quantity KC/R_θ extrapolated to $C = 0$ to give the correct molecular weight. See Figure 7.3.

Suppose we have a mixture of n macromolecular substances, of different molecular weights M_i and concentrations C_i. What result will be obtained by measuring light scattering from such a solution? Since the total intensity of scattering should be the sum of intensities from all components, then R_θ should be the sum of $R_{\theta i}$.

$$R_\theta = \sum_{i=1}^{n} R_{\theta i} = \sum_{i=1}^{n} K_i C_i M_i \tag{7.18}$$

If K_i is the same for all components (which essentially means that the refractive index increment, dn/dC_i, is the same for all i—often a good assumption), we may write

$$R_\theta = KC\frac{\sum_{i=1}^{n} C_i M_i}{\sum_{i=1}^{n} C_i} = KCM_w \tag{7.19}$$

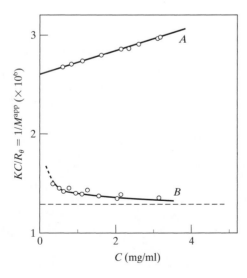

Figure 7.3 Graphs of KC/R_θ for a subunit of squid hemocyanin (A) and a dimer of that subunit (B). The subunit ($M = 385,000$) is nonideal, with a positive value of B. The dimer shows some tendency to dissociate at low concentration ($1/M^{app}$ increases). The broken horizontal line shows the value expected (770,000) for an intact dimer.

where we have multiplied and divided by the total concentration, $C = \Sigma_{i=1}^{n} C_i$. The quantity (M_w) is an *average* molecular weight for the solutes in the mixture. It is called the *weight average* molecular weight, since each component is counted according to its weight concentration in taking the average. We have already encountered this quantity in describing the sedimentation equilibrium of heterogeneous solutes (Chapter 5).

The measurement of R_θ is usually carried out in a *photometer* similar to that diagrammed in Figure 7.4. In most modern instruments, a laser is used to provide a well-collimated monochromatic beam. The intensity of light scattered at a given angle (90°, for example) is compared with the intensity of the incident light. A complication arises from the fact that the solution must be scrupulously clean; small amounts of dust will contribute as extremely large molecules to the average in Eq. 7.19, leading to serious errors. Therefore, solutions must be carefully filtered or centrifuged before scattering measurements.

In addition to R_θ values, calculation of the molecular weight requires a knowledge of dn/dC. Since the difference in refractive index between a dilute solution and pure solvent is very small, a *differential* refractometer, which directly measures the small difference $n - n_0$, is commonly used.

7.1.3   Scattering from Particles That Are Not Small Compared to Wavelength of Radiation

So far, we have assumed that the greatest dimension of the scattering particle is small compared to the wavelength of the light. There are two kinds of situations in which this will not be true. First, very large particles like viruses and immense proteins or nucleic acids are not insignificant in dimensions compared to visible light. On the other hand, we will later be discussing scattering of X-rays or neutrons from

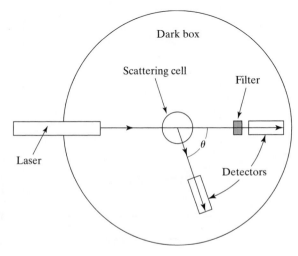

Figure 7.4 A schematic design of an instrument for light-scattering measurements. A continuous wave laser is used as the light source, and the output from the scattering detector (which can rotate about the cell) is compared with that from the direct beam detector. A filter is inserted to reduce the intensity for the latter.

solutions. Even small proteins can have dimensions much larger than the wavelengths of such radiation.

In such situations, the scattering experiment becomes more informative. In addition to the measurement of mass, it becomes possible to estimate dimensions and even obtain some information about shape or internal structure of the molecules.

Consider the example shown in Figure 7.5. The incident radiation is inducing scattering from points within a single molecule that are an appreciable fraction of a wavelength apart; the dipoles within the molecule are therefore oscillating out of phase. Furthermore, the two points shown as examples are at a significant phase difference in distance from the observer. Since the scattering centers within a given large molecule are (more or less) fixed with respect to one another, we can no longer consider them to be independent scatterers. We must therefore take into account interference between radiation from these scattering centers.

The calculation of this effect for a macromolecule must include the interference between light scattered from all pairs of scattering points within the molecule. Furthermore, because the molecules are randomly oriented in the solution, the total scattering must be averaged over all orientations.

For convenience, we express the general result by defining a function $P(\theta)$, which describes how the scattering at angle θ corresponds to that which the *same* molecule would have were its dimensions shrunk to be much smaller than λ.

$$P(\theta) = \frac{\text{scattering by real particle at angle } \theta}{\text{scattering by hypothetical point particle at angle } \theta}$$

Since a solution of the hypothetical point particles would exhibit Rayleigh scattering, we can calculate the light scattering from a solution of large particles by simply multiplying Eq. 7.12 by $P(\theta)$.

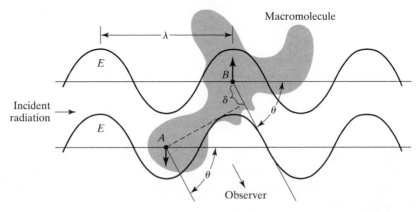

Figure 7.5 Scattering from a macromolecule of dimensions comparable to the radiation wavelength. Two points from which scattering occurs are shown at A and B. The phase of the radiation (and hence of the induced dipoles) is clearly different at the two points. Also, the two points are at different distances from the observer.

To evaluate $P(\theta)$, we can use the general formulation for the interference in scattering developed in Chapter 6. Recall from Eq. 6.45 that the structure factor for a collection of n scattering elements (atoms or electrons) is

$$F(\mathbf{S}) = \sum_{i=1}^{n} f_i e^{2\pi i \mathbf{S} \cdot \mathbf{r}_i} \tag{7.20}$$

Here $\mathbf{S}$ is the scattering vector (see Chapter 6) and $\mathbf{r}_i$ is the position of scatterer i with respect to some arbitrary reference point in the molecule. If we regard the individual scatterers within the molecule as the electrons, then all f_i are identical, and $f_i = f_0$ so that

$$F(\mathbf{S}) = f_0 \sum_{i=1}^{n} e^{2\pi i \mathbf{S} \cdot \mathbf{r}_i} \tag{7.21}$$

To calculate the intensity of the scattered light, we take the product of F with its complex conjugate, which yields

$$I(\mathbf{S}) = FF^* = f_0^2 \sum_{i=1}^{n} \sum_{j=1}^{n} e^{2\pi i \mathbf{S} \cdot \mathbf{r}_{ij}} \tag{7.22}$$

where $\mathbf{r}_{ij}$ is a vector connecting scatterers i and j. The dot product can be expressed in terms of the angle (α) between $\mathbf{r}_{ij}$ and the scattering vector $\mathbf{S}$. Since the magnitude of $\mathbf{S}$ is $2 \sin(\theta/2)/\lambda$ (see Chapter 6), we obtain

$$I(\mathbf{S}) = f_0^2 \sum_{i=1}^{n} \sum_{j=1}^{n} e^{\frac{4\pi i (\sin \theta/2) r_{ij} \cos \alpha}{\lambda}} \tag{7.23}$$

This is the numerator for $P(\theta)$. To get the denominator we consider all r_{ij} to be small compared to λ (the condition for Rayleigh scattering). Then the exponentials are all unity, the double sum equals n^2, and

$$I(0) = f_0^2 n^2$$

The ratio is

$$P(\alpha, \theta) = I(\mathbf{S})/I(0) = \frac{1}{n^2} \sum_{i=1}^{n} \sum_{j=1}^{n} e^{ihr_{ij} \cos \alpha} \tag{7.24}$$

where we have symbolized the quantity $[4\pi\sin(\theta/2)]/\lambda$ by h. Equation 7.24 is for a given orientation of a particle (as defined by the α value). If we consider that all orientations are equally possible, we may average by integrating over the angular range $0 \rightarrow \pi$.

$$P(\theta) = \frac{\langle I(\mathbf{S}) \rangle}{I(0)} = \frac{1}{n^2} \int_0^{\pi} \sum_{i=1}^{n} \sum_{j=1}^{n} e^{ihr_{ij} \cos \alpha} \sin \alpha \, d\alpha \tag{7.25}$$

With the substitution $\cos \alpha = x$, the integration readily gives

$$P(\theta) = \frac{1}{n^2}\sum_{i=1}^{n}\sum_{j=1}^{n}\left(\frac{\sin hr_{ij}}{hr_{ij}}\right) \tag{7.26}$$

This equation, often called the Debye equation, is very general, and applies to many scattering problems, utilizing different kinds of radiation.

The behavior of $P(\theta)$ at low angles can be seen by expanding the function $(\sin hr_{ij}/hr_{ij})$ in a Taylor series.

$$\frac{\sin hr_{ij}}{hr_{ij}} \cong 1 - \frac{(hr_{ij})^2}{6} + \frac{(hr_{ij})^4}{120} - \cdots \tag{7.27}$$

As either $h \to 0 (\theta \to 0)$ or $r_{ij} \to 0$ we obtain

$$P(\theta) = \frac{1}{n^2}\sum_{i=1}^{n}\sum_{j=1}^{n} 1 = \frac{n^2}{n^2} = 1 \tag{7.28}$$

Thus, at very low angles, at very long wavelengths, or for very small particles, Rayleigh scattering is approached. The situation at $\theta = 0$, in fact, can be inferred from Figure 7.5; there is no phase difference in light scattered directly in the forward direction.

At angles greater than zero, $P(\theta)$ will always be less than unity, which means that the scattering intensity at these angles will always be less for an extended particle than for a compact particle of the same weight. This can be thought of as a consequence of *internal interference* within the scattering molecule. A typical graph of the angular dependence of light scattering for a large particle is shown as the dotted line in Figure 7.2c. We have multiplied the $(1 + \cos^2\theta)/2$ factor by $P(\theta)$. Note that to obtain molecular weights for large particles from light scattering, we must in most cases extrapolate the data *both* to zero concentration and zero angle. A method to conveniently accomplish this was devised by Zimm (1948).

It may appear that this effect of particle size is merely an exasperating complication, but, in fact, it allows us to determine particle dimensions from light scattering. Suppose that the particles are in such a size range that we need to consider only the first two terms in Eq. 7.27. Then

$$P(\theta) = \frac{1}{n^2}\sum_{i=1}^{n}\sum_{j=1}^{n}(1) - \frac{h^2}{6n^2}\sum_{i=1}^{n}\sum_{j=1}^{n}r_{ij}^2 \tag{7.29}$$

The sum in the second term is related to the *radius of gyration* (R_G) of the particle, which is defined mathematically by the equation

$$R_G^2 = \frac{1}{2n^2}\sum_{i=1}^{n}\sum_{j=1}^{n}r_{ij}^2 \tag{7.30}$$

The radius of gyration can also be defined as the root mean square average of the distance of scattering elements from the center of mass of the molecule. (See Section 4.4.2.) In terms of R_G,

$$P(\theta) = 1 - \frac{h^2 R_G^2}{3} + \cdots \tag{7.31}$$

There has been no assumption made about the shape of the particle in this derivation. Thus, the angular dependence of scattering can give us unambiguously the quantity R_G. This will only be useful if the particle is fairly large, as can be seen readily from Eq. 7.31. If the radius of gyration is less than one-fiftieth of the wavelength (that is, less than about 8.0 nm for visible light in water), the deviation of $P(\theta)$ from unity will not amount to more than a small percentage at any angle. Therefore, measurement of the angular dependence of the scattering of visible light will be useful only for the larger macromolecules.

In Table 7.1, some values of M and R_G determined by light scattering are given. Although R_G itself is unambiguous, its interpretation in terms of dimensions of the particles depends on particle shape. Table 7.2 lists some values of R_G in terms of particle dimensions. It should also be emphasized that the angular dependence of scattering will depend explicitly on shape if particles are so large that higher terms in the expansion in Eq. 7.27 have to be retained. We shall encounter such cases in the later sections.

In summary, light scattering provides a powerful tool for the determination of particle weight and dimensions. Its use is limited at one extreme by the low scattering power of small molecules and at the other by the difficulties encountered in the interpretation for very large macromolecules. If the particles are very large, Eq. 7.31 no longer adequately describes the angular dependence, which then reflects the particle shape as well. If the shape is not known or if experimental difficulties prevent measurement to angles low enough for Eq. 7.31 to be valid, a clear extrapolation to $\theta = 0$ may not be possible. In practice, it is very difficult, because of diffraction effects, to measure scattering of visible light at angles less than about 5°.

Table 7.1 Representative Results from Light Scattering and Low-Angle X-Ray Scattering

Material	M_w	R_G (nm)
Ribonuclease	*12,700*	*1.48*
α-Lactalbumin	*13,500*	*1.45*
Lysozyme	*13,600*	*1.43*
β-Lactoglobulin	*36,700*	*2.17*
Serum albumin	*70,000*	*2.98*
Myosin	493,000	46.8
Tobacco mosaic virus	39×10^6	92.4

Values in *italic* are from low angle X-ray scattering.

Table 7.2   Relation Between R_G and Dimensions for Particles of Various Shapes

Shape	R_G
Sphere	$\sqrt{3/5}R$
Prolate ellipsoid	$\left(\sqrt{2 + \gamma^2/5}\right)a$
Very long rod	$L/\sqrt{12}$
Random coil	$n^{1/2}l/\sqrt{6}$

Abbreviations: R = radius of sphere; axes of ellipsoid are $2a, 2a, \gamma\, 2a$; L = length of rod; n = number of links of length l in chain.

7.2   DYNAMIC LIGHT SCATTERING: MEASUREMENTS OF DIFFUSION

In the preceding discussion of Rayleigh scattering, it was implicitly assumed that we were observing a large enough volume of solution that the number of particles therein, and hence the scattering, was essentially constant in time. However, if we follow over time the intensity of light scattering from a *small* volume element in a solution of macromolecules, we will observe that it is not constant, but fluctuates as shown in Figure 7.6. This is to be expected, for the *local* concentration of macromolecules in any element of volume is fluctuating with time, as the molecules move in and out of that volume element. The light scattering experiments we have described above measure the *average* scattering (dashed line in Figure 7.6). However there is important information to be gained by studying the fluctuations themselves, for their frequency is determined by how fast the solute molecules move.

The rate of random motion of solute molecules, as a consequence of their thermal energy is measured by the *diffusion coefficient, D.* We have already encountered D in Chapter 5, as pointed out therein it provides a direct measure of the frictional coefficient

$$D = RT/\mathcal{N}f \tag{7.32}$$

Obviously, the larger D is, the more rapid the fluctuations in scattering will be in an experiment like that shown in Figure 7.6. There are several ways in which such data can be analyzed. One way is to perform a Fourier analysis of the fluctuations. However, the method most commonly used today is called *autocorrelation analysis*. We compare the intensity $i(t)$ of the scattered light at time t with that at some later time $(t + \tau)$ and define the *autocorrelation function*

$$g^{(2)}(\tau) = \langle i(t) \cdot i(t + \tau)\rangle/\langle i\rangle^2 \tag{7.33}$$

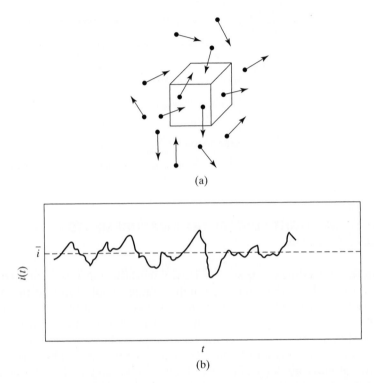

(a)

(b)

Figure 7.6 Fluctuations in light scattering are used to measure diffusion. (a) A small element of volume from which scattering is being recorded. The number of macromolecules within this volume fluctuates with time, due to their diffusional motion. (b) A recording of intensity fluctuations about the mean value, $\bar{i}$ (dashed line).

where the brackets ($\langle \rangle$) mean that we have averaged the product over time, using a fixed interval τ. The quantity $\langle i \rangle$ is the average value of scattering intensity. The correlation between i-values will decrease with time; for large τ all correlation will have been lost (see Figure 7.7). The value of $g^{(2)}(\tau)$ will then approach unity, for uncorrelated fluctuations will fluctuate about $\langle i \rangle$. In fact, it can be shown that the quantity $g^{(2)}(\tau)$ decays exponentially with τ when a single component is dominating the scattering.

$$g^{(2)}(\tau) = 1 + ce^{-2h^2 D\tau} \qquad (7.34)$$

$$\ln[g^{(2)}(\tau) - 1] = \ln c - 2h^2 D\tau \qquad (7.35)$$

where $h = 4\pi n \lambda^{-1} \sin(\theta/2)$, θ being the scattering angle and n the refractive index. Thus, D can be measured from the slope of a graph according to Eq. 7.35 (see Figure 7.8). This method is now the favored technique for measuring diffusion coefficients. It is rapid and economical of sample, for only a very small volume need be illuminated by the laser beam. The rapidity of the measurements allows the

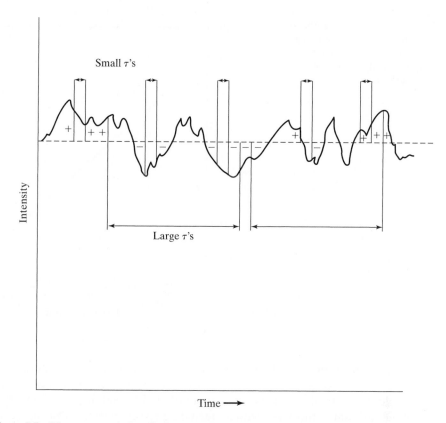

Figure 7.7 Why autocorrelation disappears at large τ. For large τ, the fluctuation about $\langle I \rangle$ becomes random, and a fluctuation is as likely to be paired with one of opposite sign as with one of the same sign. Thus, values of the product are equally likely to be negative or positive. For very short τ, they are likely to be of the same sign, giving only positive values for the product.

method to be employed for some kinetic studies involving macromolecules. If macromolecules are of dimensions comparable to the wavelength of the light, then Eq. 7.34 no longer suffices, for rotational diffusion and/or internal motion of the macromolecule will contribute to the dynamic scattering as well. For extensive discussions of multitudinous applications of dynamic light scattering see the references at the end of this chapter.

7.3   SMALL-ANGLE X-RAY SCATTERING

While the *weights* of small macromolecules can be determined from the scattering of visible light, the method fails if we seek to use the angular dependence to measure the *dimensions* of particles with R_G less than about 10 nm. Since the resolving power of the method depends on the ratio $(R_G/\lambda)^2$, an obvious solution would be to

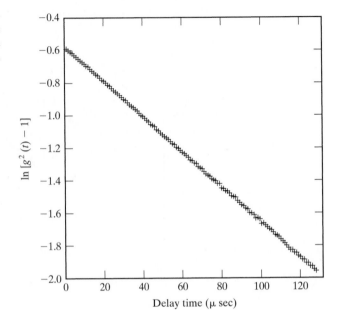

Figure 7.8 A graph of the autocorrelation function as a function of delay time (τ). [Data for TYMV virus, taken from S. E. Harding and P. Johnson (1985), *Biochem. J.* **231**, 549–555.]

use light of shorter wavelengths. But here nature has been uncooperative, for those materials that we would most like to investigate (proteins and nucleic acids, for example) begin to absorb very strongly a little below 300 nm (see Chapter 9). Below about 160 nm, almost everything, including water and air, absorbs very strongly, leaving a considerable region of the spectrum seemingly inaccessible to scattering studies. It is not until we reach the X-ray region that materials again become generally transparent. But here the scattering situation is very different; the interference effects, which were second order for the scattering of visible light, now dominate the scattering. For example, the wavelength of the Cu-α radiation (a commonly used X-radiation) is 0.154 nm, so that the distance across a ribonuclease molecule (molecular weight of 13,683) is of the order of 10λ. Clearly, an approximation such as Eq. 7.27 will fail even at quite small angles, and the complete Eq. 7.26 must be used. The situation is best illustrated by a specific example. The evaluation of the quantity $\Sigma\Sigma \sin hr_{ij}/hr_{ij}$ for electrons distributed uniformly in a solid sphere of radius R leads to the $P(\theta)$ function shown in Figure 7.9. For this example, we have chosen $\lambda = 0.154$ nm and $R = 1.54$ nm. The scattering is almost entirely confined to a very narrow angular range. The effect of the interference of waves is emphasized by the existence of maxima and minima in the $P(\theta)$ function at higher h.

Fortunately, it is possible to collimate an X-ray beam very well, so measurements can often be made to angles small enough to allow a decent extrapolation to $\theta = 0$, and hence calculation of M as by light scattering. The technique has become known as *small-angle X-ray scattering*, or SAXS. The theory looks a little different because it is most convenient to consider the individual electrons in the

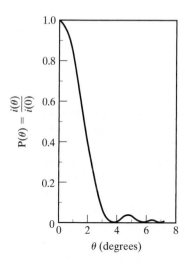

Figure 7.9 A graph of the scattering intensity versus angle for the scattering of 0.154 nm X-rays from spheres of radius 1.54 nm.

molecule as scatterers. By classical theory the intensity of X-ray scattering from a single electron is

$$i_e(\theta) = \frac{I_0}{r^2}\left(\frac{e^2}{mc^2}\right)^2 \frac{1 + \cos^2\theta}{2} \qquad (7.36)$$

where e is the magnitude of the electron charge, m the electron mass, and c is the velocity of light. For a molecule with Z electrons, we have an intensity Z^2 times as great, for the scattering from within one molecule is coherent and *amplitudes* for each electron must be added.

$$i_m(0) = \frac{I_0}{r^2}\left(\frac{e^2}{mc^2}\right)^2 Z^2 \qquad (7.37)$$

where θ has been taken as 0 degrees. To calculate the scattering from 1 cm^3 of a solution containing N molecules per cubic centimeter, we add the *intensities* (assuming the molecules to be independent of one another, so that the scattering from the collection of molecules is incoherent).

$$i(0) = \frac{I_0}{r^2}\left(\frac{e^2}{mc^2}\right)^2 Z^2 N \qquad (7.38)$$

We are interested in solutions, so we replace Z by $(Z - Z_0)$, the excess electrons in a solute molecule over the number of electrons in a volume of solvent it displaces. Also, we divide and multiply by M^2.

$$i(0) = \frac{I_0}{r^2}\left(\frac{e^2}{mc^2}\right)^2 NM^2\left(\frac{Z - Z_0}{M}\right)^2 \qquad (7.39)$$

Or, recalling that $N = \mathcal{N}C/M$ where C is the weight concentration and $\mathcal{N}$ is Avogadro's number,

$$i(0) = \frac{I_0}{r^2}\left(\frac{e^2}{mc^2}\right)^2 \mathcal{N}\left(\frac{Z - Z_0}{M}\right)^2 MC \tag{7.40}$$

The similarity to Eq. 7.12 for light scattering is obvious. The quantity $(Z - Z_0)/M$ depends on the chemical composition, since it is essentially a ratio of atomic numbers to atomic weights. As in the case of light scattering, it is necessary to extrapolate data to $C = 0$, because of nonideality of real solutions. According to Eq. 7.40, the limiting intensity at $\theta = 0$ can be used to measure the molecular weight of a macromolecule.

However, the most important applications of small-angle X-ray scattering come from studies of the angular dependence of scattering. The angular dependence of the scattering in the very small-angle range will give the radius of gyration. The extrapolation might be made by graphing $i(\theta)$ versus $\sin^2(\theta/2)$ as in light-scattering experiments, but it has become customary to attempt to extend the range of a linear graph by a slightly different method. The first two terms of Eq. 7.31 are identical with those in a Taylor's series expansion of $e^{-h^2 R_G^2/3}$; that is,

$$e^{-h^2 R_G^2/3} = 1 - \frac{h^2 R_G^2}{3} + \frac{h^4 R_G^4}{18} - \cdots \tag{7.41}$$

For spherical particles, even the third term is identical to the expansion of the exponential. Thus, to a good approximation for any particle in the small-angle range, we can write

$$\frac{i(\theta)}{I(0)} = P(\theta) = e^{-h^2 R_G^2/3} \tag{7.42}$$

or

$$\ln \frac{i(\theta)}{i(0)} = -\frac{h^2 R_G^2}{3} = -\frac{16\pi^4 R_G^2 \sin^2(\theta/2)}{3\lambda^2} \tag{7.43}$$

which means that graphing $\ln i(\theta)$ versus $\sin^2(\theta/2)$ will give a straight line with intercept $\ln i(0)$ and slope $-16\pi^4 R_G^2/3\lambda^2$. Such a graph is known as a *Guinier plot*. From the intercept $i(0)$ we can calculate M; from the slope we obtain R_G. In Table 7.1 some molecular weights and particle dimensions obtained by small-angle X-ray scattering are given. Even the smallest protein molecules can be investigated by this method.

The angular dependence of the scattering at somewhat larger angles becomes dependent on details of the shape of the particles. Under these conditions, Eq. 7.26 predicts multiple maxima and minima in the scattering curve, as shown in Figure 7.10. Note that these details occur at higher angles than the Gainier region, corresponding to the principle emphasized in Chapter 6, that higher scattering angles corresponds to

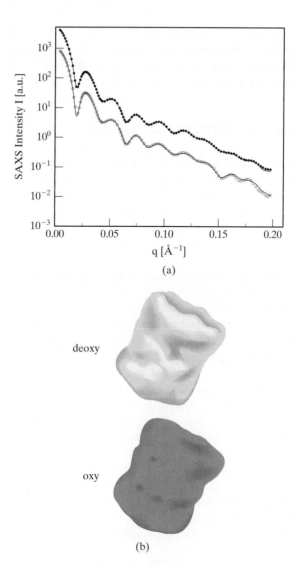

(a)

(b)

Figure 7.10   (a) Small-angle X-ray scattering from keyhole limpet hemocyanin in the oxy (*top*) and deoxy (*bottom*) states. The curves have been displaced for clarity. The two models that fit the two sets of data are shown in (b). Even though the models show only small structural differences, the accuracy of the data is so high that the conformational change is credible. [Courtesy of H. Hartmann, A. Bongers, and H. Decker.]

finer details of particle structure. Until recently, it has not been possible to analyze such high-angle data to yield details of particle structure. However, the computational power of modern computers allow techniques such as a Monte Carlo search over possible structures to provide highly accurate fits to such curves. This, in turn, can provide details of molecular structure in solution.

This capability is important, for while NMR (see Chapter 12) can provide highly resolved solution structures, it cannot at the present be used for the very large molecules that are most easily studied by small-angle X-ray scattering. On the other hand, X-ray diffraction cannot be used to study structural transitions in solution.

7.4   SMALL-ANGLE NEUTRON SCATTERING

According to the basic principles of quantum mechanics, every particle has also a wavelike character, and all radiation can show particle-like behavior. The de Broglie equation states that a particle of mass m, moving with velocity v can be considered to have an associated wavelength given by

$$\lambda = \frac{h}{mv} \tag{7.44}$$

where h is Planck's constant. Nuclear reactors can produce beams of neutrons with quite well-defined velocities, corresponding to wavelengths of the order of 0.2 to 0.4 nm. Such a *monochromatic* neutron beam can be used for scattering experiments. From such experiments, we can extract the same basic information as from small-angle X-ray scattering or Rayleigh scattering—the molecular weight from zero-angle scattering intensity and the radius of gyration from Guinier plots as Eq. 7.43 (see Zaccai and Jacrot 1983, for details). Since a nuclear reactor is much less readily available to most biochemists than a laser, or an X-ray source, it may seem surprising that neutron scattering has become an important tool in biophysical research. The reason lies in some very special features of the scattering of neutrons by matter. To give even a qualitative explanation, it is necessary to introduce some special concepts and nomenclature used in describing neutron scattering.

In Eq. 7.36, which gives the scattering by an electron, the intensity is proportional to the square of the quantity e^2/mc^2. The *amplitude* of the scattered X-ray is thus proportional to this number, which has the dimensions of length. A similar *scattering length* is hidden in Eq. 7.5, as α/λ^2 (polarizability always has the dimensions of volume). The scattering lengths (b_x) that various atoms have for X-rays are given approximately by

$$b_x = \frac{e^2}{mc^2} Z \tag{7.45}$$

where Z is the number of electrons in the atom, that is, the atomic number. There are two consequences of this relationship for X-ray scattering: (1) Since hydrogen has a very small value for b_x, hydrogen atoms do not contribute very much to the scattering; and (2) the other common elements in most biopolymers (C, N, O) are roughly the same in scattering power (see Table 7.3).

The point to this discussion is that the situation is very different for neutrons. Table 7.3 lists neutron-scattering lengths for the common elements in biopolymers. The most striking observation is the *negative* value for hydrogen. Physically, this

Table 7.3   X-Ray and Neutron-Scattering Lengths

Element	b_x	b_n
H	3.8	−3.74
D	2.8	6.67
C	16.9	6.65
N	19.7	9.40
O	22.5	5.80
P	42.3	5.10

All lengths in units of 10^{-13} cm.

means that the neutron scattering from a hydrogen atom will be 180° out of phase with that from other atoms in the molecule. Since, as in X-ray scattering, we must add the amplitudes of scattered waves from different atoms in the molecule, the presence of hydrogen atoms has a major effect. In addition, it means that the scattering from H_2O is very different from that from D_2O.

Recall that it is always the *contrast* between solvent and macromolecule that allows us to observe the scattering. For example, if we change the polarizability (refractive index) of the solvent in a light-scattering experiment (by adding sucrose, for example) the scattering intensity will change. However, the effects that can be produced in this case are small. With neutron scattering, the situation is very different. The great difference in H_2O and D_2O scattering allows matching of the solvent background to the scattering from a macromolecule. In effect, the macromolecule can be made to "disappear" by choice of a suitable H_2O–D_2O mixture for solvent.

This would be only an interesting curiosity were there not another aspect. As Table 7.4 shows, the scattering length densities of proteins, nucleic acids, lipids, and carbohydrates all differ, as a consequence of their differing elemental compositions. This can be of enormous utility in studying complex macromolecules and particles that are made up of two or more such components. It is possible, using H_2O–D_2O mixtures of different compositions, to produce solvent-matching selectively with each type of material listed in Table 7.4 in order to observe another type of material. Consider, for example, the experiments on nucleosomes described in Application 7.1. A nucleosome contains both protein and nucleic acid, but the relative disposition of

Table 7.4   X-Ray and Neutron-Scattering Densities of Biologically Important Substances

Substance	ρ (X-ray)	ρ (neutron)
H_2O	+9.3	−0.55
D_2O	+9.3	+6.36
Protein	+12.4	+3.11
Protein (deuterated)	+12.4	+8.54
Nucleic acid	+16.0	+4.44
Fatty acid	+8.2	−0.01
Carbohydrate	+14.1	+4.27

All values are in units of 10^{-4} nm^{-2}.

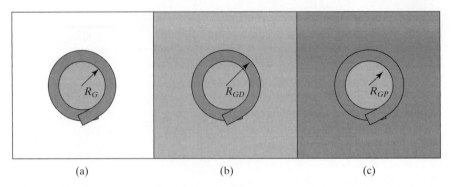

Figure 7.11   The principle of contrast matching in neutron scattering. A nucleosome, which has DNA (*dark*) wrapped about a protein core (*lighter*), is depicted schematically. (a) Solvent of low-scattering power; the entire particle contrasts, and R_G is for the whole structure. (b) Solvent matched to protein; the R_{GD} measured is that of the DNA. (c) Solvent matched for DNA; the DNA "disappears" and the R_{GP} of the protein is measured.

these two components was originally unknown. The results from neutron scattering showed unequivocally that the nucleic acid lay on the outside of the particle (see Figure 7.11 and Figure A7.1b). Similar experiments with lipoproteins have been used to demonstrate the separation between lipid and protein in such structures.

Application 7.1   Using a Combination of Physical Methods to Determine the Conformation of the Nucleosome

Chromatin is the complex of DNA and proteins (mostly histones) found in all eukaryotic nuclei. In 1973, it was discovered that micrococcal nuclease digestion of chromatin yielded a fraction (about 50%) in which the DNA was relatively resistant to further cleavage. The nuclease-resistant chromosomal particles behaved as a relatively homogeneous component, with a sedimentation coefficient of about 11 S and a protein/DNA ratio slightly greater than unity (Sahasrabuddhe and van Holde 1974). It was soon recognized that these particles must represent the *repeating units* of chromatin structure postulated by Kornberg (1974), and probably corresponded to the *beads* observed on chromatin strands in electron microscopy (Olins and Olins 1974). But the composition and structure of these objects, and the conformation of DNA they contained, remained unclear. Were they compact, globular beads, as the EM pictures implied, or was this conformation an artifact of drying on EM grids? Where was the DNA—inside, outside, or both? Earlier models of chromatin had assumed that histones coated the DNA fiber, much as insulation coats electric wire.

　　　To answer these questions, a battery of physical techniques was brought to bear over the next few years. First, a hydrodynamic study was made of the particles (now called *core particles* or *nucleosomes*) obtained by partial enzymatic digestion of the DNA in chromatin (van Holde et al. 1975). These particles were found to contain about 140 bp of DNA each, from gel electrophoresis of the DNA remaining. Sedimentation equilibrium of the core particle preparation indicated these to be homogeneous, or nearly so (Figure A7.1a). From

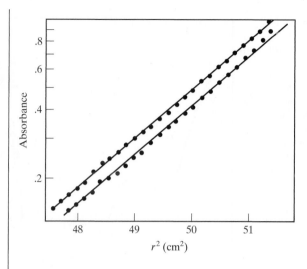

Figure A7.1a Sedimentation equilibrium experiments with two core particle preparations. The logarithm of the absorbance, as measured by the ultracentrifuge scanner, is graphed versus r^2. The upper line was obtained at 5603 rpm, 8.9°C, the lower at 5621 rpm, 4°C. The two lines give molecular weight values of 198,000 and 194,000, respectively, if $\bar{v}$ is assumed to be 0.66.

the slope of these lines, the molecular weight could be calculated, if $\bar{v}$ were known. Although the quantities of core particles available at the time made the measurement of $\bar{v}$ impractical, its value could be estimated as 0.66 cm^3/gm from the known values for histones and DNA and the histone/DNA ratio in the particles. Using Eq. 5.41, this value lead to a molecular mass of 196,000 Da for the core particle. The difference between this mass and the mass of 140 bp of DNA (~91,000 Da) leaves 105,000 Da for protein. This is almost exactly what would be expected for an octamer composed of two each of the histones H2A, H2B, H3, and H4, supporting the model proposed by Kornberg (1974).

With a molecular weight value available, it was possible to answer questions about particle shape. The sedimentation coefficient was found to be 11.0 S. This, together with M, allowed calculation of the frictional coefficient, Eq. 5.5; a value of 1.01×10^{-7} was obtained. From this we can calculate a Stokes's radius (Eq. 5.11); the value of 5.4 nm obtained was in approximal agreement with the average radii of the beads observed in electron microscopy of chromatin fibers (about 5 nm). This, then, supported the idea that the particles were compact. This could be seen in another way from a calculation of f/f_0, as described in Section 5.3.1. The value obtained (1.42) again confirms a compact object, although one that is somewhat hydrated or asymmetrical.

These results allowed a sharp focus on the question as to how the DNA was configured in the particle. Obviously, it has to be coiled or folded in some way, for 140 bp, if extended, would be about 48 nm long, much greater than any possible dimension of the particle. But was the DNA folded inside or coiled outside?

To answer this question, two groups of researchers (Pardon et al. 1975; Suau et al. 1977) turned to the use of low-angle neutron scattering. As described in this chapter, we can vary the scattering length density in D_2O–H_2O mixtures to produce contrast matching with either the DNA or the protein (see Figure 7.11). Measuring the radii of gyration for DNA and protein in this fashion provided a clear answer; R_G (protein) was about 3.5 nm, R_G (DNA) about 5.0 nm. Therefore, the DNA *had* to be on the outside. A more sophisticated analysis (Suau et al. 1977) showed that the scattering as well as all the other physical data could be well accounted for by a structure like that shown in

Figure A7.1b The best model for the nucleosome core particle, as deduced from low-angle neutron scattering. The experimental scattering intensity is shown by the open circles. Curves i, ii, and iii are calculated from the three models shown: (i) a uniform DNA helix around the histone core; (ii) a helix that is constrained to follow the assumed shape of the core; and (iii) two DNA rings about the core. Clearly, (ii) is best, as has been confirmed by X-ray diffraction studies. [From Suau et al. (1977), permission of Oxford University Press.]

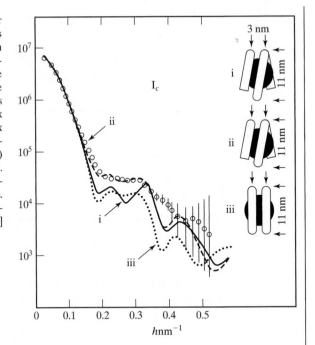

Figure A7.1b. The essential correctness of this model was finally demonstrated by X-ray diffraction studies many years later (Richmond et al. 1984).

KORNBERG, R. (1974), *Science* **184**, 868–871.

OLINS, A. L. and D. E. OLINS (1974), *Science* **183**, 330–332.

PARDON, J. F., D. L. WORCESTER, J. C. WOOLEY, K. TATCHELL, K. E. VAN HOLDE, and B. M. RICHARDS (1975), *Nucleic Acids Res.* **2**, 2163–2175.

RICHMOND, T. J., J. T. FINCH, B. RUSHTON, D. RHODES, and A. KLUG (1984), *Nature* **311**, 532–537.

SAHASRABUDDHE, C. G. and K. E. VAN HOLDE (1974), *J. Biol. Chem.* **249**, 152–156.

SUAU, P., G. G. KNEALE, G. W. BRADDOCK, J. P. BALDWIN, and E. M. BRADBURY (1977), *Nucleic Acids Res.* **4**, 3769–3786.

VAN HOLDE, K. E., B. R. SHAW, D. L. LOHR, T. M. HERMAN, and R. T. KOVACIC (1975), *Proc. 10th FEBS Meeting*, vol. 38, ed. G. Bernardi and F. Gros, 57–72.

A quantitative analysis of this contrast matching can be provided in the following way. It can be shown (see Zaccai and Jacrot 1983) that the observed radius of gyration depends in a simple way on the difference in scattering length density between macromolecule and solvent, $\langle \rho \rangle$:

$$\langle \rho \rangle = \langle \rho \text{ macromolecule} - \rho \text{ solvent} \rangle \tag{7.46}$$

where the brackets denote averaging over all elements of volume in the particle. The radius of gyration is given by the *Sturhman equation*:

$$R_G^2 = R_C^2 + \alpha/\langle\rho\rangle - \beta/\langle\rho^2\rangle \tag{7.47}$$

The constants on the right side of Eq. 7.47 have the following meanings: R_C is the radius of gyration that is observed if the particle has uniform internal structure; equivalently, it is the R_G value that would be found at infinite contrast ($1/\langle\rho\rangle = 0$). The constant α measures the inhomogeneity of the particle; it is zero if the particle is homogeneous, positive if the exterior of the molecule has a higher scattering length density than the interior, and negative if the interior is more dense. The parameter β measures the displacement of the center of scattering length density from the physical center of the molecule. Equation 7.47 can be employed by plotting R_G, measured at different solvent scattering densities, as a function of $1/\overline{\rho}$. Representative examples are shown in Figure 7.12.

There are other ways in which the phenomenon of contrast variation can be exploited. For example, if selected subunits in a multisubunit complex are deuterated, it is possible to focus attention on these particular parts of a complicated structure. Researchers have, for example, determined the distances between pairs of ribosomal proteins in a ribosome, thereby producing a three-dimensional map of their arrangement. The potentialities of neutron scattering are obviously great. The main problem at the present time is that there are only a few reactors in the world that provide sufficient neutron flux at the right wavelengths to allow such experiments to be performed on a practicable time scale.

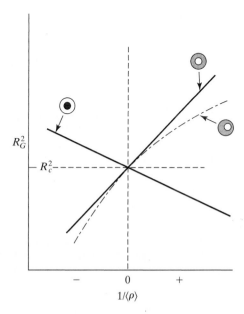

Figure 7.12   Graphs of the observed radius of gyration versus $1/\langle\rho\rangle$ for different idealized structures. In each case, high scattering density is indicated by dark shading.

Table 7.5 A Summary of Scattering Techniques

Technique Name	Acronym	Radiation Used	Basic Information Obtained
Rayleigh light scattering	—	Visible light	Molecular weight, Molecular interactions
Angular-dependent light scattering	—	Visible light	Radius of gyration
Dynamic light scattering	DLS	Visible light	Diffusion coefficient
Small-angle X-ray scattering	SAXS	X-rays	Molecular weight, radius of gyration, shape information
Small-angle neutron scattering	SANS	Neutrons	Radius of gyration, shape information, internal structure

7.5 SUMMARY

In this chapter, we have described a number of techniques, and a brief overview may be helpful. All of the methods have in common that they involve scattering of radiation from molecules in solution. They differ in the kinds of radiation employed, and the information that can be obtained (see Table 7.5). Because molecules in solution exhibit random orientations and locations, it is not possible, using these methods, to obtain the kind of high-resolution structural information that we can get from X-ray diffraction (Chapter 6). On the other hand, the fact that the molecules are free to interact with one another or the solvent environment, or to change their conformations, means that we can obtain kinds of information that a static crystal structure cannot provide. The solution-scattering methods provide another kind of tool for the physical biochemist.

EXERCISES

*7.1. At one time, it was conventional to describe light scattering in terms of the *turbidity*, τ. This is defined in analogy to the absorbance in spectroscopy. If a beam of incident intensity I_0 passes through 1 cm^3 of solution and emerges with a smaller intensity I (the loss being by scattering), we define

$$\tau = -\ln \frac{I}{I_0}$$

Show that for Rayleigh scattering of unpolarized light, $\tau = (16\pi/3)R_0$. [Hint: Integration of the scattered intensity over a sphere at distance r from the solution should give the total loss of energy, but this should also be equal to $I - I_0$ as defined above. Why?]

7.2. A complication in the absorption spectroscopy of very large molecules arises from the light lost by scattering. Calculate the apparent absorbance due to *scattering* alone, for a solution of a hemocyanin, given the following data:

1. Cell thickness in direction of beam = 1 cm.

2. Protein concentration = 0.01 g/ml, M = 3,800,000 gm/mol.

3. n_0 = 1.33, dn/dC = 0.198 ml/g, λ = 280 nm.

You may assume Rayleigh scattering. See Exercise 7.1.

7.3. To visualize the effect of dust contamination on light-scattering experiments, perform the following calculation. A solution, containing 5 mg/ml of protein of M = 100,000, is contaminated to the extent of 0.001% of the protein weight by dust. The dust particles are 0.1 μm in radius, with density = 2.00 gm/cm^3. By what percent will the 90° scattering be changed by this contamination, assuming Rayleigh scattering for all particles? Is this a good assumption? Is the true situation apt to be more or less serious than your estimate?

7.4. The data given below describe light-scattering measurements on a small globular protein. Use these to calculate the molecular weight and the second virial coefficient. Other data you will need are λ = 436 nm; (dn/dc) = 0.196 cm^3/gm,

C (mg/ml)	$(KC/R_\theta) \times 10^5$
0.5	5.50
1.5	5.60
2.5	5.79
3.5	5.86
4.5	6.05

*7.5. For a spherically symmetrical particle of radius R, the structure factor is given by

$$F = 4\pi \int_0^R \rho(r) r^2 \left(\frac{\sin hr}{hr} \right) dr$$

where $\rho(r)$ is the electron density as a function of r.

a. From this, show that the F for a sphere of uniform electron density ρ_0 is

$$F = \frac{4\pi \rho_0}{h^3} (\sin hR - hR \cos hR)$$

b. Obtain an expression for the intensity of scattering, and show that this will have maxima and minima. [Hint: Differentiate the expression for I with respect to h and set $dI/dh = 0$.]

7.6. The following data are obtained from dynamic scattering of a virus particle as a function of retardation time (τ).

$\tau(\mu\ \text{sec})$	$g^{(2)}(\tau)$
0	1.55
20	1.45
40	1.36
60	1.30
80	1.24
100	1.19
120	1.16

The scattering angle was $90°$, $n = 1.33$. Calculate the diffusion coefficient of the virus.

*7.7. If a large molecule exhibits rotational diffusion as well as translational diffusion, the dynamic light scattering autocorrelation decay is described by

$$[g^{(2)}(\tau) - 1]^{1/2} = g^1(\tau) = e^{-D_T h^2 \tau}(A_0 + A_1 e^{-6D_R \tau})$$

where D_T and D_R are translational and rotational diffusion coefficients. Devise a way to obtain both from such data.

7.8. The radius of gyration for a circle is just the radius of the circle. The expression for the R_G of a rod is given in the text. Suppose that a DNA molecule could exist either as a rigid rod 100 nm long or as a circle with this circumference. Calculate the ratio of scattering at $\theta = 90°$ for the rod form to that for the circle form. Assume that $\lambda_0 = 546$ nm. If a 1% difference is detectable, would the light-scattering measurement serve to detect the breaking of circles? Suggest a non-scattering technique that would serve better and be cheaper.

7.9. The 50S ribosomal subunits from *E. coli* have been studied by low-angle X-ray scattering. X-rays of $\lambda = 0.154$ nm were used, and the following data recorded:

θ (mrad)	ln$I$*
1.41	76
2.00	73
2.54	70
3.0	66

*Units arbitrary.

Calculate the radius of gyration. If the particles are spheres, what is their diameter?

7.10. Consider neutron scattering experiments with a hypothetical globular lipoprotein, which is composed of approximately equal masses of protein and lipid. In a mixture of 50% D_2O, 50% H_2O, the radius of gyration was found to be 12 nm. In a mixture with only 10% D_2O, R_G increased to 20 nm. Assuming a concentric structure, is there protein or lipid on the outside?

REFERENCES

Scattering Processes: General

BRUMBERGER, H. (ed.) (1995) *Modern Aspects of Small Angle Scattering,* Kluwer Academic Publishers, Dordrecht, The Netherlands.

CANTOR, C. R. and P. R. SCHIMMEL (1980) *Biophysical Chemistry,* W. H. Freeman and Company, San Francisco, chaps. 13 and 14. Rigorous and comprehensive, but somewhat dated.

FEYNMAN, R. P., R. B. LEIGHTON, and N. SANDS (1963) *The Feynman Lectures on Physics,* Addison-Wesley, Reading, MA, chaps. 28–32. These chapters contain a remarkably lucid discussion of the fundamentals of electromagnetic radiation. The entire set of three volumes is strongly recommended.

Light Scattering

BROWN, WYN (ed.) (1993) *Dynamic Light Scattering,* Clarendon Press, Oxford, U.K.

BROWN, WYN (ed.) (1996) *Light Scattering, Principles and Developments,* Clarendon Press, Oxford, U.K. Together, these two volumes provide a compendium of chapters describing light scattering theory and application.

PITTZ, E. P., J. C. LEE, B. BABLOUZIAN, R. TOWNSEND, and S. N. TIMASHEFF (1973) "Light scattering and differential refractometry," in *Methods in Enzymology,* vol. 27, ed. C. H. W. Hirs and S. N. Timasheff, Academic Press, New York, 209–256. A classic reference to static scattering.

SCHMITZ, K. S. (1990) *An Introduction to Dynamic Light Scattering by Macromolecules.* Academic Press, New York.

TANFORD, C. (1961) *Physical Chemistry of Macromolecules,* John Wiley & Sons, New York, chap. 5. Although somewhat dated, this book contains detailed and careful derivations of the fundamental equations for Rayleigh scattering.

ZIMM, B. H. (1948) "Apparatus and Methods for Measurement and Interpretation of the Angular Variation of Light Scattering: Preliminary Results on Polystyrene Solutions," *J. Chem. Phys.* **16**, 1093–1099. Development of Zimm's methods for analysis of angular dependence of light scattering.

Other Scattering Methods

PILZ, I., Q. GLATTNER, and O. KRATKY (1979) "Small Angle X-Ray Scattering," in *Methods in Enzymology,* vol. 61, ed. C. H. W. Hirs and S. N. Timasheff, Academic Press, New York, 148–249.

SERGON, D. L., V. V. VOLKOV, M. B. KOZIN, M. B. STUHRMANN, C. BARBERATO, and M. H. J. KOCH (1997) "Shape determination from solution scattering of biopolymers," *J. Appl. Crystallogr.* **30**, 798–802.

ZACCAI, G. and B. JACROT (1983) "Small Angle Neutron Scattering," *Ann. Rev. Biophys. Bioeng.* **12**, 139–157.

8

Quantum Mechanics and Spectroscopy

Spectroscopy is the observation of quantum mechanics. Light may be absorbed or emitted by a molecule undergoing a *transition* between discrete energy levels, and these energy levels, as well as other characteristics of the molecule, are predicted by quantum mechanics. Quantum mechanics is important to physical biochemists who study macromolecules because its principles underlie all of their spectroscopic observations. Whether it is the energy of a transition or the rules for the transition to occur between quantized states, we cannot understand our observations without understanding quantum mechanics.

Readers should not be intimidated by quantum mechanics because in reality it is a beautiful and particularly simple mathematics that describes the world of atoms and molecules. In our experience, confusion arises when attempts are made to relate quantum mechanics to everyday life. There is nothing obvious about atoms and molecules in quantum states to people who live in a classical world, which is the world of our everyday experience.

Most students will be familiar with the fundamentals of quantum mechanics from their undergraduate course in physical chemistry. The basic principles of quantum mechanics are briefly reviewed in this chapter to highlight the fundamentals taught in physical chemistry that are important here. We take the postulate approach and see the simplicity of the method as it attacks each problem in exactly the same way. Macromolecules are complicated, so we review the approximate methods commonly used to solve quantum chemistry problems. The hydrogen atom is briefly reviewed because its solutions are used to approximate the solutions of more complicated systems. We will see the simple progression from the hydrogen atom to the hydrogen molecule ion to the hydrogen molecule. The hydrogen molecule is particularly important to physical biochemists as the model for a bond.

Quantum mechanics permits us to predict the characteristics of a transition observed by any type of spectroscopy applied to macromolecules.

8.1   LIGHT AND TRANSITIONS

Classically, light is considered to be a wave with an electric field component and a magnetic field component. A snapshot of a linearly polarized beam of monochromatic light is depicted in Figure 8.1. Monochromatic light repeats sinusoidally in time with frequency ν and in distance with wavelength λ. Figure 8.1 can be relabeled with distance x becoming time t and λ becoming ν^{-1} to depict light as a function of time. Mathematically, the sinusoidal variation of the electric vector $\mathbf{E}$ and the magnetic vector $\mathbf{H}$ is given by

$$\mathbf{E} = \mathbf{E}_0 \cos 2\pi(\nu t - x/\lambda)$$

$$\mathbf{H} = \mathbf{H}_0 \cos 2\pi(\nu t - x/\lambda) \tag{8.1}$$

where $\mathbf{E}_0$ and $\mathbf{H}_0$ are the maximum values for the vectors, and the two vectors are in phase. Light is monochromatic when ν and λ each have a single value. These two quantities are related through the speed of light c so that

$$\nu\lambda = c \tag{8.2}$$

In the quantum mechanical view, light also has particle properties and the energy of these particles or *photons* depends on the classical frequency through Planck's law

$$E = h\nu = \frac{hc}{\lambda} \tag{8.3}$$

where Planck's constant h is 6.63×10^{-34} J · s.

Atoms and molecules exist only in discrete energy levels. Light may be absorbed or emitted for photons with an energy that matches the energy difference

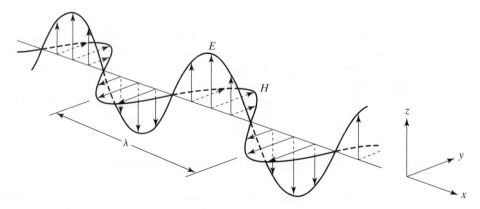

Figure 8.1   A snapshot of a linearly polarized beam of monochromatic light. The direction of propagation is in the $+x$-direction and the picture slides to the right with time.

Figure 8.2 An atom or molecule in a discrete energy level may either absorb photons with an energy that matches the difference between the occupied level and a higher-energy unoccupied level or emit photons with an energy that matches the difference between the occupied level and a lower unoccupied level.

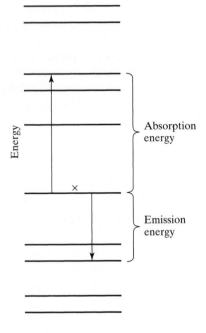

between occupied and unoccupied levels, as shown in Figure 8.2. Energy is a measurable characteristic of a transition, and as a practical matter spectrographs measure absorption or emission as a function of wavelength, so spectra are often plotted as a function of λ. When spectra are plotted as a function of energy, spectroscopists often use *wavenumbers*, $\bar{\nu} = 1/\lambda$, which are proportional to energy as stated by Eq. 8.3.

Intensity is another measurable characteristic of a transition, since not all photons impinging on a sample will be absorbed nor will all molecules in a higher energy level emit photons. After reviewing the quantum mechanical calculation of energies for various systems, we discuss intensity at various levels of sophistication. Each transition also has a characteristic direction. It is the direction of transitions that tells us about the conformation of biological molecules, and the chapter ends by discussing this important measurable.

As we shall see, quantum mechanics can be used to predict the three measurable characteristics of a transition: energy, intensity, and direction.

8.2   POSTULATE APPROACH TO QUANTUM MECHANICS

Euclidean geometry begins with a set of postulates, statements that are taken to be true and from which the properties of plane figures are derived by defining certain words and proving theorems. In quantum mechanics, six postulates are assumed to be true and a mathematics has been developed that to date has been remarkably successful in predicting the properties of atoms and molecules. This approach provides a conceptually simple recipe for solving any quantum mechanical problem. *Postulate I* says that the state of a system of n particles (for our purposes an atom or a molecule) is described as completely as possible by some function of the position coordinates for the particles q_i, and the time t, called the wavefunction and denoted by $\Psi(q_1, q_2, \ldots, q_{3n}, t)$.

The wavefunction has a physical significance given by *Postulate II:* The probability of finding the system in the differential volume element $d\tau$ with coordinates $q_1, q_2, \ldots, q_{3n}, t$ is given by

$$\Psi * \Psi d\tau \tag{8.4}$$

where the * indicates the complex conjugate, because the wavefunction may be complex. We remember from Chapter 6 that the complex conjugate of the complex number $a + ib$ is $a - ib$, where i is the imaginary square root of -1. Since the total probability of finding the system somewhere is 1, the wavefunction should be normalized so that

$$\int \Psi * \Psi d\tau = 1 \tag{8.5}$$

where the integration is over all space. Note that when Ψ are treated as functions, integration of $\Psi * \Psi d\tau$ gives us a number.

As physical biochemists, we are interested in observing the physical properties of biological molecules. *Postulate III* says that every physical observable of a system (not just the energy, but position, dipole moment, momentum, etc.) has associated with it a linear Hermitian operator. Operators change a function, for example, by taking a partial derivative. Operators and their Hermitian property are discussed below. They are denoted by a hat, such as $\hat{P}$.

We want to predict the numerical value for various physical properties of biological molecules. *Postulate IV* says that the average value of many measurements for a physical observable corresponding to operator $\hat{P}$ will be predicted by

$$\overline{P} = \frac{\displaystyle\int \Psi * \hat{P}\Psi d\tau}{\displaystyle\int \Psi * \Psi d\tau} \tag{8.6}$$

where integration again gives us a number. In some cases, an operator operating on a function may result in a number times the function, that is,

$$\hat{P}\Psi = p\Psi \tag{8.7}$$

where p is a numerical value. In this case, Ψ is said to be an *eigenfunction* of operator $\hat{P}$ with *eigenvalue* p. Then each measurement will yield the same numerical value p.

If Ψ is not an eigenfunction of $\hat{P}$, different measurements on many identical systems may give different results that we cannot predict; we can predict only the average value. The fact that the operators are Hermitian guarantees real numerical values for Eqs. 8.6 and 8.7.

What are these quantum mechanical operators that allow us to predict the results of experiments? *Postulate V* says that the operators for position, momentum, time, and total energy are those given in Table 8.1. Note that the operators for position and time merely multiply the function by the corresponding variable. The momentum and energy operators involve taking a partial derivative of the wavefunction. These are simple operations from calculus. Quantum mechanical operators for other dynamical variables are easily constructed from the classical expression for that variable by substituting the Hermitian operators from Table 8.1.

To complete the recipe for predicting the properties of a molecule, we need its wavefunction. *Postulate VI* says that the wavefunction for a system can be determined from the *Schrödinger equation*

$$\hat{H}\Psi = i\hbar\frac{\partial \Psi}{\partial t} \tag{8.8}$$

where $\hat{H}$ is the Hamiltonian operator, and $\hbar$ is $h/2\pi$. Hamilton worked out a system of classical mechanics based on the sum of the expressions for the kinetic energy K and the potential energy V of a system. This sum $(K + V)$ is symbolized as H. We can make up separate quantum mechanical operators for the kinetic energy and the potential energy of a system. While the total energy operators $\hat{H}$ and $i\hbar\partial/\partial t$ are not equal, their effect on a wavefunction should be the same. Equation 8.8 says that the kinetic energy operating on the wavefunction for a system plus the potential energy operating on the wavefunction should be equivalent to the total energy operating on the wavefunction.

Classically, the kinetic energy for a particle is given by

$$K = \frac{p^2}{2m} \tag{8.9}$$

Table 8.1 Quantum Mechanical Operators

Classical Variable	Quantum Mechanical Operator	Form of the Quantum Mechanical Operator
q (position)	$\hat{q}$	q (classical variable)
p (momentum)	$\hat{p}$	$-i\hbar\dfrac{\partial}{\partial q}$
t (time)	$\hat{t}$	t (classical variable)
E (energy)	$\hat{E}$	$i\hbar\dfrac{\partial}{\partial t}$

where p is the momentum and m is the mass of the particle. Substituting the quantum mechanical form of the momentum operator for a Cartesian coordinate system we obtain

$$\hat{K} = \frac{\left[-i\hbar\left(\dfrac{\partial}{\partial x}\mathbf{i} + \dfrac{\partial}{\partial y}\mathbf{j} + \dfrac{\partial}{\partial z}\mathbf{k}\right)\right]^2}{2m} = \frac{[-i\hbar\nabla]^2}{2m}$$

$$= \frac{-\hbar^2\left(\dfrac{\partial^2}{\partial x^2} + \dfrac{\partial^2}{\partial y^2} + \dfrac{\partial^2}{\partial z^2}\right)}{2m} = \frac{-\hbar^2\nabla^2}{2m} \tag{8.10}$$

where we have used the standard symbols ∇ for the gradient, ∇^2 for the Laplacian operator, and $\mathbf{i}, \mathbf{j}$, and $\mathbf{k}$ for the unit vectors of a Cartesian coordinate system.[1] For a given system the Hamiltonian will involve the sum over the kinetic energies of all the particles plus a potential energy that will depend on the system and will normally involve only the operators for position and time. Equation 8.8 is the *time-dependent* Schrödinger equation, a known differential equation in position and time that involves the unknown function Ψ, for which we wish to solve. Some simple systems, such as a free particle in space or a hydrogen atom, give differential equations for which the solutions have long been known. However, more complicated systems give differential equations for which the solution is unknown, so that finding the wavefunction for a given system is the central problem in quantum mechanics. Often, well-known methods are used to obtain approximate solutions and three useful approximate methods are presented in Section 8.3.

In many cases, an operator is independent of time and the system is then said to be in a *stationary state* with respect to that observable. These states that don't change with time turn out to be quantized at certain energy values. If $\hat{V}$ (and therefore $\hat{H}$) is independent of time, the stationary state solutions take the form

$$\Psi(q, t) = \psi(q)e^{-iEt/\hbar} \tag{8.11}$$

How do we know that this form is a solution? We substitute Eq. 8.11 into the differential Eq. 8.8 and show that it works. Differential equations are often solved in this way. Intelligent guesses are made about the solution; if the guess works, it *is* a solution. Substituting, we obtain

$$\hat{H}(q)\psi(q)e^{-iEt/\hbar} = i\hbar\frac{\partial}{\partial t}\psi(q)e^{-iEt/\hbar}$$

$$e^{-iEt/\hbar}\hat{H}(q)\psi(q) = i\hbar\psi(q)(-iE/\hbar)e^{-iEt/\hbar}$$

$$\hat{H}(q)\psi(q) = E\psi(q) \tag{8.12}$$

---

[1] The reader should note that the square in the numerator of Eq. 8.10 is really a dot product. Recalling that $\mathbf{i}\cdot\mathbf{i} = 1, \mathbf{i}\cdot\mathbf{j} = 0$, and so on, we see why the gradient contains only three terms.

since the time-independent $\hat{H}$ does not operate on the exponential and the total energy operator does not operate on the time-independent factor ψ. The result is the *time-independent* Schrödinger equation (Eq. 8.12) where the solution is $\psi(q)$, a time-independent eigenfunction with eigenvalue E. The complete wavefunction $\Psi(q, t)$ is *always* time dependent, although the time dependence when $\hat{H}$ is independent of time is the factor $e^{-iEt/\hbar}$. When $\hat{H}$ is time independent we usually ignore the time-dependent factor and work only with $\psi(q)$, but we must not forget that the factor exists.

8.3 TRANSITION ENERGIES

This section is a basic review of the fundamental principles of quantum mechanics as presented in many undergraduate courses on physical chemistry. Here, we consider the time-independent Schrödinger equation for a number of simple systems and discuss their wavefunctions and their corresponding energy levels. A complete discussion of the manipulations used to solve for the wavefunctions can be found in most physical chemistry texts (see references). Having obtained the energy levels for the systems, we can predict the energy for a transition between two levels.

8.3.1 The Quantum Mechanics of Simple Systems

The simplest system is a single *free particle* of given mass m that has only a given velocity (kinetic energy) and no potential energy in a one-dimensional space. Using the Cartesian coordinate x to locate this particle, we have for the Schrödinger equation

$$-\frac{\hbar^2}{2m}\left(\frac{\partial^2}{\partial x^2}\right)\psi(x) = E\psi(x) \tag{8.13}$$

There are two solutions to Eq. 8.13,

$$\psi_\pm = Ne^{\pm ikx} \tag{8.14}$$

where $k = (2mE)^{1/2}/\hbar$, and E can have any value, since there are no boundary conditions to restrict the particle; E is, of course, known from the given velocity. The two solutions are obtained from Eq. 8.14 by using either the top or bottom signs. The arbitrary constant N is usually chosen by using Eq. 8.5 to normalize the wavefunction. However, these wavefunctions are not simply normalizable because the particle is unbounded.

With this simple system it is easy to see that once we have the wavefunction for a system, we can use Postulate IV to predict the average value of any physical observable. For instance, the wavefunctions of Eq 8.14 turn out to be eigenfunctions of the linear momentum operator.

$$-i\hbar(\partial/\partial x)\psi_\pm = \pm(2mE)^{1/2}\psi_\pm \tag{8.15}$$

There are two possibilities for the momentum $\pm(2mE)^{1/2}$, corresponding to the particle traveling in the positive or negative x direction.

A linear combination of solutions to the Schrödinger equation will also be a solution, yielding in this case

$$\Psi = Ae^{+ikx} + Be^{-ikx} \tag{8.16}$$

Since $e^{\pm ikx} = \cos kx \pm i \sin kx$, Eq. 8.16 can be written in the form

$$\Psi = C \sin kx + D \cos kx \tag{8.17}$$

Sine and cosine functions are another way of expressing the solutions to Eq. 8.13 and this form will be useful for the next example.

A *particle in a one-dimensional box* is our second simple example of predicting the properties for a system through the mathematics of quantum mechanics. Inside the box of dimension a $(0 < x < a)$, the potential energy is zero, and the Schrödinger equation is just Eq. 8.13. Outside the box $(x \leq 0; x \geq a)$ the potential energy is infinite. The box constitutes a steep-sided potential well, and the Schrödinger equation for positions outside this potential well is

$$\left(-\frac{\hbar^2}{2m}\frac{\partial^2}{\partial x^2} + \infty\right)\psi(x) = E\psi(x) \tag{8.18}$$

The solution to Eq. 8.18 is $\psi(x) = 0$, and the particle cannot exist outside the box. Except for this boundary condition, the Schrödinger equation is the same as that for a free particle and we expect a similar solution. The boundary conditions and continuity at the boundaries suggest that the wavefunction should be zero at $x = 0$ and $x = a$, which in turn suggests a sine function for the way that we have set up our coordinate system, as shown in Figure 8.3. An informed guess is

$$\psi_n(x) = \left(\frac{2}{a}\right)^{1/2} \sin\left(\frac{n\pi x}{a}\right) \qquad n = 1, 2, \ldots \tag{8.19}$$

where normalization gives the multiplicative factor. Substitution into the Schrödinger equation confirms that the guess is indeed the solution and yields the corresponding energies

$$E_n = \frac{n^2 h^2}{8\,ma^2} \tag{8.20}$$

This particle is bounded and the possible energy levels E_n are quantized through the *quantum number n*. Figure 8.3 labels the first few *quantum states* as the energy proportional to n^2. Thus, we see that the energy levels diverge as n gets larger.

The quantum mechanics for a particle in a box allows us to predict transition energies for this system. For instance, to cause a particle in state $n = 1$ to undergo a

Figure 8.3 The energy levels corresponding to a particle in a one-dimensional box of dimension a. The energy difference between the levels increases with the square of the quantum number.

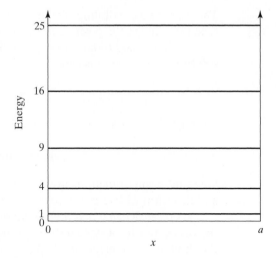

transition to state $n = 2$, we need a photon with energy

$$\Delta E = 4\left(\frac{h^2}{8\,ma^2}\right) - \frac{h^2}{8\,ma^2} = \frac{3h^2}{8\,ma^2} \tag{8.21}$$

For a box 0.05 nm (5×10^{-11} m) in length containing an electron of mass 9.11×10^{-31} kg, we can substitute into Eq. 8.21 to obtain an energy difference of 7.24×10^{-17} J.

   With the wavefunction Eq. 8.19, we can calculate the average value of any observable. Again let us consider the momentum. From Postulate V, the momentum operator for a particle in a one-dimensional box is $\hat{p} = -i\hbar(\partial/\partial x)$. In this case the wavefunctions are not eigenfunctions of the momentum operator, but the average value of the momentum for state n can be calculated using Postulate IV.

$$\begin{aligned}
\bar{p} &= -\frac{2i\hbar}{a} \int_0^a \sin\left(\frac{n\pi x}{a}\right) \frac{\partial}{\partial x} \sin\left(\frac{n\pi x}{a}\right) dx \\
&= -\frac{2n\pi i\hbar}{a^2} \int_0^a \sin\left(\frac{n\pi x}{a}\right) \cos\left(\frac{n\pi x}{a}\right) dx \\
&= \left(\frac{i\hbar}{a}\right) \cos^2\left(\frac{n\pi x}{a}\right) \Big|_0^a = 0
\end{aligned} \tag{8.22}$$

The average value turns out to be zero because the particle has an equal probability of positive or negative momentum, depending on whether it is moving in the $+x$ or $-x$ direction. Using Eq. 8.20, $p = (2mE)^{1/2} = \pm nh/2a$ for the individual directions.

   We can also calculate the average value of many position measurements $\bar{x}$ for the particle by using Postulate IV and the position operator x

$$\bar{x} = \frac{2}{a} \int_0^a \sin\left(\frac{n\pi x}{a}\right) x \sin\left(\frac{n\pi x}{a}\right) dx = \frac{a}{2} \tag{8.23}$$

Regardless of n, the average value is the center of the box, as we would expect classically.

It is also instructive for us to look at the probability distribution of finding the particle between x and $x + dx$. This is given by Postulate II as

$$\psi * \psi \, dx = \frac{2}{a} \sin^2 \left(\frac{n\pi x}{a} \right) dx \tag{8.24}$$

The probability is not constant as we would expect from classical theory, but varies at different points in the box.

The classical approach to the *simple harmonic oscillator* is discussed briefly in Chapter 3. The quantum mechanical approach presented here will prove useful when we discuss the spectroscopy of molecular vibrations. In this case the potential energy of the well is given by the parabolic curve $kx^2/2$. Figure 8.4 shows the example of a mass m stretching or compressing a spring of stiffness k. The Schrödinger equation is

$$\left(-\frac{\hbar^2}{2m} \frac{\partial^2}{\partial x^2} + \frac{1}{2}kx^2 \right)\psi(x) = E\psi(x) \tag{8.25}$$

The solutions to this Schrödinger equation are

$$\psi_n(\xi) = \left(\frac{\sqrt{\beta/\pi}}{2^n n!} \right)^{1/2} H_n(\xi)e^{-\xi^2/2} \qquad n = 0, 1, 2, \ldots \tag{8.26}$$

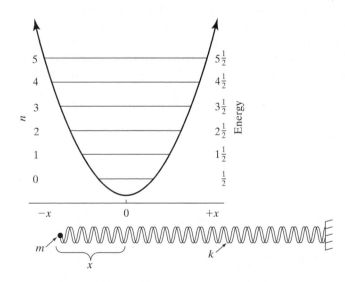

Figure 8.4 The energy levels for a simple harmonic oscillator are equally spaced. Here we also see the example of a mass m stretching and compressing a spring of stiffness k. The equilibrium position for the spring is $x = 0$.

where $\xi = \sqrt{\beta}x$, $\beta = mv_0/h$, v_0 is the basic frequency of the oscillator, and $2\pi v_0 = (k/m)^{1/2}$. The $H_n(\xi)$ are the Hermite polynomials, $H_0 = 1$, $H_1 = 2\xi$, $H_2 = 4\xi^2 - 2, \ldots$. The corresponding energies are very simple

$$E_n = \left(n + \frac{1}{2}\right)hv_0 \tag{8.27}$$

Figure 8.4 also depicts the simple harmonic potential and the corresponding energy levels. We note that even the level for $n = 0$ has a finite energy, called the *zero point energy*. Since the levels are evenly spaced, a photon with energy hv_0 could cause a transition between any two adjacent states.

The final simple example in our review of fundamental quantum mechanics is *hydrogen-like atoms*, which have a nucleus with some number of protons Z and a single electron. We separate the kinetic energy at the center of mass and assume that the center of mass is well approximated by the position of the nucleus and the reduced mass by the mass of an electron m_e. Figure 8.5 shows the nucleus at the origin of a coordinate system that defines both Cartesian and spherical coordinates for the electron. The potential energy will be the Coulomb interaction between the nucleus of charge $+Ze$, where e is the charge on a proton and the electron of charge $-e$.

$$V = \frac{-Ze^2}{4\pi\varepsilon_0 r} \tag{8.28}$$

where $\varepsilon_0 = 1/c^2\mu_0 = 8.854 \times 10^{-12}$ C^2/J$\cdot$m is the permittivity of a vacuum (the permeability of a vacuum μ_0 is defined as $4\pi \times 10^{-7}$ T$^2 \cdot$ m^3/J). For quantum mechanical calculations on atoms and molecules, it is convenient to use atomic units (au), which use dimensions of the atomic scale and greatly simplify the equations: 1 au of distance is the first Bohr radius, $a_0 = h^2\varepsilon_0/\pi m_e e^2 = 5.29 \times 10^{-11}$ m; 1 au of energy is twice the ionization energy of a hydrogen atom, $m_e e^4/4\varepsilon_0^2 h^2 = 27.21$ eV $= 4.36 \times 10^{-18}$ J. A more complete definition of atomic units is given in Table 8.2. With these units the Schrödinger equation for a hydrogen-like atom is

$$\left(-\frac{\nabla^2}{2} - \frac{Z}{r}\right)\psi(r, \theta, \phi) = E\psi(r, \theta, \phi) \tag{8.29}$$

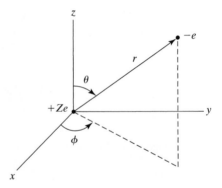

Figure 8.5 The Cartesian and spherical coordinate systems for a hydrogen-like atom.

Table 8.2   Atomic Units

Quantity	Value in SI Units*	Value in Atomic Units
Rest mass of electron, m_e	9.109×10^{-31} kg	1 au
Magnitude of electron charge, e	1.602×10^{-19} C	1 au
Unit of action (Planck's constant), h	6.625×10^{-34} J $\cdot$ sec	2π au
Unit of orbital angular momentum about any axis, $\hbar = h/2\pi$	1.055×10^{-34} J $\cdot$ sec	1 au
Permittivity, ε_0	8.854×10^{-12} C^2/J $\cdot$ m	$\dfrac{1}{4\pi}$ au
Radius of first Bohr orbit, a_0	$\dfrac{h^2\varepsilon_0}{\pi m_e e^2} = 5.292 \times 10^{-11}$ m	
	$= 0.05292$ nm	1 au
Velocity of electron in first Bohr orbit, v_0	$\dfrac{e^2}{2h\varepsilon_0} = 2.188 \times 10^6$ m/sec	1 au
Frequency of revolution in first Bohr orbit, ν_0	$\dfrac{e^2}{4\pi h\varepsilon_0 a_0} = 6.580 \times 10^{15}$/sec	$\dfrac{1}{2\pi}$ au
Time for one circuit of first Bohr orbit, τ_0	$\dfrac{1}{\nu_0} = 1.520 \times 10^{-16}$ sec	2π au
Ionization energy of hydrogen, E_H	$\dfrac{m_e e^4}{8\varepsilon_0^2 h^2} = 2.179 \times 10^{-18}$ J	
	$= 13.65$ eV	1/2 au
Electric dipole moment	$ea_0 = 8.478 \times 10^{-20}$ C $\cdot$ m	
	$= 2.542$ debye	1 au
Electric field at first Bohr orbit caused by nucleus	$\dfrac{e}{4\pi\varepsilon_0 a_0^2} = 5.142 \times 10^{11}$ J/C $\cdot$ m	1 au
Potential energy of electron in first Bohr orbit	$\dfrac{-e^2}{4\pi\varepsilon_0 a_0} = -4.359 \times 10^{-18}$ J	-1 au
Magnetic moment at nucleus due to motion electron in first Bohr orbit (Bohr magneton), μ_B	$\dfrac{e\hbar}{2m_e} = 9.273 \times 10^{-24}$ J/T	1/2 au

*Système internationale d'unités

Atoms are spherical, and the Schrödinger equation for a hydrogen-like atom is most easily solved in a spherical coordinate system. The Laplacian in spherical coordinates is

$$\nabla^2 = \frac{1}{r^2}\frac{\partial}{\partial r}\left(r^2\frac{\partial}{\partial r}\right) + \frac{1}{r^2 \sin\theta}\frac{\partial}{\partial\theta}\left(\sin\theta\frac{\partial}{\partial\theta}\right) + \frac{1}{r^2\sin^2\theta}\frac{\partial^2}{\partial\phi^2} \qquad (8.30)$$

When integrating in spherical coordinates, the volume element is $d\tau = r^2 \sin\theta\, dr\, d\theta\, d\phi$ and the limits of integration are: $r = 0, \infty$; $\theta = 0, \pi$, and $\phi = 0, 2\pi$. Solutions to Eq. 8.29 are discussed as part of undergraduate physical chemistry; here we give only the results. The radial part of the wavefunctions is the associated Laguerre functions; the angular θ part of the wavefunction is the associated Legendre polynomials; the angular ϕ part of the wavefunction is an exponential. Normalized solutions of Eq. 8.29 for the first two principal quantum numbers are

$$n = 1: \quad 1s = \frac{1}{\pi^{1/2}} Z^{3/2} e^{-Zr}$$

$$n = 2: \quad 2s = \frac{1}{(32\pi)^{1/2}} Z^{3/2} (2 - Zr) e^{-Zr/2}$$

$$2p_z = \frac{1}{(32\pi)^{1/2}} Z^{5/2} \cos\theta \, r e^{-Zr/2}$$

$$2p_x = \frac{1}{(32\pi)^{1/2}} Z^{5/2} \sin\theta \cos\phi \, r e^{-Zr/2}$$

$$2p_y = \frac{1}{(32\pi)^{1/2}} Z^{5/2} \sin\theta \sin\phi \, r e^{-Zr/2} \qquad (8.31)$$

We have written the $2p$ orbitals as the chemists notation in a Cartesian coordinate system, where $x = r \sin\theta \cos\phi$, $y = r \sin\theta \sin\phi$, and $z = r \cos\theta$. These wavefunctions correspond to the quantized negative energies of an unionized hydrogen-like atom expressed in atomic units

$$E_n = -\frac{Z^2}{2n^2} \qquad (8.32)$$

The energies are depicted in Figure 8.6, where we see that higher energies become closer and closer together. This contrasts with the harmonic oscillator, where the energies are equally spaced, and with the particle in the box, where higher energies become further apart. There is a continuum of energy levels for the hydrogen-like atom with E greater than 0, which corresponds to *ionization* or the escape of the electron.

The energies of the wavefunctions for a hydrogen-like atom depend only on the principal quantum number. We can predict the energy for the transition of an electron between any two quantum states by using Eq. 8.32. For instance, with $Z = 1$, we have for the transition from $n = 1$ to $n = 2$.

$$\Delta E_{21} = E_2 - E_1 = -\frac{1}{8} + \frac{1}{2} = \frac{3}{8} \text{ au} \qquad (8.33)$$

which corresponds to 1.635×10^{-18} J or, using Eq. 8.3, a wavelength of 121 nm.

8.3.2   Approximating Solutions to Quantum Chemistry Problems

As we turn our attention to more complicated systems, it will not be possible to find exact solutions to the Schrödinger equation. However, we can find approximate solutions that will give us a good understanding of the chemistry and other properties

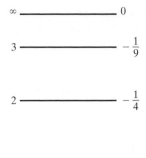

∞ ——————— 0

3 ——————— $-\dfrac{1}{9}$

2 ——————— $-\dfrac{1}{4}$

n

Energy

1 ——————— -1

Figure 8.6   The energy levels for a hydrogen-like atom. The energy difference between adjacent levels decreases with the inverse square of the quantum number.

of complicated molecules. Just three basic principles need to be understood to generate approximate wavefunctions.

Separability.   Solving the Schrödinger equation for a complicated system can be simplified considerably if the interaction terms are neglected so that the Hamiltonian becomes *separable*. Let us consider a system of individual groups i that do not interact. We denote the Hamiltonian for each group as $\hat{h}(i)$, with wavefunctions $\chi(i)$ and corresponding energies $e(i)$. Since there is no interaction between the groups, the Hamiltonian for the system is simply the sum of the Hamiltonians for the individual groups,

$$\hat{H} = \sum_{i} \hat{h}(i) \tag{8.34}$$

The exact solutions are products of the solutions for the individual groups

$$\psi_{ab\ldots}(1, 2,\ldots) = \chi^{a}(1)\chi^{b}(2)\cdots \tag{8.35}$$

where $a, b, \ldots$ index the various states for each group. The corresponding energies are a sum of the individual energies for the groups

$$E_{ab\ldots} = e^{a}(1) + e^{b}(2) + \cdots \tag{8.36}$$

We can verify that the product function is a solution for the noninteracting Hamiltonian by substitution, remembering that the Hamiltonian for group i operates only on the wavefunction for group i.

$$[\hat{h}(1) + \hat{h}(2) + \cdots]\chi^a(1)\chi^b(2) \cdots$$

$$= \chi^b(2)\cdots\hat{h}(1)\chi^a(1) + \chi^a(1)\cdots\hat{h}(2)\chi^b(2) + \cdots$$

$$= x^b(2)\cdots e^a(1)\chi^a(1) + \chi^a(1)\cdots e^b(2)\chi^b(2) + \cdots$$

$$= [e^a(1) + e^b(2) + \cdots]\chi^a(1)\chi^b(2) \cdots \qquad (8.37)$$

Students will remember from undergraduate physical chemistry that, as a good approximation, the properties of the nucleus of the hydrogen atom can be separated from the properties of the electron by putting the coordinate system for the electron at the nucleus. Furthermore, choosing a spherical coordinate system for the hydrogen atom and doing an appropriate substitution separates the r part of the wavefunction from θ and ϕ.

Perturbation theory. *Nondegenerate perturbation theory* is based on the simplification possible when a Hamiltonian is made separable. Suppose we want to solve the Schrödinger equation for some complicated system that is made up of a number of simpler groups indexed by i. A concrete example might be the groups of guanosine phosphate, cytidine phosphate, and adenosine phosphate that together make up the more complicated trimer GpCpAp. The groups must all be different and not have any energy levels at the same energy—that is, the groups must be *nondegenerate*. In general, the Hamiltonian for the system can be written as

$$\hat{\mathcal{H}} = \sum_i \hat{h}_i + \sum_i \sum_{j>i} \hat{V}_{ij} = \hat{H}_0 + \hat{V} \qquad (8.38)$$

This Hamiltonian consists of a separable part $\hat{H}_0$, and the sum of the interactions between the groups taken two at a time, $\hat{V}$. We wish to solve the Schrödinger equation for this system,

$$(\hat{H}_0 + \lambda\hat{V})\psi_k = \mathcal{E}_k\psi_k \qquad (8.39)$$

where k indexes the various states of the system and the dummy parameter λ can be used to turn the interaction on and off. With $\lambda = 0$, the Hamiltonian is separable; with $\lambda = 1$, there is a *perturbation* of the separable system due to the interactions between the groups.

Each individual group has a corresponding Schrödinger equation,

$$\hat{h}_i\chi_i^a = e_i^a\chi_i^a \qquad (8.40)$$

with normalized solutions χ_i^a and corresponding energies e_i^a, where the a indexes the various states for the group. If we set $\lambda = 0$, then we know from separability that the solution to the separable part of the Hamiltonian $\hat{H}_0$ will be the various possible products of the wavefunctions for the groups with corresponding energies that are sums of the individual energies of the groups

$$\phi_k = \prod_i \chi_i^a \tag{8.41}$$

$$E_k = \sum_i e_i^a \tag{8.42}$$

where $\prod$ is the product operator. There are many product wavefunctions ϕ_k because each group has a number of possible states indexed by a. The Schrödinger equation for the unperturbed system is then

$$\hat{H}_0 \phi_k = E_k \phi_k \tag{8.43}$$

This treatment suggests the product functions ϕ_k as approximate wavefunctions for the *perturbed* Hamiltonian. Postulate IV can be used to get the energy of the perturbed system in this approximation

$$\begin{aligned}
\overline{E} &= \int \phi_k^*(\hat{H}_0 + \hat{V})\phi_k \, d\tau \\
&= \int \phi_k^* \hat{H}_0 \phi_k \, d\tau + \int \phi_k^* \hat{V} \phi_k \, d\tau \\
&= E_k + \int \phi_k^* \hat{V} \phi_k \, d\tau
\end{aligned} \tag{8.44}$$

This was an early approximation used in solving the Schrödinger equation for a hydrogen molecule. The approximate wavefunction was taken to be the product of a $1s$ hydrogen atom wavefunction on each of the two-component hydrogen atoms. The resulting treatment is called the *valence bond* approximation.

Since ψ_k goes to ϕ_k as λ goes to zero, we are inspired to write ψ_k and its corresponding energy as a power series in λ.

$$\psi_K = \phi_k + \lambda \phi_k' + \lambda^2 \phi_k'' + \cdots \tag{8.45}$$

$$\mathcal{E}_K = E_k + \lambda E_k' + \lambda^2 E_k'' + \cdots \tag{8.46}$$

Remember, the power series form for ψ_K and $\mathcal{E}_K$ is just a guess and we do the algebra to see if the guess is useful. Substituting into Eq. 8.39, we have

$$(\hat{H}_0 + \lambda \hat{V})(\phi_k + \lambda \phi_k' + \lambda^2 \phi_k'' + \cdots)$$
$$= (E_k + \lambda E_k' + \lambda^2 E_k'' + \cdots)(\phi_k + \lambda \phi_k' + \lambda^2 \phi_k'' + \cdots)$$
$$\hat{H}_0 \phi_k + \lambda \hat{H}_0 \phi_k' + \lambda^2 \hat{H}_0 \phi_k'' + \lambda \hat{V} \phi_k + \lambda^2 \hat{V} \phi_k' + \cdots$$
$$= E_k \phi_k + \lambda E_k \phi_k' + \lambda^2 E_k \phi_k'' + \lambda E_k' \phi_k + \lambda^2 E_k' \phi_k' + \lambda^2 E_k'' \phi_k + \cdots$$
$$\hat{H}_0 \phi_k + \lambda(\hat{H}_0 \phi_k' + \hat{V} \phi_k) + \lambda^2(\hat{H}_0 \phi_k'' + \hat{V} \phi_k') + \cdots$$
$$= E_k \phi_k + \lambda(E_k \phi_k' + E_k' \phi_k) + \lambda^2(E_k \phi_k'' + E_k' \phi_k' + E_k'' \phi_k) + \cdots \tag{8.47}$$

For the last equation in 8.47 to hold for all values of λ, the equation must hold for each power of λ.

$$\hat{H}_0 \phi_k = E_k \phi_k \quad \text{for } \lambda^0 \tag{8.48}$$

$$\hat{H}_0 \phi_k' + \hat{V} \phi_k = E_k \phi_k' + E_k' \phi_k \quad \text{for } \lambda^1 \tag{8.49}$$

$$\hat{H}_0 \phi_k'' + \hat{V} \phi_k' = E_k \phi_k'' + E_k' \phi_k' + E_k'' \phi_k \quad \text{for } \lambda^2 \tag{8.50}$$

Equation 8.48 is just the Schrödinger equation for the unperturbed system, identical to Eq. 8.43. Equation 8.49 will yield the *first-order* corrections that make ψ_K and $\mathscr{E}_K$ different from ϕ_k and E_k. The *second-order* corrections can be obtained from Eq. 8.50, but we will not consider higher-order corrections. If the perturbation is small, only first-order corrections will be necessary.

How can we express ϕ_k' in a useful form? In order to understand the procedure, we need to introduce the concept of *vector spaces*. Students are undoubtedly familiar with the Cartesian coordinate system already used in this chapter, which is an example of a vector space with three axes. *Basis vectors* **i**, **j**, and **k** lie along the x-, y-, and z-axes. They are *orthonormal*—that is, they have a length of 1 and are therefore normalized in the sense of Eq. 8.5; and they are mutually orthogonal—the dot product between any two is zero. These basis vectors are *complete* since any standard distance vector **r** can be expressed as its projected components on the three axes, $\mathbf{r} = x\mathbf{i} + y\mathbf{j} + z\mathbf{k}$, as already illustrated for the hydrogen atom in Figure 8.5. Another way to express this vector is to list its components with the basis vectors understood implicitly, $\mathbf{r} = (x, y, z)$.

Most students will be familiar with digitizing a function in order to enter it into a computer or calculator. This is illustrated for positive values of $f(x) = x^2/4$ in Figure 8.7. Using unit intervals along the x-axis, digitizing represents $f(x)$ as the vector $(0.0, 0.25, 1.0, 2.25, \ldots)$ with an infinite number of components. This digitization

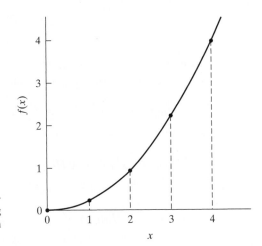

Figure 8.7 The function $f(x) = x^2/4$ for positive values of x. It can be digitized by recording $f(x)$ at various intervals along the x-axis, shown here for unit intervals.

uses the positions along the x-axis as the axes in the vector space. If the positions are taken close enough together, the representation will be exact.

For our treatment of nondegenerate perturbation theory, we have the zero-order wavefunctions ϕ_k, and these functions can also be represented in a vector space. We assume that they have been normalized. They are also orthogonal because the set of solutions to the Schrödinger equation for a given system is always orthogonal; consider the sine functions of Eq. 8.19, or the orbitals of a hydrogen atom of Eq. 8.31. Here for two different ϕ_k at least one group with wavefunction χ_i^a must be in a different state for the two ϕ_k, and the product functions are then orthogonal. For instance, let us investigate the orthogonality of the wavefunction for two groups, both in their ground state, to the wavefunction with group number 2 excited to state a. The integral of Eq. 8.5 for functions is analogous to the dot product of vectors; both result in a number. Showing orthogonality,

$$\int (\chi_1^0{}^* \chi_2^0{}^*)(\chi_1^0 \chi_2^a)\, d\tau = \int \chi_1^0{}^* \chi_1^0\, d\tau_1 \int \chi_2^0{}^* \chi_2^a\, d\tau_2 = 0 \qquad (8.51)$$

The individual wavefunctions are normalized and the integral over the volume element for group one is 1. However, the integral over the volume element for group two is zero because different states of the same system are orthogonal. The ϕ_k form a complete orthonormal basis set in the same way that $\mathbf{i}, \mathbf{j}$, and $\mathbf{k}$ form a complete basis set for the Cartesian coordinate system. We will use the orthonormality condition time and time again to simplify quantum mechanical problems. In general, it is expressed as

$$\int \psi_i^* \psi_j\, d\tau = \delta_{ij} \qquad (8.52)$$

where $\delta_{ij} = 1$ if $i = j$ and $\delta_{ij} = 0$ if $i \neq j$.

The first-order correction to the wavefunction, ϕ_k' can be expressed as a linear combination of the zero-order product functions, which are an orthonormal set of basis functions that span the space.

$$\phi_k' = \sum_\ell C_{k\ell} \phi_\ell \qquad (8.53)$$

where ℓ is an arbitrary index for ϕ, just as is k, and the $C_{k\ell}$ are the projections of vectors ϕ_k' on coordinate axes ϕ_ℓ, or the coefficients. To determine these coefficients we substitute into Eq. 8.49, obtaining,

$$\hat{H}_0 \sum_\ell C_{k\ell} \phi_\ell - E_k \sum_\ell C_{k\ell} \phi_\ell = E_k' \phi_k - \hat{V} \phi_k \qquad (8.54)$$

Working with numbers can be much simpler than working with operator equations, such as Eq. 8.54. One way to get a number is to make use of integration, as we have done for orthonormality and in Postulate IV, Eq. 8.6. Here we can turn Eq. 8.54

into an equation of numbers if we multiply both sides by ϕ_m^* (where again m is an arbitrary index for ϕ) and integrate both sides over all space.

$$\int \phi_m^* \hat{H}_0 \sum_\ell C_{k\ell}\phi_\ell \, d\tau - E_k \int \phi_m^* \sum_\ell C_{k\ell}\phi_\ell \, d\tau$$

$$= E_k' \int \phi_m^*\phi_k \, d\tau - \int \phi_m^* \hat{V}\phi_k \, d\tau \qquad (8.55)$$

where E_k and E_k' can be taken outside of the integral because they are numbers. $\hat{H}_0\phi_\ell$ is $E_\ell\phi_\ell$ from Eq. 8.43 and we have

$$\sum_\ell C_{k\ell}E_\ell \int \phi_m^*\phi_\ell \, d\tau - E_k \sum_\ell C_{k\ell} \int \phi_m^*\phi_\ell \, d\tau$$

$$= E_k' \int \phi_m^*\phi_k \, d\tau - \int \phi_m^* \hat{V}\phi_k \, d\tau \qquad (8.56)$$

The key to simplifying this equation, and many other equations in quantum mechanics, is orthonormality. The simplest case is when $m = k$ and applying the orthonormality of ϕ we obtain nonzero terms on the left side only when $\ell = k$. Equation 8.56 becomes

$$C_{kk}E_k - E_k C_{kk} = E_k' - \int \phi_k^* \hat{V}\phi_k \, d\tau$$

$$E_k' = \int \phi_k^* \hat{V}\phi_k \, d\tau \qquad (8.57)$$

We do not have to know the first-order corrections for the wavefunctions to get the first-order correction to the energy; the first-order correction to the energy involves only the zero-order wavefunctions and the perturbation operator. Indeed, the new energy is just Eq. 8.44, which we obtained by assuming that the wavefunction for separable Hamiltonian was the approximate wavefunction for the perturbed system.

We can obtain the first-order correction to the wavefunction by considering Eq. 8.56 for the case $m \neq k$. From the orthonormality of ϕ_k, the two sums on the left side of the equals sign each has a single term that occurs only when $\ell = m$. The first term to the right of the equals sign is always zero since m does not equal k. The result is

$$C_{km}E_m - E_k C_{km} = - \int \phi_m^* \hat{V}\phi_k \, d\tau \qquad (8.58)$$

which gives us each of the coefficients in the first-order correction to the wavefunction, except for C_{kk}.

$$C_{km} = - \frac{\int \phi_m^* \hat{V}\phi_k \, d\tau}{E_m - E_k} \qquad (8.59)$$

We can see why this perturbation theory works only for the nondegenerate case; the denominator in Eq. 8.59 goes to zero if the two energies are identical. However, the first-order correction to the energy is valid in the degenerate case because Eq. 8.57 involves only the zero-order wavefunctions. The coefficient C_{kk} can be determined by normalizing ψ_k corrected to the first order. Only first-order terms λ^1 are kept in the normalization in Eq. 8.5, and C_{kk} turns out to be zero. The wavefunction corrected to first order is, then,

$$\psi_k = \phi_k - \sum_{m \neq k} \frac{\int \phi_m^* \hat{V} \phi_k \, d\tau}{E_m - E_k} \phi_m \qquad (8.60)$$

Evaluating the first-order correction to the energy and the wavefunction involves evaluating integrals of the form $\int \phi_m^* \hat{V} \phi_\ell \, d\tau$. This probably seems pretty abstract at this point, but remember that we always know $\hat{V}$ because we can always write down the Hamiltonian for our system by making use of Postulate V. Also, we get to pick our groups. If we are clever, we pick them so that we have expressions for the ϕ_k. Then it is simply a matter of evaluating integrals involving a known operator and known functions, and these integrals often turn out to be rather simple. With modern computers it is not too difficult to evaluate even complicated integrals. We see specific examples in various sections of this text, including the section on the hydrogen molecule.

Variation method.   The third approximate method applies the *variation theorem*. Postulate IV says that if we have a system with Hamiltonian $\hat{H}$, then any arbitrary function ψ will have a corresponding average energy given by

$$\overline{E} = \frac{\int \psi^* \hat{H} \psi \, d\tau}{\int \psi^* \psi \, d\tau} \qquad (8.61)$$

Suppose we guess at the ground-state wavefunction ψ_0 for Hamiltonian $\hat{H}$ with the function ϕ. All of the wavefunctions of $\hat{H}$ form a complete orthonormal set that we can use to expand ϕ,

$$\phi = \sum_\ell C_\ell \psi_\ell \qquad (8.62)$$

where we assume that the coefficients normalize ϕ. Substituting Eq. 8.62 into Eq. 8.61 yields

$$\overline{E} = \sum_\ell \sum_m \int C_m^* \psi_m^* \hat{H} C_\ell \psi_\ell \, d\tau$$

$$= \sum_\ell \sum_m C_m^* C_\ell E_\ell \int \psi_m^* \psi_\ell \, d\tau$$

$$= \sum_\ell C_\ell^* C_\ell E_\ell \geq E_0 \sum_\ell C_\ell^* C_\ell = E_0 \qquad (8.63)$$

where we have made use of the fact that E_0 must be less than E_ℓ; the orthonormality of the ψ_ℓ; and normalization, which means that $\Sigma_\ell C_\ell^* C_\ell = 1$.

This proves the variation theorem: If the lowest eigenvalue of $\hat{H}$ is E_0, then

$$\overline{E} = \frac{\displaystyle\int \phi^* \hat{H} \phi \, d\tau}{\displaystyle\int \phi^* \phi \, d\tau} \geq E_0 \qquad (8.64)$$

for all ϕ. This is a powerful theorem. We might naively expect that if we made a random guess at ψ_0, we will obtain some random value for $\overline{E}$. The variation theorem says that we can make many guesses at ψ_0 and that the lowest $\overline{E}$ will be the best approximation to E_0.

The variation theorem suggests a method for obtaining good approximations to E_0. We devise a form for ϕ that contains many arbitrary coefficients. Using differential calculus to determine the coefficients that give a minimum $\overline{E}$ we will have the best approximation to the ground-state energy that can be obtained with a wavefunction of the form that we have devised. One obvious way to approximate the wavefunction for a molecule is to take a linear combination of atomic orbitals (LCAO) on each atom. The LCAO theory is very popular because the method is the same regardless of the molecule. The LCAO wavefunctions always lead to a solution of the same form and we can set up a general computer program to solve many molecular problems. We see an example of the LCAO method when we look at the hydrogen molecule in Section 8.3.3. Another approach is to take a linear combination of the zero-order product functions from perturbation theory. Degeneracy will be irrelevant if the variation method is used to determine the coefficients. This is an extended valence-bond theory that is often called *exciton theory* or *the independent systems approach*. If we know ψ_0, we can apply the variation method to wavefunctions that are orthogonal to ψ_0. The variation theorem guarantees that the lowest $\overline{E}$ obtained in this way will be greater than or equal to the energy of the next highest eigenstate corresponding to $\hat{H}$.

8.3.3 The Hydrogen Molecule as the Model for a Bond

The hydrogen molecule H_2 has two independent protons as nuclei bonded together by two electrons. As physical biochemists dealing with molecules, we need to understand bonding between atoms. Spectroscopy can be used to investigate these bonds. Figure 8.8 depicts the four particles in H_2, and defines the symbols A and B that denote the protons and 1 and 2 that denote the electrons. The electronic Hamiltonian for the H_2 molecule is given by

$$\hat{\mathcal{H}} = -\frac{\nabla^2(1)}{2} - \frac{1}{r_{1A}} - \frac{1}{r_{1B}} - \frac{\nabla^2(2)}{2} - \frac{1}{r_{2A}} - \frac{1}{r_{2B}} + \frac{1}{r_{12}} + \frac{1}{r_{AB}}$$

$$= \hat{h}(1) + \hat{h}(2) + \frac{1}{r_{12}} + \frac{1}{r_{AB}} \qquad (8.65)$$

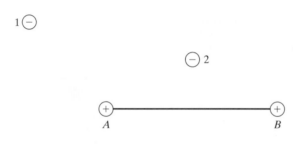

Figure 8.8 The two protons, labeled A and B, and two electrons, labeled 1 and 2, that make up a hydrogen molecule.

where r is the distance between two particles as defined by the subscripts. We see that the Hamiltonian consists of a part that depends only on electron number 1, and another part that depends only on electron number 2. If the last two terms in the potential energy are neglected, then the Hamiltonian is separable.

We can neglect the operator $1/r_{AB}$ that corresponds to the Coulomb interaction between the nuclei by invoking the *Born-Oppenheimer approximation*. The idea is that heavy nuclei in a molecule move slowly compared to the electrons, so that the electronic properties of a molecule can be calculated at any set of fixed internuclear distances. For H_2 this means that to a good approximation the electronic wavefunction will depend on $r_{1A}, r_{1B}, r_{2A}, r_{2B}$, and r_{12}, but not the *dynamics* of r_{AB} as the molecule vibrates. However, to get the total energy of a molecule, we must add the Coulomb energy due to the nuclear interaction ($1/r_{AB}$ in the case of H_2) after calculating the energy of the electrons with the electronic wavefunction.

Exact solutions to the Schrödinger equation for the H_2 molecule that include the Coulomb interaction between the electrons $1/r_{12}$ have been found. They are an infinite series and do not provide an intuitive feeling for the properties of a bond. From a physical biochemist's point of view, it is more instructive to formulate approximate solutions. These approximations simplify finding the solution and are justified when the predictions made from the approximate solutions agree well with experiments.

Separability will provide an approximate solution if we neglect the potential energy operator for the interaction between the two electrons in addition to invoking the Born-Oppenheimer approximation. Separability gives us product functions of the form

$$\psi_n(1, 2) = \chi_a(1)\chi_b(2) \tag{8.66}$$

where the χ's are solutions to the Schrödinger equation for each $\hat{h}$ of Eq. 8.65. Each $\hat{h}$ is the Hamiltonian for the hydrogen molecule ion H_2^+ that involves two protons but only one electron. The Schrödinger equation for H_2^+ can be solved in confocal elliptic coordinates to give an exact wavefunction, as is shown in most physical chemistry texts. However, we will get a better feeling for a chemical bond by approximating the solution as an LCAO and applying the variation method.

Two hydrogen atoms that are far apart will each have their electrons in a $1s$ orbital if they are in their ground state. We expect that two hydrogen atoms at their

Figure 8.9 The wavefunction for the ground state of a hydrogen molecule ion in the linear combination of atomic orbitals approximation as the sum of a $1s$ orbital on nucleus A and a $1s$ orbital on nucleus B.

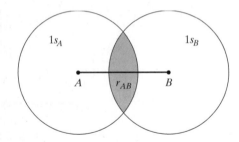

bonding internuclear distance will form a bond that is described by a wavefunction something like the sum of the $1s$ orbitals on each atom, as pictured in Figure 8.9. As an LCAO, the approximate ground-state wavefunction for the electron in H_2^+ is

$$\chi_0 = a1s_A + b1s_B \tag{8.67}$$

where the $1s$ orbital is given by Eq. 8.31 and the coefficients a and b are unknown.

We can determine the a and b that would give the lowest energy, and hence the best approximation to the ground-state energy of H_2^+, by applying the variation method. Using Postulate IV,

$$\overline{E}_0 = \frac{\int \chi_0 \hat{h} \chi_0 \, d\tau}{\int \chi_0^2 \, d\tau} \tag{8.68}$$

where we note that the $1s$ orbitals are real and assume that the coefficients will be real as well. Since the coefficients are unknown, the wavefunction is not yet normalized. On rearranging and substituting Eq. 8.67 for χ_0 we obtain

$$\overline{E}_0 \int (a1s_A + b1s_B)^2 \, d\tau = \int (a1s_A + b1s_B)\hat{h}(a1s_A + b1s_B) \, d\tau \tag{8.69}$$

and on multiplying out, we obtain

$$\overline{E}_0\left(a^2 \int 1s_A 1s_A \, d\tau + 2ab \int 1s_A 1s_B \, d\tau + b^2 \int 1s_B 1s_B \, d\tau \right)$$

$$= a^2 \int 1s_A \hat{h} 1s_A \, d\tau + 2ab \int 1s_A \hat{h} 1s_B \, d\tau + b^2 \int 1s_B \hat{h} 1s_B \, d\tau$$

$$\overline{E}_0(a^2 + b^2 + 2abS) = a^2\alpha + b^2\alpha + 2ab\beta \tag{8.70}$$

where we have made use of the fact that the $1s$ orbitals are normalized but not orthogonal, because they are on different centers. Here the coefficients a and b are the variables and the integrals S, α, and β are numbers that can be calculated.

These integrals have been found and are

$$S = \int 1s_A 1s_B \, d\tau = \frac{1}{\pi} \int e^{-r_{1A}} e^{-r_{1B}} \, d\tau$$

$$= e^{-r_{AB}}\left(1 + r_{AB} + \frac{r_{AB}^2}{3}\right)$$

$$\alpha = \int 1s_A \left(-\frac{\nabla^2}{2} - \frac{1}{r_{1A}} - \frac{1}{r_{1B}}\right) 1s_A \, d\tau$$

$$= E_H - \int 1s_A \frac{1}{r_{1B}} 1s_A \, d\tau$$

$$= -\frac{1}{2} - \frac{1}{r_{AB}}\{1 - e^{-2r_{AB}}(1 + r_{AB})\}$$

$$\beta = S E_H - \int 1s_A \frac{1}{r_{1B}} 1s_B \, d\tau$$

$$= -\frac{S}{2} - e^{-r_{AB}}(1 + r_{AB}) \tag{8.71}$$

where the Hamiltonian for a hydrogen atom operating on a $1s$ orbital gives the ground-state energy for a hydrogen atom, $E_H = -0.5$ au. We note that these integrals can be evaluated for any internuclear distance. This calculation will yield a bonding distance of 2.5 au for H_2^+, and the integrals have the following values: $S = 0.458$, $\alpha = -0.891$ au, $\beta = -0.516$ au. The integral S is called the *overlap* integral and is depicted by the shaded volume in Figure 8.9. Note however, that the $1s$ orbitals extend to infinity, so that Figure 8.9 merely denotes some volume where $1s^2$ has much of its density.

To find the coefficients that give a minimum value for $\overline{E}_0$ at the chosen r_{AB}, we take the partial derivative of both sides of Eq. 8.70 with respect to each of the coefficients, giving,

$$\overline{E}_0(2a + 2bS) = 2a\alpha + 2b\beta$$

$$\overline{E}_0(2b + 2aS) = 2b\alpha + 2a\beta \tag{8.72}$$

and on rearranging and dropping the subscript on E, we obtain

$$a(\alpha - E) + b(\beta - SE) = 0$$
$$a(\beta - SE) + b(\alpha - E) = 0 \tag{8.73}$$

Solving these equations simultaneously leads to the trivial solution $a = b = 0$. Any set of simultaneous equations from the variation method yields a trivial solution; fortunately, we have another equation that relates the coefficients, the normalization equation. Thus, we can still obtain unique values for the variables even if two equations in the set are linearly dependent. We ensure linear dependence and a nontrivial solution by setting the determinant of the coefficients equal to zero.

$$\begin{vmatrix} \alpha - E & \beta - SE \\ \beta - SE & \alpha - E \end{vmatrix} = 0 \tag{8.74}$$

The values of E that provide linear dependence in the set of equations are found by multiplying out the determinant and solving for E. In this case, the equation in E is a quadradic and there are two roots given by

$$E_\pm = \frac{\alpha \pm \beta}{1 \pm S} \tag{8.75}$$

Substituting E_+ or E_- into the original simultaneous equations gives $a = b$ and $a = -b$, respectively. With normalization, the corresponding wavefunctions are

$$\begin{aligned} E_+ \qquad \chi_+ &= (2 + 2S)^{-1/2}(1s_A + 1s_B) \\ E_- \qquad \chi_- &= (2 - 2S)^{-1/2}(1s_A - 1s_B) \end{aligned} \tag{8.76}$$

The integrals α and β are negative, and E_+ represents the best estimate of the ground-state energy for a wavefunction of the form given by Eq. 8.67. The wavefunction χ_+ is the normalized sum of the two $1s$ orbitals, which adds electron density between the two nuclei in the overlap region (see Figure 8.9). Balancing the repulsion between nuclei with electron density in the overlap region is the essence of a bond, and χ_+ is said to be a bonding orbital. In contrast, χ_- cancels electron density in the overlap region, leading to a higher-energy repulsive state, and is said to be an antibonding orbital. It is orthogonal to χ_+ and is thus an approximation to the wavefunction for the first-excited state.

The electron energies are readily calculable at any internuclear distance; for $r_{AB} = 2.5$ au they are $E_+ = -0.9648$ au and $E_- = -0.6908$ au. When the two nuclei are far apart, only one proton will possess the electron and the electronic energy of the system will be the energy of a ground-state hydrogen atom, -0.5 au. When this is taken as the new zero of energy, $E_+ = -0.465$ and $E_- = -0.191$ au. Taking the overlap at this internuclear distance into account, the wavefunctions are

$$\begin{aligned} \chi_+ &= 0.586\,\pi^{-1/2}(e^{-r_{1A}} + e^{-r_{1B}}) \\ \chi_- &= 0.961\,\pi^{-1/2}(e^{-r_{1A}} - e^{-r_{1B}}) \end{aligned} \tag{8.77}$$

The total energy of the H_2^+ molecule must take into account the Coulomb repulsion between the nuclei, $1/r_{AB}$. Figure 8.10 plots the results for various discrete values of r_{AB}. This graph gives a theoretical minimum at r_{AB} equal to 2.5 au with a total bonding energy of -0.0648 au. Experimentally, H_2^+ has an equilibrium nuclear distance of about 2.0 au and a total bonding energy of -0.1024 au.

The total energy for the antibonding state increases as the two nuclei are brought together, and this state is unstable. Using this approximate LCAO theory, the transition energy from the bonding to the antibonding state at $r_{AB} = 2.5$ is predicted to be 0.274 au and will require light with a wavelength of 166 nm.

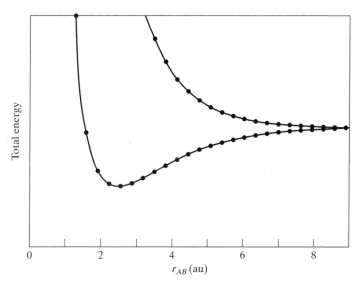

Figure 8.10   The total energy of a hydrogen molecule ion as a function of internuclear distance for the bonding (low energy) and antibonding (high energy) levels that result from a simple LCAO treatment.

We can now use the lowest energy LCAO wavefunction of H_2^+ for the χ in Eq. 8.66, and form the wavefunction for the ground state of H_2, $\psi_0(1,2)$. There is still one problem, however. The Pauli exclusion principle tells us that two particles cannot be in the same place at the same time and therefore must be represented by different wavefunctions. We know that the two electrons in a bond have different spins, and we can satisfy the Pauli exclusion principle by considering a *spin* wavefunction in addition to the *space* wavefunction χ_+ of Eq. 8.76. Essentially, we add a term to the space Hamiltonian of Eq. 8.65 that corresponds to the potential energy operator for spin and assume there is no term that corresponds to an interaction between spin and space. Then the Hamiltonian is separable in spin and space parts, and the solution to the Schrödinger equation will be a wavefunction that is a product of spin and space parts. Symbolizing the solutions to the spin part as α and β (corresponding to the two types of spin), we can write the ground-state wavefunction for H_2 as

$$\psi_0(1,2) = \chi_+(1)\alpha(1)\chi_+(2)\beta(2) \qquad (8.78)$$

This wavefunction does not take into account the indistinguishability of electrons and the principle that any electronic wavefunction must be antisymmetric, that is, must change sign on the exchange of two electron numbers. A product wavefunction with any number of factors can be antisymmetrized by forming a determinant with each electron in each factor. For the H_2 problem, the normalized determinant is

$$\psi_0(1,2) = \frac{1}{\sqrt{2}} \begin{vmatrix} \chi_+(1)\alpha(1) & \chi_+(1)\beta(1) \\ \chi_+(2)\alpha(2) & \chi_+(2)\beta(2) \end{vmatrix} \qquad (8.79)$$

This complicated-looking wavefunction is only approximate. A major deficiency is that it does not depend on r_{12}. This means that the electrons are independent and their position lacks *correlation*. The spin part will factor out for two electrons, giving

$$\psi_0(1,2) = \chi_+(1)\chi_+(2)\frac{[\alpha(1)\beta(2) - \alpha(2)(1)]}{\sqrt{2}} \tag{8.80}$$

The spin part will not be operated on by the space Hamiltonian of Eq. 8.65; it will simply factor out and, with the normalization factor, integrate to a value of unity. Thus, in this case, we do not need to consider the spin part when calculating the average ground-state energy corresponding to $\psi_0(1,2)$. Substituting the product of the space functions together with the Hamiltonian of Eq. 8.65 into Postulate IV for the electronic energy we have

$$\overline{E}_0 = \int \psi_0(1,2)\hat{\mathcal{H}}\psi_0(1,2)\,d\tau$$

$$= \int \chi_+(2)^2\,d\tau_2 \int \chi_+(1)\hat{h}(1)\chi_+(1)\,d\tau_1$$

$$+ \int \chi_+(1)^2\,d\tau_1 \int \chi_+(2)\hat{h}(2)\chi_+(2)\,d\tau_2$$

$$+ \int \chi_+(1)\chi_+(2)\left(\frac{1}{r_{12}}\right)\chi_+(1)\chi_+(2)\,d\tau_1\,d\tau_2$$

$$= 2E_+ + V \tag{8.81}$$

The integral V for the potential energy of interaction between the two electrons can be calculated, and computer programs have been written to ease the computation. In this treatment, the total ground-state energy of H_2 (including nuclear repulsion) has a minimum at $r_{AB} = 1.6$ au, and the bonding energy is calculated to be -0.0985 au. The observed value is -0.1746 au at an $r_{AB} = 1.40$ au.

An approximate wavefunction for the first-excited state of H_2 can be formed by putting one electron in χ_+ and the other in χ_-. In this example, we maintain opposite spins for the two electrons and there are two possibilities: A given spin function might multiply either χ_+ or χ_-. There are two possible determinants and the antisymmetric wavefunction will be the normalized difference between them.

$$\psi_1(1,2) = \frac{1}{\sqrt{2}}\left\{\frac{1}{\sqrt{2}}\begin{vmatrix} \chi_+(1)\alpha(1) & \chi_-(1)\beta(1) \\ \chi_+(2)\alpha(2) & \chi_-(2)\beta(2) \end{vmatrix} - \frac{1}{\sqrt{2}}\begin{vmatrix} \chi_+(1)\beta(1) & \chi_-(1)\alpha(1) \\ \chi_+(2)\beta(2) & \chi_-(2)\alpha(2) \end{vmatrix}\right\}$$

$$\tag{8.82}$$

Again, the space and spin parts separate into an antisymmetric spin factor and a symmetric space factor. We will ignore the spin factor, and the wavefunction for the first-excited state becomes

$$\psi_1(1,2) = \frac{1}{\sqrt{2}}[\chi_+(1)\chi_-(2) + \chi_+(2)\chi_-(1)] \tag{8.83}$$

The average energy for the first-excited state will be $\overline{E}_1 = E_+ + E_- + V'$ where

$$V' = \int \psi_1(1,2)(1/r_{12})\psi_1(1,2)\, d\tau \tag{8.84}$$

The electronic energy of the first-excited state for H_2 is a function of r_{AB} and the total energy including $1/r_{AB}$ is also somewhat bonding, with a minimum at an internuclear distance that is slightly larger than the minimum for the ground state.

The H_2 molecule can be considered as a model for a single bond between any two atoms and, in general, the total energies for the ground and first-excited states of a bond as a function of internuclear distance are as represented in Figure 8.11. Electronic spectroscopy measures the energy difference between the ground-electronic state and the excited states. Of course, the nuclei will vibrate and the energy of a particular vibrational state must be added to the electronic energy to get the total energy of each state. The added energy of vibration for various vibrational quantum states is represented by the horizontal lines in Figure 8.11. The potential energy well, due to the energy of the electrons and nuclei, has a harmonic shape near the minimum; vibrational energy is a simple harmonic and has equally spaced

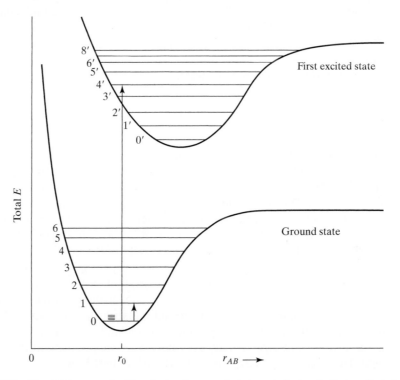

Figure 8.11   The total energy for the electronic ground state and electronic first-excited state of a hydrogen molecule that result from the approximate quantum mechanical treatment described in the text. Some vibrational energy levels and a few representative rotational energy levels are also shown.

levels similar to Figure 8.4. At higher electronic energies the potential well is anharmonic and the vibrational levels are closer together. Extra energy is added to the total for the system by quantized rotation of the molecule. Short horizontal lines represent this added energy for the lowest vibration level in Figure 8.11.

The spacing between rotational energy levels is small, about 100 cm^{-1}, and depends on the vibrational state of the molecule. Transitions can be observed between rotational energy levels, and these levels can add structure to the vibrational and electronic spectra. However, we will not consider rotational energy levels further.

The spacing of the vibrational energy levels will be different for ground and excited states, and their spacing of about 2000 cm^{-1} is large compared with the rotational spacing. Vibrational spectra are observed with light from the infrared region, and the transitions normally correspond to a transition from the ground-vibrational state of the ground-electronic level to the first-excited vibrational level of the ground-electronic state. A diatomic molecule only has one type of vibration, but a large molecule will have many different types of vibrations. Infrared spectroscopy normally observes one transition from ground to first excited state for each of the many different types of vibrations.

The spacing between ground- and first-excited electronic states varies widely but averages about 40,000 cm^{-1}. The electronic transitions require energy in the visible or ultraviolet portion of the spectrum. Vibrational and electronic spectroscopy are simplified by the fact that, for all practical purposes, molecules are in their ground-electronic and vibrational states at room temperature. Thus, vibrational or electronic spectroscopy normally observes transitions from the ground-electronic and vibrational state to some higher excited state. As we have seen, these energies are predicted by quantum mechanics.

8.4   TRANSITION INTENSITIES

The energy of a transition is determined by the spacing between the energy levels. However, not all the photons of the correct energy will be absorbed by a sample of molecules. We can investigate the fraction of photons absorbed at various levels of sophistication. This transition intensity is useful to a physical biochemist because it can give structural information about a macromolecule. Let us begin by considering the intensity of an absorption band from an experimental point of view.

Empirically, the *Beer-Lambert law* governs absorption processes at any concentration that does not change the nature of the absorbing species. Expressed in terms of differential increments, the law says that successive increments of absorbing molecules in a monochromatic light beam will absorb equal fractions of the light (Figure 8.12). An increment of absorbing molecules is $N dz$, where N is the number of molecules in the light beam per unit volume and dz is a thin slice of the path. The fraction of light absorbed is dI'/I', where I' is the intensity of the monochromatic light. The equation that expresses the Beer-Lambert Law is, then,

$$N \, dz = -\frac{1}{\sigma} \frac{dI'}{I'} \tag{8.85}$$

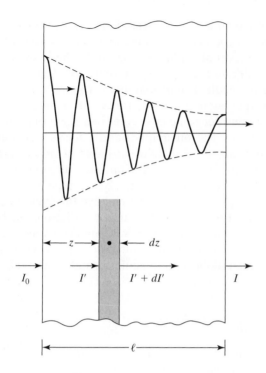

Figure 8.12 A monochromatic beam of light traversing a sample of absorbing molecules loses its intensity exponentially, according to the Beer-Lambert law. [Reprinted from W. C. Johnson, "Circular Dichroism and Its Empirical Application to Biopolymers," *Methods of Biochemical Analysis,* vol. 31, ed. D. Glick, copyright 1985 by John Wiley & Sons.]

where the minus sign indicates absorption and $1/\sigma$ is the constant of proportionality between the increment of absorbing molecules and the fraction of light absorbed. The quantity σ, in units of distance squared, is called the *absorption cross-section.* Integrating, we obtain

$$\int_{I_0}^{I} \frac{dI'}{I'} = -\sigma N \int_0^{\ell} dz \tag{8.86}$$

where the cross-section σ is the opaque area associated with each molecule in the sample. Chemists prefer to express concentrations in moles per liter c and N equals $\mathcal{N} c / 10^3$ where $\mathcal{N}$ is Avogadro's number, the number of molecules per mole. Integration yields the absorbance A, defined as

$$\log \frac{I_0}{I} = A = (-\log e) \ln \frac{I}{I_0} = \sigma N \ell \log e = \varepsilon \ell c \tag{8.87}$$

where ln symbolizes the natural logarithm, log symbolizes the logarithm to the base 10, $I/I_0 = T$ is called the *transmission,* and

$$\varepsilon = \sigma \mathcal{N} 10^{-3} \log e \tag{8.88}$$

Absorption changes as a function of frequency, and in its final form the Beer-Lambert law becomes

$$A(\nu) = \varepsilon(\nu) \ell c \tag{8.89}$$

where $\varepsilon(\nu)$ is a characteristic of the molecule and traditionally is in $l/(mol \cdot cm)$ or $1/(M \cdot cm)$. It is called the *extinction coefficient,* or sometimes the absorptivity.

The *Einstein relations* are a level of sophistication above the Beer-Lambert law. A simple model is used to derive the probability that a photon will be absorbed or emitted by a sample molecule. The probability (proportional to the intensity of a transition) is derived by assuming consistency with Planck's formula for black-body radiation. Let us consider some molecules in a beam of perturbing radiation as described by Figure 8.13. The subscripts 1 and 2 denote two states of a sample molecule, and A and B are the Einstein constants of proportionality that describe the probability of spontaneous emission and the probability of emission or absorption induced by radiation, respectively. We expect that monochromatic radiation will consist of photons, all of which have the same quantum of energy, $h\nu = E_2 - E_1$ given by Eq. 8.3. The amount of radiation can be described by the density of energy, $\rho(\nu)$. The number of molecules absorbing a quantum of energy and undergoing a transition from state 1 to 2, dN_{12}, per unit time dt, will be proportional to the number of molecules in state 1 and the radiation density:

$$\frac{dN_{12}}{dt} = B_{12}N_1\rho(\nu) \tag{8.90}$$

where this equation holds for a thin slice of molecules, so that the radiation density does not change. Similarly, perturbing radiation can cause the emission of a photon,

$$\frac{dN_{21}}{dt} = B_{21}N_2\rho(\nu) \tag{8.91}$$

In addition, there may be a spontaneous emission that is independent of the perturbing radiation.

$$\frac{dN'_{21}}{dt} = A_{21}N_2 \tag{8.92}$$

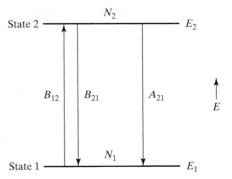

Figure 8.13   The Einstein relations consider absorption and emission for molecules in two states.

At steady state, the number of molecules per unit time that are absorbing photons will equal the number of molecules per unit time that are emitting photons, and combining Eqs. 8.90–8.92 gives

$$N_2[A_{21} + \rho(\nu)B_{21}] = N_1 B_{12}\rho(\nu) \tag{8.93}$$

The radiation density produced by a black body at temperature T is given by Planck's radiation law,

$$\rho(\nu) = \frac{8\pi h\nu^3/c^3}{e^{h\nu/k_BT} - 1} = \frac{N_2 A_{21}}{N_1 B_{12} - N_2 B_{21}} \tag{8.94}$$

where we have also used Eq. 8.93 to obtain an expression for $\rho(\nu)$. Rearranging, we have

$$\frac{1}{\rho(\nu)} = \frac{c^3}{8\pi h\nu^3}[e^{h\nu/k_BT} - 1] = \frac{N_1 B_{12}}{N_2 A_{21}} - \frac{B_{21}}{A_{21}} = e^{h\nu/k_BT}\frac{B_{12}}{A_{21}} - \frac{B_{21}}{A_{21}} \tag{8.95}$$

where we have made use of the Boltzmann distribution for molecules between two states at temperature T,

$$\frac{N_2}{N_1} = e^{-h\nu/k_BT} \tag{8.96}$$

Equating the corresponding terms, we have

$$\frac{B_{21}}{A_{21}} = \frac{c^3}{8\pi h\nu^3} = \frac{B_{12}}{A_{21}} \tag{8.97}$$

which implies that $B_{21} = B_{12}$, and the subscripts may be dropped. Furthermore, the Einstein probability for spontaneous emission is related to the Einstein probability for an induced transition by

$$A(\nu) = \frac{8\pi h\nu^3}{c^3}B(\nu) \tag{8.98}$$

where we recognize that the Einstein coefficients will depend on the frequency of the radiation.

All three processes (induced absorption, induced emission, and spontaneous emission) can always occur. The energy difference between ground and excited states for electronic and vibrational levels is large enough so that molecules, for all practical purposes, are in their ground-electronic and vibrational levels at room temperature. Electronic and vibrational absorption are normally concerned only with B. Fluorescence from electronic and vibrational excited states is normally concerned only with A because the low population of excited states means that an induced emission is much less probable than a spontaneous emission. In contrast, the energy levels explored by nuclear magnetic resonance (NMR) are quite close together and almost equally populated at room temperature; induced emission will nearly cancel the induced absorption.

The extinction coefficient of the Beer-Lambert law and the Einstein coefficients describe the intensity of a transition at a given frequency. In practice, a transition will occur over a range of frequencies and its *total probability* will be related to the area under this band. We can combine the Beer-Lambert law with the Einstein relations and relate the total probability for the transition to the experimentally measured band. Equation 8.85 can be rewritten with substitution of Eq. 8.88 to give

$$-dI(\nu) = 10^3 \varepsilon(\nu) N \, dz \, I(\nu)/\mathcal{N} \log e \tag{8.99}$$

The intensity of the radiation I is the number of photons per unit time in the slice of sample. Then, $-dI(\nu)$ will be the number of photons in the slice absorbed per unit time, which is the same as the number of molecules excited per unit time and the left side of Eq. 8.90. Moreover, Ndz is the number of molecules in the lower state in Eq. 8.90. Finally, the intensity is related to the radiation density by $I(\nu) = \rho(\nu)c/h\nu$ so that combining Eq. 8.99 with Eq. 8.90 yields

$$B(\nu) = 10^3 \varepsilon(\nu) c/\mathcal{N} \log e \, h\nu \tag{8.100}$$

The total probability for induced absorption or emission is obtained by integrating over the frequency range corresponding to the entire band.

$$B = \int B(\nu) \, d\nu = \frac{10^3 c}{\mathcal{N} h \log e} \int_{band} \frac{\varepsilon(\nu)}{\nu} d\nu \tag{8.101}$$

This relates an experimentally measured absorption band to the total probability for induced absorption or emission, as defined in the Einstein relations. In addition, Eq. 8.98 can be used to give us the total probability for spontaneous emission

$$A = \frac{8\pi 10^3 \bar{\nu}^3}{\mathcal{N} c^2 \log e} \int_{band} \frac{\varepsilon(\nu)}{\nu} d\nu \tag{8.102}$$

where we have assumed that we can use an average frequency $\bar{\nu}$ in the relationship.

We can obtain the *quantum mechanical intensity* for a transition by adding the energy of the interaction of light with the molecule to the Hamiltonian for the molecule

$$\hat{\mathcal{H}}(q, t) = \hat{H}(q) - \hat{\boldsymbol{\mu}} \cdot \hat{\mathbf{E}} = \hat{H}(q) + \hat{H}'(q, t) \tag{8.103}$$

where $\hat{H}(q)$ is a Hamiltonian for the unperturbed molecule that is time independent, and $\hat{H}'(q, t)$ is the small time-dependent perturbation due to the light interacting with the molecule. We use the classical expression for the light wave $\hat{\mathbf{E}}$, given by Eq. 8.1, and $\hat{\boldsymbol{\mu}}$ is the dipole moment induced in the molecule by the light.

$$\hat{\boldsymbol{\mu}} = \sum_i q_i \hat{\mathbf{r}}_i \tag{8.104}$$

where the sum is over all particles in the molecule and q_i is the charge on each particle. We assume that the molecule is small compared to the wavelength of light so that the intensity of the electric vector is constant over the molecule, and we fix the molecule at the origin, which eliminates the distance dependence in the expression for $\mathbf{E}$. The time-dependent perturbation becomes

$$\hat{H}'(q, t) = \hat{\boldsymbol{\mu}} \cdot \mathbf{E}_0 \cos 2\pi \nu t = -\hat{\boldsymbol{\mu}} \cdot \mathbf{E}_0 \frac{(e^{i2\pi\nu t} + e^{-i2\pi\nu t})}{2} \tag{8.105}$$

We assume that we know the wavefunctions for the unperturbed molecule, $\psi_j(q)$, that correspond to energies E_j. These stationary-state wavefunctions have a time dependence as given by Eq. 8.11. If we send light of the proper frequency into a sample of molecules so that the light induces a transition from initial state i to final state f, then the wavefunction for the perturbed molecule will be

$$\Phi(q, t) = \psi_i(q)e^{-iE_i t/\hbar} + C_f(t)\psi_f(q)e^{-iE_f t/\hbar} = \Psi_i(q, t) + C_f(t)\Psi_f(q, t) \tag{8.106}$$

where the time-dependent coefficient, $C_f(t)$, is small and the coefficient of the ground state is still 1. Substituting the wavefunction for the perturbed molecule into the time-dependent Schrödinger equation of Postulate VI yields

$$\hat{\mathcal{H}}(q, t)\Phi(q, t) = i\hbar \frac{\partial}{\partial t}\Phi(q, t)$$

$$\hat{H}(q)\Psi_i(q, t) + C_f(t)\hat{H}(q)\Psi_f(q, t) + \hat{H}'(q, t)\Psi_i(q, t)$$
$$+ \hat{H}'(q, t)C_f(t)\Psi_f(q, t)$$
$$= i\hbar \left[\frac{\partial}{\partial t}\Psi_i(q, t) + C_f(t)\frac{\partial}{\partial t}\Psi_f(q, t) + \Psi_f(q, t)\frac{\partial}{\partial t}C_f(t) \right] \tag{8.107}$$

Making use of the fact that the unperturbed functions are solutions to the time-dependent Schrödinger equation, we obtain to the first order

$$\hat{H}'(q, t)\Psi_i(q, t) = i\hbar \Psi_f(q, t)\frac{\partial}{\partial t}C_f(t) \tag{8.108}$$

since the fourth term is second order. We changed operator equations into numerical equations in nondegenerate perturbation theory, Eq. 8.55, by making use of integration in analogy to Postulate IV, Eq. 8.6. Here, we apply the same principle using the wavefunction $\Psi_f^*(q, t)$ and integrate over all space, yielding

$$\frac{\partial}{\partial t}C_f(t) = \frac{1}{i\hbar} \int \Psi_f^*(q, t)\hat{H}'(q, t)\Psi_i(q, t)\, d\tau$$

$$= -\frac{1}{i\hbar} \int \psi_f^*(q)e^{iE_f t/\hbar}(\hat{\boldsymbol{\mu}} \cdot \hat{\mathbf{E}})\psi_i(q)e^{-iE_i t/\hbar}\, d\tau$$

$$= \frac{i}{2\hbar}\mathbf{E}_0 \cdot \int \psi_f^*(q)\hat{\boldsymbol{\mu}}\psi_i(q)\, d\tau (e^{i2\pi\nu t} + e^{-i2\pi\nu t})e^{i(E_f - E_i)t/\hbar} \tag{8.109}$$

after applying the orthonormality condition. Collecting the exponentials and integrating from time zero (when the molecule is in state i) to time t gives us an expression for the perturbation coefficient

$$C_f(t) = \frac{1}{2}\mathbf{E}_0 \cdot \boldsymbol{\mu}_{fi}\left[\frac{e^{i(E_f - E_i + h\nu)t/\hbar} - 1}{E_f - E_i + h\nu} + \frac{e^{i(E_f - E_i - h\nu)t/\hbar} - 1}{E_f - E_i - h\nu}\right] \tag{8.110}$$

where the *transition dipole* is defined as

$$\boldsymbol{\mu}_{fi} = \int \psi_f^*(q)\hat{\boldsymbol{\mu}}\psi_i(q)\,d\tau \tag{8.111}$$

The coefficient $C_f(t)$ will be extremely small unless the energy of the light, $h\nu$, matches the energy difference between initial and final states. The second term in Eq. 8.110 is important for absorption when the initial state is at lower energy than the excited state; the first term is important for induced emission when the initial state is at higher energy than the final state.

We remember from Postulate II that the probability that the molecule will be in the final state is proportional to $C_f^*(t)C_f(t)$, and if we consider absorption for which the second term in Eq. 8.110 dominates

$$C_f^*(t)C_f(t) = \frac{|\mathbf{E}_0 \cdot \boldsymbol{\mu}_{fi}|^2}{4}\left[\frac{2 - e^{i(E_f - E_i - h\nu)t/\hbar} - e^{-i(E_f - E_i - h\nu)t/\hbar}}{(E_f - E_i - h\nu)^2}\right]$$

$$= E_0^2\, D_{fi} \cos^2\theta\left[\frac{\sin^2\frac{1}{2}(E_f - E_i - h\nu)t/\hbar}{(E_f - E_i - h\nu)^2}\right] \tag{8.112}$$

where θ is the angle between $\mathbf{E}_0$ and $\boldsymbol{\mu}_{if}$, and the *dipole strength* is defined as

$$D_{fi} = \boldsymbol{\mu}_{if} \cdot \boldsymbol{\mu}_{fi} = \left[\int \psi_f(q)\hat{\boldsymbol{\mu}}\psi_i(q)\,d\tau\right]^2 \tag{8.113}$$

In practice, a light source will have a distribution over a frequency interval $d\nu$, and $E_0^2 = 8\pi\rho(\nu)\,d\nu$. The probability for absorption becomes

$$C_f^*(t)C_f(t) = \frac{2D_{fi}\cos^2\theta}{\pi\hbar^2}\int \rho(\nu)\frac{\sin^2\pi(\nu_{fi} - \nu)t}{(\nu_{fi} - \nu)^2}\,d\nu$$

$$\cong \frac{2\pi D_{fi}\,\rho(\nu_{fi})\cos^2\theta}{\hbar^2}t \tag{8.114}$$

where $\nu_{fi} = (E_f - E_i)/h$. The transition probability for a single oriented molecule will be the coefficient of t in Eq. 8.114. A comparison with Eq. 8.90 for N_1 molecules relates the dipole strength to the total Einstein coefficient for induced absorption,

$$B = \frac{2\pi D_{fi}}{3\hbar^2} \tag{8.115}$$

where we have replaced $\cos^2 \theta$ with 1/3, because for N_1 molecules randomly oriented in solution, $\cos^2 \theta$ must be averaged over three dimensions.

The dipole strength D_{fi} is the quantum mechanical intensity, and it can be related to an experimentally measured absorption band by combining Eq. 8.101 with Eq. 8.115,

$$D_{fi} = \frac{3\hbar^2}{2\pi} B = \frac{2303(3hc)}{8\mathcal{N}\pi^3} \int_{\text{band}} \frac{\varepsilon(\nu)}{\nu} d\nu \tag{8.116}$$

where 2303 has been substituted for $10^3/\log e$. The transition dipole of Eq. 8.111 has atomic dimensions and, consequently, the dipole strength as given by Eq. 8.113 is conveniently expressed in atomic units

$$D = 1.421 \times 10^{-3} \int_{\text{band}} \varepsilon(\lambda) d(\ln \lambda) \tag{8.117}$$

where we have exchanged frequency dependence for wavelength dependence by making use of the relation $\nu\lambda = c$. We can convert an experimentally measured absorption band into the quantum mechanical dipole strength by plotting the extinction coefficient for the absorption versus $\ln(\lambda)$ and measuring the area under the curve. The probability that a photon will be absorbed is proportional to the integrated intensity of an absorption band measured in this special way. Spectroscopy observes the quantum mechanical intensity of a transition.

8.5   TRANSITION DIPOLE DIRECTIONS

In addition to energy and intensity, a transition has the measurable characteristic of direction, the direction of the vector called the transition dipole in Eq. 8.111. Modern spectroscopy makes extensive use of the direction of transition dipoles. These directions, predicted by quantum mechanics, allow us to use spectroscopy to learn about the conformation of macromolecules.

The electric vector of a light wave interacts with the transition dipole of a molecule according to the scalar product between the two vectors as given by the perturbation in Eq. 8.105. We see that the measured transition intensity is proportional to the square of this scalar product, the dipole strength of Eqs. 8.113 and 8.117. The scalar product will vary as the cosine squared of the angle between the electric vector and the transition dipole, as shown in Figure 8.14 and expressed in Eq. 8.112. When the electric vector of the light is aligned with the transition dipole of the molecule ($\theta = 0$ or $180°$), the probability of absorption is a maximum; when the electric vector of the light and the transition dipole of the molecule are at right angles, the probability for absorption is zero. At angles in between these limits, the probability will have an intermediate value that varies as $\cos^2 \theta$.

When a sample of molecules is randomly oriented, as is the case for a sample dissolved in a solvent, $\cos^2 \theta$ is averaged over three dimensions to give 1/3. However, if a sample of molecules is oriented, then the direction of the transition dipole can be determined by changing the angle of linearly polarized light relative to the oriented sample and finding the maximum absorption. In practice, samples may be oriented

Figure 8.14 The absorption of linearly polarized light depends on the cosine squared of the angle between the electric vector **E** and the transition dipole, **μ**.

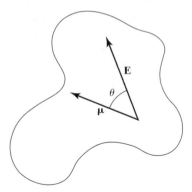

by applying a strong electric field, stretching a sample dissolved in a polymer film, making use of the shear forces in a flow, or growing a crystal.

Growing crystals has the advantage that in them the orientation is absolute and can be determined by X-ray diffraction. On the other hand, most crystals contain a high concentration of sample that absorbs all of the incident light. Stewart and Davidson (1963) measured the dipole direction for the first electronic transition in 1-methylthymine by growing a crystal and grinding it on an ultramicrotome to reduce the path length of the sample to a few microns. Their results, shown in Figure 8.15 for

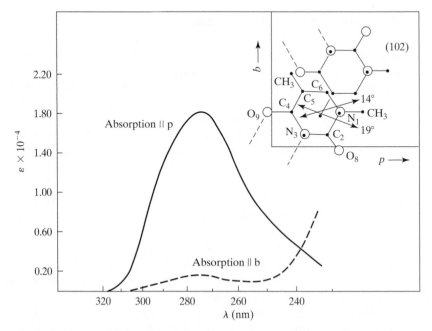

Figure 8.15 Polarized absorption spectra for crystals of 1-methylthymine. The absorption is from the first $\pi\pi^*$ transition; the two possible directions for the transition dipole are shown. [Adapted from data in R. F. Stewart and N. Davidson (1963), *J. Chem. Phys.* **39**, 255–266.]

the electric vector of the light oriented along the two crystallographic directions of the crystal face, give two possible directions for the first transition. Peterson and Simpson (1957) determined the transition dipole direction for the 190 nm electronic band in amide, the important chemical group that constitutes the backbone of proteins. They grew crystals of myristamide, a long-chain hydrocarbon with an amide group. The hydrocarbon chain served as a transparent solvent to dilute the amides and allow a measurable amount of light to be transmitted through the crystals. Their results show that the transition dipole direction is almost in line with the oxygen and nitrogen atoms of the amide.

We can determine transition dipole directions theoretically by evaluating the integral in Eq. 8.111 when the wavefunctions for the initial and final states of the molecule are known. The dipole moment operator, given by Eq. 8.104, sums over all the charged particles in the molecule. However, the position vector for the nuclei will not operate on the electron wavefunctions in Eq. 8.111, and the orthogonality of the wavefunctions for initial and final state will make the terms involving the positions of the nuclei zero. Therefore, only terms for the positions of the electrons contribute to the electronic transition dipole.

As an example, let us consider a hydrogen atom undergoing a transition from the $1s$ orbital to the $2p_z$ orbital given by Eq. 8.31. Substituting

$$\boldsymbol{\mu} = \int 2p_z q \hat{\mathbf{r}} 1s \, d\tau$$

$$= -\frac{1}{\pi(32)^{1/2}} \int \cos\theta \, re^{-r/2} r(\mathbf{i}\sin\theta\cos\phi$$

$$+ \mathbf{j}\sin\theta\sin\phi + \mathbf{k}\cos\theta)e^{-r}r^2\sin\theta \, dr \, d\theta \, d\phi \qquad (8.118)$$

where the charge on an electron is -1 au, and we have expressed the position vector in the Cartesian coordinate system. The integrals over the ϕ in the $\mathbf{i}$ and the $\mathbf{j}$ directions are zero, so the transition dipole is zero along these two coordinate axes. For the $\mathbf{k}$ direction we have

$$\mu = -\frac{1}{\pi(32)^{1/2}} \int_0^{2\pi} d\phi \int_0^\pi \cos^2\theta\sin\theta \, d\theta \int_0^\infty e^{-3r/2}r^4 \, dr$$

$$= -\frac{1}{\pi(32)^{1/2}}(2\pi)\left(\frac{2}{3}\right)\left(\frac{4!}{(3/2)^5}\right) = -0.74 \text{ au} \qquad (8.119)$$

where these definite integrals are evaluated explicitly in nearly any table of integrals. We see that the transition dipole is in the $\mathbf{k}$ direction, along the axis of the $2p_z$ orbital. Its magnitude is consistent with atomic dimensions, 0.74 au.

EXERCISES

8.1   The energy of a particle in a three-dimensional box is given by

$$E_{pqr} = \frac{h^2}{8m}\left(\frac{p^2}{a^2} + \frac{q^2}{b^2} + \frac{r^2}{c^2}\right)$$

where p is the quantum number for dimension a, and so on ($p = 1, 2, 3, \ldots$; $q = 1, 2, 3, \ldots$; $r = 1, 2, 3, \ldots$)

   a. For an electron in a 0.4 nm cubic box (an orbital of sorts) calculate E_{111}, E_{211} in joules.
   b. Calculate the same energies for a chair weighing 9.11 kg in a 4 m cubic room.
   c. Compare the energy spacing $E_{211} - E_{111}$ for (a) and (b).

8.2   Using Planck's law, calculate
   a. ν in $\sec^{-1}$,
   b. λ in nm, and
   c. the energy as $\bar{\nu} = 1/\lambda$ in cm^{-1}

   for the electron in Exercise 8.1 undergoing a transition from E_{111} to E_{211}.

8.3   Wavefunctions can be used to calculate *all* observables for a system. Calculate the expectation value of r for the $1s$ orbital of hydrogen. Remember that

$$0 \le r \le \infty, 0 \le \theta \le \pi, 0 \le \phi \le 2\pi, \text{ and } d\tau = r^2 \sin\theta\, d\phi\, d\theta\, dr$$

8.4   Prove that C_{kk} of Eq. 8.53 is zero to first order by normalizing Ψ_k. Assume the $C_{k\ell}$ are real.

8.5   For the hydrogen molecule ion, calculate the integrals S, α, and β at an internuclear distance of 2 au. Calculate E_+ and E_- at 2 au, both without and with the Coulomb repulsion of the nuclei. Write down χ_+ and χ_-. Compare your results with $r_{AB} = 2.5$ au in the text. Although the experimental bonding distance for H_2^+ is 2.0 au, this calculation gives the minimum energy at 2.5 au.

*8.6   Consider the dimer of adenosine-phosphate-adenosine (ApA). Suppose the backbone is inert and we know all the eigenfunctions χ_1^a and χ_2^b for all the states a of adenine #1 and all the states b of adenine #2.
   a. If the Hamiltonian for the dimer is $H = H_1 + H_2 + V_{12}$, what is the separable part of the Hamiltonian?
   b. What is the zero-order ground-state wavefunction ($a = b = 0$) for the separable Hamiltonian?
   c. Prove that your zero-order ground state is an eigenfunction of the separable Hamiltonian.
   d. What are the forms for the excited-state wavefunctions for the separable Hamiltonian?
   e. What is the first-order correction to the ground-state energy?

8.7   Find the absorbance, A, when
   a. $I = I_0$
   b. $I = \dfrac{I_0}{3}$

c. $I = \dfrac{I_0}{10}$

d. $I = \dfrac{I_0}{10^2}$

e. $I = \dfrac{I_0}{10^3}$

8.8 Calculate the ε for a typical IR absorption where $A = 1.0$, the path length is 1.0 mm, and the concentration is 10% (100 mg/ml, molecular weight = 200).

8.9 Find the concentration necessary for $A = 1.0$ in a 1.0 cm cell taking $\varepsilon = 6 \times 10^3$ for a typical electronic absorption. If the cell holds 3 ml and the molecular weight = 200, calculate the number of grams of material needed. Compare the answer with that of Exercise 8.8.

8.10 Evaluate the dipole strength in atomic units for the intense amide transition at about 190 nm in Figure 9.12.

8.11 Calculate the Boltzmann distributions of the excited state relative to the ground state at 27°C for the following transitions:
  a. electronic $(1/\lambda = 5 \times 10^4 \, cm^{-1})$
  b. vibrational $(1/\lambda = 2 \times 10^3 \, cm^{-1})$
  c. rotational $(1/\lambda = 10^2 \, cm^{-1})$

*8.12 Use symmetry to determine the direction (if any) of the transition dipole for a $2p_x \rightarrow 2p_y$ transition on a hydrogen atom.

REFERENCES

ATKINS, P. (2002) *Physical Chemistry,* 7th ed., W. H. Freeman, New York.

BERNATH, P. F. (1995) *Spectra of Atoms and Molecules,* Oxford University Press, Oxford, UK

DAVYDOV, A. S. (1971) *Theory of Molecular Excitons,* Plenum Press, New York. Exciton theory by its originator. Out of print, but in most university libraries.

EYRING, H., J. WALTER, and G. E. KIMBALL (1944) *Quantum Chemistry,* John Wiley & Sons, New York. A no-nonsense book on fundamental quantum mechanics. Out of print, but in most university libraries.

GRAYBEAL, J. D. (1988) *Molecular Spectroscopy,* McGraw-Hill, New York.

HALLIDAY, D., R. RESNICK, and J. WALKER (2004) *Fundamentals of Physics,* 7th ed., John Wiley & Sons, New York.

HOUSE, J. E. (2004) *Fundamentals of Quantum Chemistry,* 2nd ed., Academic Press, San Diego.

PAULING, L. and E. B. WILSON, JR. (1985) *Introduction to Quantum Mechanics with Applications to Chemistry,* Dover Publications, New York. A classic, originally published in 1935.

PETERSON, D. L. and W. T. SIMPSON (1957) "Polarized Spectra of Amides with Assignments," *JACS* **74**, 2375.

STEWART, R. F. and N. DAVIDSON (1963) "Polarized Absorption Spectra of Purines and Pyrimidines," *J. Chem. Physics* **39**, 255–266.

TINOCO, I., JR., K. SAUER, J. C. WANG, and J. D. PUGLISI (2002) *Physical Chemistry Principles and Applications in Biological Sciences,* 4th ed., Prentice Hall, Upper Saddle River, NJ.

9

Absorption Spectroscopy

Molecules are atoms held together by electrons forming bonds. The molecules as a whole can rotate, and the nuclei can vibrate because the bonds have some flexibility. As discussed in detail in Chapter 8, various quantized electronic, vibrational, and rotational states are available to a molecule, and light will be absorbed when a molecule undergoes a transition from a lower-energy state to a higher-energy state. Transitions between electronic states require the energy found in photons of the light of the visible and ultraviolet (UV) regions (Figure 9.1). Various quantized vibrational states are available to a molecule in each electronic state, and a molecule can be excited from a lower-vibrational state to a higher-vibrational state by the absorption of light. Light with the proper energy for vibrational absorption is found in the infrared (IR) region. Various quantized rotational states are available to a molecule in a given electronic and vibrational state. The rotational states are quite close together, and low energy microwaves have the proper energy to excite a molecule from a lower-rotational state to a higher-rotational state. Pure rotational spectra are rarely observed for biologically interesting substances, which are studied in solution. Because these are the substances of primary interest to physical biochemists, we will not consider rotational spectra.

Electronic absorption spectra of biological molecules can be used to monitor concentration and conformational changes. It is widely used to follow the denaturation of biopolymers. Vibrational absorption spectra can be used for molecule identification. A particular strength of this technique is that each functional group in a molecule has its own absorption band.

9.1 ELECTRONIC ABSORPTION

As we recall from Chapter 8, the energy spacing between vibrational and electronic states is large, so for all practical purposes molecules can be considered to be in their ground vibrational and electronic states at room temperature. Knowing that

Figure 9.1 The electromagnetic spectrum of light is divided into various regions. Different types of transitions are excited depending on the energy of the photons.

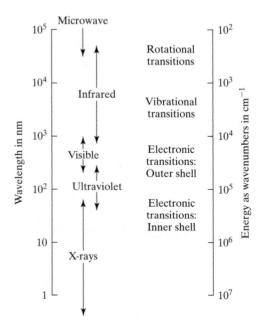

the initial state of the molecule is the ground electronic and vibrational state greatly simplifies the interpretation of electronic absorption spectra.

However, the transition to an excited electronic state can be to any of the vibrational levels, as Figure 8.11 shows. The *Franck-Condon principle* states that certain vertical transitions corresponding to no nuclear displacement during an electronic transition have the highest probability. The most probable value of the vibrational wavefunction for the ground level falls at r_0, the equilibrium bonding distance. The vibrational level of the excited electronic state with the highest probability will be the one with the largest value of the vibrational wavefunction at r_0; for the model bond shown in Figure 8.11, this corresponds to the 4′ vibrational level of the excited state. Transitions to other vibrational levels will have a lower probability, and the resulting vibrational-electronic (*vibronic*) absorption band is depicted in Figure 9.2a. When the electronic absorption spectrum is measured for molecules in solution, as is generally the case for macromolecules, individual vibronic bands are broadened by collisions with the solvent molecules, and the spectrum smears into a broad electronic absorption band (Figure 9.2b,c).

9.1.1   Energy of Electronic Absorption Bands

We will consider molecules in the orbital model where various quantized energy levels are available to electrons, either singly or as pairs with opposite spin. The lower-energy orbitals will be occupied by electron pairs and higher-energy orbitals will be unoccupied. In this approximation, electronic transitions occur when electrons are excited from an occupied orbital to an unoccupied orbital.

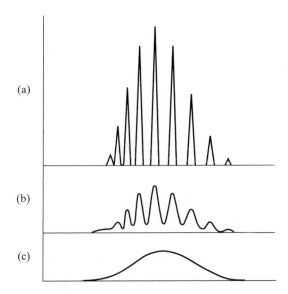

(a)

(b)

(c)

Figure 9.2 The energy of light absorbed by an electronic transition depends on the final vibrational level of the excited state (see Figure 8.11). Thus, an electronic absorption band consists of a series of vibronic bands (a). Collisions cause the vibronic bands to smear in energy (b); and in solution an electronic transition smears into a broad band (c).

What are the relative energies of these quantized electronic orbitals? We will make use of the fundamental quantum mechanical principles covered in Chapter 8. Most macromolecules will have both σ and π bonding. The Hamiltonian for the molecule can be written as the sum of the Hamiltonians for the individual σ and π parts plus the σ-π interaction.

$$\hat{H} = \hat{H}_\sigma + \hat{H}_\pi + \hat{V}_{\sigma\pi} \tag{9.1}$$

The π system in molecules is often considered to be built on a framework of σ bonds, and the interaction between π and σ parts is ignored. We remember from Chapter 8 that if $\hat{V}_{\sigma\pi} = 0$, the Hamiltonian becomes separable, and the molecular wavefunction is the product of individual wavefunctions for the σ and π systems.

In the chemist's view, individual σ bonds have an integrity, which implies a limited interaction among the individual σ bonds. We will adopt this valence-bond view and ignore the interaction between σ bonds so that $\hat{H}_\sigma$ becomes separable into the Hamiltonians for each individual σ bond. Each σ bond is then directed between two atoms and has a Hamiltonian similar to Eq. 8.65 for the hydrogen molecule, which we considered in Chapter 8 as the model for a bond. This model is pictured in Figure 9.3, where sp^3 hybrid orbitals of the bonding electron shell are directed toward each other and overlapped in preparation to form a σ bond. The inner electron shells are considered together with the nucleus. The sign of the magnitude (phases) for the two lobes of each sp^3 orbital are opposite but arbitrary, and are usually defined as in Figure 9.3. Of course, this treatment could be generalized for any type of hybrid orbital or even the $1s$ orbital of a hydrogen atom.

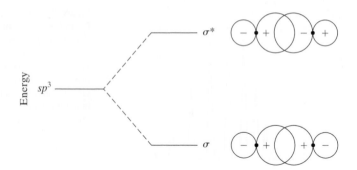

Figure 9.3 Two sp^3 hybrid orbitals aligned linearly and at the bonding distance interact to form two molecular orbitals, σ and σ^*. For the approximations used here, the bonding σ orbital is lower in energy than the sp^3 by an amount β, while the σ^* orbital is higher than the sp^3 by the energy β.

We can apply the same quantum mechanics to this problem that we did to the hydrogen molecule in Chapter 8. The Hamiltonian will be given by

$$\hat{H} = \hat{h}(1) + \hat{h}(2) + \frac{1}{r_{12}} + \frac{1}{r_{AB}} \tag{9.2}$$

where 1 and 2 number the two electrons in the bond, and A and B refer to the nuclei with the inner electron shells. Formally, this Hamiltonian is identical to the Hamiltonian for the hydrogen molecule as in Eq. 8.65. If we assume that the two atoms are the same and apply the Born-Oppenheimer approximation to eliminate $1/r_{AB}$ and neglect $1/r_{12}$, then the Hamiltonian is separable, the energy of the orbital wavefunction will have the same form as Eq. 8.75, and the corresponding wavefunctions will have the form of Eq. 8.76.

$$E_+ = \alpha + \beta: \quad \sigma = \frac{(sp_A^3 + sp_B^3)}{\sqrt{2}}$$

$$E_- = \alpha - \beta: \quad \sigma^* = \frac{(sp_A^3 - sp_B^3)}{\sqrt{2}} \tag{9.3}$$

where we have neglected the overlap integral in the normalization. This is a poor assumption, but it is commonly made, simplifies the diagrams, and does not affect the qualitative conclusions.

We remember that β is negative so that the bonding wavefunction σ corresponds to the lower energy, E_+. The initial state with $r_{AB} = \infty$ is shown on the left side of Figure 9.3 and the bonding state with $r_{AB} = r_0$, the equilibrium-bonding distance, is shown on the right side of the figure. Two electrons of opposite spin can occupy the lower energy σ orbital, and the wavefunction corresponding to the bond is the product of the wavefunctions for the two electrons with the form given by Eq. 8.80. Each of the two electrons contributes a stabilization energy of

β to the bond. An electron in the bond can undergo a transition from the bonding σ orbital to the antibonding σ^* orbital by absorbing a photon with energy $E_- - E_+ = -2\beta$.

Referring to Figure 9.3, we see that the σ bond, Eq. 9.3, corresponds to having sp_A^3 and sp_B^3 in phase, giving extra overlapping electron density between the nuclei with their inner electrons. In contrast, the σ^* orbital changes the sign of sp_B^3, putting the two hybrid orbitals out of phase and eliminating electron density in the overlap region.

In contrast to σ bonds that chemists usually think of as individual entities, π bonds are usually considered to stream over all of the overlapping $2p$ orbitals. The simplest system with a π bond is the two carbons in ethylene, pictured in Figure 9.4. The Hamiltonian for this system has the same form as Eq. 9.2 for the σ bond, and the separable Hamiltonian has solutions of the same form as well

$$E_+ = \alpha + \beta': \quad \pi = \frac{(p_A + p_B)}{\sqrt{2}}$$

$$E_- = \alpha - \beta': \quad \pi^* = \frac{(p_A - p_B)}{\sqrt{2}} \tag{9.4}$$

where we have again neglected the overlap integral. The value of the integral β' will be different from its value for the σ bond; here β' is also negative but with a lower magnitude corresponding to the smaller overlap between the $2p$ orbitals. Figure 9.4 shows the energy levels for the $2p$ orbitals when $r_{AB} = \infty$ and the energy levels for the π and π^* orbitals at $r_{AB} = r_0$. One of the paired electrons in the bonding π orbital can undergo a $\pi\pi^*$ transition to the antibonding π^* state by absorbing light of energy $-2\beta'$.

Allyl has a π streamer system involving three carbon atoms, as pictured in Figure 9.5. Three $2p$ orbitals, each on one of the three carbon atoms, combine to form three molecular orbitals of the form $ap_A + bp_B + cp_C$. Applying the variation method to the separable Hamiltonian for this system, in analogy to the way we

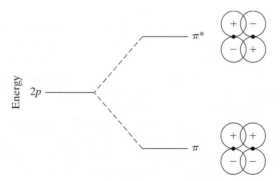

Figure 9.4 Two $2p$ orbitals aligned side by side at the bonding distance interact to form two molecular orbitals, π and π^*. With the approximations here, the bonding π orbital is lower in energy than a $2p$ by the amount β', while the π^* orbital is higher than the $2p$ by the same energy.

Figure 9.5 Three $2p$ orbitals aligned side by side at the bonding distance interact to form three molecular orbitals in a π system. Using our approximations, the π_2 orbital is at the same energy as the $2p$, the π_1 orbital is lower in energy by $\sqrt{2}\beta'$, and the π_3^* orbital is higher in energy by $\sqrt{2}\beta'$.

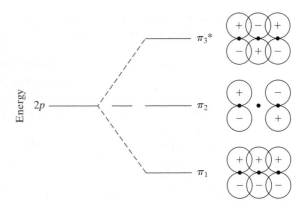

applied the method to the hydrogen molecule ion in Chapter 8, yields values for the coefficients of the three molecular orbitals and their corresponding energies.

$$\hat{H} = \sum_{i=1}^{3} \hat{h}(i)$$

$$\hat{h}(i) = \frac{-\nabla^2(i)}{2} - \frac{1}{r_{iA}} - \frac{1}{r_{iB}} - \frac{1}{r_{iC}}$$

$$E_1 = \alpha + \sqrt{2}\beta': \quad \pi_1 = \frac{(p_A + \sqrt{2}p_B + p_C)}{2}$$

$$E_2 = \alpha: \quad \pi_2 = \frac{(p_A - p_C)}{\sqrt{2}}$$

$$E_3 = \alpha - \sqrt{2}\beta': \quad \pi_3^* = \frac{(p_A - \sqrt{2}p_B + p_C)}{2} \tag{9.5}$$

where we assume interaction only between adjacent carbons.

If the equilibrium-bonding distances are the same as for ethylene, then the integral β' will have the same value as well. The wavefunction π_1 corresponds to the lowest state with a bonding energy of $\sqrt{2}\beta'$ for each of two electrons that can occupy the orbital (Figure 9.5). The overlap between adjacent $2p$ orbitals adds electron density to the bonds. The wavefunction π_2 has the same energy as an isolated $2p$ orbital in this approximation and is thus nonbonding; it has no electron density whatsoever on the center carbon. The wavefunction π_3^* changes the phase of the $2p$ orbital on the center carbon, removing electron density between each pair of atoms. The star reminds us that this is an antibonding orbital.

In this approximation, allyl obtains bonding energy in its π streamer system only from electrons occupying the π_1 orbital. Thus, this treatment predicts that an allyl cation with only two electrons in π_1, a neutral allyl radical with an additional unpaired electron in π_2, and an allyl anion with two paired electrons in π_2 will all have the same bonding energy. All three variations of allyl have been synthesized and are reasonably stable, confirming the results of our approximate treatment. Two $\pi\pi^*$-type transitions are predicted for an allyl anion. A $\pi_2\pi_3^*$ transition absorbs light with an energy $-\sqrt{2}\beta$, and a $\pi_1\pi_3^*$ transition absorbs light with an energy $-2\sqrt{2}\beta'$. Thus, the $\pi_2\pi_3^*$ transition of an allyl anion falls at a lower energy than the $\pi\pi^*$ transition of ethylene.

Wavefunctions and their energies for larger conjugate π systems can be found in a similar manner. The energy levels for some linear π-streamer systems are shown in Figure 9.6. In this approximation, we see that when there is an odd number of 2p atomic orbitals in the π streamer system, there is always a nonbonding π molecular orbital at the energy of an isolated 2p atomic orbital. As the number of 2p orbitals increases, the π orbitals get closer and closer together. For an infinitely long polyene, all the π levels are contained between $+2\beta'$ and $-2\beta'$. As the length of a polyene increases, the energy of the light required for the first transition becomes lower and lower. We see this effect in carotenoids, the biological polyenes found in carrots, which absorb light in the visible region and give carrots their orange color.

Chromophores are functional groups of a few atoms with a characteristic electronic absorption spectrum. We can make use of the qualitative quantum mechanics discussed above by constructing diagrams that will allow us to predict the electronic transitions of biologically important chromophores. Let us consider carbonyl as a simple example and introduce the additional assumption that the integral α is the same for all 2p orbitals of first-row atoms. The atomic orbitals for the bonding electrons of the carbon and oxygen of carbonyl are shown schematically in Figure 9.7. We hybridize carbon into three sp^2 orbitals oriented in the same plane and pointing their electron density in the direction of the three single bonds to be formed by carbon. The s character of the sp^2 hybrids lowers their energy with respect to the 2p orbitals. One 2p orbital on carbon remains to form the π bond, and it is perpendicular to the plane formed by the three sp^2 hybrid orbitals.

Hybridization of the oxygen atom is a little different. It will form an sp^2 hybrid, pointing electron density in the direction of the carbon sp^2 hybrid to which it

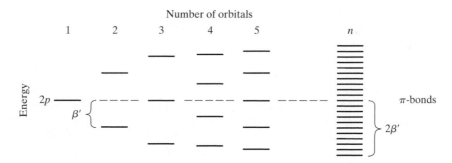

Figure 9.6 Relative energies for orbitals as linear π streamers in a simple Linear Combination of Atomic Orbitals (LCAO) approximation.

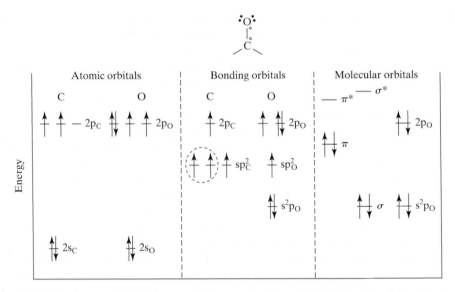

Figure 9.7 An energy-level diagram for carbonyl showing the original atomic orbitals, the hybrid orbitals formed for bonding, and the occupied and unoccupied orbitals of the carbonyl system. The carbonyl structure shows a σ bond, π electrons o and nonbonding electrons •. [Reprinted from W. C. Johnson, "Circular Dichroism and its Empirical Application to Biopolymers," in *Methods of Biochemical Analysis*, vol. 31, ed. D. Glick, copyright 1985 by John Wiley & Sons.]

will bond. A $2p$ atomic orbital on oxygen is oriented perpendicularly to the plane of the carbonyl group to form a π bond to the corresponding $2p$ orbital on carbon. We know from spectroscopy that the two nonbonding pairs on the oxygen atom (depicted by the filled circles in Figure 9.7) have quite different energies. One nonbonding pair is high energy and is well described by an atomic $2p$ orbital oriented in the plane of the carbonyl group and perpendicular to the $C=O$ bond. What remains of the original atomic orbitals on oxygen is two-thirds of an s orbital and one-third of a p orbital. We call this an s^2p hybrid to be consistent with the symbol sp^2, since the symbol sp^2 denotes a hybrid orbital consisting of one-third of a $2s$ atomic orbital and two-thirds of a $2p$ atomic orbital. The s^2p orbital has a lot of s character and is in the plane of the carbonyl group pointing away from the $C=O$ bond.

The sp^2 hybrid on oxygen and the sp^2 hybrid on carbon that point toward each other combine to form a bonding σ and an antibonding σ^* molecular orbital (Figure 9.7). The other two sp^2 hybrids on carbon form bonds with other parts of the molecule and are not shown on the molecular orbital part of the diagram. The $2p$ orbitals on carbon and oxygen that are perpendicular to the plane of the carbonyl combine to form π and π^* molecular orbitals. The splitting of the π and π^* orbitals is much less than that of the σ and σ^* orbitals. The $2p$ orbital on oxygen containing the nonbonding pair is unchanged and is denoted n. The s^2p nonbonding pair on oxygen, which is the low-energy hybrid with a lot of s character, is also unchanged and is denoted n'. The electrons of the chromophore

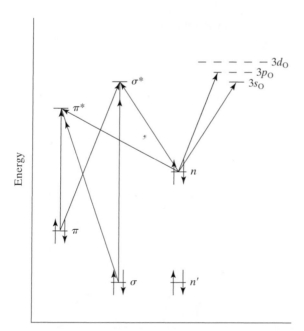

Figure 9.8   Some of the electronic orbitals for the carbonyl chromophore showing various possible electronic transitions from filled orbitals to unfilled orbitals. [Reprinted from W. C. Johnson, "Circular Dichroism and its Empirical Application to Biopolymers," in *Methods of Biochemical Analysis*, vol. 31, ed. D. Glick, copyright 1985 by John Wiley & Sons.]

occupy the lowest energy orbitals (n', σ, π, and n) as pairs with opposite spin. Students should realize that all of these orbitals in the diagram exist, whether or not they are actually occupied.

Figure 9.8 shows the transitions that we expect for this simple one-electron molecular-orbital picture. We have ignored the higher-energy unfilled atomic orbitals on the carbon and oxygen atoms in constructing the bonding picture, but now some of these are included in Figure 9.8. The lowest energy transition is predicted to be the $n\pi^*$. The $\pi\pi^*$ transition is predicted to occur at twice the energy of the $n\pi^*$. In between, we expect the $n\sigma^*$ transition, as well as transitions from the n orbital to the higher-energy unfilled atomic orbitals, called *Rydberg* transitions. Since the n orbital is located on the oxygen atom, only transitions to higher-energy oxygen orbitals are expected. There is little overlap between the n orbital on oxygen and the atomic orbitals on the other atoms, so their transition dipole magnitude (see Eq. 8.111) is negligible.

The absorption spectrum for acetone in the gas phase—an example of the carbonyl chromophore—is shown in Figure 9.9. The broad low-intensity absorption band at 280 nm is the $n\pi^*$. The fact that the lowest-energy band is the $n\pi^*$ proves that the n orbital must be a high energy $2p$. The intense $\pi\pi^*$ transition is a broad band at about 155 nm. Intermediate transitions are less certain, but the 195 nm band is probably the $n3s$ Rydberg strongly mixed with the $n\sigma^*$. The 180 nm band has been assigned to the $n3p$ Rydberg and the sharp features at 155 nm to the $n3d$ Rydberg. The $\sigma\sigma^*$ transitions are the broad features found at even higher energy.

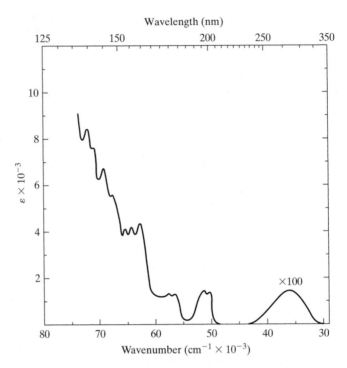

Figure 9.9 The electronic absorption spectrum of acetone, which consists primarily of the transitions of the carbonyl chromophore. [Adapted from data in Barnes and Simpson (1963), *J. Chem Phys.* **39**, 670–675. Reprinted from W. C. Johnson, "Circular Dichroism and its Empirical Application to Biopolymers," in *Methods of Biochemical Analysis,* vol. 31, ed. D. Glick, copyright 1985 by John Wiley & Sons.]

As the next example, we choose amide, an extremely important biological chromophore found in the backbone of proteins. It is diagrammed in Figure 9.10, where each σ bond is denoted by a line. The open circles denote the four electrons in the π system streaming over the oxygen, carbon, and nitrogen atoms, and the closed circles denote the electrons in the two nonbonding orbitals on the oxygen atom. The carbon atom is hybridized just as it was for carbonyl, with three sp^2 hybrids pointing electron density to form σ bonds with three other atoms and a $2p$ orbital perpendicular to the plane of the chromophore to go into the π system. The nitrogen atom is hybridized in just the same way. The oxygen atom is also hybridized as it is for carbonyl, including the two different nonbonding orbitals, $2p$ and s^2p. The bonding orbitals combine to form the molecular orbitals that bond the atoms into an amide chromophore, as shown in Figure 9.10. An sp^2 on carbon and an sp^2 on oxygen combine to form σ and σ^* orbitals between these two atoms. Similarly, an sp^2 on carbon and an sp^2 on nitrogen combine to form σ and σ^* orbitals between carbon and nitrogen. One sp^2 on carbon forms a σ bond with an atom outside the chromophore, as do two sp^2 orbitals on nitrogen. There are three $2p$ orbitals perpendicular to the

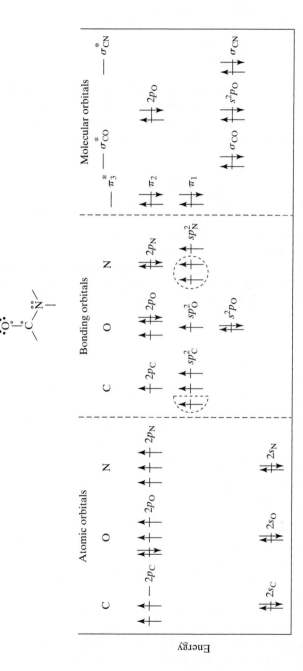

Figure 9.10 The atomic orbitals, the hybrid orbitals formed for bonding, and resulting molecular orbitals for the three atoms in the amide chromophore. [Reprinted from W. C. Johnson, "Circular Dichroism and its Empirical Application to Biopolymers," in *Methods of Biochemical Analysis*, vol. 31, ed. D. Glick, copyright 1985 by John Wiley & Sons.]

plane of the chromophore that form a π streamer system (one on each atom) and three π molecular orbitals are formed in analogy to an allyl. The two nonbonding orbitals on oxygen are unchanged in energy. The resulting qualitative energy diagram for the molecular orbitals in Figure 9.10 shows both nonbonding orbitals double-filled, the two σ bonding orbitals double-filled, and the two lowest π orbitals double-filled, in a manner that is similar to an allyl anion.

Figure 9.11 shows the lower-energy transitions that are predicted for an amide chromophore in this simple one-electron molecular-orbital treatment. The two lowest energy transitions are predicted to fall at the same energy, $n\pi_3^*$ and $\pi_2\pi_3^*$. At somewhat higher energy, we have electrons in the n or π_2 orbitals undergoing transitions to the two σ^* orbitals or the higher atomic orbitals on oxygen. At high energies we predict $n'\pi_3^*$, $\pi_1\pi_3^*$, and $\sigma\sigma^*$ transitions.

The absorption spectrum for N,N-dimethylacetamide is given in Figure 9.12. The low-energy tail between 250 and 230 nm is the red end of a very broad band corresponding to the $n\pi_3^*$ transition. The shoulder at 220 nm has been assigned to a mixture of the $n3s$ and π_23s Rydberg transitions. The broad and intense band at about 195 nm is the $\pi_2\pi_3^*$ transition. The obvious band at 175 nm has been assigned to $n3d$ Rydbergs. The $\pi_1\pi_3^*$ transition may well be the shoulder at about 140 nm. Our qualitative treatment suggests the types of transitions that we will observe for an amide chromophore, and these are in good agreement with an actual absorption spectrum.

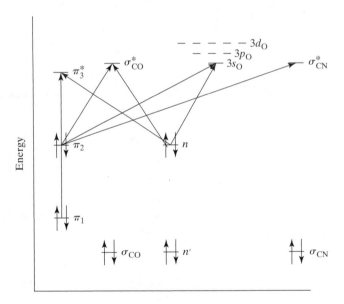

Figure 9.11   The occupied and unoccupied orbitals of the amide chromophore showing possible transitions from the filled orbitals to unoccupied orbitals. [Reprinted from W. C. Johnson, "Circular Dichroism and its Empirical Application to Biopolymers," in *Methods of Biochemical Analysis*, vol. 31, ed. D. Glick, copyright 1985 by John Wiley & Sons.]

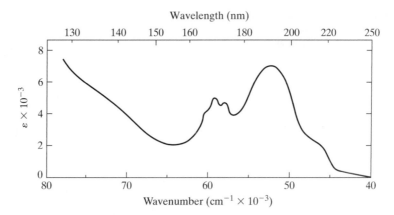

Figure 9.12   The electronic absorption spectrum of N,N-dimethylacetamide, which consists primarily of the electronic transitions for the amide chromophore. [Redrawn from data in Basch et al. (1968), *J. Chem Phys.* **49**, 5007–5018. Reprinted from W. C. Johnson, "Circular Dichroism and its Empirical Application to Biopolymers," in *Methods of Biochemical Analysis,* vol. 31, ed. D. Glick, copyright 1985 by John Wiley & Sons.]

9.1.2   Transition Dipoles

In the last section, we saw that quite simple quantum mechanical concepts predict the energy of absorption bands. But what about their intensities? In what direction will absorption be strongest? The transition dipole of Eq. 8.111 predicts the intensity and direction for a transition. If we approximate the wavefunctions for the ground and excited states of an electronic transition, then we can use Eq. 8.111 to predict these two measurable quantities. If we measure an electronic absorption band, we can calculate the magnitude of the transition dipoles by using Eqs. 8.113 and 8.117. Measuring the direction for the transition dipole in a molecule is discussed in Section 8.5.

It is often possible to deduce the direction of a transition dipole by making use of symmetry. Writing the relation for the transition dipole as

$$\boldsymbol{\mu}_{j0} = \mathbf{i} \int \psi_j^* x \psi_0 \, d\tau + \mathbf{j} \int \psi_j^* y \psi_0 \, d\tau + \mathbf{k} \int \psi_j^* z \psi_0 \, d\tau \tag{9.6}$$

we see that it can be considered as the sum of three separate integrals along each of the axes in a Cartesian coordinate system. First we will consider qualitatively the transition from a $1s$ orbital to a $2p_z$ orbital on a hydrogen atom, which is considered quantitatively in Section 8.5. Figure 9.13 pictures the two orbitals in the xz plane of the Cartesian coordinate system. The orbitals depicted in Figure 9.13 show where about 90% of the electron density is found, but students should remember that the electron density actually stretches to infinity. Four $d\tau$ volume elements are shown at symmetric positions in the four quadrants. The first term of Eq. 9.6 says that the magnitude of the transition dipole in the $\mathbf{i}$ direction can be found from the values of ψ_j^*, x, and ψ_0 in each of the volume elements $d\tau$. For each volume element, we find

Figure 9.13 The $1s$ and $2p_z$ orbitals on a hydrogen atom have the correct symmetry for a transition dipole along the z-axis for a $1s2p_z$ transition.

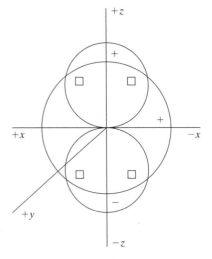

the value of $\psi_j^* x \psi_0$, multiply this times the volume $d\tau$, and sum this value over all the volume elements. In our example, ψ_j^* is the $2p_z$ orbital. By symmetry, it has the same magnitude in each of the four volume elements shown in Figure 9.13. However, the sign of the $2p_z$ orbital is positive in the upper two quadrants and negative in the lower two quadrants. Similarly, x has the same magnitude in all four volume elements, but its sign is positive in the left two quadrants and negative in the right two quadrants. The $1s$ orbital has the same magnitude and sign in all four volume elements. Clearly, the magnitude of $\psi_j^* x \psi_0 \, d\tau$ is the same in all four volume elements shown. However, the sign of this quantity is positive in the x, z and $-x, -z$ quadrants, while it is negative in the $x, -z$ and $-x, z$ quadrants. The sum of the values in the four volume elements shown in the figure is zero. This will be true for any four volume elements symmetrically arranged in the four quadrants for any value of y; and thus the first term in Eq. 9.6, which is simply a sum over all these volume elements, must also be zero. Our qualitative treatment shows that the transition dipole in the x direction for a $1s2p_z$ transition of hydrogen is zero, in agreement with the quantitative treatment in Section 8.5.

Now let us consider the third term in Eq. 9.6. The magnitudes and signs of the $1s$ and $2p_z$ are the same as discussed in the preceding paragraph for the four volume elements shown. However, the sign of z is positive in the upper two quadrants and negative in the lower two quadrants. The magnitude of $\psi_j^* z \psi_0 \, d\tau$ is the same in the symmetrically disposed volume elements and, in this case, the sign is the same for all four volume elements as well. The values in all the volume elements sum, and the transition dipole will have a nonzero value along the **k** direction. Our qualitative argument, then, shows that a $1s2p_z$ transition for a hydrogen atom will be allowed in the **k** direction, as was shown quantitatively in Section 8.5. The qualitative treatment for the **j** direction will be the same as the **i** direction, so the transition dipole is in the **k** direction only.

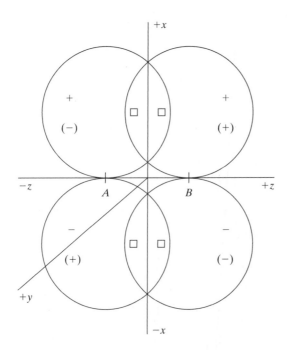

Figure 9.14   The two $2p_x$ atomic orbitals that interact to form a π bonding orbital and a π^* antibonding orbital have the correct symmetry for a transition dipole along the z-axis for a $\pi\pi^*$ transition.

As a second example, we choose the $\pi\pi^*$ transition for ethylene. The π bond made of $2p$ orbitals is pictured along the z-axis of the xz plane of a Cartesian coordinate system shown in Figure 9.14. The π^* wavefunction reverses the phase of one $2p$ function, and the signs of the $2p$ lobes for the π^* wavefunction are shown in parenthesis. Four $d\tau$ volume elements are symmetrically placed in Figure 9.14, so that the magnitude of each term in Eq. 9.6, which has contributions from both the $2p_A$ and $2p_B$, is the same in each $d\tau$. For the **i** direction, we see that the sign of $\pi^* x \pi d\tau$ is positive in the x, z and $-x, -z$ quadrants, but negative in the $x, -z$ and $-x, z$ quadrants. The sum of the values for these four $d\tau$ will be zero, and since the entire integral can be summed by choosing the $d\tau$ in this symmetric way, the transition dipole in the **i** direction is zero. For the **k** direction, the signs are positive for all four symmetrically placed $d\tau$ volume elements. Thus the $\pi\pi^*$ transition has a transition dipole along the bond axis. The integral in the **j** direction is also zero from symmetry arguments.

9.1.3   Proteins

A protein consists of an amide (peptide) backbone with various side chains attached to the α carbons between each amide. The dominant chromophore is the amide group, which has a weak $n\pi^*$ transition at about 220 nm and an intense $\pi\pi^*$ transition at about 195 nm, as we have discussed. The electronic transitions of most side chains occur below 200 nm, and are overpowered by the intense $\pi\pi^*$ of the amides. Exceptions are phenylalanine, tyrosine, tryptophan, cysteine, methionine,

Figure 9.15 The electronic absorption spectra for the aromatic side chains, in aqueous solution, pH 5 to 7: phenylalanine (—); tyrosine (--); tryptophan (---). [From data in *Practical Handbook of Biochemistry and Molecular Biology* (1989), 81–83 (compiled by Elmer Mihalyi, G. D. Fasman, ed.), CRC Press; D. B. Wetlaufer (1962) "Ultraviolet Spectra of Proteins and Amino Acids," *Adv. Protein Chem.* **17**, 303–390; R. Sussman and W. B. Gratzer, personal communication.]

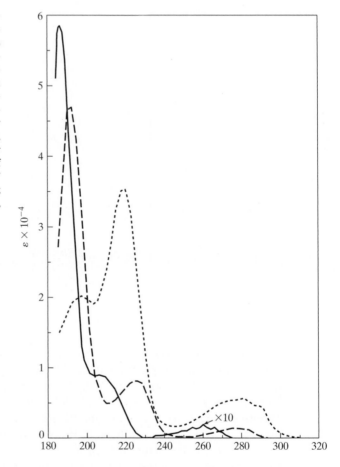

and disulfide groups, which begin their electronic transitions just below 300 nm. The π systems of phenylalanine, tyrosine, and tryptophan have the $\pi\pi^*$ transitions shown in Figure 9.15.

We are primarily interested in the spectra of polypeptide and protein backbones. As an example, let us consider the electronic absorption of poly-L-lysine hydrochloride as shown in its random coil, α-helix, and β-strand conformations (Figure 9.16). Note that spectra for biopolymers are always plotted with the intensity on a per monomer basis, in general making the intensity independent of the length of the polymer. The absorption of the random coil resembles that of N,N-dimethylacetamide in Figure 9.12, except that the second transition of moderate intensity is now blue-shifted under the very intense $\pi\pi^*$ transition. The absorption for the β strand is similar, but the band maximum of the $\pi\pi^*$ is red-shifted. However, we see that the absorption spectrum of the α helix looks quite different. The $\pi\pi^*$ transition has lost a great deal of intensity and there is a shoulder at 205 nm in addition to the obvious $\pi\pi^*$ maximum at about 190 nm.

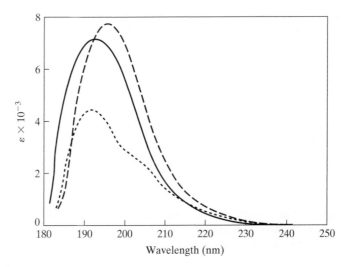

Figure 9.16   The electronic absorption spectra for poly-L-lysine hydrochloride in aqueous solution as a random coil at pH 6.0, 25°C (—); α helix at pH 10.8, 25°C (---); β strand at pH 10.8, 52°C (––). [Adapted from K. Rosenheck and P. Doty (1961), *Proc. Natl. Acad. Sci. USA* **47**, 1775–1785.]

How can we explain the marked spectral differences produced in these chromophores upon change in overall polymer conformation? *Exciton theory* is a simple quantum mechanical treatment that can predict the difference between the absorption for the random coil, which is essentially an amide absorption spectrum, and the absorption for the α helix, which has oriented amide groups interacting with one another. Exciton theory was introduced briefly in Section 8.3 as an extension of the product wavefunction suggested by nondegenerate perturbation theory. It provides a particularly simple way to predict the properties of biopolymers. As we shall see, for α helices it predicts a loss of intensity for the $\pi\pi^*$ transition of each amide because of its interaction with higher-energy transitions in the other amide chromophores. The theory also predicts that the $\pi\pi^*$ transition will have two components as a result of the interaction among the $\pi\pi^*$ transitions of the oriented amides. One will occur at lower energy with the transition dipole parallel to the helix axis, and another at higher energy with a transition dipole perpendicular to the helix axis. In Chapter 10 we find that linear dichroism confirms this prediction and demonstrates that the shoulder at 205 nm is a $\pi\pi^*$ transition with its dipole direction parallel to the helix axis.

The basic predictions of energy splitting and the rearrangement of intensity for interacting transitions can be seen by applying exciton theory to a dimer of the chromophores, with the nonchromophoric groups ignored. A dimer of amides is shown in Figure 9.17 with transition dipoles for the $\pi_2\pi_3^*$ transition. The basic tenet of exciton theory is that there is no exchange of electrons between systems, in this case the two chromophoric amides in the dimer. The interaction between systems is the Coulombic interaction between the charged particles, which in this case make

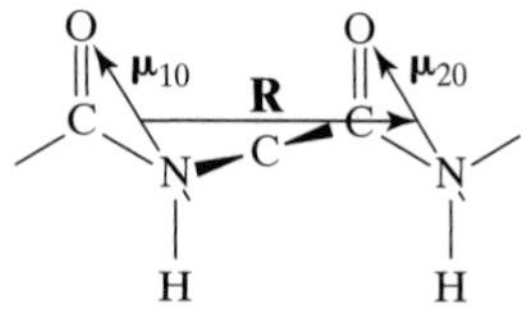

Figure 9.17 The energy of interaction between the $\pi\pi^*$ transition dipoles of an amide dimer depends on the relative orientation of the dipole vectors and the vector distance between them.

up the identical amides. The Hamiltonian for the dimer is the sum of the Hamiltonians for the individual amides plus their interaction.

$$\hat{H} = \hat{h}_1 + \hat{h}_2 + \hat{V}_{12} \tag{9.7}$$

This suggests the product wavefunctions of perturbation theory, Eq. 8.35, as the approximate wavefunctions for the system. We restrict each amide to a ground state 0, at the zero of energy, and a single excited state † at energy e. Then the possible product wavefunctions and their corresponding energies are

$$
\begin{aligned}
E_0 &= 0: & \phi_0 &= \chi_1^0 \chi_2^0 \\
E_1 &= E_2 = e: & \phi_1 &= \chi_1^\dagger \chi_2^0 \quad \phi_2 = \chi_1^0 \chi_2^\dagger \\
E_{12} &= 2e: & \phi_{12} &= \chi_1^\dagger \chi_2^\dagger
\end{aligned}
\tag{9.8}
$$

There are four possible products, a ground-state wavefunction, a doubly excited wavefunction, and two singly excited wavefunctions that fall at the same energy and thus are said to be *degenerate*. The variation theorem tells us that we can improve our approximate wavefunctions by taking a linear combination with unspecified coefficients

$$\psi = C_0 \phi_0 + C_1 \phi_1 + C_2 \phi_2 + C_{12} \phi_{12} \tag{9.9}$$

and use the variation theorem to determine the coefficients. The results of applying the variation theorem are

$$
\begin{aligned}
E_0 &= 0: & \psi_0 &= \phi_0 \\
E_+ &= e + V: & \psi_+ &= \frac{(\phi_1 + \phi_2)}{\sqrt{2}} \\
E_- &= e - V: & \psi_- &= \frac{(\phi_1 - \phi_2)}{\sqrt{2}} \\
E_{12} &= 2e: & \psi_{12} &= \phi_{12}
\end{aligned}
\tag{9.10}
$$

where many of the coefficients in Eq. 9.9 are zero. Using the dipole approximation of Eq. 3.13, V is the interaction of transition dipoles for E_0 to E_1 and E_2 (in au).

$$V = \int \phi_1^* V_{12} \phi_2 \, d\tau \cong \frac{\boldsymbol{\mu}_{10} \cdot \boldsymbol{\mu}_{20}}{R^3} - 3\frac{(\boldsymbol{\mu}_{10} \cdot \mathbf{R})(\boldsymbol{\mu}_{20} \cdot \mathbf{R})}{R^5} \tag{9.11}$$

We have assumed that the amides have no permanent dipole moment in either the ground or excited state and that the change in their respective energies will be zero.

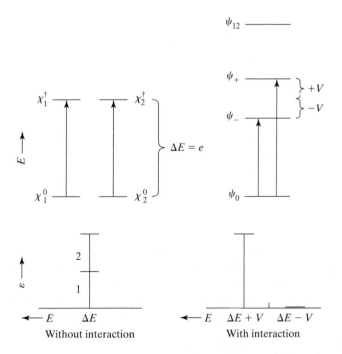

Figure 9.18 According to exciton theory, two electronic transitions on *different* chromophores that are degenerate without interaction will split in energy when their transition dipoles interact through Coulombic interaction.

Without this assumption, the interaction would mix the ground and doubly excited states. The interaction between the two singly excited states is the Coulomb interaction between their respective transition dipoles, which can be calculated using the standard equation for the Coulomb energy or can be evaluated in the dipole approximation given in Eq. 9.11. The exciton splitting between excited state ψ_+ at E_+ and excited state ψ_- at E_- is $2V$. The beauty of applying exciton theory is that we never need to know the wavefunctions to calculate the interaction; the magnitude and direction of the transition dipoles are known from experiment, and the interaction can be calculated in the dipole approximation using the geometry of the dimer.

An energy level diagram for the dimer both with and without the interaction $\hat{V}_{12}$ is given in Figure 9.18. The numerical value of the interaction V can be either positive or negative, depending on the geometry of the dimer. Figure 9.18 is drawn under the assumption that V is positive; if V is negative, then ψ_+ is the low-energy wavefunction.

The exciton wavefunctions can also be used to calculate the transition dipoles for the dimer. Considering ψ_+, we have

$$\boldsymbol{\mu}_+ = \int \psi_+{}^* \hat{\boldsymbol{\mu}} \psi_0 \, d\tau$$

$$= \frac{1}{\sqrt{2}} \int (\chi_1^\dagger \chi_2^0 + \chi_1^0 \chi_2^\dagger)^* (\hat{\boldsymbol{\mu}}_1 + \hat{\boldsymbol{\mu}}_2) \chi_1^0 \chi_2^0 \, d\tau_1 \, d\tau_2$$

$$
= \frac{1}{\sqrt{2}} \{ \int \chi_1^{\dagger*} \hat{\boldsymbol{\mu}}_1 \chi_1^0 \, d\tau_1 \int \chi_2^{0*} \chi_2^0 \, d\tau_2 + \int \chi_1^{\dagger*} \chi_1^0 \, d\tau_1 \int \chi_2^{0*} \hat{\boldsymbol{\mu}}_2 \chi_2^0 \, d\tau_2
$$

$$
+ \int \chi_1^{0*} \hat{\boldsymbol{\mu}}_1 \chi_1^0 \, d\tau_1 \int \chi_2^{\dagger*} \chi_2^0 \, d\tau_2 + \int \chi_1^{0*} \chi_1^0 \, d\tau_1 \int \chi_2^{\dagger*} \hat{\boldsymbol{\mu}}_2 \chi_2^0 \, d\tau_2 \}
$$

$$
= \frac{(\boldsymbol{\mu}_{10} + \boldsymbol{\mu}_{20})}{\sqrt{2}} \tag{9.12}
$$

where we have made use of the orthonormality of the wavefunctions for each amide. Similarly, ψ_- gives

$$
\boldsymbol{\mu}_- = \int \psi_-^* \hat{\boldsymbol{\mu}} \psi_0 \, d\tau = \frac{(\boldsymbol{\mu}_{10} - \boldsymbol{\mu}_{20})}{\sqrt{2}} \tag{9.13}
$$

Again, we do not need to know the wavefunctions explicitly; the magnitude and direction of each dimer transition dipole depends on the geometry of the dimer and is simply found from vector addition of the amide transition dipoles.

Figure 9.18 depicts the relative energies and magnitudes of the electronic transitions both with and without interaction between the amides. We see that interaction between the amides splits the transitions and rearranges the intensity, which is related to the square of the magnitude of the transition dipole. The results for a polymer are qualitatively similar to the results for a dimer. Although a helical polymer with N monomers will have N polymer transitions, when N is large only three of the N polymer transitions will have a measurable intensity. These are two mutually perpendicular transitions perpendicular to the helix axis, and a third transition parallel to the helix axis. The parallel and perpendicular polymer transitions resulting from degenerate interaction are widely split in energy, and we see the parallel $\pi\pi^*$ at 205 nm and the two nearly degenerate perpendicular $\pi\pi^*$ transitions at 190 nm for the α-helix absorption spectrum shown in Figure 9.16. Although the degenerate interaction of the transition dipoles predicts a rearrangement of intensity, there is no net change in total intensity. Of course, the $n\pi^*$ transitions undergo the same kind of interactions. However, the energy of interaction given by Eq. 9.11 depends on the magnitude of the transition dipoles. Since the magnitude of the $n\pi^*$ transition dipole is very small, the interaction is negligible.

As a concrete example of calculating the interaction energy and transition dipoles through the exciton treatment, let us consider the amide dimer of Figure 9.17. We will take the magnitude of each transition dipole to be 2 au and the distance between them to be 4 au. The angle between two vectors is found after putting them tail to tail. Since the two transition dipoles are parallel, the angle between them is zero degrees; and their scalar product is 4 au, since $\cos 0° = 1$. The angle between each transition dipole and the vector between the dipoles is the same in both cases, and we take that to be 120°. Then the scalar products $\boldsymbol{\mu}_1 \cdot \mathbf{R} = \boldsymbol{\mu}_2 \cdot \mathbf{R} = 8$ au $(\cos 120°) = -4$ au. The interaction energy of Eq. 9.11 is calculated to be $1/64 = 0.0156$ au.

Vectors are added by placing them head to tail and then drawing a new vector between the tail of the first and the head of the last. In our example, the two vectors are parallel, and the magnitude of $\boldsymbol{\mu}_+$ is $4/\sqrt{2} = 2\sqrt{2}$ au. The two amide transition dipoles cancel for $\boldsymbol{\mu}_-$ so the magnitude of $\boldsymbol{\mu}_- = 0$. The intensity of a transition is related to the dipole strength of Eq. 8.113, which is just the square of the transition dipole. Each amide has $D = 4$ au, or a total intensity of 8 au. For the dimer $D_+ = 8$ au and $D_- = 0$ au. The degenerate interaction leads to a splitting $2V$ with the dipole strength represented by the length of the vertical lines in Figure 9.18. The high-energy transition has all the intensity of 8 au, while a lower-energy transition has no intensity at all. There is a rearrangement of intensity, but no net change in total dipole strength.

There is also nondegenerate interaction between transition dipoles; the interaction of a transition dipole on one amide with transitions on other amides that fall at a different energy. These interactions are similar to the degenerate interactions detailed above, but the effects are lessened according to the energy difference between the transitions. The two transitions will be slightly split apart and there will be a rearrangement of intensity, as shown schematically in Figure 9.19. The lower-energy transition may either gain or lose intensity, depending on the geometry of the molecule and the orientation of the transition dipoles. When the low-energy transition loses intensity relative to the isolated monomer, this is called *hypochromism;* this is the effect that we observe for the $\pi\pi^*$ transition of the α helix shown in Figure 9.16.

Figure 9.20 depicts the interaction between two identical chromophores with four different transitions. With no interaction, the transitions occur as part (a) of the diagram, where the degenerate transitions are shown adding their intensity. The effect of the degenerate interaction only is shown in (b), where each pair of degenerate transitions is split substantially and the intensity is rearranged. The effect of

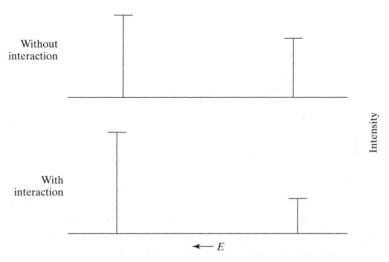

Figure 9.19 Coulombic interaction between the transition dipoles of nondegenerate transitions on *different* chromophores also cause some splitting and a rearrangement of intensity.

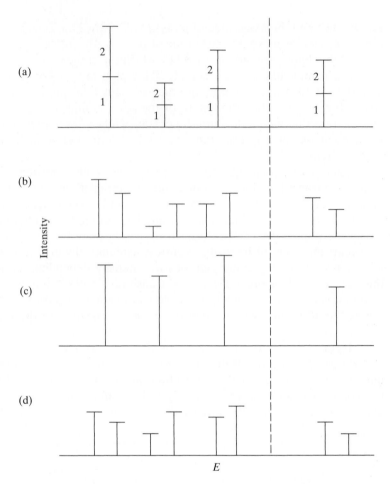

Figure 9.20 A four-part diagram for two degenerate chromophores with four electronic transitions showing (a) no interaction; (b) splitting and rearrangement of intensity due to degenerate interaction only, between the two chromophores; (c) splitting and rearrangement of intensity for the nondegenerate interactions only, with the transitions of one chromophore interacting with the nondegenerate transitions on the other chromophore; and (d) the combined results of both degenerate and nondegenerate interaction among all four transitions on the two chromophores. This diagram is consistent with the interaction between two amides in an α helix, and the dashed line represents a typical instrumental cutoff so that the higher energy transitions are not observed experimentally.

nondegenerate interactions only is shown in (c), where each pair of degenerate transitions tends to separate itself from the other pairs through the nondegenerate interaction. There is also a rearrangement of intensity among the nondegenerate transitions. Both degenerate and nondegenerate interactions are combined in (d). In this hypothetical example, the lowest-energy transition splits because of the degenerate interaction and loses overall intensity because of the nondegenerate interaction, in a manner similar to that observed for the α helix.

9.1.4   Nucleic Acids

We can also understand the electronic absorption spectra of nucleic acids by applying the same principles of exciton theory that are applied to proteins in the preceding section. The bases in nucleic acids are the chromophores, while the sugars and phosphates that make up the backbone have their transitions at higher energy. The electronic structures of the common bases (adenine, guanine, cytosine, thymine, and uracil) are given in Table 9.1. The σ skeleton is outlined; nonbonding electrons are shown as filled circles and π electrons are shown as open circles. Each of the bases has an extensive π electron system, and these give rise to intense $\pi\pi^*$ transitions that are seen in the electronic absorption spectra in Figure 9.21. Table 9.1 lists the

Table 9.1   Electronic Structure of the Bases

Compound	Structure	Number of π Orbitals	Number of π Electrons	Number of Filled π Orbitals	Number of Unfilled π Orbitals
Adenine		10	12	6	4
Guanine		11	14	7	4
Cytosine		8	10	5	3
Uracil (Thymine)		8	10	5	3

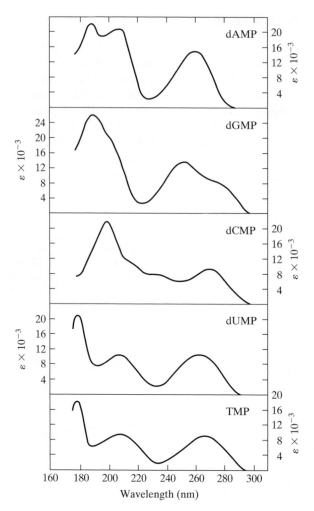

Figure 9.21 The electronic absorption spectra for five deoxyribonucleotides in aqueous solution at pH 7.0: (dAMP) deoxyadenosine-5′-phosphate; (dGMP) deoxyguanosine-5′-phosphate; (dCMP) deoxycytidine-5′-phosphate; (dUMP) deoxyuridine-5′-phosphate; (TMP) thymidine-5′-phosphate. [Adapted from data in C. A. Sprecher and W. C. Johnson, Jr. (1977), *Biopolymers* **16**, 2243–2264.]

number of π orbitals in each base, which is the same as the number of $2p$ orbitals involved in the π system. Carbon atoms, oxygen atoms, and nitrogen atoms that form two σ bonds donate one electron to the π system, while nitrogen atoms that form three σ bonds donate two electrons to the π system. There are a number of filled π orbitals and a number of unfilled π orbitals for each base, so a large number of $\pi\pi^*$ transitions is possible, and the lower-energy ones are seen in Figure 9.21. Nitrogen atoms that form two σ bonds have a nonbonding pair that is capable of undergoing an $n\pi^*$ transition predicted to have moderate intensity. These transitions have never been seen, and they are presumably underneath and overwhelmed by the $\pi\pi^*$ transitions. Similarly, each oxygen has a $2p$ nonbonding pair capable of undergoing an $n\pi^*$ transition that is predicted to have an extremely low intensity. Again, these transitions have never been seen and are presumably buried under the extremely intense $\pi\pi^*$ transitions.

The dipole directions for each of the $\pi\pi^*$ transitions have been measured, and the results from Clark's laboratory are given in Table 9.2. As we have seen for proteins above, these directions are important when making exciton calculations to predict any observable of the nucleic acid polymer from the observables of the nucleotide monomers because the base is the chromophore. Also listed in Table 9.2 are the wavelengths for the transition maxima and their transition dipole magnitude.

Since the absorption spectrum for a nucleic acid is plotted on a per monomer basis, we might expect the absorption spectrum to be the average spectrum for the

Table 9.2   Monomer Transition Dipole Data

		λ (nm)	μ (nm)	θ
Adenine	1	260	0.0513	83
	2	267	0.0701	25
	3	213	0.0700	−45
	4	204	0.0455	15
	5	186	0.0718	72
	6	179	0.0406	−45
	7	160	0.0582	6
	8	143	0.0363	−45
Cytosine	1	271	0.0591	6
	2	233	0.0254	−35
	3	224	0.0518	76
	4	198	0.0811	86
	5	161	0.0472	0
	6	152	0.0529	60
Guanine	1	272	0.0640	−4
	2	254	0.0765	−75
	3	206	0.0878	−71
	4	187	0.0914	41
	5	159	0.0541	0
	6	145	0.0517	40
Uracil	1	260	0.0674	−9
	2	213	0.0714	−53
	3	217	0.0118	⊥
	4	182	0.0721	−26
	5	167	0.0511	−31

In-plane transition dipole directions using the Tinoco-DeVoe convention. For pyrimidines, θ is positive toward N_3 from the N_1-C_4 reference axis, and for the purines θ is positive toward N_3 from the C_4-C_5 reference axis.

Source: Data from L. B. Clark (1977), *J. Am. Chem. Soc.* **99**, 3934–3938; L. B. Clark (1989), *J. Phys. Chem.* **93**, 5345–5347; L. B. Clark (1990), *J. Phys. Chem.* **94**, 2873–2879; F. Zaloudek, J. S. Novros, and L.B. Clark (1985), *J. Am. Chem. Soc.* **107**, 7344–7351; and J. S. Novros and L. B. Clark (1986), *J. Phys. Chem.* **90**, 5666–5668.

Figure 9.22 The absorption spectrum of native *E. coli* DNA (—) and the average spectrum for the four-component deoxynucleotides in aqueous solution (---).

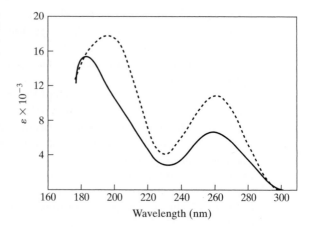

component bases, weighted by the composition of the polymer and modified by the interactions. Degenerate interactions give a splitting and a redistribution of intensity, but no net change in intensity. The splittings for nucleic acids are about 1000 cm^{-1}, which is small compared to the width of the absorption bands, and since there are seven overlapping bands for the four bases in the 260 nm region, the resulting band shape is still broad and smooth, as shown in Figure 9.22. Nondegenerate interactions can lead to a net change of intensity, and we see in Figure 9.22 that the absorption of *E. coli* DNA is about 60% of the weighted average spectrum for the component bases. This loss of intensity from nondegenerate interaction in DNA is called *hypochromism*.

Hypochromism is very useful because this change of intensity can be used to follow the melting of secondary structure for nucleic acids with temperature or a change in solvent. Figure 9.23 plots the fractional increase in intensity for the 260 nm band for DNA as a function of temperature. We see that as the DNA becomes unstacked and loses the interaction between the bases, the intensity at 260 nm increases.

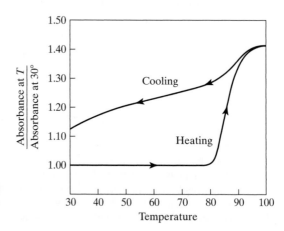

Figure 9.23 The thermal denaturation of DNA as measured through electronic absorption spectroscopy.

The maximum value upon complete melting is still less than the weighted average for the monomers (about 88%) because, even as a random coil, the DNA has interaction between the bases that causes some hypochromism. The melting temperature T_m is defined as the temperature at which half the melting has taken place. It will vary with the type of DNA, the salt concentration, and the pH. Renaturation on cooling is incomplete because in this mixture of many different sequences, a strand will have great difficulty in finding the proper partner.

9.1.5   Applications of Electronic Absorption Spectroscopy

In order to apply UV absorption to biological problems, it is important to understand how the instrument works and the practical aspects of using the instrument. A simplified diagram of a recording spectrograph is shown in Figure 9.24, where we see that the essential components are a source of light, a monochromator to disperse the light into its various wavelengths, a beam splitter so that both I and I_0 of Beer's law can be recorded simultaneously, and a detection and output system.

A spectrograph has two compartments, labeled *sample* and *reference,* and the misnaming of the reference compartment leads to confusion and misuse of the instrument. It is clear that the I of Beer's law is the amount of light that passes through the sample, which for most biological measurements consists of a cell, a solvent, and the macromolecule being studied. The parameter I_0 is then the amount of light passing through the cell and solvent without the molecule being studied. While it is tempting to put a matched cell with a solvent in the reference compartment to measure I_0 on photomultiplier 2 while I is being measured on photomultiplier 1, this procedure has two problems. First, it assumes that when the sample cell with solvent that is in the sample compartment is compared to the reference cell with solvent that is in the reference compartment it will give a spectrum (baseline) that is straight. However, it is rare that an instrument is so well balanced that a baseline is straight. Second, it assumes that the solvent absorbs very little light, so that the absorbance in the sample compartment does not exceed the maximum absorbance

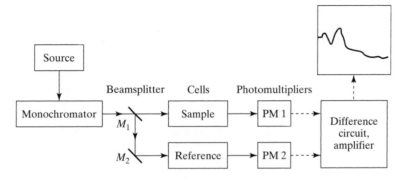

Figure 9.24   A simplified diagram for an absorption spectrograph.

that the instrument can measure accurately (usually an absorbance of 2.0, which corresponds to only 1% of the light being transmitted).

We can investigate the solvent by measuring its absorbance in the sample cell in the sample compartment versus air in the reference compartment. In general, the absorbance of the solvent should be below 1.0, and since cell path lengths are available from 10 cm to 10 μm, we can usually pick a cell path length that is appropriate for the solvent absorption, or modify the solvent if necessary. We can also monitor the total absorption of cell, solvent, and sample by following the same procedure in order to ensure that the total absorbance is less than 2.0. The imperfections of the instrument are contained in both of these spectra, so the spectrum of the macromolecule is accurately given as the absorption of cell, solvent, and sample minus the absorption of the cell and solvent, which is the baseline for this method of measurement. This procedure uses only one cell and provides the spectrum of the molecule while ensuring that the measurement is made within the limitations of the instrument. The only reason to put a second cell in the reference compartment is to straighten out the baseline of an absorbing solvent after we are familiar with our solvent and sample system.

The absorption of DNA in Figure 9.22 is clearly different from the absorption of a protein as illustrated by poly-L-lysine in Figure 9.16. In that sense, electronic absorption spectra can be used for molecule identification or for monitoring contamination of one type of macromolecule with another type of macromolecule. However, since electronic absorption bands are broad and overlapping, and are characteristic of the entire molecule, they are not as useful for molecular identification as infrared absorption or nuclear magnetic resonance, both of which have sharp nonoverlapping bands characteristic of individual bonds.

Electronic absorption spectroscopy is particularly valuable for determining the concentration of biological molecules, which are often difficult to weigh out accurately because of the presence of water and salts and because of the small quantities employed. Beer's law, Eq. 8.89, is accurate as long as the sample molecule does not aggregate, causing new interactions. Once the extinction coefficient has been determined for a particular macromolecule, the absorbance as measured by electronic absorption will give the concentration. The extinction coefficients at 190 nm are given for a variety of proteins in Table 9.3. We can see that taking $\varepsilon(190)$ equal to

Table 9.3   Extinction Coefficient at 190 nm for Some Proteins

Protein	$\varepsilon(190) \times 10^{-3}$
α-Chymotrypsin	9.69
Cytochrome c	9.72
Elastase	10.29
Hemoglobin	9.62
Lactate dehydrogenase	8.51
Lysozyme	11.46
Myoglobin	9.15
Papain	10.10
Ribonuclease	9.64

Table 9.4 Extinction Coefficients for Some DNAs

DNA	λ (nm)	GC Content	$\varepsilon \times 10^{-3}$
poly d(AT) · poly d(AT)	262	0.00	6.60
C. perfringens	258	0.31	6.30
E. coli	258	0.50	6.50
M. luteus	258	0.72	7.04
poly d(GC) · poly d(GC)	256	1.00	7.10

10,000 l/mol · cm will be sufficiently accurate for many purposes, and that it is really much more accurate than a Lowry determination of protein concentration. Furthermore, $\varepsilon(190)$ is about 100 times larger than $\varepsilon(280)$, so that much less material is necessary to monitor concentrations at 190 nm than at 280 nm. But it goes without saying that workers must use buffers that are transparent at 190 nm. The extinction coefficients for the first band of some nucleic acids are given in Table 9.4. These are closely related to the AT-GC content, and can be accurately estimated by using the Felsenfeld-Hirschman method (Felsenfeld and Hirschman 1965).

Electronic absorption is also useful for following changes in the entire biological molecule, as we have already seen for the melting of DNA in Figure 9.23. Proteins also change their electronic absorption as they are denatured since the absorption is different for the different secondary structures, as we saw in Figure 9.16.

9.2   VIBRATIONAL ABSORPTION

An atom has three degrees of freedom, which are translations along the three axes of a Cartesian coordinate system. The degrees of freedom are conserved when an atom becomes part of a molecule, although the type of freedom changes. The molecule as a whole will have three translational degrees of freedom and, if nonlinear, it will have three rotational degrees of freedom. The remaining degrees of freedom for a nonlinear molecule with n atoms will be $3n - 6$ different fundamental modes of vibration. The vibrations are *internal* degrees of freedom, usually expressed in terms of center of mass coordinates. While a macromolecule that contains a large number of atoms will have a large number of vibrations, the spectroscopy is simplified because chemical groups display characteristic vibrational bands. For example, the $C{=}O$ group has a fundamental stretching frequency (in wavenumbers, $\bar{v} = 1/\lambda$) of about 1700 cm^{-1} in many molecules, and a fundamental stretching mode of the N–H group is observed in the neighborhood of 3400 cm^{-1}. Some of the typical fundamental frequencies for the vibrational bands of chemical groups found in biological molecules are given in Figure 9.25. We see that these energies fall in the infrared region of the electromagnetic spectrum (Figure 9.1). The energy difference between ground and first excited vibrational state is large enough that the groups can be considered, for all practical purposes, to be in the ground vibrational state and this further simplifies the spectroscopy.

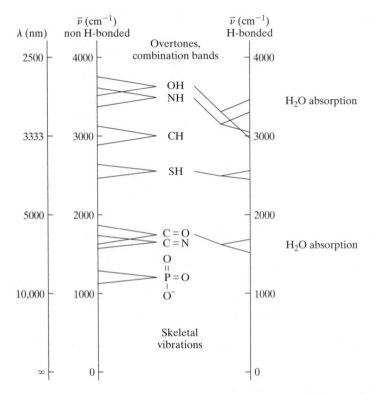

Figure 9.25 Some typical infrared absorption bands found in biological molecules. Approximate shifts on hydrogen bonding are shown and the regions of water absorption are indicated.

9.2.1 Energy of Vibrational Absorption Bands

It is the nuclei in molecules that are vibrating. To investigate the energy of the vibrational absorption bands, we must include the kinetic energy of the nuclei explicitly in the Hamiltonian,

$$\mathcal{H}(\mathbf{r}_e, \mathbf{r}_n) = K(\mathbf{r}_n) + H(\mathbf{r}_e, \mathbf{r}_n) \tag{9.14}$$

where subscripts n and e index the nuclei and electrons, respectively. The Hamiltonian $H(\mathbf{r}_e, \mathbf{r}_n)$ is just the electronic Hamiltonian of the molecule that we have treated before. If we adopt the Born-Oppenheimer approximation for the molecule, then the Hamiltonian $\mathcal{H}(\mathbf{r}_n, \mathbf{r}_e)$ is separable into nuclear and electronic parts, so the solution is the product function,

$$\Psi(\mathbf{r}_e, \mathbf{r}_n) = \psi(\mathbf{r}_e, \mathbf{r}_n)N(\mathbf{r}_n) \tag{9.15}$$

where N is the nuclear function and ψ is the electronic function that can be found for each particular value of $\mathbf{r}_n$, as we did in Section 8.3 for the hydrogen molecule. The fundamental vibrations of the molecule will be very nearly harmonic oscillations, so the $N(\mathbf{r}_n)$ will be taken as the solutions to the simple harmonic oscillator given by Eq. 8.26, with corresponding energies given by Eq. 8.27. Usually vibrational absorption spectroscopy is concerned with molecules in their ground electronic state and investigates transitions from the ground vibrational state $n = 0$ to the first excited vibrational state $n = 1$. Each of the many vibrational modes of the molecule will have a different *fundamental* frequency corresponding to the transition from $n = 0$ to $n = 1$. *Overtones* are transitions to higher excited states with n greater than 1. *Combination* bands, resulting from two or more vibrations being excited, and *difference* bands, corresponding to the difference between two vibrational energies, are also possible.

Let us consider a hypothetical chemical group with three nonlinear nuclei attached to a framework, as shown in Figure 9.26, which illustrates various types of vibrations. In-plane stretching can either be in-phase or out-of-phase. There are also in-phase and out-of-phase in-plane bending modes. Finally, out-of-plane bending can be in-phase or out-of-phase. The fundamental IR band for each of these six modes will have a characteristic frequency and intensity. This is the strength of vibrational absorption spectroscopy; a spectrum of the fundamental vibrational transitions can be used to identify the chemical groups in a molecule.

9.2.2 Transition Dipoles

It is easy for us to see the transition dipole directions for the types of vibrations illustrated in Figure 9.26. In order for the oscillations of the infrared light to excite the in-phase stretching or in-plane bending modes, the light needs to be polarized along the symmetry axis of the group. For the out-of-phase stretching or in-plane bending, the light needs to be polarized in-plane and across the symmetry axis. For the in-phase out-of-plane bending mode, the light needs to be polarized perpendicular to the plane of the group. Out-of-phase and out-of-plane bending do not have a motion that will be driven by the light, and thus will not be excited.

Stretching

In-plane bending

Out-of-plane bending

Figure 9.26 Fundamental in-phase and out-of-phase vibrations for three nonlinear atoms.

Transition dipoles can be calculated from the wavefunctions using Eq. 8.111, where we now consider the dynamics of a particular vibration explicitly by expanding the dipole operator in a Taylor series about the equilibrium position for each nucleus i

$$\hat{\boldsymbol{\mu}} = \hat{\boldsymbol{\mu}}_{q_i=0} + \sum_i \left(\frac{\partial \hat{\boldsymbol{\mu}}}{\partial q_i}\right)_{q_i=0} \hat{q}_i + \cdots \tag{9.16}$$

where q_i is the displacement from equilibrium (which is called x for the harmonic potential in Chapter 8). The operator $\hat{\boldsymbol{\mu}}$ is a sum over all particles, Eq. 8.104, and since electrons will move with the nuclei, both electrons and nuclei contribute to the second term in the series. We divide the dipole operator at the equilibrium bond distances $(q_i = 0)$ into the electronic and nuclear parts and, ignoring higher terms, we have

$$\boldsymbol{\mu}_{10} = \int \psi_0^* N_1^* \left[\boldsymbol{\mu}_{q_i=0} + \sum_i \left(\frac{\partial \hat{\boldsymbol{\mu}}}{\partial q_i}\right)_{q_i=0} \hat{q}_i\right] \psi_0 N_0 \, d\tau$$

$$= \int \psi_0^* \hat{\boldsymbol{\mu}}_{q_i=0}^e \psi_0 \, d\tau_e \int N_1^* N_0 \, d\tau_n + \int \psi_0^* \psi_0 \, d\tau_e \int N_1^* \hat{\boldsymbol{\mu}}_{q_i=0}^n N_0 \, d\tau_n$$

$$+ \sum_i \int \int \psi_0^* N_1^* \left(\frac{\partial \hat{\boldsymbol{\mu}}}{\partial q_i}\right)_{q_i=0} \hat{q}_i N_0 \psi_0 \, d\tau_n \, d\tau_e \tag{9.17}$$

The first term in Eq. 9.17 is zero because the electronic part does not operate on the nuclear wavefunctions and the harmonic wavefunction for the initial and final nuclear states are orthogonal. Similarly, the second term is zero because the nuclear part of the operator is a number for $q_i = 0$. Integrating over the electron coordinates in the third term yields

$$\boldsymbol{\mu}_{10} = \sum_i \left(\frac{\partial \hat{\boldsymbol{\mu}}}{\partial q_i}\right)_{q_i=0} \int N_1^* q_i N_0 \, d\tau_n \tag{9.18}$$

where $\partial \hat{\boldsymbol{\mu}}/\partial q_i$ at $q_i = 0$ is the change in the ground electronic state permanent dipole as the molecule vibrates, which is a number. Substituting the harmonic wavefunctions of Eq. 8.26 for the normal coordinate mode q of a *particular* fundamental vibration gives

$$\boldsymbol{\mu}_{10}^q = \left(\frac{\partial \boldsymbol{\mu}}{\partial q}\right)_{q=0} \beta \sqrt{2/\pi} \int_{-\infty}^{+\infty} e^{-\beta q^2/2} q^2 e^{-\beta q^2/2} \, dq = \frac{1}{\sqrt{2\beta}} \left(\frac{\partial \boldsymbol{\mu}}{\partial q}\right)_{q=0} \tag{9.19}$$

For a vibration to have a measurable transition dipole, $(\partial \boldsymbol{\mu}/\partial q)$ at $q = 0$ must be nonzero—that is, its permanent dipole must change as it vibrates. For instance, O_2 has no permanent dipole moment and no IR absorbance. In contrast, carbon monoxide has a dipole moment and thus an IR spectrum.

9.2.3   Instrumentation for Vibrational Spectroscopy

Scanning IR spectrographs are rare today, since they collect only one frequency at a time. Most modern IR spectrographs are Fourier transform (FT) instruments. These instruments do not use a monochromator but use a Michelson interferometer. The interferometer has both a fixed mirror and a moving mirror, and the relative positions of the two mirrors can be measured very accurately. The light from a helium-neon laser (of well-known wavelength) is split, with one-half being sent to the fixed mirror and the other half sent to the moving mirror. When the beams are recombined, there will be constructive or destructive interference, depending on the position of the moving mirror. The intensity of the recombined beam will be a sine wave as a function of position, and the nodes in the sine wave act as a ruler. Similarly, the beam of IR light, which contains all the wavelengths of experimental interest, is split, sent to the mirrors, and recombined. Different IR wavelengths will interfere at different mirror positions, and the instrument collects total intensity as a function of mirror position. The recorded function of intensity, as a function of mirror position, is Fourier transformed into intensity as a function of frequency—that is, the normal IR absorption spectrum. All IR frequencies are recorded simultaneously from the Michelson interferometer and are transformed simultaneously by a computer. Thus, the spectrum is collected without scanning, usually in a second. Fourier transforms are considered in more detail in connection with NMR spectroscopy in Chapter 12.

Unfortunately, water, the biological solvent, absorbs IR light just where the most interesting groups found in biopolymers also absorb, as shown in Figure 9.25. To some extent, this difficulty can be overcome by using dried films of the macromolecule or by working in D_2O solutions where the solvent absorption is shifted to less interesting regions around 1200 and 2500 cm^{-1}. The absorption of water can also be brought within tolerable limits by combining cells of a 6 μm pathlength with the high accuracy of a modern FTIR instrument.

9.2.4   Applications to Biological Molecules

Students will be quite familiar with the structure of the DNA bases, but at the time that Watson and Crick were seeking the structure of DNA most chemists thought the bases had the well-known benzene structure. Benzene-like structures for the bases alone are obtained in the tautomeric form where the proton is shifted from the ring nitrogen to the carbonyl oxygen to form a hydroxyl group. Watson-Crick base pairs are hydrogen bonded in the tautomeric form with carbonyl and amine substitutents, and one test of their model was to prove that these were indeed the tautomeric forms of the DNA bases. In one of the most beautiful applications of vibrational spectroscopy, Miles (1961) took up this challenge and investigated the tautomeric forms of the bases, making use of the strength of vibration spectroscopy in identifying chemical groups. He obtained nontautomerizing methyl derivatives for the

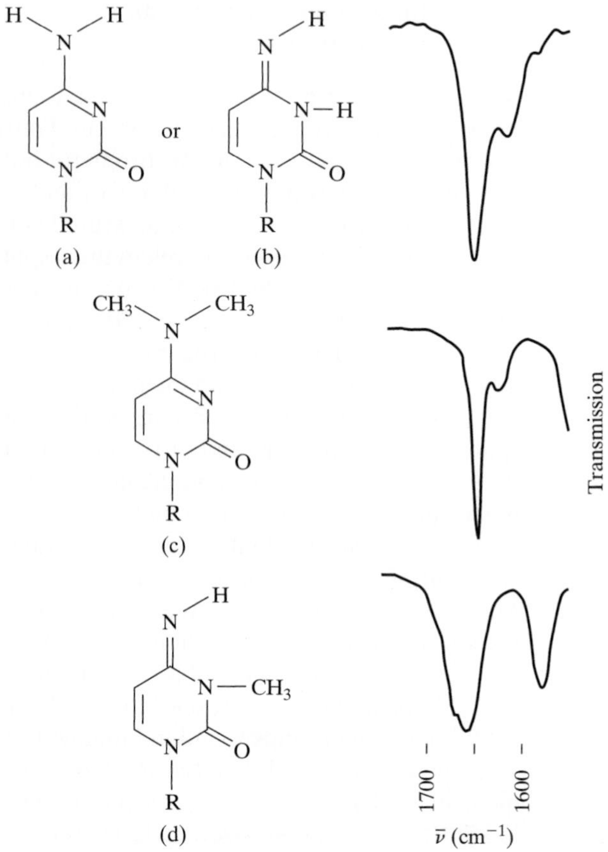

Figure 9.27 IR spectra in the 1750 to 1550 cm^{-1} region for two nontautomerizing methyl derivatives (c) and (d), and cytidine, now known to be in the first tautomeric form shown (a). [Adapted from data in H. T. Miles (1961), *Proc. Natl. Acad. Soc. USA* (Biochemistry) **47**, 793–802.]

various possible tautomers and compared them to the bases in solution. Figure 9.27 shows the amine-carbonyl and amide-imine tautomers for cytosine, along with the corresponding nontautomerizing methyl derivatives. The IR spectra in the 1750 to 1550 cm^{-1} region demonstrate that the correct tautomer for cytosine in solution is the amine-carbonyl that we take for granted today.

Another strength of vibrational spectroscopy is that hydrogen bonding shifts the characteristic frequencies of the chemical groups. Figure 9.25 also shows approximate shifts from hydrogen bonding for some typical infrared-absorption bands. Vibrational spectroscopy is therefore particularly valuable for following the hydrogen bonding of the DNA bases. Both the 1750 to 1550 cm^{-1} and the 3700 to 3100 cm^{-1} regions are useful, and the results for AU base-pairing in the latter region are shown in Figure 9.28. Bands at 3260, 3330, and 3490 cm^{-1} are indicative of hydrogen bonding. Vibrational spectra measured in deuterochloroform at various ratios for a mixture of the two bases demonstrate that the maximum hydrogen bonding occurs when a 1:1 complex of monomer derivatives is formed. This is another important early confirmation of the Watson-Crick double helix that used IR

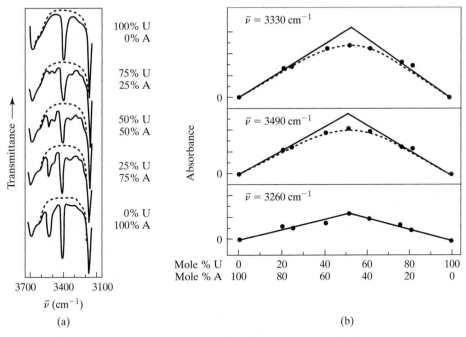

Figure 9.28 IR detection of the hydrogen-bonded association of 1-cyclohexyluracil and 9-ethyladenine in deuterochloroform solutions. (a) Spectra for different mixtures. (b) Intensity of the bands as a function of composition. The change of intensity for the bands at 3330, 3490, and 3260 cm^{-1} are indicative of hydrogen bonding, which should be maximal in equimolar mixtures. [Adapted from data in Hamlin et al. (1965), *Science* **148**, 1734–1737.]

spectroscopy. The formation of the dimer from monomers will be concentration dependent, and even at the high concentrations used in these classic experiments, only a small percentage of the monomers form the 1:1 dimer complex. The cooperative hydrogen-bonding complex with complete pairing formed by polymers of the bases shows substantial changes in the vibrational spectra.

Hydrogen bonding of the amides to form the secondary structures found in proteins has long been known to give characteristic shifts to the corresponding vibrational bands. Table 9.5 gives the vibrational frequencies for the C=O stretch (amide I), N—H bend mixed with the C—N stretch (amide II), and N—H stretch (amide A) for α helices and β strands. The vibrational spectrum for amides hydrogen

Table 9.5 Vibrational Frequencies in Wavenumbers for α Helices and β Strands in D_2O

Band	α (cm^{-1})	β (cm^{-1})
amide I	1653	1640
amide II	1545	1525
amide A	3300	3300

Figure 9.29 Infrared spectrum for a film of poly-γ-benzyl-L-glutamate in the α-helical form. The figure also illustrates linear dichroism spectroscopy to be covered in Chapter 10. A solid film of the polymer was prepared and oriented by stretching. The spectra are for infrared light linearly polarized perpendicular (---) and parallel (—) to the helix axis. [Adapted from data in Tsuboi (1962), *J. Polymer Sci.* **59**, 139–153.]

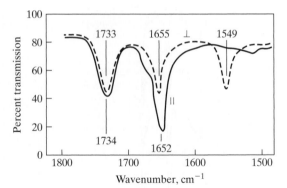

bonded into an α helix is illustrated for a film of poly-γ-benzyl-L-glutamate in Figure 9.29. The shifts in these bands on hydrogen bonding are so characteristic of secondary structure that with FTIR instrumentation the vibrational spectrum of a protein can now be successfully analyzed for the amount of each kind of secondary structure (see Application 9.1).

Application 9.1   Analyzing IR Spectra of Proteins for Secondary Structure

Byler and Susi (1986) have developed the most popular method of analyzing FTIR spectra of proteins for secondary structure. They measured the spectra for 21 proteins in the amide I region from 1600 to 1700 cm^{-1} in D_2O. These spectra consist of many overlapping bands, one for each amide. The position of each band will depend on the secondary structure in which the amide is involved. Byler and Susi used the *band-narrowing approach* of Fourier self-deconvolution to resolve the bands, as shown for ribonuclease S in Figure A9.1. When the spectra of all 21 proteins are analyzed, 11 different components are found with well-defined frequencies. Not all proteins show all components, and we see that ribonuclease S has just seven of these components.

The 11 possible component bands are assigned to α helix ($1654\ cm^{-1}$), β strand ($1624, 1631, 1637$, and $1675\ cm^{-1}$), turns ($1663, 1670, 1683, 1688$, and $1694\ cm^{-1}$), or other structure ($1645\ cm^{-1}$). Naturally, these wavenumbers are not precise and may deviate by as much as 4 cm^{-1}. The relative amounts of the four secondary structures are determined by the relative areas under the bands. For instance, ribonuclease S is predicted to have 50% α helix and 25% β strand, in excellent agreement with the results of X-ray diffraction.

The FTIR method is a rapid one for investigating the secondary structure of proteins in a variety of solvents. It requires only about 100 μg of material. The great strength of this spectroscopic technique is that, because of the long wavelength of the light, it can be applied to turbid solutions without introducing scattering artifacts. Of course, no method is perfect, and the interested student should read the critical assessment of the technique by Surewicz et al. (1993), which contains many additional references.

BYLER, D. M. and H. SUSI (1986), *Biopolymers* **25**, 469–487.

SUREWICZ, W. K., H. H. MANTSCH, and D. CHAPMAN (1993), *Biochemistry* **32**, 389–394.

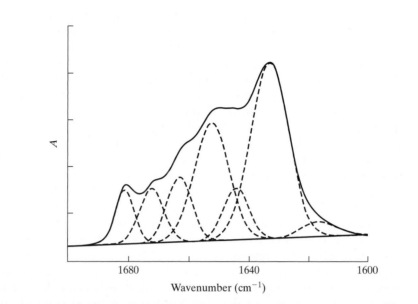

Figure A9.1   The FTIR spectrum of ribonuclease S in the amide I region deconvoluted into component bands due to α helix, β strand, turns, and other structure. [Adapted from data in Byler and Susi (1986), *Biopolymers* **25**, 469–487.]

The first step is to measure the FTIR of a number of proteins, the secondary structure of which is known from X-ray diffraction or NMR. This has been done mainly for the amide I region. Such a *basis set* of IR spectra can be used to deduce the IR patterns that correspond to each secondary structure. This has been done, but we will not discuss the *pattern recognition approach* here; it is discussed in detail in Chapter 10, which deals with circular dichroism spectroscopy. Many IR spectroscopists make use of the fact that IR bands are narrow and resolve the amide region into component bands assuming near-constant bandwidth, each band corresponding to a single secondary structure. One of these methods is discussed in Application 9.1.

9.3   RAMAN SCATTERING

In Chapter 8, we concentrated on the time-independent Hamiltonian, and showed how time independence leads to stationary energy states that are quantized with certain energy values. In the present chapter, we predicted stationary energy states for electrons in molecules by using molecular orbitals, and assumed that the vibrations of molecular bonds could be described as the stationary states of a simple harmonic oscillator. These stationary states persist and allow energy to be trapped. Molecules can also be in a *nonstationary* state, but since nonstationary states are time dependent, the molecule can exist in that state only for an instant.

In the quantum mechanical view, the scattering of light by a molecule is a two-photon process that involves nonstationary, time-dependent states, which are sometimes called *virtual states*. An incident photon excites the molecule to a nonstationary state, and then the molecule immediately emits a second photon. This second photon may be scattered in any direction, but energy must be conserved. Usually the scattered photon has the same energy as the incident photon, and the process is called *elastic* or *Rayleigh scattering*. Sometimes the molecule is excited to a higher vibrational or rotational level during the process, so the emitted photon has less energy than the incident photon. If the molecule is in an excited vibrational or rotational state, it can add energy to the emitted photon. When the emitted photon has a different energy from the incident photon, the process is called *inelastic* or *Raman scattering*. These processes are diagrammed in Figure 9.30.

Since we are interested in large biological molecules, we are concerned only with Raman scattering where a fundamental mode of vibration is excited from the ground state $n = 0$ to the first excited state $n = 1$. The wavenumber for the transition is found by subtracting the energy of the Raman-scattered light E_1 from the energy of the incident light E_0, $\bar{\nu}_{01} = (E_0 - E_1)/hc$. As we shall see, Raman scattering complements vibrational absorption in the infrared, because the selection rules for observing the transition are different. Furthermore, the experiments are done with visible light where water is transparent.

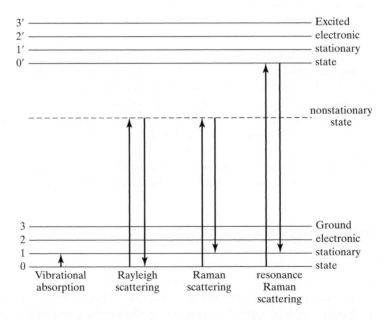

Figure 9.30 Energy-level diagram for vibrational absorption, Rayleigh scattering, Raman scattering, and resonance Raman scattering. Normal vibrational absorption in the infrared measures the energy of the $n = 0$ to $n = 1$ transition for various types of vibrations. For elastic Rayleigh scattering, the energy of the scattered light equals the energy of the exciting light. Inelastic Raman scattering that leaves the molecule in its $n = 1$ vibrational state complements vibrational absorption of biological molecules.

Since nonstationary states are not quantized, any convenient visible or UV source can be used to observe scattering. Generally, lasers are used as an intense source of monochromatic incident photons. The scattered light is usually observed at right angles to the incident beam of light, and the wavelengths of the scattered light can be separated with a monochromator or interferometer. The components of a simple Raman instrument are shown in Figure 9.31.

The transition dipole for absorption of the incident photons by the nonstationary state is the quantum mechanical analog of the classical oscillating dipole presented in Chapter 7. Analogous to Eq. 7.3, the transition dipole depends on the electric vector of the light $\mathbf{E}$, and the polarizability of the molecule $\boldsymbol{\alpha}$.

$$\boldsymbol{\mu} = \boldsymbol{\alpha}(v)\mathbf{E}(v) \tag{9.20}$$

The polarizability is a 3×3 tensor

$$\boldsymbol{\alpha} = \begin{bmatrix} \alpha_{xx} & \alpha_{xy} & \alpha_{xz} \\ \alpha_{yx} & \alpha_{yy} & \alpha_{yz} \\ \alpha_{zx} & \alpha_{zy} & \alpha_{zz} \end{bmatrix} \tag{9.21}$$

For a molecule in its ground state, its resonant components can be expressed quantum mechanically as a sum over products of the transition dipoles for all the stationary states of the molecule:

$$\alpha_{ij}(v) = \sum_n \left[\frac{(\boldsymbol{\mu}_{0n})_j (\boldsymbol{\mu}_{n0})_i}{E_{n0} - hv - i\Gamma_n} \right] \tag{9.22}$$

where $i\Gamma_n$ is the imaginary damping term that prevents the denominator from going to zero, even in the absorption $0 \to n$, where $hv = E_{n0}$.

Following the procedure we used for vibrational absorption, we can expand the polarizability operator in a Taylor series about the equilibrium position for each nucleus

$$\hat{\boldsymbol{\alpha}} = \hat{\boldsymbol{\alpha}}_{q_i=0} + \sum_i \left(\frac{\partial \hat{\boldsymbol{\alpha}}}{\partial q_i} \right)_{q_i=0} \hat{q}_i + \cdots \tag{9.23}$$

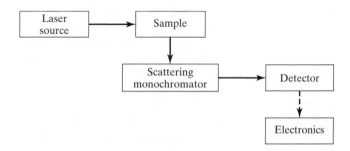

Figure 9.31   Diagram of a simple Raman scattering instrument.

Using the product function of nuclear and electronic parts defined by Eq. 9.15, we have for the polarizability corresponding to the $n = 0$ to $n = 1$ transition

$$\alpha_{10} = \int \psi_0^* N_1^* \left[\hat{\boldsymbol{\alpha}}_{q_i=0} + \sum_i \left(\frac{\partial \hat{\boldsymbol{\alpha}}}{\partial q_i} \right)_{q_i=0} \hat{q}_i \right] \psi_0 N_0 \, d\tau$$

$$= \int N_1^* N_0 \, d\tau_n \int \psi_0^* \hat{\boldsymbol{\alpha}}_{q_i=0} \psi_0 \, d\tau_e$$

$$+ \sum_i \int \psi_0^* N_1^* \left(\frac{\partial \hat{\boldsymbol{\alpha}}}{\partial q_i} \right)_{q_i=0} \hat{q}_i \psi_0 N_0 \, d\tau_n \, d\tau_e \qquad (9.24)$$

The first term in Eq. 9.24 would have a value only if the nuclear wavefunction were unchanged. This is the term that gives rise to Rayleigh scattering. The second term has a value for the mode corresponding to q_i, if the polarizability changes as the molecule vibrates. The $n = 0$ to $n = 1$ transition of a particular nuclear vibration gives

$$\alpha_{10}^q = \frac{1}{\sqrt{2\beta}} \left(\frac{\partial \boldsymbol{\alpha}}{\partial q} \right)_{q=0} \qquad (9.25)$$

and the transition dipole for the absorption of a photon that causes a corresponding Raman scattered photon becomes

$$\boldsymbol{\mu}_{10} = \frac{1}{\sqrt{2\beta}} \left(\frac{\partial \boldsymbol{\alpha}}{\partial q} \right)_{q=0} \mathbf{E} \qquad (9.26)$$

Raman scattering has about one ten thousandth the intensity of Rayleigh scattering.

The selection rule for normal vibrational absorption in the infrared is that the permanent dipole moment of the molecule must change on vibration. In contrast, the selection rule for Raman scattering is that the polarizability must change on vibration. Thus, infrared-forbidden vibrational transitions may be Raman active and vice versa.

Looking back at the polarizability operator of Eq. 9.22, we see that as the energy of the incident photon $h\nu$ approaches the energy of transition to a stationary state n, the polarizability of the molecule becomes very large and a much larger fraction of the light is also scattered. When the photon energy is such that the real part of the denominator just changes sign, then the polarizability becomes much smaller with less light scattered. Measuring Raman scattering in an absorption band where the scattering is large is called *resonance Raman scattering*. It has the advantage that the energy of excitation can be chosen to match the electronic absorption of a specific chromophore and the vibrational modes localized on that chromophore will be considerably more intense in the Raman scattering.

Application 9.2 illustrates the utility of resonance Raman spectroscopy.

Application 9.2   Using Resonance Raman Spectroscopy to Determine the Mode of Oxygen Binding to Oxygen-Transport Proteins

The resonance Raman effect has been put to use in a series of experiments designed to answer the following question: In what oxidation state and geometry do oxygen-transport proteins bind oxygen? Although for many years, study by a wide variety of techniques had answered the question pretty adequately for the heme proteins, much less was known concerning two kinds of nonheme oxygen-transport proteins found in certain invertebrates. These are *hemerythrin*, which occurs in some worms and brachiopods, and *hemocyanin*, found exclusively in arthropods and molluscs. Hemerythrins bind O_2 at a site containing two iron atoms (but no heme); hemocyanin uses a two-copper site. In both cases, oxygenation leads to pronounced color changes.

Because the O_2-binding groups in both proteins exhibit strong absorption bands in the visible and near UV, it was possible to take advantage of the resonance Raman effect to study vibrations in the bound oxygen. Although O_2 is infrared inactive because of its symmetry, it exhibits Raman scattering associated with the $O{=}O$ bond stretching. The oxygen-stretch energy has been used (Dunn et al. 1975; Freedman et al. 1976) to diagnose the oxidation state of the bond O_2. In gaseous $^{16}O_2$, the Raman band lies at 1555 cm^{-1}. However, in both proteins, the corresponding wavenumbers are found to be much lower, in the vicinity of values found for the peroxide ion O_2^- (see Table A9.2).

Thus, it was concluded that in both hemerythrins (Dunn et al. 1975) and hemocyanins (Freedman et al. 1976) binding of oxygen involves a charge transfer from the metal ligands:

$$Fe^{2+} \cdot Fe^{2+} + O_2 \rightleftharpoons {}^- Fe^{3+} \cdot O_2^{2-} \cdot Fe^{3+} \, (\text{hemerythrin})$$
$$Cu^{1+} \cdot Cu^{1+} + O_2 \rightleftharpoons {}^- Cu^{2+} \cdot O_2^{2-} \cdot Cu^{2+} \, (\text{hemocyanin})$$

Next, the same technique was employed to learn something about the geometry of the binding. A number of possible binding geometries are shown in Figure A9.2a, where we indicate either Fe^{3+} or Cu^{2+} by M^+. Note that some of these (II, III, and probably I) are symmetric; whereas IV and V are asymmetric. Structures III and V can be ruled out for a number of reasons (including the lack of a $M^+ \cdots M^+$ stretch), leaving I, II, and IV as contenders.

Table A9.2   Raman Band Wavenumbers

	Raman Wavenumbers (cm^{-1})	Sources
Free (O_2)	1555	1,2
Peroxide (O_2^{2-})	738	1
Oxyhemerythrin	845	1
Oxyhemocyanin (arthropod)	744	2
Oxyhemocyanin (mollusc)	749	2

Values are for ^{16}O, the most abundant element in natural oxygen.

1. Dunn et al. (1975).
2. Freedman et al. (1976).

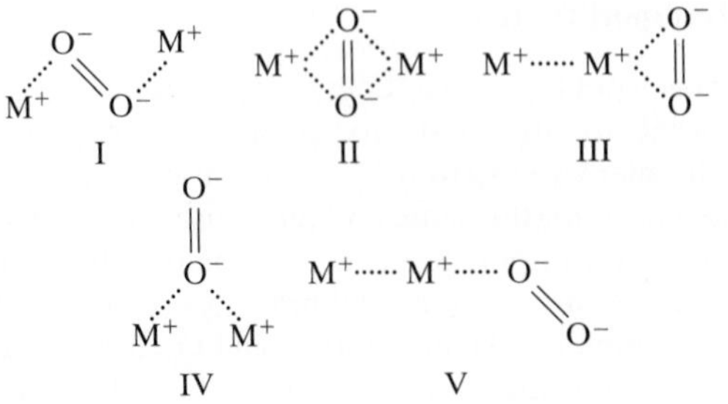

Figure A9.2a Possible models for the binding of O_2^{2-} by a metal-ion pair. M^+ can correspond to either Fe^{3+} (hemerythrin) or Cu^{2+} (hemocyanin).

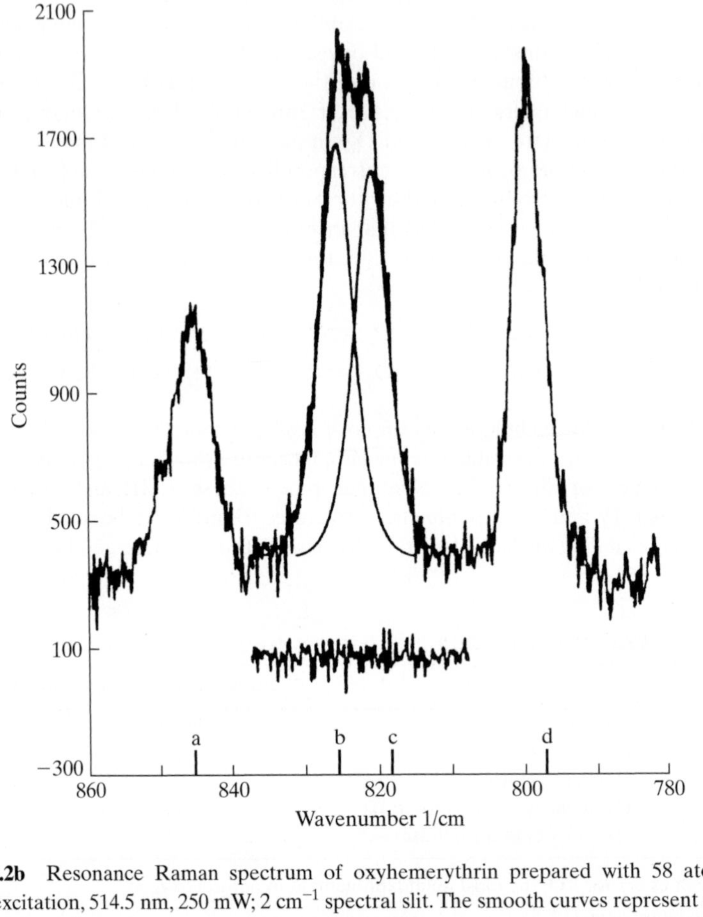

Figure A9.2b Resonance Raman spectrum of oxyhemerythrin prepared with 58 atom % ^{18}O gas: laser excitation, 514.5 nm, 250 mW; 2 cm^{-1} spectral slit. The smooth curves represent deconvolution of the 822 cm^{-1} feature into two components. The difference between observed and fitted curves is shown below the spectrum near 822 cm^{-1}. The vertical lines a, b, c, and d show the calculated peak positions for models IV and V of Fe-$^{16}O_2$ (845 cm^{-1}), Fe-^{16}O-^{18}O (825 cm^{-1}), Fe-^{18}O-^{16}O (818 cm^{-1}), and Fe-$^{18}O_2$ (797 cm^{-1}), respectively. [From Thamann et al. 1977, with permission of publisher.]

To further limit the possibilities, the same groups of researchers (Kurtz et al. 1976; Thamann et al. 1977) measured resonance Raman spectra after binding oxygen containing approximately equal amounts of isotopes ^{16}O and ^{18}O. Since the atoms combine randomly in molecular oxygen, three bands should be expected in the Raman spectrum—one corresponding to $^{16}O \cdot ^{16}O$, one to $^{18}O \cdot ^{18}O$, and one to $^{16}O \cdot ^{18}O$. If the oxygen (peroxide) is attached asymmetrically to the site, then the central ($^{16}O \cdot ^{18}O$) band should be split, because either the ^{18}O or the ^{16}O can attach to the binding site. On the other hand, symmetric binding should yield a single central band.

Data for hemerythrin (Kurtz et al. 1976) are shown in Figure A9.2b. The clear splitting of the central band (that due to $^{16}O \cdot ^{18}O$) indicates an asymmetric binding. Model IV is the likely candidate. Similar studies with hemocyanins (Thamann et al. 1977) give a quite different picture; no splitting of the central band is observed, pointing to either model I or model II. Recent X-ray diffraction studies at high resolution have shown that the latter is correct (Magnus et al. 1994).

DUNN, J. B. R., D. F. SHRIVER, and I. M. KLOTZ (1975), *Biochemistry* **14**, 2689–2694.

FREEDMAN, T. B., J. S. LOEHR, and T. M. LOEHR (1976), *J. Am. Chem. Soc.* **98**, 2809–2815.

KURTZ, D. M., D. F. SHRIVER, and I. M. KLOTZ (1976), *J. Am. Chem. Soc.* **98**, 5033–5035.

MAGNUS, K. A., B. HAZES, H. TON-THAT, C. BONAVENTURA, J. BONAVENTURA, and W. G. J. HOL (1994), *Proteins* **19**, 302–309.

THAMANN, T. J., J. S. LOEHR, and T. M. LOEHR (1977), *J. Am. Chem. Soc.* **99**, 4187–4189.

EXERCISES

9.1   Using qualitative methods, deduce the transition dipole direction for a $1s$ to sp^3 transition. Explain your reasoning.

9.2   Draw an energy-level diagram showing hybrids and molecular orbitals for the carboxylate anion. What are the three lowest energy transitions?

9.3   Calculate the splitting of the two molecular excited states, and the corresponding magnitude and direction of the transition dipoles, given the carbonyl group parameters ($\mu = 1$ au and $r = 3$ au) and the molecular geometry below.

(a)   (b)

9.4   The exciton splitting in polynucleotides appears to be of the order of 5 nm, with the absorption bands lying near 250 nm. Calculate, in atomic units, the corresponding interaction energy.

9.5   Cytosine has a molecular extinction coefficient of 6×10^3 at 270 nm at pH 7. Calculate the absorbance of 1×10^{-4} and 1×10^{-3} M cytosine solutions in a 1 mm cell.

9.6   Consider two molecules A and B that have distinguishable but overlapping absorption spectra. Outline a mathematical method, using absorption at two different wavelengths, for determining the amounts of A and B in a mixture. What will determine the choice of wavelengths?

9.7   In Table A9.2, the Raman band positions are expressed in terms of the vibrational wavenumbers (cm^{-1}). If the excitation wavelength for hemocyanin was 530.9 nm, at what *wavelengths* would one expect to observe these bands?

9.8   Using Figure 9.23, calculate the percent of DNA that is still denatured at 30°C after cooling from 100°C.

9.9   If a sample of DNA has an $\varepsilon(258)$ of 6.42×10^3, what would you estimate its GC content to be? See Table 9.4.

REFERENCES

ATKINS, P. (2002) *Physical Chemistry,* 7th ed., W. H. Freeman, New York.

DRAGO, R. S. (1992) *Physical Methods for Chemists,* Saunders College Publishing, Fort Worth, TX.

BERNATH, P. F. (1995) *Spectra of Atoms and Molecules,* Oxford University Press, Oxford, U.K.

FELSENFELD, G. and S. Z. HIRSCHMAN (1965) "A neighbor-interaction analysis of the hypochromism and spectra of DNA," *J. Mol. Biol.* **13**, 407–427.

HOLLAS, J. M. (2004) *Modern Spectroscopy,* John Wiley & Sons, New York.

MILES, H. T. (1961) "The Tautomeric Form of Cytidine," *Proc. Natl. Acad. Soc. USA* (Biochemistry) **47**, 793–802.

ROBIN, M. B. (1974, 1975) *Higher Excited States of Polyatomic Molecules,* vol. 1, vol. 2, Academic Press, New York. A compendium of electronic absorption spectra and assignment of transitions. Out of print, but in most university libraries.

SAUER, K. (ed.) (1995) *Biochemical Spectroscopy, Methods in Enzymology,* vol. 246, Academic Press, New York.

TINOCO, I., JR., K. SAUER, J. C. WANG, and J. D. PUGLISI (2002) *Physical Chemistry Principles and Applications in Biological Sciences,* 4th ed., Prentice Hall, Upper Saddle River, NJ.

10

Linear and Circular Dichroism

Dichroism is the phenomenon in which light absorption differs for different directions of polarization. Linear dichroism involves linearly polarized light where the electric vector is confined to a plane, as we have seen in Figure 8.1. The electric vector of linearly polarized light is depicted in a different representation in Figure 10.1. In Section 8.5, we discussed the use of linearly polarized light to determine transition dipole directions. Linearly polarized light oriented along the transition dipole direction for a particular transition will be absorbed maximally, while linearly polarized light oriented perpendicular to the transition dipole direction will not be absorbed at all. This property of the transition dipole is a molecular dichroism that is the basis for linear dichroism of biological polymers. Linear dichroism can tell us about the orientation of the monomers in a biopolymer.

Circularly polarized light is the antithesis of linearly polarized light. In linearly polarized light, the direction of the electric vector is constant and its magnitude is modulated; in circularly polarized light, the magnitude is constant and the direction is modulated, as we see in Figure 10.1. With both types of light, the wavelength is defined by the repeat in space, and the frequency by the repeat in time. The electric vector of circularly polarized light describes a helix, and circularly polarized light may describe either a right-handed helix, as shown in Figure 10.1, or a left-handed helix (not shown). Circular dichroism investigates the absorption of the two different kinds of circularly polarized light. The circular dichroism spectrum of a biopolymer is exquisitely sensitive to its secondary structure.

Linear and circular dichroism are special kinds of absorption spectroscopy, so they occur only at energies where normal absorption occurs. The instruments needed are similar to normal absorption instruments (Figure 9.24), although optical elements are added to produce the polarized light. As a practical matter, it is important to investigate the normal absorption first to ensure that the absorbance of the

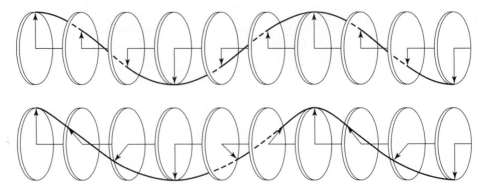

Figure 10.1 The electric vector for linearly polarized (*top*) and circularly polarized (*bottom*) light as a function of distance or time.

solution is not more than 1.0, so that at least 10% of the light is transmitted for the instrument to use in determining the dichroism.

10.1   LINEAR DICHROISM OF BIOLOGICAL POLYMERS

Many biological polymers are long, thin molecules, and can be aligned by stretching or stroking films, flowing samples in solution, or applying an electric field to polymers that are polyelectrolytes or have permanent dipole moments. Even under the most extreme conditions the alignment will not be perfect, so for quantitative interpretation the fraction of alignment must be determined or the measurements extrapolated to perfect alignment. For a helical biopolymer—like an α helix or a polynucleotide helix—the absorption of linearly polarized light parallel to the helix axis is compared to the absorption perpendicular to the helix axis. Linear dichroism is a function of wavelength λ, and is defined as the difference between parallel and perpendicular absorption.

$$LD(\lambda) = A_{\parallel}(\lambda) - A_{\perp}(\lambda) \tag{10.1}$$

The parameter $LD(\lambda)$ depends on the absorbance of the sample, so to compare measurements among various samples it is usually normalized or expressed as the reduced dichroism

$$R(\lambda) = \frac{LD(\lambda)}{A(\lambda)} \tag{10.2}$$

Both electronic and vibrational linear dichroism measurements are common. Figure 9.29 shows a portion of the vibrational absorption spectrum parallel and perpendicular to the orientation axis for a stretched film of poly-γ-benzyl-L-glutamate in the α-helical form. Hydrogen bonds are formed between C$=$O and N$-$H groups in the α helix (Figure 1.23) so that these two groups are oriented roughly along the helical axis. We see that the C=O stretch at about 1652 cm^{-1}, which has its transition dipole along the double bond, shows stronger absorption parallel to the orientation axis, as expected. The NH bend mixed with the CN stretch at 1549 cm^{-1}, which is polarized roughly perpendicular to the NH bond, shows most of its absorption perpendicular to the orientation axis, again as expected.

The electronic linear dichroism of poly-γ-methyl-L-glutamate as an α helix oriented in a stretched film is shown in Figure 10.2. This measurement confirms that the split $\pi\pi^*$ absorption of an α helix has a parallel polarized component at about 208 nm and a perpendicularly polarized component at about 190 nm. We discussed this in detail in Section 9.1.

Linear dichroism measurements can be used to determine the inclination of the bases for polynucleotides in solution. The bases of nucleic acids have both vibrational modes and strong $\pi\pi^*$ transitions with their dipoles in the plane of the bases, as we discussed in Section 9.1. If the base planes are perpendicular to the helix axis, then there is no absorption for light polarized parallel to the helix axis and a maximum absorption for light polarized perpendicular to the helix axis. As the base planes are inclined with respect to the helix axis, there will be a component of the transition dipole along the helix axis. The linear dichroism for each band i will depend on the inclination of the bases α and the angle that the transition dipole for band i makes with the axis of inclination β_i

$$LD_i(\lambda) = 1.5\, SA_i(\lambda)(3\sin^2\alpha \sin^2\beta_i - 1) \tag{10.3}$$

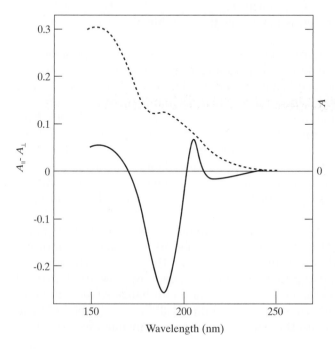

Figure 10.2 The normal isotropic absorption (--) and linear dichroism (—) of poly(γ-methyl-L-glutamate) as an α helix in an oriented film. [Adapted from data in J. Brahms, J. Pilet, H. Damany, and V. Chandrasekharan (1968), *Proc. Natl. Acad. Sci. USA* **60**, 1130–1137.]

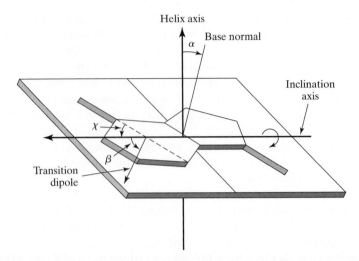

Figure 10.3 Definitions of the parameters in Eq. 10.3 that relate the linear dichroism of a polymer to its isotropic absorption and the microscopic orientation of the base. [Reprinted from X. Jin and W. C. Johnson (1995), *Biopolymers* **36**, 303–312, copyright 1995 by John Wiley & Sons.]

where the orientation factor S varies between zero and one. Figure 10.3 depicts the parameters in this equation. The derivation of this simple equation is very long and difficult, and will not be covered here. For the derivation see Norden (1978), who shows how the relation factors into a macroscopic part, S, that depends only on the orientation of the polymer, and into a microscopic part that depends only on the orientation of the transition dipoles, assuming that the long thin biopolymers behave as chains or rods.

Nucleic acids are polyanions, and electric fields have been one popular way to orient them for electronic linear dichroism measurements. The LD is usually recorded at the absorption maximum of 260 nm (Figure 9.22) as a function of field strength. Let us consider the reduced dichroism at 260 nm for a homogeneous 9200 base pair sample of DNA as a function of the reciprocal of the field shown in Figure 10.4. Extrapolated to infinite field, where presumably we have complete orientation and S is 1.0, the reduced dichroism reaches -1.41. If the bases are perpendicular to the helix axis with $\alpha = 0$, then we see from Eqs. 10.2 and 10.3 that the reduced dichroism at perfect alignment would be -1.5. If $\beta_i = 90°$, then the results in Figure 10.4 correspond to a base inclination of less than 10°.

Suppose we make linear dichroism measurements for a number of transitions. Then Eq. 10.3 applies for each transition i and gives a new piece of information for each transition dipole with a different β_i. The linear dichroism and absorbance for single-stranded poly(rA) in buffer aligned by flow are illustrated in Figure 10.5. These spectra can be decomposed into four bands, so there are four independent equations. With known transition dipole directions (Table 9.2) and only one kind of base, there are three unknowns: the base inclination, the axis of

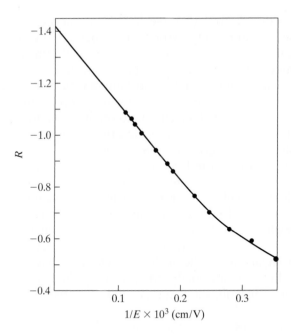

Figure 10.4 The reduced linear dichroism of a 9200 base pair DNA as a function of the inverse field strength, which was used to orient the DNA. [Adapted from data in C-H. Lee and E. Charney (1982), *J. Mol. Biol.* **161**, 289–303.]

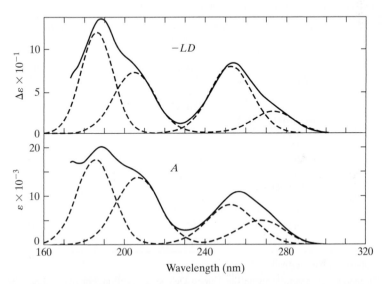

Figure 10.5 The linear dichroism and isotropic absorption of single-stranded poly(rA). [Reprinted from G. C. Causley and W. C. Johnson (1982), *Biopolymers* **21**, 1763–1780, copyright 1982 by John Wiley & Sons.]

inclination, and the orientation factor. All three unknowns can be determined from the equations, so there is no need to extrapolate to perfect alignment. The unknowns are determined from the wavelength variation of the measurement over an extended wavelength range that includes a number of transitions. One advantage of determining the parameters from the wavelength variation of the measurement is that any tertiary structure that would prevent perfect alignment does not affect the analysis. Furthermore, with electric field orientation, the appropriate extrapolation method depends on how much the orientation depends on induced or permanent dipoles. Analysis gives $\alpha = 28°$ for poly(rA).

The flow linear dichroism of B-form DNA in aqueous buffer has been measured, and the variation of LD and A over the 300 to 175 nm range has been analyzed for the average inclination angle and axis of inclination for the four bases, as well as the orientation parameter. The inclination angles are given in Table 10.1, and are larger than we might expect from the electric dichroism measurements. This is because the β_i are not 90°. However, these results are consistent with the electric dichroism measurements. When the angles of inclination, axes of inclination, and transition dipole directions are combined to calculate the reduced dichroism at 260 nm for perfect alignment, the result is -1.4, in agreement with the results in Figure 10.4.

The use of IR linear dichroism to determine base inclinations is discussed in Application 10.1.

Table 10.1   Base Inclinations in Natural Nucleic Acids

Base	Inclination (deg)[a]	Axis of Inclination (deg)[b]
10.4 B-form DNA		
dA	16.1	46.5
dT	25.0	1.8
dG	18.0	114.8
dC	25.1	215.8
10.2 B-form DNA		
dA	14.9	96.6
dT	28.1	31.9
dG	13.9	142.5
dC	27.7	201.2
A-form DNA		
dA	27.8	7.0
dT	34.7	−5.4
dG	14.3	95.3
dC	35.2	216.1

[a]Relative to base normal.

[b]In-plane axis directions using the Tinoco-DeVoe convention. For pyrimidines the angle is positive toward N_3 from the N_1-C_4 reference axis, and for the purines the angle is positive toward N_3 from the C_4-C_5 reference axis.

Application 10.1   Measuring the Base Inclinations in dAdT Polynucleotides

When are the DNA bases perpendicular to the helix axis in a B-form DNA? Baret et al. (1978) measured the vibrational LD for oriented films of poly(dAdT) · poly(dAdT) and poly(dA) · poly(dT) at high humidity and low salt. Under these conditions, the polymers are in the B-form. Figure A10.1 shows the IR transmission for light parallel and perpendicular to the orientation direction for poly(dA) · poly(dT). The results for poly(dAdT) · poly(dAdT) are very similar. For both polymers, the bands at 1696 and 1664 cm^{-1} are vibrations in the plane of dT, while bands at 1626 and 1575 cm^{-1} are vibrations in the plane of dA. The nice separation of bands due essentially to individual functional groups in vibrational spectra is an important feature of this technique. The results indicate an average inclination α of 17 to 25 degrees for the bases in both polymers.

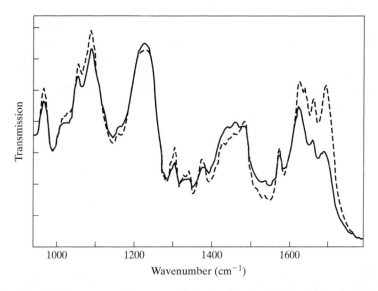

Figure A10.1   The IR transmission for an oriented film of poly(dA) · poly(dT) measured with the electric vector for the light polarized perpendicular to the helix axis (—) and parallel to the helix axis (--). [Adapted from data in J. F. Baret, G. P. Carbone, and P. Penon (1978), *Biopolymers* **17**, 2319–2339.]

BARET, J. F., G. P. CARBONE, and P. PENON (1978), *Biopolymers* **17**, 2319–2339.

10.2   CIRCULAR DICHROISM OF BIOLOGICAL MOLECULES

A solution of randomly oriented molecules will be *optically active* if the molecules are asymmetric. We know that biopolymers are asymmetric, and of course they show optical activity. The optical activity can be seen as the rotation of linearly polarized light due to the difference in refractive index for the two types of circularly polarized light (called *optical rotatory dispersion*) or by the difference in absorption for the two types of circularly polarized light (called *circular dichroism*). The two phenomena are related, and in modern work, circular dichroism is measured because it monitors the effect of absorption bands one at a time. In contrast, optical rotation is more complicated; it measures the combined effect of all the bands that gives rise to a refractive index. Both right- and left-handed circularly polarized light obey Beer's law (Eq. 8.89), and circular dichroism (CD) is defined as the difference in extinction coefficients

$$\Delta A(\lambda) = A_L(\lambda) - A_R(\lambda) = [\varepsilon_L(\lambda) - \varepsilon_R(\lambda)]\ell c = \Delta\varepsilon\ell c \qquad (10.4)$$

where the subscripts denote the type (handedness) of the light. For historical reasons, commercial spectrographs present CD as ellipticity, θ, which relates the measurement to optical rotatory dispersion. Ellipticity in degrees is related to the difference in absorbance by $\Delta A = \theta/32.98$. The corresponding expression for $\Delta\varepsilon$ is molar ellipticity in deg dl/mol dm, $[\theta] = 3298\Delta\varepsilon$, which includes an arbitrary factor of 100. We will view CD as a special kind of absorption spectroscopy and express it as $\Delta\varepsilon$.

To see how circularly polarized light is made, let us consider the optics in Figure 10.6. A linear polarizer P is oriented so the light that emerges has its electric vector at 45° to the y- and z-axes—that is, oriented like the lines on the polarizer. The linearly polarized electric vector can be considered to have two components, E_y and E_z, that are its projections on the y- and z-axes. The linearly polarized light passes through a quarter-wave retarder R, so that E_y is along a principal direction with a

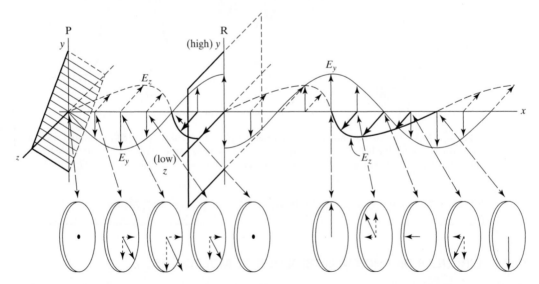

Figure 10.6 The optics for making circularly polarized light uses a linear polarizer P and a quarter-wave retarder R. Circularly polarized light can be decomposed into the sum of two mutually perpendicular, linearly polarized waves that are one-quarter of a wavelength out of phase. With E_y retarded one-quarter of a wave relative to E_z, we have right circularly polarized light as diagrammed here. If E_z were retarded one-quarter of a wave relative to E_y, then the circularly polarized light would be left-handed.

high refractive index and E_z is along a principal direction with a low refractive index. The result is that the two components emerge from the quarter-wave retarder, E_y retarded a quarter of a wave with respect to E_z, as shown in Figure 10.6. The sum of the two linearly polarized waves at the bottom of Figure 10.6 gives an electric vector that rotates as a right-handed helix, similar to Figure 10.1. If E_z were retarded a quarter of a wave with respect to E_y, then left circularly polarized light would be produced.

The difference in absorbance between the two types of circularly polarized light is very small, typically one part in a thousand. Electronic CD is illustrated for *E. coli* DNA in Figure 10.7, where we also see that CD occurs only where normal absorption occurs, and that the CD spectrum is more complicated, revealing bands that are not separated in the normal electronic absorption spectrum. The main reason for this is that CD bands, unlike absorption bands, can be either positive or negative.

In analogy with the dipole strength for normal absorption that we discussed in Chapter 8 (Eq. 8.113), the area under a CD band is related to the *rotational strength*

$$R_{fi} = Im(\boldsymbol{\mu}_{if} \cdot \mathbf{m}_{fi}) \tag{10.5}$$

where *Im* indicates that we must take the imaginary part. The quantity $\mathbf{m}_{fi}$ is the magnetic transition dipole for absorption of the magnetic component of the light

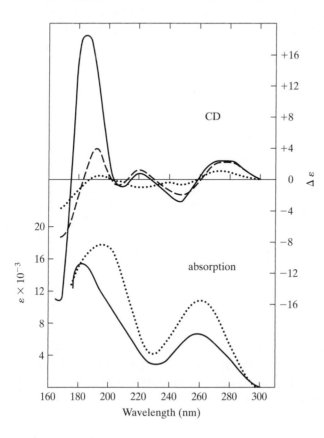

Figure 10.7 The CD of native *E. coli* DNA (—), denatured *E. coli* DNA (– –), and the average CD for the four corresponding deoxynucleotides (···) in aqueous solution. The absorption spectrum of native *E. coli* DNA (—) and the average spectrum for the four component deoxynucleotides (···) illustrates the point that there is only CD where there is normal absorption. [Reprinted from C. A. Sprecher and W. C. Johnson, Jr. (1977), *Biopolymers* **16**, 2243–2264, copyright 1977 by John Wiley & Sons.]

(Figure 8.1), a pure imaginary quantity defined in analogy to the electric transition dipole of Eq. 8.111

$$\mathbf{m}_{fi} = \int \Psi_f^*(q)\hat{\mathbf{m}}\Psi_i(q)d\tau \qquad (10.6)$$

Magnetic transition dipoles are always perpendicular to electric transition dipoles in symmetric molecules, so symmetric molecules have rotational strengths that are identically zero (Eq. 10.5) and thus no CD. The chromophores themselves in biopolymers are symmetric, and the CD bands arise because of interaction between the transition dipoles of a chromophore with transition dipoles in other parts of the molecule. The asymmetric disposition of electric and magnetic dipoles may arise because of the intrinsic asymmetry of an α carbon or sugar, or because of the super asymmetry seen in the secondary structures assumed by biopolymers. We see that interaction between transition dipoles will tell us about structure.

The rotational strength generated in a given transition by the interaction of its transition dipole with a transition dipole in some other part of the molecule is just balanced by the rotational strength of the opposite sign generated in the transition dipole from the other part of the molecule by this same interaction. Since rotational strength in symmetric chromophores is always generated with this pairwise mechanism, there is a rotational strength sum rule that says the sum of all rotational strengths in a CD spectrum must be zero. It is obvious from Eq. 10.5 that an electrically allowed transition in a chromophore, such as a $\pi\pi^*$ transition in a DNA base, can exhibit rotational strength as the scalar product with a magnetically allowed transition in some other part of the molecule, weighted by the strength of their interaction. Similarly, a magnetically allowed transition in a chromophore, such as the $n\pi^*$ transition in an amide, can exhibit a rotational strength because of its interaction with an electronically allowed transition in another part of the molecule, again weighted by the strength of their interaction. The interaction decreases with the distance between the transition dipoles and with the energy difference between the transitions. What is not so obvious is that an electrically allowed transition in a chromophore can exhibit rotational strength due to its interaction with an electronic transition dipole in another part of the molecule, because an electronic transition dipole at a distance has the properties of a magnetic transition dipole. The magnetic dipole operator is the sum not only of the magnetic dipole operators for the individual groups j but also of their electric dipole operators at a distance

$$\hat{\mathbf{m}} = \sum_j \left(\hat{\mathbf{m}}_j + i\pi\mathbf{R}_j \times \frac{\hat{\boldsymbol{\mu}}_j}{\lambda_j} \right) \qquad (10.7)$$

This means that strong electrically allowed transitions in the chromophores of biopolymers, such as the $\pi\pi^*$ transitions of the nucleic acid bases or the $\pi\pi^*$ transitions of the protein amides, can interact because of the super asymmetric structure present in biological secondary structures, and produce CD.

When the interacting transitions fall at quite different wavelengths, their rotational strengths of opposite sign do not overlap. However, when the interacting transitions are degenerate or nearly degenerate, then the resulting rotational strengths of opposite sign largely cancel, giving rise to a characteristic sigmoidal CD curve that we see in Figure 10.8. Because the CD from the degenerate transitions obviously obeys the rotational strength rule, the sigmoidal CD band is called *conservative*. The shorter wavelength CD band due to nondegenerate interactions is often beyond the range of the CD instrumentation, and the rotational strength sum rule is not obviously obeyed, so such CD bands are called *nonconservative*.

Since CD in biopolymers arises from the interaction of transition dipoles, the spectrum depends on the relative orientation of transition dipoles for the various groups in space, and thus the technique is extremely sensitive to conformation. Conservative bands in the CD spectrum of a biological polymer that arise from the degenerate interaction among the electric transition dipoles of the chromophores attest to the presence of secondary structure. Figure 10.9 illustrates this point for poly(rA). The conservative CD of poly(rA) demonstrates that this polynucleotide is helical in solution even though it is single stranded; this was an early experiment proving that hydrophobic interactions are strong enough to create a helical secondary structure in single-stranded polynucleotides. Even the dimer shows a conservative CD (Figure 10.9), indicating some helical base stacking in solution.

The rotational strength of a CD band can be measured experimentally in analogy with the measurement of dipole strength discussed in Section 8.4

$$R = \frac{2.303(3hc)}{32\mathcal{N}_0\pi^3} \int_{\text{band}} \frac{\varepsilon(\nu)}{\nu}d\nu = 0.355 \times 10^{-3} \int_{\text{band}} \varepsilon(\lambda)d(\ln \lambda) \qquad (10.8)$$

We note that the coefficient of the integral is one-fourth of the coefficient for the dipole strength in Eq. 8.116. Care must be taken to resolve the CD bands before integrating over a band, since the bands come in opposite signs and may largely cancel.

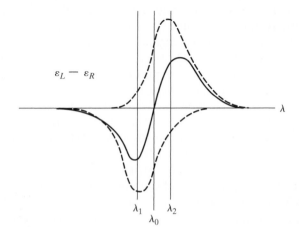

Figure 10.8 Degenerate exciton interaction in a dimer leads to two CD bands of opposite sign that are split slightly in energy. The observed sum of these two bands is a sigmoidal CD with a crossover at the energy of the original monomer transition.

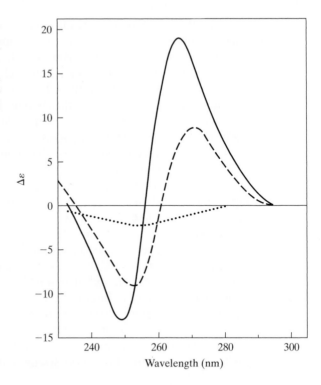

Figure 10.9 The circular dichroism of single-stranded oligo(rA) in aqueous solution at pH 7: polymer (—); dimer (– –); monomer (···). [Adapted from data in K. E. van Holde, J. Brahms, and A. M. Michelson (1965), *J. Mol. Biol.* **12**, 726–739.]

10.2.1 Electronic CD of Nucleic Acids

The circular dichroism of biopolymers is usually measured in the region of electronic absorption. When researchers talk about CD, they mean *electronic CD*. The nucleotides are the building blocks of the nucleic acids, and they have an intrinsic asymmetry due to the chiral sugar. As we expect, the interaction of the strong $\pi\pi^*$ transitions of the chromophoric bases with the higher energy transitions in the sugar yield a CD of low intensity, as shown in Figure 10.10. Formation of a helical structure is a super asymmetry that gives rise to degenerate interactions between the chromophoric bases and results in intense CD spectra, as we have already seen for poly(rA) (Figure 10.9). The CD will be characteristic of the conformation of the nucleic acid, and will depend on base composition, because each base has different transition dipoles. Figure 10.11 compares the CD of *E. coli* DNA in the B-form with 10.4 base pairs per turn, a B-form with 10.2 base pairs per turn, and the A-form. The A-form is similar to the conformation of double-stranded RNA, and is found in many alcohol solvent systems. The 10.4 base pair B-form is the normal conformation of DNA in aqueous solution at moderate salt. The 10.2 base pair B-form is found in aqueous solution at high salt, in solvent systems with a high concentration of methanol, and for DNA wound around histone cores. The structures at A- and B-form DNA are compared in Figure 1.41. We see that the three conformations are easily identified by their very different CD spectra. The positive and negative CD

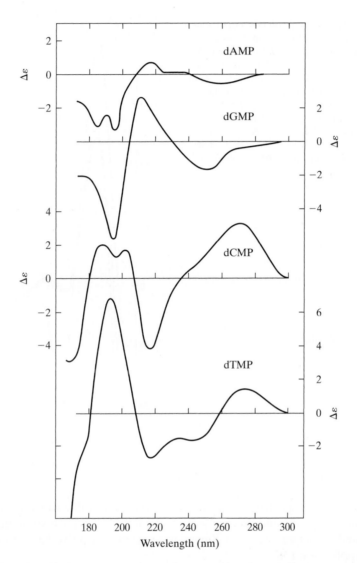

Figure 10.10 CD spectra of the four deoxyribonucleotides that are the monomers for DNA. [Adapted from data in C. A. Sprecher and W. C. Johnson (1977), *Biopolymers* **16**, 2243–2264. Reprinted from W. C. Johnson, "Circular Dichroism and Its Empirical Application to Biopolymers," in *Methods of Biochemical Analysis*, vol. 31, ed. D. Glick, copyright 1985 by John Wiley & Sons.]

couplet at about 280 and 240 nm and the intense positive CD at about 190 nm are characteristic of the 10.4 base pair B-form. Collapse of the long wavelength 280 nm CD band is the hallmark of the 10.2 base pair B-form. The more intense positive CD at 270 nm coupled with a negative CD at 210 nm, and an extremely intense positive CD at 185 nm, are characteristic of the A-form. Double-stranded RNA has a very

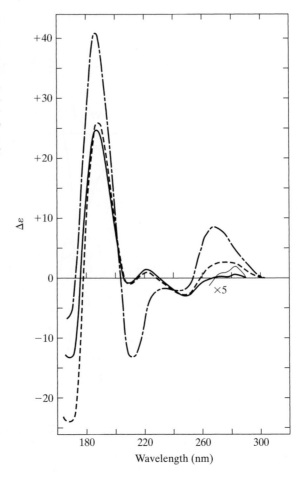

Figure 10.11 The CD of *E. coli* DNA in various secondary structures: 10.4 B-form in aqueous buffer at pH 7.0 (- - -); the 10.2 B-form in 6M NH$_4$F, aqueous buffer at pH 7 (——); the A-form in 80% 2,2,2-trifluoroethanol (·—·—·—). [Adapted from data in C. A. Sprecher, W. A. Baase, and W. C. Johnson (1979), *Biopolymers* **17**, 1009–1019. Reprinted from W. C. Johnson, "Circular Dichroism and Its Empirical Application to Biopolymers," in *Methods of Biochemical Analysis*, vol. 31, ed. D. Glick, copyright 1985 by John Wiley & Sons.]

similar CD spectrum. Poly(dGC)·poly(dGC) will form the left-handed Z-form under certain conditions of salt or alcohol. The Z-form sometimes has a negative band at 290 nm, but its hallmark is an intense negative band at about 195 nm (see Application 10.2).

Application 10.2   The First Observation of Z-form DNA Was by Use of CD

Pohl and Jovin (1972) were the first to observe the left-handed Z-form of poly(dGC)·poly(dGC), and they did this by using circular dichroism spectroscopy. The Z-form CD spectrum is compared with the A- and B-forms in Figure A10.2. Their structures are given in Figure 1.41. In spite of the different base composition, we see that the B- and A-form CD spectra resemble the spectra of *E. coli* DNA in the corresponding

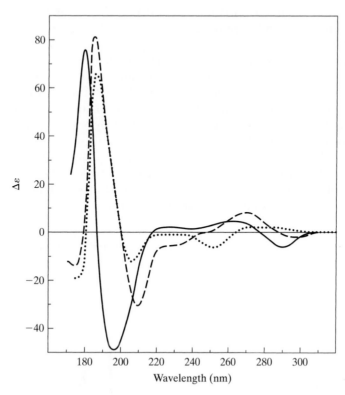

Figure A10.2   The CD of poly d(GC) · poly d(GC) as various secondary structures. The B-form in aqueous buffer at pH 7 (· · ·); the A-form in 80% 2,2,2-trifluoroethanol at pH 7 (– – –); the Z-form in 2M sodium perchlorate at pH 7 (——). [Reprinted from J. H. Riazance, W. A. Baase, W. C. Johnson, K. Hall, P. Cruz, and I. Tinoco (1985), *Nucleic Acids Res.* **13**, 4983–4989, copyright 1985 by Oxford University Press.]

conformation (Figure 10.11). The Z-form exhibits a negative band at 290 nm and a positive band at 260 nm. However, the Z-form is not the mirror image of the B-form, and this apparent reversal of the long wavelength portion of the B-form CD spectrum cannot be trusted to identify the Z-form in all polynucleotides. Instead, the blue shift of the 200 nm crossover of the B-form to about 185 nm in the Z-form appears to be the trademark of the B to Z transition. Since CD spectra have both positive and negative bands, crossovers can be as important as the bands themselves.

POHL, F. M., and T. M. JOVIN (1972) *J. Mol. Biol.* **67**, 375–396.

The fact that CD of nucleic acids depends on composition is clearly seen in the long wavelength CD of B-form poly(dA) · poly(dT) in its native form at low temperature shown in Figure 10.12, compared with B-form *E. coli* DNA (Figure 10.11) or B-form poly(dGC) · poly(dGC) (see Figure A10.2). Figure 10.12 also shows that CD is sensitive to the change in conformation that occurs when poly(dA) · poly(dT) melts with increasing temperature, as we might expect. However, Figure 10.7 demonstrates that the long wavelength bands of a natural DNA do not change much on melting. On the other hand, the 190 nm band changes substantially. In general, it is easier to follow the melting of a natural DNA by observing its hypochromism, as described in Section 9.1.

CD is also convenient for monitoring the changes in conformation of a nucleic acid as a function of solvent. In Figure 10.13, we see how the long wavelength portion of the CD of calf thymus DNA changes as methanol converts it from the 10.4 base pair B-form to the 10.2 base pair B-form.

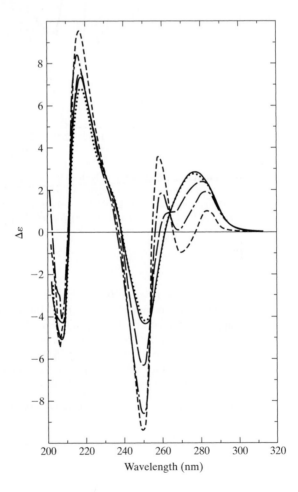

Figure 10.12 The CD of poly(dA) · poly (dT) as a function of temperature: 1.0°C (- - -); 38.8°C (−·−·); 44.7°C (− −); 48.2°C (···); 58.3°C (—). [Adapted from data in J. Greve, M. F. Maestre, and A. Levin (1977), *Biopolymers* **16**, 1489–1504. Reprinted from W. C. Johnson, "Circular Dichroism and Its Empirical Application to Biopolymers," in *Methods of Biochemical Analysis,* vol. 31, ed. D. Glick, copyright 1985 by John Wiley & Sons.]

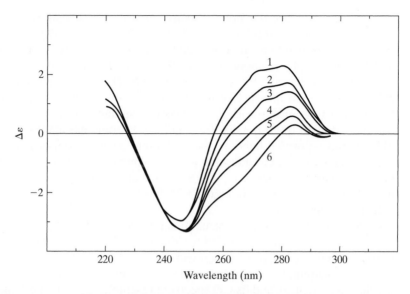

Figure 10.13   The CD of calf thymus DNA in 0.005 standard sodium citrate as a function of methanol: (1) 0%; (2) 25%; (3) 50%; (4) 65%; (5) 75%; (6) 95% methanol. [Adapted from data in J. C. Girod, W. C. Johnson, S. K. Huntington, and M. F. Maestre (1973), *Biochemistry* **12**, 5092–5096. Reprinted from W. C. Johnson, "Circular Dichroism and Its Empirical Application to Biopolymers," in *Methods of Biochemical Analysis*, vol. 31, ed. D. Glick, copyright 1985 by John Wiley & Sons.]

10.2.2 Electronic CD of Proteins

The major chromophore in proteins is the amide group that forms when two amino acids are joined. N-acetyl-L-alanine-N′-methylamide is a model amide with an asymmetric α carbon and free rotation between the two amide groups. This is the spectroscopist's chromophoric monomer for polypeptide studies, and its CD (Figure 10.14) shows a band corresponding to the amide $\pi\pi^*$ at about 195 nm. It is similar to the random coil CD spectrum of collagen shown in the same figure.

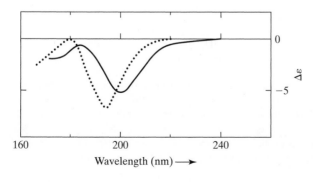

Figure 10.14   The CD of N-acetyl-L-alanine-N′-methylamide ($\cdots$). [Adapted from W. C. Johnson and I. Tinoco (1972), *J. Amer. Chem. Soc.* **94**, 4389–4390.] Collagen at 45°C in aqueous buffer at pH 3.5 (—). [Adapted from D. D. Jenness, C. A. Sprecher, and W. C. Johnson (1976), *Biopolymers* **15**, 513–521.]

We see in Figure 10.15 that the super asymmetric secondary structures formed by polypeptides and proteins have distinctive CD spectra, as expected. Typical of the α helix is a negative band at about 222 nm due to the $n\pi^*$ transition, and a negative and positive couplet at about 208 and 190 nm due to the parallel and perpendicular components of the $\pi\pi^*$ transition, respectively. The magnitude of the negative 222 nm band is a good measure of α helix content in a peptide or protein. There is a fairly linear relationship with $\Delta\varepsilon = 0$ corresponding to 0% α helix and $\Delta\varepsilon = -10$ corresponding to 100% α helix. The CD for a β strand has a negative band at about 215 nm and a positive band at about 198 nm, but the positions and intensities of these two bands vary considerably with the sample. The CD of a polypeptide that is believed to form a series of type II β turns is also shown in Figure 10.15. The structure of the α helix is given in Figure 1.23, the β strand in Figure 1.25, and the type II β turn in Figure 1.27.

Since protein secondary structures have very different CD spectra, we would expect the CD of a native protein to depend on the fractions of its component secondary structures. Figure 10.16 confirms this prediction as we compare the CD spectra of hemoglobin, EcoRI endonuclease, and tumor necrosis factor-α. Hemoglobin is primarily α helical, and its CD spectrum resembles that of an α helix, as shown in

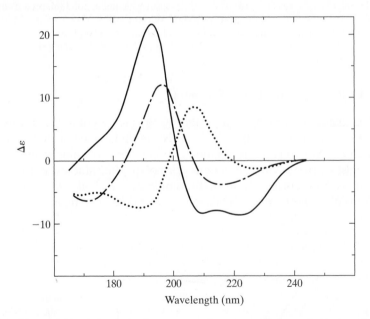

Figure 10.15 CD spectra for various secondary structures in aqueous solution: poly (L-glutamic acid) at pH 4.5 as an α-helix (—). [Adapted from W. C. Johnson and I. Tinoco (1972), *J. Amer. Chem. Soc.* **94**, 4389–4390.] Poly (L-lysine-L-leucine) in aqueous solution at pH 7 as an antiparallel β-sheet (–·–·). [Adapted from S. Brahms, G. Spach, and A. Brack (1977), *Proc. Natl. Acad. Sci. USA* **74**, 3208–3212.] Poly (L-alanine$_2$-glycine$_2$) in aqueous solution at pH 7 ($\cdots$) as a β-turn. [Adapted from S. Brahms, G. Spach, and A. Brack (1977), *Proc. Natl. Acad. Sci. USA* **74**, 3208–3212.]

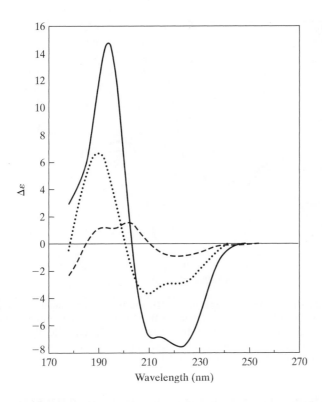

Figure 10.16 The CD of proteins hemoglobin (—), EcoRI endonuclease ($\cdots$), and tumor nectrosis factor-α (– –).

Figure 10.15. Tumor necrosis factor-α has no α helix and a fair amount of β strand. EcoRI has both α helix and β strand, and an intermediate CD spectrum.

Like nucleic acids, changes in the conformation of polypeptides and proteins due to melting, solvent change, or any other variation are easily followed with CD spectroscopy. For instance, poly(L-proline) II is normally in a 3_1 helix (see Figure 1.24), where the prolines are all *trans* across the amide bond. This single-stranded collagen-like structure is more stable than poly(L-proline) I, where the prolines are all *cis*. CD can be used to follow the slow kinetics of the transition between poly(L-proline) I and II, as we see in Figure 10.17.

In another example, Figure 10.18 shows the change in the peptide systemin as a function of temperature. Systemin is interesting because it is the only known hormone-like peptide in plants. When a plant is wounded, production of systemin starts the plant's defense systems. The CD at 5°C in Figure 10.18 demonstrates that systemin has a preference for the 3_1 helix, when we compare it with Figure 10.17. The preference for a 3_1 helix is further confirmed by the CD spectra in Figure 10.18, because the secondary structure melts out to a random coil as the temperature is raised.

Calmodulin is a protein whose interaction with a number of other proteins as well as pharmacological agents depends on calcium concentration. CD easily monitors the secondary structure of calmodulin in solution, and we see that its

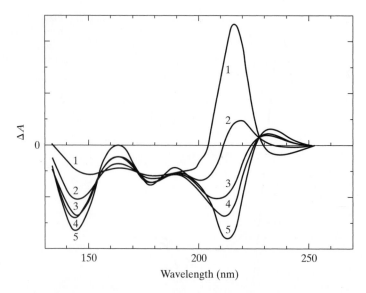

Figure 10.17 Poly(L-proline)I converts to poly(L-proline)II (the typical left-handed 3_1 helix) over time. (1) poly(L-proline)I at 3 min, (2) 88 min, (3) 179 min, (4) 291 min, (5) poly(L-proline)II at 52 hr. [Adapted from M. A. Young and E. S. Pysh (1975), *J. Amer. Chem. Soc.* **97**, 5100–5103. Reprinted from W. C. Johnson, "Circular Dichroism and Its Empirical Application to Biopolymers," in *Methods of Biochemical Analysis,* vol. 31, ed. D. Glick, copyright 1985 by John Wiley & Sons.]

conformation depends on ionic strength (Figure 10.19). We will discuss calmodulin in more detail in the next chapter that covers emission spectroscopy.

    CD was an important technique used in early studies of histones. Until the mid-1970s, scientists believed that histones coated DNA much the way insulation coats a wire. Today, we know that two copies of four types of histones (H2a, H2b, H3, and H4) form an octamer quaternary structure, similar to the quaternary structures found for larger enzymes. The DNA wraps about one and three-fourths turns around the histone quaternary structure, much the way thread is wrapped around a spool. CD was a key in determining that the histones interacted with each other, and thus behaved like complex enzymes capable of forming a quaternary structure. Changes in secondary structure often involve changes in α helix, yielding changes in the CD such as we have already seen for calmodulin (Figure 10.19). The largest change is at about 220 nm, so the CD at this wavelength can be used to monitor changes in the amount of α helix. In a series of classic experiments, D'Anna and Isenberg (1974) used CD to demonstrate that histones interact. We see their results for histone H3 interacting with H4 in Figure 10.20. If there were no interaction, the CD at 220 nm for a mixture would be the sum of contributions from H3 and H4 individually, the straight dotted line in the figure. If an interaction causes a change in α helix, the CD at 220 nm will vary in a nonlinear fashion, as observed experimentally. The maximum in the *continuous variation curve* at 50% H3 and 50% H4 demonstrates the formation of a dimer.

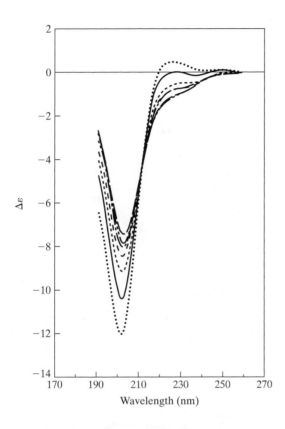

Figure 10.18   CD spectra of systemin in 10 mM phosphate buffer measured at 5°C ($\cdots$), 20°C (—), 40°C (---), 60°C (– –), 70°C (—·), 80°C (—··), and 85°C (– –). Systemin has a significant fraction of 3_1 helical structure at low temperature. [Reprinted from A. Toumadje and W. C. Johnson (1995), *J. Amer. Chem. Soc.* **117**, 7023–7024, copyright 1995 by American Chemical Society.]

10.2.3 Singular Value Decomposition and Analyzing the CD of Proteins for Secondary Structure

Theoretical mathematics can appear far removed from the practical world of experimental science. However, the power of mathematical theorems is that they are abstract, which means that they can be applied in many different practical situations. A case in point is the *singular value decomposition* (SVD) theorem, which has many uses when analyzing a set of experimental data. As we shall see, SVD reveals the number of independent variables in a set of data, is used to average among the data in a set, and monitors the experimental variation in the data as the variables are changed. We will illustrate this with a calcium titration of a polynucleotide. We will also see how the powerful SVD theorem solves many of the problems inherent in analyzing the CD of a protein for secondary structure.

Most experimentally measured data, and certainly spectra of various kinds, can be digitized into a vector as we have already shown in Figure 8.7 and described in the accompanying text. If the data are spectra, then an experiment usually consists of a set of data such as the infrared spectra of hydrogen bonded association as a function of concentration (Figure 9.28), the absorption spectrum of DNA as a

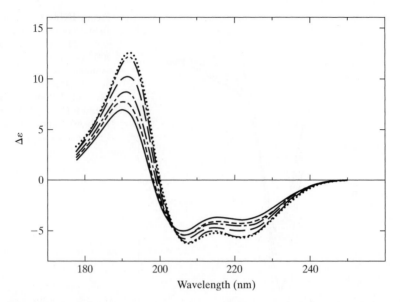

Figure 10.19 The CD of calmodulin with no calcium present at ionic strengths of 0.00 (—), 0.0075 (– –), 0.025 (–·–), 0.065 (– –), 0.115 (–··–), and 0.165 (···). [Reprinted from J. P. Hennessey, P. Manavalan, W. C. Johnson, D. A. Malencik, S. R. Anderson, M. I. Schimerlik, and Y. Shalitin (1987), *Biopolymers* **26**, 561–571, copyright 1987 by John Wiley & Sons.]

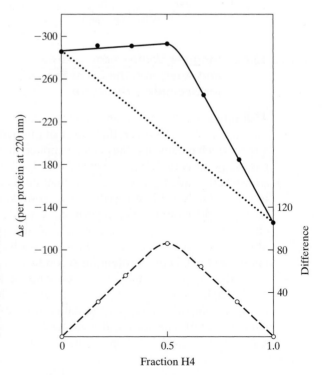

Figure 10.20 Continuous variation curve demonstrates that histones H3 and H4 interact to form a dimer. [Adapted from J. A. D'Anna and I. Isenberg (1974), *Biochemistry* **13**, 4992–4997. Reprinted from W. C. Johnson, "Circular Dichroism and Its Empirical Application to Biopolymers," in *Methods of Biochemical Analysis,* vol. 31, ed. D. Glick, copyright 1985 by John Wiley & Sons.]

function of temperature, or the CD of DNA as a function of solvent (Figure 10.13). A set of such spectra, digitized as vectors, can each be a column in a larger array of digitized data called a *matrix*. The SVD theorem says that *any* matrix **A**, and certainly a matrix of experimental data, can be decomposed into a product of three matrices.

$$\mathbf{A} = \mathbf{U}\mathbf{S}\mathbf{V}^{\mathrm{T}}$$ (10.9)

where the superscript T means the *transpose* of matrix **V**; that is, the columns of matrix **V** are written as rows in matrix **V**^{T}. Each of these three new matrices reveals important properties of the data that are useful to the experimentalist, as we will see below.

Of course it is not easy to decompose a given matrix **A** into a product of three matrices. However, we now have fast, inexpensive computers, so such a decomposition has become practical. Most statistical programs contain an SVD subroutine. A FORTRAN version of an SVD subroutine can be found in Forsythe et al. (1977).

The matrix **U** has columns that resemble the data that are column vectors in the original matrix **A**. The SVD theorem states that **U** is orthogonal and unitary, so that the column vectors can be considered a complete orthonormal set of basis vectors for the data, in the same sense that we have already discussed for nondegenerate perturbation theory in Chapter 8. The SVD theorem further states that the matrix **S** has entries only on the main diagonal, called singular values, and zeros elsewhere. The zeros ensure that each singular value in **S** corresponds to and multiplies only one particular column basis vector in **U**. Finally, the SVD theorem states that the matrix **V**^{T} is orthogonal and unitary. These are the coefficients that generate the original matrix **A** from the singular values and their corresponding basis vectors in **U**.

In order to see how the SVD theorem works for a simple transition, we will analyze 15 circular dichroism spectra for the calcium titration of poly d(Gm^{5}C) · poly d(Gm^{5}C), some of which are shown in Figure 10.21. The original raw data (smoothed to give the spectra in Figure 10.21) are digitized to form the matrix **A** in Figure 10.22. A computer decomposition of this matrix **A** results in the SVD matrices shown in Figure 10.23. Now we can see the practical value of each of these three matrices.

The singular values in the **S** matrix are arranged according to their magnitude, and we see that the first two (Figure 10.23) at 133.1 and 40.58 are much larger than the remaining singular values that range between 8.73 and 4.83. The vectors in **U** and **V**^{T} are normalized (unitary), so the singular values determine the importance of their corresponding vectors. Figure 10.24 shows the three most important basis vectors multiplied by their singular values. Only the first two vectors in **U** and **V**^{T} corresponding to the first two singular values will be important in reconstructing the original matrix **A** when performing the multiplication **USV**^{T}. Therefore, we can conclude that there are only two significant parameters (variables) in the original data, even though there is no obvious isosbestic point (Figure 10.21). Put another way, all the data in Figure 10.21, which have been digitized into matrix **A**, have an *information content* of two and correspond to only two independent equations of data. The remaining small singular values and their vectors correspond to noise, and if the data were perfect these singular values would be identically zero. The **S** matrix

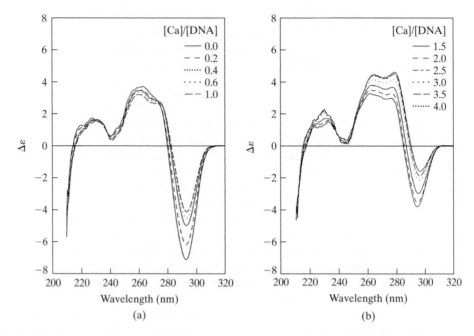

Figure 10.21 CD spectra for the calcium titration of poly d(Gm^{5}C) · poly d(Gm^{5}C) on a nucleotide basis. [Reprinted from L. Zhong and W. C. Johnson (1990), *Biopolymers* **30**, 821–828, copyright 1990 by John Wiley & Sons.]

obtained from SVD is important to us because it gives the number of independent parameters above the noise in a set of data.

We can set the insignificant singular values in **S** equal to zero to eliminate the contribution of the basis vectors that are attributed to noise. We can then regenerate the data matrix through Eq. 10.8 by using only the two significant singular values and their corresponding basis vectors and coefficients. This procedure averages among the data in matrix **A** even though they all have different shapes. This *inter*spectral smoothing removes noise at each wavelength that is uncorrelated

	r	0.0	0.1	0.2		4.0
nm						
210		−4.68	−4.59	−4.20	—	4.48
211		−2.97	−3.55	−3.21	—	−3.47
212		−2.24	−1.80	−1.77	—	−1.51
213		−1.33	−0.29	−1.43	—	−1.86
214		−0.32	0.28	−0.05	—	−0.34
—		—	—	—	—	—
—		—	—	—	—	—
320		0.03	−0.02	−0.11	—	−0.04

Figure 10.22 Measured CD spectra for the calcium titration of poly d(Gm^{5}C) · poly d(Gm^{5}C) in matrix form as **A**. The ratio r is calcium ions to nucleotides.

$$
\begin{array}{c|ccccc}
 & 1 & 2 & 3 & & 15 \\
\hline
210 & -0.145 & -0.018 & 0.330 & \text{—} & -0.056 \\
211 & -0.106 & -0.041 & 0.230 & \text{—} & 0.062 \\
212 & -0.067 & -0.008 & 0.274 & \text{—} & -0.064 \\
213 & -0.036 & -0.013 & 0.085 & \text{—} & -0.116 \\
214 & -0.001 & -0.018 & 0.021 & \text{—} & 0.097 \\
\text{—} & \text{—} & \text{—} & \text{—} & \text{—} & \text{—} \\
320 & -0.000 & -0.000 & -0.002 & \text{—} & 0.015 \\
\end{array}
\times
\begin{array}{c|ccccc}
 & 1 & 2 & 3 & & 15 \\
\hline
1 & 133.1 & 0 & 0 & \text{—} & 0 \\
2 & 0 & 40.58 & 0 & \text{—} & 0 \\
3 & 0 & 0 & 8.73 & \text{—} & 0 \\
\text{—} & \text{—} & \text{—} & \text{—} & \text{—} & \text{—} \\
15 & 0 & 0 & 0 & \text{—} & 4.83 \\
\end{array}
\times
\begin{array}{c|cccccc}
 & 0.0 & 0.1 & 0.2 & & 4.0 \\
\hline
1 & 0.32 & 0.31 & 0.29 & \text{—} & 0.23 \\
2 & -0.36 & -0.30 & -0.25 & \text{—} & 0.51 \\
3 & 0.29 & 0.21 & 0.44 & \text{—} & 0.18 \\
\text{—} & \text{—} & \text{—} & \text{—} & \text{—} & \text{—} \\
15 & 0.07 & -0.19 & 0.10 & \text{—} & 0.36 \\
\end{array}
$$

Figure 10.23 Measured CD spectra for the calcium titration of poly d(Gm⁵C) · poly d(Gm⁵C) decomposed by SVD into the product of the three matrices $\mathbf{U}$, $\mathbf{S}$, and $\mathbf{V}^{\mathrm{T}}$. Computer decomposition gives the useful parts of the matrices, not the full unitary $\mathbf{U}$ and $\mathbf{V}^{\mathrm{T}}$.

Figure 10.24 The three most important basis CD spectra from the raw data in **US**. [Reprinted from L. Zhong and W. C. Johnson (1990), *Biopolymers* **30**, 821–828, copyright 1990 by John Wiley & Sons.]

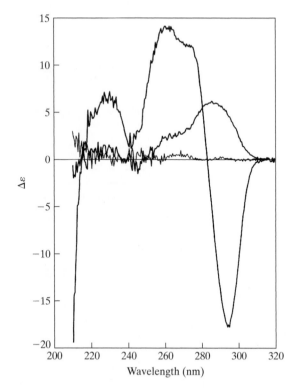

among the experimental spectra. The results for some of the data are shown in Figure 10.25. These results look noisier than the spectra presented in Figure 10.21 because those spectra were further smoothed by averaging data in a single spectrum over a range of wavelengths. The method typically used is Savitzky and Golay smoothing, which fits a wavelength range to a polynomial, and Fourier transform smoothing, which eliminates high-frequency sinusoidal components in the data. Noise that is uncorrelated in each spectrum can be removed by this *intra*spectral smoothing.

The **U** matrix generated by SVD is important to us, because its basis vectors are the shapes that are common among the data (Figure 10.24). The basis vectors in **U** that correspond to significant singular values can be used to average among the varying data.

The $\mathbf{V}^T$ matrix from SVD contains the least-squares coefficients that fit the vectors in **US** to generate the original data in **A** (Eq. 10.9). The common features in the data are given by the most significant basis vector. The most changeable features in the data above the noise are given by the least important basis vector that is still significant. One way to plot a titration curve obtained through spectroscopic measurements is to choose a particular wavelength (say, 294 nm in Figure 10.21) where the spectrum varies rapidly with the titration parameter ([Ca]/[DNA] in Figure 10.21). However, it would be much more accurate to make use of the entire

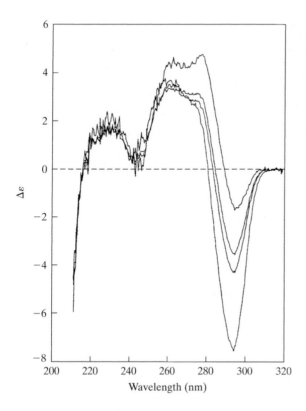

Figure 10.25 Interspectral smoothing of the raw data for four representative CD spectra resulting from SVD reconstruction using the two most important basis CD spectra. [Reprinted from W. C. Johnson (1992), *Methods in Enzymology* **210**, 426–447, ed. L. Brand and M. L. Johnson, copyright 1992 by Academic Press.]

spectrum rather than just a single wavelength, and the coefficients in $\mathbf{V}^T$ allow us to do this when there are only two parameters. Figure 10.26 plots the first three row vectors of $\mathbf{V}^T$ as a function of [Ca]/[DNA]. We see that the coefficients for the most important basis vector decrease slightly at the beginning of the titration and then remain the same. The coefficients for the second most significant basis vector form a titration curve that takes the entire spectrum into account. The coefficients for the third most significant basis vector are scattered, confirming that the third basis vector is not significant above the noise. The $\mathbf{V}^T$ matrix from SVD is important to us because it yields a titration curve that takes into account all the wavelengths in the spectra.

It should be clearly understood that the SVD basis vectors do *not* correspond to the spectra of species found in solution. These are merely an orthonormal basis for the vector space. Spectra for the individual species must be found by some other means, such as creating conditions where only a single species exists.

The simplest way to analyze the CD spectrum of a protein for secondary structure is to fit the CD spectra of pure secondary structures (Figures 10.14 and 10.15) to the measured CD spectrum for the protein. However, this procedure has a number of problems. First, there may be features other than secondary structure that contribute to the CD, such as aromatic side chains, disulfide bonds, twists in the β sheets, and the length of the α helices. Second, the CD spectra may not really

Figure 10.26 The three row vectors in $\mathbf{V}^{\mathrm{T}}$ corresponding to the first $\circ$, second $\bullet$, and third $\triangle$ most important basis CD spectra. [Reprinted from L. Zhong and W. C. Johnson (1990), *Biopolymers* **30**, 821–828, copyright 1990 by John Wiley & Sons.]

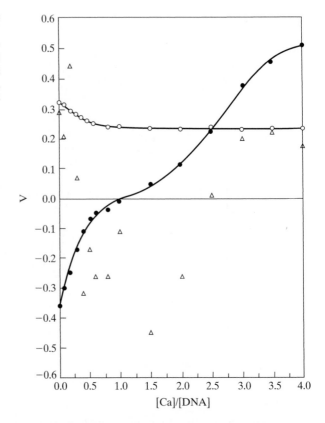

correspond to pure secondary structures. For instance, the antiparallel β sheet in Figure 10.15 may also contain β turns. Indeed, the many different antiparallel β-sheet CD spectra that have been measured are all somewhat different in the magnitude and position of the bands, although they all show a negative band around 210 nm and a positive band around 195 nm. Third, there are many types of β turns, but only a few CD spectra are available. Finally, the amides that do not fall into a particular secondary structure are also not a random coil, and their CD need not correspond to the CD of unordered structures as shown in Figure 10.14.

An alternative is to use the CD spectra of proteins with structures known from X-ray diffraction or NMR measurements to analyze the CD of proteins with unknown structure. A basis set of CD spectra containing proteins with known structure will have all the features that contribute to the CD, whether we recognize them or not. This is a pattern-recognition approach, where the CD patterns that correspond to the secondary structures are deduced from the basis set. SVD can be used to find the patterns, but first we will use SVD on the basis set to determine the information content in the CD data.

Hennessey and Johnson (1981) measured the CD spectra of 16 proteins to 178 nm, yielding data as in Figure 10.16. The secondary structures for each of these

16 proteins are known from X-ray diffraction. The data matrix can be decomposed in the normal way by an SVD subroutine to yield the $\mathbf{U}$, $\mathbf{S}$, and $\mathbf{V}^T$ matrices. The first six singular values are 116.8, 24.4, 14.9, 6.7, 4.0, and 2.7. These singular values decrease in a regular way, so it is not immediately clear how many are significant. However, if we regenerate the data matrix $\mathbf{A}$ with various numbers of singular values, we see that the first five singular values make a significant difference in the shape and magnitude of the CD curves that is above the noise level, but that further singular values make an undetectable difference in the regenerated spectra that is below the noise level of the data. This is shown in Figure 10.27 for papain. Thus, we conclude that the basis set of protein CD spectra has an information content of five, and corresponds to five independent parameters or equations. Five equations will only solve for five unknowns, so even CD data of proteins measured to 178 nm is minimal. If the basis spectra are truncated at 190 nm, then the data correspond to an information content of three or, perhaps, four.

In addition to the matrix of protein CD spectra, we can also form a matrix digitizing the fraction of the various kinds of secondary structure for each protein. We have the fractions of helix (H), antiparallel β sheet (A), parallel β sheet (P), the three most common types of β turn (type I, type II, and type III), the remaining types of β turn, and the amides that do not fall into these categories of secondary structure, which we call *other* (O). This is already eight unknowns and we have not

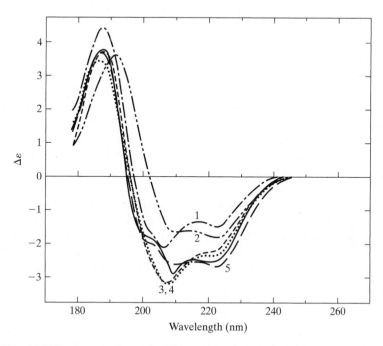

Figure 10.27 An SVD reconstruction of the CD spectrum of papain (——) by using various numbers of basis CD spectra for proteins. The numbers indicate how many basis CD spectra were used in the construction. [Adapted from J. P. Hennessey and W. C. Johnson (1981), *Biochemistry* **20**, 1085–1094.]

considered the other factors that contribute to the CD. However, the power of SVD is that it is completely general; we can apply SVD to the matrix of protein structure in the same way that we applied it to the matrix of protein CD data. SVD demonstrates that this matrix of secondary structure has an information content of only four, so it contains only four variables rather than eight. This gives us hope that we can successfully analyze the CD spectra of proteins measured to 178 nm, even though it has an information content of only five.

We will not use SVD to find the CD patterns that correspond to secondary structures, but instead implicitly use the pattern-recognition approach to solve the fundamental problem directly. We want a matrix $\mathbf{X}$ that will transform protein CD spectra into secondary structure. That is, we want the matrix $\mathbf{X}$ that is the solution to the matrix equation

$$\mathbf{F} = \mathbf{XA} \tag{10.10}$$

where $\mathbf{F}$ is the matrix of protein secondary structure. Combining the fraction of β turns under the symbol T, this matrix is shown for the 16 proteins in Figure 10.28. Making use of SVD, Eq. 10.10 can be rewritten

$$\mathbf{F} = \mathbf{XA} = \mathbf{X}(\mathbf{USV}^{\mathrm{T}}) \tag{10.11}$$

The SVD theorem indicates that the inverse of $\mathbf{U}$ is $\mathbf{U}^{\mathrm{T}}$ and the inverse of $\mathbf{V}^{\mathrm{T}}$ is $\mathbf{V}$. Therefore, we can write the transformation matrix $\mathbf{X}$ in the simple form

$$\mathbf{X} = \mathbf{F}(\mathbf{VS}^{\dagger}\mathbf{U}^{\mathrm{T}}) \tag{10.12}$$

where $\mathbf{S}^{\dagger}$ is the pseudoinverse of $\mathbf{S}$; the singular values on the main diagonal of $\mathbf{S}$ are replaced by their inverse to create $\mathbf{S}^{\dagger}$. Equation 10.12 is a particularly simple solution for matrix $\mathbf{X}$. We know the matrices $\mathbf{F}$ and $\mathbf{A}$, and we can use a singular value decomposition subroutine to create $\mathbf{U}$, $\mathbf{S}$, and $\mathbf{V}^{\mathrm{T}}$. The parameter $\mathbf{S}^{\dagger}$ is simply created using the inverses of the singular values. The parameters $\mathbf{V}$ and $\mathbf{U}^{\mathrm{T}}$ are simply created by writing the columns of $\mathbf{V}^{\mathrm{T}}$ and $\mathbf{U}$ as rows. The result for the 16 proteins is given in Figure 10.29. Multiplying the matrix $\mathbf{X}$ times the digitized CD spectrum of a protein with unknown structure will produce a vector that predicts the secondary structures.

One problem in using a basis set of 16 protein CD spectra that have an information content of five is that the analysis is overdetermined. There will be many combinations of the 16 basis protein CD spectra that will fit the measured CD spectrum of the protein with unknown structure equally well. However, SVD once again comes to the rescue. If we use only the five significant singular values in $\mathbf{S}$ and

		α-chy	cyto	elas	flav	
	H	0.10	0.38	0.10	0.38	—
	A	0.34	0.00	0.37	0.00	—
	P	0.00	0.00	0.00	0.24	—
	T	0.20	0.17	0.22	0.16	—
	O	0.36	0.45	0.31	0.22	—

Figure 10.28   The matrix containing the fractions of secondary structure for the 16 proteins.

	178	180	182	184	—	260
H	−0.0	3.0	2.2	−0.2	—	0.0
A	−19.9	−19.1	−14.4	−8.7	—	0.0
P	−18.9	−13.4	−7.0	−2.7	—	0.0
T	−13.9	−11.7	−9.2	−7.1	—	0.0
O	−42.0	−25.3	−12.2	−7.0	—	0.0

Figure 10.29 The matrix $\mathbf{X} \times 10^3$ (in units of $1/\Delta\varepsilon$) that will transform protein CD spectra into fractions of secondary structure.

set the remaining singular values equal to zero, then the problem will no longer be overdetermined. Essentially, this truncates $\mathbf{U}$ as the first five columns and $\mathbf{V}^T$ as the first five rows. The price we pay is that with truncated matrices, $\mathbf{U}^T$ will no longer be the exact inverse for $\mathbf{U}$, and $\mathbf{V}$ will no longer be the exact inverse of $\mathbf{V}^T$. However, we will obtain the best solution possible in the least-squares sense.

This method considers all the other features that contribute to the CD of a protein, even if we do not consider them explicitly in $\mathbf{F}$, because their effect is in the basis set. For instance, adding the aromatic contributions to $\mathbf{F}$ does not change the analysis for secondary structure.

The secondary structure for many proteins can be predicted successfully from their CD using this method. Three such predictions are given in Table 10.2. However, some CD spectra of proteins are not analyzed successfully; negative fractions of secondary structure may be predicted or the sum of secondary structures may not be 1.0. This arises because an information content of five is really insufficient to solve for all of the features that determine a protein CD spectrum. *Variable selection* is a statistical method to get around the problem of insufficient information content. The idea is to choose only the proteins in the basis set that have the same features as the protein being analyzed. Of course, it is impossible to know a priori which basis proteins to choose. In practice, it is found that analysis with various combinations of just eight basis CD spectra works best. We see which combinations give positive fractions of secondary structure, the amount of α helix predicted by the magnitude of the 222 nm band, and a sum of secondary structures that equals 1.0. This variable selection technique improves the success of the analyses. A further improvement introduced by Sreerama and Woody (1993) is to include the CD spectrum of the protein to be analyzed in the data matrix $\mathbf{A}$, and a guess at its secondary structure in matrix $\mathbf{F}$. The solution is then iterated until the fractions of secondary structure for the protein being analyzed are self-consistent.

One way to test the accuracy of the method is to analyze the CD spectrum for each protein in the basis set with the CD spectra of the remaining proteins. When SVD, variable selection, and self-consistency are combined, we obtain the average errors given in Table 10.3.

Table 10.2 Predictions of Secondary Structure from CD Spectra for Three Proteins

Protein	H	A + P	T	O	Total
11S Acetylcholinesterase	0.34	0.21	0.17	0.27	0.99
Interferon	0.59	0.16	0.18	0.13	1.06
*Eco*RI	0.33	0.25	0.17	0.25	1.00

Table 10.3   RMS Differences Between Predicted and X-Ray Structures Combining SVD, Variable Selection, and Self-Consistency

H	A + P	T	O
0.042	0.042	0.042	0.057

10.2.4 Vibrational CD

The CD can be measured for any type of transition, and instrumentation is now commercially available to measure *vibrational CD*. Vibrational CD (VCD) spectra have the advantage over electronic CD in that the bands are well separated and correspond to specific functional groups. VCD has been measured in the amide I and II region corresponding to the C=O stretch and N—H bend/C—N stretch modes (1750 to 1550 cm^{-1}) for both nucleic acids and proteins.

Figure 10.30 shows the vibrational absorption and CD for d(CG)$_{10}$ · d(CG)$_{10}$ in both the right-handed B-form and the left-handed Z-form. Although the usual vibrational absorption for the two conformations is very similar, the VCD is quite different. The VCD spectra for the base deformation modes of calf thymus DNA in both the A- and the B-forms have also been measured. Although the intensities of the two VCD spectra differ, in this case the shape of the VCD spectrum is not particularly characteristic of the conformation.

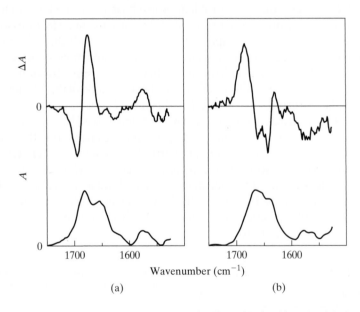

Figure 10.30   The vibrational CD and absorption of d(GC)$_{10}$ · d(GC)$_{10}$ (a) as the right-handed B-form, and (b) as the left-handed Z-form. [Adapted from T. A. Keiderling, S. C. Yasui, P. Pancoska, R. K. Dukor, and L. Yang (1989), *Biomol. Spec.* **1057**, 7–14.]

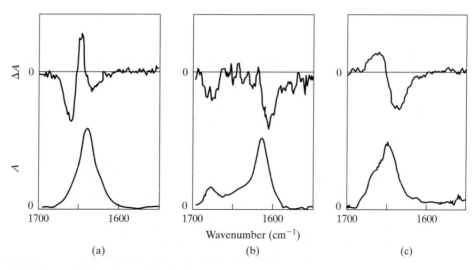

Figure 10.31   The vibrational CD and absorption of poly(L-lysine) as an (a) α helix, (b) β strand, and (c) random coil. [Adapted from T. A. Keiderling (1990) *Practical Fourier Transform Infrared Spectroscopy*, 203–284, Academic Press.]

Protein secondary structures also yield characteristic VCD spectra. Figure 10.31 shows the absorption and VCD spectra for model α helix, β strand, and random coil structures. The vibrational absorption spectra are quite different, as we expect from the discussion of vibrational absorption in Section 9.2. The secondary structures also display characteristic VCD spectra. We would expect VCD to be sensitive to the secondary structure in proteins, and this is seen to be true in Figure 10.32. Because each secondary structure has a characteristic vibrational absorption spectrum and CD spectrum, both types of protein spectra can be successfully analyzed for the fractions of component secondary structure.

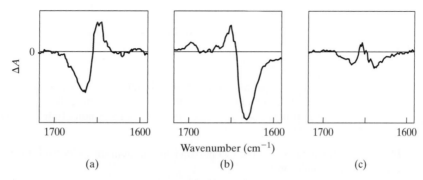

Figure 10.32   The vibrational CD of (a) myoglobin, (b) chymotrypsin, and (c) lysozyme. [Adapted from P. Pancoska, S. C. Yasui, and T. A. Keiderling (1989), *Biochemistry* **28**, 5917–5923.]

EXERCISES

10.1 If the reduced dichroism (R) for single-stranded poly (rA) at perfect alignment is -1.25 at 260 nm and the bases are inclined 28°, what is the relative angle between the axis of inclination and the transition dipole (β)?

10.2 Assume poly (dC) has the base inclination and axis of inclination found for dC as the 10.4 B-form in Table 10.1. Use the transition dipole directions from Table 9.2 to calculate the R due to the 271 nm band.

10.3 What properties of the CD spectrum of native DNA in Figure 10.7 would lead you to believe that it is helical? Compare the native CD to the CD spectrum of denatured DNA and the average spectrum of the monomers. What can you say about the denatured DNA?

10.4 Nucleosomes, which contain a length of about 150 base pairs of DNA wrapped about a protein (histone) core, have been studied by thermal denaturation, using both absorbance at 260 nm and CD. The following data were recorded as a function of temperature:

	Percent Increase in Absorbance	$\Delta\varepsilon$	
T (°C)	at 260 nm	At 273 nm	At 223 nm
20	0	+0.10	
30	0	+0.10	−11.8
35	0	+0.15	−11.8
40	0	+0.20	−11.8
45	0	+0.27	−11.8
50	0	+0.45	−11.8
55	2	+0.55	−11.8
60	6	+0.67	−11.8
65	8	+0.82	−11.2
70	12	+1.06	−7.9
75	26	+1.45	−5.5
80	33	+1.36	−4.5
85	37	+1.36	−4.2

It has been shown that only the DNA absorbs above 250 nm, and that the protein dominates the CD at 223 nm. Describe, in as much detail as possible, what happens to a nucleosome when it is heated. Under the same solvent conditions, free DNA melts at 42°C.

10.5 The following data list the CD maxima and minima for β-lactoglobulin A under two different conditions:

$\lambda_{max, min}$	$\Delta\varepsilon$
In aqueous buffer, pH 5	
215	−1.7
196	+2.6
In 99% ethanol, 0.01 M HCl	
220	−7.0
208	−8.0
192	+15.5

Describe qualitatively what happens to β lactoglobulin when it is transferred into acidic ethanol solution.

10.6 Assume that the data for synthetic polypeptides shown in Figures 10.14 and 10.15 describe accurately the contributions of α helix, β sheet, and random coil to the CD of a protein. Pick three wavelengths that would discriminate most sensitively among these forms. Show how you might analyze the data for an unknown protein in terms of CD values at these three wavelengths.

REFERENCES

General

TINOCO, I., JR., K. SAUER, J. C. WANG, and J. D. PUGLISI (2002) *Physical Chemistry Principles and Applications in Biological Sciences,* 4th ed., Prentice Hall, Upper Saddle River, NJ.

Linear Dichroism

NORDEN, B. (1978) "Applications of linear dichroism spectroscopy," *Applied Science Reviews* **14**, 157–248.

NORDEN, B., M. KUBISTA, and T. KURUCSEV (1992) "Linear Dichroism Spectroscopy of Nucleic Acids," *Quarterly Review of Biophysics* **25**, 51–72.

SAMORI, B. and E. W. THULSTRUP (eds.) (1988) *Polarized Spectroscopy of Ordered Systems,* Kluwer Academic Publishers, Dordrecht, The Netherlands.

SAUER, K. (ed.) (1995) *Biochemical Spectroscopy. Methods in Enzymology,* vol. 246, Academic Press, New York.

Circular Dichroism

D'ANNA, J. A., JR. and I. ISENBERG (1974) "A Histone Cross-Complexing Pattern," *Biochemistry* **13**, 4992–4997.

FASMAN, G. D. (ed.) (1996) *Circular Dichroism and the Conformational Analysis of Biomolecules,* Plenum Press, New York.

FORSYTHE, G. E., M. A. MALCOLM, and C. B. MOLER (1977) *Computer Methods for Mathematical Computations,* Prentice Hall, Upper Saddle River, NJ.

HENNESSEY, J. P., JR. and W. C. JOHNSON, JR. (1981) "Information Content in the Circular Dichroism of Proteins," *Biochemistry* **20**, 1085–1094.

JOHNSON, W. C., JR. (1985) "Circular Dichroism and Its Empirical Application to Biopolymers," in *Methods of Biochemical Analysis,* vol. 31, ed. D. Glick, John Wiley & Sons, New York.

JOHNSON, W. C., JR. (1999) "Analyzing Protein Circular Dichroism Spectra for Accurate Secondary Structures," *Proteins: Structure, Function, and Genetics* **35**, 307–312.

MANAVALAN, P. and W. C. JOHNSON, JR. (1987) "Variable Selection Method Improves the Prediction of Protein Secondary Structure from Circular Dichroism Spectra," *Anal. Biochem.* **167**, 76–85.

NAKANISHI, K., N. BEROVA, and R. W. WOODY (eds.) (2000) *Circular Dichroism: Principles and Applications,* 2nd ed., John Wiley & Sons, New York.

SAUER, K. (ed.) (1995) *Biochemical Spectroscopy. Methods in Enzymology,* vol. 246, Academic Press, New York.

SREERAMA, N. and R. W. WOODY (1993) "A Self Consistent Method for the Analysis of Protein Secondary Structure from Circular Dichroism," *Anal. Biochem.* **209**, 32–44.

CHAPTER **11**

Emission Spectroscopy

In discussing absorption, we have been concerned entirely with the excitation of a molecule from its ground state to a higher energy level and have given no consideration to the subsequent fate of the excited molecule. In many cases, the sequel is not very interesting; the energy is transferred as heat to the surroundings. However, in some cases light is emitted by the sample either as fluorescence or phosphorescence. We will discuss the general characteristics of emission and ignore the exceptions.

Fluorescence spectroscopy has proved particularly valuable in investigating biological molecules. It can be used to probe dynamic processes of excited electronic states, visualize samples, measure distances in biological structures, and follow the progress of reactions. Polarized fluorescence can be used to determine the relative directions of transition dipoles, investigate the shape of tertiary structures, and measure the molecular weight of biopolymers. The application of fluorescence spectroscopy to the study of single molecules will be discussed in Chapter 16.

11.1 THE PHENOMENON

Fluorescence is emission from a *singlet* state, where the electron spins in the molecule are paired. Most molecules in their ground electronic state are singlets, and we see this for the hydrogen molecule or a bond in the orbital approximation in Figure 11.1. We remember that for two electrons the space and spin parts of the wavefunction factor in Eq. 8.80. The space part of the ground state (with the spins denoted by the directions of the arrows) is a single configuration, as cartooned in Figure 11.1. It must have an antisymmetric spin part, as we saw for the ground state of the hydrogen molecule with Eqs. 8.79 and 8.80. Also qualitatively illustrated here is the electronically excited singlet, where in the orbital approximation an electron has absorbed light and occupies a higher energy orbital. There are two choices for such an excitation: either the electron with spin up or the electron with spin down can be excited. The excited

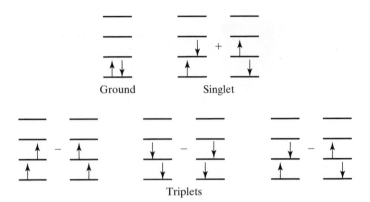

Figure 11.1 Diagrams for the space parts of the ground state, first singlet state, and first triplet states (three) for the hydrogen molecule. The signs are for the space parts with the spins denoted by the direction of the arrows.

singlet is the normalized sum of these two possibilities giving a symmetric space part of the wavefunction, as cartooned in Figure 11.1. The spin part must be antisymmetric as we saw for the hydrogen molecule in Eqs. 8.82 and 8.83.

Phosphorescence is emission from a *triplet* excited state. Although it is a rare event and thus has a low intensity, excitation to a triplet involves changing the orientation of a spin. There are three possibilities, and that is why we call this excited state a triplet. Two of the possibilities are obvious; the two unpaired electrons can either have both their spins up or both spins down. For the hydrogen molecule this would be the symmetric spin parts $\alpha(1)\alpha(2)$ or $\beta(1)\beta(2)$. Not so obvious is the third possibility, for a hydrogen molecule a symmetric spin part that is the sum of paired spin states, $\alpha(1)\beta(2) + \alpha(2)\beta(1)$. The corresponding space part must be antisymmetric as illustrated for the three possibilities in Figure 11.1, so that the wavefunction is antisymmetric.

11.2   EMISSION LIFETIMES

Following excitation by absorption of energy, a molecule loses energy as heat by cascading through the closely spaced vibronic levels of all the excited singlets taken together (*internal conversion*). The molecule hesitates at the lowest vibrational level of the first singlet, because a relatively large amount of energy must be lost to go to the ground state. During this hesitation, one of three things may happen: the molecule may fluoresce and drop to the ground state, emitting a photon; it may convert to the triplet (*intersystem crossing*); or it may go to the ground state without emitting a photon (a *nonradiative* transition). If the molecule converts to the triplet, it again hesitates before it either phosphoresces and drops to the ground state, emitting a photon, or undergoes a nonradiative transition.

Lifetimes for all these processes are illustrated in Figure 11.2 for the various states at the equilibrium nuclear distance of the ground state, in keeping with the Franck-Condon principle. We see that absorption is quite rapid, believed to take the time for a wavelength of light to pass the molecule, about 10^{-15} sec. The molecule is in its ground electronic and vibrational state at room temperature. Absorption may be to any vibrational state of any excited singlet—here we have arbitrarily chosen the third vibrational level of the second singlet. The loss of vibrational energy and transfer of electronic energy that bring the molecule to the ground vibrational level of the first excited electronic state are also rapid, with a lifetime of about 10^{-12} sec. The lifetime of the molecule in the ground vibrational state of the first excited singlet before fluorescing is fairly long, and at about 10^{-8} sec is comparable to the lifetime for losing this energy in a nonradiative manner or the lifetime for conversion of

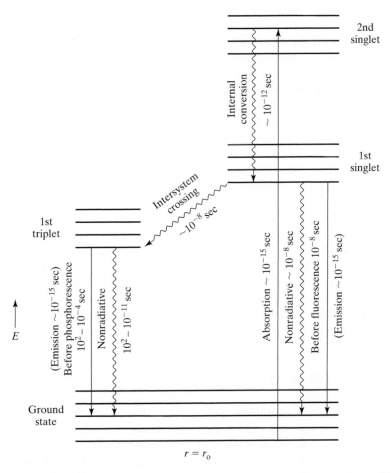

Figure 11.2   Emission lifetimes of absorption, fluorescence, and phosphorescence at the equilibrium internuclear distance of the ground state.

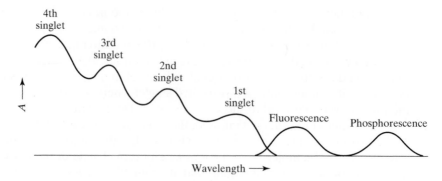

Figure 11.3   The absorption and emission spectra of a hypothetical molecule.

the first singlet to the first triplet. These three processes compete and, as we expect, the one with the shortest lifetime for a particular type of molecule will be the one that prevails. If nonradiative loss of energy is rapid for the type of molecule in question, then no emission will be seen. If the molecule emits light relatively rapidly, we can observe fluorescence. We will see that fluorescence is sensitive to dynamic processes occurring during this excited state lifetime. If the molecule converts to the triplet relatively rapidly, then there is the possibility of seeing phosphorescence. Once in the triplet state, the energy may be lost in a nonradiative manner or the molecule may emit a photon, depending on the corresponding lifetimes. Emission itself is as rapid as absorption, about 10^{-15} sec.

Figure 11.3 illustrates absorption and emission spectra on the same graph for a hypothetical molecule. Electronic absorption bands are shown that correspond to four transitions in the singlet spectrum. Triplet absorption is improbable, and such bands have very little intensity. The fluorescence emission is at longer wavelength than the singlet absorption, but there is a slight overlap of the fluorescence band with the band corresponding to the first singlet. The first triplet is generally at lower energy than the first singlet, and the corresponding phosphorescence emission is shown at an even longer wavelength. Although fluorescence is a common spectroscopic technique for studying biological molecules, their phosphorescence is rarely investigated and we will not discuss this type of emission further.

11.3   FLUORESCENCE SPECTROSCOPY

In order to understand the applications of fluorescence spectroscopy to solving biological problems, the reader must keep firmly in mind that for all practical purposes the process always takes place from the ground vibrational level of the first excited singlet. This simplifying fact is the key to understanding the results. We can see at once why the fluorescence falls at longer wavelength than the first singlet absorption but with a small overlap between the bands. As Figure 11.4 shows, absorption to the

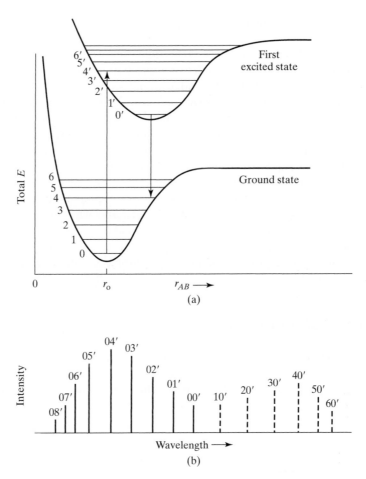

Figure 11.4   (a) The ground and first singlet electronic states of a bond with the vibrational levels. The arrows denote the most probable absorption and fluorescence. (b) The vibronic bands for the corresponding absorption ($00'-08'$) and fluorescence ($00'-60'$) are also illustrated.

first singlet is from the lowest vibrational level (quantum number 0) of the ground state to any vibrational level of the excited state. The lowest possible energy of absorption will be to the ground vibrational level (quantum number $0'$) of the first excited singlet. Excitation to higher vibrational levels of the first excited singlet require more energy and occur at shorter wavelength. Fluorescence from the ground vibrational level of the first excited singlet may be to any vibrational level of the ground electronic state, but in contrast, fluorescence to the ground vibrational level of the ground electronic state emits the most energy. Fluorescence to any other vibrational level will involve less energy and occur at a longer wavelength. Thus the fluorescence band is always at longer wavelength than absorption to the first excited singlet. The overlap between the bands occurs because both processes have the $0-0'$

band in common; the lowest energy of absorption is the same as the highest energy of fluorescence.

We will see fluorescence if this emission process has a lifetime that is shorter than the conversion to the triplet or nonradiative loss of energy. This will be related to the Einstein coefficient for induced emission B and spontaneous emission A as discussed in Section 8.4. For electronic transitions, the probability of induced emission will be small compared to the probability for spontaneous emission, which in turn is related to the probability for induced emission by Eq. 8.98. Indeed, the probability for spontaneous emission is related to the integrated intensity of an absorption band by Eq. 8.102, so that the stronger the absorption of the first singlet, the higher the probability for fluorescence. If the first singlet is a strong $\pi\pi^*$ transition, as it is for the side chains of tyrosine and tryptophan amino acids, then the lifetime for spontaneous emission is about 10^{-9} sec and we will see fluorescence. On the other hand, if the first excited singlet is a weak $n\pi^*$ transition, then the corresponding lifetime will be about 10^{-6} sec and we will not see spontaneous fluorescence because some other process will be faster. Fluorescent molecules are often referred to as *fluorophores*.

11.4 FLUORESCENCE INSTRUMENTATION

Fluorescent light is emitted from samples in an excited electronic state. Fluorescence instruments use light to excite the sample and observe the fluorescence at right angles (Figure 11.5). A monochromator before the sample chooses the

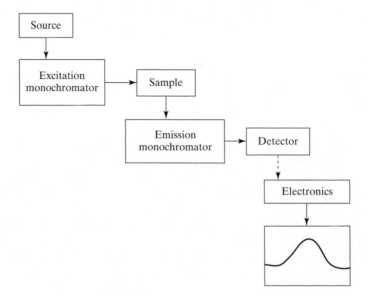

Figure 11.5   Diagram of a simple fluorescence instrument.

wavelength of exciting light, and a monochromator after the sample scans the various wavelengths of emitted light. The shape of the fluorescence spectrum as scanned by the emission monochromator will be independent of the wavelength of absorption because excited molecules decay to the ground vibrational level of the first singlet before fluorescing. The absorption and fluorescence spectra plotted in Figure 11.3 are obtained independently. There is a fluorescence spectrum for each absorption wavelength; the absorption can be scanned while observing fluorescence at some wavelength.

11.5   ANALYTICAL APPLICATIONS

Fluorescence intensity F is useful to biochemists in observing the presence of a macromolecule. For instance, biopolymers emerging from a high pressure liquid chromatograph (HPLC) might be monitored with a fluorescence detector. We see in Figure 11.6a the results for the isolation of the peptide melittin from honey bee venom on an HPLC. Melittin has a tryptophan residue that can be excited at 280 nm to fluoresce over a range of wavelengths that include 340 nm. Two peptides are eluted from the HPLC that have 340 nm fluorescence, one at 5.9 min and one at 11.7 min. The fraction at 11.7 min has melittin activity (hemolysis of red blood cells). Monitoring the elution with normal UV absorbance at 214 nm (Figure 11.6b) reveals a number of other fractions that contain biological molecules, but these do not have to be considered because they do not fluoresce. By using fluorescence for analysis, identification of the proper fraction is simplified.

Let us consider another example, in which fluorescence was used to detect the protein calmodulin and a derivative of calmodulin with its two tyrosines at positions 99 and 138 photochemically cross-linked through prolonged irradiation with UV light. When excited at 280 nm, the tyrosines in native calmodulin have a large fluorescence at 300 nm, while the dityrosine derivative has a large fluorescence at 400 nm. Figure 11.7 shows the elution profile for a mixture containing both native and cross-linked calmodulin run through a phenyl-agarose affinity column that has been equilibrated with a buffer containing 1 mM $CaCl_2$. Fluorescence detection at 400 nm shows that the dityrosine derivative elutes around fraction 85, while native calmodulin elutes along with the cross-linked dimer around fraction 124, but only on the subsequent application of 2 mM EDTA to remove bound calcium. A contaminant that fluoresces at both 300 and 400 nm elutes around fraction 38.

Green fluorescent proteins (GFPs) are very useful probes. The best characterized is from a Pacific Northwest jellyfish, *Aequorea victoria,* and has 238 amino acids that wrap around the green fluorescing fluorophore. The GFP is often attached to another protein using genetic techniques. The DNA code for GFP is spliced beside the code for the target protein, and the DNA code for both proteins is expressed as a single long amino acid sequence. The two proteins in this *fusion* sequence usually fold independently, and the proteins usually retain their function and behave independently. As an example, Greenwood and coworkers (2003) fused GFP with

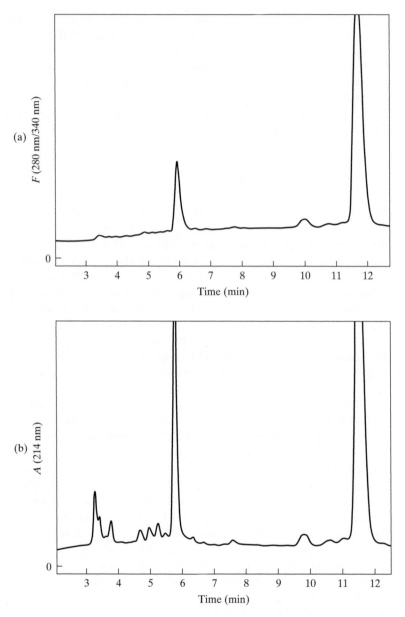

Figure 11.6 The fluorescence (a) and electronic absorption (b) for a melittin preparation emerging from a high-pressure liquid chromatograph. The melittin elutes at 11.7 min. [Courtesy of S. R. Anderson.]

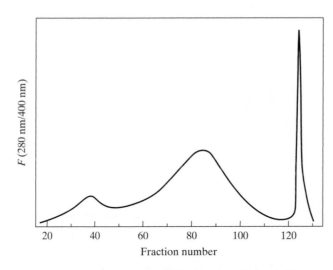

Figure 11.7 The fluorescence detected (400 nm) elution profile for a mixture containing both native and derivative calmodulin emerging from a phenyl-agarose column. The dityrosine derivative elutes around fraction 85, while native calmodulin and a cross-linked dimer elute around fraction 124. An unidentified contaminant elutes around fraction 38. [Adapted from data in D. A. Malencik and S. R. Anderson (1987) *Biochemistry* **26**, 695–704.]

α-actinin to visualize phosphoinositide inhibition of α-actinin activity. α-Actinin causes actin in cells to bundle and to adhere to the cell membrane, which is necessary for the actin to function as part of the internal structure of the cell. Phosphoinositide controls α-actinin activity, and in normal cells the GFP through its fluorescence allows visualization of normal actin bundles. However, when the α-actinin has been mutated so that it no longer binds phosphoinositide, GFP visualization shows out-of-control tangled bundling in cells, since there is no phosphoinoside control.

11.6   SOLVENT EFFECTS

Fluorescence falls at longer wavelengths than the first singlet absorption because of the vibrational levels, as we saw in Section 11.3. In addition, solvent effects will affect the position of the fluorescence band. The effects of a solvent on fluorescence can be very large, and many studies of macromolecules use this fact. *General solvent effects* depend on the polarizability of the solvent, and increasing the dielectric constant usually shifts the fluorescence to longer wavelength. *Specific solvent effects* are the result of chemical reaction of the excited state with the solvent. Important chemical reactions include hydrogen-bonding, acid-base chemistry, and the formation of charge-transfer complexes where an electron in the fluorophore is transferred to another group on excitation.

Let us consider specific solvent effects first. They often accompany general solvent effects, because solvents with a large polarizability are usually capable of hydrogen bonding. Specific solvent effects occur when the solvent reacts chemically with the fluorophore, so only a small concentration of the chemically reacting solvent is necessary to bring the effect to completion. In addition, the new species often has a new characteristic fluorescent band. An example is the hydrogen bonding

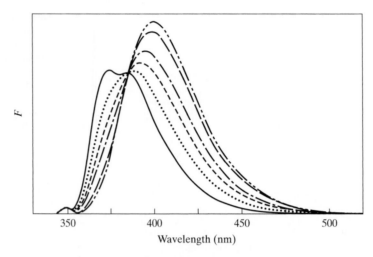

Figure 11.8 Trace amounts of ethanol hydrogen bond with 2-anilinonaphthalene. The fluorescence of 2-AN is shown in cyclohexane to which ethanol was added at 0% (—), 0.2% (···), 0.4% (– – –), 0.7% (– · – · –), 1.7% (——), and 2.7% (– · · –). [Adapted from data in Brand et al. (1971) in *Probes of Structure and Function of Macromolecules in Membranes*, ed. B. Chance et al., 17–39.]

reaction of 2-anilinonaphthalene (2-AN) with ethanol. Figure 11.8 shows that trace amounts of ethanol added to a cyclohexane bulk solvent give rise to a new fluorescent band at about 400 nm, which is presumably due to the 2-anilinonaphthalene-ethanol hydrogen-bonded complex.

General solvent effects involve the interaction of the permanent dipole moment of the molecule in both the ground and excited states with the reactive field induced in the surrounding solvent. The reactive field has two parts, the immediate reaction of the electrons of the solvent molecules and the slower reorientation reaction of the solvent molecules as a whole due to their own permanent dipole moment. We see this effect for 2-AN in Figure 11.9. Ethanol at 3% saturates the specific solvent effect and creates a new band for the hydrogen-bonded 2-AN species at about 400 nm. The large permanent dipole moment of water relative to the cyclohexane-ethanol mixture shifts the fluorescence of the hydrogen-bonded species to about 448 nm because of general solvent effects.

The *Lippert equation* predicts the energy shift for general solvent effects assuming that solvent reorientation reaches equilibrium before emission. It is a first approximation that considers fluorescence in a solvent continuum with a refractive index n and a dielectric constant ε that is unitless and relative to ε_0 ($D = 4\pi\varepsilon\varepsilon_0$). Let us consider Figure 11.10. A fluorophore in its ground state will have a permanent dipole moment μ and see an electron reactive field R_e and a reorientation reactive field R_r. Immediately after excitation to the first singlet (about 10^{-15} sec), the fluorophore will have a new (usually greater) dipole moment $\mu^{\dagger}$ and see a new electron reactive field $R_e^{\dagger}$. Reorientation of the solvent molecules due to their dipole moment is comparatively slow, so the reorientation reactive field is still R_r. However, over the

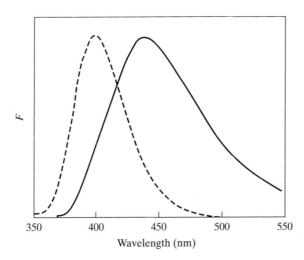

Figure 11.9 The large permanent dipole moment of water (—) relative to the cyclohexane-ethanol mixture (– – –) shifts the fluorescence of the 2-AN hydrogen-bonded species because of general solvent effects. [Adapted from data in Brand et al. (1971) and M. G. Badea, R. P. DeTomma, and L. Brand (1978) *Biophys. J.* **24**, 197–212.]

period of time the molecule remains in the excited state (about 10^{-8} sec), the solvent molecules are assumed to reorient due to their own dipole moment, so the reorientation reactive field becomes $R_r^\dagger$. This stabilizes the excited state, lowering its energy. Emission is rapid (again about 10^{-15} sec). Upon losing a photon the fluorophore returns to its ground state dipole moment and electron reactive field, but the reorientation reactive field is still $R_r^\dagger$, raising the energy of this species. With time, the ground state molecule returns to the initial conditions.

These reactive fields depend on the dipole moment of the fluorophore and the polarizability of the solvent P. The high frequency or electron polarizability is a function of the refractive index n.

$$P(n) = \frac{n^2 - 1}{2n^2 + 1} \tag{11.1}$$

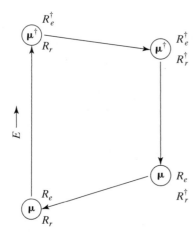

Figure 11.10 The electronic and reorientation reaction fields of a solvent have a general effect on the energy of a dipole.

The low-frequency polarizability, which includes molecular dipole reorientation as well as the electron polarizability, is a function of the relative dielectric constant.

$$P(\varepsilon) = \frac{\varepsilon - 1}{2\varepsilon + 1} \tag{11.2}$$

The polarizability due to dipole reorientation is the difference between Eqs. 11.2 and 11.1.

$$\Delta P = P(\varepsilon) - P(n) \tag{11.3}$$

The reactive fields are then given by

$$R_e = \frac{2\mu P(n)}{a^3}$$

$$R_e^\dagger = \frac{2\mu^\dagger P(n)}{a^3}$$

$$R_r = \frac{2\mu \Delta P}{a^3} \tag{11.4}$$

$$R_r^\dagger = \frac{2\mu^\dagger \Delta P}{a^3}$$

where a is the radius of the cavity occupied by the fluorophore.

    The energy shift due to general solvent effects will depend on the energy of the fluorophore dipole in the solvent, and is given by the magnitude of the dipole and the reactive field induced in the solvent by the dipole, $E = -\mu R$. We have for the energy of light absorption in the solvent,

$$E_a = E_v - \mu^\dagger R_r - \mu^\dagger R_e^\dagger - (-\mu R_r - \mu R_e) \tag{11.5}$$

where the first term is the energy of absorption for the vapor phase, and the remaining terms are the corrections for the solvent due to the reactive fields. Similarly for fluorescence from the excited state, we have

$$E_f = E_v^\dagger - \mu^\dagger R_r^\dagger - \mu^\dagger R_e^\dagger - (-\mu R_r^\dagger - \mu R_e) \tag{11.6}$$

The energy correction on fluorescence due to general solvent effects is the difference between absorption and fluorescence energies

$$
\begin{aligned}
E_a - E_f &= -\mu^\dagger(R_r - R_r^\dagger) + \mu(R_r - R_r^\dagger) \\
&= (\mu^\dagger - \mu)(R_r^\dagger - R_r) \\
&= \frac{(\mu^\dagger - \mu)(2\mu^\dagger \Delta P - 2\mu \Delta P)}{a^3} \\
&= \frac{2\Delta P(\mu^\dagger - \mu)^2}{a^3}
\end{aligned}
\tag{11.7}
$$

where we have made use of Eq. 11.4 and assumed that the energy of absorption and fluorescence are the same in the vapor phase. This is the Lippert equation, and we see that it depends on the reactive field of the solvent dipoles, but not on the reactive field of the solvent electrons. That is, it depends on the time-dependent solvent relaxation that reorients the solvent dipoles because the permanent dipole moment of the fluorophore changes in the excited state. According to the Lippert equation, the energy shift is proportional to ΔP, and linear plots of shift versus ΔP are taken as evidence that general solvent effects dominate.

11.7   FLUORESCENCE DECAY

Excitation decays by a first-order process, which can be used to relate lifetimes to probability. As we expect for a first-order decay, the number of molecules losing a quantum of energy in time dt will be proportional to the number of molecules N in the excited state. This relationship is given by

$$-\frac{dN(t)}{dt} = N(t)k \tag{11.8}$$

where k is the constant of proportionality and the probability that the molecule will lose a quantum through all processes in time dt. Rearranging the equation and integrating both sides,

$$\int_{N(0)}^{N} \frac{dN(t)}{N(t)} = \int_{0}^{t} -k\,dt \tag{11.9}$$

and we obtain

$$\ln N(t) - \ln N(0) = \ln\frac{N(t)}{N(0)} = -kt$$

$$N(t) = N(0)e^{-kt} \tag{11.10}$$

We define the *lifetime* τ as the time it takes for $N(0)$ molecules to decay to $N(t) = N(0)/e$. This occurs at time $t = 1/k = \tau$. The lifetime is measurable because the intensity of the emitted light at time t, $I(t)$, is proportional to the number of molecules in the excited state at that time. Figure 11.11 shows the fluorescence after a sample has been excited with a very short pulse of light. The intensity of emitted light decreases with time in a simple exponential manner, as we expect for a first-order process with a single lifetime. The lifetime is simply the time it takes for the maximum intensity to decrease by the factor $1/e$.

If all the quanta of energy are lost through spontaneous fluorescence, $k = A$, the Einstein coefficient (Section 8.4), and $\tau = 1/A$. This special lifetime, when all competing processes can be ignored, is called the *intrinsic lifetime* τ_0. Thus, A is

Figure 11.11 The fluorescence intensity of a sample after it has been excited by a very short pulse.

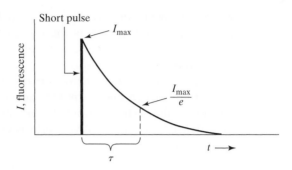

proportional to the number of molecules giving off photons through spontaneous fluorescence. The quantity k is proportional to the number of molecules losing quanta of energy through all processes. The ratio of these two numbers is the *quantum yield, q*

$$q = \frac{A}{k} = \frac{\tau}{\tau_0} \qquad (11.11)$$

We have already seen how to measure τ (Figure 11.11). We can calculate τ_0 from Eq. 8.102 for A because $A = 1/\tau_0$. Thus, we can determine the quantum yield for a given transition. This is a useful quantity because changes in the quantum yield are often diagnostic of changes in the molecular environment of a chromophore and can be used as a sensitive test for such a variation. To take a simple example: the quantum yield of a fluorophore will often increase when the molecule is taken up from solution and bound to a macromolecule such as DNA or a protein. Measurement of the quantum yield provides a simple way to measure binding (see Section 14.3).

For most biological investigations it is not important to know absolute quantum yields. We can avoid the measurement of lifetimes by simply comparing quantum yields, as shown in Figure 11.12. This example shows that the intrinsic tyrosine fluorescence of the calcium-binding protein calmodulin loses intensity when the calcium is removed by EDTA. This loss of intensity is called *quenching.* Comparing the areas under the curves, the *relative* quantum yield of calcium-free calmodulin is 50% of calmodulin with its sites saturated by bound calcium. Relative quantum yields can be made more quantitative by comparison to a standard. For instance, the fluorescence measurements of the tyrosine in calmodulin (Figure 11.12) might be compared to the fluorescence of free tyrosine. Calmodulin with its calcium sites saturated shows 46% of the fluorescence of free tyrosine, while calcium-free calmodulin shows 23%.

We remember that the quantum yield of a substance is the ratio of the total number of quanta emitted to the total number of quanta absorbed from Eq. 11.11. These quantities are in turn related to the integrated area under the fluorescence and absorption bands. However, in practice we usually measure the fluorescence

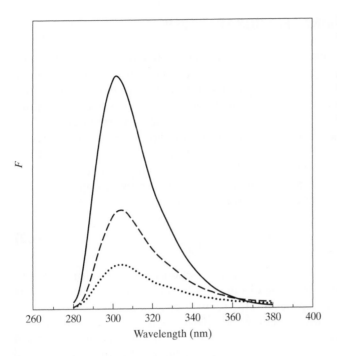

Figure 11.12 Fluorescence spectra for free tyrosine (——), the tyrosine in calmodulin with its calcium sites saturated (---), and the tyrosine in calcium-free calmodulin ($\cdots$). The relative quantum yield with free tyrosine as the basis are 46% for calcium-saturated calmodulin, and 23% for calcium-free calmodulin. [Courtesy of S. R. Anderson.]

intensity at a single wavelength; this quantity is also called the *fluorescence yield*. The fluorescence yield F will depend on the concentration of the fluorophore, the shape of the fluorescence spectrum, and the wavelength chosen for observation, as well as instrumental parameters. In contrast to normal absorbance, which is a ratio of intensities, $\log (I_0/I)$, quantitative interpretation of the fluorescence yield requires an absolute standard. Often, fluorophores are compared using the ratio of their fluorescence yields. As we see in Figure 11.13, this *relative* fluorescence yield will depend on the wavelength chosen for observation because of the shapes of the fluorescence spectra.

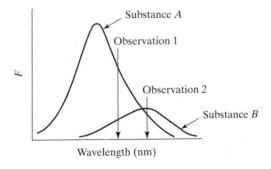

Figure 11.13 The relative fluorescence yield is defined for a single wavelength and depends on the wavelength chosen for observation.

11.8   FLUORESCENCE RESONANCE ENERGY TRANSFER

In our discussion of fluorescence so far, we have assumed that the fluorophore emitting the radiation is the same group as the absorber. This need not be the case; under favorable circumstances, excitation energy can be transferred from one fluorophore to another. This *fluorescence resonance energy transfer* or FRET is one of the more useful quenching mechanisms in experimental fluorescence. The requirements are (1) transition dipole interaction between the two fluorophores, and (2) an appreciable overlap of the fluorescence spectrum of the donor with the absorption spectrum of the acceptor. The requirement of dipole-dipole interaction between the fluorophores leads to a strong dependence of energy transfer on the distance between the participating groups. The efficiency of transfer is given by

$$\text{efficiency} = \frac{1}{1 + (r/R_0)^6} \tag{11.12}$$

where we assume that the transition dipole interaction is the same for measurements of both r and R_0. Here R_0 is the distance for 0.5 efficiency of transfer, and is characteristic of the donor-acceptor pair, the relative orientation of their transition dipoles, and the medium between them. As we see in Figure 11.14, dependence on the sixth power of the separation r means that the efficiency goes from zero to 1.0 very quickly in the region around R_0. Since R_0 is ordinarily found to be between 1 nm and 10 nm, FRET serves as a useful yardstick for the distances between groups in macromolecules such as proteins. Tyrosine and tryptophan groups in proteins often satisfy both requirements for transfer, so FRET is a common phenomenon in proteins that can be used for investigating their structure. In addition, specific amino acids can be labeled with a fluorophore such as dimethyl aminonapthalene-5-sulfonate (DNS), and FRET used to measure the distance between the dye and other aromatic groups in the protein.

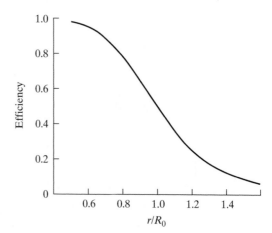

Figure 11.14   The efficiency of energy transfer changes rapidly with the distance between donor and acceptor.

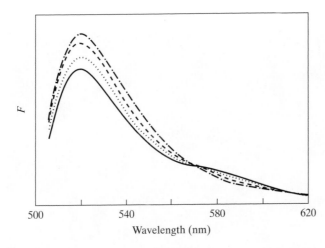

Figure 11.15 Dissociation of the four subunits in c-AMP-dependent protein kinase increases the fluorescence of fluoroscein at 520 nm at the expense of the fluorescence of rhodamine at 580 nm in the c-AMP fluorosensor. No added c-AMP (——), $0.07\,\mu$M c-AMP ($\cdots$), $0.19\,\mu$M c-AMP (– – –), and $53\,\mu$M c-AMP (–·–·–). [Adapted from data in S. R. Adams et al. (1991) *Nature* **349**, 694–697.]

The cyclic AMP (c-AMP) fluorosensor developed by Adams et al. (1991) is a particularly clever use of FRET. c-AMP-dependent protein kinase consists of two regulatory and two catalytic subunits. c-AMP causes the dissociation of these four subunits and the activation of the enzyme. To make the fluorosensor, the dye fluoroscein was attached to the catalytic subunit and the dye rhodamine attached to the regulatory subunit. The inactive tetramer exhibits FRET of the excited fluoroscein to the nearby rhodamine. Addition of c-AMP dissociates the tetramer. The dissociation eliminates FRET, which in turn increases the fluorescence of fluoroscein at 520 nm and decreases the fluorescence of rhodamine at 580 nm, as we see in Figure 11.15. The fluorescence at 520 nm relative to 580 nm is a sensitive measure of c-AMP concentration. (See Application 11.1.)

11.9   LINEAR POLARIZATION OF FLUORESCENCE

If the light used to excite fluorescence is linearly polarized, absorption will be most probable for those molecules that happen to lie with their transition dipoles parallel to the plane of polarization, as we remember from Eq. 8.114. The polarization of the emitted light, however, will depend on a number of factors, including (1) the orientation of the emitting transition dipole relative to the absorbing transition dipole, and (2) the amount of molecular rotation that takes place during the fluorescence lifetime. In general, the fluorescent light will be partially depolarized.

Application 11.1   Visualizing c-AMP with Fluorescence

Fluorescent probes have wide application. As a particularly beautiful example, DeBernardi and Brooker (1996) used the c-AMP fluorosensor discussed in Section 11.8 to detect c-AMP in live cells. Figure A11.1 shows the nondestructive video imaging of cells

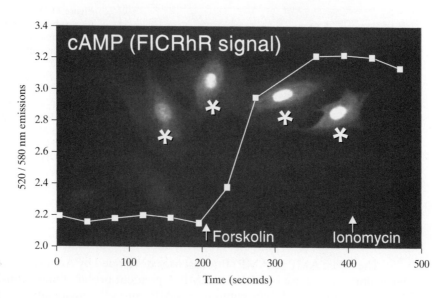

Figure A11.1 The detection of c-AMP in live cells using the fluorosensor; see Section 11.8. [Reproduced through the courtesy of Atto Instruments, Rockville, MD and Molecular Probes, Eugene, OR.]

both before and after forskolin treatment, which induces c-AMP production in the nucleus. Before forskolin treatment, we see that the cell barely fluoresces at 520 nm with 488 nm excitation. After treatment, we see marked fluorescence from the nucleus.

DeBernardi, M., and G. Brooker (1996) *Proc. Natl. Acad. Sci. USA* **93**, 4577–4582.

The experimental setup for measuring the depolarization of the fluorescent light is shown in Figure 11.16. The exciting light is sent in along the x-axis with the electric vector along the z-axis. One hypothetical molecule is at the origin of the Cartesian coordinate system and has its transition dipole for absorption at an angle θ with respect to the z-axis. The angle between the absorbing transition dipole and the emitting transition dipole of the molecule is γ. The emitted fluorescence is observed along the y-axis at right angles to the sample, and a linear polarizer is used to separate the intensity of the electric vectors parallel and perpendicular to the electric vector of the exciting light (z-axis). The depolarization is described in terms of a quantity called the *fluorescence anisotropy*.

$$r = \frac{I_\parallel - I_\perp}{I_\parallel + 2I_\perp} \tag{11.13}$$

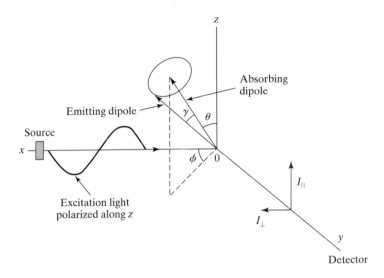

Figure 11.16   The experimental setup for measuring the depolarization of fluorescent light.

In the older literature, the *polarization* $P = (I_\parallel - I_\perp)/(I_\parallel + I_\perp)$ was used. However, this quantity was difficult to manipulate because the denominator is not proportional to the total amount of emitted light. For example, the average anisotropy measured for a mixture of components is just the sum of the individual anisotropies weighed by their fluorescence yield F,

$$r_{\mathrm{avg}} = \sum_i F_i r_i \qquad (11.14)$$

Referring to Figure 11.16 and considering the polarizations, we realize that $I_\parallel$ and $I_\perp$ are emitted along the x-axis as well as the y-axis, but that $2I_\perp$ is emitted along the z-axis. The total amount of emitted light will be proportional to the sum of the emitted light along the three mutually orthogonal Cartesian axes, or the $I_\parallel + 2I_\perp$ found in the denominator of r. The two quantities are related by $r = 2P/(3 - P)$.

First, let us concentrate on the simplest situation where the orientation of the emitting dipole is parallel to the orientation of the absorbing dipole ($\gamma = 0$), and the fluorescence lifetime is short compared to the rapidity of the molecular rotation. The two transition dipoles are always parallel for absorption in the first singlet, because $\mu_{fi} = \mu_{if}$. Molecular rotation can be slowed by freezing them in a glass, or even by dissolving them in a viscous solvent like ethylene glycol. If the molecules are all lined up parallel to the z-axis ($\theta = 0$), then $r = 1.0$. In general, each molecule will make a different angle θ with the z-axis. Then, for our hypothetical molecule in Figure 11.16, the magnitude of the electric vector absorbed, $\mathbf{E}_{\mathrm{abs}}$, will be proportional to $\cos\theta$. Furthermore, the magnitude of the electric vector emitted parallel to the linear polarization, $\mathbf{E}_\parallel$, will be proportional to $\mathbf{E}_{\mathrm{abs}} \cdot \mathbf{k}$, which is proportional to $\cos^2\theta$.

The magnitude of the electric vector perpendicular to the linearly polarized light $E_\perp$ will be proportional to $E_{abs} \cdot i$, which is proportional to $\cos\theta \sin\theta \cos\phi$. Light intensity is proportional to the electric vector squared, so

$$I_\parallel \propto \cos^4\theta$$

$$I_\perp \propto \cos^2\theta \sin^2\theta \cos^2\phi \tag{11.15}$$

A real sample will have many molecules with all values of θ and ϕ, so we must integrate over these two variables to get the emitted intensities for a solution of sample molecules

$$I_\parallel \propto \int_0^{2\pi} d\phi \int_0^\pi \cos^4\theta(\sin\theta \, d\theta) = \frac{4\pi}{5}$$

$$I_\perp \propto \int_0^{2\pi} \cos^2\phi \, d\phi \int_0^\pi \cos^2\theta \sin^2\theta(\sin\theta \, d\theta) = \frac{4\pi}{15} \tag{11.16}$$

Thus, for $\gamma = 0$ and no molecular rotation

$$r_0 = \frac{4\pi/5 - 4\pi/15}{4\pi/5 + 8\pi/15} = \frac{2}{5} \tag{11.17}$$

This is *depolarization* of the emitted light by random orientation of the molecules.

When the absorbing and emitting dipoles are not parallel, r_0 is reduced by the factor $(3\cos^2\gamma - 1)/2$. In general,

$$r_0 = \frac{1}{5}\left(3\cos^2\gamma - 1\right) \tag{11.18}$$

We see that the anisotropy when there is no molecular rotation will vary between 2/5 and −1/5; it is a further depolarization depending on the relative orientation of the absorbing and emitting transition dipoles. It is clear that the anisotropy at high viscosity, where there is no molecular rotation, can be used to determine the angle γ between absorbing and emitting transition dipoles. If the direction of the transition dipole for the first singlet is known, then we learn about the directions of the transition dipoles for the higher singlets. As an example, Figure 11.17 plots the anisotropy observed for the fluorescence of rhodamine dye as the exciting light is scanned from the beginning of the first singlet at 480 nm through shorter wavelengths to 280 nm. We see five electronic transitions. The first singlet that begins at 480 nm has an anisotropy of about +0.4, as expected, since the absorbing and emitting transition dipoles are parallel. The second singlet at about 420 nm has an anisotropy of about −0.1, which corresponds to a γ of about 66°. The third singlet at about 390 nm has an anisotropy of about +0.08, which corresponds to a γ of 47°. The next two singlets at higher energy have the same very negative anisotropy of −0.16 and these transition dipoles are nearly perpendicular to the transition dipole of the first singlet.

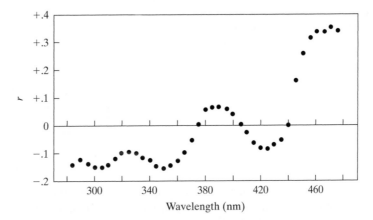

Figure 11.17   The anisotropy observed for the fluorescence of rhodamine dye as a function of the wavelength of the exciting light.

If the molecules rotate during the fluorescence lifetime, we expect even more depolarization, and the magnitude of the anisotropy will be further decreased. Most biopolymers in a nonviscous solvent can undergo a perceptible reorientation during the typical excited-state lifetime of 10^{-9} to 10^{-8} sec, so polarized fluorescence can be used to measure this dynamic property. Rotation depends on molecular volume and shape. Most proteins have a similar density and shape, so in this case rotation is sensitive to molecular weight. Polarized fluorescence is the only spectroscopic technique that is responsive to molecular weight changes, and thus is convenient for studying reactions such as monomer-dimer equilibria. Even though the molecule as a whole is large and sluggish, particular fluorescent groups (i.e., tryptophan side chains) are often sufficiently free to execute rapid rotation. These motions will also add to the intrinsic depolarization described above.

When there is no molecular rotation, the anisotropy has a maximum magnitude r_0 that depends on the wavelength of absorption as discussed above. If the molecules can randomly reorient during the fluorescent lifetime, then the emitted light will be wholly depolarized and $r = 0$. The majority of actual cases will lie between these two extremes, and molecular rotation during the fluorescence lifetime will decrease r relative to r_0.

Pulse fluorometry is a direct and intuitive way to examine molecular rotation. The fluorescent sample is excited with a very short pulse of light (about 10^{-9} sec) and then the anisotropy is measured as a function of time. Those molecules that emit immediately after the pulse will not have had time to rotate; without rotational depolarization, they will show an anisotropy value close to r_0. On the other hand, molecules that emit later will have had time to undergo rotational motion, and their anisotropy will decrease through depolarization. The decay of the anisotropy is first order (Figure 11.18), and for a molecule with a single rotational correlation time ρ is given by

$$r(t) = r_0 e^{-t/\rho} \tag{11.19}$$

Figure 11.18 Decay of the anisotropy after the sample has been excited with a very short pulse. Rotation of the molecule causes a loss in signal.

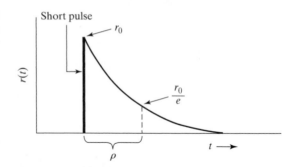

This measurement of anisotropy through *time-resolved* fluorescence provides a very convenient way to measure rotational motion of biological molecules. Two qualifications are required: (1) This relation is strictly correct only for approximately spherical molecules; for molecules with lower symmetry, the situation becomes more complex because more correlation times are involved. (2) It must be remembered that local motions of the fluorophore can contribute to the depolarization of the anisotropy; the fluorescent group must be firmly attached, so that the rotation of the molecule as a whole is being monitored.

We used calcium binding to native calmodulin to illustrate quenching in Section 11.7, and calmodulin was compared to its dityrosine derivative to illustrate the analytical applications in Section 11.5. Here we illustrate polarized time-resolved pulse fluorometry with experiments on the dityrosine derivative. Figure 11.19 shows a logarithmic plot for the fluorescence decays measured as a function of time parallel and perpendicular to the linearly polarized excitation. These experimental results were measured with calcium bound to the calmodulin sites. The anisotropies as a function of time measured for calcium-free calmodulin and calmodulin with its sites saturated by bound calcium are given in Figure 11.20. The correlation time for calcium-activated calmodulin is found to be 9.9 nsec, consistent with the rotational

Figure 11.19 Logrithmic plots of the polarized fluorescence intensity parallel $F_{\parallel}$ and perpendicular $F_{\perp}$ to the polarization of the excitation pulse E for cross-linked calmodulin. [Reprinted from E. W. Small and S. R. Anderson (1988) *Biochemistry* **27**, 419–428, copyrighted 1988 by The American Chemical Society.]

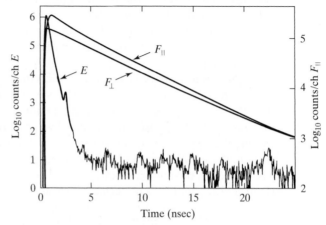

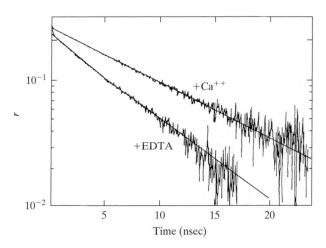

Figure 11.20 A logrithmic plot of the decay of the anisotropy of cross-linked calmodulin that is calcium-free (EDTA) and with its calcium sites saturated (Ca^{2+}). [Reprinted from E. W. Small and S. R. Anderson (1988) *Biochemistry* **27**, 419–428, copyrighted 1988 by The American Chemical Society.]

diffusion of a nonspherical elongated structure. In contrast, calcium-free calmodulin has a much shorter correlation time that indicates it is highly compact. Thus, fluorescence anisotropy demonstrates that there is a change in shape concomitant with the activation of calmodulin by calcium.

Frequency-domain fluorometry is an alternative way to measure time-resolved fluorescence. Sinusoidally modulated exciting light yields sinusoidally modulated fluorescence that is phase-shifted relative to the exciting light. This information is equivalent to the information measured in the pulse method, but the complicated analysis is not intuitive and will not be considered here.

In an older, but still frequently used method, the fluorescence depolarization is measured by determining the *steady-state* anisotropy. If a sample is continually illuminated with polarized light, we will observe an average value of the anisotropy, $\bar{r}$. This will be reduced from r_0 by the factor $1/(1 + \tau/\rho)$. The rotational correlation time for a sphere depends on its volume V, the viscosity of the solvent $\eta(T)$, and the available energy due to the temperature $k_B T$,

$$\rho = \frac{\eta(T)V}{k_B T} \tag{11.20}$$

where k_B is the Boltzmann constant. The relationship between the measured steady-state anisotropy and the intrinsic anisotropy is, then,

$$\frac{1}{\bar{r}} = \frac{1}{r_0}\left(\frac{k_B T \tau}{V\eta(T)} + 1\right) \tag{11.21}$$

For a sphere, a *Perrin plot* of $1/\bar{r}$ versus $T/\eta(T)$ should be a straight line with slope $\tau k_B/r_0 V$ and intercept $1/r_0$. The graph is usually also straight for an oblate ellipsoid. The lifetime τ could be measured as described in Section 11.7, so measurement of the steady-state anisotropy will give the volume of the biological molecule with bound solvent. If the molecule has some other shape, the graph will tend to flatten out for large values of $T/\eta(T)$. If the graph anomalously increases in slope at large $\eta(T)$, then the molecule is unfolding.

11.10   FLUORESCENCE APPLIED TO PROTEINS

Fluorescence has proved to be a particularly valuable technique for studying proteins. We have already used the fluorescence of proteins to illustrate analytical detection, changes in quantum yield, the effects of energy transfer, and changes in fluorescence polarization with time that can be used to monitor shape. Here we shall see how fluorescence can be used to study the stoichiometry of complex formation, observe the presence of intermediates, determine equilibrium constants, and ascertain the microenvironment of binding sites.

We continue to use the particularly interesting calcium-binding protein, calmodulin, as an example. It is sensitive to the calcium concentration in eucaryotic cells that in turn causes changes in shape. The changes in shape lead to the binding of small molecules and recognition by the many calmodulin-dependent enzymes.

Malencik and Anderson (1983) discovered peptide binding to calmodulin. Here we consider *Polistes* mastoparan, a toxic peptide from the social wasp that binds to calmodulin with high affinity, and contains one tryptophan fluorophore. In contrast, calmodulin contains two tyrosines as fluorophores, but no tryptophans. Excitation at 295 nm results in tryptophan fluorescence, but minimal interference from tyrosine, so it is easy to use fluorescence spectroscopy to monitor the binding of the peptide. The shift to shorter wavelength for the fluorescence of *Polistes* mastoparan on binding to calmodulin is shown in Figure 11.21. This is a solvent effect, which shows that the environment of the tryptophan on binding is hydrophobic and shielded from the aqueous solvent. There is also some quenching of the fluorescence on binding, which often accompanies solvent effects. The anisotropy measured through steady-state polarized fluorescence can be used to follow the molecular

Figure 11.21 The fluorescence spectrum of *Polistes* mastoparan free $(\cdots)$ and bound to calmodulin (—). [Adapted from data in D. A. Malencik and S. R. Anderson (1983) *Biochem. Biophys. Res. Comm.* **114**, 50–56.]

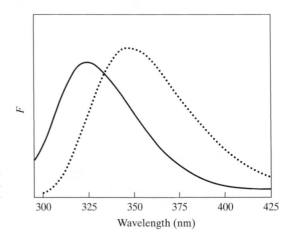

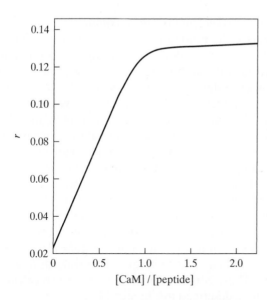

Figure 11.22 Fluorescence anisotropy for the stoichiometric titration of *Polistes* mastoparan with calmodulin. [Adapted from data in D. A. Malencik and S. R. Anderson (1983) *Biochem. Biophys. Res. Comm.* **114**, 50–56.]

weight changes that result from binding. Figure 11.22 shows the change in anisotropy that occurs when the peptide is titrated with calmodulin. The increase in anisotropy saturates at the one-to-one complex, demonstrating that there is a single binding site on calmodulin with a high affinity for the peptide (see Figure 14.5). The fluorescence can also be used to follow the calcium dependence for the binding of *Polistes* mastoparan. The effect of calcium on the fluorescence of a solution containing equal concentrations of calmodulin and the peptide is shown in Figure 11.23. The

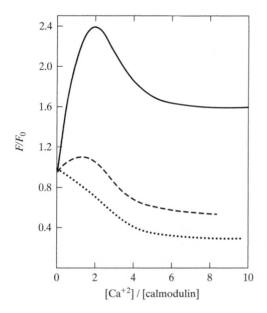

Figure 11.23 The fluorescence intensity at 320 nm (—), 340 nm (– – –), and 360 nm (···) for a solution containing equal concentrations of calmodulin and *Polistes* mastoparan titrated with calcium. [Adapted from data in D. A. Malencik and S. R. Anderson (1983) *Biochem. Biophys. Res. Comm.* **114**, 50–56.]

fluorescence was monitored at three wavelengths. The maximum in the 320 nm trace at two calciums per calmodulin (for the ratio of measured fluorescence intensity F compared to the intensity due to the peptide alone, F_0) indicates the formation of an intermediate when half of the four calcium-binding sites are saturated. The maximum occurs because the fluorescence shifts to shorter wavelength. On further addition of calcium, quenching occurs, lowering F/F_0 at all three wavelengths until all four calcium sites are saturated. The sum of these effects is clear in Figure 11.21.

Polistes mastoparan binds too tightly to calmodulin to allow determination of an equilibrium constant by this method. However, porcine glucagon shows moderate binding, and like *Polistes* mastoparan has a tryptophan that increases its fluorescence on binding. Figure 11.24 shows the ratio of fluorescence when calmodulin is added to a glucagon solution, relative to the fluorescence of the glucagon solution alone. The fluorescence increases as calmodulin is added, as we expect. Fluorescence for the complex alone without free glucagon would give a ratio of 1.71 relative to the glucagon solution alone. Therefore, with the information in Figure 11.24, it is possible to calculate the glucagon, calmodulin, and complex concentrations at each concentration of added calmodulin. These concentrations can be used to calculate the equilibrium constant for binding, as described in Chapter 14.

The properties of biopolymers are often followed by binding a dye that conveniently emits in the visible region and has a long lifetime. Some of these dyes do not fluoresce in water but only fluoresce when bound, making interpretation easier. We saw fluorescence from the dye 2-AN in Figures 11.8 and 11.9 when illustrating

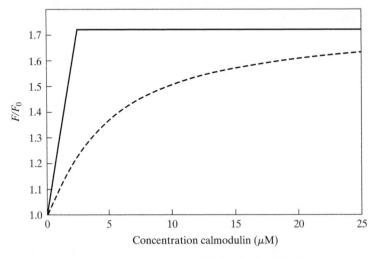

Figure 11.24   The ratio of fluorescence when calmodulin is added to 2.5 μM porcine glucagon relative to the fluorescence of glucagon alone (– – –) and the results calculated for complete binding (—). These data allow the calculation of the equilibrium constant for binding at each concentration of added calmodulin. [Adapted from D. A. Malencik and S. R. Anderson (1982) *Biochemistry* **21**, 3480–3486.]

solvent effects. In this example, we consider the closely related dye 1-anilinonaph-thalene-8-sulfonate (ANS) that binds noncovalently to the AMP-effector site of the enzyme glucogen phosphorylase. Excitation of ANS is at 360 nm, and the fluorescence monitored at 460 nm. This dye does not fluoresce appreciably in water. Binding is stronger to the more active form, phosphorylase a, than it is to the less active form, phosphorylase b. Phosphorylase is the substrate for the enzyme phosphorylase kinase, which covalently links a single phosphate to serine-14 of the subunit, causes the *N*-terminus to assume a helical conformation, and converts phosphorylase to its active form. Phosphorylase kinase is a complex enzyme with four copies of four different subunits. Interestingly, one subunit is calmodulin, which accounts for the calcium dependence of the complex enzyme. Also, protein kinase, the enzyme used to illustrate energy transfer in Section 11.8, is of interest because it activates phosphorylase kinase. The activity of phosphorylase kinase can be followed by observing the fluorescence of ANS bound to its substrate. Figure 11.25 shows the increased fluorescence of ANS as phosphorylase b is converted to phosphorylase a by the 33 kD fragment of phosphorylase kinase. This is a large fraction of the catalytic subunit, which is prepared from the complex enzyme by treatment with chymotrypsin. The fluorescence shows that the 33 kD fragment is indeed active, and can be used to follow the kinetics. Also shown is a ^{32}P radioactive assay for the incorporation of phosphate. This second assay is more time-consuming and involves the use of a dangerous isotope, so fluorescence is the assay of choice.

Several of the uses of fluorescence spectroscopy in the study of proteins discussed above are also illustrated in Application 11.2.

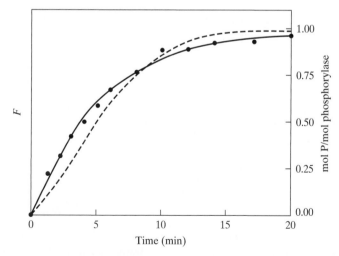

Figure 11.25   The fluorescence of the dye ANS bound to phosphorylase is used to follow the activity of the 33 kD fragment of phosphorylase kinase (——). This is less dangerous and less time-consuming than using the ^{32}P radioactive assay (– – –). [Adapted from data in D. A. Malencik, Z. Zhao, and S. R. Anderson (1991) *Biochem. Biophys. Res. Comm.* **174**, 344–350.]

Application 11.2   Investigation of the Polymerization of G-Actin

Marriott et al. (1988) used fluorescence spectroscopy in its various aspects in an elegant investigation of the muscle protein actin. Monomeric G-actin was labeled with 6-propionyl-2-(dimethylamino)naphthalene (Prodan), a fluorophore that is particularly sensitive to solvent effects and conveniently fluoresces far into the visible region with a large quantum yield. As we see in Figure A11.2a, polymerization of G-actin into F-actin blue shifts the fluorescence from 492 nm to 466 nm. This demonstrates that the Prodan has limited exposure to the solvent in the polymeric F-actin, and gives a simple method for spectral discrimination between the two forms. Time dependence of the polymerization was followed through the shift in average energy of the fluorescence (Figure A11.2b).

This fluorescence data can be used to determine the equilibrium constant and investigate the kinetics of polymerization. Figure A11.2c shows iodide-quenching of the fluorescence of labeled G-actin that is a simple straight line with the proper rate for quenching that is diffusion controlled. This indicates that the Prodan is completely exposed to the solvent in G-actin. Steady-state polarization of fluorescence was measured for labeled F- and G-actin as a function of excitation wavelength, as we see in Figure A11.2d. The high value of the polarization at long excitation wavelength means there is little depolarization due to rotational freedom of the Prodan in either form of the actin. In contrast, the low value of the polarization at short excitation wavelength indicates energy transfer from tryptophan residues, which is higher for F-actin. Taking this polarization study and the iodide-quenching data together indicates that the Prodan in G-actin is immobilized on the protein surface, with one of its faces in contact with the solvent.

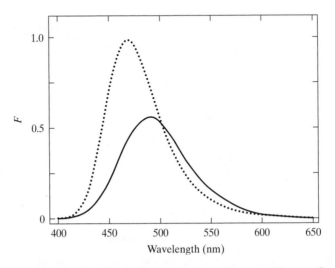

Figure A11.2a   The label Prodan shifts its fluorescence from 492 nm to 466 nm as G-actin (——) is polymerized into F-actin ($\cdots$). [Adapted from data in Marriott et al. 1988.]

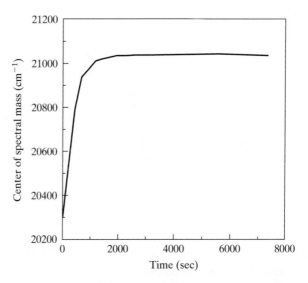

Figure A11.2b The shift in the average energy of fluorescence gives the time dependence of the polymerization of G-actin into F-actin. [Adapted from data in Marriott et al. 1988.]

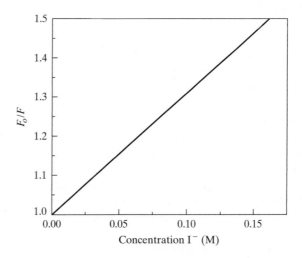

Figure A11.2c Iodide quenching of labeled G-actin. [Adapted from data in Marriott et al. 1988.]

Marriott, G., K. Zechel, and T. M. Jovin (1988) *Biochemistry* **27**, 6214–6220.

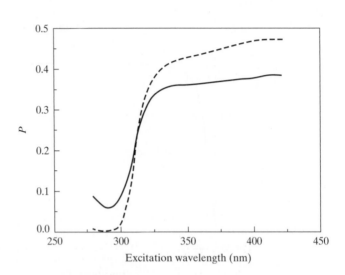

Figure A11.2d   The steady-state polarization of fluorescence for labeled G-actin (——) and F-actin (– – –) as a function of excitation wavelength. [Adapted from data in Marriott et al. 1988.]

11.11   FLUORESCENCE APPLIED TO NUCLEIC ACIDS

The DNA bases have only a very low fluorescence. However, fluorescence methods can be applied to studies of nucleic acids by substituting a fluorescent analog for a normal base or by binding a fluorophore.

The unwinding of double-stranded DNA by a helicase can be monitored by fluorescence. An oligonucleotide is synthesized substituting the fluorophore 2-aminopurine for adenine. This adenine analog hydrogen bonds to thymine without distorting the normal B-form DNA. When the oligonucleotide is hydrogen bonded to a complementary strand, the fluorescence is quenched to one-half of its normal intensity. As the helicase unwinds the oligonucleotide, the normal fluorescence returns, as we expect. Figure 11.26 shows the fraction of single-stranded oligonucleotide observed by fluorescence as T4 dda helicase unwinds the double-stranded substrate. This continuous fluorescence-based assay can be used in steady-state and kinetic studies.

Nucleic acids can also be made to fluoresce by adding a fluorophore that binds noncovalently. For instance, the most common use of fluorescence in molecular biology is to visualize DNA on gels. Ethidium is intercalated into the DNA and this fluorophore then is excited with UV light so that it in turn emits visible light. Ethidium fluorescence is strongly quenched in water, but not when shielded in the hydrophobic environment of the helix. Another simple example that also illustrates polarization of fluoroescence (see Section 11.9) is shown in Figure 11.27. In this clever early experiment, proflavin was intercalated between the bases of a double-stranded and helical

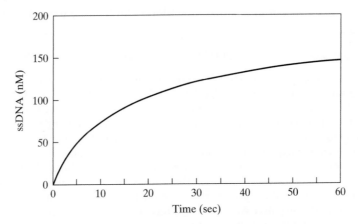

Figure 11.26 The unwinding of double-stranded DNA by T4 dda helicase is followed by monitoring the fluorescence of an adenine analog. [Adapted from data in K. D. Raney et al. (1994) *Proc. Natl. Acad. Sci. USA* **91**, 6644–6648.]

DNA. Polarization rather than anisotropy was used to describe the measurement here, and the results are graphed as a Perrin plot versus T/η. The polarization decreases slightly with temperature, until the DNA denatures (60°C), and the polarization decreases dramatically as the proflavin is released.

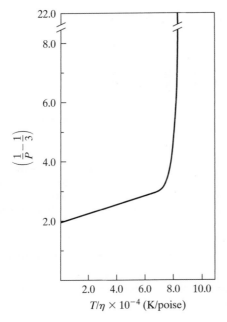

Figure 11.27 The steady-state polarization of proflavin intercalated in double-stranded DNA as a function of temperature. [Adapted from data in N. F. Ellerton and I. Isenberg (1969) *Biopolymers* **8**, 767–786.]

Application 11.3 shows how FRET can be used to study DNA.

Application 11.3   The Helical Geometry of Double-Stranded DNA in Solution

One ingenious use of FRET is the observation of the helical geometry of double-stranded DNA in solution. Clegg et al. (1993) used a series of DNAs ranging in length from 8 base pairs to 20 base pairs. They covalently attached fluorescein dye to one $5'$ end of each double-stranded DNA and rhodamine dye to the other $5'$ end, and used FRET to measure the end to end distance for each member of the series. If the DNAs were linear and not helical, then the experiment would not be very interesting, because the distance between the donor (fluorescein) and the acceptor (rhodamine) would increase by the same amount as the length of the DNA increases. But since the double-stranded DNA is helical, the donor and acceptor will be farther apart than the linear model for a complete turn of the DNA (about 10 base pairs). We expect the efficiency of the FRET to be exceptionally sensitive to the relative distance between donor and acceptor, because it goes as the sixth power of the distance (Eq. 11.12). Figure A11.3 confirms this expectation, as we see particularly low efficiency at 10 base pairs. Clegg et al. have shown that FRET has high precision for studying the conformation of DNA.

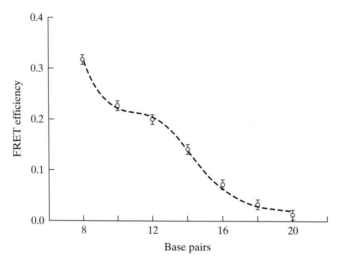

Figure A11.3   The FRET efficiency for a series of DNA oligomers with increasing numbers of base pairs is not a straight line. The efficiency undulates because the DNA is helical so that the donor and acceptor located at opposite ends of the helix are farther apart than might be expected for a complete turn of the helix. The dashed line is the calculated efficiency for the helical model, assuming 10 base pairs per turn. [Adapted from data in Clegg et al. 1993]

CLEGG, R. M., A. I. H. MURCHIE, A. ZECHEL, and D. M. J. LILLEY (1993) *Proc. Natl. Acad. Sci. USA* (Biophysics) **90**, 2994–2998.

EXERCISES

11.1   Use the absorption spectrum of tryptophan in Figure 9.15 to calculate the Einstein co-efficient A for spontaneous fluorescence in molecules sec^{-1}.

11.2   Use the answer to problem 11.1 to calculate the intrinsic fluorescence lifetime of tryptophan τ_0. If the quantum yield q for tryptophan in a protein is 0.3, what will be the corresponding lifetime τ?

11.3   Calculate the intrinsic lifetime τ_0 from the following decay curve if the quantum yield is 0.5.

t (sec) = 0	2	4	6	10	14	20	30
I = 1.0	.82	.67	.55	.37	.25	.14	.05

11.4   Explain the difference between fluorescence and phosphorescence.

11.5   Why is a fluorescence spectrum independent of the wavelength of excitation?

11.6   The following data describe the steady-state fluorescence polarization of a protein that has been labeled with a dye. The dye has a fluorescent lifetime of 7.0 nsec. Sucrose has been added to solutions at 20°C to increase η. Calculate r_0, τ, and V in H_2O.

$T/\eta = 10^{-4}$ (cgs)	$\bar{r}$
0.30	0.292
0.82	0.269
1.49	0.247
2.10	0.227
2.51	0.217
2.92 (H_2O)	0.206

11.7   It is found that there are two sites for attachment of fluorescent labels to the protein described in Problem 11.6. A pair is used for which R_0 is 2.3 nm. The energy transfer efficiency is found to be about 0.015. Estimate the distance between the labels.

11.8   Muscle contracts by sliding myosin filaments relative to actin filaments. The myosin head group hydrolyzes ATP to get the energy for its globular motor domain to move a long lever-arm domain. Suzuki et al. (1998) used GFP and its mutated relative blue fluorescent protein (BFP) in a FRET experiment to demonstrate that during the working stroke the lever-arm domain tilts against the motor domain. They constructed a fusion protein in which GFP was attached through three glycines to the N-terminal end of the myosin and BFP was attached in the same way to the C-terminal end. Exciting the BFP causes the GFP to fluoresce with an efficiency of 0.333 after the ATP is hydrolyzed to ADP, and this corresponds to a distance of 3.8 nm. Before the hydrolysis, the FRET efficiency is 0.082. How far do the two fluorophores move relative to each other during the working stroke?

REFERENCES

General

LAKOWICZ, J. R. (1999) *Principles of Fluorescence Spectroscopy,* 2nd ed., Plenum Press, New York.

LAKOWICZ, J. R. (ed.) (1992) *Topics in Fluorescence Spectroscopy,* Vol. 3, *Biochemical Applications,* Plenum Press, New York. A collection of chapters from various fluorescence spectroscopists.

SAUER, K. (ed.) (1995) *Biochemical Spectroscopy. Methods in Enzymology,* Vol. 246, Academic Press, New York. Contains articles by experts in the field on fluorescence anisotropy, resonance energy transfer, time-resolved fluorescence, and covalent labeling with fluorophores.

TINOCO, I., JR., K. SAUER, J. C. WANG, and J. D. PUGLISI (2002) *Physical Chemistry Principles and Applications in Biological Sciences,* 4th ed., Prentice Hall, Upper Saddle River, NJ.

Methods

ADAMS, S. R., A. T. HAROOTUNIAN, Y. J. BUECHLER, S. S. TAYLOR, and R. Y. TSIEN (1991) "Fluorescence Ratio Imaging of Cyclic AMP in Single Cells," *Nature* **349**, 694–697.

ANDERSON, S. R. (1991) "Time-Resolved Fluorescence Spectroscopy. Applications to Calmodulin," *J. Biol. Chem.* **266**, 11405–11408.

BRAND, L., C. J. SELISKAR, and D. C. TURNER (1971) "The Effects of Chemical Environment on Fluorescent Probes," in *Probes of Structure and Function of Macromolecules in Membranes,* ed. B. Chance, C. P. Lee, and J. K. Blaisie, Academic Press, New York, pp. 17–39.

GIULIANO, K. A., P. L. POST, K. M. HAHN, and D. L. TAYLOR (1995) "Fluorescent Protein Biosensors: Measurement of Molecular Dynamics in Living Cells," *Annu. Rev. Biophys. Biomol. Struct.* **24**, 405–434.

FRALEY, T. S., T. C. TRAN, A. M. CORGAN, C. A. NASH, J. HAO, D. R. CRITCHLEY, and J. A. GREENWOOD (2003) "Phosphoinositide Binding Inhibits α-Actinin Bundling Activity," *J. Biol. Chem.* **278**, 24039–24045.

MALENCIK, D. A., and S. R. ANDERSON (1983) "High Affinity Binding of the Mastoparans by Calmodulin," *Biochem. Biophys. Res. Comm.* **114**, 50–56.

RANEY, K. D., L. C. SOWERS, D. P. MILLAR, and S. BENKOVIC (1994) "A Fluorescence-Based Assay for Monitoring Helicase Activity," *Proc. Natl. Acad. Sci. USA* **91**, 6644–6648.

SUZUKI, Y., T. YASUNAGA, R. OHKURA, T. WAKABAYASHI, and K. SUTOH (1998) "Swing of the Lever Arm of a Myosin Motor at the Isomerization and Phosphate-Release Steps," *Nature* **396**, 380–383.

Nuclear Magnetic Resonance Spectroscopy

Nuclear magnetic resonance (NMR) results from the absorption of energy by a nucleus changing its spin orientation in a magnetic field. Protons (^{1}H) are the most commonly studied nuclei, and their resonance spectrum is characteristic of the various groups in the molecule. ^{1}H NMR is a valuable analytical technique. As we shall see, modern NMR goes far beyond analysis of groups. Two types of NMR interactions (through-bond spin-spin and the through-space nuclear Overhauser effects) can be used to determine the three-dimensional structure and dynamics of macromolecules in solution.

12.1 THE PHENOMENON

Classically, a spinning charge, such as a proton, possesses a magnetic dipole moment. Quantum mechanically, a nucleus will have a spin number I that depends on the number of protons and neutrons it contains. For many types of nuclei, the magnetic dipole vectors completely cancel and the spin number is zero. However, for some other types, the magnetic dipole vectors do not completely cancel and the spin number is different from zero. Some nuclei with a nonzero spin number are given in Table 12.1. Nuclei with a net magnetic dipole will orient the dipole axis in an external magnetic field in certain quantized orientations. The number of possible orientations is given by $2I + 1$, so that nuclei, such as protons with $I = 1/2$, can occupy two quantized orientations in the external magnetic field. This is diagrammed in Figure 12.1, where the external magnetic field H defines the z-axis, and the two orientations for the magnetic dipole of the nucleus have quantum numbers $m_s = +1/2$ or $-1/2$. The dipole axis will precess about the z-axis.

The potential energy of a magnetic dipole $\mathbf{m}$ in an external magnetic field of strength $\mathbf{H}$ is given by

$$E = -\mathbf{m} \cdot \mathbf{H} \qquad (12.1)$$

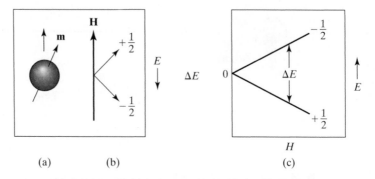

Figure 12.1   (a) A spinning charge generates a magnetic dipole **m**. (b) The dipole can take certain allowed orientations in a magnetic field. (c) The energy levels of the magnetic dipoles depend on the field strength.

The component of the magnetic dipole in the field direction m_z is given by

$$m_z = \gamma \hbar m_s \tag{12.2}$$

where γ is an empirical constant called the gyromagnetic ratio, which will differ for different nuclei. The values for some nuclei are also given in Table 12.1. For $I = 1/2$, the magnetic dipole oriented with the field ($m_s = +1/2$) has the lowest energy, $-m_z H$, while the magnetic dipole oriented against the field ($m_s = -1/2$) has energy $+m_z H$. The energy difference between the two levels is

$$\Delta E = 2m_z H = \gamma \hbar H \tag{12.3}$$

We see that this energy level difference is proportional to H, as shown in Figure 12.1.

Nuclear magnetic resonance spectroscopy involves the observation of transitions between such levels. We remember from Chapter 8 that the frequency corresponding to the energy difference is given by Planck's law, Eq. 8.3, so

$$\nu = \frac{\gamma H}{2\pi} \tag{12.4}$$

Table 12.1   Nuclei Commonly Used in Biochemical NMR

Isotope	Spin	Natural Abundance (%)	Gyromagnetic Ratio (10^7 rad/sec · T)	Relative[a] Sensitivity	Relative[b] Sensitivity in Natural Abundance	Relative NMR-Frequency
^{1}H	1/2	99.98	26.7522	1.00	1.00	100.000
^{2}H	1	1.5×10^{-2}	4.1066	9.65×10^{-3}	1.45×10^{-6}	15.351
^{13}C	1/2	1.108	6.7283	1.59×10^{-2}	1.76×10^{-4}	25.144
^{15}N	1/2	0.37	-2.7126	1.04×10^{-3}	3.85×10^{-6}	10.133
^{19}F	1/2	100	25.1815	0.83	0.83	94.077
^{31}P	1/2	100	10.8394	6.63×10^{-2}	6.62×10^{-2}	40.481
^{113}Cd	1/2	12.26	-5.9609	1.09×10^{-3}	1.33×10^{-3}	22.182

[a]At constant field for equal number of nuclei.

[b]Product of relative sensitivity and natural abundance.

The energy difference is much much less than the available energy at room temperature ($k_B T$), and the Boltzmann distribution, Eq. 8.96, predicts that the ground and excited states are almost equally populated with an excess of only about one part in 10^4 in the ground state. Since the Einstein coefficient B (Section 8.4) is identical for stimulated absorption and emission, the emission nearly cancels out the observation of absorption. This means that NMR instruments must be very sensitive and, even so, measurements require milligrams of sample. The larger the magnet the larger ΔE, Eq. 12.3, and the larger the difference in population between the ground and excited states. Typically, superconducting magnets have a field in the order of 14 Tesla with corresponding frequencies in the megahertz region.

Most NMR work on biological molecules uses nuclei with the spin number of 1/2 because the spectrum is greatly simplified when there are only two possible quantum states. Proton magnetic resonance is typical of NMR spectroscopy, and if we understand ^{1}H NMR, we will understand magnetic resonance of the other spin 1/2 nuclei. Thus, we will concentrate on protons in the remainder of the chapter.

12.2   THE MEASURABLE

When an external field is applied by a magnet, the different protons in a molecule will see different *effective* magnetic fields because of the other fields induced in the molecule itself. Equation 12.4 tells us that the frequency of electromagnetic radiation (light) needed for resonance is directly proportional to the magnetic field. This means protons that see different fields will resonant at different frequencies. Older *continuous wave* (CW) NMR instruments used a single frequency of light and varied the magnetic field to achieve resonance. Since CW instruments are still in use, they are discussed briefly below. However, we will concentrate on modern Fourier transform (FT) NMR instruments that apply the constant field from the basic magnet, and send in a pulse of light that contains many frequencies. The protons then absorb at different frequencies depending on local variations in the field. The frequency of resonance for a sample proton ν_s is given by

$$\nu_s = \nu_0(1 - \sigma) \tag{12.5}$$

where ν_0 is the frequency of excitation corresponding to the basic field of the magnet, and $-\sigma\nu_0$ is the correction necessary for excitation that compensates for local variations. The *shielding coefficient* σ relates the correction in frequency to ν_0. The ν_s needed for resonance is characteristic of chemical bonding; each equivalent group of protons will absorb a particular frequency of light at the same field. Structure also affects the frequency needed for resonance.

In practice, we use a reference compound with very shielded protons that resonate at particularly high frequencies—such as tetramethylsilane (TMS) or sodium 2,2-methyl-2-silapentane-5-sulfonate (DSS)—to calibrate the instrument. In order to make the measurements independent of the magnetic fields used by different instruments, we define the *chemical shift* δ, which compares the frequency of the sample proton to the frequency of the reference compound as measured by the particular instrument ν_r.

$$\delta = \frac{\nu_r - \nu_s}{\nu_0} \tag{12.6}$$

with units in parts per million (ppm). With this definition, δ increases when the shielding decreases.

We see the chemical shifts of some typical protons in Figure 12.2. Because chemical shifts are tabulated for various equivalent types of protons in various bonding situations, ^{1}H NMR spectroscopy is a powerful analytical tool. The chemical shift from the field expected for resonance of an isolated proton depends on the electrons in the molecule, and is called *chemical shift dispersion*. The applied magnetic field induces a magnetic field in the electron clouds that change the fields seen by the protons in the molecule. *Local positive shielding* is caused by the magnetic field induced in the spherical electron cloud around the proton that is undergoing resonance. This is positive shielding that is opposite to the applied field, as we see in Figure 12.3a. Local positive shielding tells about the kind of bond, since it depends on the electron density around the proton. Electronegative substituents withdraw electron density from the protons, giving less shielding (larger δ) in the order $H_3C-O^{(-)}$, H_3C-C, H_3C-N, H_3C-O. However, the electron cloud isn't really spherical, and the effect will be diminished according to the distortion of the electron cloud.

Interatomic shielding from neighboring atoms can augment or oppose the applied field. As an example, the electrons in the bond between two carbon atoms will have an induced magnetic field that opposes the applied field, which is shown for two representative orientations in Figure 12.3b. We see that ethane and ethylene have their protons at an angle to the bond axis, so the interatomic shielding is generally negative and augments the applied field. However, acetylene has its protons on axis, so that for some orientations of the molecule the interatomic shielding is positive and opposes the applied field. It is this effect averaged over all the orientations that causes the protons in acetylene to be more shielded than we expect, when compared to ethane and ethylene (Figure 12.2).

Ring currents can be set up in the delocalized π electrons of aromatic rings. This is a special kind of interatomic shielding that is particularly important in macromolecules. These induced electron currents of large diameter create large magnetic fields, shown in Figure 12.3c for one orientation that gives the net effect for benzene. We see that the effect may be either positive or negative, depending on the position of the proton relative to the benzene ring. As pictured in Figure 12.3c, a benzene proton would see an induced magnetic field that augments the applied field. However, in a larger biopolymer there might be a proton oriented over the center of the ring; such a proton would be highly shielded by the opposing induced field.

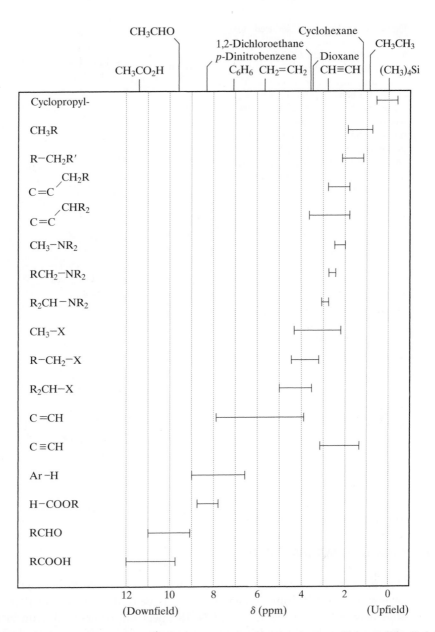

Figure 12.2 The chemical shifts of ^1H in some common functional groups. Halogen (X); alkyl (R); aromatic (Ar).

Figure 12.3 The field induced in some electron clouds for specific orientations in the magnetic field. (a) Local positive shielding for the proton in its electron cloud. (b) Interatomic shielding from a bond depends on whether the proton is on axis or off axis. (c) Interatomic shielding from a ring current depends on the position of the proton relative to the ring. This specific orientation gives the same effect as the net effect from averaging over all orientations.

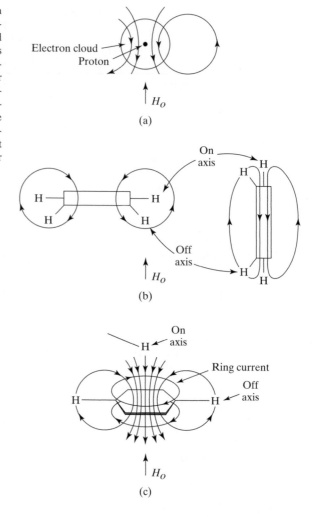

12.3 SPIN-SPIN INTERACTION

According to Figure 12.2, acetaldehyde will have an absorption due to the aldehyde proton at $\delta = 9.75$ ppm, and an absorption due to the three equivalent methyl protons at small δ. However, the ^{1}H NMR spectrum of acetaldehyde in Figure 12.4 shows us that the aldehyde resonance consists of four closely spaced peaks and the methyl resonance consists of two closely spaced peaks. *Spin-spin interactions* are responsible for the splitting of the absorption. The magnetic dipoles of one group will affect the magnetic dipoles of another group through the electrons of the bonds between them. For this reason, spin-spin interactions are often referred to as *through-bond interactions*. The splitting is independent of the field, and depends only on the nature of the bonding between the two groups. This interaction is very important, as

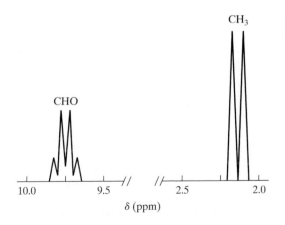

Figure 12.4   Spin-spin interaction in the ^{1}H NMR spectrum of acetaldehyde.

we shall see in Section 12.7, because it is one of the two most important effects that give crosspeaks in two-dimensional NMR.

The aldehyde proton in a particular acetaldehyde molecule may be either in its ground or excited state. Methyl protons in an acetaldehyde molecule with its aldehyde proton in the ground state will absorb at a slightly different field from methyl protons in an acetaldehyde molecule with its aldehyde proton in the excited state. A sample of molecules will have nearly equal populations of these two types of aldehyde protons, and therefore we see two absorption bands at slightly different chemical shifts for the methyl protons.

None, one, two, or three of the protons on a given methyl group can be excited. The relative populations of these possibilities are 1:3:3:1, as we see diagrammed in Figure 12.5. These four different types of methyl groups give rise to four different chemical shifts, depending on which type of methyl group a particular aldehyde proton sees. The areas under the four peaks are in the same ratio as the probabilities for

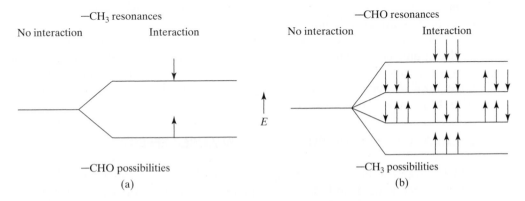

Figure 12.5   The possibilities for spin-spin interactions in acetaldehyde. (a) The two spin arrangements for the —CHO proton cause two resonances for —CH$_3$. (b) The four spin arrangements for the —CH$_3$ protons cause four resonances for —CHO.

Table 12.2 Karplus Constants for Vicinal Coupling of Selected Bonds

y	z		A	B	C
HC	CH		17	0	1.1
HC	NH		12	0	0.2
HC	OH		10	0	−1.0
HN	$C_\alpha H$	(protein, $\theta = \phi - 60°$)	6.4	−1.4	1.9

the four spin arrangements of the methyl protons. The total areas of the methyl and aldehyde absorptions are in the ratio of the number of protons involved, three to one.

The spin-spin coupling constant J measures the splitting between two different nuclei in a given molecule in Hz. Since it is independent of the applied field, this will be the splitting measured on any NMR spectrometer. Karplus (1959) has found that J depends on the relative orientation of the magnetic dipoles of the nuclei (the *torsion angle*, see Section 1.1.2 and Figure 1.5) involved in the spin-spin coupling according to

$$^xJ_{yz} = A \cos^2 \theta + B \cos \theta + C \qquad (12.7)$$

In the case of protons, θ is the angle between the protons for geminal (two-bond) coupling ($x = 2$) or the dihedral angle between protons for vicinal (three-bond) coupling ($x = 3$). The subscripts y and z define the nuclei that are interacting, and A, B, and C are empirical constants. Table 12.2 gives the values in the case of vicinal coupling for selected proton interactions. Torsion angles from spin-spin interactions can be used to learn about bond geometry.

The splittings from spin-spin interactions can be removed with the *double resonance* technique. Exciting a group of protons with one source of electromagnetic radiation removes the orientations necessary for spin-spin interaction with other protons (*decoupling*), and the resonances of the other protons can then be viewed with a second source. For a complicated molecule, the researcher may not be able to assign the resonance peaks, and the double resonance technique can be used to determine which peaks belong to protons that are close enough to interact with another proton through a bond.

The student may well ask why the protons of a particular group, such as the methyl group, do not interact among themselves and create a splitting. Of course they do interact, but only one of the split resonances has intensity when all protons are equivalent. The resonance with intensity is the one that we see, and is associated with the protons of that group.

12.4 RELAXATION AND THE NUCLEAR OVERHAUSER EFFECT

NMR instrumentation uses an oscillating magnetic field to excite protons in the external magnetic field, and make them precess in phase, as we see in Section 12.5. However, without the oscillating magnetic field, they will eventually both lose phase coherence and lose excitation energy. Two relaxation processes are involved in these losses. In *spin-lattice relaxation,* a proton in the excited state interacts with

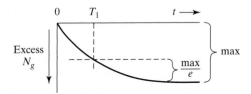

Figure 12.6 Spin-lattice relaxation returns the signal from a sample of protons that were saturated with a strong source. The return of the signal is proportional to the excess protons in the ground state, and is plotted here to emphasize its similarity to the exponential decay of fluorescence shown in Figure 11.11.

fluctuating magnetic fields generated by any other atoms in the molecule (the lattice). This interaction is an *enthalpic relaxation,* because it brings the proton to the ground state, generating heat. Enthalpic spin-lattice relaxation maintains the excess of protons in the ground state. The time for the fraction $1 - 1/e$ of the protons to relax via this process, T_1, can be measured by saturating the sample with a strong CW source. With enough energy over a period of time, the number of protons in the ground state N_g equals the number in the excited state N_e, and there is no phase coherence. Then the absorption peaks have no intensity. The return of an absorption peak is monitored with a weak CW source, as illustrated in Figure 12.6.

The *spin-spin relaxation* process involves dipole-dipole interaction between an excited state spin and a ground state spin for the same type of proton. The interaction leads to an exchange of energy, making the system more random (*entropic relaxation*) and eroding the phase coherence, as we see in Figure 12.7. However, entropic spin-spin

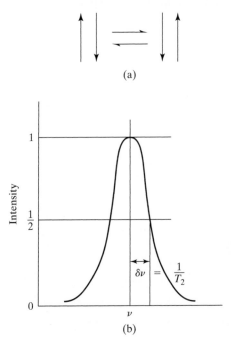

Figure 12.7 (a) Spin-spin relaxation is an exchange of energy between nuclei that makes the system more random. (b) When T_2 is the shortest relaxation time, its reciprocal is the half-width at half-height of the resonance peak.

relaxation does *not* change the net population of the excited state. The relaxation time T_2 is the time for $1/e$ of the protons to relax via this process. NMR linewidths depend on both types of relaxation processes, but often T_2 is the shortest and most important. In that case $1/T_2$ is the half-width at half-height of the absorption peak (Figure 12.7).

An important effect that is observed in NMR spectroscopy is the *nuclear Over-hauser effect* (NOE). This is a *through space interaction* where a change in signal intensity for one type of nucleus is caused by irradiation of another type of nucleus that is nearby. Exciting the nearby nucleus with a second source changes the ratio of N_g to N_e, thus changing the intensity. The NOE is a dipole-dipole interaction that depends strongly on the distance between the two types of nuclei (as r^{-6}), and can be used to measure interatomic distances up to 0.5 nm. This is the other important interaction (discussed in Section 12.7) that gives crosspeaks in two-dimensional NMR.

12.5   MEASURING THE SPECTRUM

Let us consider a conventional *continuous wave* NMR instrument as diagrammed in Figure 12.8. The magnetic field defines the z direction. An oscillating magnetic field along the y-axis is the electromagnetic radiation that provides the energy to cause the transition. The excited nuclei precess around the z-axis at a frequency equal to the frequency of absorption. These precessing spins are in phase, because the oscillating magnetic field is coherent. This in-phase precession of excited spins induces an oscillating current in a coil of wire along the x-axis, which is received and amplified to become the signal. The frequency of absorption depends on the magnetic field strength, Eq. 12.4, and rather than sweeping the frequency, continuous wave instruments generally sweep the field with a coil along the z-axis.

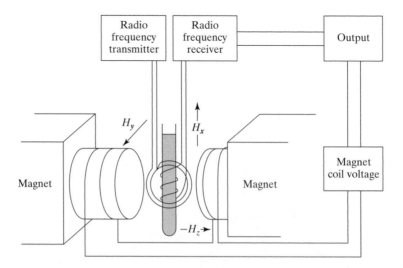

Figure 12.8   A simplified diagram of a continuous wave NMR spectrometer.

Modern NMR spectrometers are *Fourier transform* instruments that send a pulse of energy through the x-axis coil to excite all the transitions at once. The pulse has a nominal frequency, ν_0, and we remember that the functional form of the magnetic component of the light at $z = 0$ given by Eq. 8.1 is

$$\mathbf{H} = \mathbf{H}_0 \cos 2\pi\nu_0 t \tag{12.8}$$

The pulse has a short finite length τ, depicted schematically in Figure 12.9a. Because of its finite duration, Fourier analysis shows that the magnetic component contains a range of frequencies around ν_0 with amplitudes given by

$$\frac{H\tau}{2}\left[\frac{\sin \pi(\nu_0 - \nu)\tau}{\pi(\nu_0 - \nu)\tau}\right] \tag{12.9}$$

which is shown in Figure 12.9b. Although typically $10\ \mu$sec in duration, the short pulse still contains hundreds of wavelengths of each frequency, $\Delta\nu = \nu_0 - \nu$. However, the light is coherent, and the excess spins tend to precess in phase, although at different frequencies $\Delta\nu$, depending on the frequency necessary for excitation of that particular type of nucleus. Each type of precessing nucleus induces an oscillating current of the same frequency $\Delta\nu_0$ in the x-axis coil, which can be used for detection after the pulse.

We see that a sample of molecules will have the spins for a certain equivalent type of proton in both ground and excited states, precessing in random phase about the z-axis, as illustrated in Figure 12.10a. There is a slight excess of spins in the ground state, and this excess is illustrated in Figure 12.10b. Since their phases are random, the x and y components of their magnetic dipoles cancel, but their z components are all in the same direction and add to give a net magnetic dipole (*magnetization*) $\mathbf{M}$ along the z-axis. If the pulse of energy that is sent in along the x-axis has the intensity and duration to excite half the excess protons, then the net magnetization along the z-axis is zero. However, the precession of the excess spins δN is now in phase because the oscillating magnetic field is coherent (Figure 12.10c). The in-phase spins add to give a

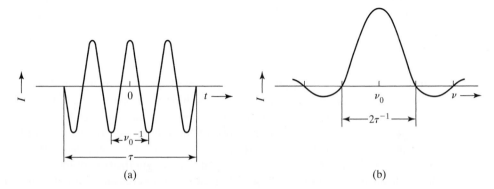

(a)   (b)

Figure 12.9   (a) A short pulse of a specific frequency ν_0 with finite length. (b) Because of its short duration, the pulse contains a range of frequencies around ν_0.

Figure 12.10   (a) A sample of molecules will have the spins for a certain equivalent type of proton in both ground and excited states precessing in random phase around the z-axis. (b) The excess of spins in the ground state sum to a net magnetic dipole along the z-axis, but the x and y components cancel because their phases are random. (c) A 90° pulse excites half of the excess protons to the excited state and causes all of the excess protons to precess in phase. This gives a net magnetization precessing in the xy-plane.

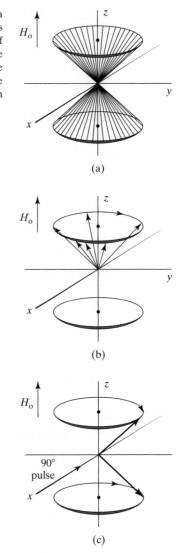

(a)

(b)

(c)

net magnetization that is precessing in the xy-plane at the excitation frequency. Precession of the magnetization in the xy-plane induces a current oscillating at the excitation frequency in the detection coil. However, the magnitude of the induced current will decrease with time because of the relaxation processes. Spin-lattice relaxation T_1 returns spins to the ground state, restoring the net magnetization along the z-axis and concomitantly decreasing the net magnetization in the xy-plane. Spin-spin relaxation T_2 degrades phase coherence, also decreasing the net magnetization in the xy-plane. The exponential decrease in the induced current with time is normally dominated by the faster spin-spin relaxation.

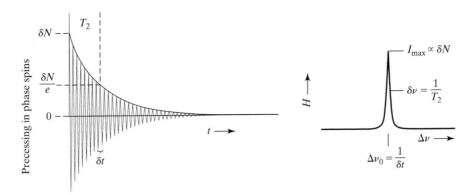

Figure 12.11   A FID with a single frequency $\Delta \nu_0 = 1/\delta t$ and spin-spin relaxation T_2 is easily transformed to the frequency domain.

We illustrate in Figure 12.11 the oscillating signal with its exponential decay, called a *free induction decay* (FID). The frequency of the FID is the frequency of the light absorbed by the particular type of equivalent protons. The magnitude of the FID at $t = 0$ is a measure of the intensity of the absorption. The exponential decay of the FID gives the relaxation time (normally T_2 predominates), whose reciprocal $\delta \nu$ is the half-width of the absorption peak at half-height. Thus, the FID can be transformed through what is known as a *Fourier transformation* to an NMR absorption peak, as we also see in Figure 12.11.

For a single type of equivalent protons, we can do the Fourier transform from the FID in the time domain to the NMR peak in the frequency domain with ease. However, a real molecule will have many different types of equivalent protons, and the measured FID will be a combination of all their frequencies, intensities, and decays. Fortunately, all the different frequencies are independent, so that it is possible to decompose the FID into the individual components. In practice, the problem is quite complicated, but modern computers have the power to carry out a Fourier transformation of measured FIDs from real molecules. Fourier transform NMR spectrometers are possible because we have the computers to transform complicated FIDs, which result from exciting all the transitions in an NMR spectrum.

As we discussed, a pulse of just the right energy to excite half the excess protons in the ground state puts the net magnetization in the xy-plane. The coherent oscillation of the magnetization produces a maximum signal, and is called a 90° or $\pi/2$ pulse. Classically, this is pictured as tipping the magnetization 90° (Figure 12.12). When the pulse contains somewhat more energy, the signal is weaker. The sample will have no detectable signal (all the excess protons in the excited state will have random phase and the net magnetization will be in the $-z$ direction) at twice the energy of a 90° pulse. This is logically called a 180° or π pulse, and is pictured classically as tipping the magnetization 180° (Figure 12.12). Adding even more energy to the pulse again gives a signal, but of opposite sign. The maximum negative magnitude occurs with a 270° or $3\pi/2$ pulse, which again equalizes the excess protons

Figure 12.12 Although (a) 90°, (b) 180°, and (c) 270° pulses redistribute the excess protons between ground and excited state and also affect their phase, the result can be viewed as a tipping of the net magnetization.

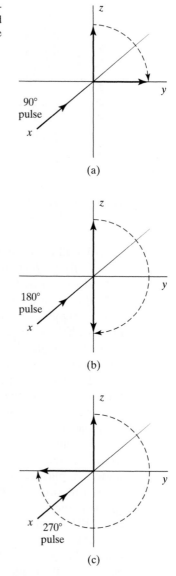

between ground and excited states. Here, the net magnetization oscillates in the *xy*-plane, but out of phase with the 90° pulse. Figure 12.12 shows the classical picture of tipping the magnetization 270°.

It is much faster to record an FID using an FT NMR instrument than to sweep a spectrum on a continuous wave instrument. This tremendous savings in time means that many FIDs can be recorded in a reasonable period of time, and then averaged to reduce noise. The *signal averaging* available with FT NMR gives the increased sensitivity necessary for studying macromolecules.

12.6   ONE-DIMENSIONAL NMR OF MACROMOLECULES

Amino acids have characteristic resonances resulting from the proton on the α carbon and the protons on the side chains. Figure 12.13a shows the CH and CH_3 resonances for alanine, which occur at low ppm. This spectrum was measured in

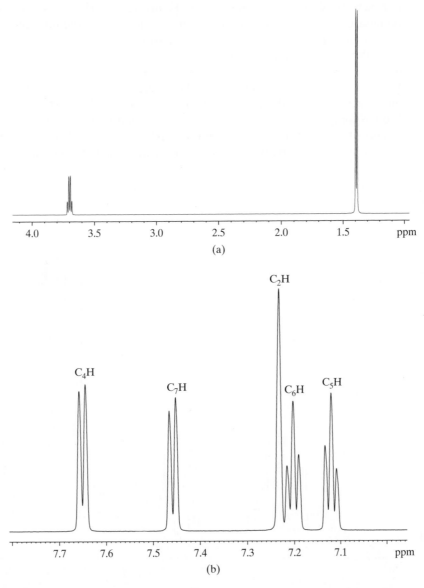

(a)

(b)

Figure 12.13   (a) The ^1H resonances for the CH and CH_3 of alanine. (b) The ^1H resonances for the nonexchangeable protons on the indole side chain of tryptophan. [Courtesy of V. L. Hsu.]

^2H_2O, so there are no resonances for the protons on the acid and amine groups, because these protons have exchanged with the deuteriums of the solvent. We see splitting due to spin-spin interaction similar to acetaldehyde, but the value of J will be different and characteristic of the bonding and the geometry between the α carbon and the methyl side chain.

The indole side chain of tryptophan has characteristic resonances for its nonexchangeable protons, as shown in Figure 12.13b. These resonances of protons on unsaturated rings occur at high ppm. Resonances for the nonexchangeable protons on the CH_2 and CH groups are not shown.

Proteins contain many different kinds of protons, so we expect their one-dimensional spectra to be quite complicated. This expectation is born out in Figure 12.14 with the spectrum of lysozyme, a relatively small protein of 129 amino acids. Even with the high field magnet of a 600 MHz instrument, Figure 12.14 is quite complicated, many resonances still overlap, and the assignment of most resonances from the one-dimensional (1D) spectrum is impossible. Still, we know the general positions of many of the resonances. Side chain protons on aliphatic carbons

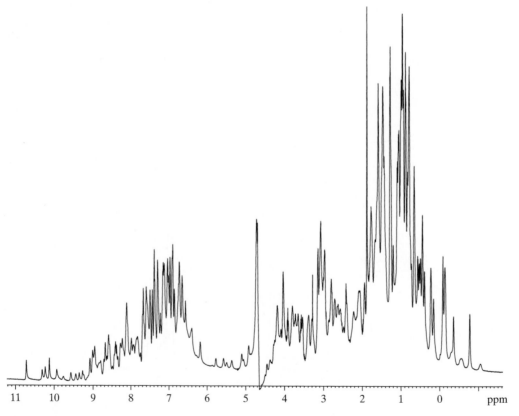

Figure 12.14 The ^1H 1D NMR spectrum of lysozyme at pH 5.3 on a 600 MHz instrument. [Courtesy of V. L. Hsu.]

usually occur at less than 4 ppm. Protons on α carbons are between 4 and 5 ppm. Aromatic protons have resonances between 6.8 and 7.8 ppm. The resonances for amide protons occur between 8 and 9 ppm. Amino and imino protons are found between 6.6 and 8.2 ppm and at high ppm.

This spectrum was taken on a Fourier transform instrument with H_2O as the solvent. H_2O has the advantage that we can see the exchangeable protons at acid pH, but this solvent is 55 molar in H_2O and normally the protons of H_2O would give such an intense peak at about 4.7 ppm that the digitizer in the instrument would saturate and the peaks due to the much lower concentration of the protons in the protein would have no value on a relative basis. The problem is minimized in this spectrum by saturating the water so that its absorption, while still broad, has a peak height that is consistent with the height of the peaks for the protein protons. Still, we see that the H_2O peak interferes with $C_\alpha H$ resonances. Also, saturation of H_2O decreases the intensity of all exchanging protons.

In principle, it is possible to separate the resonances for the individual protons of a protein by using magnets with extremely high fields, but this is really impractical. Instead, we will see that it is possible to learn about the resonances for all of the various types of nuclei in proteins by concentrating on their interactions, such as spin-spin coupling or the nuclear Overhauser effect, and spreading these interactions over two or more dimensions. We discuss this type of multidimensional NMR in Section 12.7.

One-dimensional ^1H NMR will still answer many specific questions about proteins and their substrates. As one example, let us look at the peptidyl prolyl *cis-trans* activity of cyclophilin. This 163 amino acid protein binds the important immunosuppressive drug cyclosporin A. In addition, it catalyzes the isomerization of peptidyl prolyl *cis* configuration to the *trans,* which can be the rate-determining step in protein folding. One-dimensional NMR can be used to demonstrate that cyclosporin A blocks the *cis-trans* activity of cyclophilin.

Figure 12.15a shows the CH_3 resonances for the alanines in succinyl-Ala-Ala-Pro-Phe-*p*-nitroanilide (AAPF), which is a model substrate for cyclophilin. We see major resonances at about 1.48 ppm for AAPF molecules with P in the *trans* configuration, and minor resonances at about 1.47 ppm for AAPF molecules with P in the less stable *cis* configuration. There is an isomerization between the *cis* and *trans* configuration for each proline, but these two species are in *slow exchange* compared to the NMR time scale of about two seconds to run the experiment, so we see distinct resonances for each species. If the isomerization were *fast exchange,* then we would see only one weighted average peak. Broading of the resonances occurs when two species have *intermediate exchange* at about the time scale of the NMR experiment. Figure 12.15b shows the resonances when cyclophilin is added, but AAPF is present in a 300-fold excess, so the resonances due to the cyclophilin are unimportant. We see that the resonances have broadened. The effect is most pronounced for the minor resonance, which we now see only as a shoulder. This means that cyclophilin speeds up the rate of isomerization between *cis* and *trans* proline, although the equilibrium is not changed. We see the

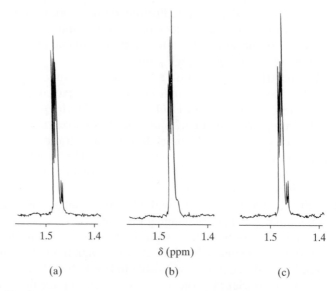

Figure 12.15   The CH$_3$ resonances for the alanines in AAPF, a model substrate of cyclophilin. (a) The major resonance is for P in the *trans* configuration and the minor resonance for P in the *cis* configuration. (b) Added cyclophilin speeds up the rate of isomerization and broadens the resonances, so the minor resonance is seen only as a shoulder. (c) The addition of cyclosporin A blocks the effect of cyclophilin, and the minor resonance returns. [Reprinted from V. L. Hsu, R. E. Handschumacher, and I. M. Armitage (1990), *J. Amer. Chem. Soc.* **112**, 6745–6747. Copyright 1990 by the American Chemical Society.]

effect of blocking the cyclophilin with cyclosporin A in Figure 12.15c, where the minor resonance has returned.

The nucleotide monomers that make up a nucleic acid strand consist of a phosphate with no protons, a sugar with nonlabile protons (carbons labeled with primed numbers), and a base with labile and nonlabile protons (carbons and nitrogens labeled with unprimed numbers). See Figures 1.36 and 1.39 for the numbering of the atoms in a nucleotide. Ranges for the chemical shift dispersion of these protons are given in Figure 12.16. The ^{1}H NMR spectrum of a nucleic acid will be the superposition of the proton resonances for the constituent monomers. Figure 12.17 shows the ^{1}H NMR of the DNA hairpin TCGCGTTTTCGCGA in ^{2}H$_2$O (all the labile imino protons on the bases exchanged for ^{2}H) as one example. Even this oligomer of 14 bases has a fairly complicated spectrum. Still, we can easily see the individual groups of resonances, as diagrammed in Figure 12.16. The intense methyl resonances are between 1.3 and 1.6 ppm, the 2',2" resonances between 1.7 and 2.6 ppm, the 3',4', and 5',5" resonances between 3.4 and 4.2 ppm, the 5 and 1' resonances between 5.0 and 6.2 ppm, and the 2, 6, and 8 resonances between 6.9 and 8.0 ppm. We will discuss the multidimensional methods usually used to assign the resonances below.

One-dimensional NMR was a very important technique for investigating the structure of DNA proposed by Watson and Crick, as we see in Application 12.1.

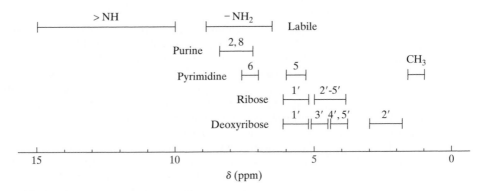

Figure 12.16   Ranges for the chemical shift dispersion in DNA and RNA nucleotide monomers.

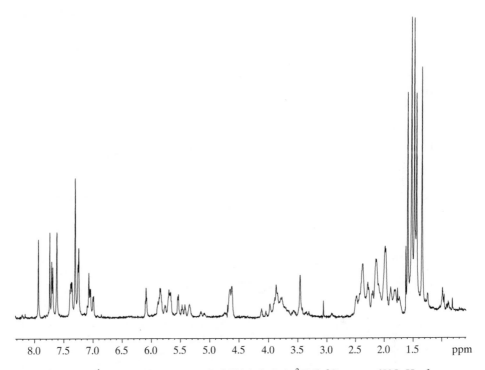

Figure 12.17   The ^{1}H 1D NMR spectrum of a DNA hairpin in ^{2}H$_2$O. [Courtesy of V. L. Hsu.]

Application 12.1   Investigating Base Stacking with NMR

We have already seen how IR absorption and UV circular dichroism played a role in confirming the Watson-Crick structure for DNA. Of course, NMR spectroscopy was also used, both to confirm base stacking and to determine the conformation of the base relative to the deoxyribose sugar. As an example we chose the early NMR investigation of the dimer rAprA by Chan and Nelson (1969), made at a time when only low-field 1D instrumentation was available. They observed only the C_2H and C_8H protons of the adenine ring (Figure A12.1), and these are far removed from the ribose resonances. The C_8H resonances were easily distinguished from the C_2H resonances because of their more rapid exchange with ^2H from the ^2H_2O solvent at higher temperatures. The Mn(II) paramagnetic ion binding to the phosphate groups and the temperature studies showed that the chemical shift dispersion for the non-equivalent adenines put the 3'-adenine at lower ppm.

Each adenine will have a large ring current that can affect the field seen by protons on the other adenine. We discussed the ring current effect in Section 12.2, and saw that protons above a ring are highly shielded, while protons beside a ring are highly deshielded. This research makes use of ring current shielding. Figure A12.1 shows the change in chemical shift of the C_2H and C_8H resonances as a function of temperature. We see that the

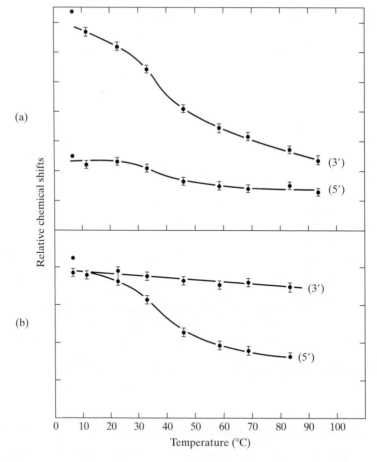

Figure A12.1 The relative chemical shifts for adenine protons in rAprA as a function of temperature (a) C_2H and (b) C_8H. [Adapted from data in Chan and Nelson 1969.]

C_2H resonance for the 5'-adenine ($C_2H(5')$) and the $C_8H(3')$ resonance are hardly affected by temperature. However, both $C_2H(3')$ and $C_8H(5')$ lose shielding with increasing temperature. The deshielding with increasing temperature must mean that the bases are stacked at low temperature. This agrees with the early CD experiments discussed at the end of Section 10.2. In addition, both adenines must be in the *anti* conformation with respect to the glycosidic bond to give the specific responses of the four protons seen in Figure A12.1. The bases are in the *anti* conformation in the Watson-Crick structure of DNA, with the bases rotated away from the deoxyribose. For the rAprA dimer, the stacked *anti-anti* conformation puts the $C_8H(3')$ and $C_2H(5')$ far from the opposing adenines, but the $C_2H(3')$ and $C_8H(5')$ over the opposing adenines.

CHAN, S. I. and J. H. NELSON (1969), *J. Amer. Chem. Soc.* **91**, 168–183.

12.7   TWO-DIMENSIONAL FOURIER TRANSFORM NMR

With its short pulses of energy that contain all the frequencies of interest, FT NMR is rapid, and therefore excellent for acquiring repeated scans that can be averaged to reduce noise. The use of pulses also introduces the possibility of simplifying spectra by looking only at the interactions between nuclei and spreading these interactions over two or more dimensions.

We have already discussed the two important interactions that are used to assign resonances and investigate the conformation of a biopolymer. First, spin-spin interactions through the bonds between nearby nuclei (that cause a splitting in the one-dimensional spectra) tell us about the bonding structure in a biopolymer. Conversely, if the bonding structure is known, the effect can be used to assign resonances. Second, the nuclear Overhauser effect (NOE), which changes the signal intensity of one nucleus when another nearby is irradiated, can be used to assign resonances to their order in the sequence of a biopolymer, and can be used to measure the distance between nuclei. Since this dipole-dipole interaction is through space, it is not only sensitive to nuclei that are close together because of bonding structure, but also to nuclei that are close together (about 0.5 nm) because of the conformational structure of the molecule. In using two-dimensional (2D) FT NMR, we plot data using two frequency axes, which will show a crosspeak at the frequencies corresponding to each pair of nuclei that interact. Usually, the 2D spectrum is either for the spin-spin interaction or for the nuclear Overhauser effect. The magnitude of the effect is represented by the intensity of the crosspeak; it is usually shown by the number of contours at the point of interaction, as diagrammed for a hypothetical molecule in Figure 12.18. The contours are like the contour map of a mountain and the more contours, the higher the peak corresponding to the interaction. The interaction may be between any two types of NMR-active nuclei: $^1H-^1H$, $^1H-^{13}C$, $^{13}C-^{13}C$, and so on. Since a nucleus will interact with itself, a 1D spectrum, as the contour lines of peaks, appears on the diagonal when the same 1D spectrum is plotted along both axes, as we see for the diagram in Figure 12.18.

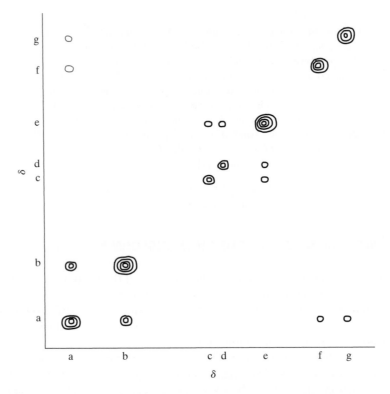

Figure 12.18   A two-dimensional FT NMR plot for a hypothetical molecule. The height of the peaks is denoted by the number of contour lines.

Two-dimensional spectra are generated by using a sequence of pulses. The pulse sequence can be divided into four periods—preparation, evolution, mixing, and detection—as shown in Figure 12.19. The preparation period is necessary to allow the excess nuclei to relax back down to the ground state. Although the FID is collected during the detection period, we remember that the decay in the FID is primarily due to spin-spin relaxation that creates randomness in the phase coherence, but does not bring the nuclei down to the ground state. The preparation period must be long enough to allow the slower spin-lattice relaxation to bring the excess excited nuclei down to the ground state. A pulse will begin the evolution period t_1, which is really the second dimension of the 2D spectrum. FIDs will be collected for different values of t_1, beginning at $t_1 = 0$ through a t_1 of up to a hundred milliseconds. Another pulse may begin a mixing period of time t_m, which would be roughly a hundred milliseconds and remains the same throughout the experiment. Finally, there is a detection period t_2, during which the FID is collected. This is the normal time dimension that we have already discussed in connection with collecting a 1D FT NMR spectra.

The two most common 2D pulse sequences are also given in Figure 12.19. The correlated spectroscopy (COSY) sequence consists of a 90° pulse, an evolution

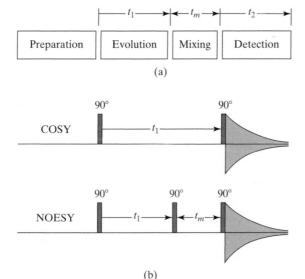

Figure 12.19 (a) A pulse sequence can be divided into four periods. (b) The two most common 2D pulse sequences.

period for the second dimension, and a 90° pulse just before detecting the FID. The mixing time is the duration of the second pulse. The COSY pulse sequence results in crosspeaks that correspond to standard short-range through-bond interactions of the spin-spin type. The nuclear Overhauser effect and exchange spectroscopy (NOESY) sequence consists of a 90° pulse that initiates the evolution period, a second 90° pulse that begins the mixing period, and a third 90° pulse that precedes the detection of the FID. The NOESY pulse sequence results in crosspeaks that are sensitive to the through-space nuclear Overhauser effect. There are many other pulse sequences that may be used, but we will limit our discussion to the fundamental COSY and NOESY experiments.

Figure 12.20 illustrates the collection and Fourier transformation of a 2D FT COSY spectrum for a hypothetical molecule. The COSY pulse sequence is shown in the upper left corner. We begin with a t_1 of 0, so that the 90° pulse at the beginning of the evolution period and the 90° pulse at the beginning of the detection period are nearly simultaneous. The FID is collected during time t_2. This is repeated a number of times, say, 64, with a preparation period in between each repeat, and the results averaged to lower the noise. Next, the evolution time is increased (for instance, a t_1 of 150 μsec), and 64 FIDs are collected, averaged, and stored in a separate location. This procedure is repeated, incrementing t_1 by 150 μsec each time, until we have a reasonable t_1 dimension. We might collect the FID in the t_2 direction for detection time of 300 msec, and 512 FIDs to correspond to 76.65 msec in the t_1 direction. The 512 FIDs, each an average of 64 scans, are transformed from the t_2 time domain into the ν_2 frequency domain to yield 512 one-dimensional spectra. The $t_1 = 0$ spectrum in the frequency domain would usually look like a typical 1D spectrum for the molecule. The other 1D spectra corresponding to the incremented t_1 time will have peaks in the same position, but their individual intensities will

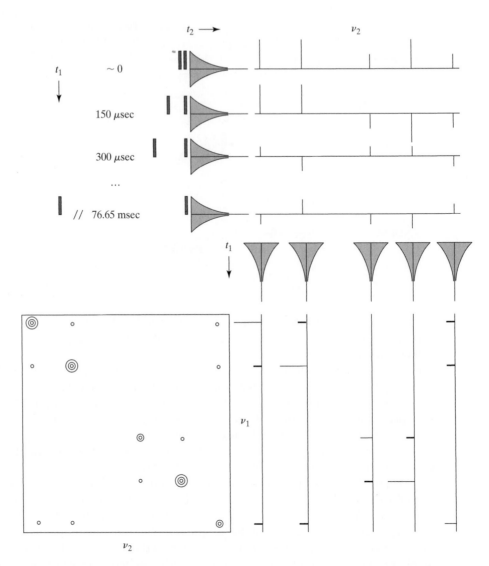

Figure 12.20 A simplified diagram of the pulse sequences for collecting a 2D FT NMR COSY and its transform to the 2D plot.

vary and may even be negative. We see this in the upper right corner of Figure 12.20. If we pick a particular frequency for the ν_2 1D spectra and plot the intensity versus t_1, the result will be a FID. This is illustrated in Figure 12.20 for the ν_2 frequencies that correspond to the five hypothetical peaks. In this example, the computer would consider 512 equally spaced frequency locations, but many of these locations would give FIDs of no intensity. The 512 FIDs in the t_1 direction are transformed to the ν_1 frequency domain. The hypothetical results for the five FIDs at the five ν_2

peaks are shown in the lower right corner of Figure 12.20. We assume that we are investigating the same nucleus in both dimensions, so there are intense resonances that correspond to the one-dimensional spectrum of that nucleus. There are also low intensity bands for nuclei that interact through the spin-spin mechanism. We can use the 512 ν_1 spectra corresponding to the different ν_2 frequencies to make a two-dimensional contour plot, as illustrated in the lower left corner of Figure 12.20. This is the COSY spectrum. A one-dimensional spectrum of the molecule is on the main diagonal. Spin-spin interactions are shown as low-intensity crosspeaks off the diagonal.

Let us consider the COSY of isoleucine in Figure 12.21 as an example of a 2D ^{1}H NMR spectrum. Only the 3.6 ppm resonance has a single crosspeak, so we know this must be due to the $C_\alpha H$. The crosspeak must be the through-bond spin-spin interaction with the $C_\beta H$, so we have identified the resonance at 1.9 ppm. We know that the $C_\beta H$ will also have interactions with the $C_\gamma H_3$ and the $C_\gamma H_2$ protons. Here, the two $C_\gamma H_2$ protons are not equivalent, and of course they interact with the $C_\delta H_3$ protons. This identifies the 1.4 and 1.2 ppm resonances as the $C_\gamma H_2$ protons as well as the 0.8 ppm resonance as the $C_\delta H_3$. We note that the two $C_\gamma H_2$ protons have a crosspeak due to their mutual interaction. We realize that the remaining 0.9 ppm resonance must be the $C_\gamma H_3$, which of course interacts with the $C_\beta H$. Thus, we see that a 2D COSY spectrum is an excellent way to identify NMR resonances.

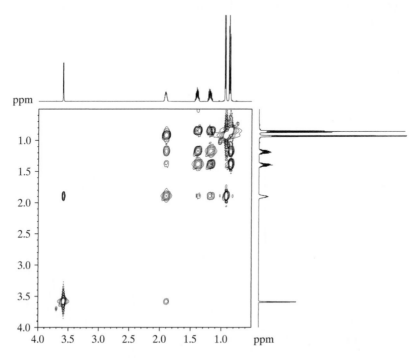

Figure 12.21   The 2D COSY spectrum of isoleucine with the corresponding 1D spectrum displayed along the axes. [Courtesy of V. L. Hsu.]

12.8   TWO-DIMENSIONAL FT NMR APPLIED TO MACROMOLECULES

Biologically active Arg-Gly-Asp (RGD) oligopeptides will provide us with a simple example of how 2D ^{1}H NMR might be used to investigate structure. The sequence RGD is interesting because it is the consensus sequence for proteins binding to membrane-spanning receptors in cells. The heptapeptide Tyr-Gly-Arg-Gly-Asp-Ser-Pro (YGRGDSP) binds tightly to membrane-spanning receptors. In contrast, the conservative change of aspartic acid for glutamic acid (E) gives a heptapeptide that does not bind to membrane-spanning receptors. Perhaps 2D ^{1}H NMR will reveal a conformational difference between the YGRGDSP and YGRGESP heptapeptides in solution, even though the rapid conversion among the various possible conformations available to this short peptide will give NMR parameters corresponding to an averaged structure.

Our first task is to assign the resonances in the 1D ^{1}H NMR spectrum of each heptapeptide, and the 1D spectrum of YGRGDSP is given in Figure 12.22. Since we are only dealing with seven amino acids, this 1D spectrum is relatively simple, with the resonances well separated. The only amino acid to appear twice is G, and the two Gs have different environments and would be expected to absorb at different frequencies (chemical shift dispersion). The spectrum was measured in H_2O at acid pH (pH 4), so that labile protons have a low exchange rate with the solvent. Thus, there are resonances due to the exchangeable protons as well as the nonexchangeable

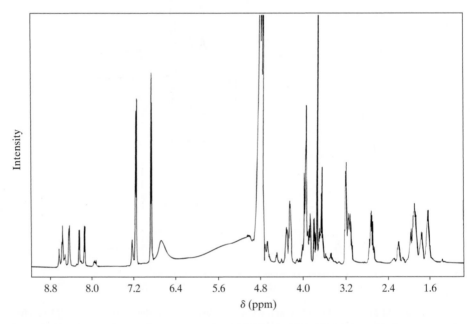

Figure 12.22 The 1D ^{1}H NMR spectrum of YGRGDSP in H_2O. The intense H_2O resonance is at 4.8 ppm and the dioxane reference at 3.75 ppm.

protons in the heptapeptide. Resonances for the labile NH protons of the amide are mandatory for NMR analysis of peptides and proteins.

As we discussed in Section 12.6, the approximate position of the resonances for the protons in each amino acid are well known from previous studies. Figure 12.23 shows the range of chemical shifts for different types of protons in peptides and Table 12.3 lists the proton chemical shifts for the 20 common amino acids. Quite obviously, different amino acids have different numbers of resonances depending on

Table 12.3 ^{1}H Chemical Shifts for the 20 Common Amino Acids (in ppm)

Residue	NH	$C_\alpha H$	$C_\beta H$	Others	
Gly	8.39	3.97			
Ala	8.25	4.35	1.39		
Val	8.44	4.18	2.13	$C_\gamma H_3$	0.97, 0.94
Ile	8.19	4.23	1.90	$C_\gamma H_2$	1.48, 1.19
				$C_\gamma H_3$	0.95
				$C_\delta H_3$	0.89
Leu	8.42	4.38	1.65, 1.65	$C_\gamma H$	1.64
				$C_\delta H_3$	0.94, 0.90
Pro(*trans*)		4.44	2.28, 2.02	$C_\gamma H_2$	2.03, 2.03
				$C_\delta H_2$	3.68, 3.65
Ser	8.38	4.50	3.88, 3.88		
Thr	8.24	4.35	4.22	$C_\gamma H_3$	1.23
Met	8.42	4.52	2.15, 2.01	$C_\gamma H_2$	2.64, 2.64
				$C_\varepsilon H_3$	2.13
Cys	8.31	4.69	3.28, 2.96		
Asp	8.41	4.76	2.84, 2.75		
Asn	8.75	4.75	2.83, 2.75	$N_\gamma H_2$	7.59, 6.91
Glu	8.37	4.29	2.09, 1.97	$C_\gamma H_2$	2.31, 2.28
Gln	8.41	4.37	2.13, 2.01	$C_\gamma H_2$	2.38, 2.38
				$N_\delta H_2$	6.87, 7.59
Lys	8.41	4.36	1.85, 1.76	$C_\gamma H_2$	1.45, 1.45
				$C_\delta H_2$	1.70, 1.70
				$C_\varepsilon H_2$	3.02, 3.02
				$N_\varepsilon H_3$	7.52
Arg	8.27	4.38	1.89, 1.79	$C_\gamma H_2$	1.70, 1.70
				$C_\delta H_2$	3.32, 3.32
				$N_\delta H$	7.17, 6.62
His	8.41	4.63	3.26, 3.20	$C_2 H$	8.12
				$C_4 H$	7.14
Phe	8.23	4.66	3.22, 2.99	$C_2 H, C_6 H$	7.30
				$C_3 H, C_5 H$	7.39
				$C_4 H$	7.34
Tyr	8.18	4.60	3.13, 2.92	$C_2 H, C_6 H$	7.15
				$C_3 H, C_5 H$	6.86
Trp	8.09	4.70	3.32, 3.19	$C_2 H$	7.24
				$C_4 H$	7.65
				$C_5 H$	7.17
				$C_6 H$	7.24
				$C_7 H$	7.50
				NH	10.22

Source: Adapted from data in K. Wüthrich (1986).

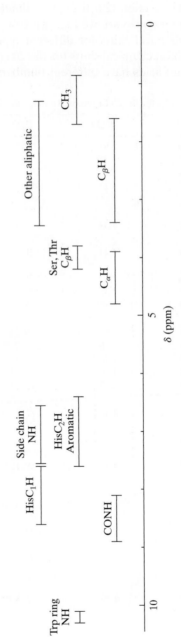

Figure 12.23 The range of chemical shifts for the different ^{1}H NMR resonances of the amino acids.

the side chain. The simplest is glycine, with only an amide NH resonance and one or two resonances for the α carbon protons. Considerably more complicated is trypto-phan with ten different resonances. Protons on adjacent atoms will have spin-spin in-teraction that give rise to crosspeaks in a COSY spectrum. Each of the 20 amino acids with their different side chains would have a characteristic pattern of COSY crosspeaks. We see the diversity of these connectivity patterns in the idealized COSY spectra for alanine, glutamic acid, and arginine of Figure 12.24. Connectivity patterns can be used to assign the resonances for each amino acid. However, some amino acids have quite similar patterns. These can often be differentiated through sequen-tial resonance assignment, since in general we know the sequence of our sample. Ser, Asp, Asn, Cys, Trp, Phe, Tyr, and His all give a three-spin (called AMX) side chain pattern with only $C_\alpha H$, and two $C_\beta H$ resonances. Other expected resonances and corresponding crosspeaks are not observed because the coupling is too weak or the protons exchange too rapidly, even at an acid pH. Similarly, Glu, Gln, and Met give a four-spin or AM(PT)X side chain pattern with only $C_\alpha H$, two $C_\beta H$, and $C_\gamma H$ reso-nances (Figure 12.24).

The COSY spectrum of YGRGDSP in Figure 12.25 shows crosspeaks be-tween protons on adjacent atoms. Great detail is present in such a spectrum, as we see in the portion shown in Figure 12.26, where we even see the spin-spin splittings of the crosspeaks. The pattern for arginine (R) is indicated in Figures 12.25 and 12.26 (compare with Figure 12.24), where we see crosspeaks for amide $NH/C_\alpha H$, $C_\alpha H/C_\beta H$, $C_\beta H/C_\gamma H$, $C_\gamma H/C_\delta H$, and $C_\delta H$ with one of the side chain NH reso-nances. Patterns such as this can be used to assign the resonances to amino acid *spin systems.*

While the COSY spectrum can be used to assign resonances to many kinds of amino acids, it cannot be used to distinguish among different amino acids with the same spin system. However, a NOESY spectrum will have additional crosspeaks be-tween the $C_\alpha H$ of amino acid i and the amide NH of $i + 1$, which can be used for *sequential assignment,* and thus distinguish between different amino acids with the same pattern. We see a portion of the NOESY spectrum in Figure 12.27, which

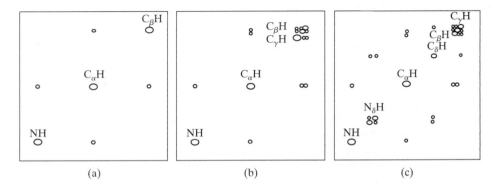

Figure 12.24   Idealized COSY spectra for (a) alanine, (b) glutamic acid, and (c) arginine.

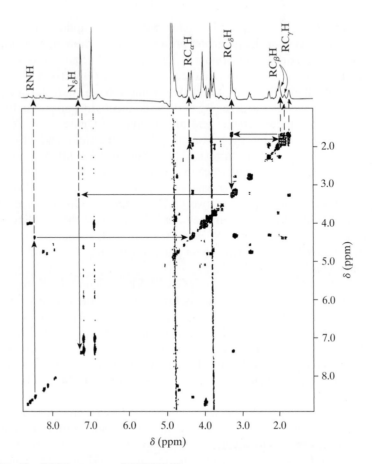

Figure 12.25   The COSY spectrum of YGRGDSP.

shows these additional crosspeaks for the RGD peptide. The NOESY peak for the amide NH of G interacting with the $C_\alpha H$ of Y1 identifies the G as the first glycine, G2. The G2 resonance for $C_\alpha H$ is identified by its NOESY interaction with the amide NH of R3. Similarly, the amide NH of G4 interacts with the $C_\alpha H$ of R3, and the $C_\alpha H$ of G4 interacts with the amide NH of D5. For this simple peptide, the NOESY peak between the $C_\alpha H$ for D5 and the amide NH for S6 provides no additional information. The P7 has no amide NH, and thus no interaction with the $C_\alpha H$ of S6. The $C_\delta H$ resonances of P are often used in sequential assignment. Sequential assignment with NOESY crosspeaks is essential for identification of each amino acid in the NMR studies of more complex proteins.

A heptapeptide is so short that it would be largely random in structure in aqueous solution, and the crosspeaks observed in the COSY and NOESY spectra are the time average of all of the conformations assumed. With the resonances assigned, we now look for unexpected NOESY crosspeaks between protons that are

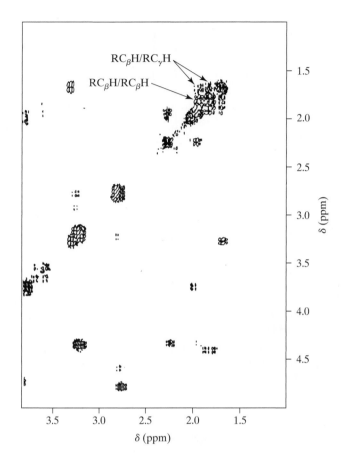

Figure 12.26 Great detail is seen in this expanded portion of the COSY spectrum for YGRGDSP, including the crosspeaks $RC_\beta H/RC_\beta H$ and $RC_\beta H/RC_\gamma H$.

normally far apart, but show up here because a conformational preference for this sequence brings them close together. We see such a crosspeak between the amide protons of aspartic acid and the glycine in the fourth position in Figure 12.28a. This crosspeak is characteristic of a type II β turn. The YGRGDSP heptapeptide is shown schematically in a type II β turn in Figure 12.29. In contrast, the YGRGESP heptapeptide has three amide NH/amide NH crosspeaks (Figure 12.28b) between R and G4, E and G4, and E and S. The R-G4 and E-G4 crosspeaks are indicative of a type I or type III β turn. A type I β turn would have the amide between G4 and R rotated 180° from the amide pictured in Figure 12.29. A type III β turn is a portion of a 3_{10} helix. The E-S crosspeak also suggests a 3_{10} helical structure. In any case, the two heptapeptides do have quite different conformational preferences in solution. Since their chemical properties are the same, it would appear that the conformational differences revealed by 2D ^{1}H NMR are relevant to the difference in the ability of these two sequences to bind to membrane-spanning receptors.

Clearly, the basic principles of 2D FT NMR spectroscopy that we have used to determine the structure of the RGD oligopeptides (assignment of spin systems from

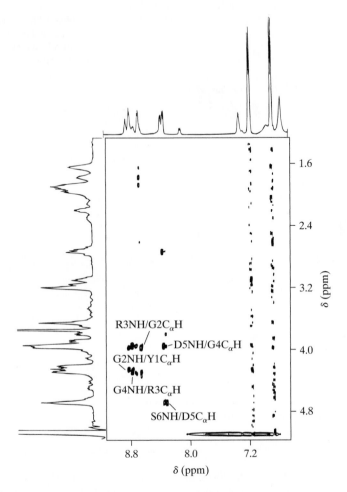

Figure 12.27   A portion of the NOESY spectrum for YGRGDSP.

a COSY, sequential assignment from a NOESY, structural information from a NOESY) can be applied to longer peptides and proteins. Like protein crystallography, the method can give a complete description of three-dimensional structure to atomic resolution, but has the important difference that it determines the structure in solution. As a practical matter, it becomes more and more tedious to assign the spin systems as the number of amino acids increases. However, construction of a NOESY-COSY connectivity diagram aids considerably in sequential assignments (Figure 12.30). The idea is simple. The NOESY crosspeaks can be used to connect the $C_\alpha H$ of residue i with the NH of residue $i + 1$. The COSY crosspeaks connect the NH with the $C_\alpha H$ of residue $i + 1$. By alternately using these COSY and NOESY connectivities, we can step through the sequential assignment along the backbone. We see the pathway for six residues in basic pancreatic trypsin inhibitor (BPTI) illustrated in Figure 12.30.

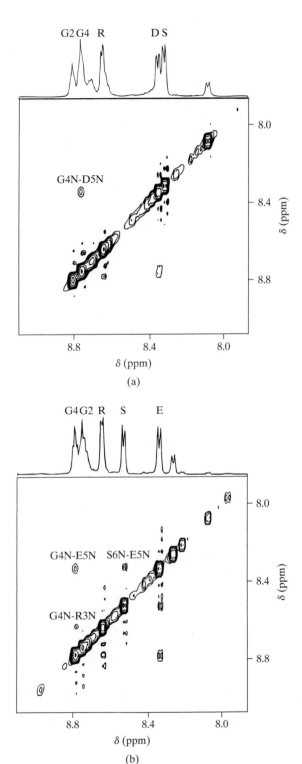

Figure 12.28 The amide portion of the NOESY spectrum for (a) YGRGDSP and (b) YGRGESP. [Reprinted from W. C. Johnson, T. G. Tagano, C. T. Basson, J. A. Madri, T. Gooley, and I. M. Armitage (1993), *Biochemistry* **32**, 268–273, copyright 1993 by The American Chemical Society.]

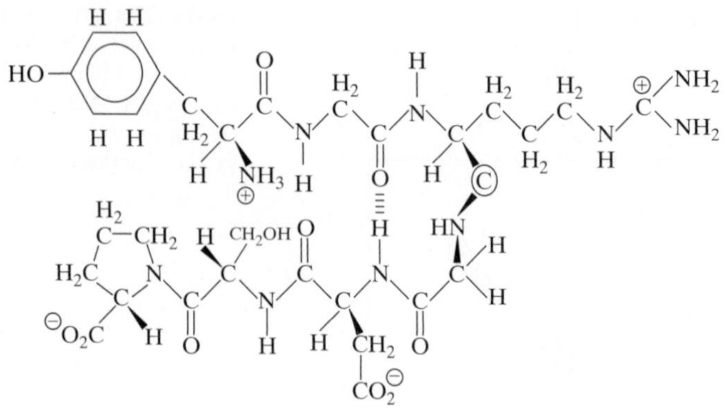

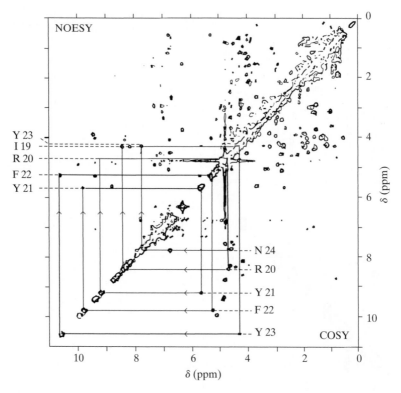

Figure 12.29 The heptapeptide YGRGDSP drawn as a type II β turn. [Reprinted from W. C. Johnson, T. G. Tagano, C. T. Basson, J. A. Madri, T. Gooley, and I. M. Armitage (1993), *Biochemistry* **32**, 268–273, copyright 1993 by The American Chemical Society.]

Figure 12.30 A NOESY-COSY connectivity diagram for sequential ¹H NMR assignments of BPTI. [Reprinted from G. Wagner, Anil-Kumar, and K. Wüthrich (1981), *Eur. J. Biochem.* **114**, 375–384, copyright 1981 by FEBS.]

Once the spin system and sequential assignments have been made, the three-dimensional structure of the protein can be investigated by using the NOESY crosspeaks between protons that are far apart in sequence but close enough in distance for the through-space interaction (up to 0.5 nm). The distance information from a ^{1}H–^{1}H NOESY is key to solving the structure. Spin-spin coupling constants are sensitive to torsion angles (Section 12.3), and provide additional constraints.

Secondary structures are easily identified early in the structure determination by the backbone crosspeaks. We have already seen that the type II β turn has an NH/NH crosspeak between residues 3 and 4, while a type I β turn has NH/NH crosspeaks between residues 2 and 3, and 3 and 4. The α helix and 3_{10} helix are regular secondary structures that have strong sequential NH/NH crosspeaks between residues i and $i + 1$ (corresponding to a distance between these protons of about 0.28 nm), and C_αH/NH crosspeaks between residues i and $i + 3$ (corresponding to a distance of about 0.32 nm). The β strands have strong sequential C_αH/NH crosspeaks between residues i and $i + 1$ (corresponding to a distance between these protons of about 0.22 nm). A complete discussion of the sequential and medium-range proton-proton distances for polypeptide secondary structures that give corresponding NOESY crosspeaks can be found in Wüthrich (1986). A diagram with the protein sequence and corresponding NOESY crosspeaks aids in the identification of secondary structure. We see this illustrated for human cyclophilin in Figure 12.31. Furthermore, secondary structures define specific torsion angles that result in specific spin-spin coupling constants. For instance, the α helix is expected to have sequential ^3J values between NH and C_αH of about 4 Hz due to the repeating ϕ angle of 60°. The β strand is expected to have sequential ^3J values between NH and C_αH of about 9 Hz due to the repeating ϕ angle of 120°. Secondary structure also affects the chemical shift dispersion of C_αH. This proton resonance is shifted to lower ppm for the α helix and higher ppm for the β strand.

The fundamental problem in using NMR to determine the structure of a protein is the assignment of the proton resonances. There exist methods for assignment that go beyond the COSY and NOESY techniques presented here. There are many pulse sequences that solve specific problems (see Kessler et al. 1988). For instance, the total correlation spectroscopy (TOCSY) is a popular 2D measurement that is sensitive to spin-spin interactions through a longer series of bonds than the COSY. This aids in resolving the ambiguities of similar spin patterns for different types of amino acids, assignments in side chains, and uncertainty due to crowding of crosspeaks. Since proteins contain carbon and nitrogen, there is the possibility of using ^{13}C and ^{15}N NMR. The natural abundance of these two isotopes is small (Table 12.1), but it is possible to do 2D ^{1}H–^{13}C *heteronuclear* experiments in the inverse mode where the ^{13}C is excited in t_1 and the ^{1}H is detected in t_2. If the protein being studied is a bacterial protein or has been cloned, it can be isotopically labeled with ^{13}C or ^{15}N. Then, NMR experiments can easily utilize ^{13}C and ^{15}N, as well as ^{1}H. It becomes possible to do 3D or 4D experiments that connect resonances and spread out crowded 2D regions over more dimensions. For instance, if the proton spin systems for two valines overlap, their α carbons might show chemical shift dispersion

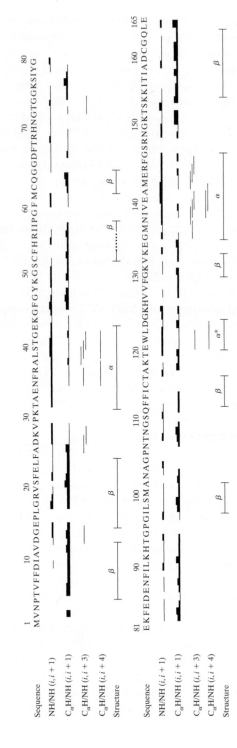

Figure 12.31 A diagram of the NOESY peaks for human cyclophilin. [Adapted from data in K. Wüthrich et al. (1991), *FEBS Lett.* **285**, 237–247.]

because of sequence effects. Spin-spin interactions of protons connected through the α carbons would be distinct (*edited* by the α carbons), and the two valines could be differentiated. By viewing ^{1}H–^{1}H interactions as edited by a heteronucleus, there is no increase in complexity and the interaction is labeled by the chemical shift of the heteronucleus. We see this in Figure 12.32 for a portion of the NOESY for interleukin 1β edited by ^{15}N resonating at 123.7 ppm. Heteronuclear experiments are mandatory for assignment purposes and for solving the structures of larger proteins by using NMR. Furthermore, large heteronuclear couplings are less sensitive to line broadening, increasing sensitivity so that larger proteins can be investigated.

The ideas behind using the NMR data to determine the three-dimensional structure of a protein are conceptually very simple. The sequence of the protein is known, as are all the distances between the covalently bonded atoms. The NMR investigation produces a table of NOE distances between backbone protons, side chain protons, and protons that are relatively far apart in sequence but close together because of

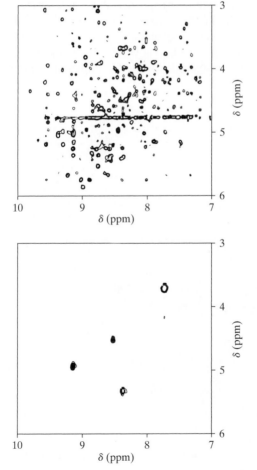

Figure 12.32   A portion of the multidimensional NOESY spectra of interleukin 1β. (top) The 2D spectrum edited for interactions with ^{15}N amide protons only. (bottom) The 3D spectrum edited for the particular ^{15}N amide protons resonating at 123.7 ppm. [Adapted from G. M. Clore, A. Bax, T. C. Driscoll, P. T. Wingfield, and A. N. Gronenborn (1990), *Biochemistry* **29**, 8172–8184; and P. C. Driscoll, G. M. Clore, D. Marion, P. T. Wingfield, and A. N. Gronenborn (1990), *Biochemistry* **29**, 3542–3556.]

structure (*distance geometry*), and perhaps a set of spin-spin coupling constants that give approximate torsion angles. Then it is simply a matter of rotating the amides about their phi-psi angles and the side chains around their single bonds to produce a packed structure that is consistent with the measured NMR data and yet does not put atoms closer together than their van der Waals distances. Of course, in practice this task is complicated and time-consuming; for a reasonably sized protein it would take more than a lifetime to do this by hand.

Fortunately, computer programs have been written to make use of the known structural parameters and the NMR data, and to produce a three-dimensional structure. The structure can be further refined by computer using *molecular mechanics,* which uses potential energy functions for all of the types of interactions (such as bond length, bond angle, hydrogen bonding, van der Waals interactions) and moves the structure around to minimize the total energy. The three-dimensional protein structure that results from molecular dynamics may be in a local potential energy minimum. Using *molecular dynamics* computer programs, each atom is assigned a velocity with magnitude and direction that is proportional to a temperature. The method uses a high temperature to shake the molecule out of a bad local minimum and then lowers the temperature in an attempt to anneal it into a deeper global potential energy minimum. If different starting conditions yield similar results from these computer programs, the structure is considered solved. As an example, Figure 12.33

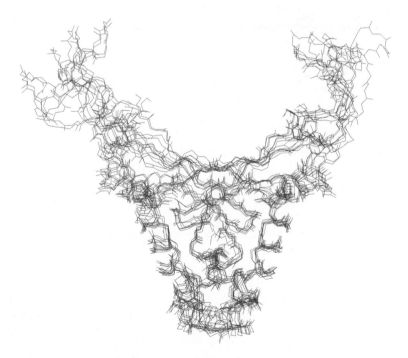

Figure 12.33   A set of eight structures for transcription factor 1 generated by a molecular mechanics program using NMR constraints. [Courtesy of V. L. Hsu.]

shows eight structures generated for transcription factor 1 by a restrained molecular dynamics computer program using NMR constraints. This is a dimer with 99 amino acids per monomer. We see that the structure is well defined except at the ends, where there are few NOEs to constrain the distances.

The methods for investigation of nucleic acids by NMR are similar to the methods we have already discussed for proteins. Although the labile imino protons of the bases are sensitive to hydrogen bonding in base pairs, they are not usually investigated for structure determination. Thus, NMR measurements of nucleic acids are often carried out in ^2H_2O where problems associated with the intense ^1H_2O peak are avoided, but the labile protons exchange with deuteriums from the solvent.

While there are 20 natural amino acids, there are only four different bases in a typical nucleic acid. Therefore, there is a lot more sequence degeneracy of near neighbors in nucleic acids, which causes overlap of the peaks. On the other hand, most proteins have more than 50 residues, while biologically important nucleic acid sequences (such as restriction sites and transcription factor binding sites) investigated by NMR are less complicated, few over 12 nucleotides long.

As we have already seen in Section 12.6, the bases and sugars of the nucleic acids have characteristic resonances resulting from their protons. The ranges for the resonances given in Figure 12.15 are conveniently wide; exceptions are the ribose protons of RNA, which are particularly crowded.

As with proteins, spin systems are identified by using through-bond interactions revealed by a COSY or related pulse sequence. However, there is no through-bond coupling between a base and its sugar, so the sugar spin systems and base spin systems are identified independently. Thymine shows a weak COSY crosspeak between the methyl protons and its C_6H, so its spin system can be unambiguously identified. Cytosine and uracil both have a C_5H/C_6H crosspeak that separates them from the purines, but have identical spin systems. The purines have no COSY crosspeaks.

We realize, however, that there are NOESY through-space crosspeaks between bases and sugars, as shown for a DNA in Figure 12.34. The C_6H of a pyrimidine or C_8H of a purine is near the base-sugar (glycosidic) bond, and interacts not only with the $C_1'H$, $C_2'H$, and $C_{2''}H$ of their own sugar, but also with these protons on the sugar linked with its 5′ carbon. These interactions give two routes for stepping through the sequential assignment of a known sequence. The $C_1'H$ route is the simplest because there are fewer possible crosspeaks, and is diagrammed for $d(CGCGAATTCGCG)_2$ in Figure 12.35. Crosspeaks may be missing or overlapped, so in practice a combination of routes may be utilized.

As with proteins, there is the possibility of using nuclei other than ^1H. The nucleus ^{31}P is a natural isotope and can be used in stepping from $C_4'H$ through P to $C_3'H$. Inverse ^{13}C experiments are also useful for obtaining geometric information.

We might expect that the NOESY can easily be used to distinguish between A- and B-DNA. In B-DNA, the C_6H/C_8H internucleotide distance of 0.22 nm and the $C_2'H/C_{2''}H$ internucleotide distance of 0.34 nm give much stronger crosspeaks than the comparable distances of 0.38 nm and 0.46 nm for A-DNA. Also, the distances in stepping through $C_2'H$ and $C_{2''}H$ are reversed in the two different secondary structures. For B-DNA, the C_6H or C_8H distance to $C_2'H$ is 0.39 nm

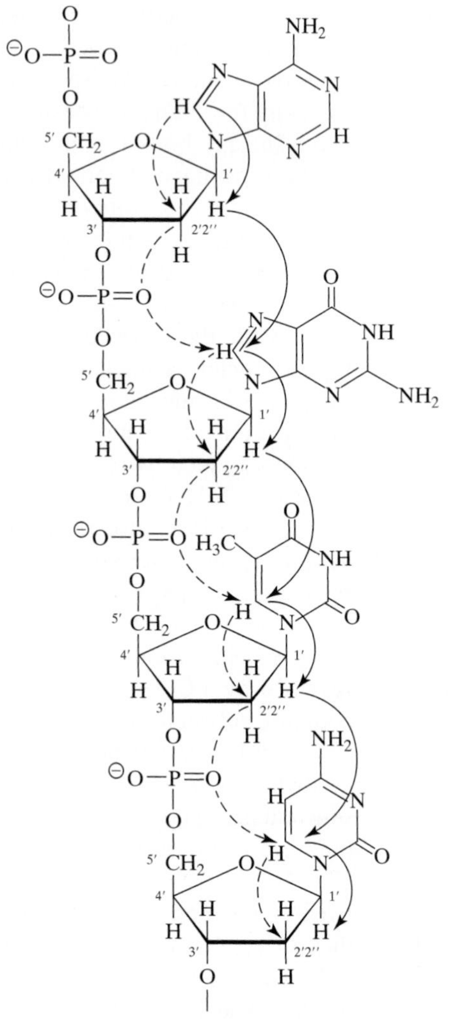

Figure 12.34 A diagram of the NOESY ^{1}H cross-peaks between the bases and sugars of DNA. The 1' route (—) and 2',2" route (- - -) for stepping through sequential assignments.

and to $C_{2''}H$ is 0.22 nm, while in A-DNA these distances are 0.17 nm and 0.32 nm. In addition, right-handedness can be inferred from the presence of a thymine methyl proton to adenine C_8H NOESY crosspeak.

Resonance assignment for RNAs is more difficult because of the crowded ribose resonances and the identical spin systems of cytosine and uracil. Deuteration of base protons and isotopic labeling are used to advantage and many structures have been solved (see the references at the end of this chapter).

As we expect from our discussion of proteins, NOESY crosspeaks that include interbase and interstrand interactions are used as distance constraints. Then the distance geometry in a restrained molecular dynamics computer program can be used to determine the structure.

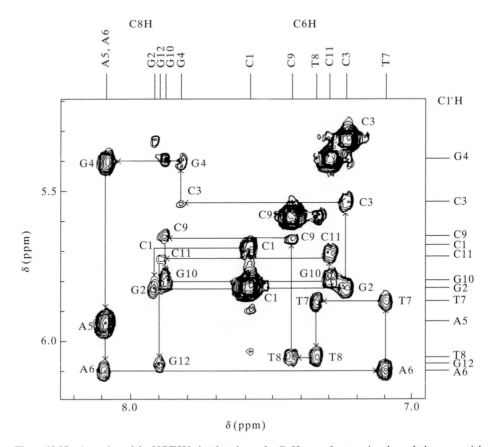

Figure 12.35 A portion of the NOESY plot that shows the $C_{1'}H$ route for stepping through the sequential assignments of the sequence d(CGCGAATTCGCG)$_2$. [Reprinted from D. R. Hare, D. W. Wemmer, S-H. Chou, G. Drobny and B. R. Reid (1983), *J. Mol. Biol.* **171**, 319–336, copyright 1983 by Academic Press.]

Nuclear magnetic resonance is an extremely versatile technique that goes far beyond the fundamental principles presented in this chapter. An entire book would be needed to cover all aspects of NMR spectroscopy, and some representative books are listed in the references.

EXERCISES

12.1 If the splitting for 2-propanol is $J = 9$ Hz, what does the Karplus equation estimate for the torsion angle between the C_2 proton and the hydroxyl proton?

12.2 Consider spin-spin (through-bond) splitting of proton peaks for CH_3—NH_2.
   a. Diagram the probabilities for the various interactions.
   b. Show the spectrum with expected splittings and relative intensities.

12.3 If the preparation time is 3 sec, t_1 varies from 0 to 51 msec at 200 μsec intervals, and the FID is collected for 1 sec, how many repetitions of each t_1 FID can one collect and complete the measurement of a COSY in about 12 hr?

12.4 Transform this FID to a band with frequency and bandwidth.

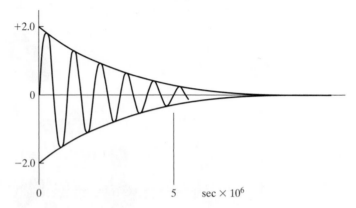

12.5 Given that Glu has the following peaks: NH = 8.4, αH = 4.3, βH = 2.1 and 1.9, γH = 2.3 for both, at what 2D coordinates do you expect COSY interactions?

12.6 In the NOESY below, the off-diagonal circles are crosspeaks that were in the corresponding COSY, while the crosses are new crosspeaks in the NOESY. Explain, using proton numbers, what each crosspeak tells us about the molecule.

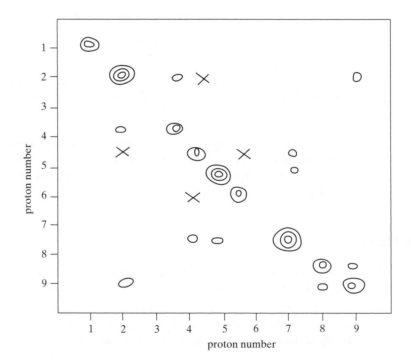

REFERENCES

General

CAVANAGH, J., W. J. FAIRBROTHER, A. G. PALMER III, and N. J. SKELTON (1996) *Protein NMR Spectroscopy, Principles and Practice,* Academic Press, New York.

EVANS, J. N. S. (1995) *Biomolecular NMR Spectroscopy,* Oxford University Press, London.

FREEMAN, R. (2003) *Magnetic Resonance in Chemistry and Medicine,* Oxford University Press, London.

HARRIS, R. K. (1986) *Nuclear Magnetic Resonance Spectroscopy, A Physicochemical View,* John Wiley & Sons, New York. A good beginning text.

JAMES, T. L. and N. J. OPPENHEIMER (eds.) "Nuclear Magnetic Resonance," in *Methods in Enzymology,* vol. 239, Academic Press, New York. Contains various topics by many experts in the field.

LEVITT, M. (2001) *Spin Dynamics: Basics of Nuclear Magnetic Resonance,* John Wiley & Sons, New York.

WILDER, G. and K. WÜTHRICH (1999) "NMR Spectroscopy of Large Molecules and Multimolecular Assemblies in Solution," *Curr. Opin. Struct. Biol.* **9**, 594–601.

WÜTHRICH, K. (1986) *NMR of Proteins and Nucleic Acids,* John Wiley & Sons, New York. An exceptionally clear description of how to investigate biopolymers by using 2D NMR.

Spin-Spin Splitting

KARPLUS, M. (1959) "Contact Electron-Spin Coupling of Nuclear Magnetic Moments," *J. Chem. Phys.* **30**, 11–15.

Pulse Sequences

BAIN, A. D. (1984) "Coherence Levels and Coherence Pathways in NMR. A Simple Way to Design Phase Cycling Procedures," *J. Magnetic Resonance* **56**, 418–427.

BODENHAUSEN, G., H. KOGLER, and R. R. ERNST (1984) "Selection of Coherence-Transfer Pathways in NMR Pulse Experiments," *J. Magnetic Resonance* **58**, 370–388.

KESSLER, H., M. GEHRKE, and C. GRIESINGER (1988) "Two-Dimensional NMR Spectroscopy: Background and Overview of the Experiments," *Angew. Chem. Int. Ed. Engl.* **27**, 490–536.

SORENSEN, O. W., G. W. EICH, M. H. LEVITT, G. BODENHAUSEN, and R. R. ERNST (1983) "Product Operator Formalism for the Description of NMR Pulse Experiments," *Progress in NMR Spectroscopy* **16**, 163–192.

2D NMR of Proteins

CLORE, G. M. and A. M. GRONENBORN (1994) "Multidimensional Heteronuclear Nuclear Magnetic Resonance of Proteins," in *Methods in Enzymology,* vol. 239, ed. T. L. James and N. J. Oppenheimer, Academic Press, New York.

WÜTHRICH, K., C. SPITZFADEN, K. MEMMERT, H. WIDMER, and G. WIDER (1991) "Protein Secondary Structure Determination by NMR. Application with Recombinant Human Cyclophilin," *FEBS Lett.* **285**, 237–247.

2D NMR of Nucleic Acids

HOSUR, R. V., K. V. R. CHARY, A. SHETH, G. GOVIL, and H. T. MILES (1988) "Refined Procedures for Accurate Determination of Solution Structures of Nucleic Acids by Two Dimensional Nuclear Magnetic Resonance Spectroscopy," *J. Biosci.* **13**, 71–86.

HOSUR, R. V., G. GOVIL, and H. T. MILES (1988) "Application of Two-Dimensional NMR Spectroscopy in the Determination of Solution Conformation of Nucleic Acids," *Magnetic Resonance in Chemistry* **26**, 927–944.

REID, B. R. (1987) "Sequence-Specific Assignments and Their Use in NMR Studies of DNA Structure," *Quart. Rev. Biophys.* **20**, 1–34.

VAN DE VEN, F. J. M. and C. W. HILBERS (1988) "Nucleic Acids and Nuclear Magnetic Resonance," *Eur. J. Biochem.* **178**, 1–38.

VARANI, G. and I. TINOCO, Jr. (1991) "RNA Structure and NMR Spectroscopy," *Quart. Rev. Biophys.* **24**, 479–532.

Others

BAX, A., M. IKURA, L. E. KAY, D. A. TORCHIA, and R. TSCHUDIN (1990) "Comparison of Different Modes of Two-Dimensional Reverse-Correlation NMR for the Study of Proteins," *J. Magnetic Resonance* **86**, 304–318. Analyzes inverse ^{1}H detected heteronuclear experiments.

CLORE, G. M. and A. M. GRONENBORN (1991) "Structures of Larger Proteins in Solution: Three- and Four-Dimensional Heteronuclear NMR Spectroscopy," *Science* **252**, 1390–1399.

13

Macromolecules in Solution: Thermodynamics and Equilibria

In the next two chapters, we describe certain aspects of the behavior of biological macromolecules in solution. We discuss membrane equilibria, provide a rigorous thermodynamic analysis of sedimentation equilibrium, and show how interactions of macromolecules with one another and with small molecules can be studied. All of these phenomena occur when the macromolecules are dissolved in solution and can be described quite accurately in thermodynamic terms. Therefore, we must begin with a short description of some of the special features of the thermodynamics of macromolecular solutions. Even students who have had a good physical chemistry course may find this section useful, for conventional courses tend to deemphasize solution thermodynamics.

The development of a true understanding of molecular biology was long inhibited by fundamental misunderstandings about the nature of solutions of proteins, polysaccharides, and other large molecules. In the early 1900s, recognition of some of the remarkable features of the behavior of such solutions led to the idea that these were not true solutions at all. Because they exhibited no measurable freezing-point depression and since the solutes could not easily be made to crystallize and exhibited very slow diffusion, such solutions were called *colloids*. This carried the unfortunate connotation that the well-known physicochemical laws for "true" solutions should not apply and that the application of conventional solution thermodynamics in this field was of doubtful validity.

Of course, we now know that proteins, polysaccharides, nucleic acids, and polymers in general form true solutions. The freezing-point depression is indeed small and diffusion is slow, but these are only consequences of the high molecular weights of the solutes. Crystallization is more difficult but possible in most cases. There is only one cause for reservation. Valid thermodynamic descriptions of solutions are dependent on there being a very large number of solute particles in the sample observed, so that macroscopic fluctuations in properties will be very unlikely. Is this condition satisfied for biopolymers? If we consider an extreme case, a solution containing 0.01 mg/ml of a virus of molecular weight 100 million, we still find approximately 10^{10} particles per milliliter. This is a number that is large enough that we need not worry about fluctuations in macroscopic volumes. It is only when we begin considering volumes comparable to that of a single cell that such questions need be raised.

Recognition that a cell may contain only a small number of molecules of some particular protein has recently increased interest in macromolecular behavior at the single molecule level. At the same time, the development of powerful techniques that allow the study of single molecules under near-physiological conditions has led to the development of a whole new field of *single molecule biophysics*. We shall describe some of these methods and their applications in Chapter 16.

13.1   SOME FUNDAMENTALS OF SOLUTION THERMODYNAMICS

13.1.1   Partial Molar Quantities: The Chemical Potential

A solution is a single-phase system containing more than one component. A *component* is an independently variable chemical substance. It should be noted that this definition of *component* is strict and that a solution will frequently contain fewer components than the molecular species that might be present. Such a situation will occur whenever chemical equilibria exist in the solution. To take an example, consider a solution containing water, the protein hemoglobin (Hb), and dissolved oxygen. Since each hemoglobin molecule can bind one, two, three, or four oxygen molecules, there will be a number of molecular species potentially or actually present: H_2O, Hb, O_2, HbO_2, $Hb(O_2)_2$, $Hb(O_2)_3$, and $Hb(O_2)_4$. Yet if the binding reactions are in equilibrium, there are only three *independently variable* substances or components. Specification of the amounts of the solvent plus any two of the others would, via the equilibrium relationships, determine the rest. Specifying these amounts, together with the temperature and pressure, will completely define the state of the system. It is important to emphasize that such simplification is possible only if the system is in equilibrium. In the case above, if the oxygenation or deoxygenation reactions were very slow, we might perturb the system (put in some more oxygen, for example) and observe the system in a nonequilibrium state. In this case, more than three components would be needed to describe the system.

The description of the state of a solution by stipulation of T, P, and the amounts of each of the n components may be thought of as a recipe. It means that if, on two or more occasions, we fix these variables at the same values, we shall find exactly the same properties, both extensive and intensive. We may logically ask, then, how some extensive property, such as the volume, will depend on the amounts of the various components. It would be extremely naive to assume simple additivity. If we imagine, for example, mixing NaCl and water, we know that the noncovalent interaction of water molecules with sodium and chloride ions is very different from the interaction that occurs in pure salt crystals or pure water. Thus, we would not expect that mixing together n_i moles of each component would necessarily give a total volume $V = \Sigma_i n_i V_i^0$, if V_i^0 is the molar volume of each pure component. In many cases, volume changes on mixing can easily be observed. The true situation may be described as follows: If we add a small amount of component i to a mixture, the change produced in an extensive property will depend not only on the amount and nature of the substance added but also on the composition of the mixture to which the addition is made. As we add salt to an increasingly concentrated solution, the environment into which these ions enter changes.

This concept leads to the definition of *partial molar* and *partial specific* quantities. Considering any extensive property X, we define the partial molar quantity $\bar{X}_i$ as

$$\bar{X}_i = \left(\frac{\partial X}{\partial n_i} \right)_{T, P, n_{j \neq i}} \tag{13.1}$$

where n_i is the number of moles of component i. The subscripts indicate that we hold T, P, and the amounts of all the components except i constant, and therefore $\bar{X}$ measures the infinitesimal change in X produced by changing *only* n_i by an infinitesimal amount. The corresponding partial specific quantity $\bar{x}_i$ is given by

$$\bar{x}_i = \left(\frac{\partial X}{\partial g_i} \right)_{T, P, g_{j \neq i}} \tag{13.2}$$

where g_i represents grams of component i. The partial specific quantities are frequently used in macromolecular chemistry, since we often do not know the molecular weight and hence the number of moles in a sample. The change dX in property X caused by adding an amount dn_i moles or dg_i grams is then

$$dX = \bar{X}_i dn_i = \bar{x}_i dg_i \tag{13.3}$$

To make these definitions concrete, let us consider the volume of a solution V as the extensive property. The partial molar volume $\bar{V}_i$ and partial specific volume $\bar{v}_i$ of component i are then given by

$$\bar{V}_i = \left(\frac{\partial V}{\partial n_i} \right)_{T, P, n_{j \neq i}} \tag{13.4a}$$

and

$$\bar{v}_i = \left(\frac{\partial V}{\partial g_i} \right)_{T, P, g_{j \neq i}} = \frac{\bar{V}_i}{M_i} \tag{13.4b}$$

respectively. Figure 13.1 provides a visualization of a quantity like $\bar{V}_i$, and shows how it is related to $\bar{V}_i^0$, the molar volume of a pure component.

These partial quantities are intensive variables, dependent on T, P, and the composition of the system, but not on its total size. This leads to a very important result, which bears on the original question as to how *extensive* properties may be calculated. Suppose that we imagine putting together a volume V of solution by

Figure 13.1 Increase in volume upon adding solute to pure solvent. The broken line represents what would be observed if $\bar{V}_2 = V_2^0$ and volumes were additive. The solid line represents a more realistic case, and the slope of the tangent line equals $\bar{V}_2$ at the indicated concentration.

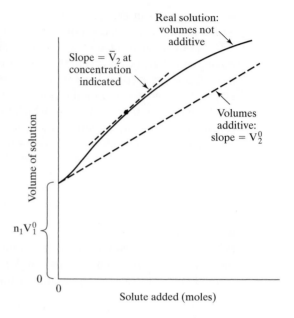

starting from nothing and adding infinitesimal increments of the n components, always in the proportions to be found in the final mixture. That is, we make $dg_1 = g_1 d\lambda$, $dg_2 = g_2 d\lambda$, ... $dg_i = g_i d\lambda$, where $g_1 g_2, \ldots, g_i$ are constants equal to the number of grams of each substance in the final mixture and λ is a dummy variable going from 0 to 1 as the solution is assembled. With this method of assembly, the composition will always be the same, and each $\bar{v}$ will therefore be constant during construction of the solution. Now

$$dV = \bar{v}_1 dg_1 + \bar{v}_2 dg_2 + \cdots + \bar{v}_i dg_i + \cdots + \bar{v}_n dg_n$$
$$= g_1 \bar{v}_1 d\lambda + g_2 \bar{v}_2 d\lambda + \cdots + g_i \bar{v}_i d\lambda + \cdots + g_n \bar{v}_n d\lambda \qquad (13.5)$$

Since both g_i and $\bar{v}_i$ are constants, we may integrate easily.

$$V = (g_1 \bar{v}_1 + g_2 \bar{v}_2 + \cdots + g_i \bar{v}_i + \cdots + g_n \bar{v}_n) \int_0^1 d\lambda \qquad (13.6a)$$

$$= \sum_{i=1}^{n} g_i \bar{v}_i \qquad (13.6b)$$

The total extensive quantity is the sum of the partial specific quantities, each multiplied by the number of grams of the appropriate component. Similarly, with partial molar quantities

$$X = \sum_{i=1}^{n} n_i \bar{X}_i \qquad (13.7)$$

These are the appropriate summation rules for real solutions.

A quantity that will be of the utmost significance in many of our problems is the partial molar Gibbs free energy, $\overline{G}_i$. We shall again and again be concerned with equilibria of multicomponent systems at constant T and P, which are normal laboratory and physiologic conditions. Therefore, the Gibbs free energy of the solution will be the property of interest, for it is this quantity that is minimized for a system at equilibrium at constant T and P. (See Chapter 2, Section 2.3.) Hence, we must know the contribution of each component to the total free energy of a solution. The partial molar Gibbs free energy is given the special symbol μ_i and is defined as

$$\overline{G}_i = \mu_i = \left(\frac{\partial G}{\partial n_i}\right)_{T,\,P,\,n_{j \neq i}} \tag{13.8}$$

The total free energy of the solution is then

$$G = \sum_{i=1}^{n} n_i\, \mu_i \tag{13.9}$$

The partial molar free energy is often called the *chemical potential*. This term is appropriate, for, as we shall see, differences in μ_i values may be regarded as the potentials driving such processes as chemical reactions or diffusion, in which changes in the amounts of chemical substances occur. It is for these reasons that we attach so much importance to this concept.

For an *open system*, in which amounts of components can vary, G is a function of T, P, and all of the n_i. The change in G for an infinitesimal change in the state of the system is then given by the general formula

$$dG = \left(\frac{\partial G}{\partial T}\right)_{P,\,n_i} dT + \left(\frac{\partial G}{\partial P}\right)_{T,\,n_i} dP + \sum_i \left(\frac{\partial G}{\partial n_i}\right)_{T,\,P,\,n_{j \neq i}} dn_i \tag{13.10}$$

or, using standard thermodynamic relationships,

$$dG = -S\,dT + V\,dP + \sum_i \mu_i\, dn_i \tag{13.11}$$

At constant T and P, we have

$$dG = \sum_i \mu_i\, dn_i \tag{13.12}$$

This equation shows how G varies with composition of an open system at constant T and P; we shall use it as a starting point for many different analyses. In addition, from the results above we can obtain an important and useful theorem, the *Gibbs-Duhem equation*. If we differentiate Eq. 13.9 in the most general way, taking into account variation in both n_i and μ_i, we obtain

$$dG = \sum_i \mu_i\, dn_i + \sum_i n_i\, d\mu_i \tag{13.13}$$

Combining this with Eq. 13.12, we have the *Gibbs-Duhem equation*

$$\sum_i n_i \, d\mu_i = 0 \tag{13.14}$$

This means that variations of μ_i are not wholly independent of one another. In a two-component system, for example, the Gibbs-Duhem equation tells us that changes causing an increase in the chemical potential of solute will result in a decrease in the chemical potential of solvent, and vice versa.

The importance of the chemical potential arises from the fact that it measures that increment of free energy accompanying an infinitesimal change in the amount of one particular component in a system. Thus, it is immediately applicable to discussions of phase equilibria and chemical equilibria. A particular example, which we shall use shortly, is the relationship between the chemical potential values of a substance in different *phases* of a multiphase system. The result can be stated simply: If a number of phases are in equilibrium, the chemical potential of a given component will have the same value in all phases into which that component can freely pass. While the general proof will not be given here, the principle involved is obvious— the transfer of an infinitesimal amount of a component between two phases (α and β) *in equilibrium* at constant T and P must involve a zero free-energy change. This can only be so if μ_i is the same in both phases, since, from Eq. 13.12

$$dG = \mu_i^\alpha \, dn_i^\alpha + \mu_i^\beta \, dn_i^\beta = 0 \tag{13.15}$$

but since $dn_i^\beta = -dn_i^\alpha$, we have

$$\left(\mu_i^\alpha - \mu_i^\beta\right) dn_i^\alpha = 0 \tag{13.16}$$

Since n_i are independent variables, Eq. 13.16 can only be true if

$$\mu_i^\alpha = \mu_i^\beta \tag{13.17}$$

This principle will be the starting point for discussion of such problems as the analysis of membrane equilibria.

13.1.2 The Chemical Potential and Concentration: Ideal and Nonideal Solutions

Equations such as 13.17 are of practical importance because the chemical potential of a substance in a mixture depends on its concentration. It is this dependence that allows us to link very general equations concerning equilibrium to specific experimental problems where we observe variations in the concentrations of substances. To make that linkage we must ask: How does μ_i depend on the concentration of i? To answer this, we must inquire a bit more into the nature of solutions.

In physical chemistry, the distinction is made between *ideal* and *nonideal* solutions. This is useful, since ideal behavior leads to exceedingly simple laws, which serve as prototypes for the more complex laws describing real nonideal solutions. There are a number of ways of approaching the definition of solution ideality. Many texts simply state that an ideal solution is one in which the chemical potential of each component is given by

$$\mu_i = \mu_i^0 + RT \ln \chi_i \tag{13.18}$$

Here χ_i is the mole fraction of i, R the gas constant, and T the absolute temperature. The quantity μ_i^0 is the *standard state* chemical potential, which in this case equals the molar free energy of pure component i, since $\mu_i = \mu_i^0$ when $\chi_i = 1$.

Equation 13.18 is correct but unilluminating. What *properties* make a solution ideal? We can show that Eq. 13.18 will obtain if the following conditions apply

1. The enthalpy change in mixing is zero,

$$\Delta H_m = 0 \tag{13.19a}$$

2. The entropy change in mixing is given by the simple statistical rule (see Chapter 2).

$$\Delta S_m = -R \sum_{i=1}^{n} n_i \ln \chi_i \tag{13.19b}$$

This definition is easy to visualize on a molecular basis; it means that there is no difference in interaction *energy* between solute and solvent molecules ($\Delta H_m = 0$) and that the entropy change arises *only* from the randomness produced by mixing the two kinds of molecules together. From Eqs. 13.19a and 13.19b, we have, for the free energy of mixing,

$$\Delta G_m = \Delta H_m - T \Delta S_m = RT \sum_i n_i \ln \chi_i \tag{13.20}$$

for an ideal solution. But in general, the free energy of mixing can also be defined by

$$\Delta G_m = G_{(\text{solution})} - \sum_i G_{i(\text{pure components})} \tag{13.21}$$

The free energy of n_i moles of a particular pure component will be given by $n_i \mu_i^0$. Then from Eqs. 13.9 and 13.21

$$dG_m = \sum_i n_i \mu_i - \sum_i n_i \mu_i^0 \tag{13.22}$$

or, setting this equal to the expression in Eq. 13.20,

$$\sum_i n_i (\mu_i - \mu_i^0) = RT \sum_i n_i \ln \chi_i \tag{13.23a}$$

or

$$\sum_i n_i(\mu_i - \mu_i^0 - RT \ln \chi_i) = 0 \qquad (13.23b)$$

Since the components of a solution are independently variable, this equation can only be generally satisfied if

$$\mu_i - \mu_i^0 = RT \ln \chi_i \qquad (13.24)$$

which is the result given in Eq. 13.18.

Although the mole fraction scale is sometimes satisfactory for describing the behavior of the solvent in a solution of macromolecules, we usually do not employ this concentration scale for a solute. We are interested in solute components that are present only at very low concentrations, and therefore the pure component as a standard state is inconvenient. Furthermore, as biochemists we are much more accustomed to molarity or weight concentration scales than mole fraction. At low concentrations, a weight or molar concentration will be proportional to the mole fraction. Since it is the logarithm of the concentration that occurs in Eq. 13.24, any proportionality constant can be absorbed in a redefined μ_i^0 (see Exercise 13.2). Thus, we may equally well write for the solute in dilute ideal solutions

$$\mu_i = \mu_i^0 + RT \ln C_i \qquad (13.25)$$

where C_i denotes concentration in grams per liter (or milligrams per milliliter). The value of μ_i^0, as well as the definition of the standard state, is understood to depend on the concentration scale used. In this case, μ_i^0 is the chemical potential of the ideal solute at unit concentration, for example, when $C_i = 1$ g/l.

Deviations from ideal behavior can be taken care of in a purely formal way by defining an activity or an activity coefficient. Thus, the analog of Eq. 13.25 for a non-ideal solution can be written as

$$\mu_i = \mu_i^0 + RT \ln C_i y_i = \mu_i^0 + RT \ln a_i \qquad (13.26)$$

where y_i is a function that describes all deviations from ideality and a_i is the *activity* or *effective* concentration, defined as $C_i y_i$. It should be noted that activity is a unitless quantity, which allows taking its logarithm to be a meaningful mathematical operation. The activity coefficient has units of inverse concentration, and we should think of it as always being present in the expression for μ_i, even though its value may be unity. In general, y_i will be a function of T, P, and *all* of the solute concentrations. We expect that, in general, solutes will approach ideal behavior in very dilute solution. Therefore, we expect y_i to approach unity as the total concentration of *all* solutes in the solution approaches zero.

In discussing some properties of macromolecular solutions (membrane equilibria, for example), we shall be interested primarily in the chemical potential of the solvent. For dilute solutions it is most practical to take the standard state for solvent as pure solvent and to use the mole fraction scale. Because we are primarily interested in the effects produced by the *solute* on the chemical potential of the solvent, it will usually be more convenient if we can express the solvent chemical potential in terms of *solute* concentration. For a two-component solution, $\chi_1 = 1 - \chi_2$ (where χ_2 is the mole fraction solute) and since $\chi_2 \ll 1$, we have $\ln(1 - \chi_2) \cong -\chi_2 - \chi_2^2/2 - \cdots$. Dropping higher terms in the expansion of $\ln(1 - \chi_2)$, we find, for dilute ideal solution

$$\mu_1 - \mu_1^0 \cong -RT(\chi_2 + \cdots) \tag{13.27}$$

It is almost always more useful to express the solute concentration in terms of C_2 rather than χ_2. For dilute solutions

$$\chi_2 \cong \frac{C_2 V_1^0}{M_2} \tag{13.28}$$

where V_1^0 is the molar volume of pure solvent. Therefore, Eq. 13.27 becomes

$$\mu_1 - \mu_1^0 \cong -\frac{RTV_1^0 C_2}{M_2} \tag{13.29}$$

The negative sign is in accord with our intuitive expectation that the chemical potential of the solvent should be reduced by adding solute.

If the solution is nonideal, we shall find that Eq. 13.29 fails at higher concentrations of the solute. This suggests that we might express that nonideality by adding terms in higher powers of the concentration to Eq. 13.29—that is, using a power series to describe the more complicated concentration dependence of μ_i for a nonideal solution. So we write what is termed the *virial expansion* of μ_1

$$\mu_1 - \mu_1^0 = -RTV_1^0\left(\frac{C_2}{M_2} + BC_2^2 + DC_2^3 + \cdots\right) \tag{13.30}$$

The quantities $1/M_2$, B, D, and so forth, are known as the first, second, third, and higher *virial coefficients*. In most cases, terms higher than quadratic in concentration are not necessary, so the second virial coefficient B serves as a convenient measure of solution nonideality. It will be zero for an ideal solution. We can make use of the Gibbs-Duhem equation to convert Eq. 13.30 into an expression for the *solute* chemical potential (see Exercise 13.1). The result, carried to the second virial coefficient, is

$$\mu_2 = \mu_2^0 + RT \ln C_2 + 2BRT\, M_2 C_2 \tag{13.31}$$

It is not difficult to make a qualitative argument to explain what features of molecular behavior should determine the sign and magnitude of the second virial coefficient. A positive value of B means that the chemical potential of the solvent decreases more rapidly, and the chemical potential of the solute increases more rapidly with the solute concentration than would be expected for an ideal solution

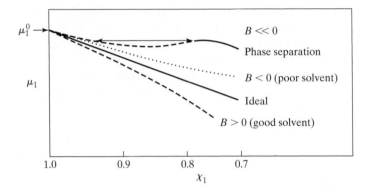

Figure 13.2   A schematic diagram of the behavior of the solvent chemical potential near $\chi_1 = 1$ for ideal and nonideal solutions. A positive value of B makes the solvent chemical potential drop even more rapidly than in an ideal solution. For $B \ll 0$, phase separation can occur, yielding the phases given by the ends of the arrow.

(Figure 13.2). This can be accounted for by either strong positive interaction between the solute and solvent (solvation) or by repulsion between solute molecules. Repulsion may be direct (as, for example, by electrostatic repulsion between macroions of like charge) or indirect (as in the simple exclusion of one molecule from space occupied by another). This *excluded volume effect* is particularly pronounced for highly extended and swollen molecules, like denatured proteins or nucleic acids. (See Table 13.1.)

A negative value of B means that the chemical potential of the solvent decreases *less* rapidly than expected for an ideal solution. This will be expected, for example, when solute-solute interactions are favored over solute-solvent interactions. The solute molecules are interacting positively with one another, rather than tying up solvent molecules. As is indicated in Figure 13.2, a very negative value of B can lead to phase separation. An example familiar to protein chemists is depicted in Figure 13.3. For many proteins, B is positive at high or low pH because the net charge on the molecule causes solute-solute repulsion. But near the isoelectric point this repulsion vanishes, and favorable protein-protein interactions become dominant. Thus, B diminishes and can even become negative in this pH region.

Table 13.1   Solution Nonideality

Particle	B^a	$(1 + BM_2C_2)^b$
Sphere, $M = 10^5$	3×10^{-5}	1.015
Rod, $L/d = 100$, $M = 10^5$	7.5×10^{-4}	1.375
Random coil, $M = 10^5$, good solvent	5×10^{-4}	1.250

[a]Order of magnitude in units of cm^3 mol/gm^2.
[b]Assuming that $C_2 = 5$ mg/ml.

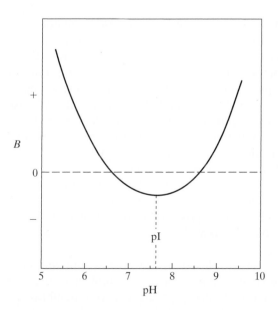

Figure 13.3 Variation of second virial coefficient with pH for a globular protein. The minimum value occurs at the isoelectric point. In some cases, the minimum will be so negative that phase separation (precipitation) may occur at this pH. This is the basis of isoelectric precipitation for the purification of proteins.

Sometimes phase separation (precipitation) will occur. This is the basis for *isoelectric precipitation* as a step in protein purification.

For the student of biochemistry, the nonideal behavior of some macromolecular solutions has another practical consequence: Erroneous answers can be obtained in physical measurements of such quantities as molecular weights and molecular dimensions, as in light-scattering (see Chapter 7) or sedimentation equilibrium (see Chapter 5, and Section 13.2), if nonideality is neglected. The best way to avoid such errors is to carry out such experiments at very low-solute concentration, where ideal behavior will be approached. If that cannot be done, methods to extrapolate data to zero-solute concentration (based on the virial expansion) can be devised. We have given an example of such an extrapolation in describing the analysis of light-scattering data (see Chapter 7), and the appropriate expression for sedimentation equilibrium is derived in Exercise 13.13.

13.2   APPLICATIONS OF THE CHEMICAL POTENTIAL TO PHYSICAL EQUILIBRIA

13.2.1   Membrane Equilibria

We now turn to applications of the principles developed above. Perhaps the simplest biochemical examples are found in equilibria across membranes.

Semipermeable membranes play a number of important roles in biochemistry. On the practical side, we employ them in dialysis and membrane filtration for the purification of macromolecular substances. Also, equilibrium across membranes is used to measure the binding of small molecules and ions to large molecules (see the section on dialysis equilibrium below) and has been historically important in determining the weights of macromolecules via osmotic pressure measurements. In the cell, membranes take on a similar role, serving to partition regions of the cell from one another, and the cell from its surroundings, with barriers that retain some substances and allow others to pass. The principal functional difference between the membranes that we employ in the laboratory and those found in the cell (aside from composition) is that the former are *passive*; that is, they act only as barriers and play no active role in the transport of materials. The *active transport* found in many cell membranes is a very different thing. Here, at a free-energy price, materials are selectively transported against concentration differences.

We shall be concerned exclusively with the equilibrium phenomena arising from the existence of passive semipermeable membranes in a system. We shall first write some general rules, then discuss dialysis equilibrium and osmotic pressure, and finally turn to the complications introduced when some of the solutes in the solutions carry an electrical charge.

Consider a solution that is separated from solvent by a semipermeable membrane, as in Figure 13.4. We shall assume that solvent molecules and some low-molecular-weight solute components can pass through the membrane, whereas some kinds of large solute molecules cannot. In practice, the membranes used most commonly in biochemical laboratories pass materials of $M < 10,000$ and retain larger molecules, but membranes can be prepared that select at much higher or much lower molecular weight.

The system must be regarded as consisting of two phases, α and β, because the large molecules are restricted to one side, which means that the system cannot be homogeneous throughout. For equilibrium to exist in this system, Eq. 13.17 requires that $\mu_i^\alpha = \mu_i^\beta$ for all substances that can pass through the membrane. For those substances that cannot pass through, we can make no corresponding statement. Equation 13.17 is the starting point for all discussions of membrane equilibria. For a

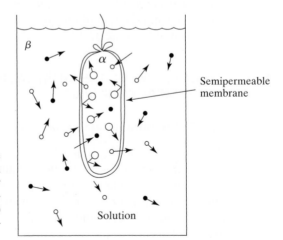

Figure 13.4   A semipermeable membrane separating one phase from another. Drawn here is the simplest arrangement commonly used for dialysis in biochemistry laboratories. Some molecules (large open circles) cannot pass the membrane; solvent and others (small circles) can.

system composed, for example, of a protein, sucrose, buffer salts, and water with a membrane through which only the protein cannot pass, Eq. 13.17 applies to each of the components except the protein.

Dialysis equilibrium.   Frequently, use is made of the membrane equilibrium to purify macromolecules or to change their solvent environment. This process is called *dialysis* and it is used in this way in almost every biochemical laboratory. In its simplest application, the macromolecule solution is placed within a sealed dialysis tubing, which is immersed in a much larger volume of the desired buffer solution. Because ions and small molecules can readily pass through the membrane, the macromolecule gradually gains the desired buffer environment. In addition, when dialysis is carried to equilibrium, it can be used to measure the binding of small molecules or ions to a macromolecule, as described in Chapter 14. This use of dialysis equilibrium provides a simple and direct application of Eq. 13.17. The principle is easily visualized. The macromolecular solution is placed inside a membrane bag (phase α) and suspended in a solution containing the small molecules whose binding is to be studied (phase β) (Figure 13.4). At equilibrium, any excess in the concentration of the low-molecular-weight substance on the macromolecular side of the membrane is taken as evidence of binding. Although this appears to be straightforward, a closer analysis is easy. Calling the low-molecular-weight substance (which we presume to be a nonelectrolyte) component 2, the inside phase α and the outside phase β, we require at equilibrium that $\mu_2^\alpha = \mu_2^\beta$. Then we obtain, from Eq. 13.26,

$$RT \ln a_2^\alpha = RT \ln a_2^\beta \qquad (13.32)$$

This means that the activity of component 2 must be the same on both sides $(a_2^\alpha = a_2^\beta)$. To this point, all the effect of the macromolecule on component 2 has been considered as simply an effect on its ideality. However, we can make a reasonable division of the activity of 2 in phase α into that of free, a_{2f}^α, and bound, a_{2b}^α, fractions

$$a_2^\alpha = a_{2f}^\alpha + a_{2b}^\alpha \qquad (13.33)$$

Because bound molecules do not contribute anything to the physical properties of the solution, we can say $a_{2b}^\alpha = 0$, and therefore

$$a_{2f}^\alpha = a_2^\alpha = a_2^\beta \qquad (13.34a)$$

or

$$C_{2f}^\alpha \gamma_{2f}^\alpha = C_2^\beta \gamma_2^\beta \qquad (13.34b)$$

If we now make the assumption that $\gamma_{2f}^\alpha = \gamma_2^\beta$, we obtain the result that is generally used in equilibrium dialysis: The concentration of *free* component 2 inside the bag is equal to the *total* concentration of 2 outside, $\left(C_{2f}^\alpha = C_2^\beta\right)$. Then, since

$$C_{2b}^\alpha = C_2^\alpha - C_{2f}^\alpha \qquad (13.35)$$

we have a result that allows us to calculate C_{2b}^α from two measurable quantities—the *total* concentration of 2 in the bag and the concentration of 2 outside the bag at equilibrium,

$$C_{2b}^\alpha = C_2^\alpha - C_2^\beta \tag{13.36}$$

It is clear that a number of somewhat arbitrary assumptions have gone into this result, the most serious of which is that $\gamma_{2f}^\alpha = \gamma_2^\beta$. It is not surprising, therefore, that the numbers obtained from such an analysis can be misleading under some circumstances. For example, there exist situations in which the data seem to indicate negative binding. Suppose that the macromolecule does not bind the solute at all, but simply excludes it from the volume occupied by the macromolecules. Then $C_2^\alpha \leq C_2^\beta$ and we interpret this as negative binding. Thus, analyses depending on the assumptions above must be treated with caution. Nevertheless, when properly employed, equilibrium dialysis is probably the most accurate and unambiguous method for binding studies (see Chapter 14).

Osmotic pressure.   The osmotic pressure difference that can exist between two solutions separated by a semipermeable membrane has long been recognized by both biologists and physical chemists. Biologists have noted, for example, that if cells are placed in distilled water instead of physiological saline solution, they will rupture from the pressure that develops within them. Physical chemists discovered early that osmotic pressure provides an easy and moderately accurate way to measure the molecular weights of large molecules.

We can show that the existence of such a pressure difference at equilibrium is a direct consequence of Eq. 13.17. Let us consider a simple system involving only a solvent (component 1) that can pass the membrane and a solute (component 2) that cannot. We shall use the arrangement shown in Figure 13.5 so that pressure can be adjusted independently on the two sides by force on the pistons. On one side of the membrane, α, we have both 1 and 2; on the other side, β, we have only the solvent, component 1. Both sides are initially assumed to be at the same T and P. The only equation we have is the statement that $\mu_1^\alpha = \mu_1^\beta$ at equilibrium. However, this seems to lead to a strange conclusion. According to Eq. 13.30, the chemical potential of a solvent in a solution will always be less than that of the pure solvent $(\mu_1 - \mu_1^0 < 0)$.

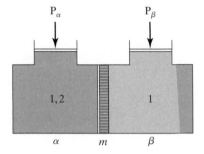

Figure 13.5   Two phases, α and β, are separated by a semipermeable membrane m. The membrane is permeable to component 1 but not to component 2. We can adjust the pressure on α and β as we like.

It would appear then that equilibrium will not be attainable, for no matter how much solvent is transferred into phase α, the inequality will still exist. This is actually the case in the system that we have set up; it is generally not possible for a solution to be in equilibrium across a membrane with a pure solvent at the same T and P. The entropy can always be increased (and the free energy decreased) by transferring some more solvent to further dilute the solution. What would happen, then, in our system is that the solvent would be transferred from β to α until phase β entirely disappeared.

To attain equilibrium, we must evidently produce some *other* difference on the two sides of the membrane, so as to produce a chemical potential difference that will compensate for the difference produced by the presence of the solute on one side. We might imagine, for example, having the two sides at different temperature. However, it seems impossible to conceive of a membrane that would transmit matter but not heat. (But see Exercise 13.6 for a system that *approximates* this idea.) This leaves us only pressure to work with.

The chemical potential, like any free-energy quantity, will depend on the pressure exerted on a system. Since a membrane can support a pressure difference, suppose we try adjusting the pressure to compensate for the effect that the solute has on the chemical potential of the solvent on side α. Using the same apparatus (see Figure 13.5), we shall keep the pressure on the solvent side β at some nominal value P_0 (1 atm, for example) and increase the pressure on side α, the solution side, to $P_0 + \pi$. What effect will this have on μ_1^α? By a standard thermodynamic relationship

$$\left(\frac{\partial G}{\partial P}\right)_T = V \tag{13.37}$$

We can then write for the partial molar free energy of solvent μ_1 the analogous result

$$\left(\frac{\partial \mu_1}{\partial P}\right)_T = \overline{V}_1 \tag{13.38}$$

where $\overline{V}_1$ is the partial molar volume of the solvent. Since on side β we have pure solvent at standard conditions, we have

$$\mu_1^\beta = \mu_1^0 \tag{13.39a}$$

On side α, we have

$$\mu_1^\alpha = \mu_1^0 - \text{effect of solute} + \text{effect of increased pressure} \tag{13.39b}$$

$$\mu_1^\alpha = \mu_1^0 - RTV_1^0\left(\frac{C_2}{M_2} + BC_2^2 + \cdots\right) + \int_{P_0}^{P_0+\pi} \overline{V}_1\,dP \tag{13.39c}$$

where we have used Eq. 13.30 to describe the effect of solute. Assuming that $\overline{V}_1 = V_1^0 = \text{constant} = \text{molar volume of pure solvent at 1 atm (a good approximation}$

for low pressures and dilute solutions), and because $\mu_1^\alpha = \mu_1^\beta$, we obtain from Eqs. 13.39a and 13.39c

$$\mu_1^0 = \mu_1^0 - RTV_1^0\left(\frac{C_2}{M_2} + BC_2^2 + \cdots\right) + V_1^0\pi \tag{13.40}$$

Solving for π, we find

$$\pi = RT\left(\frac{C_2}{M_2} + BC_2^2 + \cdots\right) \tag{13.41}$$

We have calculated the pressure difference required to equate the chemical potential of solvent on the two sides of the membrane. This we call the *osmotic pressure*. It must be applied for the system to be at equilibrium. If the solution is very dilute or if $B = 0$ (ideal solution), we obtain

$$\pi = \frac{RTC_2}{M_2} \tag{13.42}$$

This equation explains the historical importance of osmotic pressure. It allowed the measurement of the molecular weights of macromolecules by an experimentally simple technique. Indeed, the determination that hemoglobin had a molecular weight of 67,000 (Adair 1925) was a milestone in molecular biology. Even at that time, the importance of nonideality and the necessity of extrapolating data to low concentrations was appreciated. A practical equation for the calculation of M_2 for a nonideal system can be written as by dividing Eq. 13.41 by C_2,

$$\frac{\pi}{C_2} = \frac{RT}{M_2} + BRTC_2 + \cdots \tag{13.43}$$

Thus, graphing π/C_2 versus C_2 should allow determination of both M_2 and B. Graphs of osmotic pressure data for globular and unfolded proteins are shown in Figure 13.6.

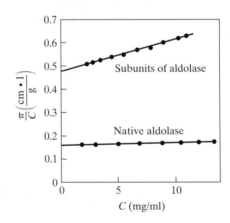

Figure 13.6 Osmotic pressure data for the protein aldolase in buffer at neutral pH (Native Aldolase) and the subunits of this protein in 6 M guanidine hydrochloride. Molecular weights are native aldolase 156,000; aldolase subunits 42,400. Thus, aldolase has 4 subunits. Note the larger virial coefficient for the unfolded subunits. [Data from F. J. Castellino and O. R. Barker (1968), *Biochemistry* **7**, 2207–2217. Reprinted by permission of the American Chemical Society.]

Table 13.2   Average Molecular Weights

	Definition[a]		
Average	**In Terms of N_i**	**In Terms of C_i**	**Examples of Methods That Yield This Average**
Number average, M_n	$\dfrac{\Sigma_i N_i M_i}{\Sigma_i N_i}$	$\dfrac{\Sigma_i C_i}{\Sigma_i (C_i/M_i)}$	Osmotic pressure, freezing point
Weight average, M_w	$\dfrac{\Sigma_i N_i M_i^2}{\Sigma_i N_i M_i}$	$\dfrac{\Sigma_i C_i M_i}{\Sigma_i C_i}$	Light scattering, sedimentation equilibrium
Z average, M_z	$\dfrac{\Sigma_i N_i M_i^3}{\Sigma_i N_i M_i^2}$	$\dfrac{\Sigma_i C_i M_i^2}{\Sigma_i C_i M_i}$	Sedimentation equilibrium

[a]C_i is the concentration in weight per volume; N_i is the concentration in numbers of moles or molecules per unit volume.

It is evident that the extrapolation to $C_2 = 0$ is necessary for accurate results, especially for the unfolded protein, for which the large volume occupied by each molecule makes B large and positive.

For a heterogeneous solute, comprising a number of macromolecular species, the osmotic pressure will be given by the *number average* molecular weight of the mixture (Table 13.2; see Exercise 13.11).

We can measure the osmotic pressure in a number of ways. The simplest *osmometer* simply fits both compartments α and β with standpipes (Figure 13.7). The solvent is allowed to flow through the membrane from β to α until the hydrostatic pressure difference equals π. Alternatively, we could adjust the pressure difference until no flow of solvent occurred. Although the latter method has been

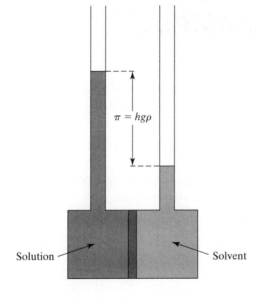

$\pi = hg\rho$

Solution

Solvent

Figure 13.7   A simple osmometer. Transfer of solvent from the pure solvent side to the solution side produces a difference in hydrostatic head. When this equals the osmotic pressure, further net transfer ceases and the system is at equilibrium.

adopted to develop accurate, automated osmometers, the technique is now rarely employed. For high-molecular-weight solutes, π becomes very small, and accurate measurement becomes difficult. Many other, more accurate methods have been developed to determine the molecular weights of macromolecules—aside from sequencing, we have already discussed sedimentation and scattering methods in earlier chapters. The most accurate physical method of all, mass spectrometry, is the subject of Chapter 15.

Even though osmotic pressure is rarely used as a technique for M determination today, understanding the phenomenon is important because of its relevance to many biological and physiological processes. Even biologists often do not appreciate the enormous pressures that can be developed in this way. The cytosol of animal cells can be considered to be approximately 0.5 M in total solutes—this would correspond to an osmotic pressure of about 13 atm, if the cell membrane were impermeable to all components except water.

Membrane potentials.   Whenever differences in the concentrations of ions that are found on two sides of a membrane, we must expect an electrical potential difference to exist across the membrane. Conversely, if we maintain such a potential difference by artificial means, we can then sustain an ionic concentration difference. The key to understanding this is to realize that in such situations the chemical potential difference must be augmented by an electrochemical term, which represents the free energy (or work) involved in moving a mole of ions between the two phases. For an ion of charge Z_i, (+ or − multiples of the proton charge) present in a phase with potential ψ (relative to some reference), the *total potential* or (as it is more often called) the *electrochemical potential* is given by

$$\mu_i' = \mu_i + Z_i F \psi \tag{13.44}$$

where F is the Faraday constant, 96.48 kJ/mol-volt or 23.0 kcal/mol-volt. If the solution can be considered to be ideal,

$$\mu_i' = \mu_i^0 + RT \ln C_i + Z_i F \psi \tag{13.45}$$

The condition for equilibrium for an ionic species across a membrane is then $\mu_i'^\alpha = \mu_i'^\beta$. If we neglect the small difference in μ_i^0 resulting from osmotic pressure, this leads to

$$RT \ln C_i^\alpha + Z_i F \psi^\alpha = RT \ln C_i^\beta + Z_i F \psi^\beta \tag{13.46}$$

which may be rearranged to yield an expression for the potential difference across the membrane, which we call the *membrane potential,* $\Delta \psi$.

$$\frac{RT}{Z_i F} \ln \frac{C_i^\alpha}{C_i^\beta} = \psi^\beta - \psi^\alpha = \Delta \psi \tag{13.47}$$

This equation can be interpreted in either of two ways: If we somehow maintain a difference in ionic concentration $(C_i^\beta \neq C_i^\alpha)$ across a membrane, a membrane potential $\Delta\psi$ will be generated. Alternatively, if we impose a potential difference $\Delta\psi$, an ion i will adopt a concentration difference given by Eq. 13.47. Note that neither of these are true equilibrium situations. If the membrane is permeable to i (which is necessary for Eq. 13.47), then the concentration should eventually equalize across the membrane, and $\Delta\psi$ should drop to zero in the absence of external influences. On the other hand, to maintain a potential gradient requires a continued expenditure of free energy. Indeed, biomembranes are frequently held away from equilibrium, by expenditure of metabolic energy. Ion pumps (which require energy) can maintain ionic concentration differences, and hence generate membrane potentials. Alternatively, a membrane potential can pump ions.

13.2.2 Sedimentation Equilibrium

Chapter 5 contains a brief description of sedimentation equilibrium, based on the concept that at the equilibrium state, opposed flows of diffusion and sedimentation must balance. But it should be clear that if a state of thermodynamic equilibrium is involved, we should be able to analyze the problem from a purely thermodynamic approach. This is the case, and the analysis that follows is more fundamental and rigorous than that given in Chapter 5. The earlier derivation left several questions unanswered: How does one take nonideality into account? Why is it the *partial* specific volume that comes into the equations? Is the molecular weight determined that of the solvated or unsolvated molecule? All of these questions can be answered unambiguously from a thermodynamic approach.

You will recall from the preceding sections that the condition for phase equilibria between solutions of uncharged particles is defined by equality of the chemical potentials in the several phases. However, when the particles carry charges, we must employ the total electrochemical potential, with an extra term that accounts for the energy that an ion has by virtue of its position in an electric field. In a completely analogous manner, the condition for equilibrium for a component i in a centrifugal field is determined by a *total potential,* defined as

$$\tilde{\mu}_i = \mu_i - \frac{1}{2}M_i\omega^2 r^2 \tag{13.48}$$

where the other symbols have the same meaning as in Chapter 5. Here, the term $1/2M_i\omega^2 r^2$ is the potential energy that 1 mole of component i (M_i grams) will have at a point r in the field. The reference zero for this energy is at $r = 0$, the center of rotation of the rotor. The term in question can also be thought of as representing the work required to raise 1 mole of i from point r to the center, working against the field.

The condition for equilibrium is that the total potential for each component of the solution be constant along the solution column, from $r = a$ (meniscus) to $r = b$

(bottom). That is, we require $d\tilde{\mu}_i/dr = 0$. Considering the solute in a two-component solution, we have

$$\frac{d\tilde{\mu}_2}{dr} = \frac{d\mu_2}{dr} - M_2\omega^2 r = 0 \tag{13.49}$$

The chemical potential μ_2 is a function of T, P, and C_2, so we can write $d\mu_2/dr$ in general as

$$\frac{d\mu_2}{dr} = \left(\frac{\partial\mu_2}{\partial T}\right)_{P,C}\frac{dT}{dr} + \left(\frac{\partial\mu_2}{\partial P}\right)_{T,C}\frac{dP}{dr} + \left(\frac{\partial\mu_2}{\partial C_2}\right)_{T,P}\frac{dC_2}{dr} \tag{13.50}$$

The terms on the right side of Eq. 13.50 are evaluated as follows.

1. The first term $= 0$, because T is constant.
2. The second term $= \bar{V}_2\rho\omega^2 r = M_2\bar{v}_2\rho\omega^2 r$ because $(\partial\mu_2/\partial P) = \bar{V}_2$, the partial molar volume, and $dP/dr = \omega^2\rho r$. The latter expression is analogous to the expression in a gravitational field, where we find that $dP/dx = g\rho$, where g is the acceleration of gravity.
3. The third term $= (RT/C_2)dC_2/dr$ if we assume the solution is ideal; for then $\mu_2 = \mu_2^0 + RT \ln C_2$.

Combining these results into Eq. 13.50 and then substituting into Eq. 13.49, we find

$$M_2(1 - \bar{v}_2\rho)\omega^2 r - \frac{RT}{C_2}\frac{dC_2}{dr} = 0 \tag{13.51}$$

This is exactly the same result we obtained in Chapter 5 (Eq. 5.41). However, the derivation presented here is far more rigorous. It identifies the volume quantity involved unambiguously as the partial specific volume, a point that involves some hand waving in the Chapter 5 derivation. Furthermore, it is now clear how solute nonideality could be incorporated into the equations for sedimentation equilibrium. If instead of the "ideal" expression for μ_2, we used Eq. 13.31, an equation analogous to Eq. 13.51 could be obtained for a nonideal solute (see Exercise 13.13).

Finally, the definition of the partial specific volume as describing the rate of volume charge upon adding anhydrous solute to solvent shows that it is the *anhydrous* molecular weight that sedimentation equilibrium measures.

13.2.3 Steady-State Electrophoresis

Recently, a technique has been devised that can be considered as approximately the electrophoretic analog to sedimentation equilibrium. In this method, also termed *membrane confined electrophoresis* (Laue et al. 1998; Durant et al. 2002), a solution of macroions is placed in a volume between two membranes, permeable to small ions but impermeable to the macroions. Electrodes are placed outside of these membranes, and a low electric field is applied. As in the case of sedimentation equilibrium, equilibrium is eventually reached between the flow produced by the applied field and the backflow due to diffusion. From the thermodynamic perspective, we can say that at this point the *electrochemical potential* must be the same at every level. If the concentration of macroions at each point in the cell is measured by an optical system, the kind of concentration gradient shown in Figure 13.8 is observed. This fits the theoretical equation

$$\frac{C(x)}{C(x_0)} = e^{(Z^*eE/k_BT)(x-x_0)}$$
(13.52)

where E is the electric field, e the proton charge, k_B is the Boltzmann constant, T is the absolute temperature, and x_0 is a reference point in the cell. The quantity Z^* is the *effective valence* of the macromolecule. The effective valence (or *reduced valence,* using the terminology of Durant et al. 2002) is related to the actual valence, Z (the number of electrons which must be acquired or lost to produce a neutral particle) by the equation

$$Z^* = Z\left\{\frac{f(\kappa a)}{1 + \kappa a}\right\}$$
(13.53)

where f is Henry's function (see van Holde, *Physical Biochemistry,* 2nd ed., p. 140 for a graphical representation of this complicated function) and a is an effective radius equal to the sum of the Stokes's radius of the macroion and the Stokes's radius of the counterion. The parameter κ is the Debye-Hückel reciprocal radius of the ion atmosphere, given by

$$\kappa = \left(\frac{8\pi \mathcal{N} e^2}{1000 D k_B T}\right)^{1/2} I^{1/2}$$
(13.54)

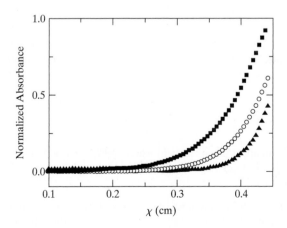

Figure 13.8 Steady-state electrophoresis of three differently charged mutants of lysozyme. The data agree within experimental error with the predicted results. See text. [From Durant et al. 2002.]

Here D is the dielectric constant of the medium, and I is the ionic strength of the solution.

An application of this method is shown in Figure 13.8. Here, mutants of lysozyme that differ only in charge have been studied. The valences determined in this way agree well with expected values from the structure.

EXERCISES

*13.1 Using the Gibbs-Duhem equation, obtain an expression for $\mu_2 - \mu_2^0$ from the equation

$$\mu_1 - \mu_1^0 = -RTV_1^0 C_2\left(\frac{1}{M_2} + BC_2\right)$$

[Hint: You must first differentiate, then integrate. You may assume $n_1 \cong V/V_1^0$, where V is solution volume.]

13.2 Starting with Eq. 13.24, derive the form given in Eq. 13.25. What is the relationship between the standard-state potential on the two scales? That is, write an equation for μ_i^0 on the C_i scale in terms of μ_i^0 on the χ_i scale.

*13.3 In practice, it is difficult to distinguish weak self-association of a solute from nonideality with a negative second virial coefficient. To show this, consider a protein molecule that dimerizes according to the reaction

$$2\,M \Leftrightarrow D \quad K = \frac{[D]}{[M]^2}$$

where the quantities in brackets are molar concentrations and K is an equilibrium constant. Derive an equation for the chemical potential of solvent in terms of the *total* concentration of solute, given by

$$[M_0] = [M] + 2[D]$$

You may assume that the contribution of the monomer and dimer are additive on a molar concentration scale. You will have to show that the free monomer concentration $[M]$ is given by

$$[M] = \frac{-1 + \sqrt{1 + 8K[M_0]}}{4K}$$

and you will have to use an approximation valid when $8K[M_0] \ll 1$. You should find that the expression for μ_1 contains a term in $[M_0]^2$, even though no explicit assumption of nonideality was made.

13.4 The osmotic pressure of a protein, as a function of concentration, is shown in the accompanying figure. The units of π/C are arbitrary. Provide a plausible explanation for this behavior.

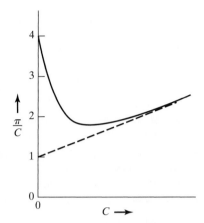

13.5 One mg of a sample of polyethylene glycol (molecular weight = 60,000) is dissolved in 1 ml of water and placed in a small open vial in a closed container. One mg of another sample of polyethylene glycol (of unknown molecular weight) is also dissolved in 1 ml H_2O, and placed in another small open vial in the same container. The entire system is kept at constant temperature.

After several days, the volume of the "reference" solution is found to be 0.5 ml, that of the "unknown" to be 1.5 ml. [Note: Polyethylene glycol is a nonionic, water-soluble polymer. You may assume it shows ideal solution behavior.]

a. What is the molecular weight of the unknown? Explain your reasoning.

b. A student attempted to use this method to measure the molecular weight of a protein. She placed a known polyethylene glycol (in water) in one container, and protein (in buffer solution) in the other. She got an absurdly low apparent molecular weight value. Why?

*13.6 A certain kind of osmometer works as follows. A droplet of solvent is suspended over a solution. Since the vapor pressure of the pure solvent (or μ_1 in the droplet) is greater than that in the solution, evaporation will be more rapid in the droplet. This cools the drop until the drop has the same vapor pressure or μ_1 as the solution. The temperature difference ΔT is measured by very sensitive probes in solvent and solution.

Derive an equation to calculate the solute molecular weight from ΔT. You may assume the solute to be nonvolatile. [Hint: You may have to go back to physical chemistry texts to figure out how μ_1 depends on T.]

13.7 The following osmotic pressure data have been obtained for lactate dehydrogenase in two different buffer systems.

a. In 0.1 M KCl/0.1 M potassium phosphate buffer at 25°C (buffer density = 1.012 g/ml), π/C had the value of 0.183 cm of solvent/(g/l), independent of concentration. Calculate M.

b. In 6.0 M guanidine hydrochloride, a denaturing and dissociating solvent, with density 1.150 g/ml, the following data were obtained:

C (g/l)	π/C (1 · cm solvent/g)
1.25	0.59
2.50	0.60
3.75	0.61
5.00	0.62

Calculate M under these circumstances, and estimate the number of subunits in a native lactate dehydrogenase molecule.

13.8 A highly supercoiled DNA molecule is cleaved at one site by a restriction endonuclease, yielding linear DNA.

Predict whether each of the following will *increase, decrease,* or remain *unchanged* as a consequence of this treatment:

1. Second virial coefficient
2. Osmotic pressure at 0.1 mg/ml
3. Limit of π/C as $C \rightarrow 0$
4. Diffusion coefficient

13.9 A protein that is isoelectric at pH = 4.0 is studied by osmotic pressure in the following three conditions:

1. pH = 8.0, 0.001 M buffer
2. pH = 8.0, 0.001 M buffer + 0.5 M NaCl
3. pH = 4.0, in 6 M guanidine hydrochloride (Note: Guanidine hydrochloride has a great ability to disrupt noncovalent bonds).

The graphs of π/c are shown in the figure below. Explain qualitatively the differences in slopes of the graphs.

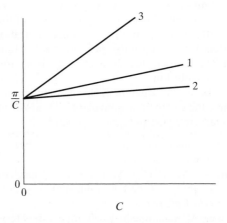

13.10 We have mentioned that dialysis equilibrium experiments can be complicated by the fact that the macromolecular substance present can occupy a significant volume of the solution. Show how you can mathematically correct for this effect.

13.11 Prove that the osmotic pressure of a heterogeneous ideal solute will be determined by the *number average* molecular weight. [Hint: π is the sum of the osmotic pressure contributions from the various solute components.]

13.12 Calculate the membrane potentials across a membrane at 37°C, if the NaCl concentration on the right is 0.1 M and on the left is 0.01 M, given each of the following conditions.
  a. Membrane is permeable only to Na^+
  b. Membrane is permeable only to Cl^-
  c. Membrane is permeable to both

13.13 For sedimentation equilibrium of a nonideal solution, obtain the expression for $d \ln C_2/d(r_2)$. [Hint: Equation 13.31 will be appropriate.] Describe the plot of $\ln C$ versus r^2 if B is positive.

*13.14 A protein preparation contains both monomer and dimer. The mixture is 50% (by weight) monomer (M = 40,000) and 50% (by weight) dimer. What molecular weight values would you expect using osmotic pressure (see Exercise 13.11) and by sedimentation equilibrium (according to Eq. 5.43)?

REFERENCES

General

TINOCO, I., JR., K. SAUER, J. WANG, and J. D. Puglisi (2004) *Physical Chemistry: Principles and Applications on Biological Sciences,* 4th ed., Prentice Hall, Upper Saddle River, NJ. An excellent physical chemistry text, with strong emphasis on biological applications. A good introduction to this text.

VAN HOLDE, K. E. (1985) *Physical Biochemistry,* 2nd ed. Prentice Hall, Upper Saddle River, NJ. Briefer than this text, and dated in portions, but also more detailed in some areas.

Membranes

ADAIR, G. S. (1925) "The osmotic pressure of haemoglobin in the absence of salt." *Proc. Roy. Soc. A.* **109**, 292–300. A pioneering application of osmotic pressure measurements. Primarily of historical interest.

KOTYK, A., K. JANACEK, and J. KORATA (1988) *Biophysical Chemistry of Membrane Functions,* John Wiley & Sons, New York. More sophisticated than the above.

YEAGLE, P. (ed.) *The Structure of Biological Membrane,* 2nd ed., CRC Press, Boca Raton, FL.

Sedimentation Equilibrium

FUJITA, H. (1975) *Foundations of Ultracentrifugal Analysis,* John Wiley & Sons, New York. The ultimate reference for questions on ultracentrifugation theory.

SCHUSTER, T. M. and T. M. LAUE (eds.) (1994) *Modern Analytical Ultracentrifugation,* Birkhäuser, Boston. Contains very useful chapters on the design and analysis of sedimentation equilibrium experiments.

Steady-State Electrophoresis

DURANT, J. A., C. CHEN, T. M. LAUE, T. P. MOODY, and S. A. ALLISON (2002) "Use of T4 lysozyme charge mutants to examine electrophoretic models." *Biophysical Chem.* **101–102**, 593–609.

LAUE, T. M., H. K. SHEPARD, T. M. RIDGEWAY, T. P. MOODY, and T. J. WILSON (1998) "Membrane-confined Analytical Electrophoresis," in *Methods in Enzymology,* vol. 295, ed. G. K. Ackers and M. L. Johnson, Academic Press, New York, 494–518.

CHAPTER 14

Chemical Equilibria Involving Macromolecules

Biochemistry is concerned with many types of chemical equilibria. Some of these involve small metabolite molecules, including those used as monomers in the construction of proteins, nucleic acids, and polysaccharides. Most of these equilibria directly involve macromolecules, either in interactions with other macromolecules or in reactions with small molecules. Almost all of these reactions occur in solution, and the usual approach of physical chemistry texts, which is to emphasize the gas phase reactions, is of little direct help to the biochemist. In this chapter, we shall first discuss solution equilibria in general; this treatment will expand upon the introduction in Chapter 2. We shall then treat in detail equilibria in reactions between macromolecules and the equilibria involving the binding of small molecules by macromolecules.

14.1 THERMODYNAMICS OF CHEMICAL REACTIONS IN SOLUTION: A REVIEW

As emphasized in Chapter 13, the discussion of the thermodynamics of solutions centers on the chemical potential. This quantity is also the key to understanding chemical equilibria in solution. Suppose that we have a general chemical reaction

$$aA + bB + \cdots \Rightarrow pP + qQ + \cdots$$

in which a moles of A, b moles of B, and so forth, at molar concentrations [A], [B], and so forth, are converted into p moles of P, q moles of Q, and so forth, at concentrations

[P], [Q], and so forth.[1] We may write the driving force of the chemical reaction, which is the free-energy change ΔG, as

$$\Delta G = G \text{ (final state)} - G \text{ (initial state)} \tag{14.1}$$

If the reaction occurs at constant temperature and pressure, Eq. 13.9 yields

$$\Delta G = p\mu_P + q\mu_Q + \cdots - a\mu_A - b\mu_B - \cdots \tag{14.2}$$

where μ_P, μ_Q, and so on are the chemical potentials at the concentrations existing in the solution. Recalling that

$$\mu_i = \mu_i^0 + RT \ln [i] y_i \tag{14.3}$$

where [i] is the molar concentration of component i and y_i is its activity coefficient, we may rewrite Eq. 14.2 as

$$\Delta G = p\mu_P^0 + q\mu_Q^0 + \cdots - a\mu_A^0 - b\mu_B^0 - \cdots + RT \ln \frac{([P]y_P)^p([Q]y_Q)^q \cdots}{([A]y_A)^a([B]y_B)^b \cdots}$$

$$= (p\mu_P^0 + q\mu_Q^0 + \cdots - a\mu_A^0 - b\mu_B^0 - \cdots)$$

$$+ RT \ln \frac{[P]^p[Q]^q \cdots}{[A]^a[B]^b \cdots} + RT \ln \frac{y_P^p y_Q^q \cdots}{y_A^a y_B^b \cdots} \tag{14.4}$$

The three terms on the right side of this equation may be distinguished as follows: The first, involving μ_i^0 values, is the standard-state free-energy change ΔG^0. It corresponds to the ΔG that would be observed if a moles of A, and so forth, in the standard state formed p moles of P, and so forth, also in the standard state. The second term gives the effect of the *actual* concentrations of reactants and products on the total free-energy change. It is corrected by the third term, which involves only the activity coefficients. For practical biochemistry, we may assume this activity coefficient term to be zero; this would be true, for example, if all components were present in such great dilution that they could be considered to behave ideally. We may then simplify Eq. 14.4 as

$$\Delta G = \Delta G^0 + RT \ln \frac{[P]^p[Q]^q \cdots}{[A]^a[B]^b \cdots} \tag{14.5}$$

which is identical to Eq. 2.61. The values of ΔG^0 for many reactions have been tabulated (see for example Fasman 1976). It should be emphasized that ΔG not ΔG^0,

---

[1] Note that we are not assuming equilibrium here; we simply take so many moles of the reactants under the stated conditions and convert them to the corresponding number of moles of the product under the given conditions. Because we will write so many equations involving molar concentrations in this chapter, we will adopt here the nomenclature [A], [B], instead of C_A, C_B, as the latter becomes awkward in complex equations.

is the driving force for a chemical reaction. The quantity ΔG^0 tells us what the relative concentrations of reactants and products will be at equilibrium, but it is the ΔG at biological concentrations that is physiologically relevant. To take a specific example: ΔG^0 for the hydrolysis of ATP to ADP and phosphate is about -31 kJ/mole. But at the concentrations of reactants and products maintained in most cells, the value of ΔG is closer to -40 kJ/mol. *This* is the free energy actually available to the cell from ATP hydrolysis.

Equation 14.5 calculates the free-energy change accompanying a given reaction involving some arbitrarily chosen initial and final conditions. Suppose, instead, that we ask that the reaction be carried out reversibly. This means that equilibrium concentrations are maintained. At constant temperature and pressure, the total free-energy change for any amount of reaction will be zero. Therefore, Eq. 14.5 becomes

$$0 = \Delta G^0 + RT \ln \left(\frac{[P]^p[Q]^q \cdots}{[A]^a[B]^b \cdots} \right)_{eq} \tag{14.6}$$

Since the concentrations involved must now satisfy equilibrium, we have the result

$$\Delta G^0 = -RT \ln \left(\frac{[P]^p[Q]^q \cdots}{[A]^a[B]^b \cdots} \right)_{eq} = -RT \ln K \tag{14.7a}$$

where K is the *equilibrium constant,* defined as

$$K = \left(\frac{[P]^p[Q]^q \cdots}{[A]^a[B]^b \cdots} \right)_{eq} \tag{14.7b}$$

Students are frequently confused at this point. Why is it that the *standard-state* free-energy change can be used to measure the *equilibrium* constant while the free-energy change at equilibrium is zero? It is helpful to write Eq. 14.5 in a different way, using Eq. 14.7a:

$$\Delta G = -RT \ln \frac{[P]_{eq}^p[Q]_{eq}^q \cdots}{[A]_{eq}^a[B]_{eq}^b \cdots} + RT \ln \frac{[P]^p[Q]^q \cdots}{[A]^a[B]^b \cdots} \tag{14.8}$$

Imagine a reaction that we begin at concentrations $[A], [B], \ldots [P], [Q], \ldots$ and follow over time, letting it approach equilibrium. The second term in Eq. 14.8 is the dynamic term that depends on the concentrations of reaction and products at any time during the course of the reaction. The first term is the static term that depends on the equilibrium constant, which in turn depends only on temperature and pressure. As the reaction approaches equilibrium, the second term approaches the first term, and ΔG approaches zero.

Physical biochemists normally make measurements at equilibrium and determine ΔG^0. Thus, they are investigating the stability at the equilibrium conditions, not the forces that caused the reaction to go to equilibrium. This distinction is particularly important in the protein-folding problem. Workers are investigating the forces that stabilize intermediates and the folded protein. They are not investigating the forces that cause folding. These forces change as the reaction proceeds and determine ΔG, not ΔG^0. Before there is any folding, the force that determines ΔG is undoubtedly the hydrophobic interaction.

For the purpose of interpreting ΔG^0 values, it is worthwhile to further examine the dependence of the equilibrium constant on ΔG^0: We can rewrite Eq. 14.7a as

$$K = e^{-\Delta G^0 / RT} \tag{14.9}$$

Since $RT \cong 2.5$ kJ/mol near room temperature, a value of $\Delta G^0 = -2.5$ kJ/mol corresponds to $e^1 = 2.7$, and a value of $\Delta G^0 = -5$ kJ/mol corresponds to $K = e^2$, about 8.7. Thus, even what seem to be small ΔG^0 values correspond to fairly large values of the equilibrium constant. If we consider an *essentially irreversible* reaction to be one in which $K > 10^4$, this will correspond to a free-energy change more negative than about 23 kJ/mol.

Another way of measuring the free-energy change in a chemical reaction is through the electrical potential of an electrochemical cell (real or hypothetical) in which the reaction occurs. It is always possible to calculate the electromagnetic force corresponding to a given reaction, since the free-energy change determines the maximum work, other than PV work, that the reaction can produce (see Chapter 2). If we interpret this other work as electrical, we equate it to the work involved in transporting n moles of electrons against a potential difference ϵ. So

$$\Delta G = -w_{\text{rev}} = -nF\epsilon \tag{14.10}$$

where n is the number of moles of electrons transferred per mole of reaction, and F is the Faraday constant (96.48 kJ/V equiv.). This leads to the *Nernst equation,* from Eq. 14.5:

$$\epsilon = \epsilon^0 - \frac{RT}{nF} \ln \frac{[P]^p [Q]^q \cdots}{[A]^a [B]^b \cdots} \tag{14.11}$$

where we have once again assumed ideality. In this equation, ϵ^0 is the standard-state potential. A table of standard potentials is equivalent to a table of standard free-energy changes, since

$$\Delta G^0 = -nF\epsilon^0 \tag{14.12}$$

and a positive value for the standard potential for a process indicates products will be favored at equilibrium, because this means that ΔG^0 will be negative. Biological oxidation-reduction reactions are often thought of in terms of standard potentials.

Just as there is a standard-state free-energy change for any reaction, there exist corresponding standard-state enthalpy and entropy changes. To obtain a full description of the energetics of a reaction, it is necessary to be able to determine the full set of thermodynamic parameters ($\Delta G^0, \Delta H^0, \Delta S^0$) involved in biochemical processes. For example, knowing the relative contributions of ΔH^0 and $T\Delta S^0$ to a process such as protein denaturation can help to sort out various contributions (such as the hydrophobic effect) to protein stability. A few words about the experimental problems are in order. If it is possible to experimentally determine the equilibrium constant for a reaction by measuring the concentrations of the reactants and products present at equilibrium, Eq. 14.7a yields ΔG^0 directly. The meaning of the value of ΔG^0 that is obtained will depend on the way in which the reaction is written and the way in which the equilibrium constant is consequently expressed. To take an example, if we express a dimerization reaction by the equation

$$2A \Leftrightarrow B$$

and consequently write

$$K = \frac{[B]}{[A]^2} \tag{14.13}$$

the ΔG^0 obtained will be for the process as written above: two moles of A (in standard state) forming 1 mole of B (in standard state). On the other hand, we could equally well write the same process as

$$A \Leftrightarrow \frac{1}{2}B$$

giving

$$K' = \frac{[B]^{1/2}}{[A]} \tag{14.14}$$

In this case, the ΔG^0 calculated from K' will be the free-energy change corresponding to the conversion of 1 mole of A into 1/2 mole of B, each in standard state. It will be, as it should, a number half as large as that given for Eq. 14.13. It is important to remember that ΔG^0 for a reaction depends upon exactly how the reaction is written.

To obtain the enthalpy change corresponding to a given reaction, two courses of action are available. The most direct is simply to let the reaction occur in a calorimeter and measure the heat evolved or absorbed. We must, of course, know the number of moles that actually react, and for traditional methods to be successful the reaction must usually be fairly rapid to allow for precise measurement; it is difficult to measure a small amount of heat that is slowly evolved. The development of *differential scanning calorimetry* has proved very useful for the study of processes like protein denaturation. This technique is described in considerable detail in Chapter 2, Section 2.5.2.

For reactions that are simply not amenable to calorimetric study, more indirect ways of determining ΔH^0 are available. Equation 14.7 shows one such way to determine ΔH^0: The equation is rewritten as

$$-RT \ln K = \Delta G^0 = \Delta H^0 - T\Delta S^0 \tag{14.15}$$

Rearranging gives

$$\ln K = \frac{-\Delta H^0}{RT} + \frac{\Delta S^0}{R} \tag{14.16}$$

As was pointed out in Chapter 2, this is called the van't Hoff equation. Equation 14.16 states that $\ln K$ should be a linear function of $1/T$, if and only if ΔH^0 and ΔS^0 are independent of T. The slope of the straight line obtained by graphing $\ln K$ versus $1/T$ (called a van't Hoff graph) should be $-\Delta H^0/R$, and the intercept at $(1/T = 0)$ should be $\Delta S^0/R$. To construct such a graph, it is only necessary to measure the equilibrium constant for the reaction of interest at a series of different temperatures.

It must be emphasized that this technique, as it is often employed, is beset with difficulties. Frequently, ΔH^0 varies with T; this will happen, for example, whenever there is a difference in heat capacity between the reactants and products (see Chapter 2, Section 2.5.2). In this case, Equation 14.16 will yield a nonlinear graph. The occurrence of the *logarithm* of K in Eq. 14.16 means that any physical quantities *proportional* to the concentrations of the reactants and products can be used to define an apparent K, which will still give the right slope. But a hazard exists if these quantities do not measure what the experimenter thinks they measure or if the reaction is a complex one. One may literally measure a standard-state enthalpy change for a completely meaningless reaction.

14.2 INTERACTIONS BETWEEN MACROMOLECULES

Many of the important biochemical processes in cells are based on or controlled by interactions between two or more macromolecules. Examples abound: the formation of multisubunit proteins from polypeptide chains; the intimate and complex association between protein and RNA molecules in the ribosome; the control of DNA expression by the binding of specific transcription factors, and so forth. Most of these interactions are based on noncovalent forces, although covalent bonds such as disulfides are sometimes involved in the stabilization of the complexes. In this section, we shall emphasize some general principles, concentrating on association reactions of simple, defined stoichiometry. The nonspecific binding of proteins to nucleic acids will be left until later in the chapter.

Those macromolecular complexes stabilized by noncovalent forces often are in thermodynamic equilibrium with their dissociated components, and in many cases it is possible to measure the thermodynamic parameters for these association-dissociation reactions. A sampling of such data is given in Table 14.1. A number of noteworthy points emerge from inspection of these data.

First, although the energies for *individual* noncovalent interactions like hydrogen bonds or salt bridges are quite small, it is clear that they can sum in the binding between macromolecules to yield quite considerable values. This is a reflection of the particular nature of such reactions; macromolecules tend to interact over binding *surfaces,* and many groups may participate simultaneously. Second, an examination

Table 14.1 Thermodynamic Parameters for Some Protein Association Reactions

Protein	Reaction Type	ΔG^0 (kJ/mol)	$\Delta G_m^{0\,a}$ (kJ/mol)	ΔH^0 (kJ/mol)	ΔS^0 (J/mol·K)
D-amino acid oxidase	$2P \Leftrightarrow P_2$	−30.5	−15.7	0	+104
Arginosuccinase	$2P \Leftrightarrow P_2$	−42.6	−21.3	+192	+791
Enolase	$2P \Leftrightarrow P_2$	−39.3	−19.7	−335	+924
Glutamate dehydrogenase	$P + P_i \Leftrightarrow P_{i+1}$	−32.6	−16.3	0	+104
Hemerythrin	$8P \Leftrightarrow P_8$	−193.7	−24.2	0	+92
Insulin	$2P \Leftrightarrow P_2$	−22.9	−11.5	−29.6	−23
	$2P_2 \Leftrightarrow P_4$	−16.5	−4.1	−68.2	−172
	$P_2 + P_4 \Leftrightarrow P_6$	−16.1	−2.7	+205	+740
β-Lactoglobulin	$4P_2 \Leftrightarrow P_8$	−63.6	−7.9	−234	−632

[a]Per monomer unit.

Source: Most data from Klotz, I. M., Darnall, D.W., and R. Langerman (1975) in *The Proteins* Vol. I, H. Neurath and R. Hill, eds., *Academic Press*, N.Y. pp. 293–411, which carries original references and further data.

of the signs and magnitudes of ΔH^0 and ΔS^0 suggests that many different patterns of noncovalent interactions can be involved. In some cases, such as those that show a significant entropy *increase* upon association, hydrophobic interactions seem to dominate; in other cases, hydrogen-bonding or electrostatic forces play the major roles. Finally, as the examples in Table 14.1 suggest, it is an inescapable fact that most biologically significant interactions between macromolecules result in the formation of defined structures of relatively simple stoichiometry. Monomer-dimer, dimer-tetramer reactions are common; indefinite aggregation is rare. Most oligomeric proteins exhibit the simpler classes of point-group symmetry described in Chapter 1.

A characteristic feature of all association-dissociation reactions is the dependence of the position of equilibrium on the *total* concentration of the subunits involved. This serves to distinguish such reactions from purely conformational changes, for which the equilibrium state will be concentration-independent. A simple case will illustrate the principle, show how such reactions can be studied, and point out some of the problems involved. Consider a simple monomer-dimer reaction,

$$2M \Leftrightarrow D$$

We have the equilibrium condition,

$$K = \frac{[D]}{[M]^2} \tag{14.17}$$

where K is the association constant in liters per mole and the concentrations of the monomer and dimer are in moles/liter. If we denote by M_0 the total molar concentration of subunits in the system (irrespective of whether they are present as monomer or incorporated into dimer), we obtain the additional equation

$$M_0 = [M] + 2[D] \tag{14.18}$$

The factor 2 appears because each mole of dimer contains 2 moles of subunits. Thus, we can write

$$[D] = \frac{M_0 - [M]}{2} \tag{14.19}$$

and substitute this into Eq. 14.17. The resulting quadratic equation can be solved for [M], as a function of the total concentration M_0, to give for the mole fraction of free monomer,

$$\frac{[M]}{M_0} = \frac{-1 \pm 1\sqrt{1 + 8\,KM_0}}{4\,KM_0} \tag{14.20}$$

Only the solution with the positive sign is physically significant. If we evaluate the limits of this expression, we find that $[M]/[M_0] \rightarrow 1$ as $M_0 \rightarrow 0$, and $[M]/M_0 \rightarrow 0$ as $M_0 \rightarrow \infty$. Thus, the reaction will tend toward total dissociation or complete dimerization, as M_0 becomes very small or very large, respectively. Graphs of monomer concentration versus total concentration are shown in Figure 14.1 for two different values of the association constant, $K = 1 \times 10^4$ l/mol and $K = 1 \times 10^8$ l/mol, corresponding to $\Delta G^0 = -23$ and $\Delta G^0 = -46$ kJ/mol, respectively. The graphs are in terms of weight concentration (mg/ml) rather than molar concentration to emphasize the following point: if $\Delta G^0 = -23$ kJ/mol, half of the material will be dissociated into the monomer at 2 mg/ml; and the reaction can be easily studied by most physicochemical techniques. On the other hand, if $\Delta G^0 = -46$ kJ/mol, the binding

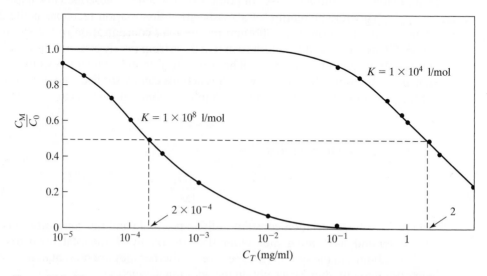

Figure 14.1 The weight fraction of monomer as a function of total protein concentration for reversible monomer-dimer association reactions. Curves for two values of the molar association constant $(1 \times 10^8$ l/mol and 1×10^4 l/mol) are shown, assuming a monomer of molecular weight 20,000.

is so strong that half-dissociation only occurs at 0.0002 mg/ml. In fact, there is very little dissociation even at 0.02 mg/ml. Such strong binding reactions are very difficult to study, since very little shift in the equilibrium occurs over the concentration range accessible to most of the methods available to the physical biochemist. The experimental techniques typically used for the analysis of such association-dissociation equilibria for protein-protein interactions are those that yield the average molecular weight values for the equilibrium mixture, such as light-scattering (see Chapter 7) or sedimentation equilibrium (see Chapter 5).

Suppose we use such a technique to measure the weight-average molecular weight of the equilibrating monomer-dimer mixture at different total weight concentrations C_T. Since the value C_T must equal $C_D + C_M$, we can write for M_w

$$M_w = f_M M_M + f_D M_D \tag{14.21}$$

where f_M and f_D are the weight fractions of monomer and dimer, and M_M and M_D their molecular weights. (See the definition of M_w in Table 13.3.) Then, since $f_D = 1 - f_M$ and $M_D = 2M_M$, we can rearrange Eq. 14.21 to yield

$$f_M = 2 - \frac{M_w}{M_M} \tag{14.22}$$

We can use this to calculate the equilibrium constant K at each concentration. First, we must express K in terms of weight concentrations,

$$K = \frac{[D]}{[M]^2} = \frac{(C_D/2M_M)}{(C_M/M_M)^2} = \frac{M_M}{2} \frac{C_D}{(C_M)^2} \tag{14.23}$$

But since $C_D = f_D C_T$, and $C_M = f_M C_T$, we get

$$K = \frac{M_M}{2} \frac{f_D C_T}{f_M^2 C_T^2} = \frac{M_M}{2C_T} \frac{(1 - f_M)}{f_M^2} \tag{14.24}$$

We can obtain the value of f_M at each total concentration, so we can calculate an apparent equilibrium constant at each concentration. The constancy of this value is a test that this is truly a monomer-dimer equilibrium. The stoichiometry of the association can also be checked by the behavior of the M_w versus $C_T(r)$ curve, as shown in Figure 14.2. As $C_T \to 0$, $M_w \to M_M$, as $C_T \to \infty$, $M_w \to 2M_M = M_D$. Other systems may be studied by the same approach, but very high precision is required if a multistep association (monomer-dimer-tetramer, for example) is to be analyzed. Sedimentation equilibrium is a very powerful technique for such studies, for a whole range of concentrations can be sampled in a single experiment. If a series of measurements like those described above are repeated at different temperatures, it may be possible to obtain ΔH^0 and ΔS^0 by using Eq. 14.16.

Systems with more than two different species of reacting macromolecules may show "order of addition" effects. For example, if both protein A and protein B bind to DNA, a different complex may form if A is added first or B is added first. Such a

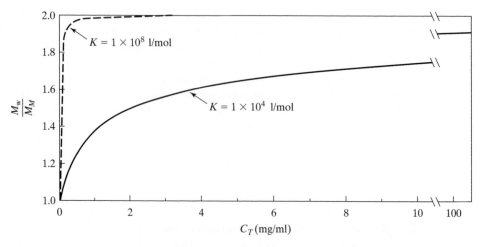

Figure 14.2   Graphs of M_w/M_M (weight average molecular weight over monomer molecular weight) versus C_0 for the two dimerization reactions shown in Figure 14.1.

system is not in equilibrium, and its behavior is dominated by kinetic effects. Dissociation of some tight macromolecular complexes may have half-lives of days. Biological processes may, in many cases, depend on kinetics rather than equilibria.

The analysis of the thermodynamics of interactions between proteins and nucleic acids presents special problems. Those proteins that bind nonspecifically to nucleic acids do so at multiple, often overlapping, sites. Thus, the study of such interactions falls in the province of multiple-binding analysis, which we shall discuss in Section 14.3. On the other hand, proteins that interact with very specific nucleic acid sequences (repressors and other regulatory proteins are examples) frequently bind very strongly. For the reasons described above, such strong interactions are hard to study by most physical techniques. To give an example, the *lac* repressor-operator complex exhibits an association constant of about 10^{14} l/mol—about a million times greater than the stronger association depicted in Figure 14.1. To study such interactions, special techniques that can be applied at very low concentrations have been developed. One such technique is *filter binding,* which takes advantage of the fact that free nucleic acids will pass through a nitrocellulose filter, whereas protein–nucleic acid complexes will not. Using radiolabeled DNA, for example, very low concentrations can be studied in this way. In the most frequently used variant of the technique, radiolabeled DNA is mixed with varying concentrations of protein and the samples are poured through the filters. Scintillation counting of the filters measures the fraction of DNA bound to protein. It is, however, very difficult to obtain precise or wholly unambiguous results. A major problem arises from washing: If the filter is not washed, adventitiously attached DNA may be counted; on the other hand, extensive washing will lead to the dissociation and release of some complexed DNA.

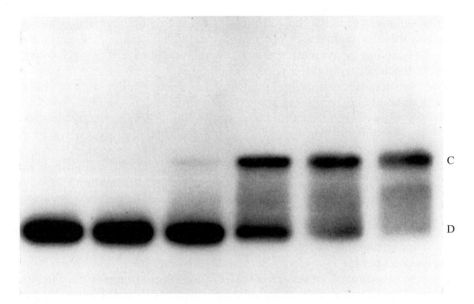

Figure 14.3   Gel shift assay for the binding of histone H1 to a cross-shaped DNA molecule. The DNA (D) has been radiolabeled and the gel subjected to autoradiography. The data represent a titration of a fixed amount of DNA with increasing amounts of the protein. Note that a single complex (C) is formed in this case. [Courtesy of Dr. Jordanka Zlatanova.]

Another, somewhat more sophisticated technique is the gel-shift assay, which depends on the fact that a protein–nucleic acid complex will migrate in gel electrophoresis more slowly than the free nucleic acid (see Chapter 5). As shown in Figure 14.3, it is possible to measure both the bound and free nucleic acids if this component has been radiolabeled. This should, in principle, allow the calculation of the equilibrium constant for the reaction. However, caution must be exercised, for the system is not truly in equilibrium and dissociation of the complex may occur during electrophoresis. This will lead to a smearing of the band corresponding to the complex, and an underestimation of the equilibrium concentration of the complex.

14.3   BINDING OF SMALL LIGANDS BY MACROMOLECULES

14.3.1   General Principles and Methods

A major biological function of many biopolymers is to bind small molecules and/or ions. Examples include enzymes, which bind substrates and effector molecules; transport proteins such as hemoglobin or storage proteins such as myoglobin, both of which bind oxygen; and the many proteins that act as buffers by binding hydrogen ions. In fact, almost all biological functions involve the interactions of those small molecules that serve as metabolites, regulators, and signals with specific sites on those macromolecules that carry out cellular processes. For this reason, an understanding of the mechanisms of such interactions is essential for comprehension of biochemistry at the molecular level.

In most cases, such binding involves the formation of some kind of noncovalent bond between the small molecule or ion (called the *ligand*) and some specific region on or near the surface of the macromolecule. This region is called the *binding site*. Most biological macromolecules possess binding sites of varying degrees of strength and specificity for a variety of ligands. Since the act of binding a particular ligand may itself induce conformational changes in the biopolymer, which may in turn modify other sites, we expect that in general the binding of several ligands to one macromolecule will be a complex process. It is, in fact, this very possibility of interdependence of binding affinities that provides a *raison d'être* for the complexity of many biopolymers, for it is in this way that one kind of metabolite can sense the concentration of another kind even though any direct interaction between them is imperceptible.

The binding of small molecules or ions to macromolecules can be so strong as to appear irreversible under ordinary conditions (as, for example, the binding of heme to hemoglobin), but unless covalent bonds are involved, we can in principle always consider it to be an equilibrium process. In fact, in most cases involving active metabolism, the binding will be found to be of such strength that an appreciable concentration of free ligand will be in equilibrium with bound ligand under physiologically significant conditions.

In investigating such phenomena, the biochemist is usually concerned with the following questions:

- What is the maximum number of moles of ligand that can be bound per mole of the macromolecule? That is, what is the number of sites n for a ligand designated A?
- What are the equilibrium constants (and other thermodynamic parameters) for binding of this ligand to each of the sites?
- Is the equilibrium constant for binding of ligand A to each of these sites independent of whether or not any of the remaining sites for A are occupied? (Obviously, this is directly related to the second question.)
- Are the equilibrium constants for the binding of ligand A modified by the binding of some other ligand B to the same macromolecule?

To show how such questions may be investigated, we shall first provide a quantitative definition of binding and describe some general ways in which it can be measured. We shall then consider several categories of binding processes, in increasing order of complexity.

Definition of binding—experimental measurements of binding.   In a sense, most measurements of equilibrium binding processes are indirect, in that we cannot observe directly which macromolecules have ligands bound to which binding sites.[2] In most cases, all we can measure is the fraction of all the ligand molecules in the system that are bound or the fraction of binding sites that are occupied. The general techniques for binding studies can be divided into classes that yield one or the other of these kinds of information.

An example of the first class is provided by equilibrium dialysis, perhaps the most direct technique employed for binding studies. This method was described in Chapter 13, where it was shown that, with certain assumptions, we can determine the concentrations of both free and bound ligands in equilibrium with the macromolecules. If the molar concentrations of free ligand and bound ligand are [A] and $[A_b]$, respectively, the total ligand concentration $[A_T]$ is obviously

$$[A_T] = [A] + [A_b] \tag{14.25}$$

If we designate by $[P_T]$ the total molar concentration of macromolecules, we can determine the number of moles of A bound per mole of P under the conditions of the particular experiment. This must be an average number since at any instant the dynamic equilibrium between the free and bound states will result in different molecules of P having different numbers of sites occupied. We define this average number as $\bar{\nu}$.

$$\bar{\nu} = \frac{[A_b]}{[P_T]} \tag{14.26}$$

If we can continue the experiments to a high enough concentration of A that the binding approaches saturation, we will observe that $\bar{\nu}$ will approach the limit n, the total number of sites per macromolecule available to A. Thus, a generalized binding curve will always behave in the manner shown in Figure 14.4; it will be a monotonic function of the free ligand concentration, approaching a finite limit as $[A] \to \infty$. Unfortunately, the limit (and hence the number n) is often difficult to deduce from the kind of graph shown in Figure 14.4. We shall show later that for certain types of binding, other graphical presentations are more useful.

The equilibrium dialysis technique, although it can yield accurate and complete data, is laborious and demands in many cases large quantities of the substances involved. Variants of this method involving gel filtration or sedimentation have been devised. These are sometimes more convenient, but to date they have not proven as accurate.

A quite different technique that is frequently employed in binding studies involves the detection of some physical change in either the macromolecule or the ligand upon binding. An example is the use of light absorption to follow the binding of

---

[2] This is not strictly true. We can observe the binding of large molecules to other large molecules using the electron microscope. However, it is very difficult to decide if such observations represent equilibrium binding because of the extreme conditions required for specimen preparation for electron microscopy. On the other hand, single-molecule techniques, such as atomic force microscopy, allow observation of some dynamic interactions under solution conditions. See Chapter 16.

Figure 14.4 A schematic graph of $\bar{\nu}$ versus [A] in some complicated binding process. Two points are general: $\bar{\nu} \rightarrow 0$ as [A] $\rightarrow 0$ and $\bar{\nu} \rightarrow n$ as [A] $\rightarrow \infty$.

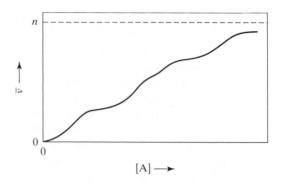

oxygen by hemoglobin. This depends on the fact that the spectrum of oxyhemoglobin is different from the spectrum of deoxyhemoglobin. If it can be demonstrated that the change with binding in some such physical property X is linear in the extent of binding, we may say that

$$\frac{\Delta X}{(\Delta X)_T} = \frac{\bar{\nu}}{n} = \theta \tag{14.27}$$

where $(\Delta X)_T$ is the total change produced when the macromolecules are saturated with A. The quantity θ is often called the *fraction saturation,* for it also represents the fraction of sites in the system that are occupied by ligand. Such techniques can be very sensitive because they can make use of a wide variety of physical measurements, including light absorption, fluorescence, nuclear magnetic resonance, and so forth (see Chapters 9 to 12).

A technique of this kind that has become quite popular in recent years is *surface plasmon resonance* (SPR). At an interface between a metal (like gold) and a dielectric (an attached protein layer, for example) reflected light can generate a surface plasma electromagnetic wave. The precise angle of incidence or wavelength that can do this depends critically on the refractive index of the dielectric layer. At the resonance angle or wavelength, reflection of the incident beam is minimized. If the protein (or other binding material) binds ligands from the solution the refractive index of the dielectric layer will be changed, and the condition for minimum reflection will change. If it can be assumed that the change in refractive index is proportional to binding, then Eq. 14.27 can be applied. The SPR technique has a number of advantages, including simplicity and versatility. Many kinds of binding can be studied with essentially the same protocol. On the other hand, it may be difficult to interpret for complicated binding processes.

Although techniques of the kind described above have become very popular, they have certain general limitations. These include:

- The change in parameter X must be linear in $\bar{\nu}$ and be the same for each site in a multisite molecule (or the precise relationship between change and $\bar{\nu}$ must be known). It often happens that binding to different sites on a single macromolecule may yield different changes in the property measured. Sometimes only a subset of the different sites for a given ligand will yield a detectable change. Of course, when combined with other studies (such as equilibrium dialysis), this kind of behavior may be advantageous, if it allows us to distinguish between classes of sites.

• In most instances n cannot be determined in this way, for we can only determine the *fraction* of total sites that are occupied. A special case occurs, however, if the binding is very strong, for in such cases a binding curve graphed as ΔX versus the *total* concentration of A in the system $[A_T]$ will be of the form shown by the solid line in Figure 14.5. Here ΔX has been graphed versus $[A_T]/[P_T]$. If the binding is very strong, a curve like the solid line will be obtained; each A molecule added to the system will be bound, changing ΔX, until the system is saturated. Then, even though more A molecules are added, they will not bind, and no more change will be observed. The break point of the curve occurs when $[A_T]/[P_T] = n$. Note, however, that if the binding is not strong, a curve such as that shown by the dashed line is observed; it would be very hazardous to attempt to deduce n from this curve. A further complication can arise from the fact that weak, nonspecific binding may be encountered at high ligand concentrations, making the determination of the n value for specific binding difficult.

On the other hand, if the value of n is known, considerable information can be extracted from a curve like the dashed line in Figure 14.5. Since $\Delta X/(\Delta X)_T = \bar{\nu}/n$, we can obtain $\bar{\nu}$. Since $\bar{\nu}$ is the average number of moles of ligands bound per mole of P, we can then obtain the concentration of bound A by $[A_b] = \bar{\nu}[P_T]$. Then the concentration of free A will be given by Eq. 14.25. This means that we know $\bar{\nu}$ as a function of $[A]$, just as we do from the equilibrium dialysis experiments and as we shall show in the following section, it is this kind of information that can be analyzed to yield binding constants and details of the mechanism of binding.

A final note: Observe that if the binding is very strong (the solid line in Figure 14.5), almost *all* A is bound until saturation is reached. This means that the concentration of free A is an indeterminably small quantity and that further analysis (to yield binding constants and so forth) will be impossible in such cases.

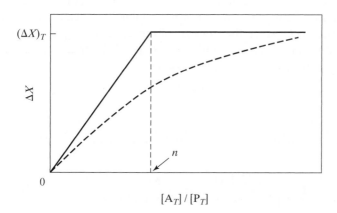

Figure 14.5 The change in some measurable parameter X graphed against *total* ligand concentration for very strong binding (solid line) and weaker binding (dashed curve). In the strong binding case, a clear break point will be observed at saturation.

With these preliminary remarks and definitions, we can now see how different kinds of binding, of increasing degrees of complexity, can be analyzed.

The simplest case: a single site per macromolecule.   In order to introduce some of the formalism and to provide a basis for the discussion of methods of data analysis, we begin with the particularly simple case in which a macromolecule P has only a single site for binding of a ligand A. Furthermore, we neglect for the moment the effects of any other ligands. The reaction is then

$$P + A \Leftrightarrow PA$$

The binding constant is then defined as

$$K = \frac{[PA]}{[P][A]} \tag{14.28}$$

(where the square brackets denote, as usual, molar concentrations). Remember that [A] and [P] in Eq. 14.28 are the concentrations of *free* ligand and *unliganded* macromolecule, respectively.

Sometimes it is more useful to think of the reaction as a *dissociation* process rather than as a binding process,

$$PA \Leftrightarrow P + A$$

in which case we can write the *dissociation constant*

$$\mathbf{K} = \frac{[P][A]}{[PA]} \tag{14.29}$$

Obviously,

$$\mathbf{K} = \frac{1}{K} \tag{14.30}$$

We now wish to express the results in terms of the experimentally accessible parameter $\bar{\nu}$, defined as the average number of bound ligand molecules per macromolecule. In the present case, since the number of moles of bound ligand per unit volume is equal to [PA], and the total number of moles of the macromolecule per unit volume is given by the sum ([P] + [PA]), we have

$$\bar{\nu} = \frac{[PA]}{[P] + [PA]} \tag{14.31}$$

This can be rewritten in a more useful form by using Eq. 14.28, in the form [PA] = K[P][A].

$$\bar{\nu} = \frac{K[P][A]}{[P] + K[P][A]} = \frac{K[A]}{1 + K[A]} \tag{14.32a}$$

since for a single site, $\bar{\nu}$ equals the fraction saturation θ we also have

$$\theta = \frac{K[A]}{1 + K[A]} \tag{14.32b}$$

Equation 14.32a describes $\bar{\nu}$ versus [A] as a rectangular hyperbola, like the curve in Figure 14.6. It is clear from the equation that $\bar{\nu} \rightarrow 1$ (saturation) only as $[A] \rightarrow \infty$. At half-saturation ($\bar{\nu} = 0.5$), we find $[A] = 1/K$. Note that if we had written Eq. 14.32 in terms of a dissociation constant, Eq. 14.30, would yield

$$\bar{\nu} = \frac{[A]/\mathbf{K}}{1 + [A]/\mathbf{K}} = \frac{[A]}{\mathbf{K} + [A]} \tag{14.33}$$

and $[A] = \mathbf{K}$ at $\bar{\nu} = 0.5$. This points out one convenience of dissociation constants: At least in simple cases, the numerical value of the dissociation constant (in moles per liter) shows immediately the ligand concentration at which the system is half-saturated and makes clear the concentration range in which dissociation occurs. The reader versed in biochemistry will note the similarity of Eq. 14.33 with the Michaelis-Menten equation for enzyme kinetics. This is because the Michaelis-Menten analysis

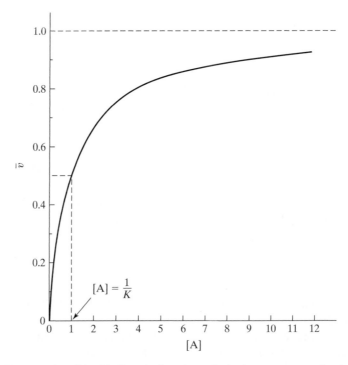

Figure 14.6   The curve describing binding of a ligand to a single site on a macromolecule. At the point where $\bar{\nu} = 0.5$, $[A] = 1/K$.

assumes an equilibrium (or at least steady state) of substrate binding to a single site on an enzyme molecule, and is framed in terms of dissociation constants.

Up to now, we have made the restrictive assumption that $n = 1$. Although there are many cases in which a macromolecule has but a single binding site for a particular ligand, most of the interesting situations arise when multiple sites are present. We turn now to such examples.

14.3.2 Multiple Equilibria

In this section, we consider those cases in which each macromolecule can bind n ligand molecules, where n can now be an integer greater than 1. We shall find that in this event, the interaction between the binding sites can produce complex and important effects. These include the whole range of *allosteric* effects that are of such great importance in protein chemistry.

Some general relationships.   If there are $n > 1$ sites per molecule, the expression for $\bar{\nu}$ becomes a bit more complicated than in the single-site case. As before, we define $\bar{\nu}$ as the average number of ligands bound per macromolecule. However, there may now be some molecules binding one ligand, some binding two, some binding three, and so forth, up to n. So

$$\text{concentration of bound ligand} = [PA] + 2[PA_2] + 3[PA_3] + \cdots$$

$$= \sum_{i=1}^{n} i[PA_i] \tag{14.34}$$

The numerical coefficients enter, of course, because each mole of $[PA_i]$ carries i moles of ligand. The total molar concentration of the macromolecule is just the sum of its molar concentrations in all forms,

$$\text{concentration of macromolecule} = \sum_{i=0}^{n} [PA_i] \tag{14.35}$$

Note that this summation starts at zero to include the term $[P]$, corresponding to unliganded macromolecule. Now $\bar{\nu}$ becomes

$$\bar{\nu} = \frac{\sum_{i=1}^{n} i[PA_i]}{\sum_{i=0}^{n} [PA_i]} \tag{14.36}$$

This statement is completely general; it includes Eq. 14.32a as the special case where $n = 1$.

In order to proceed further with the analysis, we must have some relationship among the $[PA_i]$. This can be done in a formal sense by writing a series of equilibrium equations. A number of formulations are possible; two are shown below.

Formulation I	**Formulation II**

Formulation I

$$P + A \Leftrightarrow PA \quad k_1 = \frac{[PA]}{[P][A]}$$

$$PA + A \Leftrightarrow PA_2 \quad k_2 = \frac{[PA_2]}{[PA][A]}$$

$$\vdots$$

$$PA_{n-1} + A \Leftrightarrow PA_n \quad k_n = \frac{[PA_n]}{[PA_{n-1}][A]}$$

Formulation II

$$P + A \Leftrightarrow PA \quad K_1 = \frac{[PA]}{[P][A]}$$

$$P + 2A \Leftrightarrow PA_2 \quad K_2 = \frac{[PA_2]}{[P][A]^2}$$

$$\vdots$$

$$P + nA \Leftrightarrow PA_n \quad K_n = \frac{[PA_n]}{[P][A]^n} \quad (14.37)$$

These two methods of describing the equilibria are equivalent. One set of equilibrium constants can always be expressed in terms of the other; for example, $K_i = k_1 k_2 k_3 \cdots k_i$. In any event, it should be understood that the manner in which the equilibrium constants are written says nothing about the mechanism of the reaction; for example, writing $K_n = [PA_n]/[P][A]^n$ does *not* imply that we believe n ligands bind simultaneously to a molecule of P. In many cases, Formulation II leads to simpler equations, and we shall employ it at this point.

Using the set of equations in Formulation II, we may rewrite Eq. 14.36 as

$$\bar{\nu} = \frac{\sum_{i=1}^{n} iK_i[P][A]^i}{\sum_{i=0}^{n} K_i[P][A]^i} \qquad (14.38)$$

where K_0 has been defined as equal to unity. Canceling [P], which factors out of each term in the numerator and denominator, we find that

$$\bar{\nu} = \frac{\sum_{i=1}^{n} iK_i[A]^i}{\sum_{i=0}^{n} K_i[A]^i} \qquad (14.39)$$

This very general binding equation is known as the *Adair equation*. Equation 14.32a is the special case where $n = 1$. The Adair equation will, in principle, describe almost any binding situation, for nothing has been assumed about the behavior of the individual K_i. However, if n is large (even as large as 4), there are so many adjustable parameters in the Adair equation that their determination with precision requires extensive and exact binding data. Furthermore, simply fitting the data to the Adair equation is not especially revealing about the mechanism of binding. Often, it is more useful to see if the data can be fitted by some more restrictive equation that embodies a simple model and requires fewer adjustable parameters. We shall turn now to the consideration of some such special cases.

All sites equivalent and independent.   The simplest case of a multiple equilibrium occurs when each of the n sites on a biopolymer has the same affinity for the ligand as any other, and the binding is *noncooperative;* that is, the affinity of any site is independent of whether or not other sites are occupied. At first glance, it might seem that this simply means that all of the k_i in Formulation I are equal. However, this is *not* the case, and to understand why we must look more closely at the equilibrium constants we have defined above. In writing, for example, $K_i = [PA_i]/[P][A]^i$ or $k_i = [PA_i]/[PA_{i-1}][A]$, the concentrations, such as $[PA_i]$ or $[PA_{i-1}]$, each represent the sum of concentrations of a whole class of molecules. Consider, for example, the set of molecules shown in Figure 14.7, in each of which two ligands are bound to a four-site macromolecule.

The number of such isomers of a particular liganded state is given by the number of ways in which n sites may be divided into i occupied sites and $(n - i)$ vacant sites. This is

$$N_{i,n} = \frac{n!}{(n - i)!i!} \tag{14.40}$$

In our macroscopic measurements we do not make distinctions between such isomers, and the *macroscopic* equilibrium constants we have written (the K_i or k_i) simply lump all such isomers together. But when we are talking about the *affinity* of a particular site, we are referring to a reaction in which a ligand molecule becomes bound to one particular site. The *microscopic* constant for such a reaction would be written

$$(k_i)_{kl} = \frac{[PA_{i,l}]}{[PA_{i-1,k}][A]} \tag{14.41}$$

which refers to the addition of a ligand to a particular isomer k of the class $[PA_{i-1}]$, at a specific vacant site to form a particular member l of the class $[PA_i]$. It is these microscopic constants, $(k_i)_{kl}$, that are identical for independent and equivalent sites. In physical terms, this simply means that each encounter of a ligand molecule with a site is an independent event and that all such encounters are intrinsically the same. The problem is formally very similar to the noncooperative two-state conformational transition described in Chapter 4.

Since the number of ways in which this addition can be made depends on the value of i, it turns out that even in the case of interest here, where all the microscopic constants are equal, the macroscopic constants are not equal. In order to

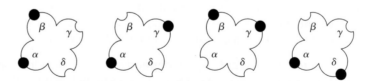

Figure 14.7 Four of the six (4!/2!2!) arrangements of two ligands over four sites. Note that the total concentration of PA_2 will be the sum of the concentrations of these six forms.

evaluate the individual terms in Eq. 14.36 for this case, we must consider first the formation of a particular species such as $[PA_{i,l}]$. We can imagine it to be made by the addition of successive ligands to P, as shown in Figure 14.8. Since the microscopic association constants for each step have been assumed to be equal, we give their value as k and can write

$$[PA_{i,l}] = k^i[P][A]^i \tag{14.42}$$

which will be valid whatever the path of additions we choose (see Figure 14.8). All of the individual species of $[PA_i]$ will have the same (average) concentration. Therefore, for the total concentration of $[PA_i]$ we need only multiply Eq. 14.42 by the statistical factors $N_{i,n}$, which represents the number of ways n sites may be arranged in i occupied sites and $(n - i)$ unoccupied sites.

$$[PA_i] = \frac{n!}{(n - i)!i!}[P]k^i[A]^i \tag{14.43}$$

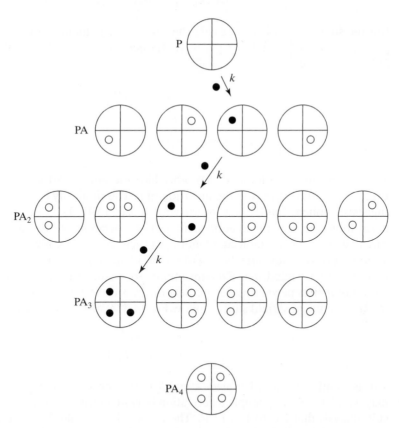

Figure 14.8 A possible path of additions of ligands to make a particular species PA_3. Each reaction in the path is described by the microscopic constant k.

This makes the Adair equation take the seemingly formidable form

$$\bar{\nu} = \frac{\sum_{i=1}^{n} i \, \dfrac{n!}{(n-i)!i!}(k[A])^i}{\sum_{i=0}^{n} \dfrac{n!}{(n-i)!i!}(k[A])^i} \tag{14.44}$$

However, Eq. 14.44 is easily simplified because the sums in the numerator and denominator are closely related to the binomial expansion. In fact, the denominator is simply the binomial expansion of $(1 + k[A])^n$, because the factor $N_{i,n}$ is just the binomial coefficient. The numerator may be evaluated by noting that

$$(1 + k[A])^{n-1} = \sum_{i=0}^{n-1} \frac{(n-1)!}{(n-1-i)!i!} k^i [A]^i$$

$$= \frac{1}{nk[A]} \sum_{i=0}^{n-1} \frac{n!(i+1)}{(n-(i+1))!(i+1)!} k^{i+1} [A]^{i+1} \tag{14.45}$$

But the sum in Eq. 14.45 is exactly the sum in the numerator of Eq. 14.44 if we replace $i + 1$ by a new index, $j = i + 1$. (It does not matter what symbol we give the index.) Therefore,

$$\bar{\nu} = \frac{nk[A](1 + k[A])^{n-1}}{(1 + k[A])^n}$$

$$\bar{\nu} = \frac{nk[A]}{1 + k[A]} \tag{14.46}$$

Equation 14.46 looks remarkably like the equation for single-site binding, Eq. 14.32a, differing only in the inclusion of n. The similarity is not accidental, and it suggests a much simpler way of arriving at Eq. 14.46. If all of the n sites on the macromolecule are indeed equivalent and independent of one another, there is no way in which a ligand molecule binding to such a site can know that it is binding to a site that is somehow attached to other sites. The fraction of all the sites in the system that are occupied, θ, is $\bar{\nu}/n$ (since there are $\bar{\nu}$ sites occupied per molecule out of a total of n sites per molecule; see Section 14.3.1). If the sites are independent, the fraction occupied should simply follow Eq. 14.32b, which we may write as

$$\theta = \frac{\bar{\nu}}{n} = \frac{k[A]}{1 + k[A]} \tag{14.47}$$

This is identical to Eq. 14.46 since k and K both represent binding to a single site. We may wonder why, if so simple a derivation is available, we resorted to the rather difficult analysis that led to Eq. 14.46. The answer is two-fold: first, the full analysis is more revealing of exactly what is meant by independent sites; and, second, the more complicated analysis will be essential for describing cooperative binding (see below).

Another advantage of the approach we have taken is that we can demonstrate how the population of various levels of ligation changes as the concentration of ligand increases. Equation 14.43 gives the individual $[PA_i]$ values, and if we divide by their sum, we obtain the fraction of protein molecules with i ligands bound. Figure 14.9a shows the distribution of site occupancy at $\bar{\nu} = 4$ for a molecule with eight binding sites. Note that for this case (binding to independent, equivalent sites) the distribution is quite broad.

When the binding of a ligand to a macromolecule is to be analyzed, we should test if all sites are equivalent and independent by determining whether or not Eq. 14.46 describes the binding with appropriate values of the parameters n and k. A simple direct plot (Figure 14.10a) is often not very satisfactory, for it is hard to obtain n as an asymptote, and the evaluation of k depends on knowing this limit. One can, of course, use nonlinear least-square analysis on a computer, with n and k as adjustable

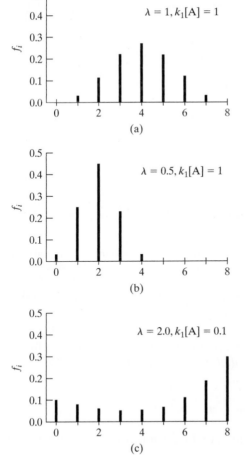

(a)

(b)

(c)

Figure 14.9 Distributions of species (PA_i) for a molecule with eight binding sites for three different binding models. The symbol f_i denotes the mole fraction of molecules with i sites occupied, $n - i$ sites empty. (a) Noncooperative binding; (b) negative cooperativity; (c) positive cooperativity. The values of λ and $k_1[A]$ used are indicated for each.

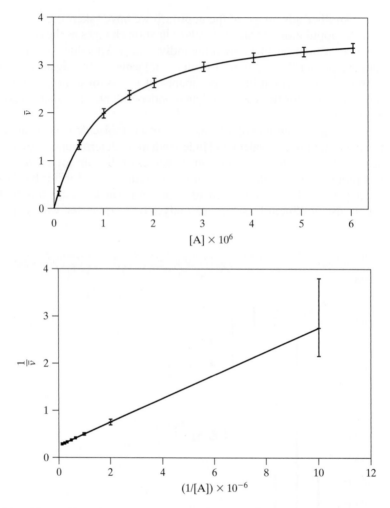

Figure 14.10   (a) A graph of Eq. 14.46 with $n = 4$ and $k = 1 \times 10^6$. An error bar is placed at each point to indicate the effect of a small error in each $\bar{\nu}$. It is clear from the data available that visual estimation of the value of n (and hence the value of k) would be difficult. (b) A double-reciprocal plot of the same data, according to Eq. 14.48. This will yield n and k. Error bars are same as in (a). Figure courtesy of Dr. M. Bruist.

parameters. This is becoming a very popular approach. Other graphical forms may at times be more advantageous. If we simply invert Eq. 14.46 and simplify, we obtain the equation for a *double reciprocal plot,* as shown in Figure 14.10b.

$$\frac{1}{\bar{\nu}} = \frac{1}{n} + \frac{1}{nk[A]} \tag{14.48}$$

Graphing $1/\bar{\nu}$ versus $1/[A]$ will give a straight line with intercept $1/n$ and slope $1/nk$.

An alternative rearrangement provides what is called a *Scatchard plot* (Figure 14.11a),

$$\frac{\bar{v}}{[A]} = nk - \bar{v}k \tag{14.49}$$

The extrapolation of a graph of $\bar{v}/[A]$ versus $\bar{v}$ will yield a slope of $-k$ and an intercept on the $\bar{v}$ axis of $\bar{v} = n$ (see Figure 14.11a). The method is widely used, but it must be employed with caution, as the extrapolation to the $\bar{v}$ axis can be somewhat

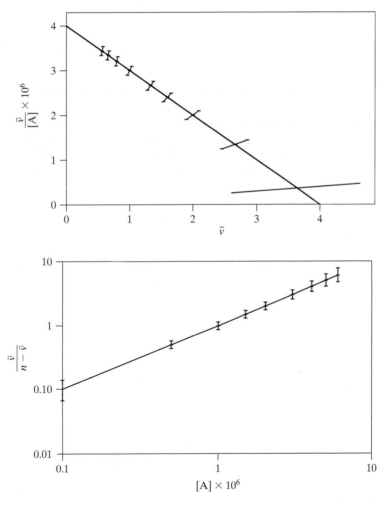

Figure 14.11   Other ways of plotting the same data as were shown in Figure 14.10. (a) A Scatchard plot (Eq. 14.49). (b) A Hill plot (as a log-log plot) (Eq. 14.50b). Again, same errors are as in Figure 14.10a. Figure courtesy of Dr. M. Bruist.

uncertain. This is particularly true in the cases where more than one type of site is present, as will be discussed in the next section.

Finally, if we rearrange Eq. 14.46 to calculate $\bar{\nu}/(n - \bar{\nu})$, we obtain a third kind of linear plot,

$$\frac{\bar{\nu}}{n - \bar{\nu}} = k[A] \tag{14.50a}$$

which is more frequently graphed in logarithmic form (Figure 14.11b)

$$\log\frac{\bar{\nu}}{n - \bar{\nu}} = \log k + \log[A] \tag{14.50b}$$

For equivalent sites this plot (called a *Hill plot*) will have a slope of unity (see Figure 14.11b). Its special utility will become evident in the following sections.

Binding to nonequivalent sites.   Considering the complexity of the structure of most proteins and other biopolymers, it is not surprising that in many cases a number of different *kinds* of sites for a given ligand may be found on one macromolecule. In this event, we should expect *a priori* that these different kinds of sites will have different binding constants for the ligand. It should be emphasized that we are *not* considering cooperative binding here; the value of the association constant for each site is still assumed to be independent of the state of occupancy of all other sites. Very often there is a large difference in affinity between the strongest sites and the weakest; and we may have to study the binding over a very wide range in ligand concentration in order to see the whole picture.

The formal analysis of such binding situations is a straightforward extension of what we have considered so far. Suppose that there are N classes of sites, numbered $s = 1 \rightarrow N$, and that there are n_s sites in each class (n_s can, of course, be as small as 1). Then $\bar{\nu}$, the total number of sites occupied by ligands per macromolecule, will simply be the sum of the $\bar{\nu}_s$ values, the number of sites of each class occupied per macromolecule.

$$\bar{\nu} = \sum_{s=1}^{N} \bar{\nu}_s \tag{14.51}$$

If each class of sites binds independently with an intrinsic association constant k_s, we have by Eq. 14.46,

$$\bar{\nu}_s = \frac{n_s k_s[A]}{1 + k_s[A]} \tag{14.52a}$$

so,

$$\bar{\nu} = \sum_{s=1}^{N} \frac{n_s k_s[A]}{1 + k_s[A]} \tag{14.52b}$$

To investigate the behavior of such systems, let us consider a simple case in which there are only two classes of sites: n_1 strong sites, each with a microscopic association constant k_1; and n_2 weak sites, with constant k_2. By definition, $k_1 > k_2$. Then Eq. 14.52b reduces to

$$\bar{\nu} = \frac{n_1 k_1 [A]}{1 + k_1 [A]} + \frac{n_2 k_2 [A]}{1 + k_2 [A]} \tag{14.53}$$

Unless $k_1 \gg k_2$, it is very difficult to distinguish this situation from the binding by $(n_1 + n_2)$ equivalent sites with k intermediate between k_1 and k_2 (see Figure 14.12a for an example). In such cases, Scatchard plots are often useful. If $k_1 \gg k_2$, the initial binding (low $\bar{\nu}$) will be dominated by k_1, so the first part of the curve will be given approximately by

$$\bar{\nu} \cong \frac{n_1 k_1 [A]}{1 + k_1 [A]} \tag{14.54}$$

Thus, the extrapolation of this region to the abscissa should give n_1, and the initial slope will yield k_1. On the other hand, the intercept of the full curve with the abscissa will yield $n_1 + n_2$, allowing in principle the calculation of n_2 and, by more indirect means, of k_2. Whatever graphical method may be used, the analysis is often ambiguous. If there are several classes of sites, the analysis becomes very difficult. In principle, the intercept with the abscissa should always give the total number of sites, but in many cases the curve approaches the $\bar{\nu}$ axis almost asymptotically and the determination of the intercept is very uncertain.

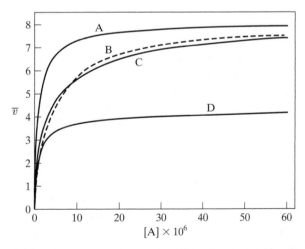

Figure 14.12a   Examples of the kinds of binding curves obtained if there are two classes of sites. It is presumed that there are eight sites in all, with four in each class 1 and 2. Curve A is obtained if $k_2 = k_1 = 1 \times 10^6 \ (M/l)^{-1}$ curve B if $k_2 = k_1 = 2.5 \times 10^5 \ (M/l)^{-1}$, curve C if $k_2 = 1 \times 10^5 \ (M/l)^{-1}$ and $k_1 = 1 \times 10^6 \ (M/l)^{-1}$, and curve D if $k_2 = 1 \times 10^3 \ (M/l)^{-1}$, $k_1 = 1 \times 10^6 \ (M/l)^{-1}$. Note that curves B and C are almost the same; it is very difficult to tell that two kinds of sites are present. Figure courtesy of Dr. M. Bruist.

Figure 14.12b The kind of Hill plot expected for a molecule with two classes of sites. Very similar curves would be observed with negative cooperativity.

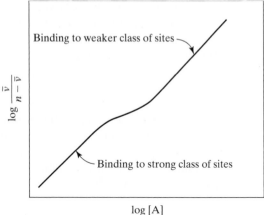

Binding to weaker class of sites

$\log \dfrac{\bar{v}}{n - \bar{v}}$

Binding to strong class of sites

log [A]

The existence of different classes of sites can also be demonstrated by the use of Hill plots. We graph $\log [\bar{v}/(n - \bar{v})]$ (or $\log [\theta/(1 - \theta)]$, where $\theta = \bar{v}/n$) versus log [A] (Figure 14.12b). The data obtained at the lowest values of [A] lie close to the line of slope unity corresponding to the strong binding sites. As [A] increases, the data move over to approach the line of slope unity corresponding to the weaker class of sites. The graph must contain a region with slope less than 1. Both k_1 and k_2 can be *estimated,* but n_1 and n_2 are not given by this method. The above examples show that in such situations a combination of two or more graphical techniques may be necessary for a full analysis.

Cooperative binding.   If binding of a ligand to one site on a macromolecule influences the affinity of other sites for the same kind of ligand, the binding is said to be *cooperative.* Such cooperativity can be *positive* (binding at one site increases the affinity at others) or *negative* (if the affinity of other sites is decreased). Such effects are a part of the general phenomenon of *allostery,* a general term that includes *homeoallostery,* which refers to the influence on ligand binding by ligands of the same kind, and *heteroallostery,* which refers to the binding of one kind of ligand modifying the binding of a second kind of ligand. For the present, we shall be concerned with homeoallosteric effects. Before proceeding to a detailed analysis of some of the models that have been proposed to explain such behavior, let us see, in a general way, just what the effects of positive or negative cooperativity will be. To do so, we will consider a simple model, which probably will not describe cooperativity in protein-ligand binding, but may describe some examples of nucleic acid–ligand interaction. We assume that the addition of each ligand changes the affinity for *all* remaining unoccupied sites by a factor λ. If λ is greater than 1, affinity increases with binding, and the cooperativity is positive. Negative cooperativity results if λ is less than 1. Of course, $\lambda = 1$ corresponds to the noncooperative case. In any event, we have for the microscopic constants $k_2 = \lambda k_1, k_3 = \lambda k_2 = \lambda^2 k_1$, etc. The consequences of $\lambda < 1$ and $\lambda > 1$ are shown in Figures 14.9b and 14.9c, respectively.

Strong positive cooperativity has the effect that the molecules tend to be *either* in the lightly occupied states *or* in the highly occupied states. Those that have, get more; those that lack, get still less. Negative cooperativity, on the other hand, means that neither extreme state is favored; the distribution of ligand occupancy is much narrower. All models of cooperative binding will behave qualitatively, like the simple example shown here.

In most cases, we cannot observe the actual distribution of binding occupancy. Rather, the average value $\bar{\nu}$ as a function of [A] (the binding curve) is more readily available by experiment (see above). How can we detect and measure cooperativity from binding curves? The differences in binding curves between noncooperative and cooperative binding are illustrated in Figure 14.13. The noncooperative curve is that given by Eq. 14.46; it is a rectangular hyperbola. The positively cooperative curve exhibits a *sigmoidal* shape. This can be qualitatively explained in the following way. At low levels of ligand the binding is very weak, because the molecule without ligands is presumed to be in a weakly binding state, and the first ligands bound are mostly going to different molecules that are in that state. But filling the first site or first few sites on any molecule somehow increases the affinity of the remaining sites on that molecule, so that binding becomes stronger as more sites are filled. Therefore, the curve turns upward, giving it the sigmoidal shape. The partial filling of binding sites somehow switches the macromolecule into a strong binding conformation. The effect of positive cooperativity is seen even more clearly on a Hill

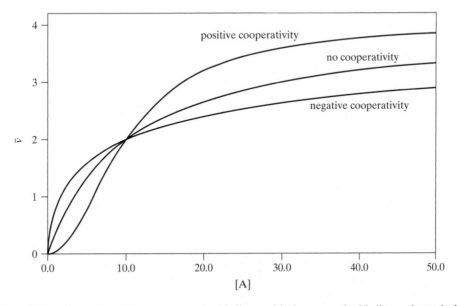

Figure 14.13 Curves describing noncooperative binding, positively cooperative binding, and negatively cooperative binding. The first follows Eq. 14.46. The second, which has a sigmoidal shape, might be described by an equation such as Eq. 14.61. The latter cannot be easily distinguished from binding to two classes of sites. Figure courtesy of Dr. M. Bruist.

Figure 14.14 A Hill plot for positively cooperative binding. The dashed lines have slopes of unity, and represent the limiting behavior at low and high ligand concentrations.

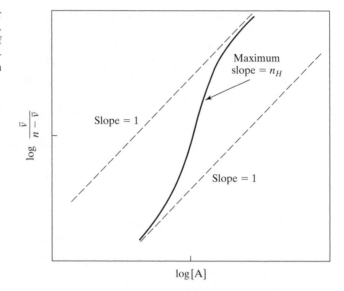

plot (Figure 14.14). The points corresponding to the binding of the first ligands in the system fall near the weak-binding-state line. As more sites are filled, the points move over toward a line corresponding to the strong binding state. By the time the last few openings are filled, almost all the molecules will be in this state, and the remaining points will lie along a line characteristic of this state.

Negative cooperativity has the opposite effect; as sites are filled, the remaining sites become weaker. A Hill plot for negative cooperativity looks like that shown in Figure 14.12b for a system with fixed strong and weak sites. Note that in all cases the curves for Hill plots will approach, at both low and high values of log [A], straight lines with slopes of unity. If *positive* cooperativity occurs, the curve must have a slope greater than unity near the middle of the binding range; if *negative* cooperativity is found, there will be a range where the slope is less than unity. Thus, Hill plots can be useful as diagnostic tests for binding type:

- If a straight line with slope of unity is found over the whole range, the binding is noncooperative and the sites are equivalent.
- If the curve has a slope greater than unity in some region, the binding must be positively cooperative.
- If the curve has a slope less than unity in some region, the macromolecule either has more than one class of sites or the binding is negatively cooperative. No matter how the data are plotted, it is almost impossible to distinguish between these two alternatives.

It is of interest to consider one idealized case that, while never really encountered, illustrates an extreme of behavior. Suppose that a system has such strong positive cooperativity that each molecule on accepting one ligand is so activated that it immediately fills the remainder of its n sites before the sites on any other molecules are filled. In this all-or-none case, as in the all-or-none conformational transition (Chapter 4), there are only two macromolecular species present in appreciable quantity, P and PA_n. This is the extreme form of the bimodal distribution shown in Figure 14.9c. In such a case, only the 0 and 8 bars would be seen. For such a model, Eq. 14.36 reduces to the simple form

$$\bar{\nu} = \frac{n[PA_n]}{[P] + [PA_n]} \tag{14.55}$$

and Eq. 14.39 becomes

$$\bar{\nu} = \frac{nK_n[A]^n}{1 + K_n[A]^n} \tag{14.56}$$

The Hill equation then takes the simple form

$$\log \frac{\bar{\nu}}{n - \bar{\nu}} = \log K_n + n \log [A] \tag{14.57}$$

which corresponds to a straight line with slope n. Since this unreal case represents the maximum possible cooperativity, the greatest slope a Hill plot can have is n, the number of cooperative sites on the macromolecule. In real systems, the maximum Hill slope n_H will always be less than this, so we can say that for a positively cooperative system the maximum Hill slope will always be greater than unity but less than the number of sites. The closer the quantity n_H approaches n, the stronger the cooperativity.

Models for cooperative binding.   In order to say anything more quantitative about cooperative binding, a model is required. A large number of models have been proposed. However, two extremely different ones essentially define their range, and have dominated thinking about cooperative binding by proteins for decades.

The theory developed by Monod, Wyman, and Changeux (1965), usually referred to as the MWC model, and all the subsequent variants of it are based on the concept of *concerted* conformational transitions in the subunits of a multisite protein. It is assumed that each subunit carries one site, and that it can exist in two states; in one state the binding site is weak and in the other the binding site is strong. A basic assumption of the MWC theory is that the symmetry of the molecule is maintained in conversion between the states; that is, although the transition may involve changes in the secondary and tertiary structures, as well as quaternary changes, all the subunits must undergo the transition in concert. Thus, there are only two conformation states for the molecules as a whole, the T state (in which all sites are weak) and the R state (in which all are strong). An equilibrium is presumed to exist between these forms, as depicted in Figure 14.15, so that at any moment a mixture of these two states of the whole molecule will exist. In the absence of the ligand, this equilibrium is described by the constant L,

$$L = \frac{[T]}{[R]} \tag{14.58}$$

Figure 14.15 The Monod-Wyman-Changeau (MWC) model for cooperative binding, using a tetramer as an example. The molecule is assumed to exist in equilibrium between two conformational states, T and R. The equilibria between liganded and unliganded forms are described in terms of two microscopic binding constants, k_T and k_R, where $k_R > k_T$. An equilibrium constant L connects the unliganded forms. Although we could write equations for other equilibria (dashed arrows), these are not necessary to define the system. Cooperativity arises from the fact that the greater liganding to the R form pulls the equilibria to the right, therefore further increasing the concentration of strong sites.

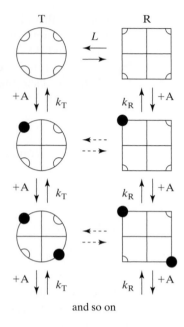

and so on

L is assumed to be much larger than 1; that is, in the absence of binding, most molecules are in the T state. The basic idea is that the binding of ligands, which occurs more strongly to the R state, will shift this equilibrium to favor the R form, in accord with Le Châtelier's principle. This provides a simple mechanism for cooperativity, as adding a ligand to the system increases the proportion of unoccupied sites in the R conformation, which then will bind with greater affinity. The R and T forms of the molecule will each, by virtue of the symmetry assumption, be a state with n identical sites. We designate the microscopic binding constants for the R and T states by k_R and k_T, respectively, and the ratio of these (k_T/k_R) as c. Since T is defined as the weak-binding state, $c < 1$, for strong cooperativity it will be a very small number. With these assumptions and the definition of $\bar{\nu}$, we may write, in analogy to Eq. 14.36,

$$\bar{\nu} = \frac{\sum_{i=1}^{n} i[RA_i] + \sum_{i=1}^{n} i[TA_i]}{\sum_{i=0}^{n}[RA_i] + \sum_{i=0}^{n}[TA_i]} \tag{14.59}$$

Using the same reasoning used in deriving Eq. 14.44, we obtain

$$\bar{\nu} = \frac{[R]\sum_{i=1}^{n} i \dfrac{n!}{(n-1)!i!} k_R^i[A]^i + [T]\sum_{i=1}^{n} i \dfrac{n!}{(n-i)!i!} k_T^i[A]^i}{[R]\sum_{i=0}^{n} \dfrac{n!}{(n-i)!i!} k_R^i[A]^i + [T]\sum_{i=0}^{n} \dfrac{n!}{(n-i)!i!} k_T^i[A]^i} \tag{14.60}$$

If we now insert $[T] = L[R]$ and $k_T = ck_R$, and evaluate the sums in the same way we did in deriving Eq. 14.46, we find that

$$\bar{\nu} = nk_R[A]\frac{(1 + k_R[A])^{n-1} + Lc(1 + ck_R[A])^{n-1}}{(1 + k_R[A])^n + L(1 + ck_R[A])^n} \tag{14.61}$$

The result is certainly not simple, but the behavior of systems obeying the MWC model can be readily visualized if we rewrite Eq. 14.61 in the form appropriate for a Hill plot. After some algebra, we find that

$$\frac{\bar{\nu}}{n - \bar{\nu}} = k_R[A]\frac{1 + Lc\left(\dfrac{1 + ck_R[A]}{1 + k_R[A]}\right)^{n-1}}{1 + L\left(\dfrac{1 + ck_R[A]}{1 + k_R[A]}\right)^{n-1}} \tag{14.62}$$

In the limits as $[A] \to 0$ and $[A] \to \infty$ we obtain

$$\frac{\bar{\nu}}{n - \bar{\nu}} \to k_R[A]\frac{1 + Lc}{1 + L} \quad \text{as } [A] \to 0 \tag{14.63a}$$

$$\frac{\bar{\nu}}{n - \bar{\nu}} \to k_R[A]\frac{1 + Lc^n}{1 + Lc^{n-1}} \quad \text{as } [A] \to \infty \tag{14.63b}$$

Each of these limiting forms corresponds to a straight line of slope equal to unity on a Hill plot. Such lines are shown in Figure 14.16. If L is large (that is, the T form is highly favored), Eq. 14.63a is given approximately by

$$\frac{\bar{\nu}}{n - \bar{\nu}} \cong k_R[A]c = k_T[A] \quad \text{as } [A] \to 0 \tag{14.64a}$$

Thus, the system begins at low $[A]$ by behaving like the pure T state. If it is also true that c is small, then, especially if n is large, we find that Eq. 14.63b will simplify to

$$\frac{\bar{\nu}}{n - \bar{\nu}} \cong k_R[A] \quad \text{as } [A] \to \infty \tag{14.64b}$$

That is, in this case the high $[A]$ limit is the R state. It should be emphasized that the Hill plot will approach lines of unit slope at very low or very high $[A]$ in any case; but only under the special conditions given above will these lines correspond to pure T or R states. The Hill plot must always lie between these limiting lines (R and T in Figure 14.16). The slope is a complicated function[3] of $[A]$, but it will always be less than n.

---

[3] The exact form was worked out by a student, in an unsolicited extrapolation of Exercise 14.10c. The result was subsequently published (see G. W. Zhou, P. S. Ho, and K. E. van Holde (1989), *Biophys. J.* **55**, 275–280).

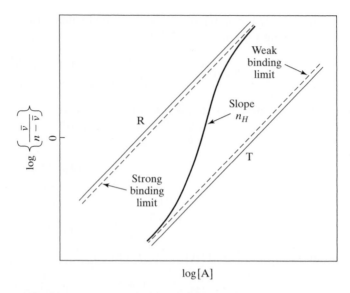

Figure 14.16   A Hill plot interpreted in terms of the MWC theory for cooperative binding. The asymptotic lines (dashed) correspond to Eqs. 14.63a,b. The solid straight lines correspond to the R and T states. Note that the system will not translate all of the way from T to R.

It is evident from Figure 14.16 that the MWC model predicts the kind of behavior observed in homeoallosteric systems that exhibit positive cooperativity. The MWC model cannot describe negative homeoallosteric cooperativity. Can you explain why? The MWC model has been extended and generalized by others to include more than two states and it can certainly describe a wide range of experimental data. But is it unique? Does it correspond to what really happens? Others have suggested that cooperativity can be explained in other ways. We will now consider a quite different alternative model.

The model proposed by Koshland, Nemethy, and Filmer (1966) (called the KNF model) takes a quite different approach. Rather than assuming symmetry in the conformational transformation of subunits, it assumes that subunits may change *one at a time* from a weak-binding form, which we shall call W to a strong-binding form, S. Thus, mixed states of the kinds shown in Figure 14.17 are to be expected. In the KNF model, the cooperativity arises because the interaction between the different pairs of subunits is assumed to depend on the states of the subunits in the pairs. That is, the stability of each pair W-W, W-S and S-S is assumed to differ. It is clear that this kind of mechanism is capable of giving rise to a wide range of behavior patterns, including both positive and negative cooperativity. One feature that makes the KNF model more complex than the MWC model is that the possible patterns of interaction depend on the geometry of the molecule or, more precisely, on the topology of the possible interactions. For example, in a tetrahedral tetramer, with D_2 symmetry, the subunit-subunit contacts are different from those in a square-planar tetramer with C_4 symmetry, and the equations describing the binding are likewise different.

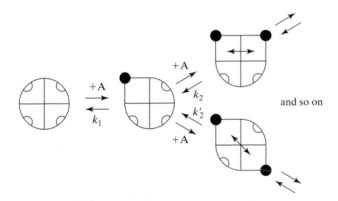

Figure 14.17 The Koshland-Nemethy-Filmer (KNF) model for cooperative binding, using a tetramer as an example. It is assumed that binding of a ligand to a site induces the subunit carrying that site to adopt a new conformation. Interactions with contiguous subunits may then promote their conversion to the new binding conformation, thus enhancing binding. Different subunit interactions may have different results, leading to considerable complexity in the theory. Note that there exists the possibility that interactions among subunits in the strong-binding conformation might be unfavorable; this would yield negative cooperativity.

In any event, the KNF model leads to much more complex expressions for the binding curve than does the MWC model. In terms of fitting individual binding curves, the two models appear to do equally well with the best data available at the present time. However, as more precise binding data are obtained and as more information becomes available concerning the intermediate states in the binding process, it is becoming evident that neither of these models is sufficiently sophisticated to mimic the real behavior of complicated proteins.

The extensive studies that have been carried out on hemoglobin provide an excellent example. Vertebrate hemoglobins are tetramers, containing two copies each of two kinds of similar subunits (α and β). The native hemoglobin molecule, therefore, should be considered as a dimer of (α/β) pairs. Each subunit can bind one O_2 molecule. Early binding studies showed that the binding must be cooperative. It was later observed that data on oxygen binding by hemoglobin could be fitted quite well by either the MWC or KNF model. The former gained credibility when early X-ray diffraction studies showed different, but symmetrical, *quaternary* structures for the deoxy- and fully oxy- forms of hemoglobin (compare Figure 14.18a and b). However, techniques like NMR have allowed us to investigate intermediate forms, and the picture appears to be more complex (see Barrick et al. 2004). As summarized by Ackers et al. (1992, 2004), the sequential binding of O_2 molecules by hemoglobin has features in common with *both* the KNF and MWC models, but agrees *exactly* with neither (Figure 14.19). Basically, individual subunits can bind O_2 with individual tertiary conformational shifts. There is a T $\rightarrow$ R quaternary change, but it occurs only after at least one of the sites on each α-β pair is occupied. Some features of this work are described in Application 14.1. We may expect that future studies will continue to reveal more complex behavior in other allosteric proteins.

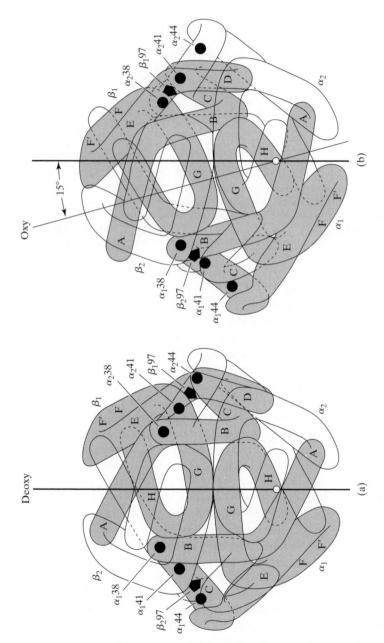

Figure 14.18 The conformational change that occurs on oxygenation of hemoglobin, as determined by X-ray diffraction. (a) Deoxyhemoglobin. (b) Oxyhemoglobin. After oxygenation the front α-β subunit pair ($\alpha_1\beta_1$) has rotated about 15° with respect to the back pair. [From Ackers et al. (1992).] Critical residues for interaction are indicated by circles.

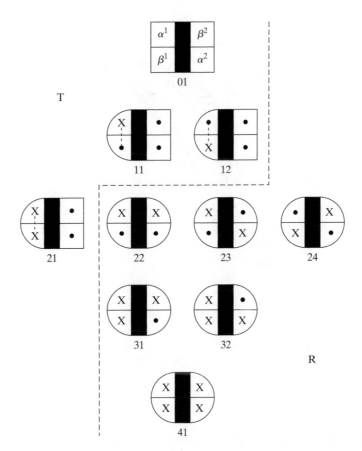

Figure 14.19 A recent model for cooperative binding of oxygen by hemoglobin. In liganded subunits, oxygen is indicated by a cross. The oxygenated tertiary structure of each subunit is depicted as curved, the deoxy as square. As oxygenation progresses, tertiary strain accumulates. Whenever there is one or more oxy subunit in both α-β pairs, the T $\rightarrow$ R quaternary transition occurs. Thus, the model shows some features in common with both KNF and MWC theories. Note that for simplicity, not all partially liganded configurations are shown. [Courtesy G. Ackers.]

Application 14.1   Thermodynamic Analysis of the Binding of Oxygen by Hemoglobin

One of the most extensive and profound analyses of binding is that of hemoglobin oxygen binding carried out in the laboratory of Dr. G. Ackers. A summary of the results is given in two reviews by this group (Ackers et al. 1992; 1994). Only a brief overview of the experiments and their analysis can be given here. One key to the analysis lies in the fact that the oxygenation process is linked to stability of the $(\alpha\beta)_2$ tetramer via the allosteric effect. The fundamental observation that both demonstrates this and allows the analysis is that the oxygen-binding curves for hemoglobin are dependent on the concentration of the protein (see Figure A14.1a). As a hemoglobin solution is diluted, the

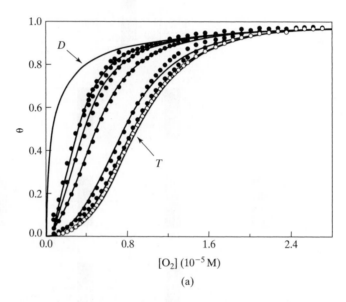

$[O_2]$ $(10^{-5}\,M)$

(a)

Figure A14.1a   Binding isotherms at decreasing hemoglobin concentrations (*right to left*). The right-most curve pertains to tetramers (T) ($n = 3.3$), and the leftmost solid curve is for dissociated $\alpha^1\beta^1$ dimers (D) ($n = 1.0$). Intermediate curves are at hemoglobin concentrations ranging from 4×10^{-8} M to 1×10^{-4} M. [From Ackers et al. (1992), reproduced with permission.]

equilibrium between $\alpha\beta$ dimers and $(\alpha\beta)_2$ tetramers is shifted toward dimers. This shift is accompanied, as Figure A14.1a shows, by both an increase in affinity and a loss in co-operativity. This is a most important observation, for it shows that the cooperativity pri-marily involves interactions *across* the interface between $\alpha\beta$ dimers. The corollary of these observations is that the association of hemoglobin dimers should be different in the deoxy state than in the fully oxygenated state. Indeed, this is true, as was shown, for example, by very careful kinetic measurements of the rates of hemoglobin dissociation and reassociation (Ip et al. 1976). The ratio of these rates then allows calculation of the equilibrium constant for tetramer formation. For deoxyhemoglobin, $\Delta G°$ for the associ-ation of *deoxy* dimers is about -14 kcal/mol heme, whereas for fully oxyhemoglobin the corresponding value is about -8 kcal/mol heme, the deoxy tetramer is more stable. On the basis of very careful measurements and data analysis, the numerical values for free-energy changes corresponding to all possible steps in the oxygenation of dimeric and tetrameric hemoglobin, and the interconversions between these forms have been de-duced (Figure A14.1b).

   The data show that the free energy involved in binding each molecule of oxygen to tetrameric hemoglobin is the algebraic sum of an intrinsic-site free energy and a cooper-ative free energy. Thus, $\Delta G°$ to bind the first O_2 to tetrameric hemoglobin, -5.4 kcal/mol,

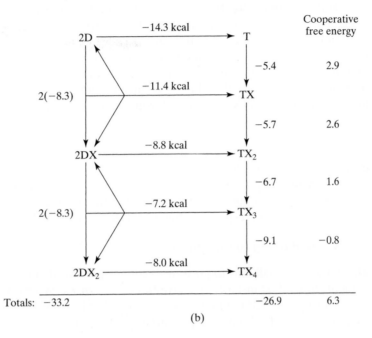

Totals: −33.2   −26.9   6.3

(b)

Figure A14.1b   Linkage diagram showing free-energy components of the cooperative mechanism. Binding of ligand species X to tetramers is depicted on the right and for dimers on the left. Dimer-tetramer assembly reactions at various stages of ligation are depicted by horizontal arrows. Standard Gibbs energies are given numerically for X = O_2 under conditions of data in Figure A14.1a. Reactions are taken left to right and top to bottom. Cooperative free energies (*right*) are the net energetic cost for binding to the assembled tetramers and reflect protein structural alterations that accompany the bindings. Quaternary assembly results in decreased affinity for the first binding step, which is paid for by 2.9 kcal/mol of free energy of protein structural rearrangement. [From Ackers et al. (1992), reproduced with permission.]

is equal to −8.3 kcal/mol +2.9 kcal/mol (see Figure A14.1b). For further developments in this analysis, see Ackers, et al. (2004).

ACKERS, G. K., M. L. DOYLE, D. MYERS, and M. A. DAUGHERTY (1992), *Science* **255**, 54–63.

ACKERS, G. K., J. M. HOLT, E. S. BURGIE, C. S. YRIAN (2004), *Methods in Enzymology* **279**, 3–28.

IP, S. H. C., M. L. JOHNSON, and G. K. ACKERS (1976), *Biochemistry* **15**, 654–660.

Heterotropic effects. In the MWC model, the behavior of the system is strongly dependent on the value of L. If L is very, very large, the macromolecules will tend to remain in the T state over the entire experimentally accessible binding range; if L is small, the molecules will yield a noncooperative binding curve close to that of the R state. Obviously, if some other factor could influence the T ⇔ R equilibrium described by L, the binding behavior would be profoundly altered. This is how the MWC theory explains heterotropic allostery. An *effector* molecule shifts the value of L. If it makes L larger, it is called an *allosteric inhibitor;* if it makes L smaller, it is an *allosteric activator.* It is often found that there are a number of molecules or ions that can thus modify the binding behavior for a particular ligand. An example is shown in Figure 14.20.

Needless to say, the KNF model can also explain such effects, often in subtle and complex ways. For example, binding of a heterotropic effector to one subunit may influence that subunit's response to binding of the major ligand to another subunit.

14.3.3 Ion Binding to Macromolecules

The binding of small ions to macromolecules can be divided into two classes of phenomena. In the first instance, we have those cases in which ions bind specifically and reversibly to a limited number of sites on a molecule such as a protein. The classical example of this is the binding of protons to weakly acidic groups on protein molecules.

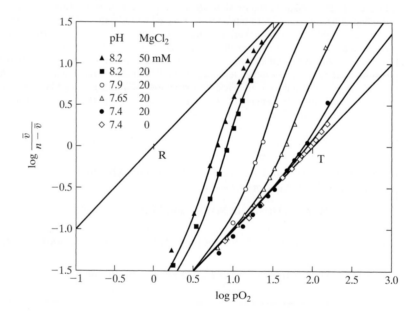

Figure 14.20 Binding of oxygen by the hemocyanin of the ghost shrimp *Callianassa*. The curves through the points have been fitted to a modification of the MWC theory, using the R and T states shown. Both hydrogen ion and divalent ions like Mg^{2+} are allosteric effectors for this protein. [From F. Arisaka and K. E. van Holde (1979), *J. Mol. Biol.* **143**, 41–73, with permission of the publisher.]

We shall use this as a paradigm for such site-specific binding of ions in general. The second case arises when there is a very high density of charged groups on a macromolecular surface, as on the surface of a DNA double helix, for example. In such cases, a rather different phenomenon, called *counterion condensation,* is observed.

Binding of protons to proteins.   The titration of proteins with protons is conventionally described in a manner rather different from that which we have employed hitherto. First, rather than *association* constants, *dissociation* constants are used. Second, instead of $\bar{\nu}$, the number of ligands bound, scientists describing protein-proton equilibria conventionally speak in terms of $\bar{r}$, the number of protons dissociated from the wholly protonated protein. If the total number of protons that can be dissociated is n, then obviously

$$\bar{r} = n - \bar{\nu} \tag{14.65}$$

We shall be interested first in the case where there are n equivalent and independent sites. Writing Eq. 14.46 in terms of a microscopic dissociation constant $\mathbf{k}$, we get

$$\bar{\nu} = \frac{n[\text{H}^+]/\mathbf{k}}{1 + [\text{H}^+]/\mathbf{k}} \tag{14.66}$$

so

$$\bar{r} = n - \bar{\nu} = n\left(1 - \frac{[\text{H}^+]/\mathbf{k}}{1 + [\text{H}^+]/\mathbf{k}}\right) \tag{14.67}$$

which can be rearranged to

$$\bar{r} = \frac{n}{1 + [\text{H}^+]/\mathbf{k}} = \frac{n\mathbf{k}/[H^+]}{1 + \mathbf{k}/[H^+]} \tag{14.68}$$

It is in this latter form that titration curves are usually expressed. Obviously,

$$\frac{\bar{r}}{n - \bar{r}} = \frac{\alpha}{1 - \alpha} = \frac{\mathbf{k}}{[\text{H}^+]} \tag{14.69}$$

where α is the fraction dissociation $\bar{r}/n$. So

$$\log\frac{\alpha}{1 - \alpha} = \log \mathbf{k} - \log[\text{H}^+] \tag{14.70a}$$

$$\log\frac{\alpha}{1 - \alpha} = \text{pH} - \text{p}K_a \tag{14.70b}$$

where $\text{p}K_a = -\log \mathbf{k}$. This is the familiar Henderson-Hasselbalch equation, which turns out to be equivalent to the Hill equation!

In the titration of any real protein, there will not be a single kind of titratable group but rather a range, from very acidic to very basic. Roughly, these can be grouped into classes, the carboxyls, the imidazoles, the lysines, and so forth. Thus, we might attempt to approximate the whole titration curve by an equation analogous to Eq. 14.52b,

$$\bar{r} = \frac{n_1 k_1/[H^+]}{1 + k_1/[H^+]} + \frac{n_2 k_2/[H^+]}{1 + k_2/[H^+]} + \cdots \tag{14.71}$$

A protein titration curve thus looks like Figure 14.21, with overlapping regions corresponding to the titration of different kinds of groups. Directly fitting curves such as the one in Figure 14.21 to Eq. 14.71 is rather unprofitable, for the number of adjustable parameters involved is too great. Obviously, more information is needed from other kinds of measurements. More seriously, Eq. 14.71 is only a very poor approximation, fundamentally because the affinity of a given group for a proton must be a function not only of the intrinsic properties of the group but also of that group's environment. The environments for a single kind of group (carboxyl, for example) in a protein molecule may vary widely, from highly hydrophobic regions to surface regions, or to strong interactions with specific other groups (lysine, for example). So a given kind of group may show a range of pK_a values. Furthermore, the environment of each binding site may change in a number of ways as the titration proceeds.

An obvious change that occurs is in the net charge on the protein molecule. At high pH, the average protein will be negatively charged (all the carboxylates will be in their anionic forms, and basic groups such as lysine will be neutral). As the pH is decreased and protons are added, the protein will go through a pH of zero charge (the isoelectric point) and then, at low pH, adopt a positive charge, since carboxylates will now be neutralized and the basic groups will have gained protons and become positive. This progressive change from a negative to a positive charge makes it more and more difficult to add further protons. This is actually a

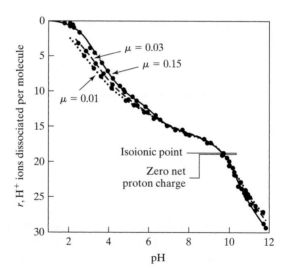

Figure 14.21 The titration of ribonuclease at 25°C, and several values of the ionic strength, μ. [From C. Tanford and J. D. Hauenstein (1956), *J. Am. Chem. Soc.* **77**, 5287–5291. Reprinted by permission of the American Chemical Society.]

kind of negative cooperativity; the more protons that are bound to a protein, the more difficult further proton-binding becomes.

As this change in charge occurs, the free-energy change involved in the dissociation of a proton from the molecule will change, for the electrical work done against the potential field of the protein will depend on the net charge. We may write the standard-state free-energy change for ionization of a particular group as

$$\Delta G^0 = \Delta G_{in}^0 + \Delta G_{el}^0 \tag{14.72}$$

where ΔG_{in}^0 is the *intrinsic* contribution, which depends only upon the intrinsic ionizability of the group, and ΔG_{el}^0 represents the extra work done against the net potential field of the protein.

Now, since $\Delta G^0 = -RT \ln \mathbf{k}$, we have

$$-RT \ln \mathbf{k} = -RT \ln \mathbf{k}_{in} + \Delta G_{el}^0 \tag{14.73a}$$

$$-\log \mathbf{k} = -\log \mathbf{k}_{in} + \frac{\Delta G_{el}^0}{2.303RT} \tag{14.73b}$$

or

$$pK_a = pK_{a,in} + \frac{\Delta G_{el}^0}{2.303RT} \tag{14.74}$$

To a first approximation, we can consider that the electrical term will be given by

$$\frac{\Delta G_{el}^0}{RT} = -w\overline{Z} \tag{14.75}$$

where w is a positive parameter whose value should depend on the dielectric constant, the ionic strength, and other properties of the medium; and $\overline{Z}$ is the average net charge on the protein. Note that if $\overline{Z}$ is positive, ΔG_{el}^0 is negative; a positive charge on the protein molecule favors the dissociation of protons. Conversely, if $\overline{Z}$ is negative, ΔG_{el}^0 is positive; protons are held more strongly by a negatively charged molecule. Using Eq. 14.75, we may now write Eq. 14.74 as

$$pK_a = pK_{a,in} - 0.434 \, w\overline{Z} \tag{14.76}$$

or, from Eq. 14.70b,

$$pH - \log \frac{a}{1-a} = pK_{a,in} - 0.434 \, w\overline{Z} \tag{14.77}$$

This result provides a method for evaluating the intrinsic pK from protein titration curves. If we can follow the titration of a single group or class of groups and evaluate both α and the net charge $\overline{Z}$ at different pH values, a graph of the left side of Eq. 14.77 versus $\overline{Z}$ will yield $pK_{a,in}$ as a limit where $\overline{Z} = 0$, and the value of w as a slope.

Methods have been developed that allow measurement of the titration of individual classes of groups, or even of individual groups in proteins and other macromolecules. The most powerful of these are spectroscopic techniques, including NMR. The details of such measurement techniques are given in Chapter 12.

As a consequence of X-ray diffraction analysis, it has become possible to know the positions of individual charged groups on many protein molecules quite exactly. With such information at hand, a much more sophisticated analysis of protein titration data can be carried out. The electrostatic free energy for any particular ionizable group i can be expressed in terms of the distances r_{ij}, between this group and all other charges on the molecule. An example of an application of this method to hemoglobin is given by Matthew et al. (1979). The results predict that the local charge environment will have widely varying effects on local pK_a values. For example, assuming an intrinsic pK_a of 6.60 for histidine led to the prediction of values ranging from 6.07 to 8.48 for different individual histidine residues. This points up the necessity for more selective experimental methods for the study of protein titration. Furthermore, it is becoming clear that proteins frequently "fine-tune" the pK_a's of important residues (as in catalytic sites) by the placement of other charged groups in their vacinity.

The description of proton binding to proteins as given above can easily be generalized to include the site-specific binding of other small ions to proteins. Such processes can be studied by the use of ion-specific electrodes to determine the free ion concentration during a titration process.

14.4   BINDING TO NUCLEIC ACIDS

14.4.1 General Principles

It is becoming evident that the binding of both small molecules and proteins to nucleic acids constitutes a group of the most important interactions in biochemistry and molecular biology. We can divide these interactions into two general classes, those in which the ligand interacts with a *specific* nucleotide sequence or binding site; and those in which the binding is nearly or completely nonspecific insofar as the nucleotide sequence is concerned. Binding in the first category, which includes, for example, many of the important interactions of transcription factors with DNA, can be analyzed by the methods described in the preceding sections; a particular DNA molecule will have one (or sometimes a few) specific binding sites for the ligand. These may be independent or interacting. An example is shown in Figure 14.22, which depicts the analysis (using quantitative footprinting) of the binding of a repressor to specific sites on λ-phage DNA. In this case, the binding to adjacent sites is cooperative.

14.4.2 Special Aspects of Nonspecific Binding

The second category, nonspecific binding, requires a somewhat different method of analysis than we have used up to this point. The situation for wholly nonspecific binding is shown in Figure 14.23; each ligand molecule can occupy a length of n

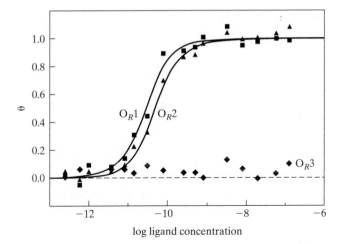

Figure 14.22   Analysis of the binding of Cro repressor protein to three specific sites (O_R1, O_R2, O_R3) on λ-bacteriophage DNA. The method used is quantitative footprinting, in which digestion of the DNA by a nuclease (DNAse I) is used to determine the positions and relative occupancy of binding sites. The binding sites show up as protected areas (footprints) on the DNA sequencing gel. Quantification of the protection offered by increasing the concentrations of the ligand (protein) allows the determination of θ for all three sites. Under the conditions shown here, only two sites bind significantly. [From Brenowitz et al. (1986).]

bases (for single-strand polynucleotides) or base pairs (for double-strand polynucleotides) *anywhere* on the nucleic acid molecules present in solution. The total
number of binding sites can best be expressed in terms of total concentration of
bases or base pairs (C_0 mol/l),

$$\text{total concentration of sites} = [S_0] = C_0/n \tag{14.78}$$

We write the equation for binding ligand A to a site as

$$A + S \Leftrightarrow AS$$

so the binding constant is defined as

$$k = \frac{[AS]}{[A][S]} \tag{14.79}$$

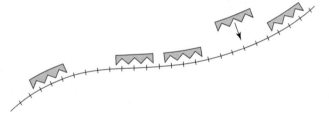

Figure 14.23   Nonspecific binding to a linear template like
DNA. Each ligand occupies *n*
bases (or base pairs) but these
can be *anywhere* on the long
DNA molecule.

The concentration of bound ligand $[A_b]$ is equal to the concentration of bound sites $[AS]$ and the concentration of free sites must be $[S_0] - [A_b]$. We then have

$$k = \frac{[A_b]}{[A]\{[S_0] - [A_b]\}} \tag{14.80}$$

Dividing by C_0, and defining $\bar{\nu}$ as $[A_b]/C_0$, we obtain

$$k = \frac{[A_b]/C_0}{[A]\left\{\dfrac{[S_0]}{C_0} - \dfrac{[A_b]}{C_0}\right\}} = \frac{\bar{\nu}}{[A]\left\{\dfrac{1}{n} - \bar{\nu}\right\}} \tag{14.81}$$

Note that $\bar{\nu}$ is here defined in terms of the number of ligand molecules bound per *base pair*. Even at saturation this will generally be a number much less than unity. In fact, at saturation $\bar{\nu} = 1/n$. Equation 14.81 may be rearranged in the form of a Scatchard plot,

$$\frac{\bar{\nu}}{[A]} = k\left\{\frac{1}{n} - \bar{\nu}\right\} \tag{14.82}$$

According to this expression, a graph of $\bar{\nu}/[A]$ versus $\bar{\nu}$ should yield a straight line with a slope $(-k)$ and an intercept $1/n$; both affinity and site size should be determinable.

Unfortunately, this simple analysis does not usually work, even in cases where it seems as if it should. Figure 14.24 shows data obtained for the binding of homogeneous oligolysines to a polynucleotide of uniform sequence. Clearly, the graph

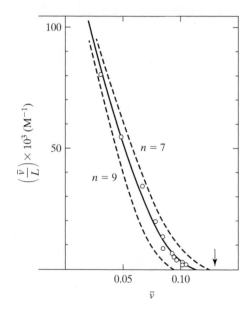

Figure 14.24 Binding of ϵ Dnp lys (lys)$_5$ to poly (rI) + poly (rC). These small molecules are expected to bind to this uniform template with a single binding constant. However, the data suggested multisite classes or negative cooperativity. This kind of anomolous behavior led to the McGhee and von Hippel theory. The data can, in fact, be fitted to Eq. 14.83 using $n = 7.8$, $K = 1.4 \times 10^5$ (M^{-1}) (the solid curve). Results with $n = 7$ or $n = 9$ are shown by dotted lines. The arrow indicates lattice saturation. [From McGhee and von Hippel (1974), *J. Mol. Biol.* **86**, 469–489, with permission from the publisher.]

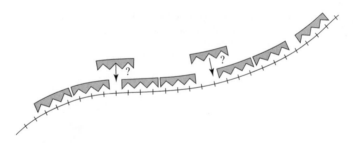

Figure 14.25   A closer look at nonspecific binding, according to the McGhee and von Hippel theory. As the lattice approaches saturation, major rearrangements may be necessary to use all of the possible sites.

exhibits curvature. But it should not be curved according to Eq. 14.82, for all sites on this simple DNA sequence must be equivalent. This problem was resolved by the realization by McGhee and von Hippel (1974) that the binding to a linear lattice exhibits a peculiar complexity, especially in its later stages. Consider Figure 14.25. As a large number of ligands become randomly bound, gaps will remain that represent the unoccupied parts of the binding matrix yet are each too small to accommodate another ligand. To completely saturate the system, massive rearrangement must occur; in fact, complete saturation of the lattice is very difficult to attain. The appropriate expression according to McGhee and von Hippel is

$$\frac{\bar{\nu}}{[A]} = k\left(\frac{1}{n} - \bar{\nu}\right)\left\{\frac{1 - n\bar{\nu}}{1 - (n - 1)\bar{\nu}}\right\}^{n-1} \tag{14.83}$$

which accounts for the curvature shown in Figure 14.24 very nicely. Note that if $n = 1$, Eq. 14.83 reduces to Eq. 14.82, as it should.

The above discussion has assumed that the binding is noncooperative. Cooperative binding to a linear lattice can be taken into account as shown in Figure 14.26a; an isolated molecule binding to a polynucleotide does so with affinity constant k, but a molecule binding adjacent to an occupied site does so with affinity ωk. If $\omega > 1$, binding is positively cooperative, if $\omega = 1$ binding is noncooperative, and $\omega < 1$ binding is negatively cooperative. Extreme positive or negative cooperativity will deform Scatchard plots in the ways shown in Figure 14.26b. The general expression obtained by an extension of McGhee and von Hippel theory, together with methods for analysis to cooperative binding are given by McGhee and von Hippel (1974).

14.4.3 Electrostatic Effects on Binding to Nucleic Acids

One aspect of the binding of proteins to polynucleotides that is both common and somewhat unusual is the major role played by the competition between the protein and condensed counterions for binding sites. Nucleic acids exhibit very high surface concentrations of charged groups that can bind counterions. An important example

Figure 14.26 Modifications of the McGhee and von Hippel theory to account for cooperative behavior. (a) A schematic to define the cooperativity parameter ω. (b) Representative Scatchard plots for $\omega > 1$, $\omega = 1$, and $\omega < 1$.

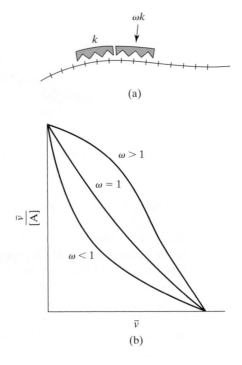

is double-strand DNA, which carries a negatively charged phosphate group on each nucleotide residue. A number of years ago, G. Manning demonstrated that a linear polyion like DNA with high charge density would exhibit unusual behavior in the presence of small ions (see Manning 1978 Record et al. 1978; see also Chapter 3). In addition to the diffuse counterion atmosphere expected with less densely charged macromolecules, counterions *condense* on a dense linear charge array. This has been termed the *counterion condensation effect*. If the average distance between polyion charges is b (see Figure 14.27), the fraction of these charges neutralized by condensed monovalent counterions is given by Manning's theory as

$$\psi = 1 - \frac{b}{(e^2/Dk_BT)} \tag{14.84}$$

ψ = fraction of charges neutralized by condensed counterions

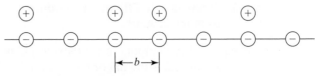

Figure 14.27 Parameters of the Manning theory of polyion condensation.

Distance between charges on chain

where e is the proton charge, D the dielectric constant, and k_B is the Boltzmann constant. The quantity (e^2/Dk_BT) has the dimension of length, and in water at room temperature, it has the value of 0.71 nm. If for double-stranded DNA, which has two charges per repeat, we take $b = 0.34$ nm$/2 = 0.17$ nm, we predict $\psi = 0.77$; that is, 77% of the charge will be neutralized by cations when a DNA molecule is immersed in a solution of a 1:1 electrolyte. This result is essentially independent of ionic strength, except at very low values. These predictions are approximately verified by experimental studies.

A consequence of this cation shielding is that when a protein binds electrostatically to DNA, it must displace some of those cations. If the protein interacts with n phosphate groups, the number of displaced cations will be $m = n\psi$, where ψ is the fraction of phosphates shielded by condensed counterions (Eq. 14.84). Thus, the binding reaction should properly be written

$$A + S \Leftrightarrow AS + mM^+$$

where M^+ stands for the monovalent cation. The equilibrium constant is then

$$K = \frac{[AS][M^+]^m}{[A][S]} \tag{14.85}$$

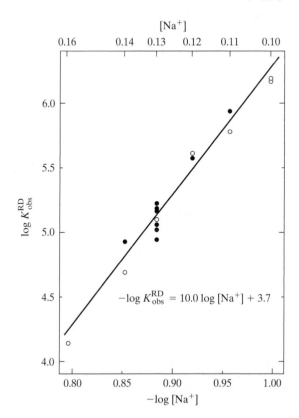

Figure 14.28 Dependence of the apparent binding constant on salt concentration for binding of lac repressor to DNA. The slope indicates that about ten monovalent ions are displaced for each protein molecule bound. [Data from P. L. de Haseth, T. M. Lohman, and M. T. Record, Jr. (1977), *Biochemistry* **16**, 4783–4790, reprinted with permission.]

But we *measure* an equilibrium constant by determining [AS], [A], and [S] at equilibrium. Therefore, the observed K is given by

$$K_{obs} = \frac{[AS]}{[A][S]} = \frac{K}{[M^+]^m} \tag{14.86a}$$

or

$$\log K_{obs} = \log K - m \log [M^+] \tag{14.86b}$$

The consequence of Eq. 14.86b is that a graph of $\log K_{obs}$ versus $\log [M^+]$ will have a slope of $-m$. Thus, a series of binding measurements at different ionic strengths will, in principle, yield the number of phosphate groups on the DNA that interact with each protein molecule, since $m = n\psi$. The second consequence of this effect is that binding of proteins to nucleic acids may be *very* strongly dependent on the ambient salt concentration (see Figure 14.28). It is this fact that lies behind the common practice of using increasing salt concentrations to dissociate proteins bound to DNA and RNA.

EXERCISES

14.1 a. The ΔG^0 for the hydrolysis of glucose-1-phosphate to glucose + phosphate, at 25°, pH 7.0, is -20.9 kJ/mol. For the corresponding reaction for glucose-6-phosphate, the value is -13.8 kJ/mol. Using these data, calculate ΔG^0 for the reaction:

glucose-1-phosphate $\Longleftrightarrow$ glucose-6-phosphate

at 25°C, and pH 7.0.
   b. What is the equilibrium constant?
   c. What is the fraction of total glucose in the form of glucose-1-phosphate at equilibrium?
   d. Suppose a mixture is prepared with equal concentrations of glucose-1-phosphate and glucose-6-phosphate. What is ΔG and what is the thermodynamically favored direction for the reaction?

*14.2 The hydrolysis of ATP to ADP + phosphate has $\Delta G^0 = -30.5$ kJ/mol at 37°C. Calculate the ratio of ADP to ATP in aqueous solution as a function of total nucleotide concentration, from 0 to 1 M. Neglect the possibility of adenosine monophosphate formation. Assume that all phosphate comes from the hydrolysis of ATP and that solutions behave ideally.

14.3 A protein, of molecular weight 14,300, dimerizes with an association constant of 1×10^5 mol/l. At what *weight* concentration will there be equal *masses* of monomer and dimer in solution?

14.4 A student believes that a protein she has isolated binds zinc. The protein has a molecular weight of 102,000. She performs a dialysis equilibrium experiment in which the concentration of protein within the bag is 1 mg/ml. The bag is equilibrated with a large volume of 0.1 mM $ZnCl_2$ solution (in appropriate buffer). Measurement of total zinc concentration inside the bag at equilibrium yields 0.12 mM. How many Zn ions are bound per molecule of protein?

14.5 It is often assumed that the folding of a globular protein is essentially a two-state reaction of the type

$$U \Leftrightarrow F$$

where U designates unfolded (random coil) and F designates the folded state, with
negligible concentrations of intermediate forms.

  a. If this is the case, express ΔG for the reaction as a function of the fraction of protein folded, (f_F). Note that the fraction unfolded (f_U) is equal to $1 - f_F$.
  b. *Sketch* the curve for ΔG as a function of f_F, from $f_F = 0$ to $f_F = 1$.
  c. What is the value of ΔG at $f_F = 0.5$?
  d. Obtain an expression for the value of f_F at $\Delta G = 0$.

*14.6 Suppose a protein is capable of *either* binding a ligand A *or* dimerizing but not
both. That is, dimerization hides the ligand-binding site. We have the reactions,

$$P + A \Leftrightarrow PA \quad k = [PA]/[P][A]$$

$$P + P \Leftrightarrow P_2 \quad K = [P_2]/[P]^2$$

  a. Obtain an expression for $\bar{\nu}$ as a function of [A] and $[P_T]$, when $[P_T]$ is total protein
  concentration.
  b. Is the result hyperbolic in [A]; that is, of the form $\bar{\nu} = \dfrac{\alpha[A]}{\beta + \gamma[A]}$?
  c. Show that in the limit as $[P_T] \to 0$, the binding curve does become hyperbolic;
  $\bar{\nu} = k[A]/(1 + k[A])$. Explain qualitatively why this is so.

14.7 The subunits of the hemocyanin of the shrimp *Callianassa* can undergo the following
association reactions:

$$2P \Leftrightarrow P_2 \quad K_{12} = \frac{[P_2]}{[P]^2}$$

$$2P_2 \Leftrightarrow P_4 \quad K_{24} = \frac{[P_4]}{[P_2]^2}$$

or

$$K_{14} = \frac{[P_4]}{[P]^4}$$

The equilibrium constants have been measured as a function of T.

$T(°C)$	K_{12}	K_{14}
4	1.3×10^7	4.5×10^{20}
10	1.2×10^7	1.0×10^{21}
15	1.5×10^7	3.0×10^{21}
30	—	4.0×10^{22}

All the concentrations are expressed in moles per liter. Estimate ΔH^0, ΔS^0, and ΔG^0
at 20°C for the reactions.

14.8 Consider a *monomeric* protein that can exist in two conformational forms (T and R).
Either form can bind one ligand molecule, A.

$$\begin{array}{ccc} P_T & \Leftrightarrow & P_R \\ +A \updownarrow & & \updownarrow +A \\ P_T A & & P_R A \end{array}$$

Here L, k_T and k_R are equilibrium constants for the $T \Leftrightarrow R$ conversion, and for binding to T and R states, respectively. We assume $k_R > k_T$.

a. Prove that this protein will exhibit noncooperative binding of A, with a binding curve of the form

$$\bar{\nu} = \frac{k[A]}{1 + k[A]}$$

b. What is the constant k, in terms of L, k_R, and k_T?

14.9 The data below show the binding of oxygen to squid hemocyanin. Determine from a Hill plot whether the binding is cooperative and estimate the *minimum* number of subunits in the molecule.

pO_2 (mm)	Percent Saturation	pO_2 (mm)	Percent Saturation
1.13	0.30	136.7	55.7
5.55	1.33	166.8	67.3
7.72	1.92	203.2	73.4
10.72	3.51	262.2	79.4
31.71	8.37	327.0	83.4
71.87	18.96	452.8	87.5
100.5	32.90	566.9	89.2
123.3	47.80	736.7	91.3

*14.10 In the original MWC model, it was assumed that the T state did not bind ligand at all.

a. Write the equation for $\bar{\nu}$ for a two-site protein according to this model. Obtain the expression for $\log [\theta/(1 - \theta)]$.

b. Compute a Hill plot for $L = 10^6$, $k_R = 10^6$ for this model. What is the maximum slope?

c. Derive an expression in terms of L for the maximum Hill slope. [Note: This is difficult, but doable.]

14.11 The following data describe binding of ligand A to a protein.

$[A](M)$	$\bar{\nu}$
1×10^{-6}	0.101
5×10^{-6}	0.381
1×10^{-5}	0.591
5×10^{-5}	1.116
1×10^{-4}	1.409
5×10^{-4}	1.813
1×10^{-3}	1.899
5×10^{-3}	1.955

Using both Scatchard plots and Hill plots, tell as much as you can about the binding reaction.

14.12 Sketch the Scatchard plots and double-reciprocal plots for:

a. positively cooperative binding.

b. negatively cooperative binding.

*14.13 We have emphasized that in the MWC model the conformational transition does not ordinarily cover the whole range from the pure T to the pure R state. Let us explore this a bit more quantitatively.

a. Show that the fraction of protein in the R state is given in the MWC model by

$$\overline{R} = \frac{(1 + \alpha)^n}{(1 + \alpha)^n + L(1 + c\alpha)^n}$$

where $\alpha = k_R[A]$.

b. Since α can only vary from 0 to ∞, there are limits to the range of values possible for $\overline{R}$. This is called the *allosteric range*. Find these limits in terms of L and c.

c. What is the allosteric range for a tetramer ($n = 4$) when $c = 10^{-2}$, $L = 10^4$; and when $c = 10^{-2}$, $L = 10^8$?

*14.14 The simplest form of a KNF-type model for cooperative binding can be described as follows: A dimeric protein, in the absence of ligand, has both subunits in the conformation W, with binding affinity k_W. But as soon as one site is occupied, *both* subunits switch to conformation S, with affinities k_S.

a. Write an expression for $\overline{\nu}$ as a function of ligand concentration [A].

b. Will this correspond to a hyperbolic (noncooperative) binding curve?

c. Can this model exhibit negative cooperativity? Under what circumstances?

*14.15 A common way of measuring DNA protein binding is by a filter-binding assay. In this method, DNA at a total concentration $[D]_T$ is mixed with protein at a total concentration $[P]_T$. (Both concentrations are in mol/l of molecules.) When these are poured through a nitrocellulose filter, only DNA molecules with protein bound to them are retained. The basic measurements are $[D]_T$, $[P]_T$, and f, the fraction of the DNA retained on the filter. For a case in which one protein-binding site is present on each DNA molecule, the equilibrium constant observed is

$$K_{obs} = \frac{[DP]}{[D][P]}$$

where [DP] is the molar concentration of complex, and [D] and [P] are molar concentrations of *free* DNA and protein, respectively. Derive an expression for calculating K_{obj} in terms of $[D]_T$, $[P]_T$, and f. (Hint: You also know that $[DP] = [D]_b = [P]_b$, where $[D]_b$ and $[P]_b$ are the concentrations of the bound DNA and the protein.)

14.16 The technique described in Exercise 14.15 is used to measure the binding of a protein to DNA at several NaCl concentrations. The results follow.

NaCl Concentration (M)	$K_{obs}(M^{-1})$
0.1	1×10^{12}
0.2	6.6×10^9
0.3	4.6×10^8
0.4	6.3×10^7
0.5	1.3×10^7
0.6	3.5×10^6

Calculate K and estimate the number of phosphate residues on the DNA interacting with the protein molecule.

REFERENCES

General

ALBERTY, R. A. (2003) *Thermodynamics of Biochemical Reactions,* Wiley Interscience, Hoboken, NJ.

DENBIGH, K. G. (1955) *The Principles of Chemical Equilibrium,* Cambridge University Press, Cambridge, U.K. An unusually clear (but not trivial) exposition of equilibrium thermodynamics.

FASMAN, G. D. (ed.) (1976) *Handbook of Biochemistry and Molecular Biology: Physical and Chemical Data,* 3rd ed., vol. 1, CRC Press, Cleveland, OH.

TINOCO, I., K. SAUER, J. WANG, and J. D. Puglisi (2004) *Physical Chemistry: Principles and Applications to the Biological Sciences,* 4th ed., Prentice Hall, Upper Saddle River, NJ. Provides an excellent introduction to the relevant thermodynamics.

WEBER, G. (1992) *Protein Interactions,* Chapman and Hall, London. A thorough description of protein thermodynamics, including both binding and association reactions.

WINZOR, D. J. and W. H. SAWYER (1995) *Quantitative Characterization of Ligand Binding.* Wiley-Liss, New York. Brief, but truly excellent.

Binding of Small Molecules and Ions to Macromolecules

BÖHM, J. and G. SCHNEIDER (eds.) (2002) *Protein-Ligand Interaction: From Molecular Recognition to Drug Design,* John Wiley and Sons, New York.

DRAPER, D. (2004) "A guide to ions and RNA structure," *RNA* **10**, 335–343. Makes distinction between Mg^{++} ions that are shelated to specific sites and those nonspecifically bound by electric field.

HOMOLA, J. (2003) "Present and future of surface plasmon resonance biosensors," *Anal. Bioanal. Chem.* **277**, 528–539.

KOSHLAND, D. E., G. NEMETHY, and D. FILMER (1966) "Comparison of Experimental Binding Data and Theoretical Models in Proteins Containing Subunits," *Biochemistry* **5**, 365–385. The primary reference to the KNF theory.

MATTHEW, J. B., G. I. H. HANANIA, and F. R. N. GURD (1979) "Electrostatic Effects in Hemoglobin: Hydrogen Ion Equilibrium in Human Deoxy- and Oxyhemoglobin A," *Biochemistry* **18**, 1919–1928. An excellent study of protein titration.

McGHEE, J. D. and P. VON HIPPEL (1974) "Theoretical Aspects of DNA-Protein Interactions: Cooperative and Non-Cooperative Binding of Large Ligands to a One-Dimensional Homogeneous Lattice," *J. Mol. Biol.* **86**, 469–489.

MONOD, J., J. WYMAN, and J. P. CHANGEUX (1965) "On the Nature of Allosteric Transition: A Plausible Model," *J. Mol. Biol.* **12**, 88–118. The original paper on the MWC theory of allostery.

PERUTZ, M. (1990) *Mechanisms of Cooperativity and Allosteric Regulation in Proteins,* Cambridge University Press, Cambridge, U.K. A compact and thoughtful discussion of a number of systems.

WEBER, G. (1975) "Energetics of Ligand Binding to Protein," *Adv. Protein Chem.* **29**, 1–83. A general treatment of the energetics of binding that does not resort to particular models.

WYMAN, J. and S. J. GILL (1990) *Binding and Linkage,* University Science Books, Mill Valley, CA. A sophisticated, but difficult treatment.

Hemoglobin and Allostery

ACKERS, G. K., M. L. DOYLE, D. MYERS, and M. A. DAUGHERTY (1992) "Molecular Code for Cooperativity in Hemoglobin," *Science* **255**, 54–63. An excellent review of hemoglobin allostery.

ACKERS, G. K., J. M. HOLT, E. S. BURGIE and C. S. Yarian (2004) "Analysing Intermediate State Cooperativity in Hemoglobin," *Methods in Enzymology* **379**, 3–28.

BARRICK, D., J. A, LUKIN, V. SIMPACANU and C. HO (2004) "Nuclear Magnetic Resonance Spectroscopy in the Study of Hemoglobin Cooperativity," *Methods in Enzymology* **379**, 28–54.

BEN-NAIM, A. (2001) *Cooperativity and Regulation in Biochemical Processes,* Kluwer Academic/ Plenum Publishers, New York.

Binding of Proteins or Small Molecules to Nucleic Acids

BRENOWITZ, M., D. F. SENEAR, M. A. SHEA, and G. K. ACKERS (1986) "Footprint Titration Yields Valid Thermodynamic Isotherms," *Proc. Natl. Acad. Sci., USA* **83**, 8462–8467. Use of footprinting to quantitate binding of proteins to DNA.

MANNING, G. (1978) "The Molecular Theory of Polyelectrolyte Solutions with Applications to the Electrostatic Properties of Polynucleotides," *Quarterly Reviews Biophys.* **11**, 179–246.

RECORD, M. T., C. F. ANDERSON, and T. M. LOHMAN (1978) "Thermodynamic Analysis of Ion Effects on the Binding and Conformational Equilibria of Proteins and Nucleic Acids: The Roles of Ion Association or Release, Screening, and Ion Effects on Water Activity," *Quarterly Reviews Biophys.* **11**, 103–178. How to analyze ionic effects on binding to nucleic acids.

STEITZ, T. A. (1990) "Structural Studies of Protein-Nucleic Acid Interaction: The Source of Sequence-Specific Binding," *Quarterly Reviews Biophys.* **23**, 205–280. Not much on binding thermodynamics, but a wonderful structural analysis as to what makes binding specific.

15

Mass Spectrometry of Macromolecules

It has long been recognized that the most precise physical method for the determination of molecular mass is mass spectrometry (MS). Furthermore, we have an appreciation for how MS can provide chemical structural information on small organic molecules from, for example, the degradation products resulting from their collisions with electrons (Figure 15.1). What has become very exciting is that the method is being adapted to solve a variety of problems in biology by applying these same basic principles to the analysis of biological macromolecules. The ability to determine accurate molecular masses, including those of fragments of large molecules such as proteins and nucleic acids, now allows us to identify molecules from a cellular mixture, monitor their changes, and even provide detailed information on their structures (from the primary sequence to potentially the tertiary fold and quaternary association of subunits) and mechanisms of folding. The basic components of MS are the same for both small and large molecules (Figure 15.1): The molecule(s) of interest as a charged ion are placed into the gas phase, then the molecular species must be separated (or resolved) according to their masses, and, finally, the molecular ions must be detected. Mass spectrometry uses an electrostatic potential to accelerate molecular ions and, thus, the method does not measure the mass of a molecule per se, but rather the mass-to-charge (m/Z) ratio of the molecular ion. We will see that molecular ions with smaller masses or larger numbers of charges will have a higher velocity, and this can be used as a basis to resolve various molecular ions according to differences in their respective masses.

The three major challenges in the mass spectrometry of biological macromolecules are essentially the same as those for small molecules, only more exaggerated. This chapter focuses on methods to generate macromolecular ions in the gas phase and to resolve their molecular masses to better than 1 part per million. We then discuss in greater detail how the resulting data are interpreted in order to identify molecules and learn something about their structures.

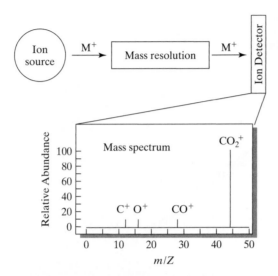

Figure 15.1 The three primary components of a mass spectrometer. The ion source generates molecular ions M^+ that can then be resolved (separated or filtered) according to their mass to charge ratios (m/Z) for detection. The mass spectrum of CO_2 after electron ionization is shown.

15.1   GENERAL PRINCIPLES: THE PROBLEM

For many years the application of mass spectrometry to study biological macromolecules was blocked by a seemingly insurmountable barrier, which is to propel such large molecules into flight. The problem has been stated as getting an elephant to fly without doing inordinate harm to the elephant or, in our case, the elephant of a molecule. The conventional method, used with low-molecular-weight materials, is simply to heat the sample. But the vapor pressures of proteins and nucleic acids are *very* low; heat sufficient to produce a detectable concentration in the vacuum leads to unacceptable degradation. The other complication is that the molecules must also be ionized. The breakthrough for this technique came from two new techniques for ion generation—each of these methods will be discussed in turn. Once we have an understanding of how to get biomolecules into flight, we will then be able to discuss how the molecular mass can be accurately determined, and see how the molecule's sequence and three-dimensional structure can be probed by mass spectrometric techniques.

Matrix-assisted laser desportion-ionization (MALDI).   The *matrix-assisted laser desportion-ionization* (MALDI) method takes advantage of the observation that pumping a lot of energy into a solid chemical matrix in which macromolecules are embedded will lead to the ionization and desorption (into a vacuum) of the macromolecule without significant degradation (Figure 15.2). The best way to pump energy in is through a laser pulse, and the matrix is chosen as a substance that absorbs strongly at the laser wavelength. The power density required to generate a significant ion current corresponds to an energy flux of ~ 20 mJ/cm^2. Aromatic molecules such as 2,5-dihydroxybenzoic acid, which absorbs in the UV, are favorite matrices because

Figure 15.2 Molecular vaporization and ionization by matrix-assisted laser desorption-ionization (MALDI). In MALDI-TOF-MS, the macromolecule to be analyzed (the *analyte*) is first embedded in a crystal of a matrix compound, typically a weak organic acid. When excited by a short laser pulse, the matrix becomes vaporized and, in the process, carries some of the analyte indirectly into the vapor phase. Ionization of the analyte results from exchange of electrons and/or protons (shown here as a proton transition) with the matrix compound. The ionized molecule then accelerates towards the mass analyzer, with the assistance of the electrostatic field.

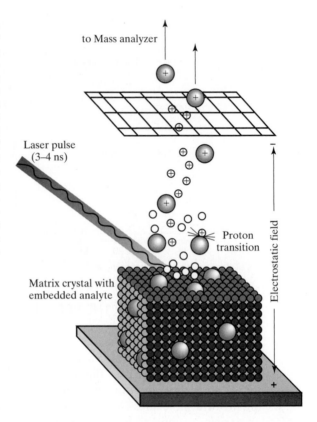

of the common use of UV lasers in the MALDI method. The pulsed nature of the excitation (from a few tens of nanoseconds to a few hundred microseconds) simplifies the data analysis, for all molecules begin their flight in a nearly synchronous fashion.

The result obtained when using MALDI is shown in Figure 15.3. Note that, as is frequently the case, more than one ionic species is obtained in the desorption process. This normally causes no difficulty in interpretation, because their m/Z ratios will be integral fractions of that for $Z = 1$. The accuracy of the method (better than 0.1%, using some techniques) exceeds that obtainable by any other physical measurement of mass. The molecular weight range that can be covered is also remarkable; protein molecules of mass up to 300,000 Da have been analyzed with accuracy.

Electrospray ionization (ESI).   An alternative technique to MALDI that may have even greater potential for the study of biological macromolecules is called *electrospray ionization* (ESI). In this technique, charged microdroplets containing the macromolecules to be studied are sprayed into the mass spectrometer through a charged nozzle (this ionizes the drops that are exiting the tip). As the droplets accelerate away from the tip, the solvent evaporates until, at some point, the concentration of charges is so high that the coulombic forces overcome the surface tension of

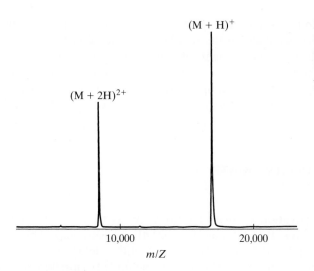

$(M + H)^+$

$(M + 2H)^{2+}$

10,000   20,000

m/Z

Figure 15.3 A mass spectrum of myoglobin. Both singly and doubly charged ions are present. [Adapted from Senko and McLafferty (1994), *Ann. Rev. Biophys. Biomol. Str.* **23**, 763–785.]

the drop, resulting in dispersion of the drop into a spray of smaller droplets. These droplets continue to evaporate and will themselves disperse into even finer sprays until all the solvent is gone, leaving the macroions they contained for analysis (Figure 15.4).

The technique is wholly nondestructive, but it produces a wide range of multiply charged macroions. This is no hindrance (and, in fact, can be an advantage) in studying pure substances, but mixtures will tend to yield a confusing multitude of peaks. The situation is further complicated by the tendency of the macroions to form adducts with small ions such as Na^+ and K^+, which often comes from the glassware used. The fact that some macromolecular adducts are stable in ESI has allowed further study of a variety of macromolecular complexes. Current ESI-MS instruments can be used to study molecules up to 500,000 Da.

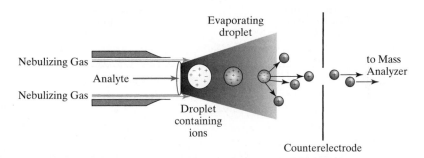

Figure 15.4 Electrospray ionization (ESI) mass spectrometry. A solution of the analyte is sprayed from an electrostatically charged nozzle (ionizing the solution and molecules in the exiting droplet) and carried toward a counterelectrode. A nebulizing gas facilitates dispersion of the droplets into a fine spray. Charged molecules eventually escape from the surface of the evaporating droplet and are accelerated toward the detector of the mass analyzer. [Adapted from J. Fenn (1984) *J. Phys. Chem.* **88**, 4451.]

Whatever technique is used, mass spectrometry provides a rapid, highly accurate way to measure masses of all but the very largest molecules. We discuss the details for determining the molecular weights of molecules by ESI-MS in Section 15.3. Furthermore, the amounts of molecules required are exceedingly small. For example, it has proved highly practical to couple ESI mass spectrometry to liquid chromatography for mass analysis of the tiny quantities (see Section 5.4.5) handled by that technique.

15.2   RESOLVING MOLECULAR WEIGHTS BY MASS SPECTROMETRY

The most obvious application of mass spectrometry is to determine molecular masses, but we should stress that MS does not directly measure mass. This principle can be understood by a more detailed consideration of the elements of operation of one simple type of mass spectrometer, the time-of-flight (TOF) instrument depicted in Figure 15.5. Positive ions of the substance to be analyzed are generated in a short *source* region, in the presence of an electrical field that imparts to each ion a kinetic energy ZEs, where Ze is the charge, E is the electric field, and s is the length of the source region. The ions emerge into the field-free drift region of length D, with each ion of a given charge having the same kinetic energy, regardless of its size. The kinetic energy then determines the velocity with which each ion moves through the drift region toward the detector, according to $K = mv^2/2$. The heavier ions, having the same energy as the lighter ones of the same charge, will move more slowly. Equating the expressions for K,

$$\frac{1}{2}mv^2 = ZEs \tag{15.1}$$

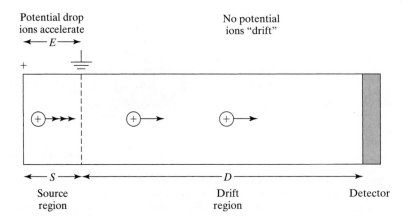

Figure 15.5   Schematic drawing of a time-of-flight mass spectrometer (TOF-MS).

which gives for the velocity v,

$$v = \left(\frac{2ZeEs}{m}\right)^{1/2} \tag{15.2}$$

The time for each ion to traverse the drift region will be

$$t = \frac{D}{v} = \left(\frac{m}{2ZeEs}\right)^{1/2} D \tag{15.3}$$

This allows us to express the mass/charge ratio for each ion in terms of t/D.

$$\frac{m}{Z} = 2eEs\left(\frac{t}{D}\right)^2 \tag{15.4}$$

You will always find mass spectra expressed in units of mass per charge. The units of the mass itself may be in atomic mass units (amu) or, for biological molecules, in daltons (Da, equivalent to 1 gm/mol) or kilodaltons (kDa equivalent to 1 kg/mol), units that are familiar to biochemists.

Although conceptually easy to understand, the simple linear TOF instrument has several drawbacks, all contributing to the reduced resolving power. Consider a situation in which a population of molecules (small or large) are ionized at the source. These molecules may all have different kinetic energies, which means that they will have different velocities (recall that $v^{1/2} = 2K/m$). Thus, even if all the ionized molecules have identical m/Z ratios, they will arrive at the detector at different times, yielding a broad distribution rather than a single peak of ions detected over time. We can see that it would be very difficult to resolve multiple molecular ions in the stream if each were to generate such broad peak profiles. To solve this resolution problem, we can force those molecules with very high kinetic energies and velocities to travel a longer distance than those with lower kinetic energies. A *reflectron* (Figure 15.6) reflects the stream of ions in the opposite direction, but a potential is applied so that faster ions penetrate further into the device, thereby forcing them to have a longer flight path and, consequently, equalizing the flight times of all the molecular ions to yield a sharper peak. These sharper peaks can now provide more accurate determinations of masses and resolution of multiple peaks from a mixture of molecular species.

The ability of MS to resolve molecular masses can be further enhanced (to as little as 1 amu mass difference) by selectively filtering the molecular ions while in flight. One simple example of such a molecular filter is to place magnets along the flight path to force the ions into a curved line of flight (magnetic sector mass spectrometer). Along this curved path, the angular momentum of the molecular ions is dependent on the m/Z of the ions and the strength of the magnetic field (Figure 15.7). It is easy to see that high molecular weight ions will be less affected by the magnetic field than lighter ones; consequently, the magnet will select for those macroions with the proper angular momentum to carry them through the curvature without colliding with the walls of the instrument before reaching the detector.

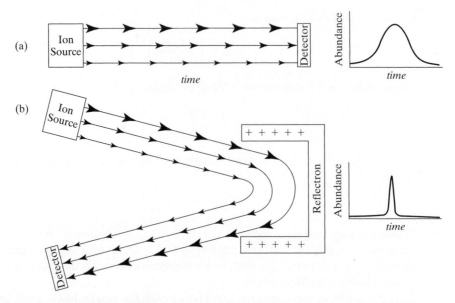

Figure 15.6   Principle of a reflectron in TOF-MS. Molecular ions exiting an ion source will have an average kinetic energy and an associated velocity (medium triangles). However, some ions will be have more kinetic energy (large triangles) and others less kinetic energy (small triangles), which will arrive at the detector at different times, even if they are the same molecular species. This generates a broad peak. By introducing a reflecting device that not only changes the direction of the molecular pathway, but also forces the faster ions to travel a longer pathway (in positive ionization mode, for example, the reflectron would have a positive potential to reflect ions, but those with more kinetic energy can penetrate farther into the field). The result is that ions with different kinetic energies will have the same approximate flight times to the detector, resulting in a sharper peak in the mass spectrum.

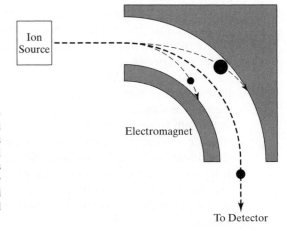

Figure 15.7   Magnetic sector mass spectrometry. This variation on the standard time of flight mass spectrometer filters molecular ions by forcing them to travel along a curved pathway. Only those ions whose angular momentum (as defined by their m/Z ratios) defines an unimpeded pathway through the magnetic field will reach the detector.

For mass spectrometry of biomolecules, the *quadrupole filter* is more routinely used as a molecular filter for mass analysis. The basic principle for the quadrupole filter is similar to that of the magnetic sector filter in that the angular momentum of the molecular ions is used to select for ions through, in this case, an electric field. The difference is that, as the name suggests, two electric potentials are applied across four electrodes or poles (Figure 15.8), with the electrodes precisely arranged around the path of the molecular ions (we will label this molecular pathway the z-axis). Unlike the magnetic filter, however, the quadrupole relies not on a fixed force to filter masses, but on a set of oscillating electrostatic fields to very selectively allow only a molecule of a specific mass to pass through.

To understand how this works, we start with the basic principle that a light molecule will have a small inertia in flight and, thus, will be greatly affected by an alternating field, while a heavy molecule with large inertia will be less susceptible to perturbations from an alternating field. We can then set up the filter by first arranging the electrodes in a square, with a distance $2r_0$ separating the electrodes (or a distance r_0 from the center to each surface of the quadrupole). A constant electric potential (dc potential) is applied across each electrode, pairing the electrodes harboring positive static potentials along an axis that we label here as the x-axis, and the electrode pair with negative potentials along the y-axis. Superimposed onto each of these static potentials is an alternating potential (ac) or a radio frequency (rf) field

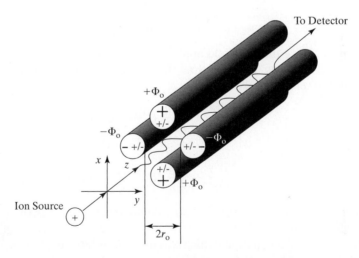

Figure 15.8 Quadrupole mass filter. A quadrupole filter for mass spectrometry is setup with four electrodes in a square arrangement (with a distance of $2r_0$ separating the surface of diametrically opposed electrodes), leaving an open pathway (z-axis) to the detector. Each electrode carries a constant voltage potential (with the positive poles for the pair along the x-axis and negative poles along the y-axis). Superimposed on the constant potential is an alternating $(+/-)$ potential. The positive potentials focus positive ions towards the center of the quadrupole (the molecular pathway), while negative potentials defocus the ions towards the electrode surfaces. Only those ions that travel through this field of constant and alternating potentials unimpeded will be detected.

with an angular frequency ($\omega = 2\pi\nu$, where ν is the frequency of the rf field). Thus, any ion coming into the quadrupole will experience a static potential (with magnitude U) and alternating potential (with a magnitude of V). The overall potential can thus be defined as

$$+\Phi_o = +(U - V \cos \omega t) \tag{15.5}$$

for each electrode with a positive static potential along the x-z plane, and

$$-\Phi_o = -(U - V \cos \omega t) \tag{15.6}$$

for each electrode with a negative static potential along the y-z plane. The potential at any point (x and y) within the quadrupole is thus

$$\Phi = \Phi_o \frac{x^2 - y^2}{r_o} = \frac{(x^2 - y^2)(U - V \cos \omega t)}{r_o} \tag{15.7}$$

For an ion with mass m and a charge Ze, the force experienced along the x and y directions, respectively, can be described by the standard equations for force ($F = ma$), or

$$F_x = m\frac{d^2x}{dt^2} = -Ze\frac{\partial\Phi}{\partial x} \tag{15.8}$$

along x, and

$$F_y = m\frac{d^2y}{dt^2} = -Ze\frac{\partial\Phi}{\partial y} \tag{15.9}$$

along y. Notice that the field applies a force of $F = 0$ along the z-axis. Substituting Eq. 15.7 into Eqs. 15.8 and 15.9, followed by derivatization and rearrangement, yields the relationships

$$\frac{d^2x}{dt^2} + \frac{2Ze}{mr_o^2}(U - V \cos \omega t)x = 0 \tag{15.10}$$

and

$$\frac{d^2y}{dt^2} + \frac{2Ze}{mr_o^2}(U - V \cos \omega t)y = 0 \tag{15.11}$$

These equations are stable if x and y do not equal r_o (that is, the ion does not collide with the electrodes). Equations 15.10 and 15.11 can be rewritten in the form of Mathieu's differential equation (which was derived in 1866 to describe the propagation of waves in membranes) to provide a quantitative prediction of the stability of ions within a quadrupole field. For this discussion, however, a qualitative description of a stable pathway through the quadrupole filter will suffice for understanding how the filter works.

Equations 15.10 and 15.11 allow us to treat the quadrupole potential in the two separate x-z and y-z planes. Consider a stream of positive ions injected into the quadrupole. A positive field applied to any two opposing electrodes would push and focus the ions toward the center axis, while a negative potential would pull and thus defocus the ions from the center. Let us first consider the electrodes along the x-axis, with their static positive potentials and alternating rf field (Figure 15.9). Very heavy ions will have a high inertia and, therefore, their paths will not be dramatically affected by the rf field, but rather will be influenced primarily by the static positive potential. Consequently, these ions will be focused toward the center axis and will pass along the x-z plane unimpeded. The paths of very small ions, however, would be dramatically affected by the rf field, and can readily be pulled away from the center

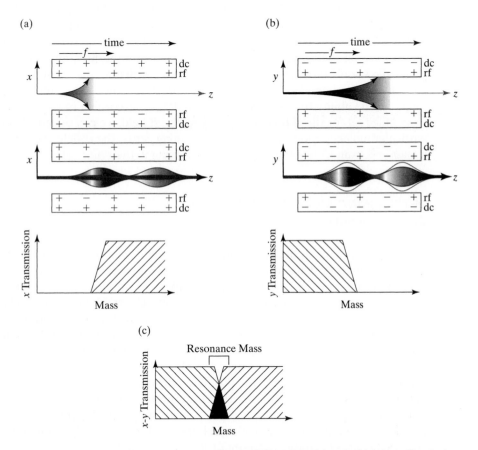

Figure 15.9 Quadrupole filter as a low mass filter in the x-z plane (a) and a high mass filter in the y-z plane (b). As ions travel through the x-z plane of the quadrupole filter, very light ions will be greatly affected by the defocusing negative potential from the rf field and will be pulled to the surface of the electrodes (a). Very heavy ions, however, feel primarily the constant focusing positive potential of the dc field and will pass through unimpeded (heavy line). In the y-z plane, the opposite is true, with the heavy ions being pulled by the negative potential of the dc field toward the electrode surfaces (b). A medium sized ion (with a specific m/Z ratio) will resonant with the rf field in both planes, and thus will pass through the filter of both planes. Thus, the x-z plane acts as a low mass filter while the y-z plane acts as a high pass filter (c), leaving only those masses that overlap between the two to be detected.

axis toward a collision with the electrode surfaces whenever the rf field presents a negative potential. This will completely defocus the light ions, and thus are filtered out along this x-z plane. Somewhere in between, ions of medium mass will fluctuate with the rf field, being focused and unfocused along their paths in resonance with the rf frequency, but these ions of medium mass will pass through the filter unimpeded. We can thus think of the electrodes along x as a low mass filter that eliminates molecular ions that are lighter than a critical mass.

The electrodes along the y-z plane harbor a static defocusing negative charge. Thus, as heavy ions pass along this plane, they are pulled toward the electrode surface, again to become entirely unfocused if the rf potential is insufficiently positive enough to change the path of these ions. The y-z plane can therefore be considered a high molecular weight filter. Putting the two sets of electrodes together, we produce the quadrupole, with the low mass filter along x and the high mass filter along y, which allows only those ions that are in resonance with both the x-z and y-z planes to pass through unimpeded toward the detector. It becomes apparent that the mass range for the ability to pass through can be defined according to the frequency and the voltage set for the rf field. By specifying the rf field, we can selectively analyze a particular molecular ion species, and by scanning across the rf field, we can generate a mass spectrum from a mixture of ions according to their m/Z ratios.

15.3   DETERMINING MOLECULAR WEIGHTS OF BIOMOLECULES

We can now discuss in detail how molecular weights can be determined from ESI-MS experiments by considering the experiment in greater detail. ESI is considered a "soft" ionization process because it is a wholly nondestructive method for introducing charges onto a biomolecule—ions are generated by adding positive charges by the addition of protons or cations (positive ionization mode) or by adding negative charges by removal of protons or the addition of anions (negative ionization mode). For small molecules ($\leq \sim 1200$ Da), only a single molecular species is observed (small molecules generally have only a single ionizable group present). That species can be written as $(M + H)^+$ for positive ionization, where M is the mass of the macromolecule, H is the added mass of the proton, and the superscript $+$ tells us that we are in positive ionization mode. We can write a similar formula for negative ionization resulting in a singly charged species. For this singly, positively charged species the mass to charge ratio (m/Z) is

$$m/Z = \frac{(\text{MW} + \text{H})}{1} = \text{MW} + 1.008 \qquad (15.12)$$

Thus, the molecular weight can be directly obtained from the ESI-MS experiment for this type of single-charged molecule. The ESI-MS of even these smaller molecules, however, is not a simple single peak. For example, the positive ionization spectrum of enkephalin (Figure 15.10) shows multiple species, each successively

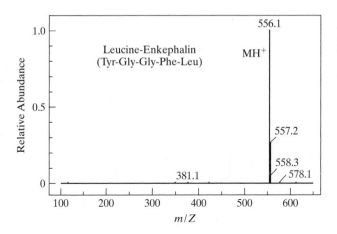

Figure 15.10 ESI-MS spectrum of leucine-enkephalin. The mass-to-charge ratio (m/Z) for the $(M + H)^+$ parent ion and the isotopes that are heavier by 1 amu are labeled.

reduced in intensity. Where do these come from? Each peak corresponds to a different isotopic form of the molecule, with the intensity related to the natural abundance of an isotope multiplied by the number of atoms of that element in the molecule. The largest peak therefore represents molecules having the "standard" (most abundant) isotopic forms of all the elements (C^{12}, H^1, O^{16}, etc.) in such low mass molecules. Notice that this peak is at 556.1 m/Z, which is the calculated mass of 555.2 g/mol for the molecule plus 1 amu for the mass of the proton added in the positive ionization mode of the experiment. The next adjacent peak corresponds to a molecule that is 1 amu higher in mass. Since the next most abundant isotope in enkephalin is C^{13}, this must correspond to a molecule having a single C^{12} replaced by a C^{13}. We don't know where this isotope is incorporated within the molecule, since all carbons have equal probability of having this heavier isotope. Each successive peak is reduced in intensity until we reach the baseline. But there is another peak at 578.1 amu, which is 23 amu greater in mass. This corresponds to the addition of an Na^+ cation rather than a proton to generate the positive ion.

For macromolecules with MW > ~1200 Da, the situation is more complicated in that there are typically numerous ionizable groups, which gives rise to multiply charged species. In proteins, they include the amines and amides, which can readily be protonated in positive ionization mode, and carboxylic acids, which can be readily deprotonated under negative ionization mode. We can use the electrospray ionization mass spectrum (ESI-MS) of lysozyme (Figure 15.11) to get a better understanding of what contributes to the spectrum and how to interpret the results of such an apparently complex spectrum. First, notice that there are multiple sets of regularly spaced peaks. These correspond to multiply charged states of the molecule that differ in charge by +1, although we do not know the actual charge of any particular species. Since this is under positive ESI-MS mode, we can write the formula for each charged species as $(M + nH)^{n+}$. Similarly, the mass to charge ratio is given as

$$\frac{m}{Z} = \frac{(MW + nH)}{n} = \frac{(MW + n1.008)}{n} \qquad (15.13)$$

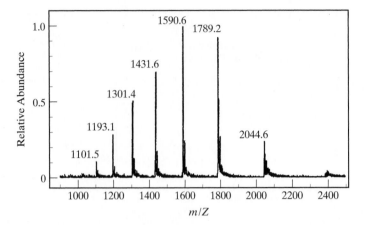

Figure 15.11   ESI-MS of lysozyme. The m/Z ratios of the $(M + nH)^{+n}$ ions are labeled.

where n is the integral number of protons added. For negative ionization, n is simply a negative integer. Given the simple relationship in Eq. 15.13, we can determine the molecular weight of lysozyme from the ESI-MS spectrum by comparing the m/Z of two adjacent primary peaks, and assuming that the molecular weight of the protein associated with each peak is identical. Thus, the peak at $m/Z = 1590.6$ can be described as the following:

$$1590.6 = \frac{(MW + nH)}{n} = \frac{(MW + n1.008)}{n} \tag{15.14}$$

while the peak at $m/Z = 1789.2$ would correspond to a species with one less charge, or $Z = n - 1$ (notice that since the m/Z ratio increases, Z must decrease for this peak if m is to remain the same), to give the equation

$$1789.2 = \frac{(MW + [n - 1]H)}{n - 1} = \frac{(MW + [n - 1]1.008)}{n - 1} \tag{15.15}$$

Solving the two simultaneous equations (15.14 and 15.15) results in $n = 9.0$, or that the $m/Z = 1590.6$ species has 9 protons added. Inserting this into Eq. 15.14, we can show that MW $= 14{,}306.4$ Da. This is compared to 14,305.1 Da calculated for lysozyme from the amino acid composition.

The pace at which MS instrumentation is being developed to improve sensitivity and extend applications has accelerated in recent years and, therefore, we will not attempt to cover all the techniques available today. We will, however, describe some of the applications of the methods for studying the structure of macromolecules using these very basic principles, which can be extended with more complex design of MS experiments.

15.4   IDENTIFICATION OF BIOMOLECULES BY MOLECULAR WEIGHTS

The obvious advantage MS has over electrophoretic and chromatographic methods (Chapter 5) for determining molecular weights is its accuracy and the small sample sizes required (which could be as little as a few picomoles, or equivalent to tens of nanograms of a protein such as lysozyme). This amount of material can be made readily available, for example, by eluting material after gel electrophoresis separation of mixtures of proteins. This has opened up a new field of study in which the entire protein composition of a biochemical system (known as the *proteome* of that system, by analogy to a *genome* for the complete spectrum of genetic material of a cell) can be analyzed (and often identifed) in terms of their mass. This *proteomic* approach has been useful, for example, in characterizing different modifications and changes in amounts of proteins present at different stages of cell's life cycle. The general strategy for an MS proteomic experiment (Figure 15.12) is conceptually simple. First, a mixture of proteins from a particular

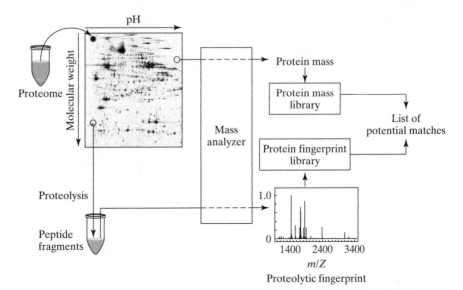

Figure 15.12 Strategy for a proteomic experiment using mass spectrometry and MS fingerprinting. A mixture of the complete spectrum of proteins from a system (the proteome) is resolved into its component proteins according to charge (horizontal axis) and molecular weight (vertical axis) by two-dimensional gel electrophoresis (see Chapter 5). Each protein "spot" can be extracted and either its mass determined by MS and compared to a library of known masses for identification, or it can be proteolyzed into fragments first, then the mass of the fragments analyzed by MS to provide a peptide fingerprint for the protein. This fingerprint can then be compared to the fingerprint predicted for each protein within a library of protein sequences for a match. The fingerprinting method provides a more reliable identification, since it is dependent on the protein sequence and not simply its composition. The reliability of identification is further improved by actually sequencing a peptide fragment from the fingerprint (see Figure 15.15).

system (potentially all of the proteins within a cell) is separated by two-dimensional gel electrophoresis, resolving the proteins according to molecular weight in one dimension, then by their charge or isoelectric points in the second (Figure 15.13). The individual protein "spots" are then extracted from the gel matrix, and their individual masses determined by MS analysis. This allows the proteins in the system to be cataloged by mass and, in cases where protein modifications are being studied, quantified according to the degree of modification. However, a single mass measurement may not be sufficient in itself to distinguish one particular protein above the rest in a system; thus, it is generally useful to obtain more detailed data on the amino acid composition or sequence of the polypeptide to definitively identify the protein.

Protein proteolytic fingerprinting by MS.   Although the overall mass of a protein may not be unique, the sequence is. The identification of a particular protein from a proteomic experiment may not, however, require a complete sequence of the polypeptide chain, but just the fragementation pattern (from proteolysis) that is associated with the sequence. This fragmentation pattern provides a "fingerprint" that can be unique for a particular protein and may be sufficient if the goal is simply to determine the protein's identity. We can consider this as a two-part problem: first, to experimentally determine the proteolytic fingerprint of an unknown protein and, second, to match the resulting fingerprint against a library of known protein patterns to find a suitable match.

The fingerprint of a protein can be generated using sequence-specific proteases (enzymes that cleave the peptide backbone at specific amino acids along the chain—examples are listed in Table 15.1). Although this method does not explicitly

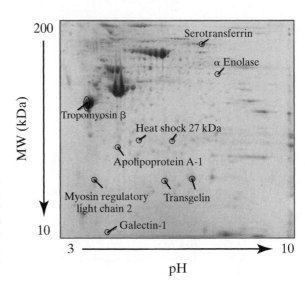

Figure 15.13 Two-dimensional (2-D) gel electrophoresis separation of the human prostrate proteome. The proteins identified by mass spectrometric fingerprinting of the tryptic peptides are labeled. [Adapted from S. Beranova-Giorgianni *Trends in Analytical Chemistry* **22**, 273–281.]

Table 15.1   Some Common Proteolytic Enzymes

$$\text{N-terminus}\cdots\!\!-\!\!N\!-\!\!\underset{\underset{H}{|}}{\overset{\overset{R_1}{|}}{C}}\!-\!\!\overset{\overset{O}{\|}}{C}\!\underset{\blacktriangle}{-}\!N\!-\!\!\underset{\underset{H}{|}}{\overset{\overset{R_2}{|}}{C}}\!-\!\!\overset{\overset{O}{\|}}{C}\!-\!\cdots\text{C-terminus}$$

Enzyme	Preference at R_1	Preference at R_2	Source
Trypsin	K, R	$\neq$P	Animal digestive systems
Chymotrypsin	Y, F, L, I, V, W	$\neq$P	Animal digestive systems
	His at high pH		
Pepsin	F, L (many others)	$\neq$P	Stomach of animals
Thrombin	R	$\neq$P	Blood (involved in coagulation)
Papain	R, K, F-X	None	Papaya latex
	(any X C-terminal of F)		
V8	E, Q (pH = 4)	None	
Thermolysin	$\neq$P	Y, F, L, I, V, W	*Bacillus proteolyticus*
		His at high pH	

Source: Excerpted from Nomenclature Committee of the International Union of Biochemistry (1984) *Enzyme Nomenclature,* Academic Press, Orlando, FL.

Note: Preferences are at standard pH = 7.0 conditions, unless specified otherwise.

determine the sequence of the polypeptide chain, the pattern of cleavage is dependent on the sequence, and the mass of each resulting fragment is dependent on the amino acid composition between each cleavage site. For example, if we treat a particular polypeptide chain with trypsin, the resulting fragments can be characterized in terms of their masses to generate a fingerprint specifically for that protein (Figure 15.14). The protein can then be identified by comparing this experimental fingerprint to one from a database of all known protein sequences. This is similar to matching a forensic fingerprint with those available from an FBI database. In our case, the fingerprint that we will try to match can be generated by predicting the fragments one would expect from treating each sequence with a particular protease. In short, we do not need to have stored the experimental fingerprints for all known proteins, just their sequences and a list of the known proteolytic cut sites. At the same time, the predicted fingerprint may not provide an exact match. There may be single amino acid differences between a particular protein and its matching protein in the database [from single or multiple replacements of amino acids that could change mass(es) of the corresponding fragment(s) or, in some instances, change the proteolytic cleavage pattern]. Consequently, a search through the sequence database will typically yield a probability for a match, and not an exact match.

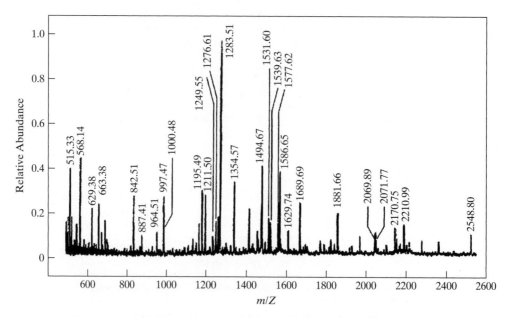

Figure 15.14   Trypsin fingerprint of seratransferrin, from the protein spot from the 2D gel of the human prostrate proteome in Figure 15.13. [Data from S. Beranova-Giorgianni *Trends in Analytical Chemistry* **22**, 273–281.]

15.5   SEQUENCING BY MASS SPECTROMETRY

How do we determine the sequence of a molecule such as a polypeptide by mass analysis? Obviously, this cannot be done with a single mass measurement, but we can systematically identify residues along a chain by comparing masses of different sized fragments of the molecule. For a very simple example of the basic technique, consider an experiment where we first measure the mass of an intact protein to be 12,000 Da. Now, we somehow fragment the polypeptide chain into two pieces with molecular weights of 4000 Da and 8000 Da. The analysis is repeated, but this time the fragments are 4147 and 7853 Da each. The difference in mass between the two smallest fragments is 147 Da, which identifies a phenylalanine residue. Assuming we know that the small fragments are at the N-terminus and that the average mass of an amino acid residue within a polypeptide chain is ~100 Da (recall that in a polypeptide chain, the mass of the amino acid residue is reduced by 18 to account for the water lost in forming the peptide bond, Table 15.2), we can now identify position 41 along the chain as Phe. If we can now generate all fragment sizes that differ by a single amino acid, then we repeat this analysis to determine the type and location of each residue along the chain, which is its sequence. Obviously, this sequencing method requires precise mass measurements (provided by MS) and methods to specifically fragment a chain (which can also be provided by the mass spectrometer). We should note that even though the masses of such fragments can be very precisely determined by MS (to 1 amu or better), there are pairs of amino acids that simply cannot be distinguished. These include leucine and isoleucine (both with

Table 15.2   Residue Masses (Within a Polypeptide Chain) of the 20
Common Amino Acids

Name	Abbreviation	Residue Mass (Da)
Glycine	G	57.02
Alanine	A	71.04
Serine	S	71.04
Proline	P	97.05
Valine	V	99.07
Threonine	T	101.05
Cysteine	C	103.01
Isoleucine	I	113.08
Leucine	L	113.08
Asparagine	N	114.04
Aspartic acid	D	115.03
Glutamine	Q	128.06
Lysine	K	128.09
Glutamic acid	E	129.04
Methionine	M	131.04
Histidine	H	137.06
Phenylalanine	F	147.07
Arginine	R	156.10
Tyrosine	Y	163.06
Tryptophan	W	186.08

Note: The masses of the isolated amino acids are the residue masses plus
18 Da.

masses of 113.08 Da), and glutamine and lysine (with residue masses of 128.06 and
128.09 Da, respectively).

A polypeptide chain can be fragmented by a number of standard biochemical
methods, including the use of site-specific proteases (Table 15.1) as we had discussed for generating a peptide fingerprint. This fragmentation is an important first
step primarily because it breaks a single, very large problem down into a large number of smaller, more manageably sized problems—instead of trying to determine
the complete amino acid sequence of a very long chain, we can simply sequence the
smaller pieces, then reassemble the entire chain by aligning the sections with overlapping fragments (Figure 15.15). This is again a classical approach to sequencing
long polypeptide chains, but can be applied to mass spectrometry (the initial fragmentation step is important since the mass spectrometric method that we will discuss for sequencing is currently applicable only to fragment sizes of up to 20–25
amino acids). To determine the sequence of the fragments, they must be broken
down to smaller pieces, each differing by single amino acids. Again, borrowing from
biochemical methods, this can be accomplished using the well-developed chemistry
in the Edman degradation, which systematically releases amino acid residues by
acid hydrolysis from the N-terminus in sequential order.

An alternative and more elegant method to generate peptide fragments with
sequential loss of amino acids from either terminus in an MS experiment is to take
advantage of the kinteic energy of the flying peptide and have it collide with another

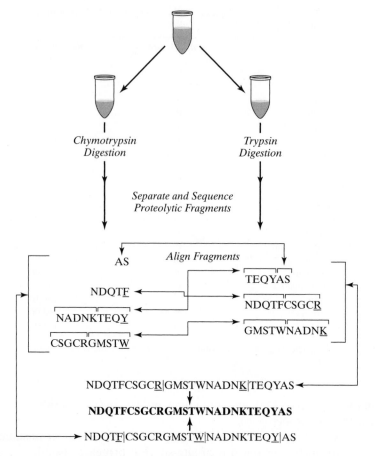

Figure 15.15 Strategy for assembling a complete polypeptide sequence from the sequences of proteolytic fragments. Two samples from a peptide solution are treated with either chymotrypsin or trypsin, each producing a unique set of peptides. These peptides are then separated and their amino acid sequences determined. By knowing the amino acids that are recognized by each protease (underlined residues), we can align regions of the chymotrypsin fragments against those from the trypsin fragments and, from the overlapping regions, can reconstruct the entire sequence of the original polypeptide.

molecule. The fragmentation is accomplished in a collision chamber, which is filled with a neutral gas (argon or xenon) that breaks the peptide backbone of the peptide—this is called collision-induced dissociation (CID) and results in a set of product or daughter ions. We place an MS instrument in front of the collision chamber that selects fragments of a very specific size from a mixture of fragments (a single peak resulting, for example, from a proteolytic fingerprint as described above) to enter the collision chamber. At the opposite end, we use an MS instrument to analyze the mass-charge ratios of the fragments resulting from the collisions. This coupling of two or more MS experiments is called *tandem mass spectrometry* (Figure 15.16).

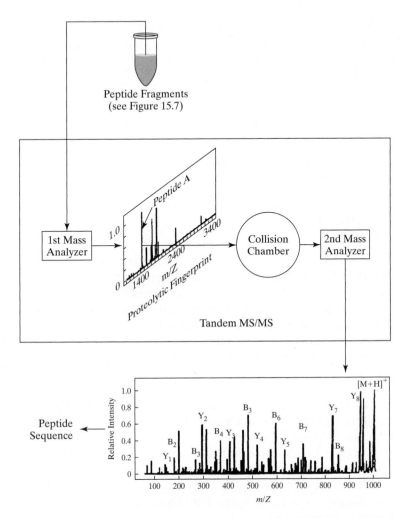

Figure 15.16 Peptide sequencing by tandem mass spectrometry (MS/MS). The sequence of a mixture of proteolytic peptide fragments is separated by the first mass spectrometer (in a manner similar to producing a peptide fingerprint, see Figure 15.12). One peptide (labeled Peptide A) is specifically sent to a collision chamber for further fragmentation into daughter ions, and the daughter ions are analyzed by the second mass spectrometer. The m/Z for the daughter ions (labeled as the B- and Y-ions) are then used to reconstruct the sequence of the Peptide A.

To see how such an experiment allows us to determine a sequence, we first need to understand the types of daughter ions that are generated by CID. Each of the three unique bonds (C_α—C, C—N, and N—C_α, where C is the carbonyl carbon and N is the amino nitrogen of the peptide bond) along a peptide backbone can be broken during CID, resulting in a set of unique products. The resulting products are not those that we expect from simple hydrolysis of a peptide bond, but will

Table 15.3 Characteristics of the CID Daughter Ions

Ion Type	Composition[a]	Change in Mass[b]	Frequency
A	+H, −CO	$\Sigma S - 27$	Common
B	+H	$\Sigma S + 1$	Common
C	+3H, +NH	$\Sigma S + 18$	Common
X	+OH, +CO	$\Sigma S + 45$	Rare
Y	+OH, +2H	$\Sigma S + 19$	Very common
Z	+OH, −NH	$\Sigma S - 27$	Very rare

[a] Composition indicates the number and types of atoms added to a peptide fragment to generate each type of daughter ion.

[b] The change in mass is the added (or subtracted) mass to the sum of the residue masses of the amino acids in a peptide fragment (ΣS) required to calculate the m/Z ratio for each type of daughter ion.

have masses and possibly additional ionic charges that will affect their m/Z ratios (Table 15.3). All of the daughter ions are positively charged, so we will use positive ionization mode mass analysis in these experiments. Notice that breaking any one of these bonds results in two complementary fragments, one representing the mass of the amino acids toward the N-terminus of the break and a second representing the mass toward the C-terminus, such that the two fragments together make up the parent peptide fragment. The daughter ions are labeled A, B, and C for the N-terminal products resulting from fragmentation at the C_α—C, C—N, and N—C_α, respectively. The C-terminal fragments are X, Y, and Z and complement A, B, and C, respectively (Figure 15.17). The most commonly observed pair of complementary daughter ions are the B- and Y-ions (Figure 15.18), so we focus the remainder of our attention on these for analysis.

Rather than starting with a description for how to analyze an actual MS/MS spectrum, let's see how such a spectrum of B- and Y-daughter ions can result from the CID of a particular peptide. This way, we get a more intuitive feel for how and why this method actually works. For this exercise, we start with the following hexapeptide sequence:

NFESGK

For simplicity, we will consider the mass of any fragment as the sum of the masses of the amino acid residues along the chain (S = masses of the amino acid residues in Table 15.2, which are 18 Da less than the natural isolated residues to account for the water lost in forming the peptide bond), plus 1 for the hydrogen at the N-terminus and 17 for the OH at the C-terminus of the parent peptide fragment. The total mass (m) of such a parent peptide is therefore

$$m = \sum_{i=1}^{n} S_i + 18 \tag{15.16}$$

From this, we would expect peaks in the mass spectrum for the parent peptide in our example at $m/Z = 681.29$ ($m + 1$ divided by 1 for positive ionization mode)

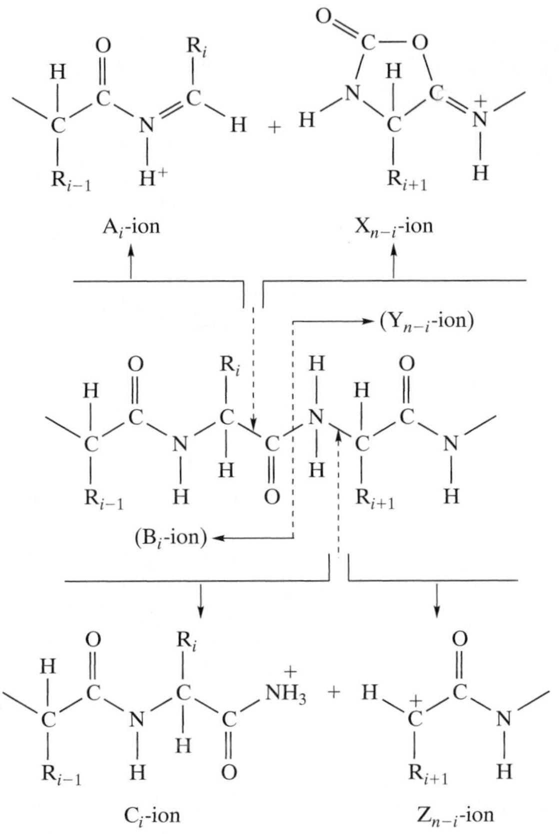

Figure 15.17 Daughter ions produced by collision-induced dispersion (CID). The complementary A- and X-ion pairs are generated by breaking the $C\alpha$—C bond, the B- and Y-ion pairs by breaking the C—N bond, and C- and Z-ion pairs by breaking the N—$C\alpha$ bond of the peptide backbone. Structures for the A/X and C/Z pairs are shown here. The structures and mechanisms for producing the ore common B/Y ion pairs are shown in Figure 15.18.

for the singly charged species, $m/Z = 341.15$ ($m + 2$ divided by 2) for the doubly charged species, and so forth. These could show up in the spectra and thus should be labeled so that they are not confused with daughter ions that result from CID. Each daughter ion from fragmentation will have very specific m/Z ratios relative to the sum of the dehydrated amino acids (Table 15.2).

As a chain of n amino acids is broken at the C—N of the individual peptide bonds, we would expect B_i-ions emerging and producing peaks in the spectrum that increase in their m/Z by the molecular mass of each sequential amino acid (counting the amino acids as $i = 1$ starting at the N-terminus), and complementary Y_{n-i}-ions with sequentially decreasing m/Z ratios, again with each decrease associated with the loss of the amino acids starting from the N-terminus. Thus, the B-ions are expected to generate a set of peaks in the spectrum that can be read from left to right, and the Y-ions peaks that can be read from right to left.

Consider the B- and Y-ions that result from fragmentation at the first peptide bond. This would theoretically yield the B_1- and Y_{n-1}-daughter ions. However, the nature of the B-ions does not allow us to observe the B_1-ion because the $i - 1$ peptide bond must remain intact to form the ring structure to carry the charge of the

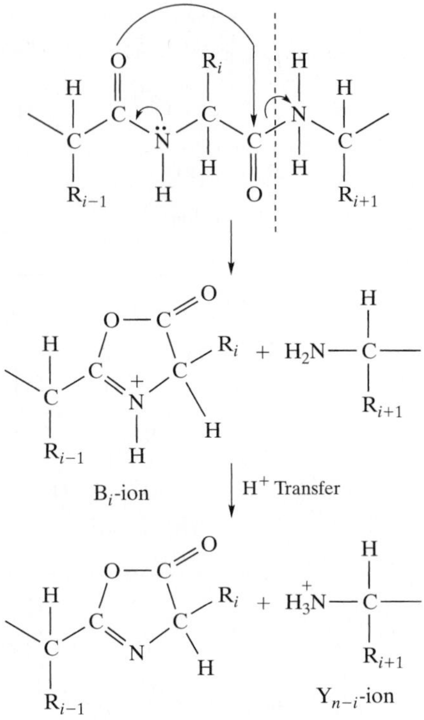

Figure 15.18 Proposed mechanism for generation of B- and Y-ions from CID fragmentation. The B-ion is generated directly from the proposed cleavage reaction, while the Y-ion requires a proton transfer. For this reason, there is a higher possibility for the exchange of the proton with protons from solvent in the Y-ions as compared to the B-ions.

B_n-ion (and for $i = 1$, there is no $i - 1$ peptide bond). In short, the B_1-ion is not observed (there will be a species associated with cleaving the first peptide bond, but it is not the same as and therefore not identified as a B-ion). There is no similar general constraint seen with the Y_{n-1}-ion and, therefore, we will start by considering the mass spectrum resulting from the Y-ions. The right-most peak for the Y-ion, therefore, should have an m/Z ratio that is the sum of S_i for all of the amino acids along the chain minus that of the asparagine residue at the N-terminus (96.04 Da) plus the mass of the hydrogen added to the amino group to develop the charge at the N-terminus and the OH of the carboxylic acid at the C-terminus. Thus, m/Z for the Y_{n-1}-ion ($i = 1$ in which the first residue is removed, or the Y_5-ion) is

$$\frac{m}{Z} = \frac{\sum_{j=2}^{n} S_j + 19}{Z} \tag{15.17}$$

or, for any Y_{n-i}-ion,

$$\frac{m}{Z} = \frac{\sum_{j=i}^{n} S_j + 19}{Z} \tag{15.18}$$

For our example, this would be 567.25 for $Z = +1$. Fragmentation of the subsequent peptide bonds would yield a Y_{n-i}-ion with an m/Z ratio of 420.18 for $i = 2$ (or the Y_4-ion), 291.14 for $i = 3$, and so on, for $Z = +1$, to give a spectrum of Y-ions shown in Figure 15.19.

The complementary B-ions would be observed first for fragmentation at the second peptide bond ($i = 2$, or the B_2-ion). The m/Z ratio of each B_i-ion is simply the sum of the masses of hydrolyzed amino acids in the fragment, plus 1 for the additional hydrogen at the N-terminus:

$$\frac{m}{Z} = \frac{\sum_{j=2}^{i} S_j + 1}{Z} \tag{15.19}$$

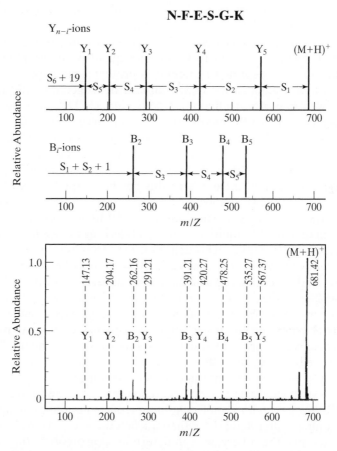

Figure 15.19   MS/MS spectrum of the peptide NFESGK. The Y- and B-ion spectra can be predicted from the molecular weights of the amino acid residue masses (Table 15.1) and the daughter ion compositions (Table 15.2). These are compared to the experimental spectrum, which includes the Y- and B-ion peaks and peaks from other daughter ions.

We can see that, for the B_2-ion, $m/Z = 262.11$ for $Z = +1$. Fragmentations farther along the polypeptide chain would result in successively larger fragments with $m/Z = 391.15$ for the B_3-ion, and so forth, to yield the B-ion spectrum.

The mass spectrum resulting from fragmentation at the C—N bonds is the sum of the B- and Y-ion spectra (Figure 15.19). Of course the actual experimental spectrum is more complicated because it will include other daughter ions (A/X- and C/Z-ion pairs) as well as multiple fragmentations.

15.6   PROBING THREE-DIMENSIONAL STRUCTURE BY MASS SPECTROMETRY

It would seem at first that information on the three-dimensional structure of a molecule such as a globular protein would be largely inaccessible to mass spectrometric analysis. However, the ability of modern MS to resolve mass differences to a single atomic mass unit allows us to study isotopic differences within a protein—we can use MS to locate and quantify the number of hydrogens and deuteriums in a peptide fragment or a protein, or in peptide fragments from proteins as we exchange between standard hydrogenated and deuterated solvent. Thus, we can perform a hydrogen/deuterium (H/D) exchange study to gain information concerning the environment of various exchangeable hydrogens in much the same way that an NMR H/D exchange study is conducted, and the results should be similar. The application of MS to studying three-dimensional structure in this way is very new and, therefore, our discussion will depend on specific examples to highlight areas where this has been applied, rather than on trying to cover all possible situations.

The advantages of using MS over NMR to monitor H/D exchange are that the amount of material required for the analysis is again much less for MS and the samples need not be in the highly concentrated forms as required for NMR. The advantage enjoyed by NMR is that each hydrogen can be assigned directly to specific amino acid residues along a chain to provide residue-level resolution and additional through-space interactions can be measured to provide a three-dimensional model for the molecule. We will see that MS can also provide residue-level measurements for H/D ratios, but the information to construct a complete model for the folded molecule is not currently available. In this section, we discuss the general methods that have been used to gather information concerning the conformation of a macromolecule, and how that information is interpreted.

The standard H/D exchange experiment in MS is very similar to that for NMR. We start with a molecule in aqueous solution, exchange it into a solution containing deuterated buffer, then quench the exchange by lowering the temperature and pH. From here, we can determine a proteolytic fingerprint to determine where protons have been exchanged with deuterons. The mass of each fragment will increase according to the number of protons that have been exchanged, although we will not know which amino acid within the fragment has undergone exchange. To resolve the H/D exchange to the amino acid level, we must perform a study similar to the sequencing experiment. In this case, the mass of the B- and Y-ions should increase according to the number of protons exchanged, and the rate can be determined by the ratio of protons and deuterons associated with each amino acid after a certain time of incubation. However, there appears to be significant scrambling of protons/deuterons during the formation of the daughter Y-ions, rendering these less useful for determining H/D exchange rates by MS. Thus, the amount of H/D exchange in a protein is most reliably determined from the B-ions, which are generally comparable to rates determined by NMR (Figure 15.20).

The rates of exchange in a particular fragment are dependent on the environment of the protons of each amino acid in that peptide fragment and the mechanism of exchange. Certain protons, including exchangeable protons of the side chains and backbone amides that are exposed to the solvent at the surface, will immediately and completely exchange with the solvent and, consequently, provide no structural information by either the MS or NMR method. The slower exchanging protons are typically the amide protons of the peptide backbone, and these can be used to study the conformation and the folding of the polypeptide chain. There can be up to a 10^8-fold difference in the rate of exchange for these hydrogens, depending on the environment of the hydrogen and the mechanism for exchange. The rates of exchange for the peptide amide protons are slower when they are involved in hydrogen bonds (for example, in α-helices and β-sheets), and generally when the residues are buried away from the protein surface. Thus, although there is currently no definitive method to assign such secondary structures by MS methods, turns can be identified in this way (Application 15.1).

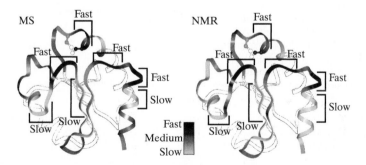

Figure 15.20   Comparison of the hydrogen/deuterium exchange rates for the protein thioredoxin. Those regions where the H/D exchange has been measured for individual amino acids by both mass spectrometry (MS) and NMR are compared in terms of regions that are in fast, medium, and slow exchange (from dark to light). Regions of the protein that have not been studied in this way are shown as open lines. [Adapted from M.-Y. Kim, C. S. Maier, D. J. Reed, P. S. Ho, and M. L. Deinzer (2001) *Biochemistry* **48**, 14413–14421.]

Application 15.1   Finding Disorder in Order

Will a protein crystallize? Only the protein itself knows for sure. There is a feeling that the more "ordered" (or structured) a protein is, the greater the chances are that it can be crystallized and thus its structure can be solved by X-ray diffraction. We know that, generally, the loops and turns of a protein are most likely to be unstructured. But, without knowing the structure of the protein beforehand, it is difficult to predict where these types of structures are and, therefore, where a protein may be ordered or disordered. The recent push in the area of structural genomics or proteomics (high throughput strategies where the entire set of proteins along a metabolic pathway, or in a biochemical system, is being crystallized for X-ray determination) makes the process of crystallization (or knowing what is crystallizable) an important process. The development of high-resolution deuterium exchange mass spectrometry (DXMS) can now provide a rapid method to identify unstructured regions of proteins (defined as having fast H/D exchange rates) and, from this, screen crystallization for proteins or variations of a protein that will have a high probability for crystallization. It was shown that proteins with large amounts of disorder as determined by DXMS generally yielded poor crystals (poor resolution, see Chapter 6), if they crystallized at all, and that the method properly located disordered regions (even for short lengths of the polypeptide chain) of proteins whose structures have already been determined (Figure A15.1).

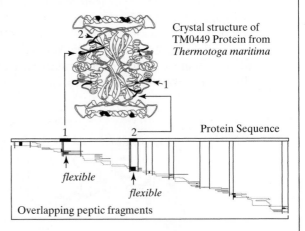

Crystal structure of TM0449 Protein from *Thermotoga maritima*

Figure A15.1 Amide hydrogens that are observed by MS to be rapidly exchanging (black bars) are mapped on the overlapping peptide fragments (bottom), resulting in consensus regions along the sequence that are considered to be highly flexible (colored black and labeled 1 and 2 along the sequence). These regions map onto loops of the protein, shown from the crystal structure to be highly disordered.

Protein Sequence

flexible

flexible

Overlapping peptic fragments

Pantazatos, D., J. S. Kim, H. E. Klock, R. C. Stevens, I. A. Wilson, S. A. Lesley, and V. L. Woods Jr. (2004), *Proc. Natl. Acad. Sci., USA* **101**, 751–756.

If the H/D exchange can be coupled with a sequencing tandem MS experiment, then each amino acid that has a hydrogen exchanged by a deuterium will be observed as having an increase in mass of +1. Thus, the actual number of deuteriums observed for any particular amino acid residue (which is a fraction between 0 and 1, and is observed as a ratio of the low and high mass peak for that residue) will be a measure of the exchange rate for that amide hydrogen at a particular time. For very fast exchanging hydrogens, we would expect to observe only the fully exchanged (high mass) peak, which provides no information about the rate of exchange.

It is more typical, however, that such amino acid level information is not readily available. However, it is still possible to glean average rate constants for proteolytic fragments. In this case, each fragment will have a low and high mass peak after a given time of exchange, with the number of deuteriums now ranging from 0 to n, for n number of exchangeable amide hydrogens within that fragment.

The discussion, to this point, is based on the assumption that the protein is fully folded and, thus, the exchange of isotopes is dependent on an intrinsic exchange rate constant (k_{int}) for this fully folded protein. The mechanism for how this exchange occurs remains unresolved, but the two current proposals are that there are small tunnels or channels for the deuterium to diffuse along in order to access the protein interior, or that the protein undergoes conformational breathing, which makes the interior accessible to deuterium exchange. In either case, we would expect that the observed rate constant for H/D exchange of a particular amino acid in a folded protein (k_f) to be dependent on some measure of its accessibility to the solvent (β), which could be the solvent accessible surface (Chapter 3) or the depth of the residue from the protein surface. This can be formulized as

$$k_f = \beta k_{int} \tag{15.20}$$

The assumption is that the closer a particular residue is to the surface (or the more exposed it is), the shorter the tunnels and channels are, or the more flexible the protein domains are, and thus the faster are the exchange rates.

As the methodologies to measure and interpret H/D exchange data are developed, there is the prospect that accurate three-dimensional models can be generated from MS studies. This will be particularly useful for characterizing proteins that are not highly abundant and thus largely inaccessible to the more common crystallographic or NMR studies. One example leading in this direction is the use of H/D exchange data on proteolytic fragments to extend the molecular model of the recombinant form of the human colony stimulating factor protein beyond what can be seen from the crystal structure (Application 15.2).

Application 15.2   When a Crystal Structure Is Not Enough

We typically think of a crystal structure as the most accurate model for a macromolecule (Chapter 6); however, for various experimental reasons, a complete structure is often simply not available. Each monomer subunit of recombinant form of the human colony stimulating factor protein dimer (rhM-CSFβ), for example, consists of 218 amino acids, but only the coordinates of residues 4–149 are available [the protein that was crystallized is a truncated form (rhM-CSFα)—this may be associated with the less-structured nature of the C-terminus; see Application 15.1]. A comparison of the number of deuteriums incorporated at the amide nitrogens within the full-length protein, as determined by mass spectrometry, to the calculated depth of the 149 amino acids into the crystal structure shows that two core helices of the crystal structure (63–76 and 114–128) have significantly less isotope exchange than predicted (Figure A15.2). This suggests that the C-terminus of the full-length protein buries the surfaces of these two helices. Using the mass spectrometric results and predictions for the most probable secondary structure for the proline-rich C-terminus, the crystal structure of rhM-CSFα was extended by 28 residues to construct a more complete model of the full length rhM-CSFβ protein (to residue 177).

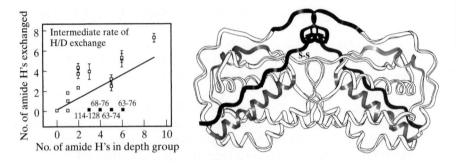

Figure A15.2   The number of amide hydrogens observed to have intermediate deuterium exchange rates by MS for proteolytic fragments of full-length rhM-CSFβ are compared (left-hand figure) to the number of hydrogens predicted in those fragments to be at intermediate depths in the crystal structure of the truncated rhM-CSFα protein. Fragments 63–76 and 114–128 (highlighted in light gray in the structural model on the right) are anomalous in that the number of intermediate exchange hydrogens is significantly lower than expected from their distances from the protein surface, suggesting that these residues are less exposed than expected from the crystal structure. These, and other H/D exchange data from MS of the protein, allowed extension of the crystal structure (dark shaded ribbons) to account for the additional buried amino acid residues in the full length protein.

YAN, X., J. WATSON, P. S. HO, and M. L. DEINZER (1991) *Molecular and Cellular Proteomics,*
3.1, 10–23.

It is also possible, however, to study the mechanism of folding and unfolding of a protein by monitoring H/D exchange with MS. In this case, we can define two possible limiting cases for the rates for unfolding and refolding relative to the H/D exchange rates, which lead to two different types of behavior for H/D exchange as observed in an MS spectrum (Figure 15.21). For both limiting cases, we will consider the folded and unfolded forms of the protein to be equilibrating with rate constants k_1 for unfolding and k_{-1} for the refolding processes. In the first limiting case (called EX1), the intrinsic exchange rate (k_{int}) is much faster than the refolding process (k_{-1}). In this situation, everytime the protein unfolds, there will be an isotope exchange. The observed rate for H/D exchange (k_{ex}) will thus be a direct measure of the rate of unfolding ($k_{ex} = k_1$). This will lead to two distinct populations, one that has not undergone the unfolding process and, therefore, will be of lower mass (all

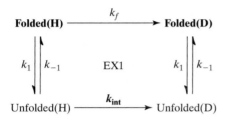

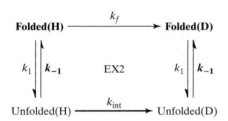

Figure 15.21 Mechanisms for hydrogen/deuterium exchange in proteins. The rate limiting steps of each mechanism are indicate by the associated rate constants in bold. In the simplest case (*top*), the exchange of deuteriums for hydrogens in a fully folded protein would have an observed rate for exchange that is dependent on the intrinsic rate constant for the particular system (k_{int}). When the exchange is dependent on unfolding and refolding of the protein, the two limiting cases can be described as EX1, where the k_{int} is faster than the refolding rate (k_{-1}), which leads to an observed rate constant that reflects the rate of unfolding (k_1). Alternatively, under the EX2 limiting case where the intrinsic exchange rate (k_{int}) is slower than refolding, (k_{-1}).the observed exchange rate (k_{ex}) is a function of k_{int} and the equilibrium between the unfolding and refolding processes (Eq. 15.21).

slow exchanging amides will remain hydrogenated), and a second population of higher mass where unfolding has led to complete deuteration of the amide protons. The EX1 mechanism will thus be invoked when a single, low mass peak in the MS spectrum gives rise to two distinct peaks over time, which then coalesce to a single high mass peak upon complete exchange (Figure 15.22).

The second limiting case (EX2) is seen when the refolding rate (k_{-1}) is much faster than the exchange rate (k_{int}). We see in this case that the observed rate for exchange is dependent on the intrinsic exchange rate and the equilibrium between the unfolded and folded states:

$$k_{ex} = k_{int}\frac{k_1}{k_{-1}} \tag{15.21}$$

Thus, as the protein unfolds, there is a possibility for exchange, but only if it occurs before the protein refolds to make the amide hydrogens inaccessible again. The result is that the mass of the population will slowly increase with time, and the resulting MS peak will reflect the average mass of the population of the deuterated protein as it converts from the unexchanged low mass form to the fully exchanged high mass form.

Interestingly, we see from Figure 15.22 that a single protein can have different domains undergoing either EX1 or EX2 exchange. In this particular case, the H/D

Figure 15.22 Time course of the electrospray mass spectra of 0% to 100% deuterated peptide fragments from oxidized thioredoxin. The fragment containing amino acids 28–39 shows a single peak that migrates from low to high mass, indicative of an EX2 exchange mechanism. In contrast, the fragment from residues 59–80 exhibits EX1 behavior, showing the evolution of two distinct peaks over time, which then collapse to a single high mass peak upon complete deuteration. [Adapted from X. Yan, J. Watson, P. S. Ho, and M. L. Deinzer (1991) *Molecular and Cellular Proteomics* **3.1**, 10–23.]

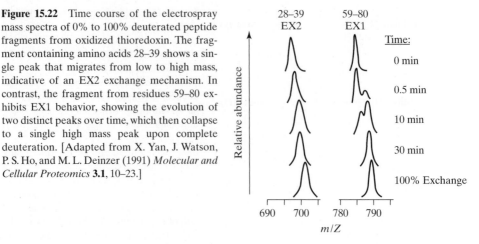

exchange rates of thioredoxin were studied, showing that the amino acids at the N-terminus (residues 28–39) are under the EX2 mechanism, while an internal set of amino acids (59–80) are under the EX1 limiting case.

Thus, modern mass spectrometry can be used to probe all levels of structure, from simple mass, to sequence, to higher-order structures. The advent of new methods to probe mass changes (through isotope exchange) has the potential to provide significant conformational information from secondary to tertiary structure, and even quaternary complexes.

EXERCISES

15.1 A supposedly homogeneous protein is studied by mass spectrometry. Rather than a single peak in the spectrum, the spectrograph records peaks arriving at times of 50 μs, 35 μs, 29 μs, and 25 μs. Show that these results are consistent with a protein homogeneous in mass.

15.2 The ESI-MS spectrum for a protein shows peaks at $m/Z = 2960.6$ and 3552.5. Calculate the number of charges associated with the 3552.5 peak and the molecular weight of the protein.

15.3 The following peptide fragments were generated after digestion of an 11 amino acid polypeptide with trypsin and separately with thermolysin.

Peptide Fragments After Digestion with	
Trypsin	Thermolysin
PA, YI, YKA, FDRH	HYI, PAYK, AFDR

What is the sequence of the parent polypeptide?

15.4 For the sequence ASTFHDKNWKQ,
   a. Calculate the masses of the fragments resulting from chymotrypsin digestion at low pH (<5) and, separately, the fragments resulting from trypsin digestion.
   b. Predict the B- and Y-ion daughter ion spectra from a tandem MS-MS experiment on this peptide sequence.

15.5 The following MS-MS spectrum was recorded after collision-induced dissociation.

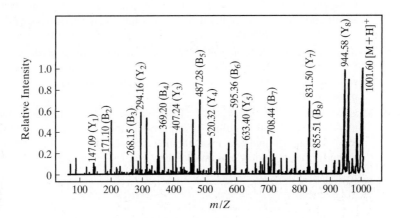

   a. What is the sequence of the parent peptide?
   b. Suggest a reason for why the Y₆-ion was not observed.

15.6 Propose mechanisms for the formation of the A-daughter ions and the X-daughter ions from CID.

REFERENCES

Mass Spectrometry

DE HOFFMAN, E. and V. STROOBANT (2001) *Mass spectrometry: Principles and Applications,* 2nd ed. Wiley and Sons, Chichester, U.K. This comprehensive text describes in detail the basic principles of mass spectrometry and its applications in the analysis of both small and large molecules.

COTTER, R. J. (1992) "Time-of-Flight Mass Spectrometry for the Structure Analysis of Biological Molecules," *Analytical Chemistry* **64**, 1027A–1039A. An excellent overall review on TOF mass spectrometry.

FENG, R. and Y. KONISHI (1992) "Analysis of Antibodies and Other Large Glycoproteins in the Mass Range of 150,000–200,000 Da by Electrospray and Ionization Mass Spectrometry," *Anal. Chem.* **64**, 2090–2095.

Two reviews that cover recent applications of mass spectrometry to both nucleic acids and proteins include the following:

FITZGERALD, M. C. and L. M. SMITH (1995) "Mass Spectrometry of Nucleic Acids," *Ann. Rev. Biophys. Biomol. Struct.* **24**, 117–140.

Senko, M. W. and F. W. McLafferty (1994) "Mass Spectrometry of Macromolecules," *Ann. Rev. Biophys. Biomol. Struct.* **23**, 763–785.

Hydrogen/Deuterium Exchange

Englander, J. J., S. W. Englander, G. Louie, H. Roder, T. Tran, and A. J. Wand (1988) in *From Proteins to Ribosomes: Protein Hydrogen Exchange Dynamics and Energetics* (R. H. Sarma and M. H. Sarma, eds.) 107–117, Adenine Press, Schenectady, NY. This work reviews the application of NMR to study protein structure by H/D exchange.

Zhang, Z. and D. L. Smith (1993) "Determination of Amide Hydrogen Exchange by Mass Spectrometry: A New Tool for Protein Structure Elucidation," *Protein Sci.* **2**, 522–531. A general procedure was described that is currently used to perform H/D exchange experiments by MS.

Miranker, A., C. V. Robinson, S. E. Radford, R. T. Aplin, and C. M. Dobson (1993) "Detection of Transient Protein Folding Populations by Mass Spectrometry," *Science* **262**, 896–900. This seminal paper was the first to describe the application of MS H/D exchange to study protein folding and unfolding.

Single-Molecule Methods

In recent years there has emerged a wholly new field of science that is often termed *single-molecule biochemistry*. Although it has long been possible to *observe* single biomacromolecules—for example, by electron microscopy—such observations have, until recently, been of fixed, static structures. But now new techniques allow us to observe individual macromolecules in dynamic interactions with one another in their normal aqueous environments, and even to manipulate them physically.

16.1 WHY STUDY SINGLE MOLECULES?

Single-molecule studies permit a much deeper analysis of molecular behavior than has been hitherto possible. If you examine Chapters 1 through 15 of this text, you will note that all of classical physical chemistry, statistical mechanics, and thermodynamics rely on the description of the *average* behavior of enormous ensembles of molecules. How any *one* molecule may behave over time is not revealed by such studies. It is quite possible, for example, that an individual macromolecule may spend periods of time in different comformation states, transient states that we could never detect by averaged measurements on a large group of molecules. (See Application 16.1.) On the other hand, individual molecules of a given type may have different conformations and different behavior. Thus, single-molecule studies may reveal nuances of molecular behavior hitherto hidden from us. The difference in approach may be compared to how an insurance actuary and a novelist each analyze human behavior. The insurance statistician bases analysis and prediction on the averaged behavior of a large number of people. The novelist is concerned with the specific behavior of one individual, or the interactions of a few. Each approach reveals truths about human behavior, but the insights are different and complementary.

Application 16.1   RNA Folding and Unfolding Observed at the Single-Molecule Level

Fluorescence resonance energy transfer (FRET) has been employed at the single-molecule level to analyze the folding and unfolding of a ribozyme (Bokinsky et al. 2003). As described in Chapter 11, energy transfer between fluorophores is very sensitive to the distance between them. The hairpin ribozyme studied by Bokinsky et al. had fluorescent donor and acceptor labels (Cy3 and Cy5 dyes, respectively) attached at the two ends (Figure A16.1a) and the fluorescence of single molecules was observed. The ribozyme exhibits fluctuations between two stable states, "docked" and "undocked" (Figure A16.1b), and the FRET transfer efficiency changes abruptly when a transition occurs. Each molecule switches erratically back and forth between the two states. Analysis of distributions of the "dwell times" in the docked and undocked states demonstrated a single rate constant for docking ($k_d = 0.018\,\text{sec}^{-1}$) but revealed a more complex, multiexponential undocking process with four different rate constants ($k_u = 0.01, 0.1, 0.8,$ and $6\,\text{sec}^{-1}$). This is interpreted that individual docked molecules can exist in any one of the four docked states corresponding to these rates. These states interconvert slowly, with a half-life of hours. A single-molecule approach was critical for this analysis, which provides new insights into RNA structural dynamics.

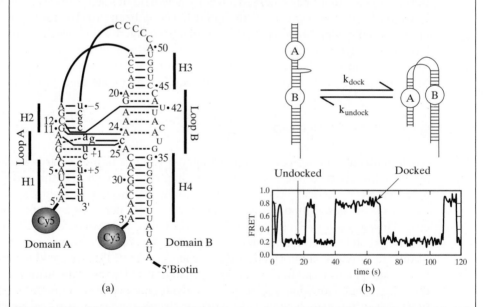

(a)   (b)

Figure A16.1   Ribozyme foldings studied by single-molecule FRET. (a) The RNA has been labeled at two ends by dye molecules (Cy3 donor, Cy5 acceptor) capable of energy transfer at short distances. (b) Schematic of the structural transition, and an experimental time trace of the FRET efficiency. When the molecule is "undocked" the sites are so far apart that FRET is negligible. But "docking" puts them in close proximity, allowing FRET. The signal oscillates between these states as indicated. [From Bokinsky et al. (2002), *Proc. Natl. Acad. Sci. USA* **100**, 9302–9307. (Copyright 2002, National Academy of Sciences, USA)]

BOKINSKY, G., D. RUEDA, V. K. MISRA, M. M. RHODES, A. GORDUS, H. P. BABCOCK, N. G. WALTER, and X. ZHUANG (2003), *Proc. Natl. Acad. Sci. USA* **100**, 9302–9307.

Two unique aspects of molecular behavior can be especially well studied by single-molecule techniques. First, specific molecules may exhibit individual *static* differences in structure, reactivity, or function, differences that cannot be easily detected in ensembles. Single-molecule techniques, when applied to a number of molecules, can provide the *distribution* of an experimental parameter, whereas classical methods yield only an *average* value. Second, individual molecules may show *dynamic* changes in function, due to spontaneous structural fluctuations. To take examples from enzymology, it has been shown that different enzyme molecules in a supposedly uniform population can function at different rates (static heterogeneity), and that a given molecule can fluctuate in its catalytic rate (dynamic heterogeneity).

In this chapter, we deal with two classes of single-molecule techniques. The first are those that allow us to *observe* dynamic processes at the single-molecule level. The second are those that allow us to *manipulate* single molecules and to measure the forces involved in molecular processes. In the following sections, we describe a number of these techniques and their applications to biochemical problems.

16.2   OBSERVATION OF SINGLE MACROMOLECULES BY FLUORESCENCE

To observe the dynamic behavior of single macromolecules by changes in their fluorescence requires overcoming some very formidable experimental barriers. First, the molecule itself must either possess a strongly fluorescent fluorophore, or have sites that can be conveniently labeled with one or more such entities. To work at the single-molecule level, dyes of high absorbance and quantum yield (see Chapter 11) are needed. Second, a way must be found in which to focus intense light of the appropriate excitation wavelength on the macromolecule. A continuous wave laser or pulsed laser is usually employed. Third, a detection system must be devised that will somehow eliminate the optical noise. This noise arises from several sources. Rayleigh scattered light (Chapter 7) is intense, but can be largely eliminated by proper filters, because the Rayleigh scattered light will have the same wavelength as the incident light, which is necessarily of shorter wavelength than the fluorescent emission. Raman scattering is a different matter, especially since water is a good Raman scatterer. Raman wavelength shifts are often such that the Raman scattered radiation is closer in wavelength to the emitted light and harder to filter out. Finally, fluorescence of even minor impurities can be a major problem.

Many of these difficulties can be circumvented by proper instrument design (see Figure 16.1). There are two general classes of techniques that have been used for most such studies. Those depicted in Figures 16.1a and 16.1b depend upon illuminating a very small volume by a focused laser beam, and then filtering out scattered or stray radiation. The total internal reflection (TIR) microscope shown in Figure 16.1c operates on a very different principle. A laser beam is directed through a prism at an angle ($>61°$) to produce total internal reflection. This creates an intense electromagnetic field very close to the reflecting surface. Fluorescence is

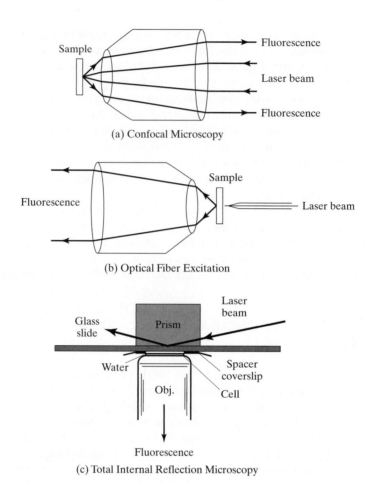

Figure 16.1 The three most frequently used techniques for observing single-molecule fluorescence. (a) Confocal microscopy: A single lens (often a microscope objective) is used both to focus the laser beam on the sample (arrows to left) and to collect the fluorescent light that is emitted (arrows to right). (b) Optical fiber excitation: The laser beam is condensed into a very fine optical fiber that illuminates the sample; fluorescence is observed in transmission (arrows to left). (c) A prism-based TIR microscope useful for single-particle FRET studies. [(a), (b) Adapted from W. E. Moerner and M. Orit (1999), *Science* **283**, 1670–1675; with permission of the publisher. (c) from S. Leuba.]

detected as normal to that surface, an image of which is focused on a 2D detector. An alternative technique that largely avoids scattering problems is *two-photon excitation*. If a very intense laser is used, producing light of half the frequency of the absorption band that will induce fluorescence, a significant number of molecules will be excited to the desired level by the absorption of two photons. This means that all scattering, Rayleigh or Raman, which will be predominately of light at or near the laser frequency, will be completely out of the frequency range of the observed fluorescent signal and thus easily eliminated by filters.

Single-molecule fluorescence can be used for a number of purposes. An example that holds great potential for biological studies is *fluorescence correlation spectroscopy* (FCS). FCS shares some features with dynamic light scattering (Chapter 7). If fluorescence is observed from a small, intensely illuminated volume (using a confocal microscope, for example), then the fluorescence fluctuation corresponding to individual molecules will be autocorrelated. An autocorrelation function of fluorescence is calculated, just as in dynamic light scattering. The autocorrelation function yields a characteristic diffusion time, τ_0, during which an average molecule resides in the observed volume. This is inversely proportional to the diffusion coefficient of the fluorescent molecule.

The technique is not at present as accurate as dynamic light scattering in the measurement of absolute values of the diffusion coefficient, mainly because of difficulties in exactly defining the zone of illumination. However, FCS has the potential to study the diffusion of specific, labeled molecules in a heterogeneous mixture, or in complex environments such as cytoplasm or the cell nucleus. This addresses a long-unrealized aim in biology—to be able to study the behavior of biological macromolecules in situ, without perturbing the environment.

A very different use of single-molecule fluorescence is in following spontaneous or induced conformational changes in macromolecules such as proteins. Figure 16.2 provides an example. The fluorescent cofactor (FAD) of the enzyme *cholesterol oxidase* loses fluorescence upon reduction to FADH at each cycle of enzymatic activity. Thus, the termination of each on/off cycle (FAD/FADH) of the enzyme is signaled by an abrupt change in fluorescence. Figure 16.3 shows the distribution of "on" times. Analysis of such distributions can give information concerning the kinetics of the process in a single molecule. If the change from the "on" state to the "off" state is first order (as we may usually expect it to be), a histogram of "on" times will decay exponentially with a relaxation time $\tau = 1/k$ where k is the first order rate constant for the on → off reaction.

The power of single-molecule fluorescence is greatly increased by the incorporation of two fluorophores, whose emission and absorption spectra overlap, thus

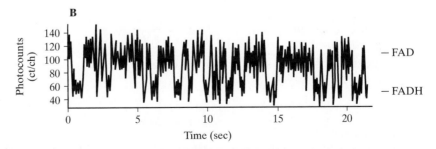

Figure 16.2 Observation of fluorescence fluctuation of an individual cholesterol oxidase molecule undergoing periodic oxidation and reduction of its fluorescent FAD cofactor. Note that the fluorescence signal switches in an erratic manner between "on" (fluorescent, oxidized) and "off" (nonfluorescent, reduced) states. [From Lu et al. (1998), *Science* **282**, 1877–1882 with permission of the publisher. Copyright 1998, AAAS.]

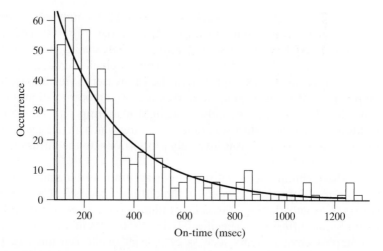

Figure 16.3   Kinetic analysis of data like that shown in Figure 16.2. A histogram of "on" times is constructed, which exhibits exponential decay (solid line). This allows calculation of the kinetic rate constant for FAD reduction. [From Lu et al. (1998), *Science* **282**, 1877–1881 with permission. Copyright 1998, AAAS.]

allowing *fluorescence resonance energy transfer* (FRET) experiments to take place. Excitation of the *donor* fluorophore can lead to nonradiative transfer of energy to the *acceptor,* followed by fluorescent emission from the latter. (See Chapter 11.) The efficiency of transfer is defined as

$$E = \frac{k_T}{k_T + \Sigma k_D} \tag{16.1}$$

where k_T measures the rate of energy transfer from donor to acceptor, and Σk_D measures the sum of all other donor transfer rates (i.e., donor fluorescence, radiationless energy loss, etc.). I_D and I_A, the fluorescent intensities observed from donor and acceptor molecules, can be taken as approximately proportional to Σk_D and k_T, respectively, which leads to an approximate expression for E in terms of these observables:

$$E \cong \frac{I_A}{I_A + I_D} \tag{16.2}$$

The efficiency depends strongly on the distance between donor and acceptor according to Eq. 16.3:

$$E = \frac{1}{1 + (R/R_0)^6} \tag{16.3}$$

The distance R_0 (called the Forster radius) typically has a value of about 5 to 6 nm, but depends on the nature of the donor and acceptor, as well as their mutual orientation. (See Figure 16.4 and Chapter 11 for further discussion.) According to Eq. 16.3, FRET can be a sensitive measure of the distance between two groups in a macromolecule.

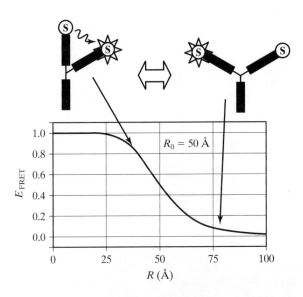

Figure 16.4 The use of FRET to follow molecular conformational changes. (a) A conformational fluctuation can change the donor-acceptor distance. (b) A graph of Eq. 16.2 for $R_0 = 50$ Å $= 5\,nm$. [Adapted from T. Ha (2001), *Methods* **25**, 78–86.]

The most accurate measurements are made between $0.5\,R_0 \leq R \leq 1.5\,R_0$ because of the asymptotic behavior of E at large and small R (see Figure 16.4).

FRET experiments with single molecules allow the possibility of observing conformational fluctuations in real time. This requires that the donor and acceptor be so positioned in the molecule that their separation changes significantly with respect to R_0 in the conformational transition. An example is shown in Application 16.1. Here, the oscillation of a single RNA molecule between two states is observed directly, through sporadic changes in the energy transfer efficiency. Furthermore, analysis of the kinetics demonstrates the existence of an intermediate transition state. Such analysis would not be possible at the many-molecule level.

16.3 ATOMIC FORCE MICROSCOPY

Atomic force microscopy (AFM) is a technique that is becoming widely used for both the observation and manipulation of biological macromolecules. The method is both remarkably simple and extraordinarily precise. A typical apparatus is depicted in Figure 16.5a. The heart of the instrument is a very fine tip (often a tungsten carbide crystal or a carbon nanotube) mounted on a springy cantilever. In the most frequently used technique, the *tapping mode,* the cantilever is made to oscillate, so the tip taps its way across the sample, which has been deposited on a very smooth (usually mica) surface. The surface is scanned as it is moved in a raster fashion by a piezoelectric driver. At the same time, a laser beam is being reflected off the cantilever onto a detector mounted some distance away, giving a very long optical lever arm. As the tip taps its way across the surface and encounters a sample molecule, the difference in height is registered (see Figure 16.5b). Because of the amplification of

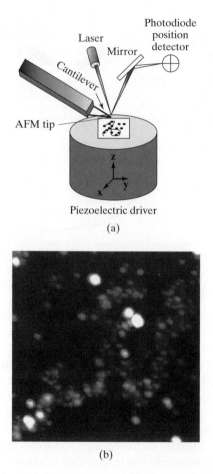

(a)

(b)

Figure 16.5   (a) Schematic diagram of an atomic force microscope. The reflected laser beam is captured by a quadrant detector, which can detect very small shifts. The sample stage is moved for scanning by a piezoelectric driver across the x-y plane as the tip of the cantilever taps across the surface. The reflected laser beam strikes a detector divided into quadrants. As the tip is deflected in the z direction, the signal difference drives a feedback circuit to the z displacement of the piezoelectric crystal. This moves the surface in the z direction to compensate for the presence of the sample, and records this displacement to form an image. (b) Scan of a chromatin sample, showing individual nucleosomes in a chromatin fiber. The nucleosomes are ~10 nm in diameter. [Both figures courtesy of S. Leuba.]

the distance by the optical lever, extremely small distances (of the order of 1 nm or less) can be measured. Although still not equal in resolution to the best electron microscopy, AFM holds a number of advantages. First, no fixation or staining of samples is required. Second, and most important, biological samples can be examined when moist, or even under aqueous conditions. For example, samples can be examined when affixed to a surface but covered by solution, allowing the effects of varying solvent compositions or the addition of substrates or other small molecules to be

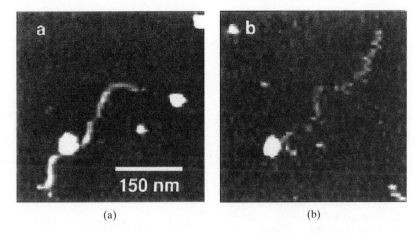

(a)   (b)

Figure 16.6 Motion of RNA polymerase on DNA observed by atomic force microscopy. (a) The polymerase is stalled by lack of nucleotide triphosphates (NTPs). After NTPs are added, and six minutes later (b) the DNA has been pulled through the polymerase. Although transcription presumably occurred, the intermediate stages could not be resolved, as the DNA was too mobile. [Reprinted with permission Kasas et al. (1997), *Biochemistry* **36**, 461–468. (Copyright 1997, American Chemical Society)]

studied. If the probe is allowed to rest on a single macromolecule, conformational changes can be observed in real time. Alternatively, consecutive scans of AFM images can be used to follow dynamic processes, such as the progress of RNA polymerase along individual DNA molecules (see Figure 16.6 and Application 16.2).

Application 16.2   Single-Molecule Studies of Active Transcription by RNA Polymerase

There has been much interest in studying the way in which individual RNA polymerase molecules move on DNA during transcription. A direct visualization of the motion can be obtained by taking sequential AFM scans of polymerases moving on DNA that is loosely affixed to a mica surface. (See Figure 16.6.) However, the technique used in most recent studies is illustrated in Figure A16.2a. The DNA molecule is affixed at one end to a microsphere that is held in an optical tweezer trap. The polymerase molecule, bound to the DNA but stalled by lack of one ribonucleotide triphosphate, is attached to the coverslip. The coverslip position is adjusted to displace the bead from the trap center enough to produce the desired force on the DNA. The solution surrounding the DNA and polymerase is then provided with the full complement of ribonucleotide triphosphates so that transcription can begin. In most applications, the bead position in the trap (and thereby force on the DNA) is held constant by a feedback curcuit that moves the coverslip to compensate for movement of the DNA through the polymerases (see Figure 16.8). Thus, the coverslip motion measures the progress of the transcription at constant force. A recent study (Adelman et al. 2002) illustrates the kind of data that can be obtained (Figure A16.2b). Several different experiments with individual molecules are shown. According to this study, the average rate for transcription (while the polymerase is moving) is about

Figure A16.2 (a) The method used by Adelman et al. to follow transcription. The polymerase was fixed to the coverslip by an antigen-antibody binding. An end of the DNA was fixed by biotin binding to a streptavidin-coated microsphere. This was held in an optical trap, while the coverslip was moved to maintain constant force as the polymerization proceeds. (b) Raw data for a number of individual transcriptions. Note that the elongation rate seems the same for all, but pausing is erratic and at different sites on different molecules. [From Adelman et al. (2002), *Proc. Natl. Acad. Sci. USA* **99**, 13538–13543. (Copyright 2002, National Academy of Sciences, USA)]

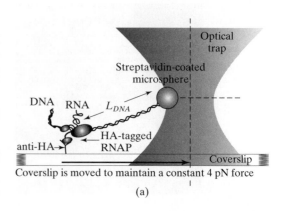

Coverslip is moved to maintain a constant 4 pN force

(a)

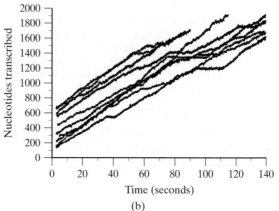

(b)

the same for all polymerase molecules; there is no static heterogeneity. However, "pausing" is highly erratic, concerning both its location on the DNA and duration. (See Davenport et al. 2000 for a somewhat different interpretation of similar experiments.) A further application of the technique of Adelman et al. involved the study of a mutant polymerase that conventional ensemble measurements had shown to have an overall slower transcription rate than the wild type. However, when individual molecules of the mutant were studied, it was shown that the mutant was indeed capable of bursts of "wild-type" elongation, but paused more often and longer than did the wild type. This is an excellent example of how single-molecule studies can provide a deeper understanding of molecular processes than can the traditional ensemble approaches.

Another application of the same technique can be used to measure the *stalling force* for polymerase—that is, the force at which the enzyme can no longer pull the DNA. To do this, the experiments described above are carried out with increasing forces. The force at which no further motion of the DNA occurs is the stalling force. Several studies indicate that transcription is stalled by a force of about 14 pN (i.e., Wang et al. 1998; Gelles and Landik 1998).

ADELMAN, K., A. LA PORTA, T. SANTANGELO, J. T. LIS, J. W. ROBERTI, and M. WANG (2002), *Proc. Natl. Acad. Sci. USA* **99**, 13538–13543.

DAVENPORT, R. I., G. J. L. WHITE, R. LANDICK, and C. BUSTAMANTE (2000), *Science* **287**, 2497–2500.

GELLES, J. and R. LANDICK (1998), *Cell* **93**, 13–18.

KASAS, S., N. H. THOMPSON, B. L. SMITH, H. G. HANSMA, X. ZHU, M. GUTHOLD, C. BUSTAMANTE, E. T. KOOT, M. KASHEV, and P. K. HANSMA (1997), *Biochemistry* **36**, 461–468.

WANG, M., M. J. SCHNITZER, H. YIN, R. LANDICK, J. GELLES, and S. BLOCK (1998), *Science* **282**, 902–907.

A very similar apparatus can be used to measure very small forces applied to individual macromolecules. If one end of a molecule is attached to the tip, and the other end is attached to a moveable support, that molecule can be stretched, and the force measured by deflection of the cantilever. There are a number of ways to make strong attachment. For example, the support can be coated with streptavidin, and a biotin molecule can be covalently linked to the molecule of intent. The very strong streptavidin-biotin interaction will then be utilized. The forces involved in such molecular processes are usually expressed in piconewtons (pN). One pN = 10^{-12} newtons; a newton is the unit of force in the mks system, with the dimension of $kg \cdot m \cdot s^{-2}$. For conversion to cgs: 1 pN = 10^{-7} dyne. For a detailed review of AFM force studies, see Fisher et al. (1999).

16.4   OPTICAL TWEEZERS

There is a powerful technique that is widely used both to manipulate macromolecules and to measure the forces involved in dynamic processes or in the stabilization of structures. This technique utilizes what are called *optical tweezers*. The principle is demonstrated in Figure 16.7a. If an intense beam of light (as from a continuous wave laser) is focused into a very small region, the intense electromagnetic field will tend to "capture" a small dielectric sphere placed in the region. The force acting on the sphere as a function of displacement from the focal point is diagrammed in Figure 16.7b: Note that there is a nearly linear restoring force in the immediate region of laser focus, pulling the bead toward the focus. Conversely, how far the bead is deflected from the center position measures any external force acting on the bead. If the bead is pulled (as by an attached macromolecule) from the center of the trap, a force will be exerted on the macromolecule that is proportional to the displacement. Dynamic processes can be studied at a constant deflection but at changing force by using a feedback circuit. This circuits holds the bead at a constant deflection by changing the laser power as the force changes (see Figure 16.8).

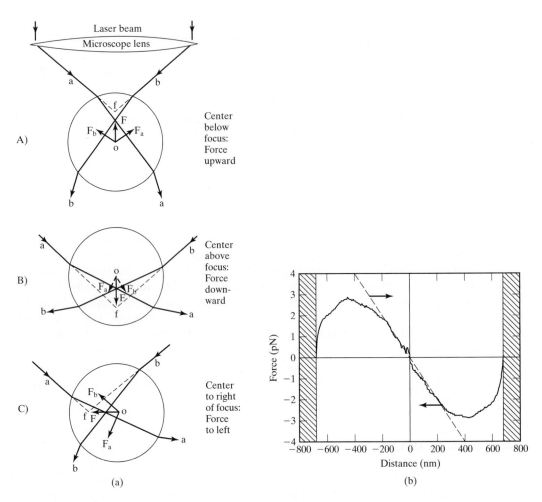

(a)   (b)

Figure 16.7 (a) The principle behind optical trapping. Two symetrically placed rays (a and b) of the light focused into a transparent dielectric sphere are depicted here. These exert forces F_a and F_b, with resultant force F. *Only* if the center of the sphere coincides with the focal point will the sum of forces be zero. Displacement of the bead up (A), down (B), or sideways (C) will lead to a restoring force that pushes the sphere back toward the focus. [Adapted from A. Ashkin (1992), *Biophys. J.* **61**, 569–582.] (b) The net restoring force as a function of lateral displacement of the bead. The force is linear in displacement up to a certain limit, beyond which the bead escapes from the trap. [Adapted from Simmons et al. (1996), *Biophys. J.* **70**, 1813–1822.]

    The ability to examine mechanical processes at the single-molecule level either with optical tweezers or with the atomic force microscope has proved very informative. For example, it has been possible to study the mechanics of individual DNA molecules by stretching them over a wide force range. The experiments reveal several stages in DNA extension (Figure 16.9). At low forces, the worm-like coil state of the DNA is straightened; this requires little increase in force. In this region of the force-extension curve, the resisting force is largely *entropic* in origin, resulting

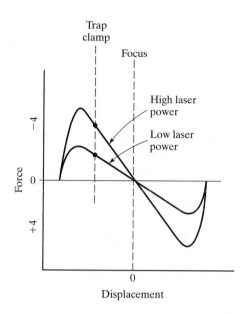

Figure 16.8 Holding a bead in an optical trap at constant displacement but varying force. Curves of displacement versus force are shown for two different laser powers. In practice, a feedback circuit, based on microscopic observation of the bead position, increases the laser power to keep the bead fixed as the force increases.

from the fact that a coiled molecule has more conformations than an extended one. However, once the coil is extended to its full length, further stretching requires a large increase in force, because the DNA helix itself is now being stretched. The resulting force is primarily *enthalpic* in this region of the curve; bonds are being bent or stretched. Then, as even greater force is applied, the DNA undergoes an abrupt conformational change and the helix converts to a more extended conformation. This is signified by a sudden increase in molecular length with almost no increase in force (see Figure 16.9). At very high forces, any nicks in the DNA will lead to unraveling of the two strands.

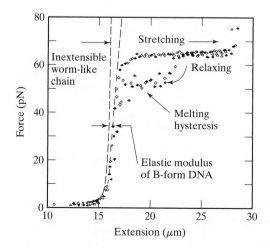

Figure 16.9 The stretching of double-stranded DNA. Up to a length of ~15 μm (the contour length of the DNA) the worm-like chain is being straightened, with little force needed. The steep part of the curve corresponds to elastic stretching of the extended chain. Stretching to this point is immediately reversible. Then, at about 17 μm, a major conformational change occurs, allowing the DNA to increase dramatically in length with little increase in force. The stretching can still be reversed, but now there is some hysteresis. [From Smith et al. (1996), *Science* **271**, 796–799, copyright 1996, AAAS.]

A number of theories have been developed to describe such behavior; all are somewhat approximate. A simple example is that of Odijk (1995), which gives for the end-to-end distance of the stretched chain (x):

$$x = L_0 \left[1 - \frac{1}{2} \left(\frac{k_B T}{F L_p} \right)^{1/2} + \frac{F}{K} \right] \tag{16.4}$$

Here L_0 is the contour length (fully extended length of the worm-like chain), L_p is the persistence length (see Chapter 4), F is force, and K is the elastic modules of the extended chain. The quantity $k_B T$ is the Boltzmann constant times temperature. In Eq. 16.4, the second term in brackets corresponds to the entropic resistance to straightening the worm-like chain, whereas the third term represents the enthalpic resistance to its stretching.

The optical tweezers and AFM techniques can be applied to the study of many other biopolymers, including multidomain proteins such as titin, and polysaccharides (see Fisher et al. 1999 for a broad review of such AFM work). The polysaccharide studies are of particular interest, for stretching such molecules can induce conformational changes (such as boat-chair transitions) in the individual sugars, allowing their energetics to be measured.

These techniques have also revealed quite a bit concerning the mechanical stability of more complex macromolecular structures. Individual chromatin fibers, when subjected to mechanical stress, display a stepwise force-distance curve (Figure 16.10), which corresponds to the disruption of individual nucleosomes. (See Application 7.1 for a description of the nucleosome.) The extension steps are of a size that indicates that the DNA is pulled off the histone core in a segmental fashion.

The optical tweezers technique has been especially useful in the study of dynamic processes such as transcription by RNA polymerase. (See Application 16.2

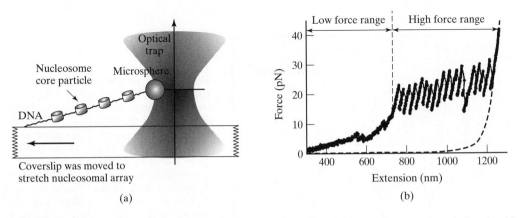

Figure 16.10   Stretching of a single chromatin fiber by optical tweezers reveals a stepwise increase in length, with intervals comparing to the dissociation of individual nucleosomes. [From Brower-Toland et al. (2002), *Proc. Natl. Acad. Sci (USA)* **99**, 1950–1955. (Copyright 2002, National Academy of Science, USA)]

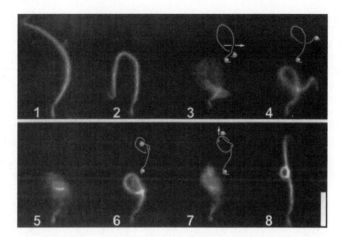

Figure 16.11 Tying a knot in an actin fiber, using optical tweezers. The fiber has been attached to beads at both ends and both are held in optical traps. One trap is moveable in three dimensions, the other is fixed. One end of the fiber is then moved to form the knot. [From Arai et al. (1999), *Nature* **399**, 446–448 with permission from *Nature*.]

for examples.) Not only can the progress of the DNA through the enzyme be studied in real time, but the forces involved can be measured as well.

The delicacy and precision with which optical tweezers can be manipulated are strikingly exhibited in Figure 16.11, which depicts the use of this technique to tie a knot in a single macromolecule. The molecule has microspheres attached to both ends, and is held by optical tweezers at both ends. Then one tweezer is manipulated to move one end to tie the knot. If the knot is pulled too tightly, the molecule breaks at a force much less than that required to break by directly stretching the molecule. The likely explanation is that the local bending at the knot is responsible.

16.5   MAGNETIC BEADS

The possibility of covalently attaching minute magnetic beads to a biopolymer molecule has opened up another entire range of possible studies. As Figure 16.12 shows, a nearby magnet can impose a precisely regulatable tension on the fiber (by adjusting the distance of the magnet from the bead). However, the unique contribution that magnetic beads can contribute is the measurement and manipulation of torsion. By rotating the magnet, one can twist the fiber to any desired degree. This means, for example, that the twist of a DNA molecule can be monitored and changed by external control. Such experiments have demonstrated that at a critical degree of twisting, the DNA molecule will begin plectonemic writhing to reduce the strain. For the first time, it becomes possible to examine the dynamics of such conformational changes.

Figure 16.12 The application of the magnetic bead technique to study DNA rotation during transcription. The polymerase, bound to the DNA in an arrested state, is attached to a surface. A magnetic bead, carrying small fluorescent beads, is attached to the end of the DNA. Transcription is begun by providing a full complement of nucleoside triphosphate. The force on the DNA can be regulated, and rotation observed by a periodicity in the fluorescent signal. [Adapted from Harada et al. (2001), *Nature* **409**, 113–115.]

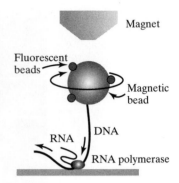

On the other hand, the technique can be used to measure molecular rotation. An important example is found in studies by Harada et al. (2001) (see Figure 16.12) in which the rotation of DNA produced by the action of a transcribing polymerase has been directly observed and measured. In this application, fluorescent beads were attached to the magnetic bead, so that the rotational motion of the bead could be directly observed.

EXERCISES

16.1 The R_0 values for the fluorophores used in many biological FRET experiments range around 6 nm. If this is the case, how far apart do the donor and acceptor have to be for the efficiency to drop to 10% of its maximum value?

16.2 An RNA molecule that can exist in either a compact, folded conformation or an open conformation is studied by single-molecule FRET. For the fluorophores used, $R_0 = 6.5$ nm.
    a. The efficiency of transfer is observed to fluctuate erratically between 0.9 and 0.2. Estimate distances between donor and acceptor in these two states.
    b. Data on "dwell time" in the high-transfer (folded) state have been accumulated for long-time observation on a number of molecules:

Average Dwell Time (sec)	Number of Examples
0.1	33
0.3	23
0.5	18
0.7	14
0.9	9
1.1	6
1.3	4

Calculate the rate of unfolding from these data. Is this a first-order process?

16.3 Stretching of the muscle protein titin has been studied by AFM. [Tskhovrebeva et al. (1997), *Nature* **387**, 308–312.] When the molecule was stretched, the force exerted was found to decay in the peculiar manner shown here. Each peak and stepwise decrease represents a separate stretching and relaxation cycle. Suggest an explanation for such behavior.

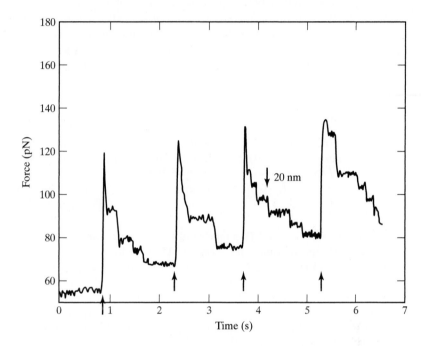

16.4 How would you expect the force-elongation curve for single-strand DNA would compare with that for double-strand DNA as shown in Figure 16.8? A qualitative answer will suffice.

REFERENCES

Methods

ARAI, Y., R. YASUDA, K. AHACHI, Y. HARADA, H. MYATA, K. KINOSITA, and H. ITOH (1999) "Tying a Molecular Knot with Optical Tweezers," *Nature* **399**, 446–448.

ASHKIN, A. (1992) "Forces of a Single-Beam Gradient Laser Trap on a Dielectric Sphere in the Ray-Optics Regime," *Biophys. J.* **61**, 569–582. A rigorous analysis of the optical tweezers trap.

FISHER, T. E., P. E. MARSZALEK, A. OBERHAUSER, M. CARRION-VAZQUEZ, and J. M. FERNANDEZ (1999) "The Micromechanics of Single Molecules Studied with Atomic Force Microscopy," *The Journal of Physiology* **520.1**, 5–14.

HA, T. (2001) "Single Molecule Fluorescence Resonance Energy Transfer," *Methods* **25**, 78–86.

HESS, S. T., S. HUANG, A. A. HEIKAL, and W. W. WEBB (2002) "Biological and Chemical Applications of Fluorescence Correlation Spectroscopy: A Review," *Biochemistry* **41**, 697–705.

SIMMONS, R. M., J. T. FINER, S. CHU, and J. SPUDICH (1996) "Quantitative Measurements of Force and Displacement Using an Optical Trap," *Biophys. J.* **70**, 1813–1822. A very useful and clear description of the optical tweezers technique.

WEISS, S. (1999) "Fluorescence Spectroscopy of Single Biomolecules," *Science* **283**, 1676–1682. Contains descriptions of both present and potential applications.

ZLATANOVA, J. and S. H. LEUBA (2003) "Magnetic Tweezers: A Sensitive Tool to Study DNA and Chromatin at the Single-Molecule Level," *Biochem. Cell. Bio.* **81**, 151–159.

Applications

BROWER-TOLAND, B. D., C. L. SMITH, R. C. YEH, J. T. LIS, C. PETERSON, and M. WANG (2002) "Mechanical Disruption of Individual Nucleosomes Reveals a Multistage Release of DNA," *Proc. Natl. Acad. Sci. USA* **99**, 1950–1955.

BUSTAMANTE, C., S. B. SMITH, J. LIPHARDT, and D. SMITH (2000) "Single Molecule Studies of DNA Mechanics," *Curr. Opin. Struct. Biol.* **10**, 279–285. A concise overview of the DNA studies.

HA, T. (2004) "Structural Dynamics and Processing of Nucleic Acids Revealed by Single-Molecule Spectroscopy," *Biochemistry* **43**, 4056–4063.

HARADA, Y., O. OHARA, A. TAKATSUKI, M. ITOH, N. SHIMAMOTO, and K. KINOSITA (2001) "Direct Observation of DNA Rotation During Transcription by *Escherichia coli* RNA Polymerase," *Nature* **409**, 113–115. Application of magnetic bead technique.

LEUBA, S., and J. ZLATANOVA, eds. (2001) *Biology at the Single Molecule Level* Pergamon, Amsterdam. An excellent collection of papers describing a variety of applications of single-molecule techniques.

LU, H. P., L. XUN, and S. XIE (1998) "Single-Molecule Enzymanic Dynamics," *Science* **282**, 1877–1882. Enzyme turnover, one step at a time.

ODIJK, T. (1995) "Stiff Chains of Filament Under Tension," *Macromolecules* **28**, 7016–7018.

STRICK, T., J. F. ALLEMAND, D. CROQUETTE, and D. BENSIMON (2000) "Twisting and Stretching Single DNA Molecules," *Progress in Biophys. & Molec. Biol.* **74**, 115–140. Includes important material on twisting. More detail than Bustamante et al. on stretching.

TSKHOVREBOVA, L., J. TRINIK, J. A. SLEEP, and R. M. SIMMONS (1997) "Elasticity and Unfolding of Single Molecules of the Giant Muscle Protein Titin," *Nature* **387**, 308–312. An AFM application to protein structure.

ZLATANOVA, J. and S. LEUBA (2003) "Chromatin Fibers, One At-a-Time," *J. Mol. Biol.* **331**, 1–19. An excellent review of the applications of single molecule techniques to the study of chromatin.

Answers to Odd Numbered Problems

CHAPTER 1

1.1 **a.** The rotation angle for $8_1, 8_2,$ and 8_4 helical symmetry is $360°/8 = 45°$ (or 2π radians/8 $= 0.25\pi$ radians). The translational component is P/8 for 8_1, P/4 for 8_2 and P/2 for 8_4, where P is the helical pitch.

  b. For 8_1, the repeating motif is a monomer, for 8_2 it is a dimer, and for 8_4 it is a tetramer.

  c. In the scheme of $8_1, 8_2$ and 8_4 helical symmetry, 8_3 symmetry would be difficult to define (in terms of the repeating motif). However, if we recall that an α-helix has 18_3 symmetry because it is a right-handed helix with a non-integral number of residues per turn (3.6 amino acids per turn, which is equivalent to 18 residues per 5 turns), we can imagine that a structure with 8_3 symmetry would have 8 residues in 3 turns, or that it is a right-handed helix with 2.67 residues per turn.

1.3 **a.** Recalling that the extended conformation has 2_1 symmetry with a rise of 0.36 nm, we can define the symmetry operator for an extended polypeptide chain along the y-axis as

$$\begin{vmatrix} x_{i+1} \\ y_{i+1} \\ z_{i+1} \end{vmatrix} = \begin{vmatrix} -1 & 0 & 0 \\ 0 & 1 & 0 \\ 0 & 0 & -1 \end{vmatrix} \times \begin{vmatrix} x_i \\ y_i \\ z_i \end{vmatrix} + \begin{vmatrix} 0 \\ 0.36 \text{ nm} \\ 0 \end{vmatrix}$$

If the C_α-carbon of the first amino acid is at $(0, 0.23, 0.05)$ nm, the coordinates of the second is at $(0, 0.59, -0.05)$ nm, the third at $(0, 0.95, 0.05)$ nm, the fourth at $(0, 1.31, -0.05)$ nm, the fifth at $(0, 1.67, 0.05)$ nm and the sixth at $(0, 2.03, -0.05)$ nm.

  b. The symmetry operator for an α-helix along the z-axis is

$$\begin{vmatrix} x_{i+1} \\ y_{i+1} \\ z_{i+1} \end{vmatrix} = \begin{vmatrix} -0.174 & -0.985 & 0 \\ 0.985 & -0.174 & 0 \\ 0 & 0 & 1 \end{vmatrix} \times \begin{vmatrix} x_i \\ y_i \\ z_i \end{vmatrix} + \begin{vmatrix} 0 \\ 0 \\ 0.154 \text{ nm} \end{vmatrix}$$

For this structure, with the first C_α-carbon at $(0, 0.23, 0.05)$ nm, the coordinates of the second is at $(-0.227, -0.04, 0.204)$ nm, the third at $(0.079, -0.216, 0.358)$ nm, the fourth at $(0.199, 0.115, 0.512)$ nm, the fifth at $(-0.148, 0.176, 0.666)$ nm and the sixth at $(-0.148, -0.176, 0.820)$ nm.

1.5 **a.** The plot shows contacts parallel to the main diagonal indicative of α-helices from 1–7 and 46 to 57 (recall that for the α-helix, all of the residues along both the x and y axes are included in the structure). The continuous set of contacts perpendicular to the main diagonal is indicative of a pair of antiparallel β-strands (residues 10–24 are aligned antiparallel to 26–40).

b. The schematic should have the amino (N) and carboxyl (C) termini in close proximity (as indicated by the contacts points in the upper right-hand corner).

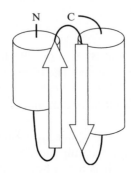

1.7 **a.** Assuming that the DNA remains entirely B-form:

$$Tw = 5250 \text{ bp}/10.5 \text{ bp/turn} = 500 \text{ turns}$$
$$Wr = -10 \text{ turns (for the 10 negative supercoils)}$$
$$Lk = Wr + Tw = -10 \text{ turns} + 500 \text{ turns} = 490 \text{ turns}$$
$$\sigma = Wr/Tw = -10 \text{ turns}/500 \text{ turns} = -0.02$$

b. First, to answer this question, recall that ΔTw, ΔWr, and ΔLk is relative to a reference state (relaxed closed circular B-DNA) which, for this length of DNA is $Tw° = 5250/10.5 = 500$ turns, $Wr° = 0$ turns (no supercoils), and $Lk° = Wr° + Tw° = 0$ turns $+ 500$ turns $= 500$ turns. Next, we need to understand that unless a topoisomerase is present, the linking number (Lk) does not change. Thus, even though the number of negative supercoils is reduced from -10 to -8, Lk for the plasmid remains 490 turns, so that $\Delta Lk = Lk - Lk° = 490$ turns $- 500$ turns $= -10$ turns and $\Delta Wr = Wr - Wr° = -8$ turns $- 0$ turns $= -8$ turns. Since $\Delta Lk = \Delta Wr + \Delta Tw$, we can calculate $\Delta Tw = \Delta Lk - \Delta Wr = -10$ turns $- (-8 \text{ turns}) = -2$ turns. This means that the DNA has unwound by, on average, 2 turns (that is, the equivalent of 2 turns of B-DNA has been lost). Since the intercalator reduces the number of negative supercoils, it will migrate slower in a gel during electrophoresis.

c. To separate a mixture of topoisomers (with $\Delta Lk = \Delta Wr = +5$ and -5 supercoils), we recall that supercoiled closed circular DNAs migrate faster in a gel during electrophoresis. Under normal conditions, these two topoisomers ($\Delta Wr = +5$ and -5) would migrate identically in the gel under normal conditions. Now, if an intercalator such as ethidium bromide is added to the running gel, for example, sufficient enough unwind 2 turns of B-DNA ($\Delta Tw = -2$ turns), then the topoisomer with $\Delta Lk = +5$ turns will have now a $\Delta Wr = \Delta Lk - \Delta Tw = +5$ turns $- (-2 \text{ turns}) = +7$ turns, while the topoisomer with $\Delta Lk = -5$ will have a

$\Delta Wr = -3$ turns. Now, the topoisomer with $\Delta Lk = +5$ turns will migrate more than twice (7/3) as fast as the topoisomer with $\Delta Lk = -5$ turns.

d. If the entire DNA is A-form (helical repeat = 11 bp/turns), Tw = 5250 bp/11 bp/turn = 477.3 turns. If $Wr = -10$ turns, then $Lk = Wr + Tw = -10$ turns + 477.3 turns = 467.3 turns.

1.9 **a.** Since both the secondary (helical) and tertiary structures of RNAs are held together by base pairing, the most useful contact distance is one that will be indicative of such interactions, but also that excludes the immediately adjacent nucleotides along the sequence. Therefore, we should set the contact distance to be greater than 0.75 nm and ~1 nm on average.

b. Setting the contact distance between 0.8 to 1.2 nm (to allow for some mispairings that would have shorter and longer C1′ to C1′ distance) yields the following contact plot for the tRNA in Figure 1.45 (closed circles represent double-stranded contacts, while open circles are tertiary structure contacts).

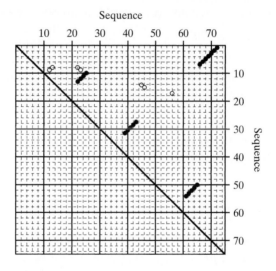

CHAPTER 2

2.1 $q = 7.95$ kJ, $w = -7.95$ kJ, $\Delta E = 0$, and $\Delta H = 0$.

2.3 **a.** 0.16 j/mol   **b.** 21.7 J/mol K   **c.** $-0.008°C$

2.5 If the α-helix represents a single conformation, the entropy for that form is defined explicitly as $S_{helix} = R\ln(1) = 0$ J/mol K. If each amino acid residue (ignoring side-chains) has on average 3 possible orientations in the random coil state, then the entropy for a 100 amino acid chain is $S_{coil} = R\ln(3)^{100} = 100R\ln(3) = 9.13$ J/mol K. The change in entropy from the helix to coil is thus $\Delta S = S_{coil} - S_{helix} = 9.13$ J/mol K. At the melting point for the helix, $\Delta G = 0 = \Delta H - T\Delta S$, or that $\Delta H = T\Delta S = (323$ K$)(9.13$ J/mol K$) = 2.95$ kJ/mol.

2.7

Number of supercoils (i)	i^2	$\Delta E = Ki^2$	Relative number of molecules ($n/n_o = e^{-\Delta E/RT}$)
0	0	0	1
1	1	0.4	0.856
2	4	1.6	0.537
10	100	40	1.81×10^{-7}

2.9 The correlation between the van't Hoff enthalpy and the calorimetric enthalpy from ITC indicates that the binding reaction is truly a two-state process between the unbound and bound forms of, for example, a protein and its ligand (there are no intermediates). One very simple potential explanation for the difference between the van't Hoff and DSC enthalpy for this system would be that there is a temperature dependent change in the population of one or both components of the reaction. Notice, however, that if, for example, the protein were to undergo a temperature folding and unfolding equilibrium, this should also be evident in the ITC experiment (the binding process would no longer be a simple two-state system, but would involve an additional equilibrium that would affect the enthalpy). An alternative would be that there is a third component in the solution that is not involved in the binding reaction, but that can undergo a temperature dependent transition. For example, if there were a contaminating protein in the solution, it can be observed to undergo a folding-unfolding equilibrium that is temperature dependent. This additional equilibrium would contribute to the overall enthalpy measured in a DSC study. However, with ITC, since we only measure the change in heat associated with the binding step, this contaminating protein would be unobserved in the change in heat at each titration point.

CHAPTER 3

3.1 **a.** -1.65×10^{-20} J $= -9.93$ kJ/mol. Treating this energy as a perturbation to the ionization energy, $\Delta pK_a = 1.73$. Histidine should become a better base in the presence of aspartic acid. The pK_a of an isolated His in water is 6.0, so its pK_a in the ion pair would be predicted to be 7.73.

3.3 **a.** $F = -dV/dr = 0$ at $r = r_0$. For a 6-12 potential, $F_{vdw} = 12A/r^{13} - 6B/r^7 = 0$. Treating this and Equation 3.17 as two simultaneous equations, $A = 7.78 \times 10^{-5}$ kJ $\cdot$ nm^{12}/mol, and $B = 8.04 \times 10^{-2}$ kJ $\cdot$ nm^6/mol.

   b. -15.6 kJ/mol, -19.2 kJ/mol.

3.5 From Equation 3.38, $\Delta S = -28$ J/mol K. From Equation 3.33, the unfolded state is shown to represent only about 29 unique conformations, while the compact intermediate state represents about 7 conformations.

3.7 $K \geq 12$ kJ/mol. $T \geq 2020$ K.

3.9 Assuming an ideal hydrogen bond (N-H aligned with the carbonyl oxygen), we can use the Equation $V_{dd} = -2|\mu|^2/D|r|^3$ for the head-to-tail alignment of dipole moments to estimate the hydrogen bond strength. This requires that we define $|\mu|$ and $|r| \cdot |\mu| = 3.7$ debye $= 1.23 \times 10^{-29}$ C·m. $|r|$ is estimated using 0.12 nm for a C=O bond, 0.147 nm for a C-N bond, 120° for O=C—N and C-N-H angles, and 0.29 nm for the N—H···O hydrogen bond. This gives a value $|r| \approx 0.5$ nm between the dipole moments. In water $(D = 78.5 \times (4\pi \times 8.85 \times 10^{-12}$ C²/J·m$))$, $V_{dd} = -3.47$ kJ/mol, while in a protein interior $(D = 3.5 \times (4\pi \times 8.85 \times 10^{-12}$ C²/J·m$))$, $V_{dd} = -3.47$ kJ/mol.

CHAPTER 4

4.1 Degeneracy for the all-or-none model $= 1$; noncooperative model $= \dfrac{n!}{j!(10 - j)!}$ for each state j; for the zipper model $= (8 - j)$ for each state j. Recall that the general form of degeneracy for the zipper model is $g = (n - j + 1)$, but for the coil to helix transition, 3 amino acids are fixed in the nucleation step, or that for $j = 0$, three amino acids must already be fixed in the helix conformation. Thus, for $j = 0, g = 8$. The partition functions for

All-or-none model: $Q = 1 + s^n = 1 + 1^{10} = 2$;

Noncooperative model: $Q = 1 + \displaystyle\sum_{j=1}^{n} \dfrac{n!}{j!(n - j)!} s^n = 1 + \sum_{j=1}^{10} \dfrac{10!}{j!(10 - j)!} 1^{10} = 1022$

Zipper model: $Q = 1 + \sigma \displaystyle\sum_{j=1}^{7} (8 - j) 1^{10} = 1 + \sigma(7 + 6 + 5 + 4 + 3 + 2 + 1)$

$$= 1 + 2 \times 10^{-4}(28) = 1.0056$$

4.3 For all heads, $g = 1$, or that $S = R \ln g = 0$. For half heads, half tails, the degeneracy is the coefficient for $j = 5$ of the polynomial expansion $(1 + s)^{10}$, or that $g = \dfrac{n!}{j!(10 - j)!}$, which for $n = 10$ and $j = 5, g = 252$. The entropy of this state is $S = R \ln(252) = 45.9$ J/mol K.

4.5 If σ is dependent only on the conformational entropy for fixing n number of amino acids prior to formation of the first hydrogen bond, then ΔS will reflect the difference of fixing 3 amino acids for the α-helix and 2 amino acids for the 3_{10} helix or that $\dfrac{g_{3_{10}}}{g_{\alpha\text{-}helix}} = \dfrac{2}{3}$. From the relationship $\sigma = e^{-T\Delta S/RT}$, we can show that $\sigma \propto \dfrac{1}{g}$ and, therefore,

$$\sigma_{3_{10}} = \sigma_{\alpha\text{-}helix} \left(\dfrac{g_{\alpha\text{-}helix}}{g_{3_{10}}} \right) = 1.3 \times 10^{-4}.$$

4.7 Recall that the probability of melting the middle versus the ends of a duplex is $\dfrac{P_{middle}}{P_{end}} = \dfrac{(n-2)\sigma}{2}$, or, for $\dfrac{P_{middle}}{P_{end}} = 1$, $n = \dfrac{2}{\sigma} + 2$. For $\sigma = 10^{-7}$, $n \approx 2 \times 10^{7}$ base pairs.

4.9 To calculate the propagation and nucleation terms, first, we need to calculate the free energies (ΔG^0) for each insert length (n). Recall that the nucleation term is independent of length of base pairs undergoing the transition and, therefore, the terms can be calculated by determining the slope and y-intercept of the relationship between ΔG^0 at the midpoint of transition and n. For this calculation, we must first calculate the change in twist (ΔTw). Recall that this is at the midpoint of the transition, which means that ΔTw will be half of the maximum expected for the number of base pairs (n). From this, we can calculate the writhe (ΔWr), which can then be used to determine ΔG^0 at the midpoints for each transition.

n (bp)	ΔLk (turns)	$\Delta Tw = \dfrac{n*(0.179) - 0.4}{2}$	$\Delta Wr = \Delta Lk - \Delta Tw$ (turns)	$K = \dfrac{1100RT}{n}$	$\Delta G^0 = K\Delta Wr^2$ (kcal/mol)
12	−12.5	−1.27 turns	−11.3 turns	636	80.2 kcal/mol
24	−15.5	−2.35 turns	−13.2 turns	634	110 kcal/mol
48	−22.0	−4.50 turns	−17.5 turns	631	193 kcal/mol

The relationship between ΔG^0 and n is linear:

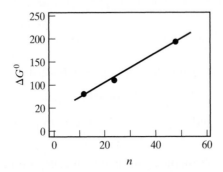

A linear regression analysis shows yields the relationship $\Delta G^0 = 3.19n + 38.4$ kcal/mol. The slope is the propagation free energy (ΔG_P^0) associated with the propagation parameter (s), while the y-intercept is nucleation free energy (ΔG_N^0) associated with the nucleation parameter (s). Thus, $s = e^{-\Delta G_P^0/RT} = 0.276$ and $\sigma = e^{-\Delta G_N^0/RT} = 1.84 \times 10^{-7}$.

4.11 The unfolded forms are:

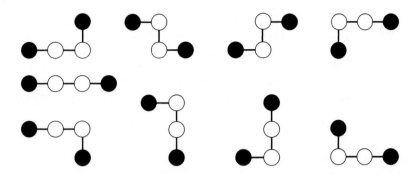

The folded form is, where the dotted line indicates a stabilizing H-H interaction:

b. 2

c. The lowest energy state entropy is $S_{folded} = R \ln(1) = 0$, while highest energy state entropy is $S_{unfolded} = R \ln(9) = 18.3$ J/mol K; therefore, $\Delta S_{unfolded\text{-}folded} = 18.3$ J/mol K.

d. The free energy of the folded state is $\Delta G_{folded} = \epsilon$, while for the unfolded state is $\Delta G_{unfolded} = -T\Delta S$, where ϵ is the enthalpy of the H-H interaction.

CHAPTER 5

5.1 $x_{1/2} = \pm 2 (D \ln 2)^{1/2} t^{1/2}$

5.2 $D_{20,w} \leq 8.7 \times 10^{-7} \ cm^2/sec$

5.3 **a.** $S_{20,w} = 1.96 \times 10^{-13}$ sec
   b. A minimum estimate from S is 1.4×10^5 g/mol
   c. $f/f_o = 3.7$ This strongly suggests that the subunits are uncoiled, that is, denatured.

5.7 $S_2/S_1 = 1.50$ (Remember you must include both terms for R_{12} and R_{21} in the sum.)

5.9 At high pH, both S and D decrease, but their ratio remains about the same. This suggests unfolding, with no change in M. At low pH, S drops but D increases, suggesting dissociation.

5.11 Label growing bacteria with radioactive amino acids. Look for the appearance of label in the polysome fraction in a sucrose gradient.

5.13 a. For concanavalin, 6240 RPM, for myosin 2888 RPM.
 b. For concanavalin, 34 hours, for myosin 159 hours.

5.15 a. M = 65,300
 b. Probably homogeneous, for the data lie pretty well on a straight line in lnc vs r^2 plot. (This was the *first* evidence that proteins were homogeneous in size.)

5.17 The s and D values give a "native" molecular weight of about 66,000 Da. In the absence of BME, the gel data show two components—one of about 32,000 Da, the other about 18,000 Da. To make the complete molecule, one of the former and two of the latter are needed (sum = 68,000). The 32,000 Da subunit is an S-S bridged dimer of two 16,000 Da units, as shown by SDS gel in the presence of BME.

5.19 Bands correspond to (roughly) 200, 400, 600, and 800 bp. This clearly indicates a repeat of about 200 bp.

CHAPTER 6

6.1 a. From Equation 6.1, the limit of resolution (LR) is approximately half of the wavelength (λ). Therefore, if $LR \approx 0.7$ nm, we would want $\lambda \approx 1.4$ nm. The energy of this wavelength, from $E = hv = \dfrac{hc}{\lambda}$, where h is Plank's constant and c is the speed of light, we can show that $E = 885$ eV for this wavelength of light.
 b. With 0.7 nm resolution, we could not resolve, for example, individual turns of an α-helix (Pitch is 0.54 nm) or distinguish between base pairs in DNA or RNA (spacing of ~0.34 nm), but can potentially revolve one helix from another, as in the case of the structure of the ribosome.

6.3 a. In an orthorhombic or tetragonal unit cell, the fractional unit cell coordinates (x, y, z) are simply the Cartesian coordinates (x, y, z) divided by the unit cell lengths $(|\mathbf{a}|, |\mathbf{b}|, |\mathbf{c}|)$, such that $x = x/|\mathbf{a}| = 0.0755$, $y = y/|\mathbf{b}| = 0.0175$, and $z = z/|\mathbf{c}| = 0.0167$.
 b. If there is a single atom in the unit cell, the phase angle is simply $\alpha = 2\pi \mathbf{S} \cdot \mathbf{r} = 2\pi(hx + ky + lz) = 2\pi(0.27) = 97.2°$.

6.5 a. If each contour is normalized to one contour representing a single vector, then the number of contours at the origin is the number of atoms in the molecule that contributes to the Patterson map. In this case, we can show that this is 4. For most macromolecules, it is impossible to determine this number simply because of the large numbers of scattering atoms contributing to the intensity at the origin.

b. To solve this structure, we can consider the four atoms of the molecule essentially in the x, y-plane. Then, we would expect that the Patterson map is generated by simply repeating the structure in the x,y-plane, placing each atom at the origin. The structure is thus:

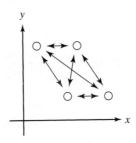

where the double-headed arrows represent two distance vectors pointing in opposite directions. The coincidence of two such sets of vectors in the horizontal directions, for example, account for the two sets of contours in the Patterson map along the x-axis. We do not, however, know the actual x, y, z-coordinates.

6.7 **a.** The Harker section with $y = 0.5$ and $z = 0.5$ indicate either 2_1 or 4_2 symmetry axes parallel to the **b** and **c** crystal axes. Assuming the simplest case of 2_1 axes, then the Cartesian x and y-coordinates would be calculated as the associate coordinates in the x, y-planes of the Patterson map divided by 2 (for the two-fold), and the Cartesian x and y coordinates would be calculated as the associate coordinates in the x, z-planes of the Patterson map divided by 2. Thus, the Cartesian atomic coordinates are $x = 0.215$, $y = 0.11$, and $z = 0.175$.

b. For this set of reflections, $\mathbf{S} \cdot \mathbf{r} = 1$. Thus, $F_H = f_H$.

6.9 **a.** For $l = 0$, $n = 0$; for $l = 5$, $n = 5$.

b. 0.329 nm^{-1}.

c. 0.26 nm^{-1}; 0.45 nm^{-1}, 0.53 nm^{-1}; 0.91 nm^{-1}.

d. Recall that Z-DNA is a left-handed double-helix with a helical repeat of 12 bp/turn, but it is really composed of 6 repeating dinucleotide repeats (e.g., the helical symmetry is actually 6_5, see Table 1.7). Thus, the repeating pattern is at the 6th and 12th layer lines, with the spacing of the 12th layer line at $1/0.37$ nm^{-1} = 2.7 nm^{-1}, while the (each layer line spaced at 0.225 nm^{-1}) in reciprocal space. The spacing of reflections along the equatorial axis would be 0.289 nm^{-1}, 0.5 nm^{-1}, 0.577 nm^{-1}, 1 nm^{-1}, etc., as shown here:

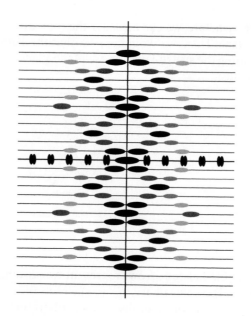

CHAPTER 7

7.1 Defining an element of surface area, $2\pi r^2 \sin\theta\, d\theta$ over which scattering intensity is constant, we have

$$\int_0^\pi 2\pi r^2 i \sin\theta\, d\theta = I - I_0$$

Substitution of i from Eq. 7.13, followed by integration gives

$$I - I_0 = (16\pi/3) R_\theta I_0$$

In most cases, $I - I_0 \ll I_0$, so we can use the approximation

$$-\tau = \ln\frac{I}{I_0} = \ln\left(1 - \frac{I - I_0}{I_0}\right) \cong -\frac{I - I_0}{I_0}$$

Therefore $\tau = (16\pi/3) R_\theta$.

7.3 Calculating the weight-average molecular weight of the mixture, and taking

$$\% \text{ error} = \frac{M_w - M_p}{M_p} \times 100$$

where M_p = protein molecular weight, we obtain about 50% for the error.

7.5 **a.** With $\rho(r) = \rho_0$, integration by parts immediately gives the result

$$F = \frac{4\pi\rho_0}{h^3}\{\sin(hR) - (hR)\cos(hR)\}$$

b. Because $I = KF^2$, we obtain

$$dI/dh = -2khR^2[\sin(hR) - (hR)\cos(hR)]\sin(hR)$$

For this to equal zero, either

$$h = 0 \text{ (the maximum value)}$$

or

$$\sin(hR) - (hR)\cos(hR) = 0$$

This means that

$$\tan(hR) = hR$$

which will be satisfied for multiple values of hR.

7.7 Note that the first term is a function of h, the second is not:

$$ln\ g^1 = -D_T h^2 \tau + ln(A_0 + A_1 e^{-6D_R\tau})$$

Plot $ln\ g^1$ vs h^2 at constant τ (several τ values would be good). Slope will give D_T. Intercept will be $ln\ g^1(0) = ln(A_0 + A_1 e^{-6D_R\tau})$ or $g^1(0) = A_0 + A_1 e^{-6D_R\tau}$ which can be fit by exponential fitting to give D_R.

7.9 $R_G = 51\ nm$.

CHAPTER 8

8.1 **a.** 1.13×10^{-18} J, 2.26×10^{-18} J
     b. 1.13×10^{-69} J, 2.26×10^{-69} J
     c. 1.13×10^{-18} J, 1.13×10^{-18} J. The spacing is 10^{51} times smaller for the chair.

8.3 The expectation value of r is $3a_0/2$. The most probable value of r is a_0.

8.5 $S = 0.586$ au, $\alpha = -0.973$ au, $\beta = -0.699$ au, $E_+ = -1.05$ au,
     $E_- = -0.662$ au, $E_+ + 1/R = -0.505$ au, $E_- + 1/R = -0.162$ au,
     $\chi_+ = 0.561\pi^{-1/2}(e^{-r_{1A}} + e^{-r_{1B}})$, $\chi_- = 1.099\pi^{-1/2}(e^{-r_{1A}} - e^{-r_{1B}})$

8.7 **a.** 0.0   **b.** 0.478   **c.** 1.0   **d.** 2.0   **e.** 3.0

8.9 1.67×10^{-4} M, 10^{-4} g

8.11 **a.** 0.0 for all practical purposes   **b.** 6.83×10^{-5} **c.** 0.383

CHAPTER 9

9.1 Along the axis of the 2p part of the sp^3 orbital.

9.3 **a.** $2|V| = 2/27$ au, $\mu_+ = 2$ au along the CO bond, $\mu_- = 0$ au Note that all the intensity falls at $+|V|$.

b. $2|V| = 4/27$ au, $\mu_+ = 0$ au, $\mu_- = 2$ au along the CO bond Note that all the intensity falls at $-|V|$.

9.5 $0.06, 0.6$

9.7 The reciprocal of 530.9 nm gives an energy of 18,836 cm^{-1}. Then the energies of the Raman bands can be subtracted and the reciprocals give 578.7, 552.5, 555.8, 552.7, and 552.9 nm, respectively.

9.9 GC content of about 42%.

CHAPTER 10

10.1 30.1 deg

10.3 The CD of native DNA from 160 to 200 nm shows an intense exciton interaction that can only come from asymmetrically oriented transition dipoles in stacked bases. The denatured CD is nearly superimposable with the native CD above 200 nm, but shows lower exciton intensity below 200 nm, as expected. Still, there must be some base-base interaction in the denatured DNA, since the CD is more intense than the average spectrum of the monomers.

10.5 It becomes very α helical.

CHAPTER 11

11.1 Use Eqs. 8.116 and 8.98 to obtain A in terms of the area under the absorption curve. Using an average wavelength of 272 nm the integral is about 770 l cm^{-1} mol^{-1}. Be careful of units when evaluating the factors in front of the integral, and remember the factor of 10^3 from Eq. 8.88, which has units cm^{3}l^{-1}. $A = 1.1 \times 10^8$ molecules sec^{-1}.

11.3 20 sec

11.5 In almost all cases, the excitation undergoes internal conversion to the ground vibrational level of the first excited singlet before fluorescing.

11.7 4.62 nm

CHAPTER 12

12.1 0 deg

12.3 42 repetitions

12.5 8.4–4.3, 4.3–8.4, 4.3–2.1, 2.1–4.3, 4.3–1.9, 1.9–4.3, 2.1–1.9, 1.9–2.1, 2.1–2.3, 2.3–2.1, 1.9–2.3, 2.3–1.9

CHAPTER 13

13.1 $\mu_2 - \mu_2^0 = RT \ln C_2 + 2BRT\, M_2 C_2$. Note that the standard state is a hypothetical state where $C_2 = 1$, but the solution behaves ideally.

13.3 We write π in terms of molar concentrations:

$$\pi = RT([M] + [D])$$

but since $[D] = \dfrac{M_0 - [M]}{2}$

$$\pi = RT\left(\frac{M_0}{2} + \frac{[M]}{2}\right)$$

the expression for $[M]$, obtained by solving the quadratic equation for $[M]$, can be written, for small $K M_0$ as:

$$[M] \cong \left\{-1 + \left(1 + \frac{8KM_0}{2} - \frac{1}{8}(8KM_0)^2 + \cdots\right)\right\}/4K$$
$$= M_0 - 2KM_0^2$$

so

$$\pi = RT\left(\frac{M_0}{2} + \frac{M_0}{2} - KM_0^2 \cdots\right) = RT(M_0 - KM_0^2 + \cdots)$$

writing $M_0 = \dfrac{C_0}{M_1}$, where M_1 is molecular weight, of monomer, and C_0 is weight concentration,

$$\pi = \frac{RT}{M_1}\left(C_0 - \frac{K}{M_1}C_0^2 + \cdots\right)$$

which is of the form expected.

13.5 **a.** $M_{\text{unknown}} = 20{,}000$ Da
b. The presence of buffer salts, which cannot pass from one vessel to the other, will yield a very low number average molecular weight.

13.7 **a.** 1.35×10^5 gm/mol
b. 0.35×10^5 gm/mol. Since the ratio is 3.9, the protein is probably a tetramer of its subunits.

13.9 Line 1 shows a moderate + virial coefficient, which is reduced in 2 by presence of salt. Thus, non-ideality in 1 is likely due to electrostatic repulsion. Line 3 shows a much greater non-ideality, likely due to the unfolding action of the denaturant.

13.11 Summing π_i:

$$\pi = \sum_i \pi_i = RT \sum_i C_i/M_i$$

$$= RTC\left(\sum_i C_i/M_i\right)\Big/\left(\sum_i C_i\right) = RTC/\overline{M}n$$

(see Table 13.2)

13.13 **a.** The result is:

$$\frac{d\ln C_2}{dr^2} = \frac{M_2(1 - \bar{v}_2\rho)\omega^2}{2RT(1 + 2BC_2)}$$

b. Since C_2 increases with r, the term in the denominator on the right side indicates that $d\ln C_2/dr^2$ will not be constant but will decrease with increasing r. Thus, the slope of the $\ln C_2$ vs r^2 graph decreases with r, and the curve turns downward from a straight line.

CHAPTER 14

14.1 **a.** -7.11 KJ/mol
b. 17.6
c. 0.054
d. -7.11 KJ/mol; toward $G - 6 - P$

14.3 Using eqn 14.24, we find $C = 0.143$ g/L

14.5 **a.** $\Delta G = \Delta G^0 + RT \ln(f/(1 - f))$
b. The curve is sigmoidal, asymptotic (going toward $-, +\infty$) as $f \to 0, 1$ respectively.
c. ΔG^0
d. $f = \dfrac{1}{1 + e^{\Delta G^0/RT}}$ (one of many possible forms)

14.7 For reaction (1-2): $\Delta G^0 = -39.7$ kJ/mol, $\Delta H^0 \cong 0$, $\Delta S^0 \cong 138$ J/mol·K. For reaction (1–4): $\Delta G^0 = -44.8$ kJ/mol, $\Delta H^0 = 117$ kJ/mol, $\Delta S^0 = 548$ J/mol·K.

14.9 A Hill plot is definitely sigmoidal, with a maximum slope of about 3.5. This means there are at least four sites in the molecule. In fact, there are 70 sites in this giant structure, so the allosteric unit must be much less than the whole molecule.

14.11 The Scatchard plot clearly reveals two sites. However, they must be either of different affinities or exhibit negative cooperativity. The Hill plot supports this, indicating further that the stronger and weaker sites have affinity constants of about 10^5 $(\text{mol/l})^{-1}$ and 10^4 $(\text{mol/l})^{-1}$ respectively.

14.13 a. One begins by noting that

$$\overline{R} = \left(\sum_{0}^{n} [\text{RX}_i] \right) \Big/ \left(\sum_{0}^{n} [\text{RX}_i] + \sum_{0}^{n} [\text{TX}_i] \right)$$

and then note that both sums are just binomial expansion. The result follows

b. $\lim\limits_{\alpha \to 0} \overline{R} = 1/(1 + L)$; $\quad \lim\limits_{\alpha \to \infty} \overline{R} = 1/(1 + Lc^n)$

c. when $c = 10^{-2}$, $L = 10^4$; $\quad \lim\limits_{\alpha \to 0} \overline{R} \cong 10^{-4}$; $\lim\limits_{\alpha \to \infty} \overline{R} \cong 1$

when $c = 10^{-2}$, $L = 10^8$; $\quad \lim\limits_{\alpha \to 0} \overline{R} \cong 10^{-8}$; $\lim\limits_{\alpha \to \infty} \overline{R} \cong 0.5$

14.14 $K_{obs} = \dfrac{f}{(1 - f)} \dfrac{1}{([P]_T - f[D]_T)}$

CHAPTER 15

15.1 From Equation 15.3, we can show that $t^2 = K\dfrac{m}{Z}$, where K is a constant to take into account the parameters that are specific for the experiment. If we make the simplifying assumption that the mass (m) is large and therefore not affected by the addition of protons ($m + 1 \approx m$), then we can also show that $\dfrac{t_1^2}{t_2^2} = \dfrac{m/Z_1}{m/Z_2} = \dfrac{Z_2}{Z_1}$. Let's consider the case where the slowest peak (at 50 μs) has a charge $Z = 1$, the second slowest has $Z = 2$, and so forth. Then, we would expect the ratio $\dfrac{t_1^2}{t_2^2} = \dfrac{Z_2}{Z_1} = 2$, $\dfrac{t_1^2}{t_3^2} = \dfrac{Z_3}{Z_1} = 3$, and $\dfrac{t_1^2}{t_4^2} = \dfrac{Z_4}{Z_1} = 4$, or that $\dfrac{t_1^2}{t_Z^2} = Z$ The actual values are the following

time (t)	Charge (Z)	$\dfrac{t_1^2}{t_Z^2}$ Predicted	$\dfrac{t_1^2}{t_Z^2}$ Measured
50 μs	+1	1	1
35 μs	+2	2	2.04
29 μs	+3	3	2.97
25 μs	+4	4	4.00

15.3 The sequence is constructed by arranging fragments according the overlapping regions of each sequence. Recall that thermolysin recognizes the aromatic amino acids tyrosine (Y), phenylalanine (F), tryptophan (W), along with leucine (L), isoleucine (I), valine (V), and, at high pHs, histidine (H) then cleaves the peptide bond just in front of one of these. Trypsin recognizes the basic amino acids lysine (K) and arginine (R) and cleaves the peptide bond just after those (Table 15.1).

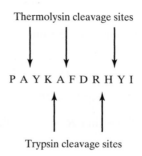

15.5 With eight Y-ions, there must be nine amino acids in the peptide (see Fig. 15.19). If we start at the $[M + H]^+$ peak at 1001.60 and work towards the left along the Y_n-ions, we see that the mass of the $[M + H]^+ - Y_8$ ion = 57.02 (associated with the amino acid G), $Y_8 - Y_7 = 113.08$ (S_2 which is I or L), $Y_7 - Y_5 = 198.1$ (S_3 and S_4 could not be assigned because the Y_6 ion is absent), $Y_5 - Y_4 = 113.08$ ($S_5 = $ I or L), $Y_4 - Y_3 = 113.08$ ($S_6 = $ I or L), $Y_3 - Y_2 = 113.08$ ($S_7 = $ I or L), $Y_2 - Y_1 = 147.07$ ($S_8 = $ F), and $Y_1 = 147.09 = S_9 + 19$ ($S_9 = $ K). From this, we can show that the sequence at this point is:

$$G\text{-}(I, L)\text{-}S_3\text{-}S_4\text{-}(I, L)\text{-}(I, L)\text{-}(I, L)\text{-}F\text{-}K$$

S_2, S_5, S_6, and S_7 are either I or L, but cannot be resolved. However, S3 and S4 can be identified from the B-ions. Starting with $B_2 = 147.09 = S_1(G) + S_2(I \text{ or } L) + 19$, we can confirm the first two amino acids. The mass difference $B_3 - B_2 = 123.06$ ($S_3 = $ P), $B_4 - B_3 = 101.05$ ($S_4 = $ T), $B_5 - B_4 = 113.08$ ($S_5 = $ I or L), $B_6 - B_5 = 113.08$ ($S_6 = $ I or L), $B_7 - B_6$ ($S_7 = 113.08$), and $B_8 - B_7 = 147.07$ ($S_8 = $ F). Thus, the sequence of the peptide is:

$$G\text{-}(I, L)\text{-}P\text{-}T\text{-}(I, L)\text{-}(I, L)\text{-}(I, L)\text{-}F\text{-}K$$

b. Cleavage of the peptide bond between the proline residue at S_3 and the threonine at S_4 to produce first the B_3-ion, followed by a proton transfer from the B_3-ion to the amino group of the remaining C-terminal peptide fragment are the two steps required to generate the Y_6-ion (see Fig. 15.18). The lack of the Y_6-ion in this spectrum would suggest that either the Pro-Thr peptide bond could not be cleaved or that the proton could not be transferred from the resulting B_3-ion. Since the B_3-ion is seen in the spectrum, it is clear that the Pro-Thr peptide bond can be cleaved in the CID chamber. We can therefore conclude that the proton transfer step must be restricted or missing. Notice that the B_3-ion would place the Pro amino acid at the charge five-membered ring, but with Pro at this position, the side-chain would be covalently linked to the amino nitrogen of the ion and, consequently, this nitrogen could not carry a proton. Thus, since there is no proton to transfer, the remaining peptide at the C-terminus cannot become charged.

CHAPTER 16

16.1 R = 8.64 nm

16.3 The protein has multiple, similar "domains" that can be unravelled from their globular conformation under tension. When tension is increased, domains "pop open" one by one, yielding the stepwise relaxation.

Index

A

Absorbance, definition of, 409
Absorption bands, infrared, 450
Absorption cross-section, 409
Absorption spectroscopy, 421–464
    electronic absorption, 421–449
    Raman scattering, 457–463
    special kinds of, 465
    vibrational absorption, 449–457
Acetone, absorption spectrum, 429, 430
Acid-base chemistry, solvent effects
    and, 509
α-Actinin, 509
Activity concentration, 586
Adair equation, 623, 626
Adenosine diphosphate (ADP), 80
Adenosine triphosphate (ATP), 80
A-DNA, structure of, 57
ADP, *see* Adenosine diphosphate
Aequorea victoria, 507
AFM, *see* Atomic force microscopy
Alanine residue, potential energy
    profiles for, 125, 126
All-or-none transition, 169
Allosteric activator, 644
Allosteric inhibitor, 644
Allostery, 632
Amide
    hydrogen bonding of, 455
    wavefunction orthonormality,
      440
Amino acid
    characteristic resonances, 549
    common, 29
    composition, residue volumes,
      231
    enzyme cofactors, 28
    ^{1}H chemical shifts for, 561
    α-helix formation and, 181
    hydropathy of, 30
    hydrophilic, 28
    hydrophobic, 28, 30
    L-configuration of, 6
    residue masses, 677
    spin systems, 563
    structure, 27
Amphipathic molecules, 16, 17
Analytical ultracentrifuge, 225, 226
Anisotropy
    fluorescence, 518, 525
    steady-state, 523
Anomalous dispersion, 333
Argand diagram, 312, 315, 333
ASP, *see* Atomic solvation parameter
Association constant, intrinsic, 630
Asymmetric unit, 280

Atom

Atom
    atomic scattering components
      of, 314
    degrees of freedom, 449
    discrete energy level, 382
    disorder of, 337
    hydrogen, 380, 434
    kinetic energy of, 147
    occupancy of, 337
    united, 149
Atomic force microscopy (AFM), 699
Atomic orbitals
    amide chromophore, 431
    carbonyl, 428
Atomic position, 309
Atomic resolution, 276, 277
Atomic scattering factor, 313
Atomic solvation parameter (ASP),
    156, 157
Atomic units, 391
ATP, *see* Adenosine triphosphate
Autocorrelation
    analysis, 363
    function, 363
Autoradiography, 264
Average molecular weights, 246
Avogadro's number, 356, 409

B

Bacterial cell, applying thermodynamic
    ideas to, 73
Base pair, 137
    energies of hydrogen bonds in,
      139
    melting of homoduplexes, 185
    potentials, 138
    stacking energies, 141
    statistical weight for melting,
      186
    Watson-Crick, 133, 453
Base stacking, NMR and, 553
Basic pancreatic trypsin inhibitor
    (BPTI), 566
Basis vectors, 396
B-DNA
    to A-DNA transition, 189
    duplex, modeling of, 143
    fiber diffraction photograph
      of, 58, 339
    model, 137, 141
    structure of, 57
    sugar conformations, 134
    Watson-Crick model for, 59
Beer-Lambert law, 408, 412
Bessel functions, 342, 345

Binding

Binding
    constant, 649
    cooperative, 632, 635
    experimental measurements
      of, 616
    ion, 644
    isotherms, 642
    noncooperative, 623–624
    nonspecific, 648
    occupancy, 633
    site, 616
Biochemically pure sample, 286
Biological macromolecules, 1–71
    cell environment, 10–19
    general principles, 1–8
    molecular interactions, 8–10
    structure of nucleic acids,
      52–68
    structure of proteins, 27–52
    symmetry elements, 25
    symmetry relationships, 19–27
Biopolymer, 1
    average linear dimension of, 201
    behavior, 204
    circular dichroism spectrum
      of, 465
    definition of, 3
    folding of, 151
    molecule, magnetic beads
      attached to, 707
    neutron-scattering lengths for
      elements in, 370
    two-state models for structural
      transitions in, 169
Blotting, 264
    Northern, 266
    Southern, 264, 265
    Western, 266
Bohr radius, 390
Boltzmann constant, 82, 147, 220, 706
Boltzmann distribution, 79, 88, 168, 411
Bomb calorimeter, 94
Bonding energy, 112, 426
Born-Oppenheimer approximation,
    401, 450
Bovine serum albumin (BSA), 229
BPTI, *see* Basic pancreatic trypsin
    inhibitor
Bragg reflection plan, 302
Bragg's angle, 297
Bragg's law, 292, 300, 302, 335
Bravais lattices, 281, 282
Brownian motion, 149, 214
BSA, *see* Bovine serum albumin
Buoyancy factor, 224
B-Z junction, nucleation parameter
    for, 193

B-Z transition, 191
  partition function for, 194
  two-dimensional gel
    electrophoresis analysis
    of, 196
  zipper model, 194

C

Calmodulin
  CD monitoring of, 483
  relative quantum yield of, 514
Calorimetry, 94–102
  differential scanning calorimetry,
    95–98
  isothermal titration calorimetry,
    99–102
cAMP, *see* Cyclic AMP
Capillary electrophoresis, 266, 267
Carbonyl, atomic orbitals, 428
ccDNA, *see* Closed circular DNA
CD, *see* Circular dichroism
Cell
  diffusion within, 221
  membranes, active transport
    in, 590
Cell, unit, 279, 280
  cylindrical, 340
  dimensions and shape of, 310
  Patterson maps, 324
  reciprocal, 300
  unknown location of, 320
Cell environment, 10–19
  interaction of molecules with
    water, 15–16
  nonaqueous environment of
    biological molecules, 16–19
  water structure, 11–15
Centrifugal force, 223
Centrifugation, sucrose gradient, 239, 240
Chain tracing, 331
Charge-charge interactions, 115
Charge-transfer complexes, solvent
  effects and, 509
Chemical equilibria involving
  macromolecules, 605–659
  binding to nucleic acids, 648–654
  binding of small ligands, 615–648
  interactions between
    macromolecules, 610–615
  thermodynamics, 605–610
Chemical potential, 583, 584
Chemical shift dispersion, 538
Chiral molecule, chemical groups, 5–6
Cholesterol oxidase, fluorescent cofactor
  of, 697
Chromatin structure, repeating units
  of, 372
Chromophores, degenerate, 442
CID, *see* Collision-induced dissociation
Circular dichroism (CD), 58, 471–497
  continuous variation curve, 484
  nucleic acids, 476–480
  proteins, 481–484
  singular value decomposition,
    485–496

spectra, important basis, 490
spectroscopy, 457
variable selection, 495
vibrational, 496–497
Circular dichroism band
  conservative, 475
  nonconservative, 475
  rotational strength, 473, 475
Closed circular DNA (ccDNA), 66, 190
  free energy of superhelicity
    of, 192
  topoisomers, 191
  topology of, 67
Coil-helix transitions, 175, 177, 178,
  182, 187
Collagen, random coil CD spectrum
  of, 481
Collision-induced dissociation
  (CID), 678
Colloids, 579
Combination bands, 451
Component, definition of, 580
Computer programs, molecular
  dynamics, 572
Conformational entropy, 150–152,
  175, 178
Conservation of mass, 216
Contact plot
  globular protein, 46
  α-helix, 47
  protein from NMR, 48
  β-sheets, 47
Continuity equation, 216, 217
Continuous helices, fiber diffraction
  of, 341
Continuous variation curve, CD, 484
Continuous wave (CW) NMR
  instruments, 537, 544
Contrast matching, neutron scattering,
  372
Cooperative binding, 632, 635
Cooperativity coefficient, 174
Correlated spectroscopy (COSY)
  sequence, 556, 557, 563
COSY sequence, *see* Correlated
  spectroscopy sequence
Coulomb energy, 439
Coulomb's law, 115, 132, 143, 249
Counterion
  condensation effect, 652
  screening, 132
  Stokes's radius of, 599
CPK models, 44
Cross vectors, 324
Crystal
  atoms, in isomorphous
    replacement, 321
  definition of, 279
  growth of, 285
  heavy atom derivatives for
    macromolecular, 328
  isomorphous, 284, 327
  lattice formation, nucleation
    of, 287
  lysozyme, 305
  1-methylthymine, 416

morphology of, 281
one-dimensional, 339
possible space groups in, 284
space group of, 283
symmetry, rotational
  components of, 281
Crystallization, 579
  solutions, 287
  vapor diffusion methods of, 289
Crystallography, phase problem in, 319
CW NMR instruments, *see* Continuous
  wave NMR instruments
Cyclic AMP (cAMP), 517
Cysteine, electronic transition, 435–436

D

D-amino acids, 28
Debye equation, 361
Debye-Hückel screening parameter,
  143
Degeneracy factors, 171
Degenerate state, molecular, 169
Degenerate wavefunctions, 438
Density gradient, 238, 246, 247
Deoxyoligonucleotide single crystals,
  X-ray diffractions studies of, 134
Deoxyribonucleic acid, *see* DNA
Depolarization, 520
Deuterium exchange mass spectrometry
  (DXMS), 686
Diagonal plot, globular protein, 46
Dialysis equilibrium, 591
Dielectric constant, 11, 17, 116, 600
Difference bands, 451
Difference electron density map, 322
Differential refractometer, 358
Differential scanning calorimetry
  (DSC), 95–98, 609
Diffraction, 292, *see also* X-ray
  diffraction
  angle, reflecting pairs and, 309
  Bragg's law of, 293, 335
  data, recording of, 298
  electron, 19
  fiber, 338, 341
  Huygen's principle of, 291
  intensity, isocitrate lyase
    crystals, 290
  von Laue conditions for, 294,
    295, 296
Diffraction pattern, 299
  continuous helices, 343
  discontinuous helices, 343
  systematic absences in, 308
Diffusion
  coefficient, 215, 227, 363, 364
  -controlled kinetics, 18
  data, 230
  description of, 215
  differential equation for, 219
  examples of, 216
  narrow zone, 218
  smearing, 236
  within cells, 221
Dimerization, 51, 138

Dipole
 -dipole interactions, 117, 516
 magnetic transition, 474
 moment, 352, 438, 511
 oscillating, 459
 strength, 414
 transition, 433, 441, 451
Direct radioactivity scanning, 264
Discrete energy level, atom in, 382
Dispersion interactions, 9
Dissociation
 constant, 620
 energies, chemical bonds in
   organic molecules, 112
Distance geometry, 571–572
Disulfide groups, electronic transition,
   435–436
DNA (deoxyribonucleic acid), 52, 53
 absorption of, 448
 base pairs found in, 137
 bases, 530
 closed circular, 66, 190, 192
 cruciform, 60
 crystal, electron density of, 321
 displacement of sodium ions
   from, 144
 double-helix, annealing and
   melting of, 184
 electrophoresis, 259
 force-elongation curve, 709
 fragments, gel electrophoresis
   of, 254
 G-quartet, 61
 helical transitions, 189
 helix, stretched, 705
 Holliday junctions, 60, 61
 kinetoplast, 258
 melting of, 185
 motion of RNA polymerase
   on, 701
 negatively charged phosphates
   and, 142
 persistence length of, 202
 polymorphic double helices,
   57–58
 possible conformations, 54
 recombination, 60
 /RNA hybrids, 248
 rotation, 708
 satellite, 248
 structure, polymorphic, 133
 supercoiled, 66, 67, 255–256
 tertiary structure, effects of
   on electrophoretic
   mobility, 255
 thermal denaturation of, 446
 topology, 65
 transitions, supercoil-dependent,
   190
 use of FRET to study, 532
 X-ray diffraction studies, 58
DNA duplex, 56, 60, 154
 melting of, 139
 models for melting and
   annealing of, 188
 stability of, 161

DNA molecules
 measuring diffusion of small, 222
 mechanics of, 704
Dot product, 201, 360
Double reciprocal plot, 628
Double resonance technique, spin-spin
   interactions and, 542
Drug discovery, protein inhibition and,
   102
DSC, see Differential scanning
   calorimetry
DXMS, see Deuterium exchange mass
   spectrometry
Dyad axis, 23
Dyad symmetry, 23
Dye
 fluorescent, 221–222
 labels, photobleaching of, 221
 mobility, 251–252
Dynamic light scattering, 221, 363

E

Effective concentration, 586
Eigenvalue, 383
Einstein coefficient, 414
Einstein relations, 410
Elastic Rayleigh scattering, 458
Electrochemical potential, 596, 599
Electromagnetic spectrum, 279
Electron
 amplitudes, 367
 diffraction, 19
 energies, 404
 indistinguishability of, 405
 ionization, 392
Electron density, 291, 313
 DNA crystal, 321
 map, 316, 317
Electronic absorption, 449
 deoxyribonucleotides, 444
 N,N-dimethylacetamide, 433
 melittin, 508
 spectroscopy, applications of, 447
Electronic circular dichroism, 476
Electrophoresis, 248–269
 capillary, 266, 267
 DNA, 259
 free, 250–251
 general principles, 249–253
 idealized model for, 249
 isoelectric focusing, 266–269
 methods for detecting and
   analyzing components on
   gels, 264–266
 moving boundary, 250–251
 nucleic acids, 253–259
 one-dimensional, 256
 pulsed field, 255, 256
 SDS gel, 259–264
 steady-state, 598–600
Electrophoresis, gel, 251, 257
 apparatus, 252
 of DNA fragments, 254
 two-dimensional, 195
Electrophoretic mobility, 250, 255

Electrophoretic transfer techniques, 266
Electrospray ionization (ESI), 662
Electrospray ionization mass spectrum
   (ESI-MS), 671
Electrostatic interactions, 115, 131
Emission lifetimes, 502–504
Emission spectroscopy, 501–534
 analytical applications, 507–509
 emission lifetimes, 502–504
 fluorescence applied to nucleic
   acids, 530–532
 fluorescence applied to proteins,
   524–529
 fluorescence decay, 513–515
 fluorescence instrumentation,
   506–507
 fluorescence resonance energy
   transfer, 516–517
 fluorescence spectroscopy,
   504–506
 linear polarization of
   fluorescence, 517–524
 phenomenon, 501–502
 solvent effects, 509–513
Energy
 base-stacking, 140
 Boltzmann distribution of, 79
 bonding, 112, 404, 426
 Coulomb, 439
 distribution of, 76
 electron, 404
 electronic absorption bands, 422
 first-order correction to, 399
 Gibbs free, 87, 92
 Helmholtz free, 88
 individual noncovalent
   interactions, 610
 internal, 74
 levels, number of particles in, 79
 potential, hydrogen-hydrogen
   bond, 113
 profile, 110
 self-, 19, 131
 shift, general solvent effects and,
   512
 spectra plotted as function of, 382
 splitting, exciton theory and, 437
 standard free, 91
 total, 109
 transfer, fluorescence resonance,
   516–517
 transition, 386–408
 wavefunctions, 427
 zero point, 390
Energy, free, 88
 ccDNA superhelicity, 192
 determination of
   biochemical system, 93
 mixing, 585
 partial molar, 583
 perturbation, 161
 simulation of, 158
 topoisomer, 192
Energy, kinetic, 384
 atom, 147
 nuclei, 450

Energy, potential
  dipole-dipole interactions, 118
  force field, 148
  hydrogen bond, 122
  macromolecule, 110
Energy minimization, 111
  goal in, 146
  total energy and, 147
Energy states
  distributions of particles over, 78
  protein, 90
Enthalpic relaxation, 543
Enthalpy, 74
  change in, 75
  determination of for
    biochemical system, 93
  van't Hoff, 97
Entropic relaxation, 543
Entropy, 82, 149–153
  change, 83, 585
  conformational, 150–152, 175, 178
  decrease in, 91
  determination of for
    biochemical system, 93
  increased, 86
  loss in, 139
  mixing, 85
  normal mode analysis, 152–153
  statistical definition of, 84
Enzyme
  -catalyzed reactions, proton
    transfers and, 15
  fluorescence cofactor of, 697
  functional domains, 44
  proteolytic, 675
Equilibrium
  constants, macroscopic, 624
  dialysis, 617
  states, 74, 81
ESI, see Electrospray ionization
ESI-MS, see Electrospray ionization
  mass spectrum
Ethylene, equilibrium-bonding
  distances, 426
Eukaryotic transcription factor, 34
Ewald sphere, 303, 305
Exciton theory, 400, 437
Excluded volume effect, 588
Experimental thermochemistry
  calorimetry, 94–102
  van't Hoff relationship, 93–94
Extinction coefficient, 410, 412, 448

F

Faraday constant, 596
FCS, see Fluorescence correlation
  spectroscopy
FDPB, see Finite difference solution to
  PB relationship
Ferguson plots, 252, 253
Fermi-Dirac permutation, 34
Fiber diffraction, 338, 341
  discontinuous helices, 344
  history, 346
  pattern, 345, 346

Fiber unit cell, 340
Fibrinogen, 231
Fick's first law, 217, 241
Fick's second law, 217
FID, see Free induction decay
Filter binding, 614
Finite difference solution to PB
  relationship (FDPB), 132
First law for reversible processes, 87
First law of thermodynamics, 73, 77
Flotation, sedimentation versus, 224
Flow
  concept of, 214
  linear dichroism, 470
  relation between molecular
    velocity and, 242
Fluorescence, 222
  anisotropy, 518, 525
  correlation spectroscopy
    (FCS), 697
  definition of, 501
  ground vibrational level, 505
  instrumentation, 506–507
  linear polarization of, 517
  melittin, 508
  observation of single
    macromolecules by, 695
  polarized, 501
  Polistes mastoparan, 524
  resonance energy transfer
    (FRET), 516, 694, 698
  spectroscopy, 501, 504–506,
    528–530
  time-resolved, 522
  yield, 515
Fluorescence decay, 513–515
  intrinsic lifetime, 513
  lifetime, 513
  quenching, 514
Fluorometry, frequency-domain, 523
Fluorophore, 506, 512
  dipole-dipole interaction
    between, 516
  emissions and absorption
    spectra, 697–698
  tryptophan, 524
Folding funnel, 208, 209
Force field
  limitations of, 145
  molecular mechanics, 111
  potential energy, 148
Forster radius, 698
Fourier self-deconvolution, band-
  narrowing approach, 456
Fourier transform (FT), 316, 547
  instruments, 453, 537, 545
  NMR, 555
Fractional cell coordinates, 309
Fraction saturation, 618
Franck-Condon principle, 422, 503
Free electrophoresis, 250–251
Free energy, 88
  ccDNA superhelicity, 192
  determination of for
    biochemical system, 93
  mixing, 585

partial molar, 583
perturbation, 161
simulation of, 159
topoisomer, 192
Free energy change
  calculation of, 607
  essentially irreversible reaction
    and, 608
  standard state, 606
Free induction decay (FID), 547, 556
Free particle, 386
Freezing-point depression, 579
Frequency-domain fluorometry, 523
FRET, see Fluorescence resonance
  energy transfer
Frictional coefficient
  ratios, 220
  unhydrated sphere of radius, 230
Frictional force, 223
Friedel pairs, 318
Friedel's law, 307, 333
FT, see Fourier transform

G

Gaussian distribution, 151–152, 217
Gel
  glycine-chloride boundary, 263
  running, 262, 263
  shift assay, binding of histone,
    615
  stacking, 262, 263
  two-dimensional, 258
Gel electrophoresis, 251, 257
  apparatus, 252
  DNA fragments, 254
  two-dimensional, 195
General solvent effects, 509, 512
GFP, see Green fluorescent protein
Gibbs-Duhem equation, 583, 584, 587
Gibbs free energy, 87, 92, 583
Globular protein
  behavior of, 231
  contact plot, 46
  prediction of secondary
    structures of, 181
  side chains at surface of, 131
Globular proteins, structure, 42–52
  domains, 43–44
  protein folds, 48–49
  quaternary structure, 49–52
  supersecondary structures, 42–43
  tertiary structure, 44–48
Gramicidin, 18, 28
Greek key motif, 43
Green fluorescent protein (GFP), 507, 509
Guinier plot, 368

H

β-Hairpin, 43
Hamiltonian
  hydrogen molecule, 424
  operator, 384
  perturbed, 395

Hard sphere approximation, 119
Harker planes, 325
Harker sections, 325
H/D exchange, *see* Hydrogen/deuterium
    exchange
H-DNA, triple-stranded, 60
Heat, definition of, 74
Heisenberg uncertainty principle, 335
Helical angle, 36
Helical symmetry, 26, 36
    hexokinase, 51
    left-handed, 38
    macromolecular, 38
    right-handed, 38
Helical twist, 36
Helix
    axis, 36
    -coil reverse transition, 180
    dipole, 129–130
    model for discrete steps of, 36
    structure, 38
    symmetry matrix, 37
Helmholtz free energy, 88
Hemocyanin
    MAD and, 334
    sedimentation, 227, 243
    small-angle X-ray scattering
        from, 369
    symmetry in, 51
Hemoglobin
    binding of oxygen by, 641
    concentrations, binding
        isotherms at decreasing, 642
    quaternary structures, 50, 639
Henderson-Hasselbach equation, 645
Henry's function, 599
Hermitian operator, 383
Heteroallostery, 632
Heterodimer, 49
Heteronuclear experiments, 569
Hexamer toy model (HTM), 205
    distinct states, 206
    statistical-mechanics analysis
        of, 206
Hexokinase, helical symmetry of, 51
HH interaction, *see* Hydrophobic-
    hydrophobic interaction
High pressure liquid chromatograph
    (HPLC), 507
HIV, *see* Human immunodeficiency
    virus
Homeoallostery, 632, 638
Homodimer, 49
Homopolymer, 175
Host-guest peptides, 179
HPLC, *see* High pressure liquid
    chromatograph
HTM, *see* Hexamer toy model
Human immunodeficiency virus (HIV),
    28, 101
Hybrid orbitals, amide chromophore,
    431
Hydration free energy, 158
Hydrogen atom
    quantum mechanics and, 380
    transition dipole, 434

Hydrogen bond, 119
    acceptor, 12, 13
    donors, 12, 13, 34
    energies in, 139
    formation of, 466
    interactions, 123
    potential energies, 113
Hydrogen/deuterium (H/D) exchange,
    684, 685
Hydrogen molecule
    composition of, 401
    Hamiltonian, 424
    as model for bond, 400
Hydrophilic compounds, 15
Hydrophobic bonding, 91
Hydrophobic compounds, 16
Hydrophobic effect, 16, 153
Hydrophobic-hydrophobic (HH)
    interaction, 205, 208
Hypochromism, 441, 446, 447
Hysteresis, DNA annealing and, 184

I

Imaginary numbers, 311
Independently variable substances,
    580
Independent systems approach, valence
    bond, 400
Infrared spectra, basis set of, 457
Interatomic shielding, 538
Internal conversion, 502
Internal energy, 74
Intersystem crossing, 502
Intrinsic lifetime, 513
Ion
    binding, 644
    mass/charge ratio for, 665
Ionization
    electron, 392
    soft, 670
Irreversible equilibrium states, 74
Isocitrate lyase, diffraction intensity
    from crystals of, 290
Isoelectric focusing, 248, 250, 266–269
Isoelectric precipitation, 589
Isokinetic gradient, 240
Isomer counting, 150
Isomerization, fast exchange, 551
Isomorphous replacement, 321, 326
Isothermal titration calorimetry (ITC),
    99–102
Isozymes, 253
ITC, *see* Isothermal titration
    calorimetry

K

Kinetic energy
    atom, 147
    nuclei, 450
    particle, 384
Klebsiella aerogenes, 328
KNF model, 638, 639

L

Laguerre functions, 392
Lamm equation, 236
Laplacian operator, 385, 391
Lasers, 459
Lattice motif, 280
LCAO, *see* Linear combination of
    atomic orbitals
Legendre polynomials, 392
Lennard-Jones potential, 120, 136
Levinthal's paradox, 151, 207, 209
Light
    absorption of, 380
    circularly polarized, 472
    electric vector of, 416, 417
    energy for vibrational
        absorption, 421, 423
    monochromatic, 381
    particle properties, 381
Light microscope, 277
Light scattering, 351–363
    dynamic, 363
    fluctuations in, 364
    fundamental concepts, 351–355
    measurements, instrument for,
        358
    Rayleigh scattering, 355–358
    scattering from particles that are
        not small, 358–363
Limit of resolution, optical method, 278
Linear combination of atomic orbitals
    (LCAO), 400, 427
Linear dichroism, 466–471
    flow, 470
    isotropic absorption and, 468,
        469
    measurements, 467, 468
LINUS, 202–203
Lippert equation, 510
Local positive shielding, 538
London dispersion forces, 119
Low-angle X-ray scattering, 362
Lysozyme
    ESI-MS of, 672
    tetragonal crystal of, 305

M

Macromolecular helices, helical
    symmetry of, 38
Macromolecular structure, 3
    hierarchical organization of, 3, 4
    modeling of, 107
    simplest method for solving, 321
Macromolecular systems, simplifying
    of, 108
Macromolecule, *see also* Biological
    macromolecules
    conformations, 7
    counterion atmosphere
        surrounding, 250
    crystallizing, 288
    definition of, 3
    effective valence of, 599
    folding problem, 107

Macromolecule (*continued*)
　hydrogen-bond donors and
　　acceptors, 13
　interactions between, 610
　ion binding to, 644
　liganded, 622
　measurement of dimensions
　　of, 365
　one-dimensional NMR of, 549
　order of addition effects, 613
　potential energy of, 110
　structures of at atomic
　　resolution, 277
　torsion angle, 7, 569
　two-dimensional FT NMR
　　applied to, 560
　unliganded, 620
　weak interactions, 8
Macroscopic dipole, 129–130
MAD, *see* Multiple-wavelength
　anomalous dispersion
Magnetic beads, 707
Magnetic dipole moment, 535
Magnetic transition dipoles, 474
Magnetization, 545
MALDI, *see* Matrix-assisted laser
　desorption-ionization
Manning's theory, 652
Mass spectrometry (MS), 660–692
　challenges in, 660
　determining molecular weights
　　of biomolecules, 670–672
　general principles, 661–664
　identification of biomolecules
　　by molecular weights,
　　673–675
　probing of three-dimensional
　　structure, 684–690
　protein proteolytic
　　fingerprinting by, 674
　resolving molecular weights,
　　664–670
　sequencing, 676–684
　tandem, 678
Matrix-assisted laser desorption-
　ionization (MALDI), 661
Maxwell-Boltzmann relationship, 148
Melittin, fluorescence and electronic
　absorption, 508
Membrane equilibria, 589–597
　dialysis equilibrium, 591–592
　membrane potentials, 596–597
　osmotic pressure, 592–596
Membrane potentials, 596–597
Meselson-Stahl experiment, 248
Metalloprotein, 334
Methionine, electronic transition,
　435–436
1-Methylthymine, polarized absorption
　spectra for crystals of, 416
Michaelis-Menten analysis, 621–622
Microcalorimeters, 95
Microequilibrium constant, 167
Miller indices, 297, 305, 316
Mirror symmetry, 20, 21
Mixing, free energy of, 585

Model
　cooperative binding, 635, 636, 638
　CPK, 44
　DNA duplex, 188
　electrophoresis, 249
　hexamer toy, 205
　KNF, 638, 639
　nucleosome core particle, 374
　simple exact, 205
　structural transition, 169
　Watson-Crick, 59, 137
Model, zipper
　application of, 174, 175
　B-Z transition, 194
　partition function of, 172, 173
　propagation parameter of, 179
Molecular chaperones, 151
Molecular dynamics, 111, 572
Molecular interactions
　energies of, 9
　intramolecular, 11
Molecular mechanics, 110, 111
Molecular partition function, 80
Molecular potentials, 111
Molecular scattering factor, 313–314
Molecular simulation, 107
Molecular structures, structure
　refinement, 331
Molecular thermodynamics, 107–165
　complexities in modeling,
　　107–109
　molecular mechanics, 109–124
　simulating macromolecular
　　structure, 145–161
　stabilizing interactions, 124–145
Molecular trajectory, 110
Molecular velocity, flow and, 242
Molecular weight
　anhydrous, 598
　identification of biomolecules
　　by, 673
　number average, 595
　separation of nucleic acids
　　by, 253
Molecule
　absorption and emission
　　spectra, 504
　amphipathic, 16, 17
　chiral, stereochemistry of, 5–6
　definition of, 2
　degenerate state of, 169
　denatured conformation, 8
　in discrete energy level, 382
　electronegativities of elements
　　in, 12
　enantiomer, 6
　geometry, conformation, 5, 6, 7
　identification, 421
　infrared absorption bands, 450
　light absorbed by, 380
　melting of, 149
　monomer units, 2–3
　noncovalent interactions
　　between, 75
　nonstationary state, 457
　normal mode analysis of, 152

　precipitated, 285
　scattering, 361
　solvent-accessible surface of, 155
　stereoisomers, 6
　structural purity, 286
　subunit of, 2
　symmetry relationships, 19
　three-dimensional structure
　　of, 34
　topology of, 3
Molecule, hydrogen
　composition of, 401
　as model for bond, 400
Molten globules, 4, 151
Monochromatic incident photons,
　intense source of, 459
Monochromatic light, 381
Monomer
　building blocks, configuration
　　of, 7
　-dimer reactions, 611, 613
　stereochemistry of, 6
　transition dipole data, 445
Monte Carlo simulation, 160, 204
Motif, 19, 24
　Greek key, 43
　lattice, 280
Moving boundary
　electrophoresis, 250–251
　sedimentation, 225–237
MS, *see* Mass spectrometry
Multiexponential curve fitting, 245
Multiple-wavelength anomalous
　dispersion (MAD), 334, 335
MWC theory, 635, 638
Myoglobin, 49
Myosin, 231

N

Nernst equation, 608
Neutron scattering
　contrast matching in, 372
　lengths, 371
　studies, 259
Newtonian physics, classical, 107
Newton's laws of motion, 146
Newton's second law of motion, 110
NMA, *see* N-Methyl acetamide
N-Methyl acetamide (NMA), 122, 123
　energies for dimerization of, 124
　hydrogen-bonding interactions
　　of, 123
NMR, *see* Nuclear magnetic resonance
NOE, *see* Nuclear Overhauser effect
NOESY, *see* Nuclear Overhauser effect
　and exchange spectroscopy
NOESY-CPSY connectivity diagram,
　568
Nonbonding potentials, 115
Noncooperative binding, 623–624
Noncooperative transition model, 170, 172
Nondegenerate perturbation theory, 394,
　413, 437
Nonideality, measure of, 357

Nonradiative transition, 502
Normal mode analysis, chemical bond in, 153
Northern blotting, 266
Nuclear magnetic resonance (NMR), 213
    base stacking and, 553
    contact plot of protein from, 48
    energy levels explored by, 411
    Fourier transform, 555
    one-dimensional, 551, 552
    two-dimensional, 555
Nuclear magnetic resonance instruments
    continuous wave, 537, 544
    Fourier transform, 537, 545
Nuclear magnetic resonance spectroscopy, 10, 46, 318, 535–578
    Fourier transforms and, 453
    measurable, 537–539
    one-dimensional NMR of macromolecules, 549–555
    phenomenon, 535–537
    relaxation and nuclear Overhauser effect, 542–544
    spectrum measurement, 544–548
    spin-spin interaction, 540–542
    two-dimensional Fourier transform NMR, 555–559
Nuclear Overhauser effect (NOE), 542, 544, 555
Nuclear Overhauser effect and exchange spectroscopy (NOESY), 557, 564
Nucleation parameter, 172, 186
    B-Z junction, 193
    estimated, 177
Nucleic acid
    amphipathic, 16
    association of peptides and proteins to, 144
    atomic solvation parameters, 157
    bases of, 140
    binding to, 648
    crystals of, 326
    electronic absorption spectra, 443
    electronic circular dichroism of, 476–481
    electrophoresis of, 253
    fluorescence applied to, 530
    separation of by molecular weight, 253
    time scale of molecular processes in, 149
Nucleic acid, structure of, 52–68
    helical structure, 55–61
    higher-order structures, 61–68
    principles determining, 133
    torsion angles, 54–55
Nucleobase, 52
    dimerization of, 138
    energies for base-stacking between, 140
Nucleosomes, 372, 374
Nucleotide
    conformations, potential energy profiles, 135
    sugar pucker of, 135
Number average molecular weight, 595

O

O'Farrell technique, 268–269
Oligolysine, binding of, 145
Oligomers, 3
Oligonucleotides, double-helical, 322
Omit map, 322
One-dimensional box, particle in, 387, 388
One-dimensional electrophoresis, 256
Optically active solution, 471
Optical trapping, 703, 704
Optical tweezers, 703, 707
Orbitals, relative energies for,
Oscillating charges, 352
Osmometer, 595
Osmotic pressure, 72, 592–596
Overtones, 451
Oxygen
    binding of by hemoglobin, 641
    -transport proteins, 461
Oxyhemerythrin, resonance Raman spectrum of, 463

P

Partial molar quantities, 580, 581
Partial specific quantities, definition of, 581
Particle
    kinetic energy for, 384
    weight, determination of, 362
Partition function, 168, 172
Patterson function, 323
Patterson map, 324, 325
Patterson method, 320, 321
Pauli exclusion principle, 405
Pauling, Linus, 34, 141, 338
Peptide
    bond, 32, 40
    composition, 33
    trans-conformation of, 38, 39
    helical versus nonhelical, 130
    host-guest, 179
    interactions between dipole moments of, 126
    sequencing, 679
Perrin plot, 523
Persistence length, 202
Perturbation theory, 394
Phenylalanine, electronic transition, 435–436
Phosphoimagers, 264
Phosphorescence, definition of, 502
Photobleaching, 221
Photograph
    fiber diffraction, 339
    precession, 305
    still, 304
Photometer, 358
Physical equilibria
    membrane equilibria, 589–597
    sedimentation equilibrium, 597–598
    steady-state electrophoresis, 598–600

Planck's law, 381, 411, 536
Point groups, 20, 25, 51
Point symmetry, 20
Polarization, 459, 519, 531
Polarized fluorescence, 501
Polyacids, 248
Polyampholytes, 248
Polyanions, 468
Polybases, 248
Polynucleic acids, helical structures of, 55
Polynucleotide
    chain, protein binding to, 144
    D-sugars in, 6
Polynucleotide duplex
    hydrogen bonding of, 185
    melting and annealing of, 184
Polypeptide
    coil-helix transition in, 174
    compact denatured forms of, 206
    -heme complex, 2
    L-amino acids in, 6
    macroscopic dipole of α-helix in, 129
Polypeptide chain
    conformation space, 128
    domains, 43
    fragmented, 677
Polysaccharides
    carbohydrate building blocks in, 6
    D-sugars in, 6
Potential energy
    barrier, 128
    dipole-dipole interactions, 118
    force field, 148
    hydrogen bond, 113, 122
    macromolecule, 110
    map, sugar pucker, 136
    profiles, nucleotide conformations, 135
    single bond, 114
Precession photography, 304
Prodan, 528
Proline, 40
Propagation parameter, 173, 178, 179
Protein
    amphipathic, 16
    association reactions, thermodynamic parameters, 611
    binding, 144, 645
    chemistry, allosteric effects, 622
    circular dichroism spectra of, 492, 493
    complex, communication among subunits in, 50
    composition, analysis of, 673
    conformations, peptide bond and, 40
    crystals, growth of under microgravity conditions, 289
    denaturation, 72
    dominant chromophore, 435
    dye, 264
    electronic circular dichroism of, 481–484

Protein (continued)
  energy states of, 90
  examples of point group
    symmetry, 51
  fluorescence applied to, 524
  folds, 48, 49, 182, 687
  FRET and, 516
  FTIR spectra of, 456
  hydrogen/deuterium exchange
    in, 689
  infrared spectra analysis, 456
  inhibition, drug discovery and, 102
  macromolecule, structure of, 45
  macroscopic dipole of α-helix
    in, 129
  molecule, denatured, 86
  nature of solutions of, 579
  –nucleic acid complexes, 244
  overall charge of, 31
  oxygen-transport, 461
  prosthetic groups, 28
  proteolytic fingerprinting, 674
  recombinant, crystallized, 330
  ribosomal, 375
  RNA-binding, 35
  SDS gel electrophoresis of,
    259, 261
  secondary structures of, 34
  sequences, 33, 209
  stability, electrostatic
    interactions and, 131
  structure, intramolecular
    interactions, 125
  TATA-binding, 10
  time scale of molecular
    processes in, 149
Protein, globular, 42–52
  behavior of, 231
  contact plot, 46
  domains, 43–44
  prediction of secondary
    structures of, 181
  protein folds, 48–49
  quaternary structure, 49–52
  side chains at surface of, 131
  supersecondary structures, 42–43
  tertiary structure, 44–48
Protein, secondary structure of
  CD spectra, 482
  matrix of, 494
  methods for predicting, 184
  statistical methods for
    predicting, 181
  VCD spectra, 497
Protein, structure of, 27–52
  amino acids, 27–31
  effect of peptide bond, 40–41
  globular proteins, 42–52
  helical symmetry, 36–40
  hydrogen bonds in, 141–142
  secondary structures, 34–36
  unique protein sequence, 31–34
Protein-folding
  pathways, 151
  problem, 4
Proteolytic enzymes, 675

Proton
  binding of to proteins, 645
  dissociation of, 647
  spin-spin interactions of, 571
  transfers, enzyme-catalyzed
    reactions and, 15
Pseudorotation angle, 54
Pseudosymmetry, 22
Pulsed field electrophoresis, 255, 256
Pulse fluorometry, 521

Q

Quadrupole filter, 667
Quantum chemistry problems,
    approximating solutions to, 392–400
  perturbation theory, 394–399
  separability, 393–394
  variation method, 399–400
Quantum mechanical intensity, 412
Quantum mechanical operators, 384
Quantum mechanics and spectroscopy,
    380–420
  light and transitions, 381–382
  postulate approach, 382–386
  transition dipole directions,
    415–417
  transition energies, 386–408
  transition intensities, 408–415
Quantum number, 387
Quantum states, 387
Quantum yield, 514
Quenching, 514

R

Radial dilution effect, 227
Radiation
  density, 411, 412
  scattering of, linearly polarized,
    352
  unpolarized, 353
  wavelength, scattering from
    macromolecule and, 359
Radius of gyration, 361
Ramachandran plot, 41
Raman scattering, 457, 458, 695
  energy-level diagram for, 458
  inelastic, 458
Raman wavelength shifts, 695
Random coil, 175
Random walk, 200
  biopolymer behavior and, 204
  pathway of, 199
Rayleigh ratio, 356
Rayleigh scattering, 355, 356, 370, 458
Reciprocal space, 299
Recombinant proteins, crystallized, 330
Reduced valence, 599
Reflection
  equatorial, 347
  exclusionary condition, 307
  scattering vector for, 303
Reflectron, 665
Relative buffer viscosity, 228

Relative mobility, 251
Repulsive potential, 119
Residue volumes, 231
Resonance Raman scattering
  advantage of, 460
  energy-level diagram for, 458
  oxygen-transport proteins and,
    461
Reverse transition
  coil-to-helix transition and, 187
  helix to coil, 180
Reversible equilibrium states, 74
Reversible processes, first law for, 87
R factors, 331
Ribonuclease
  denatured, 107
  molecule, distance across, 366
  renatured, 107
Ribonucleic acid, see RNA
Ribosomal proteins, distances between
    pairs of, 375
Ribosomal RNA (rRNA), 62
Ring currents, 538
RNA (ribonucleic acid), 52, 53
  base pairs found in, 137
  -binding protein, 35
  differences, 64
  double-helix, annealing and
    melting of, 184
  folding, 62
  negatively charged phosphates
    and, 142
  polymerase, active transcription
    by, 701, 706
  possible conformations, 54
  ribosomal, 62
  structural dynamics, 694
  transfer, 62, 204
  variations in sequence, 64
Rotational symmetry, 20, 22
rRNA, see Ribosomal RNA
Rubredoxin, mirror images of, 30
Running gel, 262, 263
Rydberg transitions, 429

S

SAS, see Solvent-accessible surface
SAXS, see Small-angle X-ray scattering
Scanning absorption optical system, 225
Scatchard plot, 629
Scattered waves, 291, 292
Scattering
  angular dependence of, 368
  experiments, monochromatic
    neutron beam used for, 370
  length, 370, 374
  low-angle X-ray, 362
  molecule, internal interference
    within, 361
  neutron, 372
  Raman, 457, 458, 695
  Rayleigh, 355, 356, 458
  techniques, 376
  vector, 302, 303, 360

Scattering intensity
    angle versus, 367
    unpolarized incident light, 354
Scattering from solutions of
    macromolecules, 351–379
    dynamic light scattering, 363–365
    light scattering, 351–363
    small-angle neutron scattering,
      370–375
    small-angle X-ray scattering,
      365–370
Schrödinger equation, 384, 389
    first-order corrections, 396
    linear combination of solutions
      to, 387
    second-order corrections, 396
    simplified solution, 393
    time-dependent, 385, 413
Screw symmetry, 20, 26
SDS, *see* Sodium dodecyl sulfate
SDS gel electrophoresis, 259
    acid-soluble nuclear proteins,
      261
    analysis of multisubunit
      structures by, 260
Second law of thermodynamics, 80, 87
Second viral coefficient, 357
Sedimentation, 223–248
    behavior, alteration of, 228
    data, 230
    density gradient, 246–248
    moving boundary, 225–237
    sedimentation equilibrium,
      241–246
    velocity, 237, 238
    zonal, 237–241, 251
Sedimentation coefficient, 224, 233, 234
    interpretation of, 229
    ratios, predicted, 234
Sedimentation equilibrium, 241–246,
    597–598
    concentration curves at, 245
    density gradient, 246, 247
    experiments, 373
    hemocyanin, 243
Self-avoiding walk, 205
Self-energy, 19, 131
Self-vectors, 324
Separation and characterization of
    molecules, 213–275
    diffusion, 214–223
    electrophoresis and isoelectric
      focusing, 248–269
    general principles, 213–214
    sedimentation, 223–248
β-Sheet
    antiparallel, 47
    contact plot, 47
    parallel, 43, 47
Shielding coefficient, 537
Side chain–side chain interactions, 131
Signal averaging, FT NMR and, 548
Simple exact models, 205
Simple harmonic oscillator, 389
Simulated annealing, 149, 150
Single-molecule biochemistry, 693

Single-molecule biophysics, 580
Single-molecule methods, 693–710
    atomic force microscopy,
      699–703
    magnetic beads, 707–708
    observation by fluorescence,
      695–699
    optical tweezers, 703–707
    reason for studying, 693–695
Singular value decomposition
    (SVD), 485
    basis vectors, 491
    least-squares coefficients, 490
    matrix, 487
Small-angle neutron scattering, 370
Small-angle X-ray scattering (SAXS),
    365, 366
    hemocyanin, 369
    particle dimensions given by, 368
Sodium dodecyl sulfate (SDS), 259
Soft ionization process, 670
Solute concentration, 587
Solution nonideality, 356, 588
Solution thermodynamics, 579–604
    applications of chemical
      potential to physical
      equilibria, 589–600
    fundamentals, 580–589
Solvent
    absorbance, 448
    -accessible surface (SAS), 155
    addition of solute to pure, 582
    interactions, modeling of, 158
    transfer, in vapor diffusion
      methods, 288
Solvent effects
    general, 509, 512
    specific, 509
Southern blotting, 264, 265
Space group, crystal, 283
Space wavefunction, 405
Specific refractive index increment, 355
Specific solvent effects, 509
Spectrograph, compartments, 447
Spin factor, 406
Spin-lattice relaxation, 542
Spin-spin interactions, 540, 571
Spin-spin relaxation, 543
SPR, *see* Surface plasmon resonance
Stacking gel, 262, 263
Standard free energy, 91
Standard state, definition of, 91, 92
Stationary state system, 385
Statistical thermodynamics, 166–212
    general principles, 166–175
    nonregular structures, 198–209
    structural transitions in
      polypeptides and proteins,
      175–184
Statistical weights, 167
Steady-state anisotropy, 523
Steady-state electrophoresis, 598–600
Still photograph, 304
Stirling's approximation, 82, 85
Stokes's law, value for spherical
    molecule, 220, 223

Stokes's radius, 19, 132, 233
Streptococcus bacteria, 35
Structural purity, 286
Structural transitions, two-state models
    for, 169
Sturhman equation, 375
Succinic acid, helical versus nonhelical
    peptides, 130
Sucrose gradient centrifugation, 239, 240
Sugar pucker, 54, 59, 134, 135
    change in, 149
    potential energy map of, 136
Supersaturation, 288
Surface plasmon resonance (SPR), 618
Surface tension, 74
SVD, *see* Singular value decomposition
Svedberg equation, 235
Symmetry
    dyad, 23
    helical, 26, 36, 51
    hemoglobin quaternary
      structure, 50
    mirror, 20, 21
    octahedral, 25
    point, 20
    pseudosymmetry, 22
    -related objects, 19
    relationships, multiple, 25
    rotational, 20, 22
    screw, 20, 26
    symbols for, 24
    tetrahedral, 25
    two-fold, 23

T

Tandem mass spectrometry, 678
TATA-binding protein, 10
Temperature factor, 337
Tetramethylsilane (TMS), 537
TFE, *see* 2,2,2-Trifluoroethanol
Thermodynamics
    first law of, 73
    second law of, 80, 87
Thermodynamics and biochemistry,
    72–106
    experimental thermochemistry,
      93–103
    first law of thermodynamics,
      73–76
    molecular interpretation of
      thermodynamic quantities,
      76–80
    second law of thermodynamics,
      80–91
    standard state, 91–93
Through-bond interactions, 540
Through-space interaction, 544
TIM, *see* Triose isomerase
Time-average conformation, modeling
    of, 166
Time-of-flight (TOF) instrument,
    664, 666
Time-resolved fluorescence, 522
TIR microscope, *see* Total internal
    reflection microscope

TMS, *see* Tetramethylsilane
TOF instrument, *see* Time-of-flight
    instrument
Topoisomer
    bacterial plasmid, 68
    ccDNA, 191
    free energy of, 192
    separation of by one-dimensional
        electrophoresis, 256
Torsion angles, 54–55, 569
Torsion force constant, 114
Total energy, 109
Total internal reflection (TIR)
    microscope, 695
Total potential, 596, 597
Transfer RNA (tRNA), 62, 204
    structure of, 63
    synthetase, functional
        domains, 44
    tertiary structure, 64
Transition
    all-or-none, 169
    B-DNA to A-DNA, 189
    coil-helix, 175, 177, 178, 182, 187
    cooperative length of, 170
    nonradiative, 502
    nucleation point, 172
    reverse, 187
Transition, B-Z, 191
    partition function for, 194
    two-dimensional gel
        electrophoresis analysis
        of, 196
    zipper model, 194
Transition dipoles, 414, 433, 451
    calculation of, 452
    directions, 415
    nondegenerate interaction
        between, 441
Transition energies, 386–408
    approximating solutions to
        quantum chemistry
        problems, 392–400
    hydrogen molecule as model for
        bond, 400–408
    quantum mechanics of simple
        systems, 386–392
Transition model
    noncooperative, 170, 172
    two-state, 169
Transmission, 409
Transport processes, 213
2,2,2-Trifluoroethanol (TFE), 175
Triose isomerase (TIM), 49
Triplet absorption, 504
Triplet excited state, 502
tRNA, *see* Transfer RNA
Trypsin fingerprint, 675, 676
Tryptophan
    electronic transition, 435–436
    fluorophore, 524
    indole side chain of, 550
Two-dimensional gels, 258, 264

Two-fold rotational axis, 23
Two-fold symmetry, 23
Two-photon excitation, 696
Tyrosine, electronic transition, 435–436

U

Unit cell, 279, 280
    cylindrical, 340
    dimensions and shape of, 310
    fiber, 340
    Patterson maps, 324
    reciprocal, 300
    unknown location of, 320
United atom, 149
Urease, crystal structure of, 328, 329

V

Valence bond
    approximation, 395
    independent systems approach,
        400
Valinomycin, 28
van der Waals interactions, 118
van der Waals potential, 121
van der Waals radius, 9
van't Hoff relationship, 93
Vapor diffusion methods
    crystallization, 289
    solvent transfer in, 288
Variable selection, 495
Variation theorem, 399
VCD, *see* Circular dichroism, vibrational
Vibrational absorption, 449–457
    applications to biological
        molecules, 453–457
    energy-level diagram for, 458
    energy of vibrational absorption
        bands, 450–451
    instrumentation for vibrational
        spectroscopy, 453
    transition dipoles, 451–452
Vibrational circular dichroism
    (VCD), 496
Vibrational modes, fundamental
    frequency of, 451
Vibrational transitions, infrared-
    forbidden, 460
Virial coefficients, 587
Virtual states, 458
von Laue conditions for diffraction,
    294, 295 296

W

Water
    dipole moment of, 511
    interaction of molecules with, 15
    pairing of isolated nucleotides
        in, 138
    phase diagram for, 14

    self-dissociation of, 14
    structures formed by
        amphipathic molecules
        in, 17
    -water hydrogen bond, 12
Watson-Crick base pairs, 56, 133
Wavefunction, 388
    degenerate, 438
    doubly excited, 438
    energies, 427
    exciton, 439
    ground-state, 438
    hydrogen molecule ion, 402
    orthonormality of, 440
    space, 405
    zero-order, 398
Wavenumbers, 382, 461
Wave propagation, 311
Wave vector, Argand diagram for, 312
Weight average molecular weight,
    246, 358
Western blotting, 266
Work, definition of, 74

X

X-ray, amplitude of scattered, 370
X-ray diffraction, 19, 46, 213, 276–350
    crystal morphology, 305–308
    deoxyoligonucleotide single
        crystals, 134
    DNA, 58
    experiment, reflections of, 294
    fiber diffraction, 338–347
    pattern, photographic film, 304
    resolution in, 334
    solving macromolecular
        structures, 309–338
    structures at atomic resolution,
        277–279
    theory, 291–305
X-ray radiation and resolution, 336
X-ray scattering
    intensity, 367
    lengths, 371

Z

Z-DNA, 59
    genomic sequence analyzed
        for, 197
    left-handed, 134, 153, 184
    structure of, 57
Zero-order wavefunctions, 398
Zero point energy, 390
Zipper model
    application of, 174, 175
    B-Z transition, 194
    partition function of, 172, 173
    propagation parameter of, 179
Zonal electrophoresis, media for, 251
Zonal sedimentation, 237–241, 251

PHYSICAL CONSTANTS

Name	Symbol	SI Units	cgs Units
Avogadro's number	$\mathcal{N}$	$6.022137 \times 10^{23}/\text{mol}$	$6.022137 \times 10^{23}/\text{mol}$
Boltzmann constant	k_B	$1.38066 \times 10^{-23} \text{ J/K}$	$1.38066 \times 10^{-16} \text{ erg/K}$
Debye		$3.336 \times 10^{-30} \text{ C} \cdot \text{m}$	$1.000 \times 10^{-18} \text{ esu} \cdot \text{cm}$
Electron charge (magnitude)	e	$1.602177 \times 10^{-19} \text{ C*}$	$4.80321 \times 10^{-10} \text{ esu}$
Rest mass electron	m_e	$9.109 \times 10^{-31} \text{ kg}$	$9.109 \times 10^{-28} \text{ g}$
Faraday constant	$\mathcal{F}$	$96485 \text{ J/V} \cdot \text{mol}$	$9.6485 \times 10^{11} \text{ erg/V} \cdot \text{mol}$
Permittivity (vacuum)	ε_0	$8.854 \times 10^{-12} \text{ C}^2/\text{J} \cdot \text{m}$	$1/(4\pi) \text{ esu}$
Gas constant	$R = \mathcal{N}k_B$	$8.31451 \text{ J/K} \cdot \text{mol}$	$8.31451 \times 10^7 \text{ erg/K} \cdot \text{mol}$
Gravity acceleration	g	9.80665 m/sec^2	980.665 cm/sec^2
Light speed (vacuum)	c	$2.99792 \times 10^8 \text{ m/sec}$	$2.99792 \times 10^{10} \text{ cm/sec}$
Planck's constant	h	$6.626075 \times 10^{-34} \text{ J} \cdot \text{sec}$	$6.626075 \times 10^{-27} \text{ erg} \cdot \text{sec}$

* 1 C = 1 J/V

CONVERSION FACTORS

Energy: 1 Joule $= 10^7$ ergs $= 0.239$ cal
        1 cal $= 4.184$ Joule

Length: 1 nm $= 10 \text{ Å} = 1 \times 10^{-7} \text{ cm} = 1 \times 10^{-9} \text{ m}$

Pressure: 1 atm $= 760$ torr $= 14.696$ psi
        1 torr $= 1$ mm Hg

Temperature: K $= {}^\circ\text{C} + 273$

ONE- AND THREE-LETTER SYMBOLS FOR THE AMINO ACIDS

Alanine	Ala	A	Leucine	Leu	L
Arginine	Arg	R	Lysine	Lys	K
Asparagine	Asn	N	Methionine	Met	M
Aspartic acid	Asp	D	Phenylalanine	Phe	F
Cysteine	Cys	C	Proline	Pro	P
Glutamic acid	Glu	E	Serine	Ser	S
Glutamine	Gln	Q	Threonine	Thr	T
Glycine	Gly	G	Tryptophan	Trp	W
Histidine	His	H	Tyrosine	Tyr	Y
Isoleucine	Ile	I	Valine	Val	V